Graduate Texts in Mathematics

Graduate Texts in Mathematics bridge the gap between passive study and creative understanding, offering graduate-level introductions to advanced topics in mathematics. The volumes are carefully written as teaching aids and highlight characteristic features of the theory. Although these books are frequently used as textbooks in graduate courses, they are also suitable for individual study.

More information about this series at http://www.springer.com/series/136

Konrad Schmüdgen

The Moment Problem

Springer

Konrad Schmüdgen
Mathematisches Institut
Universität Leipzig
Germany

ISSN 0072-5285   ISSN 2197-5612   (electronic)
Graduate Texts in Mathematics
ISBN 978-3-319-87817-1   ISBN 978-3-319-64546-9   (eBook)
DOI 10.1007/978-3-319-64546-9

Mathematics Subject Classification (2010): 44A60, 14P10, 47A57

Printed on acid-free paper

This Springer imprint is published by Springer Nature
The registered company is Springer International Publishing AG
The registered company address is: Gewerbestrasse 11, 6330 Cham, Switzerland

To KATJA

Thomas Jean Stieltjes (1856 – 1894)

Pafnuty Lvovich Chebyshev
(1821 – 1894)

Andrey Andreyevich Markov
(1856 – 1922)

Contents

Preface and Overview

Let μ be a positive Radon measure on a closed subset $\mathcal{K}$ of $\mathbb{R}^d$ and $\alpha = (\alpha_1, \ldots, \alpha_d)$ a multi-index of nonnegative integers α_j. If the integral

$$s_\alpha(\mu) := \int_\mathcal{K} x_1^{\alpha_1} \ldots x_d^{\alpha_d} \, d\mu(x)$$

is finite, the number $s_\alpha \equiv s_\alpha(\mu)$ is called the α-*th moment of the measure* μ. If all α-th moments exist, the sequence $(s_\alpha)_{\alpha \in \mathbb{N}_0^d}$ is called the *moment sequence* of μ.

Moments appear at many places in physics and mathematics. For instance, if $\mathcal{K}$ is a solid body in $\mathbb{R}^3$ with mass density $m(x_1, x_2, x_3)$, then the number

$$s_{(0,2,0)} + s_{(0,0,2)} = \int_\mathcal{K} (x_2^2 + x_3^2) \, m(x_1, x_2, x_3) dx_1 dx_2 dx_3$$

is the moment of interia of the body with respect to the x_1-axis. Or if X is a random variable with distribution function $F(x)$, the expectation value of X^k is defined by

$$E[X^k] = s_k = \int_{-\infty}^{+\infty} x^k dF(x)$$

and the variance of X is $\text{Var}(X) = E[(X - E[X])^2] = E[X^2] - E[X]^2 = s_2 - s_1^2$ (provided that these numbers are finite).

The moment problem is a classical mathematical problem. In its simplest form, the *Hamburger moment problem* for the real line, it is the following question:

Let $s = (s_n)_{n \in \mathbb{N}_0}$ be a real sequence. Does there exist a positive Radon measure μ on $\mathbb{R}$ such that for all $n \in \mathbb{N}_0$ the integral $\int_{-\infty}^{+\infty} x^n d\mu$ converges and satisfies

$$s_n = \int_{-\infty}^{+\infty} x^n \, d\mu \, ? \tag{1}$$

1

That is, the moment problem is the inverse problem of "finding" a representing measure μ when the moment sequence s is given.

For a real sequence $s = (s_n)_{n\in\mathbb{N}_0}$ let L_s denote the linear functional on the polynomial algebra $\mathbb{R}[x]$ (or on $\mathbb{C}[x]$) defined by $L_s(x^n) = s_n$, $n \in \mathbb{N}_0$. By the linearity of the integral it is clear that (1) holds for all $n \in \mathbb{N}_0$ if and only if we have

$$L_s(p) = \int_{-\infty}^{+\infty} p(x)\,d\mu \quad \text{for } p \in \mathbb{R}[x]. \tag{2}$$

Thus, the moment problem asks whether or not the functional L_s on $\mathbb{R}[x]$ admits an integral representation (2) with respect to some positive measure μ. To get some flavour of what this book is about, we sketch without giving proofs some cornerstones from the theory of the one-dimensional Hamburger problem.

Assume that $s = (s_n)_{n\in\mathbb{N}_0}$ is the moment sequence of some positive measure μ on $\mathbb{R}$. Then, for any polynomial $p(x) = \sum_{k=0}^{n} a_k x^k \in \mathbb{R}[x]$ we obtain

$$L_s(p^2) = \int p(x)^2\,d\mu = \int \left(\sum_{k,l=0}^{n} a_k a_l x^{k+l}\right)d\mu = \sum_{k,l=0}^{n} a_k a_l s_{k+l} \geq 0.$$

Therefore, $L_s(p^2) \geq 0$ for all $p \in \mathbb{R}[x]$, that is, the functional L_s is positive, and the Hankel matrix $H_n(s) := (s_{k+l})_{k,l=0}^{n}$ is positive semidefinite for each $n \in \mathbb{N}_0$. These are two (equivalent) necessary conditions for a sequence to be a moment sequence. Hamburger's theorem (1920) says that each of these necessary conditions is also sufficient for the existence of a positive measure. That is, the *existence* problem for a solution is easily answered in terms of positivity conditions.

The question concerning the *uniqueness* of representing measures is more subtle. A moment sequence is called *determinate* if it has only one representing measure. For instance, the lognormal distribution

$$F(x) = \frac{1}{\sqrt{2\pi}}\chi_{(0,+\infty)}(x)x^{-1}e^{-(\log x)^2/2}$$

gives a probability measure whose moment sequence is not determinate.

Let us assume that s has a representing measure μ supported on the bounded interval $[-a, a]$, $a > 0$. It is not difficult to show that the support of any other representing measure, say $\tilde{\mu}$, is also contained in $[-a, a]$. Then (2) implies that

$$\int_{-a}^{a} f(x)d\mu = \int_{-a}^{a} f(x)\,d\tilde{\mu} \tag{3}$$

for $f \in \mathbb{R}[x]$. By Weierstrass' theorem each continuous functions on $[-a, a]$ can be approximated uniformly by polynomials. Therefore (3) holds for all continuous functions f on $[-a, a]$. Hence $\mu = \tilde{\mu}$, so that s is determinate.

A useful sufficient criterion for determinacy is the Carleman condition

$$\sum_{n=1}^{\infty} s_{2n}^{-\frac{1}{2n}} = +\infty.$$

Now suppose s is a moment sequence such that $L_s(p\bar{p}) > 0$ for $p \in \mathbb{C}[x], p \neq 0$, or equivalently, s has a representing measure with infinite support. Then

$$\langle p, q \rangle_s = L_s(p\bar{q}), \quad p, q \in \mathbb{C}[x],$$

defines a scalar product on the vector space $\mathbb{C}[x]$. Let X denote the multiplication operator by the variable x on $\mathbb{C}[x]$, that is, $(Xp)(x) = xp(x)$. Applying the Gram–Schmidt procedure to the sequence $(x^n)_{n \in \mathbb{N}_0}$ of the unitary space $(\mathbb{C}[x], \langle \cdot, \cdot \rangle_s)$ yields an orthonormal sequence $(p_n)_{n \in \mathbb{N}_0}$ of polynomials $p_n \in \mathbb{C}[x]$ such that each p_n has degree n and a positive leading coefficient. Then there exist numbers $a_n > 0$ and $b_n \in \mathbb{R}$ such that the operator X acts on the orthonormal base (p_n) by

$$Xp_n(x) = a_n p_{n+1}(x) + b_n p_n(x) + a_{n-1}p_{n-1}(x), \ n \subset \mathbb{N}_0, \ \text{where} \ p_{-1} := 0.$$

That is, X is unitarily equivalent to the *Jacobi operator* T_J for the Jacobi matrix

$$J = \begin{pmatrix} b_0 & a_0 & 0 & 0 & \cdots \\ a_0 & b_1 & a_1 & 0 & \cdots \\ 0 & a_1 & b_2 & a_2 & \cdots \\ 0 & 0 & a_2 & b_3 & \cdots \\ & \ddots & \ddots & \ddots & \ddots \end{pmatrix}$$

If A is a self-adjoint extension of T_J on a possibly larger Hilbert space and E_A is the spectral measure of A, then $\mu = s_0 \langle E_A(\cdot)1, 1 \rangle$ is a solution of the moment problem for s. Each solution is of this form. Further, the moment sequence s is determinate if and only if the closure of the symmetric operator X on the Hilbert space completion of $(\mathbb{C}[x], \langle \cdot, \cdot \rangle_s)$ is self-adjoint. All these facts relate the moment problem to the theory of orthogonal polynomials and to operator theory.

Let s be an indeterminate moment sequence. Nevanlinna's theorem yields a parametrization of the solution set. Let $\mathfrak{P}$ denote the set of holomorphic functions on the upper half-plane with nonnegative imaginary part and set $\overline{\mathfrak{P}} := \mathfrak{P} \cup \{\infty\}$. Then Nevanlinna's theorem states that there is a bijection $\phi \to \mu_\phi$ of $\overline{\mathfrak{P}}$ to the set of all solutions of the moment problems for s given by

$$\int_{-\infty}^{\infty} \frac{1}{x-z} d\mu_\phi(x) = -\frac{A(z) + \phi(z)C(z)}{B(z) + \phi(z)D(z)}, \quad z \in \mathbb{C}, \ \text{Im} \, z > 0.$$

Here A, B, C, D are certain entire functions depending only on the sequence s. Thus, a moment sequence is either determinate or it has a "huge" family of representing measures. All this and much more is developed in detail in Part I of the book.

There are a number of other variants of the moment problem. The *Stieltjes moment problem* asks for representing measures on the half-line $[0, +\infty)$ and the *Hausdorff moment problem* for measures on the interval $[0, 1]$. If only the first moments $s_0, \dots, s_m$ are prescribed, we have a *truncated moment problem*. In this case there exist finitely atomic representing measures and one is interested in measures with "small" numbers of atoms.

The passage from one-dimensional to multidimensional moment problems leads to fundamental new difficulties. As already observed by Hilbert, there are positive polynomials in $d \geq 2$ variables that are not sums of squares. As a consequence, there exists a linear functional L on $\mathbb{R}_d[x] \equiv \mathbb{R}[x_1, \dots, x_d]$ which is positive (that is, $L(p^2) \geq 0$ for $p \in \mathbb{R}_d[x]$), but L is not a moment functional (that is, it cannot be written in the form (2)). For this reason the $\mathcal{K}$-*moment problem* is invented. It asks for representing measures with support contained in a given closed subset $\mathcal{K}$ of $\mathbb{R}^d$.

Let $\{f_1, \dots, f_k\}$ be a finite subset of $\mathbb{R}_d[x]$. Then the closed set

$$\mathcal{K} = \{x \in \mathbb{R}^d : f_1(x) \geq 0, \dots, f_k(x) \geq 0\}$$

is called a semi-algebraic set. For such sets methods from real algebraic geometry provide powerful tools for the study of the $\mathcal{K}$-moment problem. Our main result for a compact semi-algebraic set $\mathcal{K}$ is the following: A linear functional L on $\mathbb{R}_d[x]$ is a moment functional with representing measure supported on $\mathcal{K}$ if and only if

$$L(f_1^{e_1} \cdots f_k^{e_k} p^2) \geq 0 \quad \text{for } p \in \mathbb{R}_d[x], \ e_1, \dots, e_k \in \{0, 1\}.$$

If $\mathcal{K}$ is only closed but not compact, this is no longer true. But if there exist nontrivial bounded polynomials on $\mathcal{K}$, there is a fibre theorem that allows one to reduce the $\mathcal{K}$-moment problem to "lower dimensional" cases. The multidimensional determinacy question is much more complicated than its one-dimensional counterpart.

It turns out that most methods that have been successfully applied to the moment problem in dimension one either fail in higher dimensions or at least require more involved additional technical considerations.

The study of moment problems and related topics goes back to the late nineteenth century. The Russian mathematicians P.L. Chebychev (1974) and A.A. Markov (1984) applied moments in their "theory of limiting values of integrals" and invented important notions; a survey of their ideas and further developments was given by M.G. Krein [Kr2]. The moment problem itself as a problem in its own was formulated for the first time by the Dutch mathematician T.J. Stieltjes (1894) in his pioneering memoir [Stj]. Stieltjes treated this problem for measures supported on the half-line and developed a far reaching theory. The cases of the real line and of bounded intervals were studied only later by H. Hamburger (1920) and F. Hausdorff

(1920). Important early contributions have been made by R. Nevanlinna, M. Riesz, T. Carleman, M.H. Stone, and others.

A surprising feature of the moment problem theory is the connections and the close interplay with many branches of mathematics and the broad range of applications. H.J. Landau wrote in the introduction of an article in the AMS volume "Moments in Mathematics" [L2]: "The moment problem is a classical question in analysis, remarkable not only for its own elegance, but also for its extraordinary range of subjects theoretical and applied, which it has illuminated. From it flow developments in function theory, in functional analysis, in spectral representation of operators, in probability and statistics, in Fourier analysis and the prediction of stochastic processes, in approximation and numerical methods, in inverse problems and the design of algorithms for simulating physical systems." Looking at the developments of the multidimensional moment problem over the last two decades I would like to add real algebraic geometry, optimization, and convex analysis to Landau's list.

This book is an advanced text on the moment problem on $\mathbb{R}^d$ and its modern techniques. It is divided into four main parts, two devoted to one-dimensional moment problems and two others to multidimensional moment problems. In each group we distinguish between full and truncated moment problems. Though our main emphasis is on real moment problems we include short treatments of the moment problem on the unit circle and of the complex moment problem.

Here is a brief description of the four parts.

Part I deals with the *one-dimensional full moment problem* and develops important methods and technical tools such as orthogonal polynomials, Jacobi operators, and Nevanlinna functions in great detail. Basic existence and uniqueness criteria are obtained, but also a number of advanced results such as the Nevanlinna parametrizations for indeterminate Hamburger and Stieltjes problems, finite order solutions, Nevanlinna–Pick interpolation, Krein and Friedrichs approximants of Stieltjes solutions, and others are included.

Part II is about *one-dimensional truncated moment problems*. The truncated Hamburger and Stieltjes moment problems are treated and Gauss' quadrature formulas are derived. In the case of bounded intervals the classical theory of Markoff, Krein, and Akhiezer on canonical and principal measures, maximal masses, and the moment cone are studied. Part II also contains a self-contained digression to the trigonometric moment problem and some highlights of this theory (Schur algorithm, Verblunsky and Geronimus' theorems).

In Part III the *multidimensional full moment problem* on closed semi-algebraic subsets of $\mathbb{R}^d$ is investigated. Here real algebraic geometry and operator theory on Hilbert space are the main tools. For compact semi-algebraic sets the interplay between strict Positivstellensätze and the moment problem leads to satisfactory existence results for the moment problem. In the case of closed semi-algebraic sets existence criteria and determinacy are much more subtle. The fibre theorem and its applications and the multidimensional determinacy problem are investigated in detail. Further, we derive basic existence and determinacy results for the complex moment problem. The two main Chaps. 12 and 13 are the heart of Part III and also of the book. At the end of Part III we touch very briefly upon semidefinite programming and applications of moment methods to polynomial optimization.

Part IV gives an introduction to the *multidimensional truncated moment problem*. Existence theorems in terms of positivity and the flat extension theorem are derived. Fundamental technical tools (Hankel matrices, evaluation polynomials, the apolar scalar product for homogeneous polynomials) are developed and discussed in detail. A number of important special topics (the core variety, maximal masses, determinacy, Carathéodory numbers) are also studied. The multidimensional truncated moment problem is an active topic of present research. It is expected that convex analysis and algebraic geometry will provide new powerful methods and that the status of this area might essentially change in the coming decades. For this reason, we have not treated all recent developments; instead we have concentrated on basic results and concepts and on selected special topics.

All moment problems treated in this book deal with integral representations of linear functionals on a commutative unital algebra or on a (in most cases finite-dimensional) vector space of continuous functions on a locally compact space. Most of them are moment problems on certain ∗-semigroups. In two introductionary chapters, general results on integral representations of positive functionals are obtained and notions concerning moment problems for ∗-semigroups are developed. These results and notions will play an essential role throughout the whole book.

As mentioned above, the main focus of this book is on the four versions of the *scalar classical moment problem*. In the course of this we develop fundamental concepts and technical tools and we derive deep classical theorems and very recent results as well. Also, we present a number of new results and new proofs. In this book, we do *not* treat matrix moment problems, operator moment problems, infinite-dimensional moment problems, or noncommutative moment problems.

Apart from the two introductory chapters the parts of the book and a number of chapters can be read (almost) independently from each other. Sometimes a technical fact from another part is used in a proof; it can be filled easily. In order to be independent from previous chapters we have occasionally repeated some notation.

Several courses and seminars on moment problems can be built on this book by choosing appropriate material from various parts. Each course or seminar should probably start with the corresponding results from Sects. 1.1 and 1.2. For a one semester basic course on the one-dimensional moment problem this could be followed by Sects. 2.1–2.2, Chap. 3, and the core material of the first sections of Chaps. 4–6. A one semester advanced course on the multidimensional moment problem could be based on Sects. 11.1–11.3, 11.5–11.6, and selected material from Chaps. 12 and 13, avoiding technical subtleties. Here applications to optimization from Chap. 15 are optional. Chapters 8, 9, 10, and 16 are almost self-contained and could be used for special seminars on these topics. Most exercises at the end of each chapter should be solvable by active students; some of them contain additional results or information on the corresponding topics.

As the title of the book indicates, the real moment problem is the central topic. Our main emphasis is on a rigorous treatment of the moment problem, but also on important methods and technical tools. Often several proofs or approaches to fundamental results are given. For instance, Hamburger's theorem 3.8 is derived in Sect. 3.2 from Haviland's theorem, in Sect. 6.1 from the spectral theorem, and in Sect. 9.2 from the truncated moment problem. This is also true for a number of other results such as Stieltjes' theorem, Carleman's theorem, Markov's theorem, and the Positivstellensätze and moment problem theorems in Chap. 12. At various places explicit formulas in terms of the moments are provided even if they are not needed for treating the moment problem. Large parts of the material, especially in Parts III and IV, appear for the first time in a book and a number of results are new.

Necessary prerequisities for this book are a good working knowledge of measure and integration theory and of polynomials, but also the basics of holomorphic functions in one variable, functional analysis, convex sets, and elementary topology. With this background about two thirds of the book should be readable by graduate students. In the remaining third (more precisely, in parts of Chaps. 4–7 and Part III) Hilbert space operator theory and real algebraic geometry play an essential role. The short disgression into real algebraic geometry given in Sect. 12.1 covers all that is needed; for more details we refer to the standard books [Ms1] and [PD]. Elementary facts on unbounded symmetric or self-adjoint operators and the spectral theorem (multidimensional versions in Sect. 12.5 and 15.3) are used at various places; necessary notions and facts are collected in Appendix A.7. In Chap. 8 Friedrichs and Krein extensions of positive symmetric operators occur; they are briefly explained in Sect. 8.1. All operator-theoretic notions and results needed in this book can be found (for instance) in the author's Graduate Text [Sm9].

In large parts of the book results and techniques from other mathematical fields are used. In most cases we state or develop such results with reference to the literature at the places where they are needed. Further, I have added six appendices: on measure theory, on Pick functions and Stieltjes transforms, on positive semidefinite matrices, on locally convex topologies, on convex sets, and on Hilbert space operators. These appendices collect notions and facts that are used often and at different places of the text. For some results we have included proofs.

Some general notation is collected after the table of contents. Though I tried to retain standard terminology in most cases, occasionally I have made some changes, we hope for the better. For instance, instead of the term "moment matrix" I preferred "Hankel matrix" and denoted it by $H_n(L)$ or $H(L)$. Also there is some overlapping notation. While the symbol A is used for one of the four Nevanlinna functions in Chap. 7 (following standard notation), it denotes a matrix or a Hilbert space operator at other places. The meaning will be always clear from the context. The underlying algebras are usually denoted by sanserif letter such as A, B.

Continued fractions are avoided in this book (in Sect. 6.7 only the notion is briefly explained). Instead I have put my emphasis on operator-theoretic approaches, because I am convinced that these methods are more promising and powerful, in particular concerning the multidimensional case.

In writing this book I benefited very much from N.I. Akhiezer's classic [Ak] and B. Simon's article [Sim1], but also from standard books such as [BCR1], [KN], [KSt], [Ms1], [Sim3], [Chi1] and from the surveys [La2], [AK]. Applications and ramifications of moment problems and related topics are discussed and developed in the AMS volume [L2] and in the books [Ls2], [BW].

I feel unable to give precise credits for all results occuring in this book. In the Notes at the end of each chapter I have given (to the best of my knowledge) credits for some results, including the main theorems, and some hints to the literature and for further reading. In the bibliography I have listed some key classical papers.

I am indebted to B. Reznick, C. Scheiderer, and J. Stochel for valuable comments on parts of the manuscripts. I am grateful to Ph. di Dio for reading the whole text and for his helpful suggestions. Also, I should like to thank R. Lodh and A.-K. Birchley-Brun from Springer-Verlag for their help in publishing this book.

Leipzig, Germany   Konrad Schmüdgen
May 7, 2017

General Notations

Polynomials

$x^\alpha = x_1^{\alpha_1} \cdots x_d^{\alpha_d}$ for $\alpha = (\alpha_1, \cdots, \alpha_d) \in \mathbb{N}_0^d$

$\mathbb{R}_d[\underline{x}] = \mathbb{R}[x_1, \ldots, x_d]$

$\mathbb{R}_d[\underline{x}]_n = \{p \in \mathbb{R}_d[\underline{x}] : \deg(p) \le n\}$

$\mathcal{H}_{d,m}$ homogeneous polynomials from $\mathbb{R}_d[\underline{x}]$ of degree m

q_h homogenization of q

$\mathrm{Pos}(K) = \{f \in E : f(x) \ge 0, \ x \in K\}$, where $E \subseteq C(\mathcal{X}; \mathbb{R})$, $K \subseteq \mathcal{X}$

$\mathrm{Pos}(\mathsf{A}, \mathcal{K}) = \{p \in \mathsf{A} : p(x) \ge 0, \ x \in \mathcal{K}\}$

$\mathcal{Z}(p) = \{x \in \mathbb{R}^d : p(x) = 0\}$ for $p \in \mathbb{R}_d[\underline{x}]$

d is the number of variables and n, m denote the degrees of polynomials!

Moments, Moment Sequences, and Measures

$s_\alpha = \int x^\alpha \, d\mu$ α-the moment of μ

$s = (s_\alpha)$ moment sequence

$\mathfrak{s}(x), \mathfrak{s}_N(x)$ moment vector of the delta measure δ_x

$\mathcal{S}_{m+1}, \mathcal{S}, \mathcal{S}(\mathsf{A}, \mathcal{K})$ moment cones

$\mathcal{M}_+(\mathbb{R}^d) = \{\mu \in M_+(\mathbb{R}^d) : \int |x^\alpha| \, d\mu < +\infty$ for $\alpha \in \mathbb{N}_0^d\}$ Radon   measures with finite moments

$\mathcal{M}_s = \{\mu \in M_+(\mathbb{R}^d_\cdot) : s_\alpha = \int x^\alpha \, d\mu$ for $\alpha \in \mathbb{N}_0^d\}$ representing measures of s

L_s Riesz functional associated with s defined by $L_s(x^\alpha) = s_\alpha$

$H_n(s)$ finite Hankel matrix associated with s

$H(s)$ infinite Hankel matrix associated with s

$D_n(s) = \det H_n(s)$ Hankel determinant

Orthogonal Polynomials and Functions

p_n orthonormal polynomial of the first kind

P_n monic orthogonal polynomial of the first kind

q_n orthogonal polynomial of the second kind

Q_n monic orthogonal polynomial of the second kind

$\mathfrak{p}_z = (p_0(z), p_1(z), p_2(z), \ldots)$

$\mathfrak{q}_z = (q_0(z), q_1(z), q_2(z), \ldots)$

$\mathfrak{P}$ Pick functions

γ_s Friedrichs parameter

$A(z), B(z), C(z), D(z)$ entire functions in the Nevanlinna parametrization

Operators

X multiplication operator by the variable x

J infinite Jacobi matrix

$T = T_J$ Jacobi operator for the Jacobi matrix J

$\mathsf{d} = \{(c_0, c_1, \ldots, c_n, 0, 0, \ldots) : c_j \in \mathbb{C}, n \in \mathbb{N}_0\}$ finite complex sequences

$\langle p, q \rangle_s = L_s(p\overline{q})$ scalar product on $\mathbb{C}[x]$ associated with s

$\mathcal{H}_s$ Hilbert space completion of $(\mathbb{C}[x], \langle \cdot, \cdot \rangle_s)$

$X_1, \ldots, X_d$ multiplication operators by the variables $x_1, \ldots, x_d$

$\langle p, q \rangle_L = L(p\bar{q})$	scalar product on $\mathbb{C}_d[\underline{x}]$ associated with L
$\mathcal{H}_L$	Hilbert space completion of $(\mathbb{C}_d[\underline{x}], \langle \cdot, \cdot \rangle_L)$
π_L	GNS representation associated with L
T_F	Friedrichs extension of a positive symmetric operator T
T_K	Krein extension of a positive symmetric operator T
$\sigma(T)$	spectrum of T
$\rho(T)$	resolvent set of T
$\mathcal{D}(T)$	domain of T
$\mathcal{N}(T)$	null space of T
E_T	spectral measure of T

Real Algebraic Geometry

$T(\mathfrak{f})$	preordering generated by $\mathfrak{f}$, where $\mathfrak{f}$ is a finite subset of $\mathbb{R}_d[\underline{x}]$
$Q(\mathfrak{f})$	quadratic module generated by $\mathfrak{f}$
$\mathcal{K}(\mathfrak{f})$	semi-algebraic set defined by $\mathfrak{f}$
$\mathbb{R}[V]$	algebra of regular functions on a real algebraic set V
$\sum_n^2$	sum of squares p^2 of polynomials $p \in \mathbb{R}_d[\underline{x}]$, where $\deg p \leq n$
$\hat{\mathsf{A}}$	characters of a real unital algebra A
$\sum \mathsf{A}^2$	sum of squares a^2 of elements a of a real algebra A
$\mathsf{A}_+ = \{f \in \mathsf{A} : \chi(f) \geq 0, \ \chi \in \hat{\mathsf{A}}\}$	

Moment Functionals and Related Sets

$\mathcal{L}, \mathcal{L}(\mathsf{A}, \mathcal{K})$	cones of moment functionals
l_x	point evaluation functional at x
$L_g = L(g\cdot)$	localization of L at g
$\mathcal{M}_L = \{\mu \in M_+(\mathbb{R}^d) : L(p) = \int p(x)\, d\mu \text{ for } p \in \mathbb{R}_d[\underline{x}]\}$	representing measures of L
$\rho_L(t)$	maximal mass of representing measures of L at t
$H(L)$	Hankel matrix of L
$W(L), \mathcal{W}(L)$	set of atoms of representing measures of L
$V(L), \mathcal{V}(L)$	core variety of L
$\mathcal{N}_L = \{f \in \mathsf{A} : L(fg) = 0, \ g \in \mathsf{A}\}$	
$\mathcal{V}_L = \{t \in \mathbb{R}^d : f(t) = 0, \ f \in \mathcal{N}_L\}$	
$\mathcal{N}_+(L, \mathcal{K}) = \{f \in \mathrm{Pos}(\mathsf{A}, \mathcal{K}) : L(f) = 0\}$	
$\mathcal{V}_+(L, \mathcal{K}) = \{t \in \mathbb{R}^d : f(t) = 0, \ f \in \mathcal{N}_+(L, \mathcal{K})\}$	
$\mathcal{N}_+(L) = \mathcal{N}_+(L, \mathbb{R}^d)$	
$\mathcal{V}_+(L) = \mathcal{V}(L, \mathbb{R}^d)$	
$N_+(L) = \{f \in E_+ : L(p) = 0\}$	
$V_+(L) = \{x \in \mathcal{X} : f(x) = 0, \ f \in N_+(L)\}$	

Sets are denoted by braces such as $\{x_i : i \in I\}$, while sequences are written as $(x_n)_{n \in \mathbb{N}}$ or (x_n).

Chapter 1
Integral Representations of Linear Functionals

All variants of moment problems treated in this book deal with following problem:

Given a linear functional L on a vector space E of continuous functions on a locally compact Hausdorff space $\mathcal{X}$ and a closed subset K of $\mathcal{X}$, when does there exist a (positive) Radon measure μ supported on K such that

$$L(f) = \int_{\mathcal{X}} f(x)\, d\mu(x) \quad \text{for } f \in E?$$

Functionals of this form are called K-moment functionals or simply moment functionals when $K = \mathcal{X}$. In this chapter, we develop the underlying basic setup and introduce a number of general notions.

In Sect. 1.1, we prove various integral representation theorems for functionals on adapted spaces (Theorems 1.8, 1.12, and 1.14) and derive properties of the set of representing measures (Theorems 1.19, 1.20, and 1.21). Our existence theorems for *full* moment problems derived in Parts I and III are based on these results.

Section 1.2 is devoted to the case when E has finite dimension. Then, by the Richter–Tchakaloff theorem (Theorem 1.24), each moment functional has a finitely atomic representing measure. Strictly positive linear functionals (Theorem 1.30), determinate moment functionals (Theorem 1.36) and the cone of moment functionals are investigated. Further, we study the set of possible atoms of representing measures (Theorem 1.45) and prove that it coincides with the core variety (Theorem 1.49). The last subsection deals with extreme values of integrals $\int h\, d\mu$, where the measure μ has fixed moments (Theorems 1.50 and 1.52). The results obtained in Sect. 1.2 are useful for *truncated* moment problems treated in Parts II and IV, but most of them are also of interest in themselves.

Throughout this chapter, $\mathcal{X}$ denotes a **locally compact topological Hausdorff space**, E is a linear subspace of the space $C(\mathcal{X}; \mathbb{R})$ of real-valued continuous functions on $\mathcal{X}$, and L is a linear functional on E.

© Springer International Publishing AG 2017
K. Schmüdgen, *The Moment Problem*, Graduate Texts in Mathematics 277,
DOI 10.1007/978-3-319-64546-9_1

1.1   Integral Representations of Functionals on Adapted Spaces

Recall that $M_+(\mathcal{X})$ denotes the set of Radon measures on $\mathcal{X}$ and that in our terminology Radon measures are always nonnegative (see Appendix A.1).

1.1.1   Moment Functionals and Adapted Spaces

If C is a subset of E, the functional L is called *C-positive* if $L(f) \geq 0$ for $f \in C$. Set

$$E_+ := \{f \in E : f(x) \geq 0 \quad \text{for all} \quad x \in \mathcal{X}\}.$$

If μ is a measure from $M_+(\mathcal{X})$ such that $E \subseteq \mathcal{L}^1(\mathcal{X}, \mu)$, we define an E_+-positive linear functional L^μ on E by

$$L^\mu(f) = \int_{\mathcal{X}} f(x) d\mu(x), \quad f \in E. \tag{1.1}$$

The following two definitions introduce basic notions that will be used throughout this book. The terminology "moment functionals" will be clear later when we study moment functionals on examples of $*$-semigroups, see Sect. 2.3.1.

Definition 1.1 A linear functional L on E is a *moment functional* if there exists a measure $\mu \in M_+(\mathcal{X})$ such that $L = L^\mu$. Any such measure μ is called a *representing measure* of L. The set of all representing measures of L is

$$\mathcal{M}_L = \{\mu \in M_+(\mathcal{X}) : L = L^\mu\}.$$

A moment functional L is called *determinate* if it has a unique representing measure, or equivalently, if the set $\mathcal{M}_L$ is a singleton.

Definition 1.2 Let K be a closed subset of $\mathcal{X}$. A functional L on E is a *K-moment functional* if there exists a measure $\mu \in M_+(\mathcal{X})$ supported on K such that $L = L^\mu$. The set of such measures is

$$\mathcal{M}_{L,K} = \{\mu \in M_+(\mathcal{X}) : L = L^\mu \quad \text{and} \quad \operatorname{supp} \mu \subset K\}.$$

A K-moment functional L is said to be *K-determinate* if the set $\mathcal{M}_{L,K}$ is a singleton.

The aim of this section is to apply Choquet's concept of *adapted spaces* to the study of moment functionals.

Definition 1.3 For $f, g \in C(\mathcal{X}; \mathbb{R})$ we say that g *dominates* f if for any $\varepsilon > 0$ there exists a compact subset K_ε of $\mathcal{X}$ such that $|f(x)| \leq \varepsilon |g(x)|$ for all $x \in \mathcal{X} \backslash K_\varepsilon$.

Roughly speaking, g dominates f means that $|f(x)/g(x)| \to 0$ as $x \to \infty$.

We give a slight reformulation of the domination property and set

$$\mathcal{U} := \{\eta \in C_c(\mathcal{X}; \mathbb{R}) : 0 \le \eta(x) \le 1 \quad \text{for} \quad x \in \mathcal{X}\}. \tag{1.2}$$

Lemma 1.4 *For any* $f, g \in C(\mathcal{X}; \mathbb{R})$ *the following statements are equivalent:*

(i) *g dominates f.*
(ii) *For* $\varepsilon > 0$ *there is an* $\eta_\varepsilon \in \mathcal{U}$ *such that* $|f(x)| \le \varepsilon|g(x)| + |f(x)|\eta_\varepsilon(x)$, $x \in \mathcal{X}$.
(iii) *For* $\varepsilon > 0$ *there is an* $h_\varepsilon \in C_c(\mathcal{X}; \mathbb{R})$ *such that* $|f(x)| \le \varepsilon|g(x)| + h_\varepsilon(x)$, $x \in \mathcal{X}$.

Proof
    (i)→(ii) Choose $\eta_\varepsilon \in \mathcal{U}$ such that $\eta_\varepsilon(x) = 1$ on K_ε.
    (ii)→(iii) is clear by setting $h_\varepsilon := |f|\eta_\varepsilon$.
    (iii)→(i) Since $h_\varepsilon \in C_c(\mathcal{X}; \mathbb{R})$, the set $K_\varepsilon := \operatorname{supp} h_\varepsilon$ is compact and we have $|f(x)| \le \varepsilon|g(x)|$ for $x \in \mathcal{X}\backslash K_\varepsilon$.   □

Definition 1.5 A linear subspace E of $C(\mathcal{X}; \mathbb{R})$ is called *adapted* if the following conditions are satisfied:

(i) $E = E_+ - E_+$.
(ii) For each $x \in \mathcal{X}$ there exists an $f \in E_+$ such that $f(x) > 0$.
(iii) For each $f \in E_+$ there exists a $g \in E_+$ such that g dominates f.

Lemma 1.6 *If* E *is an adapted subspace of* $C(\mathcal{X}; \mathbb{R})$, *then for any* $f \in C_c(\mathcal{X}; \mathbb{R})_+$ *there exists a* $g \in E_+$ *such that* $g(x) \ge f(x)$ *for all* $x \in \mathcal{X}$.

Proof Let $x \in \mathcal{X}$. By Definition 1.5(ii) there exists a function $g_x \in E_+$ such that $g_x(x) > 0$. Multiplying g_x by some positive constant we get $g_x(x) > f(x)$. This inequality remains valid in some neighbourhood of x. By the compactness of $\operatorname{supp} f$ there are finitely many $x_1, \ldots, x_n \in \mathcal{X}$ such that $g(x) := g_{x_1}(x) + \cdots + g_{x_n}(x) > f(x)$ for $x \in \operatorname{supp} f$ and $g(x) \ge f(x)$ for all $x \in \mathcal{X}$.   □

1.1.2   Existence of Integral Representations

In the proof of Theorem 1.8 below we use the following extension theorem.

Proposition 1.7 *Let* E *be a linear subspace of a real vector space* F *and let* C *be a convex cone of* F *such that* $F = E + C$. *Then each* $(C \cap E)$-*positive linear functional* L *on* E *can be extended to a* C-*positive linear functional* $\tilde{L}$ *on* F.

Proof Let $f \in F$. We define

$$q(f) = \inf\{L(g) : g \in E, g - f \in C\}. \tag{1.3}$$

Since $F = E + C$, there exists a $g \in E$ such that $-f + g \in C$, so the corresponding set in (1.3) is not empty. It is easily seen that q is a sublinear functional and $L(g) = q(g)$

for $g \in E$. Therefore, by the Hahn–Banach theorem, there is an extension $\tilde{L}$ of L to F such that $\tilde{L}(f) \le q(f)$ for all $f \in F$.

Let $h \in C$. Setting $g = 0, f = -h$ we have $g - f \in C$, so that $q(-h) \le L(0) = 0$ by (1.3). Hence $\tilde{L}(-h) \le q(-h) \le 0$, so that $\tilde{L}(h) \ge 0$. Thus, $\tilde{L}$ is C-positive.   □

Most existence results on the moment problem derived in this book have their origin in the following theorem.

Theorem 1.8 *Suppose that E is an adapted subspace of $C(\mathcal{X}; \mathbb{R})$. For any linear functional $L : E \to \mathbb{R}$ the following are equivalent:*

(i) *The functional L is E_+-positive, that is, $L(f) \ge 0$ for all $f \in E_+$.*
(ii) *For each $f \in E_+$ there exists an $h \in E_+$ such that $L(f + \varepsilon h) \ge 0$ for all $\varepsilon > 0$.*
(iii) *L is a moment functional, that is, there exists a measure $\mu \in M_+(\mathcal{X})$ such that $L = L^\mu$.*

Proof The implications (i)→(ii) and (iii)→(i) are clear.

(ii)→(i) Let $f \in E_+$. Letting $\varepsilon \to 0$ in the inequality $L(f) + \varepsilon L(h) = L(f + \varepsilon h) \ge 0$, we get $L(f) \ge 0$.

(i)→(iii) We begin by setting

$$\tilde{E} := \{f \in C(\mathcal{X}, \mathbb{R}) : |f(x)| \le g(x), x \in \mathcal{X}, \quad \text{for some} \quad g \in E\}$$

and claim that $\tilde{E} = E + (\tilde{E})_+$. Obviously, $E + (\tilde{E})_+ \subseteq \tilde{E}$. Conversely, let $f \in \tilde{E}$. We choose a $g \in E_+$ such that $|f| \le g$. Then we have $f + g \in (\tilde{E})_+, -g \in E$ and $-f = -g + (g + f) \in E + (\tilde{E})_+$. That is, $\tilde{E} = E + (\tilde{E})_+$.

By Proposition 1.7, L can be extended to an $(\tilde{E})_+$-positive linear functional $\tilde{L}$ on $\tilde{E}$. We have $C_c(\mathcal{X}; \mathbb{R}) \subseteq \tilde{E}$ by Lemma 1.6. From the Riesz representation theorem (Theorem A.4) it follows that there is a measure $\mu \in M_+(\mathcal{X})$ such that $\tilde{L}(f) = \int f \, d\mu$ for $f \in C_c(\mathcal{X}; \mathbb{R})$. By Definition 1.5(i), $E = E_+ - E_+$. To complete the proof it therefore suffices to show that $f \in \mathcal{L}^1(\mathcal{X}, \mu)$ and $L(f) \equiv \tilde{L}(f) = \int f \, d\mu$ for $f \in E_+$.

Fix $f \in E_+$. Let $\mathcal{U}$ be the set defined by (1.2). For $\eta \in \mathcal{U}, f\eta \in C_c(\mathcal{X}; \mathbb{R})$ and hence $\tilde{L}(f\eta) = \int f\eta \, d\mu$. Using this fact and the $(\tilde{E})_+$-positivity of $\tilde{L}$, we derive

$$\int f d\mu = \sup_{\eta \in \mathcal{U}} \int f\eta \, d\mu = \sup_{\eta \in \mathcal{U}} \tilde{L}(f\eta) \le \tilde{L}(f) = L(f) < \infty, \qquad (1.4)$$

so that $f \in \mathcal{L}^1(\mathcal{X}, \mu)$.

By (1.4) it suffices to prove that $L(f) \le \int f d\mu$. From Definition 1.5(iii), there exists a $g \in E_+$ that dominates f. Then, by Lemma 1.4, for any $\varepsilon > 0$ there exists a function $\eta_\varepsilon \in \mathcal{U}$ such that $f \le \varepsilon g + f\eta_\varepsilon$. Since $f\eta_\varepsilon \le f$, we obtain

$$L(f) = \tilde{L}(f) \le \varepsilon \tilde{L}(g) + \tilde{L}(f\eta_\varepsilon) = \varepsilon L(g) + \int f\eta_\varepsilon d\mu \le \varepsilon L(g) + \int f d\mu.$$

Note that g does not depend on ε. Passing to the limit $\varepsilon \to +0$, we get $L(f) \leq \int f d\mu$. Thus, $L(f) = \int f d\mu$ which completes the proof of (iii).   □

If $\mathcal{X}$ is compact, then $C(\mathcal{X}; \mathbb{R}) = C_c(\mathcal{X}; R)$, so condition (iii) in Definition 1.5 is always fulfilled and can be omitted. But in this case we can obtain the desired integral representation of L more directly, as the following proposition shows.

Proposition 1.9 *Suppose that $\mathcal{X}$ is a compact Hausdorff space and E is a linear subspace of $C(\mathcal{X}; \mathbb{R})$ which contains a function e such that $e(x) > 0$ for $x \in \mathcal{X}$.*

Then each E_+-positive linear functional L on E is a moment functional, that is, there exists a measure $\mu \in M_+(\mathcal{X})$ such that $L(f) = \int f d\mu$ for $f \in E$.

Proof Set $F = C(\mathcal{X}; \mathbb{R})$ and $C = C(\mathcal{X}; \mathbb{R})_+$. Let $f \in F$. Since $\mathcal{X}$ is compact, f is bounded and e has a positive infimum. Hence there exists a $\lambda > 0$ such that $f(x) \leq \lambda e(x)$ on $\mathcal{X}$. Since $\lambda e - f \in C$ and $-\lambda e \in E$, $f = -\lambda e + (\lambda e - f) \in E + C$. Thus, $F = E + C$. By Proposition 1.7, L extends to a C-positive linear functional $\tilde{L}$ on F. By the Riesz representation theorem, $\tilde{L}$, hence L, can be given by a measure $\mu \in M_+(\mathcal{X})$.   □

Remark 1.10 If $\mathcal{X}$ is compact, the assumption $e(x) > 0$ on $\mathcal{X}$ in Proposition 1.9 implies that e is an interior point of the cone E_+. This is a standard assumption of the theory of ordered vector spaces which will be used in Theorem 1.26 below as well. In applications in this book we usually take $e = 1$.   ○

In the proof of Theorem 1.8 condition (iii) of Definition 1.5 was crucial. We give a simple example where this condition fails and L has no representing measure.

Example 1.11 Set $E := C_c(\mathbb{R}; \mathbb{R}) + \mathbb{R} \cdot 1$ and define a linear functional on E by

$$L(f + \lambda \cdot 1) := \lambda \quad \text{for} \quad f \in C_c(\mathbb{R}; \mathbb{R}), \ \lambda \in \mathbb{R},$$

where 1 denotes the constant function equal to 1 on $\mathbb{R}$. Then L is E_+-positive, but it is not a moment functional. (Indeed, since $L(f) = 0$ for $f \in C_c(\mathbb{R}; \mathbb{R})$, (1.1) would imply that the measure μ is zero. But this is impossible, because $L(1) = 1$.)   ○

The next result is called *Haviland's theorem*. For a closed subset K of $\mathbb{R}^d$ we set

$$\text{Pos}(K) = \{p \in \mathbb{R}_d[\underline{x}] : p(x) \geq 0 \quad \text{for all} \ x \in K\}.$$

Theorem 1.12 *Let K be a closed subset of $\mathbb{R}^d$ and L a linear functional on $\mathbb{R}_d[\underline{x}]$. The following statements are equivalent:*

(i) *$L(f) \geq 0$ for all $f \in \text{Pos}(K)$.*

(ii) *$L(f + \varepsilon 1) \geq 0$ for $f \in \text{Pos}(K)$ and $\varepsilon > 0$.*

(iii) *For any $f \in \text{Pos}(K)$ there is an $h \in \text{Pos}(K)$ such that $L(f + \varepsilon h) \geq 0$ for all $\varepsilon > 0$.*

(iv) *L is a K-moment functional, that is, there exists a measure $\mu \in M_+(\mathbb{R}^d)$ supported on K such that $f \in \mathcal{L}^1(\mathbb{R}^d, \mu)$ and $L(f) = \int_K f d\mu$ for all $f \in \mathbb{R}_d[\underline{x}]$.*

Proof Set $\mathcal{X} = K$. Then $E = \mathbb{R}_d[x]$ is an adapted subspace of $C(K, \mathbb{R})$. Indeed, condition (i) in Definition 1.5 follows from the relation $4p = (p+1)^2 - (p-1)^2$. Condition (ii) is trivial. If $p \in E_+$, then $g = (x_1^2 + \cdots + x_d^2)f$ dominates f, so condition (iii) is also fulfilled.

Note that the implications (iv)→(i)→(ii)→(iii) in Theorem 1.12 are obvious. The other assertions follow from Theorem 1.8. $\square$

Now suppose that A is a *finitely generated commutative real unital algebra*. We develop some notation and facts that will be used in Chaps. 12 and 13.

Definition 1.13 A *character* of A is an algebra homomorphism $\chi : \mathsf{A} \to \mathbb{R}$ satisfying $\chi(1) = 1$. The set of characters of A is denoted by $\hat{\mathsf{A}}$.

We fix a set $\{f_1, \ldots, f_d\}$ of generators of A. Then there exists a unique surjective unital algebra homomorphism $\pi : \mathbb{R}_d[x] \to \mathsf{A}$ such that $\pi(x_j) = f_j, j = 1, \ldots, d$. If $\mathcal{J}$ denotes the kernel of π, then $\mathcal{J}$ is an ideal of $\mathbb{R}_d[x]$ and A is isomorphic to the quotient algebra $\mathbb{R}_d[x]/\mathcal{J}$, that is,

$$\mathsf{A} \cong \mathbb{R}_d[x]/\mathcal{J}.$$

Each character χ of A is uniquely determined by the point $x_\chi := (\chi(f_1), \ldots, \chi(f_d))$ of $\mathbb{R}^d$. We identify χ with x_χ and write $f(x_\chi) := \chi(f)$ for $f \in \mathsf{A}$. Under this identification, $\hat{\mathsf{A}}$ becomes the real algebraic set

$$\hat{\mathsf{A}} = \mathcal{Z}(\mathcal{J}) := \{x \in \mathbb{R}^d : p(x) = 0 \text{ for } p \in \mathcal{J}\}. \tag{1.5}$$

Since $\mathcal{Z}(\mathcal{J})$ is closed in $\mathbb{R}^d$, $\hat{\mathsf{A}}$ is a locally compact Hausdorff space in the induced topology of $\mathbb{R}^d$ and elements of A can be considered as continuous functions on $\hat{\mathsf{A}}$. In the case $\mathsf{A} = \mathbb{R}_d[x]$ we can take $p_1 = x_1, \ldots, p_d = x_d$ and obtain $\hat{\mathsf{A}} = \mathbb{R}^d$.

For a closed subset K of $\hat{\mathsf{A}}$, we define

$$\mathrm{Pos}(K) = \{f \in \mathsf{A} : f(x) \geq 0 \text{ for } x \in K\}.$$

Then we have the following generalized version of Haviland's theorem for A.

Theorem 1.14 *Let* A *be a finitely generated commutative real unital algebra and* K *a closed subset of* $\hat{\mathsf{A}}$. *For a linear functional* L *on* A, *the following are equivalent:*

(i) $L(f) \geq 0$ *for all* $f \in \mathrm{Pos}(K)$.

(ii) $L(f + \varepsilon 1) \geq 0$ *for* $f \in \mathrm{Pos}(K)$ *and* $\varepsilon > 0$.

(iii) *For any* $f \in \mathrm{Pos}(K)$ *there is an* $h \in \mathrm{Pos}(K)$ *such that* $L(f + \varepsilon h) \geq 0$ *for all* $\varepsilon > 0$.

(iv) L *is a* K-*moment functional, that is, there exists a measure* $\mu \in M_+(\hat{\mathsf{A}})$ *supported on* K *such that* $\mathsf{A} \subseteq \mathcal{L}^1(\hat{\mathsf{A}}, \mu)$ *and* $L(f) = \int_{\hat{\mathsf{A}}} f \, d\mu$ *for all* $f \in \mathsf{A}$.

Proof (iv)→(i)→(ii)→(iii) are obviously satisfied. (iii)→(i) follows as in the proof of Theorem 1.8. It remains to prove the main implication (i)→(iv).

We define a linear functional $\tilde{L}$ on $\mathbb{R}_d[x]$ by $\tilde{L} = L \circ \pi$. Let $p \in \mathbb{R}_d[x]$ and set $f := \pi(p)$. Since π is an algebra homomorphism and $\pi(x_j) = f_j$, we have $p(f_1,\ldots,f_d) = p(\pi(x_1),\ldots,\pi(x_d)) = \pi(p(x)) = f$. Hence, for $x \in K$,

$$f(x) = \chi_x(f) = \chi_x(p(f_1,\ldots,f_d)) = p(\chi_x(f_1),\ldots,\chi_x(f_d)) = p(x_1,\ldots,x_d) = p(x).$$

Thus, if $p(x) \geq 0$ on K, then $\pi(p) = f \geq 0$ on K and hence $\tilde{L}(p) = L(f) \geq 0$ by assumption (i). Therefore, by Theorem 1.12 there exists a measure $\mu \in M_+(\mathbb{R}^d)$ such that $\operatorname{supp}\mu \subseteq K$ and $\tilde{L}(p) = \int_K p\,d\mu$ for all $p \in \mathbb{R}_d[x]$.

Let $f \in \mathsf{A}$. Then $f = \pi(p)$ for some $p \in \mathbb{R}[x]$. Using the definition of $\tilde{L}$ and the equality $p(x) = f(x)$ for $x \in K$ we obtain

$$L(f) = L(\pi(p)) = \tilde{L}(p) = \int_K p(x)\,d\mu(x) = \int_K f(x)\,d\mu(x).$$

This proves (iv). □

Remark 1.15 Let A and K be as in Theorem 1.14. Let $\mathcal{L}(K)$ denote the K-moment functionals on A. Then $\operatorname{Pos}(K)$ and $\mathcal{L}(K)$ are cones in A resp. in its dual satisfying

$$\operatorname{Pos}(K)^\wedge = \mathcal{L}(K) \quad \text{and} \quad \mathcal{L}(K)^\wedge = \operatorname{Pos}(K), \tag{1.6}$$

where the dual cones $\operatorname{Pos}(K)^\wedge$ and $\mathcal{L}(K)^\wedge$ are defined by (A.19). Thus, the cones $\operatorname{Pos}(K)$ and $\mathcal{L}(K)$ are dual to each other. Indeed, the first equality of (1.6) is Theorem 1.14(i)↔(ii), while the second follows from the bipolar theorem [Cw, Theorem V.1.8]. ∘

Example 1.16 Let A be the unital real algebra of functions on $\mathbb{R}$ generated by the two functions $f_1 = \frac{1}{1+x^2}$ and $f_2 = \frac{x}{1+x^2}$. Then the identity $(f_1 - \frac{1}{2})^2 + f_2^2 = \frac{1}{4}$ holds in A and the set $\hat{\mathsf{A}}$ is given by the points of the circle in $\mathbb{R}^2$ with center $(\frac{1}{2},0)$ and radius $\frac{1}{2}$. Note that the character $\chi \in \hat{\mathsf{A}}$ with $\chi(f_1) = \chi(f_2) = 0$ cannot be obtained by a point evaluation on $\mathbb{R}$. ∘

1.1.3 The Set of Representing Measures

In this subsection we use some facts concerning the vague convergence of measures, see Appendix A.1.

Definition 1.17 A *directed net* $(L_i)_{i\in I}$ *of linear functionals* on E is a net of linear functionals L_i defined on vector subspaces E_i, $i \in I$, of E such that $E = \cup_{i\in I} E_i$ and $E_i \subseteq E_j$ and $L_j\lceil E_i = L_i$ for all $i,j \in I, j \geq i$.

Clearly, for such a net $(L_i)_{i\in I}$ there is a unique well-defined linear functional L on E such that $L(f) = L_i(f)$ if $f \in E_i, i \in I$; we shall write $L = \lim_i L_i$.

Lemma 1.18 *Let E be an adapted subspace of $C(\mathcal{X};\mathbb{R})$ and let $(\mu_i)_{i\in I}$ be a net of measures $\mu_i \in M_+(\mathcal{X})$ which converges vaguely to $\mu \in M_+(\mathcal{X})$. Suppose that* *for each $i \in I$ there is a linear subspace $E_i \subseteq \mathcal{L}^1(\mathcal{X}, \mu_i)$ of E such that $(L^{\mu_i})_{i \in I}$ is a directed net of linear functionals on E, where $L^{\mu_i}(f) := \int f \, d\mu_i$ for $f \in E_i$. Then $\lim_i L^{\mu_i} = L^\mu$.*

Proof Let $L := \lim_i L^{\mu_i}$. Take $h \in E_+$. Since $E = \cup_{i \in I} E_i$, $h \in E_{i_0}$ for some $i_0 \in I$. Let $\eta \in \mathcal{U}$, where $\mathcal{U}$ is defined by (1.2). Then, by the definition of L,

$$\int h\eta \, d\mu_j \leq \int h \, d\mu_j = L^{\mu_j}(h) = L(h) \quad \text{for} \quad j \in I, j \geq i_0.$$

Since $h\eta \in C_c(\mathcal{X}, \mathbb{R})$, $\lim_i \int h\eta \, d\mu_i = \int h \, d\mu$ by the vague convergence, so that

$$\int h \, d\mu = \sup_{\eta \in \mathcal{U}} \int h\eta \, d\mu = \sup_{\eta \in \mathcal{U}} \lim_i \int h\eta \, d\mu_i \leq \int h \, d\mu_j = L(h), \quad j \geq i_0. \quad (1.7)$$

Thus, $h \in \mathcal{L}^1(\mathcal{X}; \mu)$. Therefore, since $E = E_+ - E_+$, it follows that $E \subseteq \mathcal{L}^1(\mathcal{X}; \mu)$.

Let $f \in E_+$. There exists a $g \in E_+$ which dominates f, that is, for any $\varepsilon > 0$ there is an $h_\varepsilon \in C_c(\mathcal{X}; \mathbb{R})$ such that $f \leq \varepsilon g + h_\varepsilon$, so that $f - h_\varepsilon \leq \varepsilon g$. Since the index set I is directed, there is an $i_0 \in I$ such that f and g are in E_{i_0}. Suppose that $i \in I, i \geq i_0$. Then, $L(f) = \int f \, d\mu_i$ by the definition of L and similarly $L(g) = \int g \, d\mu_i$. Using these facts and (1.7), applied first with $h = f$ and then twice with $h = g$, we derive

$$\left| L(f) - \int f \, d\mu \right| = L(f) - \int f \, d\mu = \int f \, d\mu_i - \int f \, d\mu$$

$$= \int (f - h_\varepsilon) \, d\mu_i - \int (f - h_\varepsilon) \, d\mu + \int h_\varepsilon \, d\mu_i - \int h_\varepsilon \, d\mu$$

$$\leq \varepsilon \left(\int g \, d\mu_i + \int g \, d\mu \right) + \int h_\varepsilon \, d\mu_i - \int h_\varepsilon \, d\mu$$

$$\leq \varepsilon (L(g) + L(g)) + \int h_\varepsilon \, d\mu_i - \int h_\varepsilon \, d\mu, \quad i \geq i_0.$$

Passing to the limit $\lim_i$ the preceding inequality yields $|L(f) - \int f \, d\mu| \leq 2 \, \varepsilon L(g)$. Now, letting $\varepsilon \to +0$, we get $L(f) = \int f \, d\mu$. Since $E = E_+ - E_+$, it follows that $L(h) = \int h \, d\mu$ for all $h \in E$. $\qquad\qquad\qquad\square$

Theorem 1.19 *Suppose that L is a moment functional on an adapted subspace E of $C(\mathcal{X}; \mathbb{R})$. The set $\mathcal{M}_L$ of representing measures is convex and vaguely compact.*

Proof It is clear that $\mathcal{M}_L$ is convex. Let $f \in C_c(\mathcal{X}, \mathbb{R})$. By Lemma 1.6, there exists a $g \in E$ such $|f(x)| \leq g(x)$ for $x \in \mathcal{X}$. Then

$$\sup_{\mu \in \mathcal{M}_L} \left| \int f \, d\mu \right| \leq \int g \, d\mu = L(g) < \infty.$$

Hence, by Theorem A.6, $\mathcal{M}_L$ is relatively vaguely compact in $M_+(\mathcal{X})$. It therefore suffices to show that $\mathcal{M}_L$ is closed in $M_+(\mathcal{X})$ with respect to the vague topology.

For let $(\mu_i)_{i \in I}$ be a net from $\mathcal{M}_L$ which converges vaguely to $\mu \in M_+(\mathcal{X})$. From Lemma 1.18, applied with $E_i = E$ for all i, it follows that $L^\mu = \lim_i L^{\mu_i}$. Since $\mu_i \in \mathcal{M}_L$, we have $L^{\mu_i} = L$. Hence $L^\mu = L$, that is, $\mu \in \mathcal{M}_L$. $\qquad\qquad\square$

Theorem 1.20 *Suppose that $\mathcal{X}$ is a locally compact Hausdorff space such that $C_0(\mathcal{X}; \mathbb{R})$, equipped with the supremum norm, is separable. Let E be an adapted subspace of $C(\mathcal{X}; \mathbb{R})$ and let $(\mu_n)_{n \in \mathbb{N}}$ be a sequence of measures $\mu_n \in M_+(\mathcal{X})$. Suppose for $n \in \mathbb{N}$ there is a linear subspace $E_n \subseteq \mathcal{L}^1(\mathcal{X}, \mu_i)$ of E such that $(L^{\mu_n})_{n \in \mathbb{N}}$ is a directed sequence of functionals on E according to Definition 1.17, where $L^{\mu_n}(f) = \int f \, d\mu_n, f \in E_n$.*

Then there exists a subsequence $(\mu_{n_k})_{k \in \mathbb{N}}$ that converges vaguely to a measure $\mu \in M_+(\mathcal{X})$ and we have $\lim_{k \to \infty} L^{\mu_{n_k}} = L^\mu$. If the functional L^μ is determinate, then the sequence $(\mu_n)_{n \in \mathbb{N}}$ itself converges vaguely to μ.

Proof We first show that the set $M := \{\mu_n : n \in \mathbb{N}\}$ is relatively compact in the vague topology. Let $f \in C_c(\mathcal{X}, \mathbb{R})$. By Lemma 1.6, $|f(x)| \leq g(x)$ for some $g \in E_+$. By Definition 1.17, there is a $k \in \mathbb{N}$ such that $g \in E_k$ and $\int g \, d\mu_n = \int g \, d\mu_k$ for $n \geq k$. Hence

$$\sup_{n \in \mathbb{N}} \left| \int f \, d\mu_n \right| \leq \sup_{n \in \mathbb{N}} \int g \, d\mu_n = \max_{j=1,\dots,k} \int g \, d\mu_j < \infty,$$

so M is relatively vaguely compact by Theorem A.6. Further, since $C_0(\mathcal{X}; \mathbb{R})$ is separable, so is its subset $C_c(\mathcal{X}; \mathbb{R})$ and the vague topology on $M_+(\mathcal{X})$ is metrizable by Proposition A.7. Therefore, $(\mu_n)_{n \in \mathbb{N}}$ has a subsequence $(\mu_{n_k})_{k \in \mathbb{N}}$ that converges vaguely to some measure $\mu \in M_+(\mathcal{X})$. Then $L^\mu = \lim_{k \to \infty} L^{\mu_{n_k}}$ by Lemma 1.18.

Suppose that L^μ is determinate. Let $(\mu_{m_k})_{k \in \mathbb{N}}$ be another subsequence which converges vaguely, say to $\tilde{\mu}$. From Definition 1.17 it follows that $\lim_{k \to \infty} L^{\mu_{m_k}}$ is independent of the subsequence. Therefore, by Lemma 1.18, $L^{\tilde{\mu}} = \lim_{k \to \infty} L^{\mu_{n_k}} = L^\mu$ and hence $\tilde{\mu} = \mu$, because L^μ is determinate. Thus, since each convergent subsequence has the same limit, the sequence $(\mu_n)_{n \in \mathbb{N}}$ itself converges vaguely. $\qquad\square$

The next result characterizes the extreme points of the convex set $\mathcal{M}_L$.

Theorem 1.21 *Let E be a linear subspace of $C(\mathcal{X}; \mathbb{R})$. Suppose that E contains a function e such that $e(x) \geq 1, x \in \mathcal{X}$. Let L be a moment functional on E. Then a measure $\mu \in \mathcal{M}_L$ is an extreme point of $\mathcal{M}_L$ if and only if E is dense in $L^1(\mathcal{X}; \mu)$.*

Proof First we assume that $\mu \in \mathcal{M}_L$ is not an extreme point of $\mathcal{M}_L$. Then there are measures $\mu_1, \mu_2 \in \mathcal{M}_L$, $\mu_1 \neq \mu_2$, such that $\mu = \frac{1}{2}(\mu_1 + \mu_2)$. Since $\mu_1 \leq 2\mu$, the Radon–Nikodym theorem (Theorem A.3) implies that $d\mu_1 = g \, d\mu$ for some $g \in L^\infty(\mathcal{X}; \mu)$. By $\mu_1 \neq \mu_2$, we have $1 - g \neq 0$. Using that $\mu_1, \mu \in \mathcal{M}_L$ we obtain

$$\int f(1-g) d\mu = \int f d\mu - \int fg d\mu = \int f d\mu - \int f d\mu_1 = 0, \quad f \in E.$$

Thus, $1-g \neq 0$ defines a nontrivial continuous linear functional on $L^1(\mathcal{X}; \mu)$ which annihilates E. Hence E is not dense in $L^1(\mathcal{X}; \mu)$.

Conversely, assume that E is not dense in $L^1(\mathcal{X}, \mu)$. Since $e \in E \subseteq L^1(\mathcal{X}, \mu)$, the measure μ is finite, so $L^\infty(\mathcal{X}, \mu)$ is the dual of $L^1(\mathcal{X}, \mu)$. Hence there is a function $g \in L^\infty(\mathcal{X}, \mu)$, $\|g\|_\infty = 1$, which annihilates E. Define $\mu_\pm$ by $d\mu_\pm = (1 \pm g)d\mu$. Since $\|g\|_\infty = 1$, μ_+ and μ_- are positive measures. From the relation $\int fg d\mu = 0$ for $f \in E$ it follows that μ_+ and μ_- are in V_L. Since $\mu = \frac{1}{2}(\mu_+ + \mu_-)$ and $\mu_+ \neq \mu_-$ (by $g \neq 0$), μ is not an extreme point of $\mathcal{M}_L$. $\qquad\square$

Remark 1.22

1. The proof and the assertion of Proposition 1.21 remain valid for Borel functions rather than continuous functions.
2. If $\mu \in \mathcal{M}_L$ and E is dense in $L^2(\mathcal{X}; \mu)$ in Theorem 1.21, E is also dense in $L^1(\mathcal{X}, \mu)$, so μ is an extreme point of $\mathcal{M}_L$. The converse is not true, that is, there are extreme points μ of $\mathcal{M}_L$ for which E is not dense in $L^2(\mathcal{X}, \mu)$. $\qquad\circ$

The following simple fact will often be used to localize the support of representing measures. We will apply it mainly to semi-algebraic sets and polynomials.

Proposition 1.23 *Let $f \in C(\mathcal{X}; \mathbb{R})$ and $\mu \in M_+(\mathcal{X})$. Suppose that $f(x) \geq 0$ for $x \in \mathcal{X}$ and $\int f(x)\, d\mu = 0$. Then*

$$\operatorname{supp}\mu \subseteq \mathcal{Z}(f) := \{x \in \mathcal{X} : f(x) = 0\}.$$

Proof Let $x_0 \in \mathcal{X}$. Suppose that $x_0 \notin \mathcal{Z}(f)$. Then $f(x_0) > 0$. Since f is continuous, there are an open neighbourhood U of x_0 and an $\varepsilon > 0$ such that $f(x) \geq \varepsilon$ on U. Then

$$0 = \int_{\mathcal{X}} f(x)\, d\mu \geq \int_U f(x)\, d\mu \geq \varepsilon\mu(U) \geq 0,$$

so that $\mu(U) = 0$. Therefore, since U is an open set containing x_0, it follows from the definition of the support that $x_0 \notin \operatorname{supp}\mu$. $\qquad\square$

1.2 Integral Representations of Functionals on Finite-Dimensional Spaces

In this section we suppose that E is a **finite-dimensional** linear subspace of $C(\mathcal{X}; \mathbb{R})$. We denote by l_x the point evaluation at $x \in \mathcal{X}$, that is, $l_x(f) = f(x)$ for $f \in E$.

For some results we will use the following condition:

There exists a function $e \in E$ such that $e(x) \geq 1$ for $x \in \mathcal{X}$. $\qquad$ (1.8)

1.2.1 Atomic Measures

Since E is finite-dimensional, it is natural to look for integral representations of functionals by finitely atomic measures. To simplify the formulation of the results we consider the zero measure as a 0-atomic measure.

Our first main result is the following *Richter–Tchakaloff theorem*.

Theorem 1.24 *Suppose that $(\mathcal{Y}, \mu)$ is a measure space, V is a finite-dimensional linear subspace of $\mathcal{L}^1_{\mathbb{R}}(\mathcal{Y}, \mu)$, and L^μ denotes the linear functional on V defined by $L^\mu(f) = \int f \, d\mu$, $f \in V$. Then there is a k-atomic measure $\nu = \sum_{j=1}^k m_j \delta_{x_j} \in M_+(\mathcal{Y})$, where $k \le \dim V$, such that $L^\mu = L^\nu$, that is,*

$$\int f d\mu = \int f d\nu \equiv \sum_{j=1}^k m_j f(x_j), \quad f \in V.$$

Proof Let C be the convex cone in the dual space V^* of all nonnegative linear combinations of point evaluations l_x, where $x \in \mathcal{Y}$, and let $\overline{C}$ be the closure of C in V^*. We prove by induction on $m := \dim V$ that $L^\mu \in C$.

First let $m = 1$ and $V = \mathbb{R} \cdot f$. Set $c := \int f d\mu$. If $c = 0$, then $\int (\lambda f) d\mu = 0 \cdot l_{x_1}(\lambda f)$, $\lambda \in \mathbb{R}$, for any $x_1 \in \mathcal{Y}$. Suppose now that $c > 0$. Then $f(x_1) > 0$ for some $x_1 \in \mathcal{Y}$. Hence $m_1 := cf(x_1)^{-1} > 0$ and $\int (\lambda f) d\mu = m_1 l_{x_1}(\lambda f)$ for $\lambda \in \mathbb{R}$. The case $c < 0$ is treated similarly.

Assume that the assertion holds for vector spaces of dimension $m-1$. Let V be a vector space of dimension m. By standard approximation of $\int f d\mu$ by integrals of simple functions it follows that $L^\mu \in \overline{C}$. We now distinguish between two cases.

Case 1: L^μ is an interior point of $\overline{C}$.

Since C and $\overline{C}$ have the same interior points (by Proposition A.33(ii)), we have $L^\mu \in C$.

Case 2: L^μ is a boundary point of $\overline{C}$.

Then there exists a supporting hyperplane F_0 for the cone $\overline{C}$ at L^μ (by Proposition A.34(ii)), that is, F_0 is a linear functional on V^* such that $F_0 \ne 0$, $F_0(L^\mu) = 0$ and $F_0(L) \ge 0$ for all $L \in \overline{C}$. Because V is finite-dimensional, there is a function $f_0 \in V$ such that $F_0(L) = L(f_0)$, $L \in V^*$. For $x \in \mathcal{Y}$, we have $l_x \in C$ and hence $F_0(l_x) = l_x(f_0) = f_0(x) \ge 0$. Clearly, $F_0 \ne 0$ implies that $f_0 \ne 0$. We choose an $(m-1)$-dimensional linear subspace V_0 of V such that $V = V_0 \oplus \mathbb{R} \cdot f_0$. Let us set $\mathcal{Z} := \{x \in \mathcal{Y} : f_0(x) = 0\}$. Since $0 = F_0(L^\mu) = L^\mu(f_0) = \int f_0 d\mu$ and $f_0(x) \ge 0$ on $\mathcal{Y}$, it follows that $f_0(x) = 0$ μ-a.e. on $\mathcal{Y}$, that is, $\mu(\mathcal{Y} \backslash \mathcal{Z}) = 0$. Now we define a measure $\tilde{\mu}$ on $\mathcal{Z}$ by $\tilde{\mu}(M) = \mu(M \cap \mathcal{Z})$. Then

$$L^\mu(g) = \int_{\mathcal{Y}} g \, d\mu = \int_{\mathcal{Z}} g \, d\mu = \int_{\mathcal{Z}} g \, d\tilde{\mu} = L^{\tilde{\mu}}(g) \quad \text{for } g \in V_0.$$

We apply the induction hypothesis to the functional $L^{\tilde{\mu}}$ on $V_0 \subseteq \mathcal{L}^1(\mathcal{Z}, \tilde{\mu})$. Since $L^{\mu} = L^{\tilde{\mu}}$ on V_0, there exist $\lambda_j \geq 0$ and $x_j \in \mathcal{Z}, j = 1, \ldots, n$, such that for $f \in V_0$,

$$L^{\mu}(f) = \sum_{j=1}^{n} \lambda_j l_{x_j}(f). \tag{1.9}$$

Since $f_0 = 0$ on $\mathcal{Z}$, hence $f_0(x_j) = 0$, and $L^{\mu}(f_0) = 0$, (1.9) holds for $f = f_0$ as well and so for all $f \in V$. Thus, $L^{\mu} \in C$. This completes the induction proof.

The set C is a cone in the m-dimensional real vector space V^*. Since $L^{\mu} \in C$, Carathéodory's theorem (Theorem A.35(ii)) implies that there is a representation (1.9) with $n \leq m$. This means that L^{μ} is the integral of the measure $\nu = \sum_{j=1}^{n} \lambda_j \delta_{x_j}$. Clearly, ν is k-atomic, where $k \leq n \leq m$. (We only have $k \leq n$, since some numbers λ_j in (1.9) could be zero and the points x_j are not necessarily different.)   □

The next corollary will be crucial for the study of truncated moment problems.

Corollary 1.25 *Each moment functional on a finite-dimensional linear subspace E of $C(\mathcal{X}; \mathbb{R})$ has a k-atomic representing measure ν, where $k \leq \dim E$. Further, if μ is a representing measure of L and $\mathcal{Y}$ is a Borel subset of $\mathcal{X}$ such that $\mu(\mathcal{X} \backslash \mathcal{Y}) = 0$, then all atoms of ν can be chosen from $\mathcal{Y}$.*

Proof Apply Theorem 1.24 to the measure space $(\mathcal{Y}, \mu \lceil \mathcal{Y})$ and $V = E$.   □

Let $\mathcal{L}$ denote the cone of all moment functionals on E. The first assertion of the following theorem is the counterpart of Proposition 1.9 for atomic measures.

Theorem 1.26 *Suppose that $\mathcal{X}$ is compact and condition (1.8) is satisfied.*

(i) *For each E_+-positive linear functional L_0 on E there exists a k-atomic measure $\mu \in M_+(\mathcal{X})$, $k \leq \dim E$, such that $L_0(f) = \int f \, d\mu$ for $f \in E$.*

(ii) *The cone $\mathcal{L}$ is closed in the norm topology of the dual space E^* of E.*

Proof Set $m := \dim E$. As in the preceding proof, let C denote the cone in the dual space E^* of all nonnegative linear combinations of point evaluations l_x at $x \in \mathcal{X}$. By Carathéodory's theorem (Theorem A.35(i)), each $L \in C$ is a combination of at most m point evaluations l_x, that is, L is of the form

$$L = \sum_{j=1}^{m} \lambda_j l_{x_j}, \quad \text{where} \quad \lambda_j \geq 0, \ x_j \in \mathcal{X}. \tag{1.10}$$

We prove that C is closed in E^*. Let $(L^{(n)})_{n \in \mathbb{N}}$ be a sequence from C converging to $L \in E^*$. Let $\lambda_j^{(n)}$ and $x_j^{(n)}$ be the corresponding numbers resp. points in (1.10). Now we use the function e occurring in condition (1.8) and obtain

$$L^{(n)}(e) = \sum_i \lambda_i^{(n)} e(x_i^{(n)}) \geq \lambda_j^{(n)}, \quad j = 1, \ldots, m.$$

Hence, since the converging sequence $(L^{(n)}(e))_{n\in\mathbb{N}}$ is bounded, so are the sequences $(\lambda_j^{(n)})_{n\in\mathbb{N}}, j = 1, \dots, m$. Because $\mathcal{X}$ is compact, there are subsequences $(\lambda_j^{(n_k)})_{k\in\mathbb{N}}$ and $(x_j^{(n_k)})_{k\in\mathbb{N}}$ which converge in $\mathbb{R}$ and $\mathcal{X}$, respectively. Passing to the limit in the representation (1.10) for $L^{(n_k)}$ yields $L \in C$. Thus, C is closed in E^*.

Next we show that $L_0 \in C$. Assume the contrary. Since C is closed, by the separation of convex sets (Theorem A.26 (ii)) there exists a linear functional F_0 on E^* such that $F_0(L_0) < 0$ and $F_0(L) \geq 0$ for $L \in C$. Because E is finite-dimensional, there is an $f_0 \in E$ such that $F_0(L) = L(f_0), L \in E^*$. For $x \in \mathcal{X}, l_x \in C$ and hence $F_0(l_x) = l_x(f_0) = f_0(x) \geq 0$, so that $f_0 \in E_+$. But $F_0(L_0) = L_0(f_0) < 0$ which contradicts the assumption that L_0 is E_+-positive. This proves that $L_0 \in C$.

Hence L_0 is of the form (1.10) and so the integral of the measure $\mu = \sum_{j=1}^m \lambda_j \delta_{x_j}$. Since some λ_j may be zero, μ is k-atomic with $k \leq m$. This proves (i).

Each functional of $\mathcal{L}$ is obviously E_+-positive and hence in C by (i). Thus $\mathcal{L} \subseteq C$. Since trivially $C \subseteq \mathcal{L}$, we have $\mathcal{L} = C$. Hence $\mathcal{L}$ is closed in E^*. This proves (ii). □

Let $\overline{\mathcal{L}}$ denote the closure of the cone $\mathcal{L}$ of moment functionals in E^*. The next proposition reformulates some results in terms of dual cones (see (A.19)).

Proposition 1.27 $\mathcal{L} \subseteq (E_+)^\wedge = \overline{\mathcal{L}}$ and $\mathcal{L}^\wedge = E_+ = (E_+)^{\wedge\wedge}$.
If $\mathcal{K}$ is compact and condition (1.8) is satisfied, then $\mathcal{L} = (E_+)^\wedge$.

Proof First we prove that $(E_+)^\wedge \subseteq \overline{\mathcal{L}}$. Assume to the contrary that there exists a functional $L_0 \in (E_+)^\wedge$ such that $L_0 \notin \overline{\mathcal{L}}$. Then, by the separation of convex sets (Theorem A.26(i)) applied to the closed (!) cone $\overline{\mathcal{L}}$ in E^*, there is a linear functional F on E^* such that $F(L_0) < 0$ and $F(L) \geq 0$ for $L \in \mathcal{L}^\wedge$. Since E is finite-dimensional, there exists an element $f \in E$ such that $F(L) = L(f), L \in E^*$. For $x \in \mathcal{X}, l_x \in \mathcal{L}$, so that $F(l_x) = l_x(f) = f(x) \geq 0$. Hence $f \in E_+$. Therefore, since $L_0 \in (E_+)^\wedge$, we get $F(L_0) = L_0(f) \geq 0$, a contradiction. Thus we have shown that $(E_+)^\wedge \subseteq \overline{\mathcal{L}}$.

Since $\mathcal{L} \subseteq (E_+)^\wedge \subseteq \overline{\mathcal{L}}$ as just proved and $(E_+)^\wedge$ is obviously closed, $(E_+)^\wedge = \overline{\mathcal{L}}$.

Because E_+ is closed in E, it follows from the bipolar theorem (Proposition A.32) that $E_+ = (E_+)^{\wedge\wedge}$. Hence $E_+ = ((E_+)^\wedge)^\wedge = (\overline{\mathcal{L}})^\wedge = \mathcal{L}^\wedge$.

Clearly, if $L \in \mathcal{L}$ and $p \in E_+$, then $L(p) \geq 0$. Therefore, $\mathcal{L} \subseteq (E_+)^\wedge$.

Now suppose that $\mathcal{K}$ is compact and (1.8) holds. Let $L \in (E_+)^\wedge$. Because L is E_+-positive, we have $L \in \mathcal{L}$ by Proposition 1.9 (or by Theorem 1.26(i)). That is, $(E_+)^\wedge \subseteq \mathcal{L}$. Since $\mathcal{L} \subseteq (E_+)^\wedge$ as noted in the preceding paragraph, $\mathcal{L} = (E_+)^\wedge$. □

1.2.2 Strictly Positive Linear Functionals

In this subsection we derive a number of results on strictly E_+-positive functionals that do not hold for E_+-positive functionals in general.

Definition 1.28 A linear functional L on E is called *strictly E_+-positive* if

$$L(f) > 0 \quad \text{for all } f \in E_+, f \neq 0. \tag{1.11}$$

The following simple lemma gives some equivalent conditions.

Lemma 1.29 *For a linear functional L on E the following are equivalent:*

(i) *L is strictly E_+-positive.*
(ii) *Let $\|\cdot\|$ be a norm on E. There exists a number $c > 0$ such that*

$$L(f) \geq c\|f\| \quad \text{for } f \in E_+. \tag{1.12}$$

(iii) *L is an inner point of the cone $(E_+)^\wedge$ in E^*.*

Proof (i)→(ii) Let c be the infimum of $L(f)$ on the set $U_+ = \{f \in E_+ : \|f\| = 1\}$. For $x \in \mathcal{X}$, the linear functional l_x is continuous on the finite-dimensional normed space $(E, \|\cdot\|)$, so there exists a number $C_x > 0$ such that

$$|l_x(f)| = |f(x)| \leq C_x\|f\| \quad \text{for } f \in E. \tag{1.13}$$

This implies that E_+ is closed. Therefore, U_+ is bounded and closed, hence compact, in the normed space $(E, \|\cdot\|)$. Since L is also continuous on $(E, \|\cdot\|)$, the infimum is attained at $f_0 \in U_+$. From $f_0 \in U_+$ we have $f \neq 0$ and $f \in E_+$, so that $L(f_0) = c > 0$ by (i). Hence $L(f) \geq c$ for $f \in U_+$. By scaling this yields (1.12).

(ii)→(iii) Let $\|\cdot\|^*$ denote the the dual norm of $\|\cdot\|$ on E^*. Suppose that $L' \in E^*$ and $\|L - L'\|^* < c$. Let $f \in E_+$ Then, using (1.12) we obtain

$$c\|f\| - L'(f) \leq L(f) - L'(f) \leq \|L - L'\|^* \|f\| \leq c\|f\|$$

so that $L'(f) \geq 0$. Thus, $L_1 \in (E_+)^\wedge$. This shows that L is an inner point of $(E_+)^\wedge$.

(iii)→(i) Let $f \in E_+, f \neq 0$. Then there exists an $x \in \mathcal{X}$ such that $f(x) > 0$. Since L is an inner point of $(E_+)^\wedge$, there exists an $\varepsilon > 0$ such that $(L - \varepsilon l_x) \in (E_+)^\wedge$. Hence $L(f) \geq \varepsilon f(x) > 0$. □

In general, E_+-positive functionals are not moment functionals (see Example 1.32), but *strictly E_+-positive* functionals are, as shown by the next theorem.

Theorem 1.30 *Let L be a strictly E_+-positive linear functional on E.*

(i) *L is a moment functional.*
(ii) *For each $x \in \mathcal{X}$, there is a finitely atomic measure $\nu \in \mathcal{M}_L$ such that $\nu(\{x\}) > 0$.*
(iii) *For each $x \in \mathcal{X}$, the infimum*

$$\kappa_L(x) := \inf\{L(f) : f \in E_+, f(x) = 1\} \tag{1.14}$$

is a minimum.

Proof

(i) The strictly E_+-positive functional L is an inner point of $(E_+)^\wedge = \overline{\mathcal{L}}$ by Lemma 1.29 and Proposition 1.27. Since the convex set $\mathcal{L}$ and its closure $\overline{\mathcal{L}}$ have the same inner points by Proposition A.33, L is also an inner point of $\mathcal{L}$. In particular, L belongs to $\mathcal{L}$, that is, L is a moment functional which proves (i).

Let $\|\cdot\|$ be a norm on E and $x \in \mathcal{X}$. By Lemma 1.29 and its proof, there exist positive numbers c and C_x such that (1.12) and (1.13) hold.

(ii) Choose $\varepsilon > 0$ such that $\varepsilon C_x < c$. Let $f \in E_+, f \neq 0$. By (1.12) and (1.13),

$$(L - \varepsilon l_x)(f) \geq c\|f\| - \varepsilon f(x) \geq (c - \varepsilon)\|f\| > 0.$$

Therefore, $L - \varepsilon l_x$ is also strictly E_+-positive and hence a moment functional by (i). Corollary 1.25 implies that $L - \varepsilon l_x$ has a finitely atomic representing measure μ. Then $\nu := (\mu + \varepsilon \delta_x) \in \mathcal{M}_L$ is finitely atomic and $\nu(\{x\}) \geq \varepsilon > 0$.

(iii) By (1.14), there is a sequence $(f_n)_{n \in \mathbb{N}}$ of functions $f_n \in E_+, f_n(x) = 1$, such that $\lim_n L(f_n) = \kappa_L(x)$. Since $\|f_n\| \leq c^{-1}L(f_n)$ by (1.12), $(f_n)_{n \in \mathbb{N}}$ is a bounded sequence in the finite-dimensional normed space $(E, \|\cdot\|)$. Hence it has a convergent subsequence (f_{n_k}). Set $f = \lim_k f_{n_k}$. From (1.13) it follows that $f \in E_+$ and $f(x) = 1$. Clearly, $L(f) = \lim_k L(f_{n_k}) = \kappa_L(x)$, so the infimum in (1.14) is attained at f. $\qquad\square$

Corollary 1.31 *Let L be a moment functional on E. Suppose that there exist a closed subset $\mathcal{U}$ of $\mathcal{X}$ and a measure $\mu \in \mathcal{M}_L$ such that $\operatorname{supp}\mu \subseteq \mathcal{U}$ and the following holds: If $f(x) \geq 0$ on $\mathcal{U}$ and $L(f) = 0$ for some $f \in E$, then $f = 0$ on $\mathcal{U}$.*

Then each $x \in \mathcal{U}$ is an atom of some finitely atomic representing measure of L.

Proof Being a closed subset of $\mathcal{X}$, $\mathcal{U}$ is a locally compact Hausdorff space. Since $\operatorname{supp}\mu \subseteq \mathcal{U}$, there is a well-defined (!) moment functional $\tilde{L}$ on the linear subspace $\tilde{E} := E{\restriction}\mathcal{U}$ of $C(\mathcal{U}; \mathbb{R})$ given by $\tilde{L}(f{\restriction}\mathcal{U}) = L(f), f \in E$. Clearly, $\tilde{L}$ is $(\tilde{E})_+$-positive on $\tilde{E}$. The condition on $\mathcal{U}$ implies that $\tilde{L}$ is even strictly $(\tilde{E})_+$-positive. Hence the assertion follows from Theorem 1.30(i), applied to $\tilde{L}$ and $\tilde{E} \subseteq C(\mathcal{U}, \mathbb{R})$. $\qquad\square$

Example 1.32 Set $\mathcal{X} = \mathbb{R}$ and $E = \operatorname{Lin}\{1, x, x^3, \dots, x^{2n+1}\}$, where $n \in \mathbb{N}_0$. Let L be a linear functional on E. Then $E_+ = \mathbb{R}_+ \cdot 1$. Therefore, if $L(1) > 0$, then L is strictly E_+-positive, so that $L \in \mathcal{L}$ by Theorem 1.30(i). If $L(1) = 0$ and $L \neq 0$, then L is E_+-positive, but $L \notin \mathcal{L}$.

In the case $E = \operatorname{Lin}\{x, x^3, \dots, x^{2n+1}\}$ we have $E_+ = \{0\}$, so each linear functional on E is strictly E_+-positive and hence $E^* = \mathcal{L}$. $\qquad\circ$

1.2.3 Sets of Atoms and Determinate Moment Functionals

In this subsection, L denotes a moment functional on E and we suppose that:

For each $x \in \mathcal{X}$ there exists a function $f_x \in E_+$ such that $f_x(x) > 0$. $\qquad$ (1.15)

Clearly, condition (1.8) implies (1.15).

Definition 1.33 $W(L) := \{x \in \mathcal{X} : \mu(\{x\}) > 0 \text{ for some } \mu \in \mathcal{M}_L \}.$

That is, $W(L)$ is the set of points of $\mathcal{X}$ that occur as an atom of some representing measure of L. This is an important set for the moment functionals L.

Lemma 1.34

(i) *If $L = 0$ and $\mu \in \mathcal{M}_L$, then $\mu = 0$.*
(ii) *The set $W(L)$ is not empty if and only if $L \neq 0$.*

Proof

(i) Since $L = 0$, we have $L(f_x) = 0$, where $f_x \in E_+$ is the function from assumption (1.15). Hence, by Proposition 1.23, $\operatorname{supp}\mu \subseteq \mathcal{Z}(f_x)$ for all $x \in \mathcal{X}$. But $\cap_{x \in \mathcal{X}} \mathcal{Z}(f_x)$ is empty by (1.15), hence is $\operatorname{supp}\mu$. Thus, $\mu = 0$.
(ii) By Corollary 1.25, L has a finitely atomic representing measure μ. If $L \neq 0$, then $\mu \neq 0$, so $W(L)$ is not empty. If $L = 0$, then $\mu = 0$ by (i), so $W(L)$ is empty. □

Lemma 1.35

(i) *Suppose that M is a Borel subset of $\mathcal{X}$ containing $W(L)$. Then $\mu(\mathcal{X}\backslash M) = 0$ for each measure $\mu \in \mathcal{M}_L$.*
(ii) *If $W(L)$ is finite, there exists a measure $\mu \in \mathcal{M}_L$ such that $\operatorname{supp}\mu = W(L)$.*
(iii) *If $W(L)$ is infinite, then for any $n \in \mathbb{N}$ there exists a measure $\mu \in \mathcal{M}_L$ such that $|\operatorname{supp}\mu| \geq n$.*

Proof The proofs of all three assertions make essentially use of Theorem 1.24.

(i) Assume $\mu(\mathcal{X}\backslash M) > 0$ to the contrary. We define functionals L_1 and L_2 by

$$L_1(f) = \int_M f(x)d\mu \quad \text{and} \quad L_2(f) = \int_{\mathcal{X}\backslash M} f(x)d\mu, \quad f \in E.$$

We apply Theorem 1.24 to the functional L_2 on the measure space $\mathcal{X}\backslash M$ with measure induced from μ. Therefore, L_2 has a finitely atomic representing measure μ_2 with atoms in $\mathcal{X}\backslash M$. The measure μ_1 on M which is induced from μ is a representing measure of L_1. Since $\mu \in \mathcal{M}_L$, we have $L = L_1 + L_2$ and hence $\tilde{\mu} := (\mu_1 + \mu_2) \in \mathcal{M}_L$. From $\mu(\mathcal{X}\backslash M) > 0$ and Lemma 1.34(i) it follows that $L_2 \neq 0$. Hence $\mu_2 \neq 0$, so there exists an atom $x_0 \in \mathcal{X}\backslash M$ of μ_2. Then, $\tilde{\mu}(\{x_0\}) \geq \mu_2(\{x_0\}) > 0$, that is, $x_0 \in W(L) \subseteq M$. This contradicts $x_0 \in \mathcal{X}\backslash M$.
(ii) Let $W(L) = \{x_1, \ldots, x_n\}$. By the definition of $W(L)$, for each x_i there is a measure $\mu_i \in \mathcal{M}_L$ such that $\mu_i(\{x_i\}) > 0$. Then $\mu := \frac{1}{n}(\mu_1 + \cdots + \mu_n) \in \mathcal{M}_L$ and $\mu(\{x_i\}) \geq \frac{1}{n}\mu_i(\{x_i\}) > 0$, $i = 1, \ldots, n$. Thus, $W(L) \subseteq \operatorname{supp}\mu$. Since $\operatorname{supp}\mu \subseteq W(L)$ by (i) applied with $M = W(L)$, we have $\operatorname{supp}\mu = W(L)$.
(iii) is proved by a similar reasoning as (ii). □

By Theorem 1.49 below, $W(L)$ is a closed subset, hence a Borel subset, of $\mathcal{X}$.

Recall from Definition 1.1 that a moment functional L is called *determinate* if it has a unique representing measure, or equivalently, if the set $\mathcal{M}_L$ is a singleton. The following theorem characterizes determinacy in terms of the size of the set $W(L)$.

Let $\{f_1, \dots, f_m\}$ be a vector space basis of E. We abbreviate

$$\mathfrak{s}(x) = (f_1(x), \dots, f_m(x))^T \in \mathbb{R}^m. \tag{1.16}$$

Note that $\mathfrak{s}$ is the moment vector of the delta measure δ_x at $x \in \mathcal{X}$.

Theorem 1.36 *For each moment functional L on E the following are equivalent:*

(i) *L is not determinate.*
(ii) *The set $\{\mathfrak{s}(x) : x \in W(L)\}$ is linearly dependent in $\mathbb{R}^m$.*
(iii) *$|W(L)| > \dim(E\lceil W(L))$.*
(iv) *L has a representing measure μ such that $|\operatorname{supp}\mu| > \dim(E\lceil W(L))$.*

Proof (i)→(iii) Assume to the contrary that $|W(L)| \leq \dim(E\lceil W(L))$ and let μ_1 and μ_2 be representing measures of L. Then, since $\dim E$ is finite, so is $W(L)$, say $W(L) = \{x_1, \dots, x_n\}$ with $n \in \mathbb{N}$. In particular, $W(L)$ is a Borel set. Hence, from Lemma 1.35(i) applied to $M = W(L)$, we deduce that $\operatorname{supp}\mu_i \subseteq W(L)$ for $i = 1, 2$, so there are numbers $c_{ij} \geq 0$ for $j = 1, \dots, n, i = 1, 2$, such that

$$L(f) = \int f(x)\, d\mu_i = \sum_{j=1}^{n} f(x_j) c_{ij} \quad \text{for } f \in E.$$

From the assumption $|W(L)| \leq \dim(E\lceil W(L))$ it follows that there are functions $f_j \in E$ such that $f_j(x_k) = \delta_{jk}$. Then $L(f_j) = c_{ij}$ for $i = 1, 2$, so that $c_{1j} = c_{2j}$ for all $j = 1, \dots, n$. Hence $\mu_1 = \mu_2$, so L is determinate. This contradicts (i).

(iii)→(ii) Since the cardinality of the set $\{\mathfrak{s}(x) : x \in W(L)\}$ exceeds the dimension of $E\lceil W(L)$ by (iii), the set must be linearly dependent.

(ii)→(i) Since the set $\{\mathfrak{s}(x) : x \in W(L)\}$ is linearly dependent, there are pairwise distinct points $x_1, \dots, x_k \in W(L)$ and real numbers $c_1, \dots, c_k$, not all zero, such that $\sum_{i=1}^{k} c_i \mathfrak{s}(x_i) = 0$. Then, since $\{f_1, \dots, f_m\}$ is a basis of E, we have

$$\sum_{i=1}^{k} c_i f(x_i) = 0 \quad \text{for } f \in E. \tag{1.17}$$

We choose for $x_i \in W(L)$ a representing measure μ_i of s such that $x_i \in \operatorname{supp}\mu_i$. Clearly, $\mu := \frac{1}{k}\sum_{i=1}^{k} \mu_i$ is a representing measure of s such that $\mu(\{x_i\}) > 0$ for all i. Let $\varepsilon = \min\{\mu(\{x_i\}) : i = 1, \dots, k\}$. For each number $c \in (-\varepsilon, \varepsilon)$,

$$\mu_c = \mu + c \cdot \sum_{i=1}^{k} c_i \delta_{x_i}$$

is a positive (!) measure which represents L by (1.17). By the choice of x_i, c_i, the signed measure $\sum_i c_i \delta_{x_i}$ is not the zero measure. Therefore, $\mu_c \neq \mu_{c'}$ for $c \neq c'$. This shows that L is not determinate.

(iii)↔(iv) If $W(L)$ is finite, by Lemma 1.35(ii) we can choose a $\mu \in \mathcal{M}_L$ such that supp $\mu = W(L)$. If $W(L)$ is infinite, Lemma 1.35(iii) implies that there exists a $\mu \in \mathcal{M}_L$ such that $|\text{supp}\,\mu| > \dim(E\lceil W(L))$. Thus, in both cases, (iii)↔(iv).   □

An immediate consequence of Theorem 1.36 is the following.

Corollary 1.37 *Let* $\mathcal{X}$*,* E*, and* L *be as in Theorem 1.36. If* $|W(L)| > \dim E$ *or if there is a measure* $\mu \in \mathcal{M}_L$ *such that* $|\text{supp}\,\mu| > \dim E$*, then* L *is not determinate. In particular,* L *is not determinate if* $W(L)$ *is an infinite set or if* L *has a representing measure of infinite support.*

Corollary 1.38 *Suppose that* L *is a strictly* E_+*-positive moment functional on* E*. Then* L *is determinate if and only if* $|\mathcal{X}| \leq \dim E$*.*

Proof From Theorem 1.30(ii), $\mathcal{X} = W(L)$. Therefore, $\dim E = \dim (E\lceil W(L))$. Hence the assertion follows from Theorem 1.36(iii) ↔(i).   □

1.2.4   Supporting Hyperplanes of the Cone of Moment Functionals

In this subsection, L is a moment functional on E.

Now we introduce two other important sets for the moment functional L.

Definition 1.39

$$N_+(L) := \{f \in E_+ : L(f) = 0\},$$

$$V_+(L) := \{x \in \mathcal{X} : f(x) = 0 \text{ for } f \in N_+(L)\}.$$

The next proposition contains some properties of these sets.

Proposition 1.40

(i) *For each measure* $\mu \in \mathcal{M}_L$ *we have* supp $\mu \subseteq V_+(L)$*.*
(ii) $W(L) \subseteq V_+(L)$*.*
(iii) *If* L *is strictly* E_+*-positive, then* $N_+(L) = \{0\}$ *and* $V_+(L) = W(L) = \mathcal{X}$*.*

Proof

(i) Since $E \subseteq C(\mathcal{X}; \mathbb{R})$, Proposition 1.23 implies that supp $\mu \subseteq V_+(L)$.
(ii) Let $x \in W(L)$. Then, by definition, $\mu(\{x\}) > 0$ for some $\mu \in \mathcal{M}_L$. Thus $x \in$ supp μ and hence $x \in V_+(L)$ by (i).
(iii) By Definition 1.28, $N_+(L) = \{0\}$. Hence $V_+(L) = \mathcal{X}$. From Theorem 1.30(ii) we obtain $W(L) = \mathcal{X}$.   □

Example 1.41 (An example for which $W(L) \neq V_+(L)$*)* Let $\mathcal{X}$ be the subspace of $\mathbb{R}^2$ consisting of three points $(-1,0), (0,0), (1,0)$ and two lines $\{(t,1): t \in \mathbb{R}\}$ and $\{(t,-1): t \in \mathbb{R}\}$ and let $E = \mathbb{R}[x_1, x_2]_2\lceil \mathcal{X}$. We easily verify that the restriction map

$f \mapsto f \lceil \mathcal{X}$ on $\mathbb{R}[x_1, x_2]_2$ is injective; for simplicity we write f instead of $f \lceil \mathcal{X}$ for $f \in \mathbb{R}[x_1, x_2]_2$.

We consider the moment functional L defined by

$$L(f) = f(-1, 0) + f(1, 0), f \in E. \tag{1.18}$$

We show that $N_+(L) = \{x_2(bx_2 + c) : |c| \leq b, \, b, c \in \mathbb{R}\}$. It is obvious that these polynomials are in $N_+(L)$. Conversely, let $f \in N_+(L)$. Then $f(-1, 0) = f(1, 0) = 0$, so that $f = x_2(ax_1 + bx_2 + c) + d(1 - x_1^2)$, with $a, b, c, d \in \mathbb{R}$. Further, $d = f(0, 0) \geq 0$. From $f(t, \pm 1) \geq 0$ for all $t \in \mathbb{R}$ we conclude that $d = 0$ and $|c| \leq b$.

The zero set $V_+(L)$ of $N_+(L)$ is the intersection of $\mathcal{X}$ with the x_1-axis, that is, $V_+(L) = \{(-1, 0), (0, 0), (1, 0)\}$. Let μ be an arbitrary representing measure of L. Then, since μ is supporting on $V_+(L)$, there are numbers $\alpha, \beta, \gamma \geq 0$ such that $\mu = \alpha \delta_{(-1,0)} + \beta \delta_{(0,0)} + \gamma \delta_{(1,0)}$. By (1.18), we have $L(x_1) = 0 = \int x_1 d\mu = -\alpha + \gamma$ and $L(x_1^2) = 2 = \int x_1^2 d\mu = \alpha + \gamma$, which implies that $\alpha = \gamma = 1$. Therefore, since $L(1) = 2 = \int 1 d\mu = \alpha + \beta + \gamma$, it follows that $\beta = 0$. Hence, $\mu(\{(0, 0)\}) = 0$, so that $(0, 0) \notin W(L)$. Thus, $W(L) \neq V_+(L)$. The functional L on E is determinate. ∘

If $L = 0$, then $\mathcal{N}_+(L) = E$ and $V_+(L) = \emptyset$. If $L \neq 0$ is a moment functional, then $V_+(L) \neq \emptyset$, since it contains the support of all representing measures.

Proposition 1.42

(i) *Let* $p \in N_+(L), p \neq 0$. *Then* $\varphi_p(L') = L'(p), L' \in E^*$, *defines a supporting functional* φ_p *of the cone* $\mathcal{L}$ *at* L. *Each supporting functional of* $\mathcal{L}$ *at* L *is of this form.*

(ii) *L is a boundary point of the cone* $\mathcal{L}$ *if and only if* $N_+(L) \neq \{0\}$.

(iii) *L is an inner point of the cone* $\mathcal{L}$ *if and only if* $N_+(L) = \{0\}$.

Proof

(i) Let $p \in N_+(L), p \neq 0$. Since $p \in E_+$, the functional φ_p is $\mathcal{L}$-positive. Further, $\varphi_p(L) = L(p) = 0$. Since $p \neq 0$, there exists an $x \in \mathcal{X}$ such that $p(x) \neq 0$. Then $\varphi_p(l_x) = l_x(p) = p(x) \neq 0$, so that $\varphi_p \neq 0$. This shows that φ_p is a supporting functional of $\mathcal{L}$ at L.

Conversely, let φ be a supporting functional of $\mathcal{L}$ at L. Since E is finite-dimensional, we have $(E^*)^* \cong E$, so $\varphi = \varphi_p$ for some $p \in E$. For $x \in \mathcal{X}, l_x \in \mathcal{L}$ and hence $\varphi(l_x) = l_x(p) = p(x) \geq 0$. That is, $p \in E_+$. From $\varphi(L) = L(p) = 0$ we obtain $p \in N_+(L)$. Clearly, $p \neq 0$, because $\varphi \neq 0$.

(ii) By Proposition A.34(ii), L is a boundary point of $\mathcal{L}$ if and only if there is a supporting functional of $\mathcal{L}$ at L. Hence (i) implies the assertion of (ii).

(iii) follows from (ii), since L is inner if and only if it is not a boundary point. □

A nonempty *exposed face* (see Definition A.36) of a cone C in a finite-dimensional real vector space is a subcone $F = \{f \in C : \varphi(f) = 0\}$ for some functional $\varphi \in C^\wedge$.

By Proposition 1.27, $\mathcal{L}^\wedge = E_+$, that is, each functional $\varphi \in \mathcal{L}^\wedge$ is of the form $\varphi_p(L') = L'(p), L' \in E^*$, for some $p \in E_+$. Hence the nonempty exposed faces of the cone $\mathcal{L}$ in E^* are exactly the sets

$$F_p := \{L' \in \mathcal{L} : \varphi_p(L') \equiv L'(p) = 0\}, \quad \text{where } p \in E_+. \tag{1.19}$$

Let $L \in \mathcal{L}$. Since $\mathcal{L} \subseteq (E_+)^\wedge$, $N_+(L)$ is an exposed face of the cone E_+. If $\mathcal{X}$ is compact and condition (1.8) holds, then $(E_+)^\wedge = \mathcal{L}$ by Proposition 1.27 and hence each exposed face of E_+ is of this form. Thus, in this case the subcones $N_+(L)$ are precisely the nonempty exposed faces of the cone E_+.

Proposition 1.43 *Let $L \in \mathcal{L}$ and $x \in \mathcal{K}$. Suppose that $\mathcal{X}$ is compact and condition (1.8) is fulfilled. Then the supremum*

$$c_L(x) := \sup \{c \in \mathbb{R} : (L - cL_x) \in \mathcal{L}\} \tag{1.20}$$

is attained and we have $c_L(x) \le e(x)^{-1}L(e)$. Further, $L - c_L(x)l_x$ is a boundary point of $\mathcal{L}$ and there exists a $p \in E_+, p \ne 0$, such that $L(p) = c_L(x)p(x)$.

Proof If $(L - cl_x) \in \mathcal{L}$, then $(L - cl_x)(e) \ge 0$, so that $c \le e(x)^{-1}L(e)$. Therefore, $c(x) \le e(x)^{-1}L(e)$.

Since $\mathcal{K}$ is compact and (1.8) holds, the cone $\mathcal{L}$ is closed in E^* by Theorem 1.26. There is a sequence $(c_n)_{n\in\mathbb{N}}$ such that $(L - c_n l_x) \in \mathcal{L}$ for all n and $\lim_n c_n = c_L(x)$. Then $(L - c_n l_x) \to (L - c_L(x)l_x)$. Since $\mathcal{L}$ is closed, $(L - c_L(x)l_x) \in \mathcal{L}$, that is, the supremum (1.20) is attained.

The definition of $c_L(x)$ implies that $L - c_L(x)l_x$ is a boundary point of $\mathcal{L}$. Therefore, by Proposition 1.42(ii), there exists a $p \in \mathcal{N}_+(L - c_L(x)l_x), p \ne 0$. Then $L(p) = c_L(x)p(x)$. $\quad\square$

Proposition 1.43 is a tool to reduce problems on moment functionals to boundary functionals. If L is an inner point of $\mathcal{L}$, it is clear that $c_L(x) > 0$ for all $x \in \mathcal{X}$.

Proposition 1.44 *For each moment functional L there exists a $p \in N_+(L)$ such that*

$$V_+(L) = \mathcal{Z}(p) := \{x \in \mathcal{X} : p(x) = 0\}.$$

Proof First let L be an inner point of $\mathcal{L}$. Then, $N_+(L) = \{0\}$ by Proposition 1.42(iii), hence $\mathcal{V}_+(L) = \mathcal{X}$; so we can set $p = 0$.

Now let L be a boundary point of $\mathcal{L}$. Then, $N_+(L) \ne \{0\}$. Let $p_1, \ldots, p_k$ be a maximal linearly independent subset of $N_+(L)$. We prove that $p := p_1 + \cdots + p_k$ has the desired properties. Obviously, $p \in N_+(L)$ and $V_+(L) \subseteq \mathcal{Z}(p)$ by definition. Suppose that $x \in \mathcal{Z}(p)$. Let $q \in N_+(L)$. By the choice of $p_1, \ldots, p_k$, the function q is a linear combination $q = \sum_i \lambda_i p_i$ with $\lambda_i \in \mathbb{R}$. From $p(x) = p_1(x) + \cdots + p_k(x) = 0$ and $p_j(x) \ge 0$ (by $q_j \in N_+(L)$) it follows that we have $p_i(x) = 0$ for all i and therefore $q(x) = 0$. Since $q \in N_+(L)$ was arbitrary, $x \in V_+(L)$. $\quad\square$

For inner points of $\mathcal{L}$ we have $W(L) = \mathcal{X}$ (by Lemma 1.29 and Theorem 1.30(ii)) and hence $W(L) = V_+(L)$. In general, $W(L) \ne V_+(L)$ as shown by Example 1.41. The next theorem characterizes those boundary points for which $W(L) = V_+(L)$.

Theorem 1.45 *Let L be a boundary point of $\mathcal{L}$. Then $W(L) = V_+(L)$ if and only if L lies in the relative interior of an exposed face of the cone $\mathcal{L}$.*

Proof First assume that $W(L) = V_+(L)$. By Proposition 1.44, there exists a $p \in N_+(L)$ such that $\mathcal{Z}(p) = V_+(L)$. Since $\varphi_p \in \mathcal{L}^\wedge$ and $L(p) = 0$, the set F_p defined by (1.19) is an exposed face of $\mathcal{L}$ containing L. We will prove that L is an inner point of F_p.

If $x \in \mathcal{Z}(p)$, then $l_x(p) = p(x) = 0$, so that $l_x \in F_p$. We choose a maximal number of points $x_1, \ldots, x_k \in \mathcal{Z}(p)$ such that the point evaluations $l_{x_1}, \ldots, l_{x_k}$ on E are linearly independent. Since $\mathcal{Z}(p) = V_+(L) = W(L)$, we have $x_i \in W(L)$. Therefore, for each i there exists a representing measure μ_i of L such that x_i is an atom of μ_i. Then $\mu := \frac{1}{k} \sum_{i=1}^k \mu_i$ is also a representing measure of L and each x_i is an atom of μ.

Suppose that $L' \in F_p$. Let $\mu' = \sum_j c_j \delta_{y_j}$ be a finitely atomic representing measure of L'. Since $L'(p) = 0$, we have $\operatorname{supp} \mu' \in \mathcal{Z}(p)$, so that $y_j \in \mathcal{Z}(p)$ for all j. Hence, by the choice of the points x_i, $L' = \sum_j c_j l_{y_j}$ is in the span of $l_{x_1}, \ldots, l_{x_k}$. That is, there are reals $\lambda_1, \ldots, \lambda_k$ such that $L' = \sum_{i=1}^k \lambda_i l_{x_i}$. Since μ has positive masses at all points x_i, there exists an $\varepsilon > 0$ such that $\mu + c \cdot \mu'$ is a positive (!) measure for $c \in (-\varepsilon, \varepsilon)$. Its moment functional is $(L + cL') \in \mathcal{L}$. Therefore, $(L + cL') \in F_p$, since $L, L' \in F_p$. This shows that L is an inner point of F_p.

Conversely, suppose that L is an inner point of an exposed face F of $\mathcal{L}$. Then F is of the form (1.19) for some $p \in N_+(L)$. Let $x \in V_+(L)$. Then, since $p \in N_+(L)$, $p(x) = 0$ and hence $l_x \in F$. Since L in an inner point of F, there is a $c > 0$ such that $L' := L - c \cdot l_x \in F$. If μ' is a representing measure of L', then $\mu = \mu' + c \cdot \delta_x$ is a representing measure of L and $\mu(\{x\}) \geq c > 0$, so that $x \in W(L)$. Since $W(L) \subseteq V_+(L)$ always holds by Proposition 1.40(ii), we get $W(L) = V_+(L)$. $\square$

Most results of this section are stated in terms of moment functionals. Sometimes it is convenient to work instead with moment sequences and the moment cone.

Fix a vector space basis $\{f_1, \ldots, f_m\}$ of E. Let S denote the cone of moment sequences $s = (s_j)_{j=1}^m$, that is, of sequences of the form $s_j = \int f_j(x) \, d\mu$ for some Radon measure μ on $\mathcal{X}$. The linear map $s \mapsto L_s$ is a homeomorphism of $S \subseteq \mathbb{R}^m$ onto $\mathcal{L} \subseteq E^*$. Using this simple fact notions and results on $\mathcal{L}$ can be reformulated in terms of the cone S and vice versa. We encourage the reader to carry this out. As a sample, we describe the supporting hyperplanes and exposed faces of the cone S.

The vector $\mathfrak{s}(x) = (f_1(x), \ldots, f_m(x))^T \in \mathbb{R}^m$ from (1.16) is just the moment vector of the delta measure $\delta_x, x \in \mathcal{X}$. Let $\langle \cdot, \cdot \rangle$ be the standard scalar product on $\mathbb{R}^m$. For $v = (v_1, \ldots, v_m)^T \in \mathbb{R}^m$ we abbreviate

$$f_v := v_1 f_1 + \cdots + v_m f_m.$$

Then $E = \{f_v : v \in \mathbb{R}^m\}$. Since $f_v(x) = \langle v, \mathfrak{s}(x) \rangle$ for $x \in \mathcal{X}$, we have

$$E_+ = \{f_v : v \in \mathbb{R}^m, \langle v, \mathfrak{s}(x) \rangle \geq 0 \text{ for } x \in \mathcal{X}\}.$$

It is easily verified that $L_s(f_v) = \langle v, s \rangle$ for $v, s \in \mathbb{R}^m$. This implies that for each linear functional h on E there is a unique vector $u \in \mathbb{R}^m$ such that

$$h_u(f_v) := h(f_v) = \langle v, u \rangle, \quad v \in \mathbb{R}^m. \tag{1.21}$$

Set $N_+(s) := N_+(L_s)$. For $u \in \mathbb{R}^m$ we define

$$h_u(t) := \langle u, t \rangle, \ t \in \mathbb{R}^m, \quad \text{and} \quad H_u := \{x \in \mathbb{R}^m : \langle u, x \rangle = 0\}.$$

Lemma 1.46 *Let $u \in \mathbb{R}^m$ and $s \in \mathcal{S}$. Then $f_u \in N_+(s)$ if and only if $h_u(s) = 0$ and $h_u(t) \geq 0$ for $t \in \mathcal{S}$.*

Proof Let $t \in \mathcal{S}$. Since t has a finitely atomic representing measure, we can write $t = \sum_i c_i \mathfrak{s}(x_i)$, where $x_i \in \mathcal{X}$ and $c_i \geq 0$ for all i. Using (1.21) we compute

$$L_t(f_u) = \langle u, t \rangle = h_u(t) = \sum_i c_i \langle u, \mathfrak{s}(x_i) \rangle = \sum_i c_i f_u(x_i).$$

Using the definition of $N_+(s)$ the assertions follow at once from this equality. $\square$

Let $s \in \mathcal{S}$. By Lemma 1.46, the functional h_u, $u \in \mathbb{R}^m$, is a supporting hyperplane of $\mathcal{S}$ at s if and only if $f_u \in N_+(s)$ and $u \neq 0$. Each supporting hyperplane of $\mathcal{S}$ at s is of the form. Thus, s is a boundary point of $\mathcal{S}$ if and only if $N_+(s) \neq \{0\}$.

Further, $H_u \cap \mathcal{S}$ is an exposed face of $\mathcal{S}$ if and only if $f_u \in N_+(s)$. All nonempty exposed faces of $\mathcal{S}$ are of this form.

1.2.5   The Set of Atoms and the Core Variety

Throughout this section, L is a moment functional on E such that $L \neq 0$.

We define inductively linear subspaces $N_k(L)$, $k \in \mathbb{N}$, of E and subsets $V_j(L)$, $j \in \mathbb{N}_0$, of $\mathcal{X}$ by $V_0(L) = \mathcal{X}$,

$$N_k(L) := \{p \in E : L(p) = 0, \ p(x) \geq 0 \ \text{for} \ x \in V_{k-1}(L)\}, \tag{1.22}$$

$$V_j(L) := \{x \in \mathcal{X} : p(x) = 0 \ \text{for} \ p \in N_j(L)\}. \tag{1.23}$$

Definition 1.47 The *core variety* $V(L)$ of the moment functional L on E is

$$V(L) := \bigcap_{j=0}^{\infty} V_j(L). \tag{1.24}$$

Since $L \neq 0$, it follows from Proposition 1.48(ii) that $V_k(L) \neq \emptyset$ for all $k \in \mathbb{N}$. Note that $N_1(L) = N_+(L)$ and $V_1(L) = V_+(L)$, where $N_+(L)$ and $V_+(L)$ have been introduced in Definition 1.39.

Some properties of these sets are contained in the next proposition. Assertion (1.25) is the crucial step in the proof of Theorem 1.49 below.

Proposition 1.48

(i) $N_{j-1}(L) \subseteq N_j(L)$ and $V_j(L) \subseteq V_{j-1}(L)$ for $j \in \mathbb{N}$.
(ii) If μ is representing measure of L, then $\operatorname{supp} \mu \subseteq V(L)$.
(iii) There exists a $k \in \mathbb{N}_0$ such that

$$\mathcal{X} = V_0(L) \supsetneqq V_1(L) \supsetneqq \ldots \supsetneqq V_k(L) = V_{k+j}(L) = V(L), \quad j \in \mathbb{N}. \tag{1.25}$$

Proof

(i) follows easily by induction; we omit the details.

Let $j \in \mathbb{N}_0$. We denote by $E^{(j)} := E \lceil V_j(L) \subseteq C(V_j(L); \mathbb{R})$ the vector space of functions $f_j := f \lceil V_j(L)$, where $f \in E$, and by $\mathcal{L}^{(j)}$ the corresponding cone of moment functionals on $E^{(j)}$. Clearly, $E^{(0)} = E$ and $L = L^{(0)}$. Note that in general dim $E^{(j)}$ is smaller than dim E.

(ii) We prove by induction that $\operatorname{supp} \mu \subseteq V_j(L)$ for $j \in \mathbb{N}_0$. For $j = 0$ this is obvious. Assume that this holds for some j. Then

$$L^{(j)}(f_j) = \int_{V_j(L)} f(x)\, d\mu, \quad f \in E, \tag{1.26}$$

defines a moment functional on $E^{(j)}$. Then $\operatorname{supp} \mu \subseteq V_+(L^{(j)}) = V_{j+1}(L)$ by Proposition 1.40(i) which completes the induction proof. Thus $\operatorname{supp} \mu \subseteq \cap_j V_j(L) = V(L)$.

(iii) Fix $\mu \in \mathcal{M}_L$. Let $L^{(j)} \in \mathcal{L}^{(j)}$ be given by (1.26). Then, $N_{j+1}(L) = N_+(L^{(j)})$ and $V_{j+1}(L) = V_+(L^{(j)})$. From Proposition 1.44, applied to $L^{(j)}$, we conclude that there exists a $p_{j+1} \in E$ such that $p_{j+1} \lceil V_j(L) \in N_+(L^{(j)}) = N_{j+1}(L)$ and

$$V_+(L^{(j)}) = V_{j+1}(L) = \mathcal{Z}(p_{j+1} \lceil V_j(L)) = \{x \in V_j(L) : p_{j+1}(x) = 0\}. \tag{1.27}$$

Suppose that L is an inner point of $\mathcal{L}$. Then, by Proposition 1.42(iii), $N_+(L) = N_1(L) = \{0\}$, so that $V_1(L) = \mathcal{X}$. Hence it follows from the corresponding definitions that $N_j(L) = \{0\}$ and $V_j(L) = \mathcal{X}$ for all $j \in \mathbb{N}$, so the assertion holds with $k = 0$.

Now let L be a boundary point of $\mathcal{L}$. Then $N_1(L) \neq \{0\}$ and hence $V_1(L) \neq \mathcal{X}$. Assume that $r \in \mathbb{N}$ and

$$V_0(L) \supsetneqq \ldots \supsetneqq V_r(L). \tag{1.28}$$

We show that the functions $p_1, \ldots, p_r$ are linearly independent. Assume the contrary. Then $\sum_{j=1}^r \lambda_j p_j = 0$, where $\lambda_j \in \mathbb{R}$, not all zero. Let n be the largest index such that $\lambda_n \neq 0$. Then $p_n(x) = \sum_{j<n} \lambda_j \lambda_n^{-1} p_j$. (The sum is to set zero if $n = 1$.) Since $V_i(L) \subseteq V_j(L)$ if $j \leq i$ and p_j vanishes on $V_j(L)$ by

(1.27), it follows that $p_n = 0$ on $V_{n-1}(L)$. Hence $V_n(L) \subseteq V_{n-1}(L)$ by (1.27), a contradiction to (1.28).

In the preceding two paragraphs we have shown that there is a $k \in \mathbb{N}_0$, $k \leq \dim E$, such that $V_k(L) = V_{k+1}(L)$. Then $N_{k+1}(L) = N_{k+2}(L)$ and hence $V_{k+1}(L) = V_{k+2}(L)$. By induction, $V_{k+j}(L) = V_k(L)$ for $j \in \mathbb{N}$, so that $V(L) = V_k(L)$. ∎

The following result says that the core variety is just the set of possible atoms. It implies that $W(L)$ is a Borel subset of $\mathcal{X}$.

Theorem 1.49 *Suppose that L is a moment functional on E and $L \neq 0$. Then $W(L) = V(L)$. In particular, $W(L)$ is a closed subset of $\mathcal{X}$.*

Proof From Proposition 1.48(ii) it follows that $W(L) \subseteq V(L)$.

By Proposition 1.25, there exists a $k \in \mathbb{N}_0$ such that (1.25) holds. We show that the set $\mathcal{U} := V(L)$ fulfills the assumptions of Corollary 1.31. By Proposition 1.48(ii), we have $\operatorname{supp}\mu \subseteq V(L)$. Further, if $f \in E$ satisfies $f(x) \geq 0$ on $\mathcal{U} = V_k(L)$ and $L(f) = 0$, then $f \in N_{k+1}(L)$ and hence $f(x) = 0$ on $V_{k+1}(L) = V(L) = \mathcal{U}$. Now Corollary 1.31 applies and gives the converse inclusion $\mathcal{U} = V(L) \subseteq W(L)$. Thus, $W(L) = V(L)$.

Since $V(L) = V_k(L)$ is closed by its definition, so is $W(L)$. ∎

1.2.6   Extremal Values of Integrals with Moment Constraints

In this subsection, we investigate the supremum and infimum of the integral $\int h \, d\mu$ of some measurable function h under the constraint that the measure μ has given "moments" $\int f_j \, d\mu = s_j, j = 1, \ldots, n$.

Let $\mathcal{M}(E)$ denote the set of Radon measures μ on $\mathcal{X}$ for which all functions of E are μ-integrable. We fix a vector space basis $\{f_1, \ldots, f_n\}$ of the finite-dimensional subspace E of $C(\mathcal{X}; \mathbb{R})$ and define the moment cone

$$\mathcal{S} = \left\{ s = (s_1, \ldots, s_n) : s_j = \int f_j(x) d\mu, j = 1, \ldots, n, \text{ where } \mu \in \mathcal{M}(E) \right\}.$$

For $s \in \mathcal{S}$ let $\mathcal{M}_s$ denote the set of representing measures of s, that is,

$$\mathcal{M}_s = \{\mu \in \mathcal{M}(E) : s_j = \int f_j(x) d\mu, j = 1, \ldots, n\}.$$

Let h be a fixed real-valued Borel function on $\mathcal{X}$ such that the integral $\int h \, d\mu$ is finite for all $\mu \in \mathcal{M}_s$. For instance, if the function h is bounded on $\mathcal{X}$ and condition (1.8) holds, then each measure $\mu \in \mathcal{M}_t, t \in \mathcal{S}$, is finite and hence $\int h(x) d\mu$ is finite.

For an interior point s of $\mathcal{S}$, we are interested in the upper bound $I_{\sup}(h; s)$ and the lower bound $I_{\inf}(h; s)$ of the integral $\int h(x) \, d\mu$ under the constraints $\mu \in \mathcal{M}_s$:

$$I_{\sup}(h;s) = \sup\left\{\int h(x)d\mu : \mu \in \mathcal{M}_s\right\}, \tag{1.29}$$

$$I_{\inf}(h;s) = \inf\left\{\int h(x)d\mu : \mu \in \mathcal{M}_s\right\}. \tag{1.30}$$

If h is the indicator function χ_A of a Borel set A, $I_{\sup}(\chi_A;s)$ is the supremum of masses $\mu(A)$ for measures $\mu \in \mathcal{M}_s$. This is an important quantity in moment theory.

Let $\mathcal{S}_{ext}$ denote the moment cone obtained by adding the function h to the sequence $\{f_1,\ldots,f_n\}$. That is,

$$\mathcal{S}_{ext} = \left\{(t,t_{n+1}) \in \mathbb{R}^{n+1} : t \in \mathcal{S}, \ t_{n+1} = \int h(x)d\mu \quad \text{for } \mu \in \mathcal{M}_t\right\}.$$

If $h \in E$, then $\int h\,d\mu = L_s(h)$ for $\mu \in \mathcal{S}_s$, so that $I_{\sup}(h;s) = I_{\inf}(h;s) = L_s(h)$, so we are done in this case. Further, we set

$$E_{\geq}(h) = \{f \in E : f(x) \geq h(x) \quad \text{for } x \in \mathcal{X}\}, \tag{1.31}$$

$$E_{\leq}(h) = \{f \in E : f(x) \leq h(x) \quad \text{for } x \in \mathcal{X}\}. \tag{1.32}$$

The following two results relate the problems (1.29) and (1.30) to two other problems (1.33) and (1.34) and give existence criteria for these problems.

Theorem 1.50 *Let s be an interior point of $\mathcal{S}$ such that $I_{\sup}(h;s)$ and $I_{\inf}(h;s)$ are finite. Further, assume that $\int h\,d\mu$ is finite for all $\mu \in \mathcal{M}_t$ and $t \in \mathcal{S}$. Then*

$$I_{\sup}(h;s) = \inf\{L_s(f) : f \in E_{\geq}(h)\}, \tag{1.33}$$

$$I_{\inf}(h;s) = \sup\{L_s(f) : f \in E_{\leq}(h)\}, \tag{1.34}$$

and there exist functions $f_+ \in E_{\geq}(h)$ and $f_- \in E_{\leq}(h)$ such that the infimum in (1.33) is attained at f_+ and the supremum in (1.34) is attained at f_-.

Further, if the supremum in (1.29) is attained at $\mu_+ \in \mathcal{M}_s$ and $f_+ \in E_{\geq}(h)$ satisfies $I_{\sup}(h;s) = L_s(f_+)$, then μ_+ is supported on $\{x \in \mathcal{X} : h(x) = f_+(x)\}$.

Likewise, if the infimum in (1.30) is attained at some $\mu_- \in \mathcal{M}_s$ and $f_- \in E_{\leq}(h)$ is such that $I_{\inf}(h;s) = L_s(f_-)$, then μ_- is supported on $\{x \in \mathcal{X} : h(x) = f_-(x)\}$.

Proof As stated before the theorem, the assertion holds if $h \in E$. Thus we can assume that $h \notin E$.

Since $I_{\sup}(h;s) \in \mathbb{R}$ is the supremum (1.29), $(s, I_{\sup}(h;s))$ is in the closure of $\mathcal{S}_{ext}$ and $(s, I_{\sup}(h;s) + \varepsilon) \notin \mathcal{S}_{ext}$ for all $\varepsilon > 0$. Thus $(s, I_{\sup}(h;s))$ is a boundary point of the convex cone $\mathcal{S}_{ext}$ of $\mathbb{R}^{n+1}$, so there exists a supporting hyperplane through this point at $\mathcal{S}_{ext}$. That is, there are $a \in \mathbb{R}^n$ and $a_{n+1} \in \mathbb{R}$ such that $(a, a_{n+1}) \neq 0$ and

$$a \cdot t + a_{n+1}t_{n+1} \geq 0 \quad \text{for all } (t, t_{n+1}) \in \mathcal{S}_{ext}, \tag{1.35}$$

$$a \cdot s + a_{n+1}I_{\sup}(h;s) = 0. \tag{1.36}$$

For any $x \in \mathcal{X}$, the delta measure δ_x is in $\mathcal{M}(E)$. Letting t be the moment sequence of δ_x in (1.35) we obtain

$$a_1 f_1(x) + \cdots + a_n f_n(x) + a_{n+1} h(x) \geq 0 \quad \text{for } x \in \mathcal{X}. \tag{1.37}$$

Next we prove that $a_{n+1} < 0$. Since $h \notin E$, it is easily seen that $\mathcal{S}_{\text{ext}}$ is not contained in a hyperplane. Hence there is a $(t, t_{n+1}) \in \mathcal{S}_{\text{ext}}$ such that $a \cdot t + a_{n+1} t_{n+1} > 0$. Since s is an inner point of $\mathcal{S}$, we have $t' := s + \varepsilon(s - t) \in \mathcal{S}$ for small $\varepsilon > 0$. Put $t'_{n+1} := \int h d\mu$ for some $\mu \in \mathcal{M}_{t'}$. Then $(t', t'_{n+1}) \in \mathcal{S}_{\text{ext}}$. Therefore, setting $t''_{n+1} := (1 + \varepsilon)^{-1} t'_{n+1} + \varepsilon(1 + \varepsilon)^{-1} t_{n+1}$,

$$(s, t''_{n+1}) = (1 + \varepsilon)^{-1}(t', t'_{n+1}) + \varepsilon(1 + \varepsilon)^{-1}(t, t_{n+1})$$

is a convex combination of points from $\mathcal{S}_{\text{ext}}$, so that $(s, t''_{n+1}) \in \mathcal{S}_{\text{ext}}$. The inequalities $a \cdot t + a_{n+1} t_{n+1} > 0$ and $a \cdot t' + a_{n+1} t'_{n+1} \geq 0$ imply that $a \cdot s + a_{n+1} t''_{n+1} > 0$. Combining the latter with (1.36) we obtain $a_{n+1} t''_{n+1} > a_{n+1} I_{\sup}(h; s)$. Therefore, since $(s, t''_{n+1}) \in \mathcal{S}_{\text{ext}}$ and hence $I_{\sup}(h; s) \geq t''_{n+1}$ by the definition of $I_{\sup}(h; s)$, it follows that $a_{n+1} < 0$.

Set $f_+(x) := -a_1 a_{n+1}^{-1} f_1 - \cdots - a_n a_{n+1}^{-1} f_n$. Dividing (1.37) by $-a_{n+1} > 0$ we get $f_+(x) - h(x) \geq 0$ for $x \in \mathcal{X}$, so that $f_+ \in E_\geq(h)$. From (1.36) and (1.37) we derive $L_s(f_+) = I_{\sup}(h; s)$. Thus we have shown that the infimum in (1.33) is attained at f_+ and equal to $I_{\sup}(h; s)$.

The assertion concerning $I_{\inf}(h; s)$ follows either by a similar reasoning or directly from the result on $I_{\sup}(h; s)$, applied with h replaced by $-h$ and f_i by $-f_i$.

Finally, let us suppose that the supremum (1.29) is attained for some $\mu_+ \in \mathcal{M}_s$ and let $I_{\sup}(h; s) = L_s(f_+)$, where $f_+ \in E_\geq(h)$. Then

$$\int f_+(x) d\mu_+ = L_s(f_+) = I_{\sup}(h; s) = \int h(x) \, d\mu_+,$$

so that $\int (f_+(x) - h(x)) \, d\mu_+ = 0$. Since $f_+ - h \geq 0$ on $\mathcal{X}$, it follows that $f_+ - h = 0$ μ_+-a.e.. This means that μ_+ is supported on the set $\{x \in \mathcal{X} : h(x) = f_+(x)\}$. The proof for μ_- is similar. $\square$

Remark 1.51 Theorem 1.50 and its proof remain valid if E consists of measurable functions for some σ-algebra instead of continuous functions. $\circ$

Since $h(x) \geq f(x)$ on $\mathcal{X}$ for $f \in E_\leq(h)$, we have

$$\inf_{f \in E_\leq(h)} \int |h - f| d\mu = \inf_{f \in E_\leq(h)} \left(\int h d\mu - \int f d\mu \right) = \int h d\mu - \sup_{f \in E_\leq(h)} L_s(f).$$

Thus, finding the supremum in (1.34) is equivalent to the problem of finding the best approximation of h in $L^1(\mathcal{X}, \mu)$ by functions from $E_\leq(h)$. In other words, the supremum in (1.34) is attained at $f \in E_{\leq}(h)$ if and only if f is the best approximation of h in $L^1(\mathcal{X}, \mu)$ by functions of E from *below*.

Recall that a real-valued function h on $\mathcal{X}$ is called *upper semicontinuous* if for each $a \in \mathbb{R}$ the set $\{x \in \mathcal{X} : f(x) < a\}$ is open in $\mathcal{X}$. Obviously, the characteristic function $h = \chi_A$ of some subset A is upper semicontinous if and only if A is closed.

The next theorem is the main result of this section. It contains sufficient conditions ensuring that the supremum in (1.29) is attained.

Theorem 1.52 *Suppose that $\mathcal{X}$ is a compact topological space and E is a finite-dimensional subspace of $C(\mathcal{X}; \mathbb{R})$ satisfying condition (1.8). Let s be an interior point of S and h an upper semicontinuous bounded real-valued function on $\mathcal{X}$.*

Then there exist a k-atomic measure $\mu_+ = \sum_{j=1}^{k} m_j \delta_{x_j} \in \mathcal{M}_s$, $k \leq \dim E + 1$, and a function $f_+ \in E_{\geq}(h)$ such that the supremum (1.29) and the infimum (1.33) are attained at f_+ and μ_+, respectively, and both numbers coincide. That is, we have $f_+(x) \geq h(x)$ on $\mathcal{X}$, $h(x_j) = f_+(x_j)$ for $j = 1, \dots, k$, and for each $v \in \mathcal{M}_s$,

$$\int f(x) dv = \int f(x) d\mu_+ = \sum_{j=1}^{k} m_j f(x_j) \quad for \quad f \in E,$$

$$\sup_{\mu \in \mathcal{M}_s} \int h(x) d\mu = \int h(x) d\mu_+ = \sum_{j=1}^{k} m_j h(x_j) = \int f_+(x) dv = \inf_{f \in E_{\geq}(h)} \int f dv.$$

Proof From the definition of the supremum (1.29) it follows that there is a sequence $(\mu_n)_{n \in \mathbb{N}}$ of measures $\mu_n \in \mathcal{M}_s$ such that $\lim_n \int h d\mu_n = I_{\sup}(h; s)$. By condition (1.8), we have $\mu_n(\mathcal{X}) \leq \int e d\mu_n = L_s(e)$ for $n \in \mathbb{N}$. Therefore, Theorem A.6 applies and implies that the set $\{\mu_n : n \in \mathbb{N}\}$ is relatively vaguely compact. Then there exist a Radon measure $\mu_+ \in M_+(\mathcal{X})$ and a subnet $(\mu_{n_i})_{i \in I}$ which converges vaguely to μ_+. Since $\mathcal{X}$ is compact, $E \subseteq C_c(\mathcal{X}, \mathbb{R})$ and hence $\int f d\mu_+ = \lim_i \int f d\mu_{n_i} = L_s(f)$ for $f \in E$. Thus, $\mu_+ \in \mathcal{M}_s$.

Further, the function 1 is in $C_c(\mathcal{X}, \mathbb{R})$ again by the compactness of $\mathcal{X}$. Hence $\lim \mu_{n_i}(\mathcal{X}) = \lim \int 1 d\mu_{n_i} = \int 1 d\mu = \mu(\mathcal{X})$, so condition (i) in Proposition A.8 is fulfilled. From $\mu_n(\mathcal{X}) \leq L_s(e)$ and $\mu(\mathcal{X}) \leq L_s(e)$ we get $\mu_n, \mu \in M_+^b(\mathcal{X})$. Since h is upper semicontinuous, it follows from the implication (i)$\rightarrow$(iii) of Proposition A.8 that $\int h d\mu_+ \geq \limsup_i \int h d\mu_{n_i} = I_{\sup}(h; s)$. Obviously, $I_{\sup}(h; s) \geq \int h d\mu_+$ by definition. Thus, $\int h d\mu_+ = I_{\sup}(h; s)$, so the supremum (1.29) is attained at μ_+.

Applying Theorem 1.24 to the functional $L(f) = \int f d\mu$ on $E \oplus \mathbb{R} \cdot h$ we conclude that μ_+ can be chosen k-atomic, say $\mu_+ = \sum_{j=1}^{k} m_j \delta_{x_j}$, with $k \leq \dim E + 1$.

All remaining assertions are contained in Theorem 1.50. Because μ_+ is supported on the set $\{x \in \mathcal{X} : h(x) = f_+(x)\}$, we have $h(x_j) = f_+(x_j)$ for all $j = 1, \dots, k$. $\qquad\square$

In particular, Theorem 1.52 holds if h is the characteristic function of a closed subset of $\mathcal{X}$. If we assume that the function h is *lower* semicontinuous (for instance, the characteristic function of an open set), then the counterpart of Theorem 1.52 remains valid almost verbatim for the infimum in (1.30) and the supremum in (1.34).

1.3 Exercises

1. Decide whether or not the following subspace E of $C(\mathbb{R}; \mathbb{R})$ is an adapted space:

   a. E is the span of functions $(x^2 + n)^{-1}$, $n \in \mathbb{N}$.
   b. E is the span of functions $e^{\alpha x}$, $\alpha > 0$.
   c. $E = C_c(\mathbb{R}; \mathbb{R})$.

2. Are the polynomials $\mathbb{R}[x]_n$ of degree at most n an adapted space of $C(\mathbb{R}; \mathbb{R})$ or of $C([-1, 1]; \mathbb{R})$?

3. Let E be the vector space of bounded continuous real functions on $\mathbb{R}$. Each $f \in E$ has a unique continuous extension $\hat{f}$ to the Stone–Čech compactification $\beta(\mathbb{R})$ of $\mathbb{R}$ (see e.g. [Cw, Chapter V, §6]). Let x_0 be a point of $\beta(\mathbb{R}) \backslash \mathbb{R}$ and define $L(f) := \hat{f}(x_0), f \in E$. Show that L is an E_+-positive linear functional on E which is not a moment functional.

4. Let μ_n, $n \in \mathbb{N}$, and μ be positive Radon measures on a locally compact Hausdorff space $\mathcal{X}$ such that the sequence $(\mu_n)_{n \in \mathbb{N}}$ converges vaguely to μ and $\sum_{n=1}^{\infty} \mu_n(\mathcal{X}) < \infty$. Show that $\lim_n \int f d\mu_n = \int f d\mu$ for $f \in C_0(\mathcal{X}; \mathbb{R})$.

5. Let $a, b \in \mathbb{R}$, where $a < b$. Determine the character set $\hat{A}$ for the $*$-algebras $A = C(\mathbb{R}; \mathbb{R}) + \mathbb{R} \cdot \chi_a$ and $A = C(\mathbb{R}; \mathbb{R}) + \mathbb{R} \cdot \chi_{[a,b]}$.
   Hint: Look for a topological space $\mathcal{X}$ such that A is isomorphic to $C(\mathcal{X}, \mathbb{R})$.

In the following exercises, E is a finite-dimensional subspace of $C(\mathcal{X}; \mathbb{R})$, $\mathcal{S}$ is the moment cone, and $\mathcal{S}_j$ is the set of $s \in \mathcal{S}$ which have a k-atomic representing measure with $k \leq j$.

6. Give an example such that $\mathcal{S}_1$ is not closed.
7. Prove that $\mathbb{R}^m = \mathcal{S} - \mathcal{S}$, where $m = \dim E$.
8. Let C be the smallest number such that each $s \in \mathcal{S}$ has a k-representing measure with $k \leq$ C. Show that $\mathcal{S}_{j-1} \neq \mathcal{S}_j$ for $j = 1, \ldots,$ C.
9. Assume that condition (1.15) holds.

   (i) Prove that the cone $\mathcal{S}$ is pointed, that is, $\mathcal{S} \cap (-\mathcal{S}) = \{0\}$.
   (ii) Prove that if $\mathcal{S}_1$ is closed, so is $\mathcal{S}_k$ for all $k \in \mathbb{N}$.

10. Assume that $\mathcal{X}$ is compact and (1.8) holds. Prove $\mathcal{L}_k$ is closed for all $k \in \mathbb{N}$.
11. Suppose that L is a strictly E_+-positive moment functional on E and $|\mathcal{X}| > \dim E$. Prove that L is not determinate by using the Hahn–Banach theorem.
    Hints: Take $h \in C_c(\mathcal{X}; \mathbb{R})$, $h \notin E$. Show that $S_h < I_h$, where

    $$S_h := \sup \{L(f); f \in E \text{ and } f(x) \leq h(x), x \in \mathcal{X}\}, \qquad (1.38)$$

    $$I_h := \inf \{L(f) : f \in E \text{ and } h(x) \leq f(x), x \in \mathcal{X}\}. \qquad (1.39)$$

Choose $\gamma \in \mathbb{R}$, $S_h < \gamma < I_h$. Show that the functional L_γ on $F := E + \mathbb{R} \cdot h$ defined by $L_\gamma(f + \lambda h) = L(f) + \lambda\gamma, f \in E$, is strictly F_+-positive. Apply Theorem 1.30.

1.4 Notes

Choquet's theory of adapted spaces was elaborated in [Chq]. Haviland's Theorem 1.12 goes back to [Ha]. Since the one-dimensional case was noted by M. Riesz [Rz2], Theorem 1.12 is often called Riesz–Haviland theorem in the literature. Theorem 1.21 (for polynomials) is due to M.A. Naimark [Na]; the general version is from [Do].

The important Theorem 1.24 was first proved in full generality by H. Richter [Ri] in 1957, see also W.W. Rogosinsky [Rg]. V. Tchakaloff [Tch] treated the simpler compact case about the same time. Richter's paper has been ignored in the literature and a number of versions of his result have been reproved even recently.

Theorem 1.26 is due to [FN2]. Assertion (i) of Theorems 1.30 is based on an idea from [FN1]. Theorems 1.30, 1.36, 1.45, and 1.49 were proved in [DSm1]. The core variety (for polynomials) was introduced in [F2]. More results on the moment cone can be found in [DSm1], [DSm2]; proofs of Exercises 1.7–1.10 are given in [DSm2].

The results of Sect. 1.2.6 were obtained in [Ri], [Rg], [Ii], [Kp1]; see [Kp2] for a survey. They will not be used later in this book.

Chapter 2
Moment Problems on Abelian ∗-Semigroups

In this chapter we collect a number of general concepts and simple facts on moment problems on commutative ∗-semigroups that will be used throughout the text, often without mention. Section 2.1 is about positive functionals on ∗-algebras and positive semidefinite functions on ∗-semigroups. In Sect. 2.2 we specialize to commutative ∗-algebras and ∗-semigroups and introduce moment functionals, moment functions, K-determinate moment functions, and generalized Hankel matrices. Some standard examples of commutative ∗-semigroups are given in Sect. 2.3.

Throughout this chapter, $\mathbb{K}$ is either the real field $\mathbb{R}$ or the complex field $\mathbb{C}$.

2.1 ∗-Algebras and ∗-Semigroups

In this section we discuss the one-to-one correspondence between positive semidefinite functions on ∗-semigroups and positive functionals on semigroup ∗-algebras.

Let us begin with some basic definitions.

Definition 2.1 A *∗-algebra* A over $\mathbb{K}$ is an algebra over $\mathbb{K}$ equipped with a mapping $* : \mathsf{A} \to \mathsf{A}$, called *involution*, such that for $a, b \in \mathsf{A}$ and $\alpha, \beta \in \mathbb{K}$,

$$(\alpha a + \beta b)^* = \overline{\alpha}\, a^* + \overline{\beta}\, b^*, \quad (ab)^* = b^* a^*, \quad (a^*)^* = a.$$

The *Hermitian part* A_h of a ∗-algebra A is $\mathsf{A}_h := \{a \in \mathsf{A} : a = a^*\}$.

Our standard examples of real or complex ∗-algebras in this book are the polynomial algebras $\mathbb{R}[x_1, \dots, x_d]$ and $\mathbb{C}[x_1, \dots, x_d]$, respectively, with involutions determined by $(x_j)^* = x_j, j = 1, \dots, d$.

© Springer International Publishing AG 2017
K. Schmüdgen, *The Moment Problem*, Graduate Texts in Mathematics 277,
DOI 10.1007/978-3-319-64546-9_2

Definition 2.2 Let A be a $\mathbb{K}$-linear subspace of a some ∗-algebra over $\mathbb{K}$. We define

$$A^2 = \text{Lin}\,\{b^*a : a, b \in A\}, \tag{2.1}$$

$$\sum A^2 = \left\{ \sum_{i=1}^{k} (a_i)^* a_i : a_i \in A, k \in \mathbb{N} \right\}. \tag{2.2}$$

A linear functional $L : A^2 \to \mathbb{K}$ is called *positive* if L is $(\sum A^2)$-positive, that is, if

$$L(a^*a) \geq 0 \quad \text{for} \quad a \in A. \tag{2.3}$$

Lemma 2.3 *Let* A *be a linear subspace of a* ∗*-algebra* B *over* $\mathbb{K}$ *and* $L : A^2 \to \mathbb{K}$ *a positive linear functional. Then the Cauchy–Schwarz inequality holds:*

$$|L(b^*a)|^2 \leq L(a^*a)L(b^*b) \quad \text{for } a, b \in A. \tag{2.4}$$

Further, if B *is unital and the unit element of* B *is contained in* A*, then*

$$L(a^*) = \overline{L(a)} \quad \text{for } a \in A. \tag{2.5}$$

Proof We carry out the proof in the case $\mathbb{K} = \mathbb{C}$; the case $\mathbb{K} = \mathbb{R}$ is even simpler. For $\alpha, \beta \in \mathbb{K}$ and $a, b \in A$ we have

$$L((\alpha + \beta b)^*(\alpha a + \beta b))$$

$$= \overline{\alpha}\alpha L(a^*a) + \overline{\alpha}\beta L(a^*b) + \alpha\overline{\beta}L(b^*a) + \overline{\beta}\beta L(b^*b) \geq 0. \tag{2.6}$$

Hence $\overline{\alpha}\beta L(a^*b) + \alpha\overline{\beta}L(b^*a)$ is real. Letting $\overline{\alpha}\beta = 1$ and $\overline{\alpha}\beta = i$, we derive $L(a^*b) = \overline{L(b^*a)}$. If B has a unit element 1 and $1 \in A$, we set $b = 1$ and get (2.5).

The expression in (2.6) is a positive semidefinite quadratic form. Hence its discriminant has to be nonnegative. Since $L(a^*b) = \overline{L(b^*a)}$ as just shown, this yields

$$L(a^*a)L(b^*b) - L(b^*a)L(a^*b) = L(a^*a)L(b^*b) - |L(b^*a)|^2 \geq 0. \quad \square$$

If the linear subspace A in Definition 2.2 is itself a ∗-algebra, then (2.3) is just the definition of a positive functional on the ∗-algebra A.

In this book we deal mainly with *commutative real* algebras. Each such algebra is a ∗-algebra over $\mathbb{R}$ if we take the identity map as involution. In this case, $\sum A^2$ is the set of finite sums of squares a^2 of elements $a \in A$ and the Cauchy–Schwarz inequality (2.4) has the following form:

$$L(ab)^2 \leq L(a^2)L(b^2) \quad \text{for } a, b \in A. \tag{2.7}$$

By a *semigroup* $(S, \circ)$ we mean a nonempty set S with an associative composition $\circ$ (that is, a mapping $S \times S \ni (s_1, s_2) \mapsto s_1 \circ s_2 \in S$ such that $s_1 \circ (s_2 \circ s_3) = (s_1 \circ s_2) \circ s_3$ for all $s_1, s_2, s_3 \in S$) and a neutral element $e \in S$ (that is, $e \circ s = s \circ e = s$ for $s \in S$).

Definition 2.4 A *-semigroup $(S, \circ, *)$ is a semigroup $(S, \circ)$ endowed with a mapping $* : S \to S$, called an *involution*, such that

$$(s \circ t)^* = t^* \circ s^* \quad \text{and} \quad (s^*)^* = s, \quad s, t \in S.$$

If no confusion can arise we write simply S instead of $(S, \circ, *)$.

Any *abelian* semigroup becomes a *-semigroup if we take the identity map as involution. Each group S is a *-semigroup with involution $s^* := s^{-1}$, $s \in S$.

Definition 2.5 A function $\varphi : S \to \mathbb{K}$ on a *-semigroup S is *positive semidefinite* if for arbitrary elements $s_1, \ldots, s_n \in S$, numbers $\xi_1, \ldots, \xi_n \in \mathbb{K}$ and $n \in \mathbb{N}$,

$$\sum_{i,j=0}^{n} \varphi(s_i^* \circ s_j) \, \overline{\xi_i} \, \xi_j \geq 0.$$

The set of positive semidefinite functions $\varphi : S \to \mathbb{K}$ on S is denoted by $\mathcal{P}_{\mathbb{K}}(S)$.

Suppose that S is *-semigroup. We define the semigroup *-algebra $\mathbb{K}[S]$. A vector space basis of $\mathbb{K}[S]$ is given by the elements of S and product and involution of $\mathbb{K}[S]$ are determined by the corresponding operations of S. That is, $\mathbb{K}[S]$ is the vector space of all sums $\sum_{s \in S} \alpha_s s$, where $\alpha_s \in \mathbb{K}$ and only finitely many numbers α_s are nonzero, with pointwise addition and scalar multiplication. The vector space $\mathbb{K}[S]$ becomes a unital *-algebra over $\mathbb{K}$ with product and involution defined by

$$\left(\sum_{s \in S} \alpha_s s \right) \left(\sum_{t \in S} \beta_t t \right) := \sum_{s, t \in S} \alpha_s \beta_t (s \circ t),$$

$$\left(\sum_{s \in S} \alpha_s s \right)^* := \sum_{s \in S} \overline{\alpha_s} \, s^*.$$

Since the elements of S form a basis of $\mathbb{K}[S]$, there is a one-to-one correspondence between functions $\varphi : S \to \mathbb{K}$ and linear functionals $L_\varphi : \mathbb{K}[S] \to \mathbb{K}$ given by

$$L_\varphi(s) := \varphi(s), \quad s \in S.$$

Definition 2.6 The unital *-algebra $\mathbb{K}[S]$ over $\mathbb{K}$ is the *semigroup *-algebra* of S and the functional L_φ is called the *Riesz functional* associated with the function φ.

Proposition 2.7 *For a function $\varphi : S \to \mathbb{K}$ the following are equivalent:*

  (i) *φ is a positive semidefinite function.*
 (ii) *L_φ is a positive linear functional on the *-algebra $\mathbb{K}[S]$.*
(iii) *$H(\varphi) = (a_{st} := \varphi(s^* \circ t))_{s,t \in S}$ is a positive semidefinite Hermitian matrix.*

Proof For arbitrary $a = \sum_{s \in S} \alpha_s s \in \mathbb{K}[S]$, we have

$$L_\varphi(a^* a) = \sum_{s,t \in S} \varphi(s^* \circ t)\overline{\alpha}_s \alpha_t = \sum_{s,t \in S} a_{st}\overline{\alpha}_s \alpha_t.$$

Comparing Definitions 2.5 and 2.2 the first equality implies the equivalence of (i) and (ii), while the second equality yields the equivalence of (i) and (iii). □

Corollary 2.8 *If* $\varphi : S \to \mathbb{K}$ *is a positive semidefinite function on S, then*

$$\varphi(s^*) = \overline{\varphi(s)} \quad and \quad \varphi(s^* \circ s) \geq 0 \quad for \ s \in S, \tag{2.8}$$

$$|\varphi(s^* \circ t)|^2 \leq \varphi(s^* \circ s)\varphi(t^* \circ t) \quad for \ s, t \in S. \tag{2.9}$$

In particular, if $\varphi(e) = 0$, *then* $\varphi(t) = 0$ *for all* $t \in S$.

Proof By Proposition 2.7, L_φ is a positive linear functional on the unital ∗-algebra $\mathbb{K}[S]$, that is, $L_\varphi(s^* s) = \varphi(s^* \circ s) \geq 0$. Therefore, (2.8) and (2.9) follow at once from (2.5) and (2.4), respectively. If $\varphi(e) = 0$, then it follows that $\varphi \equiv 0$ on S by setting $s = e$ in (2.9). □

Definition 2.9 The positive semidefinite matrix

$$H(\varphi) = (\varphi(s^* \circ t))_{s,t \in S}$$

with (s, t)-entry $a_{st} := \varphi(s^* \circ t)$ is called the *generalized Hankel matrix* associated with the positive semidefinite function $\varphi : S \to \mathbb{K}$.

2.2 Commutative ∗-Algebras and Abelian ∗-Semigroups

Throughout this section, we assume that A is a **commutative** unital ∗-algebra over $\mathbb{K}$ and that S is an **abelian** ∗-semigroup. As is common, we write $+$ for the composition $\circ$ of S and denote the neutral element of S by 0.

Definition 2.10 A *character* on A is linear functional $\chi : A \to \mathbb{K}$ satisfying

$$\chi(1) = 1, \quad \chi(ab) = \chi(a)\chi(b), \quad \chi(a^*) = \overline{\chi(a)}, \quad a, b \in A. \tag{2.10}$$

If A is a real algebra with identity involution, this coincides with Definition 1.13.

The set of characters of A is denoted by $\hat{A}$. We equip $\hat{A}$ with the topology of pointwise convergence and assume that $\hat{A}$ is then a locally compact topological Hausdorff space. The latter is always fulfilled if the algebra A is finitely generated.

The following definition restates Definitions 1.1 and 1.2 in the present setting.

Definition 2.11 Let K be a closed subset of $\hat{A}$. A linear functional $L : A \to \mathbb{K}$ is called a K-*moment functional* if there exists a Radon measure μ on $\hat{A}$ such that

$$\operatorname{supp}\mu \subseteq K, \tag{2.11}$$

the function $\chi \mapsto \chi(a)$ is μ-integrable on $\hat{A}$ and satisfies

$$L(a) = \int_{\hat{A}} \chi(a)\, d\mu(\chi) \quad \text{for all} \quad a \in A. \tag{2.12}$$

A K-moment functional L is said to be K-*determinate* if there is only one Radon measure μ on $\hat{A}$ for which (2.11) and (2.12) holds.

In the case $K = \hat{A}$ we call K-moment functionals simply *moment functionals* and K-determinate K-moment functionals *determinate*.

Lemma 2.12 *Each K-moment functional is a positive linear functional on* A.

Proof Let $a \in A$. For $\chi \in \hat{A}$ we have $\chi(a^*a) = \chi(a^*)\chi(a) = |\chi(a)|^2$ by (2.10) and therefore

$$L(a^*a) = \int_{\hat{A}} \chi(a^*a)\, d\mu(\chi) = \int_{\hat{A}} |\chi(a)|^2\, d\mu(\chi) \geq 0. \qquad \square$$

Remark 2.13 A positive linear functional L on A satisfying $L(1) = 1$ is called a *state*. Let S(A) denote the set of states of A. Each character of A is an extreme point of the convex set S(A). (The reasoning of the proof of Lemma 18.3(ii) below gives a proof of this well-known fact.) In general not all extreme point of S(A) are characters. (Indeed, by Proposition 13.5, there exists a state L on $\mathbb{R}[x_1, x_2]$ which is not a moment functional. From the decomposition theory of states on *-algebras [Sm4, Section 12.4] it follows that L is an integral of extreme points of S(A). Since L is not a moment functional, not all extreme points of S(A) can be characters; see also the discussion in [Sm4, Remark 12.4.6].)   ∘

Next we turn to characters on the abelian *-semigroup S.

Definition 2.14 A *character* of S is a function $\chi : S \to \mathbb{K}$ satisfying

$$\chi(0) = 1, \quad \chi(s+t) = \chi(s)\chi(t), \quad \chi(s^*) = \overline{\chi(s)}, \quad s, t \in S. \tag{2.13}$$

The set S^* of characters of an abelian semigroup S is also an abelian *-semigroup, called the *dual semigroup* of S, with pointwise multiplication as composition, complex conjugation as involution and the constant character as neutral element.

Let us assume that S^* equipped with the topology of pointwise convergence is a locally compact Hausdorff space. This holds if the *-semigroup is finitely generated. The following is the counterpart of Definition 2.11 for *-semigroups.

Definition 2.15 Let K be a closed subset of S^*. We say that a function $\varphi : S \to \mathbb{K}$ is a *K-moment function* if there exists a Radon measure μ on S^* such that

$$\operatorname{supp} \mu \subseteq K, \tag{2.14}$$

the function $\chi \mapsto \chi(s)$ is μ-integrable on S^* and

$$\varphi(s) = \int_{S^*} \chi(s) d\mu(\chi) \quad \text{for all } s \in S. \tag{2.15}$$

A K-moment function φ is called *K-determinate* if the measure μ satisfying (2.14) and (2.15) is uniquely determined.

If $K = S^*$ we call a K-moment function simply a *moment function* and a K-determinate moment function *determinate*.

Comparing the preceding definitions and facts we see that we have a one-to-one correspondence between notions on a *-semigroup S and its semigroup *-algebra $\mathbb{K}[S]$. By (2.13) and (2.10), a function $\chi : S \to \mathbb{K}$ is a character on the *-semigroup S if and only if its Riesz functional on $\mathbb{K}[S]$ is a character on the *-algebra $\mathbb{K}[S]$. Comparing Definitions 2.11 and 2.15, it follows that φ is a K-moment function on S if and only if the Riesz functional L_φ is a K-moment functional on $\mathbb{K}[S]$. Further, by these definitions, φ is K-determinate if and only if L_φ is. That is, the moment problems for the semigroup *-algebra $\mathbb{K}[S]$ and for the *-semigroup S are equivalent. We shall use these fact throughout the book without mention.

From Proposition 2.7 (i)$\leftrightarrow$(ii), and Lemma 2.12 we obtain the following.

Corollary 2.16 *Each moment function $\varphi : S \to \mathbb{K}$ is positive semidefinite.*

By Corollary 2.16 and Lemma 2.12, each moment function $\varphi : S \to \mathbb{K}$ is positive semidefinite and the Riesz functional L_φ on $\mathbb{K}[S]$ is positive. We shall show later (Proposition 13.5) that the converse is not true for *-semigroup $S = \mathbb{N}_0^d$ when $d \geq 2$. Even more, the converse is only true in rare cases! Finding sufficient conditions on a positive linear functional L on $\mathbb{R}[\mathbb{N}_0^d]$ ensuring that L is a moment functional will be one of our main tasks in this book.

Next let us suppose that A is a *commutative real* algebra. We want to define its complexification $A_\mathbb{C}$. The direct sum $A_\mathbb{C} := A \oplus iA$ of vector spaces A and iA becomes a commutative *complex* *-algebra with multiplication, involution and scalar multiplication defined by

$$(a + ib)(c + id) = ac - bd + i(bc + ad), \quad (a + ib)^* := a - ib,$$

$$(\alpha + i\beta)(a + ib) := \alpha a - \beta b + i(\alpha b + \beta a),$$

where $a, b, c, d \in A$ and $\alpha, \beta \in \mathbb{R}$. This complex *-algebra $A_\mathbb{C}$ is called the *complexification* of A. The real algebra A is then the Hermitian part $(A_\mathbb{C})_h$ of $A_\mathbb{C}$.

Recall that $\sum (A_\mathbb{C})^2$ denotes all finite sums $\sum_j x_j^* x_j$ and $\sum A^2$ is the set of finite sums $\sum_i a_i^2$, where $x_j \in A_\mathbb{C}$ and $a_i \in A$.

Lemma 2.17 $\sum (\mathsf{A}_{\mathbb{C}})^2 = \sum \mathsf{A}^2.$

Proof Let $x \in \mathsf{A}_{\mathbb{C}}$. Then $x = a + \mathrm{i}b$ with $a, b \in \mathsf{A}$. Using that the algebra A is commutative (!) we obtain

$$x^*x = (a + \mathrm{i}b)^*(a + \mathrm{i}b) = a^2 + b^2 + \mathrm{i}(ab - ba) = a^2 + b^2 \in \sum \mathsf{A}^2. \quad \square$$

Each $\mathbb{R}$-linear functional L on A has a unique extension $L_{\mathbb{C}}$ to a $\mathbb{C}$-linear functional on $\mathsf{A}_{\mathbb{C}}$. An immediate consequence of Lemma 2.17 is the following corollary.

Corollary 2.18 *L is positive on the real $*$-algebra A (with the identity map as involution) if and only if $L_{\mathbb{C}}$ is positive on the complex $*$-algebra $\mathsf{A}_{\mathbb{C}}$.*

At the end of this section we briefly discuss the choice of the field $\mathbb{K}$. A large part of this book deals with the K-moment problem for the $*$-semigroup $S = \mathbb{N}_0^d$ with identity involution. By (2.13), all characters on $\mathbb{N}_0^d$ are real-valued, $(\mathbb{N}_0^d)^* \cong \mathbb{R}^d$ and we have $\mathbb{R}[\mathbb{N}_0^d] \cong \mathbb{R}[x_1, \ldots, x_d]$, see Example 2.3.1 below. That is, we can work with the real field and real algebras. In Chaps. 12 and 13 we will apply powerful methods from real algebraic geometry to the real algebra $\mathsf{A} = \mathbb{R}[x_1, \ldots, x_d]$.

But operator-theoretic treatments require complex Hilbert spaces. In this case it is more convenient to use the *complex* semigroup $*$-algebra $\mathbb{C}[\mathbb{N}_0^d] \cong \mathbb{C}[x_1, \ldots, x_d]$. Since $\mathbb{C}[x_1, \ldots, x_d]$ is just the complexification of $\mathbb{R}[x_1, \ldots, x_d]$, it is easy from the preceding discussion and Corollary 2.18 to pass from one algebra to the other.

In Chaps. 11 and 15 we will study the moment problem on the unit circle and the complex moment problem, respectively. In these cases it is unavoidable to work with the complex field, since otherwise we would not have enough characters.

2.3 Examples

In this section we discuss four important examples that will be crucial for this book.

2.3.1 *Example 1:* $\mathbb{N}_0^d, \mathfrak{n}^* = \mathfrak{n}$

The additive semigroup $\mathbb{N}_0^d$ with identity involution is a $*$-semigroup and the map

$$(n_1, \ldots, n_d) \mapsto x_1^{n_1} \cdots x_d^{n_d}$$

is a $*$-isomorphism of the semigroup $*$-algebra $\mathbb{K}[\mathbb{N}_0^d]$ on the $*$-algebra $\mathbb{K}[x_1, \ldots, x_d]$ of polynomials with involution determined by $x_j^* = x_j, j = 1, \ldots, d$. By identifying $(n_1, \ldots, n_d)$ and $x_1^{n_1} \cdots x_d^{n_d}$ we obtain

$$\mathbb{K}[\mathbb{N}_0^d] \cong \mathbb{K}[x_1, \ldots, x_d].$$

Clearly, for any $t \in \mathbb{R}^d$ there is a character χ_t given by the point evaluation $\chi_t(p) = p(t), p \in \mathbb{K}[x_1, \ldots, x_d]$. Conversely, if χ is a character, we set $t_j := \chi(x_j)$ for $j = 1, \ldots, d$. Then $\chi(x_j) = \chi((x_j)^*) = \overline{\chi(x_j)}$, so $t = (t_1, \ldots, t_d) \in \mathbb{R}^d$ and

$$\chi(p(x_1, \ldots, x_d)) = p(\chi(x_1), \ldots, \chi(x_d)) = p(t_1, \ldots, t_d) = \chi_t(p).$$

Hence $\chi = \chi_t$. Thus we have shown that the character space of $S = \mathbb{N}_0^d$ is

$$(\mathbb{N}_0^d)^* = \{\chi_t : t \in \mathbb{R}^d\} \cong \mathbb{R}^d, \quad \text{where} \quad \chi_t(p) = p(t). \tag{2.16}$$

A function on the semigroup $\mathbb{N}_0^d$ is just a multisequence $s = (s_\alpha)_{\alpha \in \mathbb{N}_0^d}$. Its Riesz functional L_s is given by $L_s(x^\alpha) = s_\alpha, \alpha \in \mathbb{N}_0^d$. By the definition of the involution, positive semidefinite functions on $\mathbb{N}_0^d$ are real-valued. By Definition 2.5, a real sequence $s = (s_\alpha)_{\alpha \in \mathbb{N}_0^d}$ is a positive semidefinite function on $\mathbb{N}_0^d$ if and only if

$$\sum_{\alpha, \beta \in \mathbb{N}_0^d} s_{\alpha + \beta} \, \xi_\alpha \, \xi_\beta \geq 0$$

for all finite real multisequences $(\xi_\alpha)_{\alpha \in \mathbb{N}_0^d}$.

From Definition 2.15 and Eq. (2.16) it follows that s is a moment function on $\mathbb{N}_0^d$, briefly a *moment sequence*, if and only if there is a Radon measure μ on $\mathbb{R}^d \cong (\mathbb{N}_0^d)^*$ such that $x^\alpha \in \mathcal{L}^1(\mathbb{R}^d, \mu)$ and

$$s_\alpha = \int_{\mathbb{R}^d} x^\alpha \, d\mu \quad \text{for} \quad \alpha \in \mathbb{N}_0^d, \tag{2.17}$$

or equivalently, $p(x) \in \mathcal{L}^1(\mathbb{R}^d, \mu)$ and $L_s(p) = \int_{\mathbb{R}^d} p(x) \, d\mu$ for $p \in \mathbb{R}[x_1, \ldots, x_d]$. Equation (2.17) means that s_α is the α-th moment of the measure μ. Thus, s is a moment sequence if and only if there is a Radon measure μ such that each s_α is the α-th moment $s_\alpha(\mu$ of μ, or equivalently, L_s is a moment functional according to Definition 2.11. This explains and justifies the names "moment sequence" and "moment functional".

By Definition 2.9, the corresponding generalized Hankel matrix $H(s)$ has the (α, β)-entry $s_{\alpha + \beta}$. In the case $d = 1$ the matrix $H(s)$ is a "usual" one-sided infinite *Hankel matrix* which is constant on cross-diagonals:

$$H(s) = \begin{pmatrix} s_0 & s_1 & s_2 & \ldots s_n & \ldots \\ s_1 & s_2 & s_3 & \ldots s_{n+1} \ldots \\ s_2 & s_3 & s_4 & \ldots s_{n+2} \ldots \\ \ldots \ldots & \ldots & \ldots \ldots & \ldots \\ s_n & s_{n+1} & s_{n+2} & \ldots s_{2n} & \ldots \\ \ldots \ldots & \ldots & \ldots \ldots & \ldots \end{pmatrix}. \tag{2.18}$$

Remark 2.19

1. It should be emphasized that notions such as positive semidefinite sequence, moment sequence, and Hankel matrix depend essentially on the underlying $*$-semigroup.
2. In the literature the Hankel matrix is often called the *moment matrix*, because its entries are moments if s is a moment sequence. We will not use this terminology. The reason is that we will work with Hankel matrices as technical tools even if we do not know whether or not s is a moment sequence. In algebraic geometry the Hankel matrix appears under the name *catalecticant matrix*.   ○

2.3.2 Example 2: $\mathbb{N}_0^{2d}$, $(\mathfrak{n}, \mathfrak{m})^* = (\mathfrak{m}, \mathfrak{n})$

The additive semigroup $\mathbb{N}_0^{2d}$ with involution

$$(n_1, \ldots, n_d, m_1, \ldots, m_d)^* = (m_1, \ldots, m_d, n_1, \ldots, n_d)$$

is a $*$-semigroup and the map

$$(n_1, \ldots, n_d, m_1, \ldots, m_d) \mapsto z_1^{n_1} \cdots z_d^{n_d} \bar{z}_1^{m_1} \cdots \bar{z}_d^{m_d}$$

is a $*$-isomorphism of $\mathbb{C}[\mathbb{N}_0^{2d}]$ onto the $*$-algebra $\mathbb{C}[z_1, \ldots, z_d, \bar{z}_1, \cdots, \bar{z}_d]$ of complex polynomials with involution given by $(z_j)^* := \bar{z}_j, j = 1, \ldots, d$. That is,

$$\mathbb{C}[\mathbb{N}_0^{2d}] \cong \mathbb{C}[z_1, \bar{z}_1, \ldots, z_d, \bar{z}_d].$$

Arguing as in the preceding example it follows that the character space is given by the evaluations at points of $\mathbb{C}^d$, that is,

$$S^* = \{\chi_z : z \in \mathbb{C}^d\} \cong \mathbb{C}^d, \quad \text{where} \quad \chi_z(p) = p(z_1, \ldots, z_d, \bar{z}_1, \ldots, \bar{z}_d).$$

Positive semidefinite functions on this $*$-semigroup S are complex multisequences $s = (s_{\alpha, \beta})_{\alpha, \beta \in \mathbb{N}_0^d}$ for which

$$\sum_{\alpha, \alpha', \beta, \beta' \in \mathbb{N}_0^d} s_{\alpha + \alpha', \beta + \beta'} \, \overline{\xi_{\alpha, \beta}} \, \xi_{\alpha', \beta'} \geq 0$$

for all finite complex multisequences $(\xi_{\alpha, \beta})_{\alpha, \beta \in \mathbb{N}_0^d}$.

The $((\alpha, \beta), (\alpha', \beta'))$-entry of the corresponding generalized Hankel matrix $H(s)$ is $s_{\alpha + \alpha', \beta + \beta'}$. The first equality of (2.8) yields $\overline{s_{\alpha + \alpha', \beta + \beta'}} = s_{\beta + \beta', \alpha + \alpha'}$ for

$\alpha, \beta, \alpha', \beta' \in \mathbb{N}_0^d$. In the case $d = 1$ the matrix $H(s)$ has the form

$$H(s) = \begin{pmatrix} s_{00} & s_{01} & s_{02} & \cdots & s_{0n} & \cdots \\ \overline{s_{01}} & s_{11} & s_{12} & \cdots & s_{1n} & \cdots \\ \overline{s_{02}} & \overline{s_{12}} & s_{22} & \cdots & s_{2n} & \cdots \\ \cdots & \cdots & \cdots & \cdots & \cdots & \cdots \\ \overline{s_{0n}} & \overline{s_{1n}} & \overline{s_{2n}} & \cdots & s_{nn} & \cdots \\ \cdots & \cdots & \cdots & \cdots & \cdots & \cdots \end{pmatrix}. \tag{2.19}$$

2.3.3 Example 3: $\mathbb{Z}^d, \mathfrak{n}^* = -\mathfrak{n}$

The additive group $\mathbb{Z}^d$ equipped with the involution $(n_1, \ldots, n_d)^* = (-n_1, \ldots, -n_d)$ is a ∗-semigroup. There is a ∗-isomorphism

$$(n_1, \ldots, n_d) \mapsto z_1^{n_1} \cdots z_d^{n_d}$$

of the group ∗-algebra $\mathbb{C}[\mathbb{Z}^d]$ onto the ∗-algebra of trigonometric polynomials in d variables, or equivalently, of complex polynomials in $z_1 \in \mathbb{T}, \ldots, z_d \in \mathbb{T}$. Thus,

$$\mathbb{C}[\mathbb{Z}^d] \cong \mathbb{C}[z_1, \overline{z_1}, \ldots, z_d, \overline{z_d} : z_1\overline{z_1} = \overline{z_1}z_1 = 1, \ldots, z_d\overline{z_d} = \overline{z_d}z_d = 1].$$

It is easily verified that the character space $(\mathbb{Z}^d)^*$ consists of point evaluations at points of the d-torus $\mathbb{T}^d = \{(z_1, \ldots, z_d) \in \mathbb{C}^d : |z_1| = \cdots = |z_d| = 1\}$, that is,

$$(\mathbb{Z}^d)^* = \{\chi_z : z \in \mathbb{T}^d\} \cong \mathbb{T}^d, \quad \text{where} \quad \chi_z(p) = p(z).$$

The $(\mathfrak{n}, \mathfrak{m})$-entry of the generalized Hankel matrix $H(s)$ is $s_{\mathfrak{m}-\mathfrak{n}}$ and we have $\overline{s_{\mathfrak{m}-\mathfrak{n}}} = s_{\mathfrak{n}-\mathfrak{m}}$ for $\mathfrak{n}, \mathfrak{m} \in \mathbb{Z}^d$. In the case $d = 1$ the matrix $H(s)$ is given by

$$H(s) = \begin{pmatrix} \cdots & \cdots & \cdots & \cdots & \cdots & \cdots \\ \cdots & s_0 & s_1 & s_2 & s_3 & \cdots \\ \cdots & \overline{s_1} & s_0 & s_1 & s_2 & \cdots \\ \cdots & \overline{s_2} & \overline{s_1} & s_0 & s_1 & \cdots \\ \cdots & \overline{s_3} & \overline{s_2} & \overline{s_1} & s_0 & \cdots \\ \cdots & \cdots & \cdots & \cdots & \cdots & \cdots \end{pmatrix}, \tag{2.20}$$

where $\underline{s_0}$ stands at the $(0,0)$-entry. A matrix of this form is called a *Toeplitz matrix*. This matrix is constant on each descending diagonal from left to right.

2.3.4   Example 4: $\mathbb{Z}, n^* = n$

The additive group $\mathbb{Z}$ with identity involution is a $*$-semigroup. The map $n \mapsto x^n$ is a $*$-isomorphism of the group $*$-algebra $\mathbb{R}[\mathbb{Z}]$ on the $*$-algebra $\mathbb{R}[x, x^{-1}]$ of Laurent polynomials with involution given by $x^* = x$. The character space $\mathbb{Z}^*$ is

$$\mathbb{Z}^* = \{\chi_t : t \in \mathbb{R}\backslash\{0\}\} \cong \mathbb{R}\backslash\{0\}, \quad \text{where } \chi_t(p) = p(t, t^{-1}).$$

The (n, m)-entry of the generalized Hankel matrix $H(s)$ is s_{n+m}, so that

$$H(s) = \begin{pmatrix} \cdots \cdots \cdots & \cdots \cdots & \cdots \cdots \cdots \\ \cdots \; s_{-2} \; s_{-1} \; s_0 \; \cdots \\ \cdots \; s_{-1} \; s_0 \; s_1 \; \cdots \\ \cdots \; s_0 \; s_1 \; s_2 \; \cdots \\ \cdots \; s_1 \; s_2 \; s_3 \; \cdots \\ \cdots \cdots \cdots & \cdots \cdots & \cdots \cdots \cdots \end{pmatrix}. \tag{2.21}$$

Here again s_0 is located at the $(0,0)$-entry of the matrix. That is, $H(s)$ is a "usual" two-sided infinite *Hankel matrix* which is constant on cross-diagonals.

2.4   Exercises

1. Let S be an abelian $*$-semigroup. Show that $S \times S$ is an abelian $*$-semigroup with product $(s_1, s_2) \circ (s_1', s_2') = (s_1 s_1', s_2 s_2')$ and involution $(s_1, s_2)^* = (s_2, s_1)$, where $s_1, s_1', s_2 s_2' \in S$. Which examples of Sect. 2.3 fit into this scheme?

2. Let $s = (s_n)_{n \in \mathbb{N}_0}$ be a complex positive semidefinite sequence for the $*$-semigroup $\mathbb{N}_0$ with involution $n^* = n$. Prove the following:

   a. $s_n \in \mathbb{R}$ and $s_{2n} \geq 0$ for $n \in \mathbb{N}_0$.
   b. $(s_{m+n})^2 \leq s_{2m} s_{2n}$ for $m, n \in \mathbb{N}_0$.
   c. $(s_n)^{2^k} \leq (s_0)^{2^k - 1} s_{n2^k}$ for $n \in \mathbb{N}_0$, $k \in \mathbb{N}$. In particular, $s_0 = 0$ implies that $s_n = 0$ for all $n \in \mathbb{N}$.

3. (*Schur's theorem*) Show that if $A = (a_{ij})_{i,j=1}^n$ and $B = (b_{ij})_{i,j=1}^n$ are positive semidefinite matrices over $\mathbb{R}$, then so is the matrix $C := (a_{ij} b_{ij})_{i,j=1}^n$.

4. Show that if $s = (s_n)_{n \in \mathbb{N}_0}$ and $t = (t_n)_{n \in \mathbb{N}_0}$ are positive semidefinite sequences for the $*$-semigroup $\mathbb{N}_0$, then so is the pointwise product sequence $(s_n t_n)_{n \in \mathbb{N}_0}$.

5. Let φ and ψ be positive semidefinite functions on the additive group $\mathbb{R}$. Show that $\overline{\varphi}$, $\varphi + \psi$, and $\varphi \psi$ are also positive semidefinite functions on $\mathbb{R}$.

6. Show that $\varphi(t) = \cos t$ is a positive semidefinite function on $\mathbb{R}$.

7. Let $\mu \in M_+(\mathbb{R})$ be a finite measure. Prove that

$$\varphi(t) = \int_{\mathbb{R}} e^{-itx} \, d\mu(x), \quad t \in \mathbb{R},$$

is a continuous positive semidefinite function on $\mathbb{R}$.

(Bochner's theorem (see e.g. [RS2]) states that each continuous positive semidefinite function φ on $\mathbb{R}$ is of this form with μ uniquely determined by φ.)

2.5 Notes

Basics on positive functionals and general ∗-algebras can be found (for instance) in [Sm4]. The notion of a ∗-semigroup appeared first in the Appendix written by B. Sz.-Nagy of the functional analysis textbook [RzSz]. The standard monograph about harmonic analysis on semigroups is [BCR1].

Part I
The One-Dimensional Moment Problem

Chapter 3
One-Dimensional Moment Problems on Intervals: Existence

In this chapter we begin the study of one-dimensional (full) moment problems:

Given a real sequence $s = (s_n)_{n \in \mathbb{N}_0}$ and closed subset K of $\mathbb{R}$, the K-moment problems asks: When does there exist a Radon measure μ on $\mathbb{R}$ supported on K such that $s_n = \int_{\mathbb{R}} x^n d\mu(x)$ for all $n \in \mathbb{N}_0$?

Our main aims are the solvability theorems for $K = \mathbb{R}$ (Hamburger's Theorem 3.8), $K = [0, +\infty)$ (Stieltjes' Theorem 3.12), and $K = [a,b]$ (Hausdorff's Theorems 3.13 and 3.14). They are derived in Sect. 3.2 from Haviland's theorem 1.12. To apply this result representations of positive polynomials in terms of sums of squares are needed. In Sect. 3.1 we develop such descriptions that are sufficient for the applications in Sects. 3.2 and for the truncated moment problems treated in Sects. 9.4 and 10.1.

In Sect. 3.3 we establish a one-to-one correspondence between the Stieltjes moment problem and the symmetric Hamburger moment problem. In Sect. 3.4 we derive *unique* representations of nonnegative polynomials on intervals (Propositions 3.20–3.22). These results are stronger than those obtained in Sect. 3.1 and they are of interest in themselves.

3.1 Positive Polynomials on Intervals

Suppose that $p(x) \in \mathbb{R}[x]$ is a fixed nonconstant polynomial. Since p has real coefficients, it follows that if λ is a non-real zero of p with multiplicity l, so is $\bar{\lambda}$. Clearly, $(x - \lambda)^l (x - \bar{\lambda})^l = ((x - u)^2 + v^2)^l$, where $u = \operatorname{Re} \lambda$ and $v = \operatorname{Im} \lambda$. Therefore, by the fundamental theorem of algebra, each nonzero real polynomial p factors as

$$p(x) = a(x - \alpha_1)^{n_1} \cdots (x - \alpha_r)^{n_r} (x - \lambda_1)^{j_1} (x - \bar{\lambda}_1)^{j_1} \ldots (x - \lambda_k)^{j_k} (x - \bar{\lambda}_k)^{j_k} \qquad (3.1)$$

$$= a(x - \alpha_1)^{n_1} \cdots (x - \alpha_r)^{n_r} ((x - u_1)^2 + v_1^2)^{j_1} \ldots ((x - u_k)^2 + v_k^2)^{j_k}, \qquad (3.2)$$

© Springer International Publishing AG 2017
K. Schmüdgen, *The Moment Problem*, Graduate Texts in Mathematics 277,
DOI 10.1007/978-3-319-64546-9_3

where

$$n_1, \ldots, n_r, j_1, \ldots, j_k \in \mathbb{N}, \ a, \alpha_1, \ldots, \alpha_r \in \mathbb{R},$$

$$\lambda_1 = u_1 + iv_1, \ldots, \lambda_k = u_k + iv_k, \ u_1, \ldots, u_k \in \mathbb{R}, \ v_1 > 0, \ldots, v_k > 0,$$

$$\alpha_i \neq \alpha_j \text{ if } i \neq j, \text{ and } \lambda_i \neq \lambda_j, \ \lambda_i \neq \overline{\lambda}_j \text{ if } i \neq j.$$

Thus, Eq. (3.2) expresses p as a product of a constant a, of powers of pairwise different linear polynomials $x - \alpha_i$ with real zeros α_i, and of powers of pairwise different quadratic polynomials $(x - u_j)^2 + v_j^2$ with no real zeros. Note that linear factors or quadratic factors may be absent in (3.2). Up to the numeration of factors *the representation (3.2) (and likewise the representation (3.1)) of p is unique.*

If a, b, c, d are elements of a commutative ring, there is the *two square identity*

$$(a^2 + b^2)(c^2 + d^2) = (ac - bd)^2 + (ad + bc)^2. \tag{3.3}$$

This implies that each product of sums of two squares is again a sum of two squares.

Recall that $\mathrm{Pos}(M)$ is the set of $p \in \mathbb{R}[x]$ that are nonnegative on $M \subseteq \mathbb{R}$ and $\sum \mathbb{R}[x]^2$ is the set of finite sums of squares p^2, where $p \in \mathbb{R}[x]$. We denote by $\mathbb{R}[x]_n$ and $\mathrm{Pos}(M)_n$ the corresponding subsets of polynomials p such that $\deg(p) \leq n$ and by $\sum_n^2$ the set of finite sums of squares p^2, where $\deg(p) \leq n$.

The following three propositions contain all results on positive polynomials needed for solving the moment problem on intervals. The formulas containing polynomial degrees will be used later for the truncated moment problems.

Proposition 3.1

(i) $\mathrm{Pos}(\mathbb{R}) = \sum \mathbb{R}[x]^2 = \{f^2 + g^2 : f, g \in \mathbb{R}[x]\}.$
(ii) $\mathrm{Pos}(\mathbb{R})_{2n} = \sum_n^2 = \{f^2 + g^2 : f, g \in \mathbb{R}[x]_n\}.$

Proof

(i) Let $p \in \mathrm{Pos}(\mathbb{R})$, $p \neq 0$. Since $p(x) \geq 0$ on $\mathbb{R}$, it follows that $a > 0$ and the numbers $k_1, \ldots, k_r$ in (3.2) are even. Hence p is a product of squares and of sums of two squares. Therefore, by (3.3), p is of the form $f^2 + g^2$, where $f, g \in \mathbb{R}[x]$. The other inclusions are obvious.
(ii) follows at once from (i), because $\deg(f^2 + g^2) = 2\max(\deg(f), \deg(g))$. $\square$

Proposition 3.2

$$\mathrm{Pos}([0, +\infty)) = \{f + xg : f, g \in \textstyle\sum \mathbb{R}[x]^2\}, \tag{3.4}$$

$$\mathrm{Pos}([0, +\infty))_{2n} = \{f + xg : f \in \textstyle\sum_n^2, \ g \in \textstyle\sum_{n-1}^2\}, \ n \in \mathbb{N}, \tag{3.5}$$

$$\mathrm{Pos}([0, +\infty))_{2n+1} = \{f + xg : f, g \in \textstyle\sum_n^2\}, \ n \in \mathbb{N}_0. \tag{3.6}$$

Proof Clearly, (3.4) implies (3.5) and (3.7), since

$$\deg(\sum_i f_i^2 + x g_i^2) = \max_i(2\deg(f_i), 1 + 2\deg(g_i)).$$

Let us abbreviate $Q := \sum \mathbb{R}[x]^2 + x \sum \mathbb{R}[x]^2$. It is obvious that $Q \subseteq \mathrm{Pos}([0, +\infty))$. Thus, it suffices to prove that $\mathrm{Pos}([0, +\infty)) \subseteq Q$.

Next we note that the set Q is closed under multiplication. Indeed, for arbitrary $f_1, f_2, g_1, g_2 \in \sum \mathbb{R}[x]^2$, we have

$$(f_1 + xg_1)(f_2 + xg_2) = (f_1f_2 + x^2g_1g_2) + x(f_1g_2 + g_1f_2) \in Q.$$

Let $p \in \mathrm{Pos}([0, +\infty)), p \neq 0$, and consider the representation (3.2). Since Q is closed under multiplication, it suffices to show that all factors from (3.2) are in Q. Products of quadratic factors and even powers of linear factors are obviously in Q. It remains to handle the constant a and the linear factor $x - \alpha_i$ for each real zero α_i of odd multiplicity. Since $p(x) \geq 0$ on $[0, +\infty)$, we have $a > 0$ by letting $x \to +\infty$ and $\alpha_i \leq 0$, because $p(x)$ changes its sign in the neighbourhood of a zero with odd multiplicity. Hence $a \in Q$ and $x - \alpha_i = (-\alpha_i + x) \in \sum \mathbb{R}[x]^2 + x \sum \mathbb{R}[x]^2 = Q$. □

Proposition 3.3 *Suppose that $a, b \in \mathbb{R}$, $a < b$. Then:*

$$\mathrm{Pos}([a, b]) = \{f + (x - a)g : f, g \in \sum \mathbb{R}[x]^2\}, \tag{3.7}$$

$$\mathrm{Pos}([a, b])_{2n} = \{f + (b - x)(x - a)g : f \in \Sigma_n^2, g \in \Sigma_{n-1}^2\}, \tag{3.8}$$

$$\mathrm{Pos}([a, b])_{2n+1} = \{(b - x)f + (x - a)g : f, g \in \Sigma_n^2\}. \tag{3.9}$$

Proof The equality (3.7) is an immediate consequence of (3.8) and (3.9).

All polynomials on the right-hand sides of (3.8) and (3.9) belong to the corresponding left-hand sides. We prove the converse inclusions of (3.8) and (3.9) by induction on n. Both (3.8) and (3.9) are trivial for $n = 0$. Assume that (3.8) and (3.9) hold for n. Let $p \in \mathrm{Pos}([a, b])_{2n+2}$ or $p \in \mathrm{Pos}([a, b])_{2n+3}$.

Suppose that p has a quadratic factor q without real zeros in (3.2). Multiplying by -1 if necessary we can assume that $q \geq 0$ on $\mathbb{R}$. Then $p = qp_0$ with $p_0 \in \mathrm{Pos}([a, b])$ and $\deg(p_0) \leq \deg(p) - 2$. Applying the induction hypothesis to p_0 it follows that p is in the corresponding set on the right-hand side of (3.8) or (3.9).

Now we treat the case when p has a real zero, say α. Upon a linear transformation we can assume without loss of generality that $a = 0$ and $b = 1$. Then $(b-x)(x-a) = x(1-x)$. First let $\alpha \in (0, 1)$. Considering $p(x)$ in a neighbourhood of α, we conclude that α has even multiplicity. Hence we can factorize $p = (x - \alpha)^2 p_0$ and argue as in the preceding paragraph. Thus we can assume that $\alpha \notin (0, 1)$.

Case 1: $p \in \mathrm{Pos}([0, 1])_{2n+2}$.
First suppose that $\alpha \leq 0$. Then $x - \alpha \geq 0$ on $[0, 1]$, so we can write $p = (x - \alpha)p_0$ with $p_0 \in \mathrm{Pos}([0, 1])_{2n+1}$. By the induction hypothesis we have

$$p_0 \in (1 - x)f + xg \quad \text{with } f, g \in \Sigma_n^2. \tag{3.10}$$

Therefore, from the identity

$$p = (x - \alpha)p_0 = x(1 - x)f + x^2g - \alpha[x(1 - x)(f + g) + (1 - x)^2f + x^2g]$$

we conclude that $p \in \sum_{n+1}^2 + x(1 - x)\sum_n^2$.

Now suppose that $\alpha \geq 1$. Then $p = (\alpha - x)p_0$ with $p_0 \in \mathrm{Pos}([a, b])_{2n+1}$. The assertion $p \in \sum_{n+1}^2 + x(1 - x)\sum_n^2$ follows from the induction hypotheses (3.10) combined with the identity

$$p = (\alpha - x)p_0 = (1 - x)p_0 + (\alpha - 1)p_0$$

$$= (1 - x)^2f + x(1 - x)g + (\alpha - 1)[x(1 - x)(f + g) + (1 - x)^2f + x^2g].$$

Case 2: $p \in \mathrm{Pos}([a, b])_{2n+3}$.

First let $\alpha \leq 0$. Then we write $p = (x - \alpha)p_0$ with $p_0 \in \mathrm{Pos}([a, b])_{2n+2}$. Hence, by the induction hypothesis,

$$p_0 = f + x(1 - x)g \quad \text{with } f \in \sum_{n+1}^2, \ g \in \sum_n^2.$$

Then the identity

$$p = (x - \alpha)p_0 = xf + (1 - x)x^2g - \alpha/2\,[x\,(2f + (1 - x)^2g) + (1 - x)(2f + x^2g)]$$

implies that $p \in (1 - x)\sum_{n+1}^2 + x\sum_{n+1}^2$.

Now let $\alpha \geq 1$. Then it follows from

$$p = (\alpha - x)p_0 = (1 - x)p_0 + (\alpha - 1)p_0$$

$$= (1 - x)f + x(1 - x)^2g + (\alpha - 1)/2\,[x\,(2f + (1 - x)^2g) + (1 - x)(2f + x^2g)]$$

that $p \in (1 - x)\sum_{n+1}^2 + x\sum_{n+1}^2$.

This completes the induction proof of (3.8) and (3.9).   □

Another proof of formulas (3.8) and (3.9) is sketched in Exercise 3.2. Descriptions of $\mathrm{Pos}(K)$ for some other sets K are given in Exercise 3.7. In Sect. 3.4 we give stronger forms of representations of positive polynomials.

The next proposition is a classical result due to *S. Bernstein*. It enters into the solution of Hausdorff's moment problem given by Proposition 3.14 below.

Proposition 3.4 *Suppose that $p \in \mathbb{R}[x]$ and $p(x) > 0$ for all $x \in [-1, 1]$. Then there are numbers $m \in \mathbb{N}$ and $a_{kl} \geq 0$ for $k, l = 1, \ldots, m$ such that*

$$p(x) = \sum_{k,l=0}^m a_{kl}(1 - x)^k(1 + x)^l. \tag{3.11}$$

The proof of Proposition 3.4 is based on two classical lemmas which are of interest in themselves. The following lemma is due to E. *Goursat*.

Lemma 3.5 *Suppose that $p \in \mathbb{R}[x]$, $p \neq 0$, and $m = \deg(p)$. The* Goursat transform *of p is the polynomial $\tilde{p} \in \mathbb{R}[x]$ defined by*

$$\tilde{p}(x) = (1+x)^m p\left(\frac{1-x}{1+x}\right). \tag{3.12}$$

Then we have:

(i) $\deg(\tilde{p}) \leq m$ *and we have* $\deg(\tilde{p}) = m$ *if and only if* $p(-1) \neq 0$.
(ii) $p \in \text{Pos}([-1, 1])$ *if and only if* $\tilde{p} \in \text{Pos}([0, +\infty))$.
(iii) $p(x) > 0$ *on* $[-1, 1]$ *if and only if* $\tilde{p}(x) > 0$ *on* $[0, +\infty)$ *and* $\deg(\tilde{p}) = m$.

Proof

(i) Let $p(x) = \sum_{k=0}^{m} a_k x^k$. It is obvious that

$$\tilde{p}(x) = \sum_{k=0}^{m} a_k (1+x)^{m-k}(1-x)^k$$

is a polynomial and $\deg(\tilde{p}) \leq m$. Its coefficient of x^m is $\sum_{k=0}^{m} a_k (-1)^k = p(-1)$ Thus $\deg(\tilde{p}) = m$ if and only if $p(-1) \neq 0$.

(ii) For $x \neq -1$ we set $t = \frac{1-x}{1+x}$. Then $t \neq -1$ and $x = \frac{1-t}{1+t}$. Further, $x \in (-1, 1]$ if and only if $t \in [0, +\infty)$. Therefore, we have $p(x) \geq 0$ on $(-1, 1]$, or equivalently $p(x) \geq 0$ on $[-1, 1]$, if and only if $\tilde{p}(t) \geq 0$ on $[0, +\infty)$.

(iii) Clearly, $p(x) > 0$ for $x \in (-1, 1]$ if and only if $\tilde{p}(t) > 0$ on $[0, +\infty)$. If this holds, then $p(-1) \geq 0$, so that $p(-1) > 0$ if and only if $\deg(\tilde{p}) = m$ by (i). $\square$

Remark 3.6 Let us note the following interesting facts:
The inverse of the mapping $x \mapsto t = \frac{1-x}{1+x}$ is given by the same formula $t \mapsto x = \frac{1-t}{1+t}$. If p and its Goursat transform $\tilde{p}$ have degree m, then the Goursat transform of $\tilde{p}$ is just $2^m p$. ∘

The next lemma is the one-dimensional version of a classical result of G. *Polya*; a multivariate version is given by Proposition 12.51 below.

Lemma 3.7 *Suppose that $p \in \mathbb{R}[x]$ and $p(x) > 0$ for $x \in [0, +\infty)$. Then there exists an $N \in \mathbb{N}$ such that $(1 + x)^N p(x)$ has only positive coefficients, that is,*

$$(1+x)^N p(x) = \sum_{k=0}^{m} b_k x^k \quad \text{with } b_k > 0, \ k = 0, \dots, m.$$

Proof Let us introduce the notation $(z)_t^j := z(z-t) \dots (z - (j-1)t)$. Then

$$\left(\frac{k}{m}\right)^j_{1/m} = \frac{k}{m}\left(\frac{k}{m} - \frac{1}{m}\right) \cdots \left(\frac{k}{m} - \frac{j-1}{m}\right) = \frac{k!}{(k-j)! \, m^j}. \tag{3.13}$$

Let $p(x) = \sum_{j=0}^{n} a_j x^j$, where $a_n \neq 0$, and define

$$P(x,y) := \sum_{j=0}^{n} a_j x^j y^{n-j} \quad \text{and} \quad P_t(x,y) := \sum_{j=0}^{n} a_j (x)^j_t (y)^{n-j}_t. \tag{3.14}$$

Since $p(x) > 0$ on $[0, +\infty)$ and $\deg(p) = n$, the homogenous polynomial P is positive on $\Delta = \{(x,y) : x \geq 0, y \geq 0, x + y = 1\}$, so P has a positive minimum, say c, on the compact set Δ. For $N \in \mathbb{N}$ we have

$$(x + y)^N P(x,y) = \sum_{j=0}^{n} \sum_{i=0}^{N} a_j \binom{N}{i} x^{i+j} y^{N+n-i-j}. \tag{3.15}$$

Fix $k \in \mathbb{N}_0$ such that $k \leq m$. Set $m := N + n$ and $l := m - k$. The coefficient b_k of $x^k y^l$ in (3.15) is

$$
\begin{aligned}
b_k &= \sum_{j=0}^{k} a_j \binom{N}{k-j} \sum_{j=0}^{k} a_j \frac{N!}{(k-j)!(N-(k-j))!} \\
&= \frac{N! \, m^n}{k! \, l!} \sum_{j=0}^{k} a_j \frac{k!}{(k-j)! \, m^j} \frac{l!}{(l-(n-j))! \, m^{n-j}} \\
&= \frac{N! \, m^n}{k! \, l!} \sum_{j=0}^{k} a_j \left(\frac{k}{m}\right)^j_{1/m} \left(\frac{l}{m}\right)^{n-j}_{1/m} = \frac{N! \, m^n}{k! \, l!} \, P_{1/m}\left(\frac{k}{m}\right)\left(\frac{l}{m}\right).
\end{aligned}
$$

Here the equality before last holds by (3.13) and the last equality is the definition of $P_{1/m}$. Since $P_t(x,y) \to P(x,y)$ uniformly on Δ as $t \to +0$ and $P(x,y) \geq c > 0$ on Δ, it follows that $b_k > 0$ for all k if N, hence $m = N + n$, is sufficiently large. $\qquad \square$

Proof of Proposition 3.4 Let $n = \deg(p)$. Since $p(x) > 0$ on $[-1, 1]$, Lemma 3.5(iii) implies that the Goursat transform $\tilde{p}$ has degree n and $\tilde{p}(x) > 0$ on $[0, +\infty)$. Thus, by Lemma 3.7, there are numbers $N \in \mathbb{N}$ and $a_0 > 0, \ldots, a_{N+n} > 0$ such that

$$(1 + t)^N \tilde{p}(t) = \sum_{j=0}^{N+n} a_j t^j. \tag{3.16}$$

Set $m := N + n$ and $t = \frac{1-x}{1+x}$ for $x \neq -1$. Then $x = \frac{1-t}{1+t}$ and $(1 + t)^{-1} = \frac{1+x}{2}$, so that $\tilde{p}(t) = (1 + t)^n p(x)$. Inserting these facts and using Eq. (3.16) we derive

$$p(x) = (1 + t)^{-n}\tilde{p}(t) = (1 + t)^{N-m}\tilde{p}(t) = \sum_{j=0}^{N+n} a_j t^j (1 + t)^{-m}$$

$$= \sum_{j=0}^{m} a_j \left(\frac{1-x}{1+x}\right)^j \left(\frac{1+x}{2}\right)^m = \sum_{j=0}^{m} 2^{-m} a_j (1-x)^j (1+x)^{m-j}. \qquad \square$$

3.2   The Moment Problem on Intervals

In this section, we solve the moment problem for closed intervals J by combining Haviland's theorem with the descriptions of Pos(J) given in the preceding section.

Let $\mathcal{P}(\mathbb{N}_0)$ denote the set of real sequences $s = (s_n)_{n \in \mathbb{N}_0}$ which are positive semidefinite, that is, for all $\xi_0, \xi_1, \ldots, \xi_n \in \mathbb{R}$ and $n \in \mathbb{N}$ we have

$$\sum_{k,l=0}^{n} s_{k+l} \xi_k \xi_l \geq 0. \tag{3.17}$$

Let $s = (s_n)_{n \in \mathbb{N}_0}$ be a real sequence. Recall that L_s is the Riesz functional on $\mathbb{R}[x]$ defined by $L_s(x^n) = s_n$, $n \in \mathbb{N}_0$. Let Es denote the shifted sequence given by

$$(Es)_n = s_{n+1}, \quad n \in \mathbb{N}_0.$$

Clearly, $L_{Es}(p(x)) = L_s(xp(x))$ for $p \in \mathbb{R}[x]$.

Further, we define the *Hankel matrix* $H_n(s)$ and the *Hankel determinant* $D_n(s)$ by

$$H_n(s) = \begin{pmatrix} s_0 & s_1 & s_2 & \ldots & s_n \\ s_1 & s_2 & s_3 & \ldots & s_{n+1} \\ s_2 & s_3 & s_4 & \ldots & s_{n+2} \\ \ldots & \ldots & \ldots & & \ldots \\ s_n & s_{n+1} & s_{n+2} & \ldots & s_{2n} \end{pmatrix}, \quad D_n(s) = \det H_n(s). \tag{3.18}$$

The following result is *Hamburger's theorem*.

Theorem 3.8 (Solution of the Hamburger Moment Problem) *For a real sequence $s = (s_n)_{n \in \mathbb{N}_0}$ the following are equivalent:*

(i) *s is a Hamburger moment sequence, that is, there is a measure $\mu \in M_+(\mathbb{R})$ such that $x^n \in \mathcal{L}^1(\mathbb{R}, \mu)$ and*

$$s_n = \int_{\mathbb{R}} x^n d\mu(x) \quad \text{for} \quad n \in \mathbb{N}_0. \tag{3.19}$$

(ii) *$s \in \mathcal{\dot{P}}(\mathbb{N}_0)$, that is, the sequence s is positive semidefinite.*

(iii) *All Hankel matrices $H_n(s)$, $n \in \mathbb{N}_0$, are positive semidefinite.*

(iv) *L_s is a positive linear functional on $\mathbb{R}[x]$, that is, $L_s(p^2) \geq 0$ for $p \in \mathbb{R}[x]$.*

Proof From Proposition 2.7 we know that (i) implies (ii) and that (ii) and (iv) are equivalent. The Hankel matrix $H_n(s)$ is just the symmetric matrix associated with the quadratic form in (3.17); hence (ii) and (iii) are equivalent. The main implication (iv)→(i) follows from Haviland's Theorem 1.12 combined with Proposition 3.1.   □

The next proposition deals with representing measures of *finite* support.

Proposition 3.9 *For a positive semidefinite sequence s the following are equivalent:*

(i) *There is a number $n \in \mathbb{N}_0$ such that*

$$D_0(s) > 0, \ldots, D_{n-1}(s) > 0 \quad and \quad D_k(s) = 0 \quad for \ k \geq n. \tag{3.20}$$

(ii) *s is a moment sequence with a representing measure μ supported on n points.*

Proof By Theorem 3.8 the sequence s has a representing measure μ. For $c = (c_0, c_1, \ldots, c_k)^T \in \mathbb{R}^{k+1}$ we define $p_c(x) := \sum_{j=0}^k c_j x^j$. Then, by (3.19) we derive

$$c^T H_k(s) c = \sum_{j,l=0}^k s_{j+l} c_j c_l = \int \left| \sum_{j=1}^k c_j x^j \right|^2 d\mu(x) = \int |p_c(x)|^2 d\mu(x). \tag{3.21}$$

The proof is based on the following two facts.

I. First suppose that supp μ consists of n points. Then, for $k \geq n$ we can choose $c \in \mathbb{R}^{k+1}, c \neq 0$, such that the polynomial $p_c(x)$ vanishes on supp μ. Then (3.21) is zero, so $H_k(s)$ is not positive definite and hence $D_k(s) = 0$.

II. Suppose that $D_k(s) = 0$. Then $H_k(s)$ is not positive definite, so there exists a $c \neq 0$ such that the expression in (3.21) is zero. Therefore, by Proposition 1.23, supp $\mu \subseteq \mathcal{Z}(p_c)$. Since $\deg(p) \leq k$, supp μ contains at most k points.

(i)→(ii) Since $D_n(s) = 0$ by (i), supp μ has at most n points by II. If supp μ had fewer than n points, then we would have $D_{n-1}(s) = 0$ by I, which contradicts (i).

(ii)→(i) Then $D_k(s) = 0$ for $k \geq n$ by I. If $D_k(s)$ were zero for some $k < n-1$, then supp μ would have at most $n-1$ points by II. This contradicts (ii).   □

Remark 3.10 It was recently proved in [BS3] that the assumption "s is positive semidefinite" in Proposition 3.9 can be omitted. That is, if s is an arbitrary real sequence satisfying condition (i), then s is a Hamburger moment sequence (and has an n-atomic representing measure by Proposition 3.9).   ∘

Many considerations in subsequent chapters require the stronger assumption that the moment sequence $s = (s_n)_{n \in \mathbb{N}_0}$ s is *positive definite*, that is,

$$\sum_{k,l=0}^n s_{k+l} c_k c_l > 0 \quad \text{for all} \ c = (c_0, c_1, \ldots, c_n)^T \in \mathbb{R}^{n+1}, c \neq 0, n \in \mathbb{N}_0.$$

The following proposition characterizes this property.

Proposition 3.11 *For a Hamburger moment sequence* $s = (s_n)_{n \in \mathbb{N}_0}$ *the following statements are equivalent:*

(i) *Each representating measure μ of s has infinite support.*
(ii) *s is positive definite.*
(iii) *$H_n(s)$ is positive definite for all $n \in \mathbb{N}_0$.*
(iv) *$D_n(s) > 0$ for all $n \in \mathbb{N}_0$.*

Proof The equivalence (ii)↔(iii) and the implication (iii)→(iv) are clear from elementary linear algebra. Proposition 3.9 yields (i)↔(iv).

It suffices to prove (i)→(iii). Assume that (3.21) vanishes for some c. Then the *infinite* set $\mathrm{supp}\mu$ is contained in the zero set of the polynomial p_c. Hence $p_c = 0$, so that $c = 0$. Thus, $H_k(s)$ is positive definite for each $k \in \mathbb{N}_0$. This proves (iii). □

The second main result is *Stieltjes' theorem.*

Theorem 3.12 (Solution of the Stieltjes Moment Problem) *For any real sequence s the following statements are equivalent:*

(i) *s is a Stieltjes moment sequence, that is, there is a measure $\mu \in M_+([0, +\infty))$ such that $x^n \in \mathcal{L}^1([0, +\infty), \mu)$ and*

$$s_n = \int_0^\infty x^n d\mu(x) \quad \text{for} \quad n \in \mathbb{N}_0. \tag{3.22}$$

(ii) *$s \in \mathcal{P}(\mathbb{N}_0)$ and $Es \in \mathcal{P}(\mathbb{N}_0)$.*
(iii) *All Hankel matrices $H_n(s)$ and $H_n(Es)$, $n \in \mathbb{N}_0$, are positive semidefinite.*
(iv) *$L_s(p^2) \geq 0$ and $L_s(xq^2) \geq 0$ for all $p, q \in \mathbb{R}[x]$.*

Proof The proof is almost the same as the proof of Theorem 3.8; instead of Proposition 3.1 we apply formula (3.5) in Proposition 3.3. □

Combining Haviland's theorem with (3.7) the same reasoning used in the proofs of Theorems 3.8 and 3.12 yields the following result.

Theorem 3.13 (Solution of the Moment Problem for a Compact Interval) *Let $a, b \in \mathbb{R}$, $a < b$. For a real sequence s the following are equivalent:*

(i) *s is an $[a, b]$-moment sequence.*
(ii) *$s \in \mathcal{P}(\mathbb{N}_0)$ and $((a + b)Es - E(Es) - ab\, s) \in \mathcal{P}(\mathbb{N}_0)$.*
(iii) *$L_s(p^2) \geq 0$ and $L_s((b - x)(x - a)q^2) \geq 0$ for all $p, q \in \mathbb{R}[x]$.*

Bernstein's theorem (Proposition 3.4) allows us to derive two solvablity criteria of moment problems which are not based on squares of polynomials.

Theorem 3.14 *Let $s = (s_n)_{n \in \mathbb{N}_0}$ be a real sequence and let L_s be its Riesz functional on $\mathbb{R}[x]$. Then s is a $[-1, 1]$-moment sequence if and only if*

$$L_s((1 - x)^n(1 + x)^k) \geq 0 \quad \text{for all} \quad k, n \in \mathbb{N}_0. \tag{3.23}$$

Proof The only if part is obvious, since all polynomials $(1 - x)^n(1 + x)^k$ are nonnegative on $[-1, 1]$. To prove the only if part we assume that condition (3.23) holds. Then, by Proposition 3.4, $L_s(p) \geq 0$ for all strictly positive polynomials on $[-1, 1]$. Therefore, by Haviland's theorem 1.12 (ii)→(iv), L_s is a $[-1, 1]$-moment functional. Hence s is a $[-1, 1]$-moment sequence. □

Theorem 3.14 leads to the following criterion for the Hausdorff moment problem.

Theorem 3.15 *A real sequence s is a $[0, 1]$-moment sequence if and only if*

$$((I - E)^n s)_k \equiv \sum_{j=0}^n (-1)^j \binom{n}{j} s_{k+j} \geq 0 \quad for \quad k, n \in \mathbb{N}_0. \tag{3.24}$$

Proof By applying the bijection $x \mapsto \frac{1}{2}(x + 1)$ of the intervals $[-1, 1]$ onto $[0, 1]$ we conclude from Theorem 3.14 that s is a $[0, 1]$-moment sequence if and only if $L_s((1 - x)^n x^k) \geq 0$ for $k, n \in \mathbb{N}_0$. But the latter is equivalent to (3.24), since

$$((I - E)^n s)_k = \sum_{j=0}^n (-1)^j \binom{n}{j} (E^j s)_k = \sum_{j=0}^n (-1)^j \binom{n}{j} s_{k+j}$$

$$= \sum_{j=0}^n (-1)^j \binom{n}{j} L_s(x^{k+j}) = L_s((1 - x)^n x^k). \qquad □$$

Condition (3.24) is an important property in the context of $*$-semigroups. Let S be an abelian unital semigroup with identity map as involution. For $y \in S$ and a function φ on S we define the shift E_y and a mapping Δ_y by

$$(E_y \varphi)(z) := \varphi(z + y), \quad z \in S, \quad \text{and} \quad \Delta_y := E_y - I.$$

A function $\varphi : S \to \mathbb{R}$ is called *completely monotone* if $\varphi(z) \geq 0$ and

$$(-1)^n (\Delta_{y_1} \ldots \Delta_{y_n} \varphi)(z) = ((I - E_{y_1}) \ldots (I - E_{y_n}) \varphi)(z) \geq 0 \quad \text{for } z \in S$$

and $y_1, \ldots, y_n \in S$. Completely monotone functions are moment functions, see e.g. [BCRl, Chapter 4, Theorem 6.4]. It can be shown that condition (3.24) implies that the function $\varphi(n) = s_n, n \in \mathbb{N}_0$, on the semigroup $S = \mathbb{N}_0$ is completely monotone, so Theorem 3.15 becomes a special case of this general result.

We close this section by treating the moment problem for the $*$-semigroup $\mathbb{Z}$ with involution $n^* = n$. The corresponding moment problem is called the *two-sided Hamburger moment problem* or *strong Hamburger moment problem*.

Clearly, the map $n \mapsto x^n$ yields an isomorphism of the semigroup algebra $\mathbb{R}[\mathbb{Z}]$ and the algebra $\mathbb{R}[x, x^{-1}]$ of real Laurent polynomials. It is easily checked that the characters of $\mathbb{R}[x, x^{-1}]$ are precisely the evaluations at points of $\mathbb{R}^\times := \mathbb{R} \backslash \{0\}$.

Theorem 3.16 (Solution of the Two-Sided Hamburger Moment Problem) *For a real sequence $s = (s_n)_{n \in \mathbb{Z}}$ the following statements are equivalent:*

(i) *s is a moment sequence for the $*$-semigroup $\mathbb{Z}$, that is, there exists a positive Radon measure μ on $\mathbb{R}^\times$ such that the function x^n on $\mathbb{R}^\times$ is μ-integrable and*

$$s_n = \int_{\mathbb{R}^\times} x^n \, d\mu \quad \text{for all} \ \ n \in \mathbb{Z}. \tag{3.25}$$

(ii) *$s \in \mathcal{P}(\mathbb{Z})$, that is, s is positive semidefinite on $\mathbb{Z}$.*

(iii) *L_s is a positive functional, that is, $L_s(f^2) \geq 0$ for all $f \in \mathbb{R}[x, x^{-1}]$.*

Proof (ii)↔(iii) and (i)→(ii) follow from Proposition 2.7 and Corollary 2.16, respectively. We prove the main implication (iii)→(i).

Let $p \in \mathbb{R}[x, x^{-1}]_+$, that is, $p(x) \geq 0$ for all $x \in \mathbb{R}^\times$. Because p is a Laurent polynomial, $x^{2k}p \in \mathbb{R}[x]$ for some $k \in \mathbb{N}$. Since $x^{2k}p(x) \geq 0$ on $\mathbb{R}^\times$ and hence on $\mathbb{R}$, by Proposition 3.1 there are polynomials $f, g \in \mathbb{R}[x]$ such that $x^{2k}p = f^2 + g^2$. Then $p = (x^{-k}f)^2 + (x^{-k}g)^2 \in \sum \mathbb{R}[x, x^{-1}]^2$. Hence $L_s(p) \geq 0$ by (iii). Therefore, by Theorem 1.14,(i)→(iv), L_s is a moment functional on $\mathbb{R}[x, x^{-1}] \cong \mathbb{R}[\mathbb{Z}]$, so s is a moment sequence on $\mathbb{Z}$. This proves (i). $\qquad\square$

3.3 The Symmetric Hamburger Moment Problem and Stieltjes Moment Problem

A Radon measure μ on $\mathbb{R}$ is called *symmetric* if $\mu(M) = \mu(-M)$ for all Borel sets M. Let $M_+^{sym}(\mathbb{R})$ denote the symmetric measures of $M_+(\mathbb{R})$. Set $\mathbb{R}_+ := [0, +\infty)$.

We define mappings $\tau : \mathbb{R} \to \mathbb{R}_+$ and $\kappa : \mathbb{R}_+ \to \mathbb{R}$ by $\tau(x) = x^2$ and $\kappa(x) = \sqrt{x}$. For $\mu \in M_+^{sym}(\mathbb{R})$ and $\nu \in M_+(\mathbb{R}_+)$ let $\mu_+ := \tau(\mu) \in M_+(\mathbb{R}_+)$ and $\kappa(\nu) \in M_+(\mathbb{R})$ denote the corresponding images of μ and ν under τ and κ, respectively. That is, $\mu_+(M) = \mu(\tau^{-1}(M))$ and $\kappa(\nu)(M) = \nu(\kappa^{-1}(M))$. Further, we set

$$\nu^{sym} := \big(\kappa(\nu) + (-\kappa)(\nu)\big)/2. \tag{3.26}$$

Then we have

$$\int_0^\infty f(y) \, d\mu_+(y) = \int_{\mathbb{R}} f(x^2) \, d\mu(x), \tag{3.27}$$

$$\int_{\mathbb{R}} g(x) \, d\nu^{sym}(x) = \frac{1}{2} \int_0^\infty \big(g(\sqrt{y}) + g(-\sqrt{y})\big) \, d\mu(y) \tag{3.28}$$

for Borel functions f on $\mathbb{R}_+$ and g on $\mathbb{R}$ if the integrals on one side exist.

Lemma 3.17 *The map $\mu \mapsto \mu_+$ is a bijection of $M_+^{sym}(\mathbb{R})$ onto $M_+(\mathbb{R}_+)$ with inverse given by $\nu \mapsto \nu^{sym}$.*

Proof The proof is given by simple verifications. As samples we show that ν^{sym} is symmetric for $\nu \in M_+(\mathbb{R}_+)$ and that $(\mu_+)^{sym} = \mu$ for $\mu \in M_+(\mathbb{R})$.

Let $M \subseteq \mathbb{R}$ be a Borel set. Inserting the corresponding definitions we derive

$$
\begin{aligned}
2\nu^{sym}(M) &= \nu(\kappa^{-1}(M)) + \nu((-\kappa)^{-1}(M)) \\
&= \nu(\{t \in \mathbb{R}_+ : t^2 \in M\}) + \nu(\{t \in \mathbb{R}_+ : -t^2 \in M\}) \\
&= \nu(\{t \in \mathbb{R}_+ : -t^2 \in -M\}) + \nu(\{t \in \mathbb{R}_+ : t^2 \in -M\}) = 2\nu^{sym}(-M).
\end{aligned}
$$

Let μ^+ and μ^- denote the restrictions of μ to $(0, +\infty)$ and $(-\infty, 0)$, respectively. Clearly, $\mu = \mu(\{0\})\delta_0 + \mu^+ + \mu^-$. We easily verify that

$$
\kappa\tau(\mu) = \mu(\{0\})\delta_0 + 2\mu^+, \quad (-\kappa)\tau(\mu) = \mu(\{0\})\delta_0 + 2\mu^-,
$$

so that

$$
2(\mu_+)^{sym} = \kappa\tau(\mu) + (-\kappa)\tau(\mu) = 2\mu(\{0\})\delta_0 + 2\mu^+ + 2\mu^- = 2\mu. \qquad \square
$$

Proposition 3.18 *Suppose that $s = (s_n)_{n \in \mathbb{N}_0}$ is a Stieltjes moment sequence. Set $\hat{s} = (\hat{s}_n)_{n \in \mathbb{N}_0}$, where $\hat{s}_{2n} = s_n$ and $\hat{s}_{2n+1} = 0$ for $n \in \mathbb{N}_0$. The map $\nu \mapsto \nu^{sym}$ is a bijection of the solutions of the Stieltjes moment problem for s on the set of symmetric solutions of the Hamburger moment problem for $\hat{s}$ and the inverse of this map is given by $\mu \mapsto \mu_+$.*

Proof If μ solves the Hamburger moment problem for $\hat{s}$, then by (3.27),

$$
\int_0^\infty y^n d\mu_+(y) = \int_\mathbb{R} x^{2n} d\mu(x) = \hat{s}_{2n} = s_n, \quad n \in \mathbb{N}_0,
$$

that is, μ_+ solves the Stieltjes moment problem for s.

Conversely, let ν be a solution of the Stieltjes moment problem for s. By (3.28),

$$
\int_\mathbb{R} x^n d\nu^{sym}(x) = \frac{1}{2}\int_0^\infty \left((\sqrt{x})^n + (-\sqrt{x})^n\right) d\nu(x).
$$

This number vanishes if n is odd. If n is even, say $n = 2k$, then it is equal to $s_{n/2} = s_k = \hat{s}_n$. Thus ν^{sym} is a symmetric solution of the moment problem for $\hat{s}$. The remaining assertions are already contained in Lemma 3.17. $\qquad \square$

A Hamburger (resp. Stieltjes) moment sequence is called *determinate* if it has only one representing measure in $M_+(\mathbb{R})$ (resp. in $M_+([0, +\infty))$).

Proposition 3.19 *A Stieltjes moment sequence s is determinate if and only if the Hamburger moment sequence $\hat{s}$ is determinate.*

Proof By Proposition 3.18, there is a one-to-one correspondence between solutions of the Stieltjes moment problem for s and symmetric solutions of the Hamburger moment problem for $\hat{s}$. To complete the proof it therefore suffices to show that if the symmetric sequence $\hat{s}$ is indeterminate, its moment problem has at least two symmetric (!) solutions. This will be achieved by Corollary 6.26 in Sect. 6.   □

3.4 Positive Polynomials on Intervals (Revisited)

The representation of elements of $\mathrm{Pos}(\mathbb{R})$ as sums of two squares given in Propositions 3.1 is far from being unique. For instance, we have

$$x_1^2 + x_2^2 = (ax_1 + bx_2)^2 + (bx_1 - ax_2)^2 \quad \text{for} \quad a, b \in \mathbb{R}, \ a^2 + b^2 = 1.$$

In this section, we develop *unique* representations of nonnegative polynomials on intervals. The Markov–Lukacs theorem (Corollary 3.24) enters into the proof of Theorem 10.29 below. Except for this, these results are not used in the rest of the book.

Throughout this section, suppose that $p(x)$ is a **nonconstant** polynomial in $\mathbb{R}[x]$.

We consider the representation (3.2) and set $p_{\mathrm{rz}}(x) := a(x - \alpha_1)^{n_1} \cdots (x - \alpha_r)^{n_r}$. Then $p_{\mathrm{nrz}}(x) := p(x)p_{\mathrm{rz}}(x)^{-1}$ is a polynomial with leading term 1 which has no real zero. The factorization

$$p(x) = p_{\mathrm{rz}}(x)p_{\mathrm{nrz}}(x) \tag{3.29}$$

decomposes p into a polynomial $p_{\mathrm{rz}}(x)$ which captures all real zeros of p and a polynomial $p_{\mathrm{nrz}}(x)$ which has no real zero.

Now let $p \in \mathrm{Pos}(\mathbb{R})$. Then the leading coefficient a of p is positive and the multiplicity of each real zero α_j of p is even, say $n_j = 2k_j$ with $k_j \in \mathbb{N}$. Thus,

$$p_{\mathrm{rz}}(x) = a \prod_{j=1}^{r}(x - \alpha_j)^{2k_j}, \tag{3.30}$$

so that $p_{\mathrm{rz}} \in \mathrm{Pos}(\mathbb{R})$ and hence $p_{\mathrm{nrz}} \in \mathrm{Pos}(\mathbb{R})$.

If $p_{\mathrm{rz}} = a$ or $p_{\mathrm{nrz}} = 1$, then the formulas in Propositions 3.20–3.22 should be interpreted in the obvious manner by omitting the corresponding factors.

Proposition 3.20 *The polynomial p is in $\mathrm{Pos}(\mathbb{R})$ if and only if there are integers $k_1, \dots, k_r \in \mathbb{N}$ and reals $a > 0, c > 0, \alpha_1 < \cdots < \alpha_r, x_1 < x_2 < \cdots < x_{2n-1}$ such that*

$$p(x) = a \prod_{j=1}^{r}(x - \alpha_l)^{2k_l}\left[\prod_{j=1}^{n}(x - x_{2j-1})^2 + c \prod_{j=1}^{n-1}(x - x_{2j})^2 \right]. \tag{3.31}$$

(One of the two polynomial factors of p in (3.31) might be absent.) The numbers $a, c, \alpha_1, \ldots, \alpha_r, k_1, \ldots, k_r, x_1, \ldots, x_{2n-1}$ are uniquely determined by these requirements.

Proof The if part is easily checked. We carry out the proof of the only if part.

As noted above, the assumption $p \in \text{Pos}(\mathbb{R})$ implies that p_{rz} has the form (3.30).

The polynomial p_{nrz} has no real zero, leading coefficient 1 and even degree, say $2n$. In the case $n = 0$ the second main factor in (3.31) is absent. Assume now that $n \in \mathbb{N}$. By (3.1), the polynomial p_{nrz} has precisely n zeros, say $z_1, \ldots, z_n$, with positive imaginary parts. Setting

$$f(x) = (x - z_1) \cdots (x - z_n) \text{ and } g(x) = (x - \bar{z}_1) \cdots (x - \bar{z}_n),$$

we have $p_{nr}(x) = f(x)g(x)$. Then $u(x) := \frac{1}{2}(f(x)+g(x))$ and $v(x) := \frac{1}{2i}(f(x)-g(x))$ are in $\mathbb{R}[x]$ and satisfy

$$u(x)^2 + v(x)^2 = f(x)g(x) = p_{nrz}(x). \tag{3.32}$$

Now we consider the rational functions

$$\varphi_j(x) = (x - z_j)(x - \bar{z}_j)^{-1}, j = 1, \ldots, n, \text{ and } \varphi(x) = \varphi_1(x) \ldots \varphi_n(x).$$

Clearly, φ_j is an injective map of the real line to the unit circle. Since $\text{Im}\, z_j > 0$, $\arg \varphi_j(x)$ strictly increases on $(0, 2\pi)$ as x increases on $\mathbb{R}$. Therefore,

$$\arg \varphi(x) = \arg \varphi_1(x) + \cdots + \arg \varphi_n(x)$$

strictly increases on $(0, 2\pi n)$ as x increases on $\mathbb{R}$. Hence there exist real numbers $x_1 < x_2 < \cdots < x_{2n-1}$ such that $\varphi(x_k) = (-1)^k$ for $k = 1, \ldots, 2n - 1$. Since

$$\varphi(x) = \frac{f(x)}{g(x)} = \frac{u(x) + iv(x)}{u(x) - iv(x)} = \frac{u(x)^2 - v(x)^2 - 2iu(x)v(x)}{u(x)^2 + v(x)^2}$$

and $u(x), v(x) \in \mathbb{R}$ for $x \in \mathbb{R}$, $x_1, x_3, \ldots, x_{2n-1}$ are zeros of the real part $u(x)$ and $x_2, \ldots, x_{2n}$ are zeros of the imaginary part $v(x)$. Since $\deg(u) = n$ and $\deg(v) = n - 1$, these numbers exhaust the zero sets of u and v, respectively. The leading term of u is 1. Put $c := b^2$, where b is the leading term of v. Then, by (3.32),

$$p_{nrz}(x) = u(x)^2 + v(x)^2 = \prod_{j=1}^{n}(x - x_{2j-1})^2 + c\prod_{j=1}^{n-1}(x - x_{2j})^2. \tag{3.33}$$

Since $p = p_{rz}p_{nrz}$, (3.31) follows by combining (3.30) and (3.33).

To prove the uniqueness assertion we assume that $a', c', \alpha'_l, k'_l, x'_j, r', m'$ is another set of numbers satisfying the above conditions. From (3.31) it is clear that p has the leading term $a = a'$ and the real zeros $\alpha_l = \alpha'_l$ with multiplicities $2k_l = 2k'_l$. Hence $n = \deg(p) - r = \deg(p) - r' = n'$. Thus, it follows from (3.31) and (3.33) that

$$p_{nrz}(x) = \prod_{j=1}^{n}(x - x'_{2j-1})^2 + c' \prod_{j=1}^{n-1}(x - x'_{2j})^2. \tag{3.34}$$

Comparing (3.33) and (3.34) we obtain

$$q(x) := \prod_{j=1}^{n}(x - x_{2j-1})^2 - \prod_{j=1}^{n}(x - x'_{2j-1})^2 = c' \prod_{j=1}^{n-1}(x - x'_{2j})^2 - c \prod_{j=1}^{n-1}(x - x_{2j})^2.$$

The proof is complete once we have shown that $q(x) \equiv 0$. Assume the contrary. Without loss of generality, let $x'_1 \leq x_1$. We denote by l_i the number of roots of q which are equal to x_i and by r_i the number of roots of q in the open interval (x_i, x_{i+1}). Then the number of zeros of q in the interval $[x_1, x_{2n+1}]$ is

$$m := l_1 + \cdots + l_{2n+1} + r_1 + \cdots + r_{2m-2}. \tag{3.35}$$

If $q(x_{2j}) \neq 0$, then $q(x_{2j}) > 0$, and if $q(x_{2j+1}) \neq 0$, then $q(x_{2j+1}) < 0$ by the definition of q. Hence, if $l_i = l_{i+1} = 0$, there is a zero in (x_i, x_{i+1}), so that $r_i \geq 1$. Further, if $l_i > 0$, then x_i is a zero of multiplicity at least 2 and so $l_i \geq 2$. The preceding implies that $r_i + (l_i + l_{i+1})/2 \geq 1$ for each $i = 1, \ldots, 2n - 2$ and hence $m \geq l_1/2 + 2n - 2$ by (3.35). If $l_1 \neq 0$, then $m > 2n - 2$. If $l_1 = 0$, then $x'_1 < x_1$ and hence q has a zero in (x'_1, x_1), since $q(x'_1) > 0$ and $q(x_1) > 0$. In both cases q has at least $2n - 1$ zeros. Since $\deg(q) = 2n - 2$, this is the desired contradiction. $\qquad\square$

Next we consider the half-axis. Let $p_{rz}^{[0,+\infty)}(x)$ denote the product of the constant a and all factors $(x - \alpha_j)^{n_j}$ in (3.2), where $\alpha_j \in [0, +\infty)$. Then the polynomial $p_{nrz}^{[0,+\infty)}(x) := p(x)p_{rz}^{[0,+\infty)}(x)^{-1}$ has leading term 1, no zero in $[0, +\infty)$, and we have

$$p(x) = p_{rz}^{[0,+\infty)}(x)p_{nrz}^{[0,+\infty)}(x). \tag{3.36}$$

Let $p \in \mathrm{Pos}([0, +\infty))$. Then we have $a > 0$ and the multiplicity n_l of each zero $\alpha_l \in (0, +\infty)$ of p is even, $n_l = 2k_l$ with $k_l \in \mathbb{N}$. Thus

$$p_{rz}^{[0,+\infty)}(x) = ax^{k_0} \prod_{l=1}^{r}(x - \alpha_l)^{2k_l}, \tag{3.37}$$

where $k_0 \in \mathbb{N}_0$. In particular, $p_{rz}^{[0,+\infty)}(x)$ and $p_{nrz}^{[0,+\infty)}(x)$ are in $\mathrm{Pos}([0, +\infty))$.

Proposition 3.21 Let $m := \deg(p_{nrz}^{[0,+\infty)})$. Then $p \in \mathrm{Pos}([0, +\infty))$ if and only if there are integers $k_0, k_1, \ldots, k_r \in \mathbb{N}$, $r \in \mathbb{N}_0$, and real numbers $a > 0, c > 0$,

$$0 < \alpha_1 < \cdots < \alpha_r, \quad 0 < x_1 < x_2 < \cdots < x_{2n-1}, \tag{3.38}$$

such that

$$p(x) = ax^{k_0} \prod_{l=1}^{r}(x - \alpha_l)^{2k_l} \left[\prod_{j=1}^{n}(x - x_{2j-1})^2 + cx \prod_{j=1}^{n-1}(x - x_{2j})^2 \right] \ for \ m = 2n,$$

$$p(x) = ax^{k_0} \prod_{l=1}^{r}(x - \alpha_l)^{2k_l} \left[x \prod_{j=1}^{n}(x - x_{2j})^2 + c \prod_{j=1}^{n}(x - x_{2j-1})^2 \right] \ for \ m = 2n+1.$$

These numbers are uniquely determined by p and the above requirements.

Proof It is enough to prove the only if part. For simplicity we drop the upper index $[0, +\infty)$. By the formula (3.43) and the factorization (3.42) it suffices to prove that p_{nrz} has the form given in square brackets.

Put $P(x) := p_{\mathrm{nrz}}(x^2)$. Since p_{nrz} has no zero on $[0, \infty)$, P has no real zeros and $\deg(P) = 2m$. By Proposition 3.20, P can be represented as

$$P(x) = \prod_{j=1}^{m}(x - t_{2j-1})^2 + c \prod_{j=1}^{m-1}(x - t_{2j})^2, \tag{3.39}$$

where $t_1 < t_2 < \cdots < t_{2m-1}$ and $c > 0$. Because P is even, we also have

$$P(x) = \prod_{j=1}^{m}(x + t_{2j-1})^2 + \acute{c} \prod_{j=1}^{m-1}(x + t_{2j})^2, \tag{3.40}$$

where $-t_{2m+1} < \cdots < -t_2 < -t_1$. Comparing (3.39) and (3.40) it follows from the uniqueness assertion of Proposition 3.20 that

$$t_1 = -t_{2m-1}, \quad \ldots, \quad t_{m-1} = -t_{m+1}, \ t_m = 0. \tag{3.41}$$

Hence, setting $x_j := t_{m+j}^2$ for $j = 1, \ldots, m-1$, the inequalities in (3.38) hold. Inserting (3.41) into (3.39) we obtain

$$p_{\mathrm{nrz}}(x^2) = P(x) = \prod_{j=1}^{n}(x^2 - x_{2j-1})^2 + cx^2 \prod_{j=1}^{n-1}(x^2 - x_{2j})^2,$$

for $m = 2n$ and

$$p_{\mathrm{nrz}}(x^2) = P(x) = x^2 \prod_{j=1}^{n}(x^2 - x_{2j})^2 + c \prod_{j=1}^{n}(x^2 - x_{2j-1})^2,$$

for $m = 2n + 1$. Replacing x^2 by x, we obtain the formulas in square brackets. The uniqueness assertion is easily reduced to that of Proposition 3.20.   □

Finally, we turn to the interval $[-1, 1]$. We denote by $p_{rz}^{[-1,1]}(x)$ the product of $|a|$, all factors $(x - \alpha_j)^{n_j}$ with $\alpha_j \in (-1, 1)$, $(x + 1)^{n_i}$ if $\alpha_i = -1$, and $(1 - x)^{n_i}$ if $\alpha_i = 1$ in the representation (3.2) of p. As above, $p_{nrz}^{[-1,1]}(x) := p(x) p_{rz}^{[-1,1]}(x)^{-1}$ is a polynomial which has no zero in $[-1, 1]$ and

$$p(x) = p_{rz}^{[-1,1]}(x) p_{nrz}^{[-1,1]}(x). \qquad (3.42)$$

Let $p \in \mathrm{Pos}([-1, 1])$. Then the multiplicity n_l of all zeros $\alpha_l \in (-1, 1)$ of p is even, that is, $n_l = 2k_l$ with $k_l \in \mathbb{N}$. Thus we have

$$p_{rz}^{[-1,1]}(x) = |a|(1 + x)^{k_0} (1 - x)^{k_{r+1}} \prod_{l=1}^{r} (x - \alpha_l)^{2k_l}, \qquad (3.43)$$

where $k_0, k_{r+1} \in \mathbb{N}_0$. Further, $p_{rz}^{[-1,1]}(x)$ and $p_{nrz}^{[-1,1]}(x)$ are in $\mathrm{Pos}([-1, 1])$.

Proposition 3.22 *Set $m = \deg(p_{nrz}^{[-1,1]})$. The polynomial p is in $\mathrm{Pos}([-1, 1])$ if and only if there there are numbers $k_0, k_{r+1} \in \mathbb{N}_0$, $k_1, \ldots, k_r \in \mathbb{N}$, $a > 0, b > 0, c > 0$,*

$$-1 < \alpha_1 < \cdots < \alpha_r < 1, \quad -1 < x_1 < x_2 < \cdots < x_{2n-1} < 1, \qquad (3.44)$$

such that (3.42) and (3.43) hold and

$$p_{nrz}^{[-1,1]}(x) = b \prod_{j=1}^{n} (x - x_{2j-1})^2 + c(1 - x^2) \prod_{j=1}^{n-1} (x - x_{2j})^2, \qquad m = 2n,$$

$$p_{nrz}^{[-1,1]}(x) = b(1 + x) \prod_{j=1}^{n} (x - x_{2j})^2 + c(1 - x) \prod_{j=1}^{n} (x - x_{2j-1})^2, \quad m = 2n + 1.$$

The corresponding numbers are uniquely determined by p and these conditions.

Proof Again the if part is easily verfied. To prove the formulas for $p_{nrz}^{[-1,1]}$ we abbreviate $p = p_{nrz}^{[-1,1]}$. We set $t = \frac{1+x}{1-x}$ and define

$$P(t) = p(1)^{-1}(1 + t)^m p\left(\frac{t - 1}{t + 1}\right). \qquad (3.45)$$

(In fact, P is just the Goursat transform of the polynomial $p(1)^{-1}p(-x)$.) Then we have $x = \frac{t-1}{t+1}$ and hence

$$p(x) = p(1)(1 + t)^{-m} P(t) = p(1) \left(\frac{1 - x}{2}\right)^m P(t). \qquad (3.46)$$

Note that $t \in [0, +\infty)$ if and only if $x \in [-1, 1)$. Therefore, by (3.45), $p(x) > 0$ on $[-1, 1]$ implies that $P(t) > 0$ on $[0, +\infty)$. The factor $p(1)^{-1}$ in (3.45) ensures that the polynomial P has the leading term 1. The preceding implies that $P = P_{\mathrm{nrz}}^{[0,+\infty)}$. Clearly, $\deg(P) = m$. Thus, by Proposition 3.21, P has the form

$$P(t) = \prod_{j=1}^{n}(t - t_{2j-1})^2 + \gamma t \prod_{j=1}^{n-1}(t - t_{2j})^2 \quad \text{for} \ \ m = 2n, \tag{3.47}$$

$$P(t) = t \prod_{j=1}^{n}(t - t_{2j})^2 + \gamma \prod_{j=1}^{n}(t - t_{2j-1})^2 \quad \text{for} \ \ m = 2n + 1, \tag{3.48}$$

where $\gamma > 0$ and $0 < t_1 < t_2 < \cdots < t_{2n+1}$. Setting $x_j := \frac{t_j - 1}{t_j + 1}$, (3.44) holds and

$$t - t_l = \frac{1 + x}{1 - x} - \frac{1 + x_l}{1 - x_l} = \frac{2(x - x_l)}{(1 - x)(1 - x_l)}, \quad l = 1, \ldots, 2n + 1.$$

Inserting this into (3.47) and (3.48) by using the equality $t = \frac{1+x}{1-x}$ and (3.46) we get

$$p(x) = p(1)\left(\frac{1-x}{2}\right)^m P(t) = b\prod_{j=1}^{n}(x - x_{2j-1})^2 + c(1 - x)(1 + x)\prod_{j=1}^{n-1}(x - x_{2j})^2,$$

for $m = 2n$ and

$$p(x) = p(1)\left(\frac{1-x}{2}\right)^m P(t) = b(1 + x)\prod_{j=1}^{n}(x - x_{2j})^2 + c(1 - x)\prod_{j=1}^{n-1}(x - x_{2j-1})^2,$$

for $m = 2n + 1$, where $b, c \in [0, +\infty)$. This proves the formulas for $p_{\mathrm{nrz}}^{[-1,1]}$.

The uniqueness can be shown either by repeating the corresponding reasoning from the proof of Proposition 3.20 or by tracing it back to the uniqueness statement in Proposition 3.21. We do not carry out the details.   □

Remark 3.23 Since $p_{\mathrm{nrz}}^{[-1,1]}$ has leading term 1, $|b - c| = 1$ in Proposition 3.22.   ∘

From the preceding Propositions 3.21 and 3.22 we easily derive nice and useful descriptions of positive polynomials of degree at most $m = 2n$ resp. $m = 2n + 1$.

The following result is usually called the *Markov–Lukacs theorem*.

Corollary 3.24 *For $a, b \in \mathbb{R}$, $a < b$, and $n \in \mathbb{N}_0$, we have*

$$\mathrm{Pos}([a, b])_{2n} = \{p_n(x)^2 + (b-x)(a-x)q_{n-1}(x)^2 : p_n \in \mathbb{R}[x]_n, q_{n-1} \in \mathbb{R}[x]_{n-1}\}, \tag{3.49}$$

$$\mathrm{Pos}([a, b])_{2n+1} = \{(b - x)p_n(x)^2 + (a - x)q_n(x)^2 : p_n, q_n \in \mathbb{R}[x]_n \}. \tag{3.50}$$

Proof The right-hand sides are obviously contained in the left-hand sides.

We prove the converse inclusion. By a linear transformation we can assume that $a = -1, b = 1$. Let $p \in \mathrm{Pos}([-1, 1])_m$ for $m = 2n$ resp. $m = 2n + 1$. Collecting the factors of $p = p_{\mathrm{r}}^{[-1,1]} p_{\mathrm{nr}}^{[-1,1]}$ in the formulas of Propositions 3.22 it follows that p belongs to the corresponding sets on the right.   □

In a similar manner Proposition 3.21 yields at once the following corollary.

Corollary 3.25 *For any $n \in \mathbb{N}_0$, we have*

$$\mathrm{Pos}([0, +\infty))_{2n} = \{p_n(x)^2 + x q_{n-1}(x)^2 : p_n \in \mathbb{R}[x]_n, q_{n-1} \in \mathbb{R}[x]_{n-1}\}, \quad (3.51)$$

$$\mathrm{Pos}([0, +\infty))_{2n+1} = \{x p_n(x)^2 + q_n(x)^2 : p_n, q_n \in \mathbb{R}[x]_n\}. \quad (3.52)$$

Remark 3.26 All three Propositions 3.20–3.22 give *unique* representations of nonnegative polynomials on the corresponding intervals. Propositions 3.21 and 3.22 are stronger than Corollaries 3.25 and 3.24, while Corollaries 3.25 and 3.24 are stronger than Propositions 3.2 and 3.3, respectively. However, as already mentioned earlier, Propositions 3.1–3.3 are sufficient for solving the moment problems on intervals. ○

3.5 Exercises

1. Let A be a commutative ring. Find the counterpart of the two square identity (3.3) for $n = 4$ and $n = 8$: Given elements $a_1, \dots, a_n, b_1, \dots, b_n \in A$, there are elements $c_1, \dots, c_n \in A$ which are bilinear functions of the a_i and b_j such that

$$(a_1^2 + \cdots + a_n^2)(b_1^2 + \cdots + b_n^2) = c_1^2 + \cdots + c_n^2.$$

   Remark: As shown by A. Hurwitz, $n = 1, 2, 4, 8$ are the only natural numbers for which there is such an n square identity, see [Hu] for precise formulation.
2. Use the Goursat transform (3.12) to derive the formulas (3.8) and (3.9) for the interval $[-1, 1]$ from the corresponding formulas (3.5) and (3.7) for $[0, +\infty)$.
3. Show that $\mathrm{Pos}([-1, 1]) = \{f + (1 + x)g + (1 - x)h : f, g, h \in \sum \mathbb{R}[x]^2\}$ and use this description to formulate a solvability criterion for the $[-1, 1]$-moment problem.
4. Let $f \in \mathbb{R}[x]$, $a, b \in \mathbb{R}$, $a < b$, and set $\mathcal{T}_f = \{p + fq : p, q \in \sum \mathbb{R}[x]^2\}$. Suppose that $[a, b] = \{x \in \mathbb{R} : f(x) \geq 0\}$. Let k and l be the multiplicities of the zeros a and b of f, respectively.

   a. Find a polynomial $q \in \mathrm{Pos}([-1, 1])$ such that $q \notin \mathcal{T}_f$ for $f(x) = (1 - x^2)^3$.
   b. Show that $\mathrm{Pos}([a, b]) = \mathcal{T}_f$ if and only if $k = l = 1$.

   Details for b. can be found in the proof of [PR, Corollary 11].
5. Let $p(x) = x^2 + c$, where $0 < c < 1$. Show that p cannot be written in the form $p(x) = \sum_{k,l=0}^{2} a_{kl}(1 - x)^k(1 + x)^l$ with $a_{kl} \geq 0$ for $k, l = 0, 1, 2$.

6. Let $a, b, c \in \mathbb{R}$, $a < b < c$, and set $K = [a, b] \cup \{c\}$. Describe $\mathrm{Pos}(K)$ and find solvability conditions for the K-moment problem.

7. Let $a, b, c, d \in \mathbb{R}$, $a < b < c < d$. Show that

   a. $\mathrm{Pos}((-\infty, a] \cup [b, +\infty)) = \{f + (x - a)(x - b)g : f, g \in \sum \mathbb{R}[x]^2\}$,
   b. $\mathrm{Pos}([a, b] \cup [c, d]) = \{f + (x-a)(b-x)(x-c)(x-d)g : f, g \in \sum \mathbb{R}[x]^2\}$.

8. Use Exercise 7 to give solvability criteria for the K-moment problem, where

   a. $K = (-\infty, a] \cup [b, +\infty)$,
   b. $K = [a, b] \cup [c, d]$.

9. Suppose that $-\infty < a_1 < b_1 < a_2 < \cdots < a_n < b_n < +\infty$. Define $K_1 := \cup_{k=1}^{n}[a_k, b_k]$ and $K_2 := \mathbb{R} \backslash \cup_{k=1}^{n}(a_k, b_k)$. Describe $\mathrm{Pos}(K_j)$ and give necessary and sufficient conditions for K_j-moment sequences, where $j = 1, 2$.

10. Show that the sequence $s = (0, 1, 0, 0, \dots)$ satisfies $D_k(s) = 0$ for all $k \in \mathbb{N}_0$, but s is not a moment sequence.

11. Give an alternative proof of Proposition 3.4 by showing following steps:

    a. It suffices to prove the result for linear and for quadratic polynomials.
    b. The assertion holds for linear and for quadratic polynomials.

    (This proof was given by F. Hausdorff [Hs], p. 98–99, see e.g [PSz], p. 276–277.)

12. Suppose that s is a moment sequence. Prove that $\sum_{k=0}^{2n} \frac{s_k}{k!} \geq 0$ for $n \in \mathbb{N}$.

13. Let K be a closed subset of $\mathbb{R}$. Prove that the following statements are equivalent:

    (i) If $s = (s_n)$ and $t = (t_n)$ are K-moment sequences, then so is $st := (s_n t_n)$.
    (ii) If $x, y \in K$, then $xy \in K$.

    Hint: For (ii)$\Rightarrow$(i), use Haviland's theorem and $L_{st,z}(f(z)) = L_{s,x}(L_{t,y}(f(xy)))$.

14. "Guess" representing measures for the following sequences $(s_n)_{n=0}^{\infty}$:

    a. $s_n = \frac{a^{n+1}}{n+1} + cb^n$, where $b \in \mathbb{R}$ and $a, c \geq 0$,
    b. $s_n = n!$,
    c. $s_n = \frac{1}{(n+1)(n+2)}$.

15. (*Solution of the two-sided Stieltjes moment problem*)
    Show that for a real sequence $s = (s_n)_{n \in \mathbb{Z}}$ the following are equivalent:

    (i) s is a two-sided Stieltjes moment sequence, that is, there exists a Radon measure μ on $(0, +\infty)$ such that x^n is μ-integrable and $s_n = \int_0^{+\infty} x^n \, d\mu$ for $n \in \mathbb{Z}$.
    (ii) $s \in \mathcal{P}(\mathbb{Z})$ and $Es \in \mathcal{P}(\mathbb{Z})$.
    (iii) L_s and L_{Es} are positive functionals on $\mathbb{R}[x, x^{-1}]$, that is, $L_s(f^2) \geq 0$ and $L_s(xf^2) \geq 0$ for all $f \in \mathbb{R}[x, x^{-1}]$.

3.6 Notes

The results on positive polynomials have a long and tricky history, see [PR] for some discussion. Polya's theorem appeared in [P]; we reproduced his proof. The Markov–Lukacs theorem is due to A.A. Markov [Mv1] for $m = 2n$ and F. Lukacs [Lu]. Proposition 3.22 is due to S. Karlin and L.S. Shapley [KSh, p. 35], while Proposition 3.21 can be found in [KSt, p. 169]. The proof of Proposition 3.20 follows [Ml]; our proof of Proposition 3.22 seems to be new. Other proofs of the Markov–Lukacs theorem are given in [Sz],[KSt] (see also [Ka]) and [KN]; Szegö [Sz, p. 4] derived it from the Fejér–Riesz theorem. Bernstein's theorem was proved in [Bn].

The existence criteria for moment problems on intervals were obtained in the classical papers by T.J. Stieltjes [Stj], H. Hamburger [Hm] and F. Hausdorff [Hs]. The results on symmetric Hamburger moment problems are from [Chi2].

Chapter 4
One-Dimensional Moment Problems: Determinacy

The main aim of this chapter is to develop some very useful results concerning the uniqueness of solutions of one-dimensional Hamburger and Stieltjes moment problems. These are Carleman's conditions (4.2) and (4.3) in Theorem 4.3, which are sufficient for determinacy, and Krein's conditions (4.19) and (4.23) in Theorems 4.14 and 4.17, which provide necessary criteria.

4.1 Measures Supported on Bounded Intervals

The following proposition contains a number of characterizations of measures supported on an interval $[-c, c]$, $c > 0$, in terms of their moment sequences.

Proposition 4.1 *Suppose that* $s = (s_n)_{n \in \mathbb{N}_0}$ *is a Hamburger moment sequence. Let* μ *be representing measure of* s *and let* $c \in \mathbb{R}, c > 0$. *The following are equivalent:*

(i) μ *is supported on* $[-c, c]$.
(ii) *There exists a number* $d > 0$ *such that* $|s_n| \leq dc^n$ *for* $n \in \mathbb{N}_0$.
(iii) *There exists a number* $d > 0$ *such that* $s_{2n} \leq dc^{2n}$ *for* $\in \mathbb{N}_0$.
(iv) $S := \liminf_{n \to \infty} s_{2n}^{\frac{1}{2n}} \leq c$.

Further, if $s_0 = \mu(\mathbb{R}) \leq 1$, *then the following statements are equivalent:*

(v) μ *is supported on* $[-1, 1]$.
(vi) $\liminf_{n \to \infty} s_{2n} \leq 1$.
(vii) $\liminf_{n \to \infty} s_{2n} < +\infty$.

Proof The implications (i)→(ii)→(iii)→(iv) are obviously true with $d = s_0$. Therefore, for the equivalence of (i)–(iv) it suffices to prove that (iv) implies (i).

© Springer International Publishing AG 2017
K. Schmüdgen, *The Moment Problem*, Graduate Texts in Mathematics 277,
DOI 10.1007/978-3-319-64546-9_4

For $\alpha > 0$, let χ_α denote the characteristic function of the set $\mathbb{R}\backslash(-\alpha, \alpha)$ and put $M_\alpha := \int \chi_\alpha d\mu$. Then

$$M_\alpha \alpha^{2n} = \int_\mathbb{R} \alpha^{2n} \chi_\alpha d\mu \leq \int_\mathbb{R} x^{2n} \chi_\alpha d\mu \leq \int_\mathbb{R} x^{2n} d\mu = s_{2n}$$

and hence $(M_\alpha)^{\frac{1}{2n}} \alpha \leq s_{2n}^{\frac{1}{2n}}$ for $n \in \mathbb{N}$. Therefore, if $M_\alpha > 0$, by passing to the limits we obtain $\alpha \leq S$. Thus, $M_\alpha = \mu(\mathbb{R}\backslash(-\alpha, \alpha)) = 0$ when $\alpha > S$. Since $S \leq c$ by (iv), this implies that $\operatorname{supp} \mu \subseteq [-S, S] \subseteq [-c, c]$, which proves (i).

We verify the equivalence of (v)–(vii). Since $\mu(\mathbb{R}) \leq 1$, (v) implies $s_{2n} \leq 1$ and hence (vi). The implication (vi)→(vii) is trivial, so it remains to prove (vii)→(v).

Assume to the contrary that (v) does not hold. Then we can find an interval $[a, b] \subseteq \mathbb{R}\backslash[-1, 1]$ such that $\mu([a, b]) > 0$. Set $A = a$ if $a > 1$ and $A = -b$ if $b < -1$. Then $s_{2n} \geq A^{2n} \mu([a, b])$ for $n \in \mathbb{N}$. Since $A > 1$ and $\mu([a, b]) > 0$, we deduce that $\lim_n s_{2n} = +\infty$. This contradicts (vii). $\qquad \square$

Corollary 4.2 *If a Hamburger moment sequence s has a representing measure with compact support, then s is determinate.*

Proof Let $\mu_1, \mu_2 \in \mathcal{M}_s$. By Proposition 4.1 (iv)→(i), μ_1 and μ_2 are supported on $[-S, S]$. Then, for all $f \in \mathbb{R}[x]$,

$$\int_{-S}^S f(x) \, d\mu_1 = \int_{-S}^S f(x) \, d\mu_2. \tag{4.1}$$

Since the polynomials $\mathbb{R}[x]$ are dense in $C([-S, S]; \mathbb{R})$ by the Weierstrass theorem, (4.1) holds for all continuous functions f. This in turn implies that $\mu_1 = \mu_2$. $\qquad \square$

4.2 Carleman's Condition

Recall that a Hamburger moment sequence is determinate if it has a unique representing measure, while a Stieltjes moment sequence is called determinate if it has only one representing measure supported on $[0, +\infty)$.

The *Carleman theorem* contains a powerful *sufficient* condition for determinacy.

Theorem 4.3 *Suppose that $s = (s_n)_{n \in \mathbb{N}_0}$ is a positive semidefinite sequence.*

(i) *If s satisfies the* Carleman condition

$$\sum_{n=1}^\infty s_{2n}^{-\frac{1}{2n}} = +\infty, \tag{4.2}$$

then s is a determinate *Hamburger moment sequence.*

(ii) *If in addition $Es = (s_{n+1})_{n \in \mathbb{N}_0}$ is positive semidefinite and*

$$\sum_{n=1}^{\infty} s_n^{-\frac{1}{2n}} = +\infty, \tag{4.3}$$

then s is a determinate Stieltjes moment sequenc..

The main technical ingredient of the proof of Theorem 4.3 given in this section is a result on quasi-analytic functions (Corollary 4.5). Another proof of Theorem 4.3 based on Jacobi operators can be found at the end of Sect. 6.4.

Let us begin with some notions on quasi-analytic functions. Suppose $(m_n)_{n \in \mathbb{N}_0}$ is a positive sequence and $J \subseteq \mathbb{R}$ is an open interval. Let $C\{m_n\}$ denote the set of functions $f \in C^{\infty}(J)$ for which there exists a constant $K_f > 0$ such that

$$\sup_{t \in J} |f^{(n)}(t)| \leq K_f^n \, m_n \quad \text{for } n \in \mathbb{N}_0. \tag{4.4}$$

We say $C\{m_n\}$ is a *quasi-analytic class* if the following holds: if $f \in C\{m_n\}$ and there is a point $t_0 \in J$ such that $f^{(n)}(t_0) = 0$ for all $n \in \mathbb{N}_0$, then $f(t) \equiv 0$ on J. In this case the functions of $C\{m_n\}$ are called *quasi-analytic*.

Quasi-analyticity is characterized by the following *Denjoy–Carleman theorem.*

Theorem 4.4 *$C\{m_n\}$ is a quasi-analytic class if and only if*

$$\sum_{n=1}^{\infty} (\inf_{k \geq n} m_k^{1/k})^{-1} = \infty. \tag{4.5}$$

Proof [Hr, Theorem 1.3.8]. For log convex sequences $(m_n)_{n \in \mathbb{N}_0}$ a proof is contained in [Ru2, Theorem 19.11]. □

For our proof of Theorem 4.3 the following corollary is sufficient.

Corollary 4.5 *Suppose that $(m_n)_{n \in \mathbb{N}_0}$ is a positive sequence such that*

$$\sum_{n=1}^{\infty} m_n^{-1/n} = \infty. \tag{4.6}$$

Suppose that $f \in C^{\infty}(J)$ and there is a constant $K_f > 0$ such that (4.4) is satisfied. If there exists a $t_0 \in J$ such that $f^{(n)}(t_0) = 0$ for all $n \in \mathbb{N}_0$, then $f(t) \equiv 0$ on J.

Proof Since obviously $m_n^{1/n} \geq \inf_{k \geq n} m_k^{1/k}$, (4.6) implies (4.5). Hence $C\{m_n\}$ is a quasi-analytic class by Theorem 4.4. This proves the assertion. □

The simplest examples of quasi-analytic functions are analytic functions.

Example 4.6 ($(m_n = n!)_{n \in \mathbb{N}_0}$) Since $n! \leq n^n$, the sequence $(n!)_{n \in \mathbb{N}_0}$ satisfies (4.5) and (4.6). Hence $C\{n!\}$ is a quasi-analytic class. It is well-known (see e.g. [Ru2, Theorem 19.9]) that each function $f \in C\{n!\}$ has a *holomorphic extension* to a strip

$\{z : \operatorname{Re} z \in J, |\operatorname{Im} z| < \delta\}$, $\delta > 0$. Therefore, if $f \in C\{n!\}$ and $f^{(n)}(t_0) = 0$ for some $t_0 \in J$ and all $n \in \mathbb{N}_0$, then we have $f(t) \equiv 0$ on J. That is, for the special class $C\{n!\}$ the assertion of Corollary 4.5 holds by the uniqueness theorem for analytic functions without refering to the Denjoy–Carleman theorem.   ∘

Remark 4.7

1. Let $(m_n)_{n \in \mathbb{N}_0}$ be a positive sequence such that $m_0 = 1$ and

$$m_n^2 \le m_{n-1} m_{n+1} \quad \text{for } n \in \mathbb{N}. \tag{4.7}$$

Condition (4.7) implies that $(\ln m_n)_{n \in \mathbb{N}_0}$ is a convex sequence. Indeed, it can be shown that then $m_n^{1/n} \le m_k^{1/k}$ for $n \le k$, so that $m_n^{1/n} = \inf_{k \ge n} m_k^{1/k}$. Therefore, by Theorem 4.4, in this case $C\{m_n\}$ is quasi-analytic if and only if $\sum_{n=1}^{\infty} m_n^{-1/n} = \infty$, that is, if (4.6) is satisfied.

2. Let $s = (s_n)_{n \in \mathbb{N}_0}$ be a Hamburger moment sequence such that $s_0 = 1$. Then the Hölder inequality implies that (4.7) holds for $m_n = s_{2n}$. Therefore, since $m_0 = 1$, it follows from the preceding remark that $(\ln s_{2n})_{n \in \mathbb{N}_0}$ is a convex sequence.   ∘

The following simple lemma is used in the proofs of Theorems 4.3 and 4.14. Let $\mathcal{M}_+(\mathbb{R})$ denote the Radon measures μ on $\mathbb{R}$ satisfying $\int |x|^n \, d\mu < \infty$ for all $n \in \mathbb{N}_0$. Recall that $\mathcal{M}_s$ is the set of representing measures of a moment sequence s.

Lemma 4.8 *Suppose that $\mu \in \mathcal{M}_+(\mathbb{R})$ and $\xi \in L^\infty(\mathbb{R}, \mu)$. Then the function*

$$g(t) := \int_{\mathbb{R}} e^{itx} \xi(x) \, d\mu(x) \tag{4.8}$$

is in $C^\infty(\mathbb{R})$ and satisfies

$$g^{(n)}(t) = \int_{\mathbb{R}} (ix)^n e^{itx} \xi(x) \, d\mu(x) \quad \text{for } n \in \mathbb{N}_0, \ t \in \mathbb{R}. \tag{4.9}$$

Proof We proceed by induction on n. For $n = 0$ the assertion holds by definition. Suppose that (4.9) is valid for n and all $t \in \mathbb{R}$. Fix $t \in \mathbb{R}$ and put

$$\varphi_h(x) := h^{-1}(e^{ihx} - 1), \quad \psi_h(x) := \varphi_h(x)(ix)^n e^{itx} \xi(x) \quad \text{for } h \in \mathbb{R}, \ h \ne 0.$$

Then $\psi_h(x) \to (ix)^{n+1} e^{itx} \xi(x)$ on $\mathbb{R}$ as $h \to 0$. By the complex version of the mean value theorem, $|\varphi_h(x)| = |\varphi_h(x) - \varphi_h(0)| \le |x| \sup \{ |\varphi_h'(y)| : |y| \le |x|, y \in \mathbb{R} \}$. Therefore, since $|\varphi_h'(y)| = |e^{ihy}| = 1$, we get

$$|\psi_h(x)| = |\varphi_h(x)(ix)^n e^{itx} \xi(x)| \le |x^{n+1}| \, \|\xi\|_{L^\infty(\mathbb{R},\mu)} \quad \text{a.e. on } \mathbb{R}.$$

Therefore, since $\mu \in \mathcal{M}_+(\mathbb{R})$ and hence $x^{n+1} \in \mathcal{L}^1(\mathbb{R}, \mu)$, Lebesgue's dominated convergence theorem applies and yields

$$\lim_{h\to 0}\frac{g^{(n)}(t+h)-g^{(n)}(t)}{h}=\lim_{h\to 0}\int_{\mathbb{R}}\psi_h(x)d\mu(x)=\int_{\mathbb{R}}(\mathrm{i}x)^{n+1}e^{\mathrm{i}tx}\xi(x)d\mu(x),$$

which gives (4.9) for $n+1$. $\qquad\square$

An immediate consequence of Lemma 4.8 is the following

Corollary 4.9 *Let* $\mu \in \mathcal{M}_+(\mathbb{R})$. *Then the Fourier transform* $f_\mu(t) := \int_{\mathbb{R}} e^{-\mathrm{i}tx}d\mu(x)$ *of* μ *is in* $C^\infty(\mathbb{R})$ *and*

$$f_\mu^{(n)}(t)=\int_{\mathbb{R}}(-\mathrm{i}x)^n e^{\mathrm{i}tx}d\mu(x)\quad \text{for } n\in\mathbb{N}_0,\ t\in\mathbb{R}. \tag{4.10}$$

In particular,

$$s_n(\mu)=\int_{\mathbb{R}}x^n d\mu=(-\mathrm{i})^n f_\mu(0)\quad \text{for } n\in\mathbb{N}_0. \tag{4.11}$$

Proof of Theorem 4.3 By Hamburger's theorem 3.8 and Stieltjes' theorem 3.12 it suffices to prove the the assertions about the determinacy.

(i) Suppose that $\mu_1,\mu_2\in\mathcal{M}_s$ and set $f:=f_{\mu_1}-f_{\mu_2}$. Then $f\in C^\infty(\mathbb{R})$ by Corollary 4.9. Define $m_n=\sup_{t\in\mathbb{R}}|f^{(n)}(t)|$ for $n\in\mathbb{N}_0$. By (4.10),

$$m_{2n}\le\sup_{t\in\mathbb{R}}(|f_{\mu_1}^{(2n)}(t)|+|f_{\mu_2}^{(2n)}(t)|)\le\int_{\mathbb{R}}x^{2n}d\mu_1(x)+\int x^{2n}d\mu_2(x)=2s_{2n}$$

for $n\in\mathbb{N}$. Hence condition (4.6) is fulfilled, since

$$\sum_{n=1}^\infty m_n^{-1/n}\ge\sum_{n=1}^\infty m_{2n}^{-1/2n}\ge\sum_{n=1}^\infty 2^{-1/2n}s_{2n}^{-1/2n}\ge\frac{1}{2}\sum_{n=1}^\infty s_{2n}^{-1/2n}=\infty.$$

Applying again (4.10) we obtain for $n\in\mathbb{N}_0$,

$$f^{(n)}(0)=f_{\mu_1}^{(n)}(0)-f_{\mu_2}^{(n)}(0)=\int(\mathrm{i}x)^n d\mu_1(x)-\int(\mathrm{i}x)^n d\mu_2(x)=\mathrm{i}^n s_n-\mathrm{i}^n s_n=0.$$

Thus the assumptions of Corollary 4.5 are satisfied. Therefore, $f(t)\equiv 0$, hence $f_{\mu_1}(t)\equiv f_{\mu_2}(t)$, on $\mathbb{R}$. Since the Fourier transform uniquely determines a finite measure, we get $\mu_1=\mu_2$. This shows that s is determinate.

(ii) Let v_1 and v_2 be two solutions of the Stieltjes moment problem for s. Define symmetric measures $\mu_j=v_j^{sym}, j=1,2$, on $\mathbb{R}$ by (3.26). Then, by Proposition 3.18, $\mu_j\in\mathcal{M}_+(\mathbb{R})$ has the moment sequence $\hat{s}=(\hat{s}_n)_{n\in\mathbb{N}_0}$, where $\hat{s}_{2n}=s_n$ and $\hat{s}_{2n+1}=0$ for $n\in\mathbb{N}_0$. In particular, μ_1 and μ_2 are representing measures of $\hat{s}$. Since $\hat{s}_{2n}=s_n$, it follows from assumption (4.3) that $\hat{s}$ satisfies (4.2). Hence, by (i), the moment problem for $\hat{s}$ is determinate. Therefore, $\mu_1 = \mu_2$ and hence $\nu_1 = \nu_2$. $\qquad\qquad\qquad\qquad\qquad\qquad\square$

Corollary 4.10 *Suppose that $s = (s_n)_{n\in\mathbb{N}_0}$ is a positive semidefinite sequence.*

(i) *If there is a constant $M > 0$ such that*

$$s_{2n} \le M^n (2n)! \quad \text{for } n \in \mathbb{N}, \tag{4.12}$$

then Carleman's condition (4.2) holds and s is a determinate moment sequence.

(ii) *If $Es = (s_{n+1})_{n\in\mathbb{N}_0}$ is also positive semidefinite and there is an $M > 0$ such that*

$$s_n \le M^n (2n)! \quad \text{for } n \in \mathbb{N}, \tag{4.13}$$

then s is a determinate Stieltjes moment sequence.

Proof

(i) For $n \in \mathbb{N}$ we have $(2n)! \le (2n)^{2n}$. It follows that $[(2n)!]^{1/2n} \le 2n$ and hence $\frac{1}{2n} \le [(2n)!]^{-1/2n}$, so that

$$M^{-1/2} \frac{1}{2n} \le M^{-1/2} [(2n)!]^{-1/2n} \le s_{2n}^{-1/2n}, \quad n \in \mathbb{N}.$$

Therefore, Carleman's condition (4.2) is satisfied, so that Theorem 4.3(i) applies. (As noted in Example 4.6, in this case the Denjoy–Carleman theorem is not needed.)

(ii) follows from Theorem 4.3(ii) by the same reasoning. $\qquad\qquad\qquad\qquad\square$

Corollary 4.11

(i) *Let $\mu \in M_+(\mathbb{R})$. If there exists an $\varepsilon > 0$ such that*

$$\int_{\mathbb{R}} e^{\varepsilon|x|} \, d\mu(x) < \infty, \tag{4.14}$$

then $\mu \in \mathcal{M}_+(\mathbb{R})$, condition (4.12) holds, and the Hamburger moment problem for μ determinate.

(ii) *Suppose that $\mu \in M_+([0, +\infty))$. If there exists an $\varepsilon > 0$ such that*

$$\int_{\mathbb{R}} e^{\varepsilon\sqrt{|x|}} \, d\mu(x) < \infty, \tag{4.15}$$

then $\mu \in \mathcal{M}_+([0, +\infty))$, condition (4.13) is satisfied, and the Stieltjes moment problem for μ is determinate.

Proof

(i) Let $n \in \mathbb{N}_0$ and $x \in \mathbb{R}$. Clearly, we have $e^{\varepsilon|x|} \geq (\varepsilon x)^{2n} \frac{1}{(2n)!}$. Hence

$$x^{2n} e^{-\varepsilon|x|} \leq \varepsilon^{-2n}(2n)! \tag{4.16}$$

and therefore

$$\int_{\mathbb{R}} x^{2n} d\mu(x) = \int_{\mathbb{R}} x^{2n} e^{-\varepsilon|x|} e^{\varepsilon|x|} d\mu(x) \leq \varepsilon^{-2n}(2n)! \int_{\mathbb{R}} e^{\varepsilon|x|} d\mu(x). \tag{4.17}$$

Thus, by the assumption (4.14), we have $\int x^{2n} d\mu(x) < \infty$ for $n \in \mathbb{N}_0$ and hence $\int |x^k| d\mu(x) < \infty$ for all $k \in \mathbb{N}_0$, so that $\mu \in \mathcal{M}_+(\mathbb{R})$. Further, (4.17) implies that $s_{2n} \leq M^n (2n)!$ for $n \in \mathbb{N}_0$ and some constant $M > 0$. Thus (4.12) is satisfied and the assertion follows from Corollary 4.10(i).

(ii) follows in a similar manner with $x \in [0, +\infty)$ and using Corollary 4.10(ii). □

In probability theory the sufficient determinacy conditions (4.14) and (4.15) are called *Cramer's condition* and *Hardy's condition*, respectively.

The examples treated below indicate that Carleman's condition (4.2) is an extremely powerful sufficient condition for determinacy. Nevertheless, this condition is not necessary, as shown by Example 4.18 and also by Remark 7.19.

Remark 4.12 L.B. Klebanov and S.T. Mkrtchyan [KlM] proved the following:

Let $s = (s_n)_{n \in \mathbb{N}_0}, s_0 = 1$, be a Hamburger moment sequence. Set $C_m := \sum_{n=1}^{m} s_n^{-\frac{1}{2n}}$. If μ and ν are representing measures of s, then

$$L(\mu, \nu) \leq c(s_2) \, C_m^{-1/4} \log(1 + C_m) \quad \text{for } m \in \mathbb{N}, \tag{4.18}$$

where $c(s_2) > 0$ is a constant depending only on s_2 and $L(\mu, \nu)$ denotes the Levy distance of μ and ν (see e.g. [Bl]).

If Carleman's condition (4.2) holds, then $\lim_{m \to \infty} C_m = +\infty$, hence (4.18) implies $L(\mu, \nu) = 0$, so that $\mu = \nu$. This is another proof of Carleman's Theorem 4.3(i).   ∘

Remark 4.13 C. Berg and J.P.R. Christensen [BC1] proved that Carleman's condition (4.2) implies the denseness of $\mathbb{C}[x]$ in $L^p(\mathbb{R}, \mu)$ for $p \in [1, +\infty)$, where μ is the unique representing measure of s. We sketch a proof of this result in Exercise 4.7.   ∘

4.3 Krein's Condition

The following theorem of Krein shows that, for measures given by a density, the so-called *Krein condition* (4.19) is a sufficient condition for *indeterminacy*.

Theorem 4.14 *Let f be a nonnegative Borel function on $\mathbb{R}$. Suppose that the measure μ defined by $d\mu = f(x)dx$ is in $\mathcal{M}_+(\mathbb{R})$, that is, μ has finite moments*

$s_n := \int x^n d\mu(x)$ *for all* $n \in \mathbb{N}_0$. *If*

$$\int_{\mathbb{R}} \frac{\ln f(x)}{1 + x^2} \, dx > -\infty, \tag{4.19}$$

then the moment sequence $s = (s_n)_{n \in \mathbb{N}_0}$ *is indeterminate. Moreover, the polynomials* $\mathbb{C}[x]$ *are not dense in* $L^2(\mathbb{R}, \mu)$.

Proof The proof makes essential use of some fundamental results on boundary values of analytic functions in the upper half plane. (All facts needed for this proof can be found e.g. in [Gr]). Recall that the Hardy space $H^1(\mathbb{R})$ consists of all analytic functions h in the upper half-plane satisfying

$$\sup_{y>0} \int_{\mathbb{R}} |h(x + iy)| dx < \infty.$$

Each $h(z) \in H^1(\mathbb{R})$ has a nontangential limit function $h(x) \in \mathcal{L}^1(\mathbb{R})$ [Gr, Theorem 3.1]. From assumption (4.19) it follows that there is an $h \in H^1(\mathbb{R})$ such that $|h(x)| = f(x)$ a.e. on $\mathbb{R}$ [Gr, Theorem 4.4]. (In fact, (4.19) implies that the Poisson integral

$$u(z) = u(x + iy) = \frac{1}{\pi} \int_{\mathbb{R}} \frac{y}{(x - t)^2 + y^2} \ln f(t) \, dt$$

of the function $\ln f(x)$ exists. The corresponding function is $h(z) = e^{u(z)+iv(z)}$, where v is harmonic conjugate to u.) Since $h \in H^1(\mathbb{R})$, it follows from a theorem of Paley–Wiener [Gr, Lemma 3.7 or p. 84] that

$$\int_{\mathbb{R}} e^{itx} h(x) dx = 0 \quad \text{for} \quad t \geq 0. \tag{4.20}$$

Set $\xi(x) = h(x) f(x)^{-1}$ if $f(x) \neq 0$ and $\xi(x) = 0$ if $f(x) = 0$. Then $|\xi(x)| \leq 1$ on $\mathbb{R}$ and $d\mu = f dx$, so Lemma 4.8 applies to the function

$$g(t) := \int_{\mathbb{R}} e^{itx} h(x) dx = \int_{\mathbb{R}} e^{itx} \xi(x) d\mu(x).$$

Recall that $g(t) = 0$ for $t \geq 0$ by (4.20). Therefore, by formula (4.9) in Lemma 4.8,

$$(-i)^n g^{(n)}(0) = \int_{\mathbb{R}} x^n \xi(x) d\mu(x) = \int_{\mathbb{R}} x^n h(x) dx = 0 \quad \text{for} \quad n \in \mathbb{N}_0. \tag{4.21}$$

Let $h_1(x) := \operatorname{Re} h(x)$ and $h_2(x) := \operatorname{Im} h(x)$. From (4.21) we obtain

$$\int_{\mathbb{R}} x^n h_1(x) dx = \int_{\mathbb{R}} x^n h_2(x) dx = 0 \quad \text{for} \quad n \in \mathbb{N}_0. \tag{4.22}$$

Since $f \neq 0$ by (4.19) and $|h(x)| = f(x)$ a.e. on $\mathbb{R}$, h_1 or h_2 is nonzero, say h_j, and we have $f(x) - h_j(x) \geq 0$ on $\mathbb{R}$. Hence the positive Radon measure ν on $\mathbb{R}$ given by $d\nu := (f(x) - h_j(x))dx$ has the same moments as μ by (4.22). But ν is different from μ, because $h_j \neq 0$.

Since h is nonzero, so is ξ. By (4.21), $\xi \in L^2(\mathbb{R}, \mu)$ is orthogonal to $\mathbb{C}[x]$ in $L^2(\mathbb{R}, \mu)$. Hence $\mathbb{C}[x]$ is not dense in $L^2(\mathbb{R}, \mu)$. $\qquad\square$

Let us briefly discuss the Krein condition (4.19). First we note that (4.19) implies that $f(x) > 0$ a.e. (Indeed, if $f(x) = 0$, hence $\ln f(x) = -\infty$, on a set with nonzero Lebesgue measure, then the integral in (4.19) is $-\infty$.)

We set $\ln^+ x := \max(\ln x, 0)$ and $\ln^- x := -\min(\ln x, 0)$ for $x \geq 0$. Then $\ln^- x \geq 0$ and $\ln x = \ln^+ x - \ln^- x$. Since $f(x) \geq 0$ and hence $\ln^+ f(x) \leq f(x)$, we have

$$0 \leq \int_{\mathbb{R}} \frac{\ln^+ f(x)}{1 + x^2} \, dx \leq \int_{\mathbb{R}} \frac{f(x)}{1 + x^2} \, dx \leq \int_{\mathbb{R}} f(x) \, dx = s_0 < +\infty.$$

Therefore, (4.19) is equivalent to

$$\int_{\mathbb{R}} \frac{\ln^- f(x)}{1 + x^2} \, dx < +\infty.$$

Remark 4.15 In the literature, the integral

$$\frac{1}{\pi} \int_{\mathbb{R}} \frac{\ln f(x)}{1 + x^2} \, dx$$

is often called the *entropy integral* or *logarithmic integral*, see [Ks]. $\qquad\circ$

Remark 4.16 An interesting converse of the preceding theorem was proved by J.-P. Gabardo [Gb]. Suppose that s is an indeterminate moment sequence. Then there exists a solution of the moment problem for s given by a density $f(x)$ such that (4.19) holds and the entropy integral is maximal among all densities of absolutely continuous solutions of the moment problem. $\qquad\circ$

The next theorem is about Krein's condition for the Stieltjes moment problem.

Theorem 4.17 *Let f, μ, and s be as in Theorem 4.14. If the measure $\mu \in \mathcal{M}_+(\mathbb{R})$ is supported on $[0, +\infty)$ and*

$$\int_{\mathbb{R}} \frac{\ln f(x^2)}{1 + x^2} \, dx \equiv \int_0^\infty \frac{\ln f(x)}{(1 + x)} \frac{dx}{\sqrt{x}} > -\infty, \tag{4.23}$$

then the Stieltjes moment problem for s is indeterminate.

Proof Note that $d\mu = f(y)dy$ and $\mu \in M_+(\mathbb{R}_+)$. We define a symmetric measure $\nu \in M_+^{\text{sym}}(\mathbb{R})$ by $d\nu = |x|f(x^2)dx$ and we compute for $h \in C_c(\mathbb{R}_+)$,

$$\int_0^\infty h(y)\,d\mu(y) = 2\int_0^\infty h(x^2)xf(x^2)\,dx = \int_{\mathbb{R}} h(x^2)|x|f(x^2)\,dx = \int_{\mathbb{R}} h(x^2)\,d\nu(x).$$

Comparing this equality with (3.27) we conclude that $\nu = \mu^{\text{sym}}$, that is, μ^{sym} has the density $|x|f(x^2)$. By assumption (4.23), we have

$$\int_{\mathbb{R}} \frac{\ln|x|f(x^2)}{1+x^2}\,dx = \int_{\mathbb{R}} \frac{\ln|x|}{1+x^2}\,dx + \int_0^\infty \frac{\ln f(y)}{(1+y)}\frac{dy}{\sqrt{y}} > -\infty.$$

Therefore, by Theorem 4.14, the Hamburger moment sequence of μ^{sym} is indeterminate, so the Stieltjes moment sequence s is indeterminate by Proposition 3.19.   □

We apply the preceding criteria to treat a number of examples.

Example 4.18 (The Hamburger moment problem for $d\mu = e^{-|x|^\alpha}dx,\ \alpha > 0$)
   Clearly, $\mu \in \mathcal{M}_+(\mathbb{R})$. If $0 < \alpha < 1$, Krein's condition (4.19) is satisfied, since

$$\int_{\mathbb{R}} \frac{\ln e^{-|x|^\alpha}}{1+x^2}\,dx = \int_{\mathbb{R}} \frac{-|x|^\alpha}{1+x^2}\,dx > -\infty.$$

Therefore, the Hamburger moment problem for μ is indeterminate.
   If $\alpha \geq 1$, then

$$s_n = \left(\int_{-1}^1 + \int_{|x|\geq 1}\right)x^n e^{-|x|^\alpha}dx \leq 2 + \int_{\mathbb{R}} x^n e^{-|x|}dx = 2 + 2n! \leq 2^n n!. \qquad (4.24)$$

Hence, by Corollary 4.10(i), the Hamburger moment problem for μ is determinate.
                                                                                                                     ○

Example 4.19   (The Stieltjes moment problem for $d\mu = \chi_{[0,\infty)}(x)e^{-|x|^\alpha}dx,\ \alpha > 0$)
   If $0 < \alpha < 1/2$, then (4.23) holds, so the Stieltjes moment problem is indeterminate. If $\alpha \geq 1/2$, then $2\alpha \geq 1$ and hence by (4.24),

$$s_n = \int_0^\infty x^n d\mu(x) = \int_0^\infty (x^2)^n d\mu(x^2) = \int_0^\infty x^{2n} e^{-|x|^{2\alpha}}dx \leq 4^n(2n)!.$$

Thus, by Corollary 4.10(ii), the Stieltjes moment problem is determinate. Since μ has no atom at 0, it follows from Corollary 8.9 proved in Chap. 8 that the Hamburger moment problem for s is also determinate if $\alpha \geq 1/2$.
   By some computations it can be shown that the moments of μ are

$$s_n = \Gamma((n+1)/\alpha) \quad \text{for} \quad n \in \mathbb{N}_0.$$

The Gamma function has an asymptotics $\Gamma(x) \sim \sqrt{2\pi}\, e^{-x} x^{x+1/2}$ as $|x| \to \infty$, see e.g. [RW, p. 279]. From this it follows that we have an asymptotics $s_n^{1/n} \sim c n^{1/\alpha}$ for some $c > 0$. Hence $\sum_{n=1}^{\infty} s_{2n}^{-\frac{1}{2n}} < \infty$ for $0 < \alpha < 1$.

Therefore, by the preceding, if $1/2 \le \alpha < 1$, then the Hamburger moment sequence s is determinate, but Carleman's condition (4.2) is not satisfied! Another example of this kind can be found in Remark 7.19 in Sect. 7.   ∘

Example 4.20 $dv = \frac{1}{2} e^{-|x|} dx$ and $d\mu(x) = \frac{1}{2}\chi_{[0,+\infty)}(x)\, e^{-\sqrt{x}} dx$.

The moments of the measures v and μ are easily computed. They are

$$s_n = \frac{1}{2}\int_{\mathbb{R}} x^n e^{-|x|} dx = (2n)!\,, \quad t_n = \frac{1}{2}\int_0^{\infty} x^n e^{-\sqrt{x}} dx = (2n+1)!\,, \quad n \in \mathbb{N}_0,$$

so the sequences s and t satisfy conditions (4.12) and (4.13), respectively. Therefore, by Corollary 4.10 (i) and (ii), the Hamburger moment problem for v is determinate and the Stieltjes moment problem for μ is determinate. Note that in both cases the corresponding Krein conditions (4.19) and (4.23) are violated.   ∘

Example 4.21 $d\mu(x) = \chi_{(0,+\infty)}(x) x^{\alpha} e^{-x^2} dx$, where $\alpha > -1$.

In this case we calculate

$$s_n = \int_0^{\infty} x^n x^{\alpha} e^{-x^2} dx = \Gamma((n+\alpha+1)/2), \quad n \in \mathbb{N}_0. \qquad (4.25)$$

Then Corollary 4.11 applies and implies that s is determinate.   ∘

Example 4.22 (Lognormal distribution) The first examples of indeterminate measures were given by T. Stieltjes in his memoir [Stj]. He showed that the *log-normal distribution* $d\mu = f(x)dx$ with density

$$f(x) = \frac{1}{\sqrt{2\pi}}\, \chi_{(0,+\infty)}(x) x^{-1} \exp(-(\ln x)^2/2)$$

is indeterminate. We carry out his famous classical example in detail.

Let $n \in \mathbb{Z}$. Substituting $y = \ln x$ and $t = y - n$, we compute

$$s_n = \int_{\mathbb{R}} x^n \, d\mu(x) = \frac{1}{\sqrt{2\pi}}\int_0^{\infty} x^{n-1} e^{-(\ln x)^2/2}\, dx = \frac{1}{\sqrt{2\pi}}\int_{\mathbb{R}} e^{ny} e^{-y^2/2}\, dy$$

$$= \frac{1}{\sqrt{2\pi}}\int_{\mathbb{R}} e^{-(y-n)^2/2} e^{n^2/2}\, dy = e^{n^2/2}\frac{1}{\sqrt{2\pi}}\int_{\mathbb{R}} e^{t^2/2}\, dt = e^{n^2/2},$$

so that $s = (e^{n^2/2})_{n\in\mathbb{N}_0}$. This proves μ that finite moments, that is, $\mu \in \mathcal{M}_+(\mathbb{R})$.

For arbitrary $c \in [-1, 1]$ we define a positive (!) measure μ_c by

$$d\mu_c(x) = [1 + c \sin(2\pi \ln x)]\, d\mu(x).$$

Clearly, $\mu_c \in \mathcal{M}_+(\mathbb{R})$, since $\mu \in \mathcal{M}_+(\mathbb{R})$. For $n \in \mathbb{Z}$, a similar computation yields

$$\int_{\mathbb{R}} x^n \sin(2\pi \ln x) \, d\mu(x) = \frac{1}{\sqrt{2\pi}} \int_{\mathbb{R}} e^{ny} (\sin 2\pi y) \, e^{-y^2/2} dy$$

$$= \frac{1}{\sqrt{2\pi}} \int_{\mathbb{R}} e^{-(y-n)^2/2} e^{n^2/2} \sin 2\pi y \, dy = \frac{1}{\sqrt{2\pi}} e^{n^2/2} \int_{\mathbb{R}} e^{-t^2/2} \sin 2\pi (t+n) \, dt = 0,$$

where we used that the function $\sin 2\pi(t+n)$ is odd. From the definition of μ_c and the preceding equality it follows at once that μ_c has the same moments as μ. (This was even shown for all moments s_n with $n \in \mathbb{Z}$.) Hence μ is indeterminate.   o

Example 4.23 (Lognormal distributions (continued)) Let $\alpha \in \mathbb{R}$ and $r > 0$ be arbitrary. Then the function

$$f(x) = \frac{1}{r\sqrt{2\pi}} \, \chi_{(0,\infty)}(x) \, x^{-1} \, e^{-\frac{(\ln x - \alpha)^2}{2r^2}}$$

defines a probability measure μ on $\mathbb{R}$ by $d\mu = f(x)dx$. This measure μ is indeterminate and it has the moments $s_n = e^{n\alpha + n^2 \frac{r^2}{2}}$ for $n \in \mathbb{N}_0$.   o

4.4 Exercises

1. Show that each Stieltjes moment sequence $s = (s_n)$ satisfying

$$\sum_{n=1}^{\infty} s_n^{-\frac{1}{2n}} = +\infty$$

   is determinate as a Hamburger moment sequence.
2. Suppose that $(a_n)_{n \in \mathbb{N}_0}$ is a sequence of positive numbers such that

$$2 \ln a_{2n} \leq \ln a_{n+1} + \ln a_{n-1} \quad \text{for} \quad n \in \mathbb{N}.$$

   a. Show that $\sqrt[n]{a_n} \, {}^{n+1}\!\sqrt{a_0} \leq {}^{n+1}\!\sqrt{a_{n+1}} \sqrt[n]{a_0}$ for $n \in \mathbb{N}_0$.
   b. Suppose that $a_0 = 1$. Show that $(\sqrt[n]{a_n})_{n \in \mathbb{N}_0}$ is monotone increasing.
3. Show that for a moment sequence $s = (s_n)$ the following are equivalent:

   (i) s satisfies Carleman's condition (4.2).
   (ii) $\sum_{n=1}^{\infty} s_{4n}^{-1/(4n)} = \infty$.
   (iii) $\sum_{n=1}^{\infty} s_{4n+2}^{-1/(4n+2)} = \infty$.
   (iv) $\sum_{n=1}^{\infty} s_{2kn+2l}^{-1/(2kn+2l)} = \infty$ for some (and then for all) $k \in \mathbb{N}, l \in \mathbb{N}_0$.
4. Prove that the moment sequence (4.25) in Example 4.21 is determinate.

5. Prove that the measure in Example 4.23 is indeterminate.
6. Let $\mu \in M_+(\mathbb{R})$. Suppose that $\int_{\mathbb{R}} x^{2n} e^{\varepsilon x^2} d\mu < +\infty$ for some $\varepsilon > 0$ and $n \in \mathbb{N}_0$. Show that $\mu \in \mathcal{M}_+(\mathbb{R})$ and the moment sequence of μ is determinate.
7. (*Carleman's condition implies denseness of* $\mathbb{C}[x]$ *in* $L^p(\mathbb{R}, \mu), p \in [1, +\infty)$ [BC2])
   Let $\mu \in \mathcal{M}_+(\mathbb{R})$. Suppose the moment sequence of μ satisfies Carleman's condition (4.2). Prove that $\mathbb{C}[x]$ is dense in $L^p(\mathbb{R}, \mu)$ for any $1 \le p < +\infty$.

   Sketch of proof It suffices to prove the denseness for $p = 2k, k \in \mathbb{N}$. We mimic the proof of Lemma 4.8. Let $\xi \in L^p(\mathbb{R}, \mu)' \cong L^q(\mathbb{R}, \mu), \frac{1}{p} + \frac{1}{q} = 1$, be such that $\int \xi(x) f(x) d\mu = 0$ for all $f \in \mathbb{C}[x]$. Define $g(t)$ by (4.8). As in the proof of Lemma 4.8, we show that $g \in C^\infty(\mathbb{R})$ and that Eq. (4.9) holds. Since $\int \xi(x) f(x) d\mu = 0$ for $f \in \mathbb{C}[x]$, $g^{(n)}(0) = 0$ for $n \in \mathbb{N}_0$ by (4.9). Applying the Hölder inequality to (4.9) we obtain $|g^{(n)}(t)| \le s_{2kn}^{-1/(2k)} \|\xi\|_{L^q(\mathbb{R}, \mu)}$. By Exercise 4.3, (4.2) implies $\sum_{n=1}^{\infty} s_{2kn}^{-1/(2kn)} = +\infty$. Thus, Corollary 4.5 applies with $m_n = s_{2kn}^{1/(2k)}$, $t_0 = 0$, and yields $g(t) \equiv 0$. Hence $\xi = 0$, so $\mathbb{C}[x]$ is dense in $L^p(\mathbb{R}, \mu)$.

8. (*Moment generating function*)
   Let $\mu \in M_+(\mathbb{R})$ and $c > 0$. Suppose that the function $x \mapsto e^{tx}$ is μ-integrable for $|t| < c$. Then $g(t) := \int_{\mathbb{R}} e^{tx} d\mu(x)$ is called the *moment generating function* of μ.

   a. Show that $\mu \in \mathcal{M}_+(\mathbb{R})$, that is, μ has finite moments $s_n = \int x^n d\mu$ for $n \in \mathbb{N}_0$.
   b. Show that $s_n = g^{(n)}(0)$ for $n \in \mathbb{N}_0$.
   c. Show that $g(t) = \sum_{n=0}^{\infty} \frac{t^n}{n!} s_n$ for $t \in (-c, c)$.

4.5 Notes

Carleman's condition and Theorem 4.3(i) are due to T. Carleman [Cl].

Theorem 4.14 is stated in [Ak, Exercise 14 on p. 87] where it is attributed to M.G. Krein. It follows from Krein's results in [Kr1]. Our proof based on boundary values of analytic functions is from [Lin1] and [Sim1]. Another proof using Jensen's inequality is given in [Be, Theorem 4.1]. A generalization of Krein's condition and a discrete analogue were obtained by H.L. Pedersen [Pd2]. The denseness of polynomials in $L^p(\mathbb{R}, \mu)$ and Carleman's condition are studied in [BC1], [BC2], [Ks], [KMP], [Bk1]. [BR], [If]. An index of determinacy for determinate measures is defined in [BD]. Further elaborations of the determinacy problem can be found in [Lin2], [Stv], [SKv].

An interesting characterization of determinacy was discovered by C. Berg, Y. Chen, and M.E.H. Ismail [BCI], see also [BS1] for more refined results: Let λ_n denote the smallest eigenvalue of the Hankel matrix $H_n(s)$ of a moment sequence s. Then s is determinate if and only if $\lim_{n \to \infty} \lambda_n = 0$.

Chapter 5
Orthogonal Polynomials and Jacobi Operators

In the preceding chapters we derived basic existence and uniqueness results for moment problems. In this chapter we develop two powerful tools for a "finer" study of one-dimensional moment problems: orthogonal polynomials and Jacobi operators.

Throughout this chapter we assume that $s = (s_n)_{n \in \mathbb{N}_0}$ is a fixed **positive definite moment sequence**. The positive definiteness is crucial for the construction of orthogonal polynomials and subsequent considerations. By Proposition 3.11 a moment sequence s is positive definite if and only if all Hankel determinants

$$D_n \equiv D_n(s) \begin{vmatrix} s_0 & s_1 & s_2 & \dots & s_n \\ s_1 & s_2 & s_3 & \dots & s_{n+1} \\ s_2 & s_3 & s_4 & \dots & s_{n+2} \\ \dots\dots & \dots & \dots\dots & & \\ s_n & s_{n+1} & s_{n+2} & \dots & s_{2n} \end{vmatrix}, \; n \in \mathbb{N}_0, \tag{5.1}$$

are positive. We shall retain the notation (5.1) in what follows.

In Sect. 5.1 we define and study orthogonal polynomials associated with s. There are two distinguished sequences of orthogonal polynomials, the sequence $(p_n)_{n \in \mathbb{N}_0}$ of orthonormal polynomials with positive leading coefficients (5.3) and the sequence $(P_n)_{n \in \mathbb{N}_0}$ of monic orthogonal polynomials (5.6). In Sect. 5.2 we characterize these sequences in terms of three term reccurence relations and derive Favard's theorem (Theorem 5.10). The three term reccurence relation for $(p_n)_{n \in \mathbb{N}_0}$ implies that the multiplication operator X is unitarily equivalent to a Jacobi operator (Theorem 5.14). The interplay of moment problems and Jacobi operators is studied in Sect. 5.3.

Orthogonal polynomials of the second kind are investigated in Sect. 5.4. In Sect. 5.5 the Wronskian is defined and some useful identities on the orthogonal polynomials are derived. Section 5.6 contains basic results about zeros of orthogonal polynomials, while Sect. 5.7 deals with symmetric moment problems.

© Springer International Publishing AG 2017
K. Schmüdgen, *The Moment Problem*, Graduate Texts in Mathematics 277,
DOI 10.1007/978-3-319-64546-9_5

The study of orthogonal polynomials is an important subject that is of interest in itself. We therefore develop some of the beautiful classical results and formulas in this chapter, even if not all of them are used for the moment problem.

5.1 Definitions of Orthogonal Polynomials and Explicit Formulas

Since the sequence s is positive definite, the equation

$$\langle p,q \rangle_s := L_s(p\,\overline{q}), \quad p,q \in \mathbb{C}[x], \tag{5.2}$$

defines a scalar product $\langle \cdot, \cdot \rangle_s$ on the vector space $\mathbb{C}[x]$. (Indeed, it is obvious that $\langle \cdot, \cdot \rangle_s$ is a positive semidefinite sesquilinear form. If $\langle p,p \rangle_s = \sum_{k,l} s_{k+l} \overline{c}_k c_l = 0$ for some $p(x) = \sum_k c_k x^k \in \mathbb{C}[x]$, then $c_k = 0$ for all k and hence $p = 0$, since s is positive definite. This proves that $\langle \cdot, \cdot \rangle_s$ is a scalar product.)

Note that $\langle p,q \rangle_s$ is real for $p,q \in \mathbb{R}[x]$, because the sequence $s = (s_n)$ is real.

The following orthonormal basis of the unitary space $(\mathbb{C}[x], \langle \cdot, \cdot \rangle_s)$ will play a crucial role in what follows.

Proposition 5.1 *There exists an orthonormal basis* $(p_n)_{n \in \mathbb{N}_0}$ *of the unitary space* $(\mathbb{C}[x], \langle \cdot, \cdot \rangle_s)$ *such that each polynomial p_n has degree n and a positive leading coefficient. The basis* $(p_n)_{n \in \mathbb{N}_0}$ *is uniquely determined by these properties. Moreover,* $p_n \in \mathbb{R}[x]$.

Proof For the existence it suffices to apply the Gram–Schmidt procedure to the basis $\{1, x, x^2, \dots\}$ of the unitary space $(\mathbb{C}[x], \langle \cdot, \cdot \rangle_s)$. Since the scalar product is real on $\mathbb{R}[x]$, we obtain an orthonormal sequence $p_n \in \mathbb{R}[x]$ such that $\deg(p_n) = n$. Upon multiplying by -1 if necessary the leading coefficient of p_n becomes positive. The uniqueness assertion follows by a simple induction argument.   □

That the sequence $(p_n)_{n \in \mathbb{N}_0}$ is *orthonormal* means that

$$\langle p_k, p_n \rangle_s = \delta_{k,n} \quad \text{for} \quad k,n \in \mathbb{N}_0.$$

Definition 5.2 The polynomials p_n, $n \in \mathbb{N}_0$, are called *orthogonal polynomials of the first kind*, or *orthonormal polynomials*, associated with the positive definite sequence s.

The existence assertion from Proposition 5.1 will be reproved by Proposition 5.3. We have included Proposition 5.1 in order to show that proofs are often much shorter if no explicit formulas involving the moments are required. This is true for many other results as well, such as the recurrence relations (5.9) and (5.11). But our aim in this book is to provide explicit formulas for most quantities if possible.

Proposition 5.3 *Set $D_{-1} = 1$. Then $p_0(x) = \frac{1}{\sqrt{s_0}}$ and for $n \in \mathbb{N}$ and $k \in \mathbb{N}_0$,*

$$p_n(x) = \frac{1}{\sqrt{D_{n-1}D_n}} \begin{vmatrix} s_0 & s_1 & s_2 & \cdots & s_n \\ s_1 & s_2 & s_3 & \cdots & s_{n+1} \\ s_2 & s_3 & s_4 & \cdots & s_{n+2} \\ \cdots & & & & \\ s_{n-1} & s_n & s_{n+1} & \cdots & s_{2n-1} \\ 1 & x & x^2 & \cdots & x^n \end{vmatrix}, \tag{5.3}$$

$$\langle x^n, p_n \rangle_s = \sqrt{D_n/D_{n-1}} \quad and \quad \langle x^k, p_n \rangle_s = 0 \quad if \quad k < n. \tag{5.4}$$

The leading coefficient of p_n is $\sqrt{D_{n-1}/D_n}$. In particular, $p_1(x) = \frac{s_0 x - s_1}{\sqrt{s_0(s_0 s_2 - s_1^2)}}$.

Proof Obviously, $p_0(x) = \frac{1}{\sqrt{s_0}}$. In this proof let p_n denote the polynomial (5.3).

First we verify (5.4). The polynomial $x^k p_n(x)$ is obtained by multiplying the last row of the determinant in (5.3) by x^k. Applying the functional L_s to $x^k p_n$ means that each terms x^{k+j} in the last row has to be replaced by s_{k+j}. Thus,

$$\langle x^k, p_n \rangle_s = L_s(x^k p_n) = \frac{1}{\sqrt{D_{n-1}D_n}} \begin{vmatrix} s_0 & s_1 & s_2 & \cdots & s_n \\ s_1 & s_2 & s_3 & \cdots & s_{n+1} \\ s_2 & s_3 & s_4 & \cdots & s_{n+2} \\ \cdots & & & & \\ s_{n-1} & s_n & s_{n+1} & \cdots & s_{2n-1} \\ s_k & s_{k+1} & s_{k+2} & \cdots & s_{k+n} \end{vmatrix}. \tag{5.5}$$

If $k < n$, the last row coincides with the $(k+1)$-th row in (5.5), so that $\langle x^k, p_n \rangle_s = 0$. If $k = n$, the determinant in (5.5) is just the Hankel determinant D_n, that is,

$$\langle x^n, p_n \rangle_s = \frac{1}{\sqrt{D_{n-1}D_n}} D_n = \sqrt{D_n/D_{n-1}},$$

which completes the proof of (5.4).

Next we prove that $\langle p_k, p_n \rangle_s = \delta_{kn}$. First let $k < n$. Since $\deg p_k = k < n$, we conclude from (5.4) that $\langle p_k, p_n \rangle_s = 0$. Similarly, $\langle p_k, p_n \rangle_s = 0$ for $k > n$. Now let $k = n$. Since $\langle x^j, p_n \rangle_s = 0$ for $j < n$, $\langle p_n, p_n \rangle_s$ is equal to $\langle x^n, p_n \rangle_s$ multiplied by the leading coefficient of p_n. From (5.3) it follows that p_n has the leading coefficient $\frac{1}{\sqrt{D_{n-1}D_n}} D_{n-1} = \sqrt{D_{n-1}/D_n}$. Since $\langle x^n, p_n \rangle_s = \sqrt{D_n/D_{n-1}}$ by (5.4), this yields $\langle p_n, p_n \rangle_s = 1$.

From the uniqueness assertion of Proposition 5.1 it follows that the polynomial p_n defined by (5.3) concides with the polynomial p_n in Proposition 5.1.   □

We now define the general notion of orthogonal polynomials.

Definition 5.4 A sequence $(R_n)_{n \in \mathbb{N}_0}$ is called a sequence of *orthogonal polynomials*, briefly an OPS, with respect to s if $R_n \in \mathbb{R}[x]$, $\deg R_n = n$, and

$$\langle R_k, R_n \rangle_s = 0 \quad \text{for} \quad k \neq n, \ k,n \in \mathbb{N}_0.$$

Let $(R_n)_{n \in \mathbb{N}_0}$ be an OPS. Then $\|R_n\|_s \neq 0$, since $\langle \cdot, \cdot \rangle_s$ is a scalar product, and $\tau_n R_n$ has positive leading term for $\tau_n = +$ or $\tau_n = -$. Hence, by the uniqueness of the orthonormal sequence (p_n), we have $\tau_n \|R_n\|_s^{-1} R_n = p_n$ for all $n \in \mathbb{N}_0$.

While there are many OPS for a given s, there is a *unique* OPS consisting of monic polynomials. Such an OPS will be called *monic*. Recall that a polynomial P of degree n is *monic* if its coefficient of x^n is 1. Since p_n has the leading term $\sqrt{D_{n-1}/D_n}$, the polynomial

$$P_n(x) := \sqrt{D_n/D_{n-1}}\, p_n(x) = \frac{1}{D_{n-1}} \begin{vmatrix} s_0 & s_1 & s_2 & \dots & s_n \\ s_1 & s_2 & s_3 & \dots & s_{n+1} \\ s_2 & s_3 & s_4 & \dots & s_{n+2} \\ \dots & & & & \\ s_{n-1} & s_n & s_{n+1} & \dots & s_{2n-1} \\ 1 & x & x^2 & \dots & x^n \end{vmatrix}, \quad n \in \mathbb{N}, \qquad (5.6)$$

is monic. Set $P_0(x) = 1$. Then $(P_n)_{n \in \mathbb{N}_0}$ is the *unique monic OPS* for s. Thus, there are two distinguished sequences of orthogonal polynomials associated with s, the orthonormal sequence $(p_n)_{n \in \mathbb{N}_0}$ and the monic OPS $(P_n)_{n \in \mathbb{N}_0}$ given by the above formulas (5.3) and (5.6), respectively.

We close this section with a beautiful classical formula for the polynomial P_n. It will be not used later in this book. Another formula for P_n is given in Lemma 6.27(i).

Proposition 5.5 *(Heine Formulas) Suppose that $\mu \in \mathcal{M}_s$. Then, for $n \in \mathbb{N}$, $n \geq 2$, we have*

$$P_n(x) = \frac{1}{n!\, D_{n-1}} \int_{\mathbb{R}^n} \prod_{j=1}^{n} (x - x_j) \prod_{1 \leq k < l \leq n} (x_k - x_l)^2 \, d\mu(x_1) \dots d\mu(x_n), \qquad (5.7)$$

$$D_{n-1} = \frac{1}{n!} \int_{\mathbb{R}^n} \prod_{1 \leq k < l \leq n} (x_k - x_l)^2 \, d\mu(x_1) \dots d\mu(x_n). \qquad (5.8)$$

Proof Let us abbreviate $\tilde{P}_n(x) := D_{n-1} p_n(x)$.

If σ is a permutation of $\{0, 1, \ldots, n-1\}$, we compute

$$\tilde{P}_n(x) = \int_{\mathbb{R}^n} d\mu(x_{\sigma(0)}) \ldots d\mu(x_{\sigma(n-1)}) \begin{vmatrix} 1 & x_{\sigma(0)}^1 & \cdots & x_{\sigma(0)}^n \\ x_{\sigma(1)}^1 & x_{\sigma(1)}^2 & \cdots & x_{\sigma(1)}^{n+1} \\ \cdots & \cdots & \cdots & \cdots \\ x_{\sigma(n-1)}^{n-1} & x_{\sigma(n-1)}^n & \cdots & x_{\sigma(n-1)}^{2n-1} \\ 1 & x & \cdots & x^n \end{vmatrix}$$

$$= \int d\mu(x_{\sigma(0)}) \ldots d\mu(x_{\sigma(n-1)}) \begin{vmatrix} 1 & x_{\sigma(0)}^1 & \cdots & x_{\sigma(0)}^n \\ 1 & x_{\sigma(1)}^1 & \cdots & x_{\sigma(1)}^n \\ \cdots & \cdots & \cdots & \cdots \\ 1 & x_{\sigma(n-1)}^1 & \cdots & x_{\sigma(n-1)}^n \\ 1 & x & \cdots & x^n \end{vmatrix} x_{\sigma(0)}^0 x_{\sigma(1)}^1 \ldots x_{\sigma(n-1)}^{n-1}$$

$$= \int d\mu(x_0) \ldots d\mu(x_{n-1}) \begin{vmatrix} 1 & x_0^1 & \cdots & x_0^n \\ 1 & x_1^1 & \cdots & x_1^n \\ \cdots & \cdots & \cdots & \cdots \\ 1 & x_{n-1}^1 & \cdots & x_{n-1}^n \\ 1 & x & \cdots & x^n \end{vmatrix} (\text{sign } \sigma) x_{\sigma(0)}^0 x_{\sigma(1)}^1 \ldots x_{\sigma(n-1)}^{n-1}.$$

Here the first equality follows from formula (5.6) by replacing the moments s_l in row $j+1$ by $\int_{\mathbb{R}} x_{\sigma(j)}^l \, d\mu(x_{\sigma(j)})$ and using the multilinearity of the determinant. For the third equality the rows in the determinant are permuted and the integration variables are changed to $x_0, \ldots, x_{n-1}$.

Summing over all $n!$ permutations σ of $\{0, 1, \ldots, n-1\}$ and inserting the determinant definition it follows that the polynomial $n! \tilde{P}_n(x)$ is equal to

$$\int_{\mathbb{R}^n} d\mu(x_0) \ldots d\mu(x_{n-1}) \begin{vmatrix} 1 & x_0^1 & \cdots x_0^n \\ 1 & x_1^1 & \cdots x_1^n \\ \cdots & \cdots & \cdots \\ 1 & x_{n-1}^1 & \cdots x_{n-1}^n \\ 1 & x & \cdots x^n \end{vmatrix} \cdot \begin{vmatrix} 1 & x_0^1 & \cdots x_0^{n-1} \\ 1 & x_1^1 & \cdots x_1^{n-1} \\ \cdots & \cdots & \cdots \\ 1 & x_{n-1}^1 & \cdots x_{n-1}^{n-1} \end{vmatrix}.$$

Both determinants under the integral are Vandermonde determinants. The formula

$$\begin{vmatrix} 1 & y_0^1 & \cdots & y_0^m \\ 1 & y_1^1 & \cdots & y_1^m \\ \cdots & \cdots & \cdots & \cdots \\ 1 & y_m^1 & \cdots & y_m^m \end{vmatrix} = \prod_{0 \le k < l \le m} (y_k - y_l)$$

for a Vandermonde determinant implies that these determinants are

$$\prod_{j=0}^{n-1}(x-x_j)\prod_{0\le k<l\le n-1}(x_k-x_l)\quad\text{and}\quad\prod_{0\le k<l\le n-1}(x_k-x_l),$$

respectively. Changing the variables from $x_0,\dots,x_{n-1}$ to $x_1,\dots,x_n$ and inserting the preceding expressions shows that $n!\tilde{P}_n(x)=n!D_{n-1}p_n(x)$ is equal to the integral in (5.7). This proves (5.7). Comparing the coefficients of x^n in (5.7) gives (5.8). □

5.2 Three Term Recurrence Relations

Orthogonal polynomials can be characterized and studied by means of three term recurrence relations. We begin with the orthonormal sequence $(p_n)_{n\in\mathbb{N}_0}$.

Proposition 5.6 *Set* $a_n=\sqrt{D_{n-1}D_{n+1}}\,D_n^{-1}$ *and* $b_n=L_s(xp_n^2)$ *for* $n\in\mathbb{N}_0$. *Then we have* $a_n>0$ *and* $b_n\in\mathbb{R}$ *for* $n\in\mathbb{N}_0$, *and*

$$xp_n(x)=a_np_{n+1}(x)+b_np_n(x)+a_{n-1}p_{n-1}(x),\ n\in\mathbb{N}_0,\qquad(5.9)$$

where $a_{-1}:=1$ *and* $p_{-1}(x):=0$. *In particular*, $p_0(x)=s_0^{-1/2}$,

$$p_1(x)=s_0^{-1/2}a_0^{-1}(x-b_0),\quad p_2(x)=s_0^{-1/2}(a_0^{-1}a_1^{-1}(x-b_0)(x-b_1)-a_0a_1^{-1}).$$

Proof Since $D_k>0$ and $p_k\in\mathbb{R}[x]$ for all k, we have $a_n>0$ and $b_n\in\mathbb{R}$.

Since $xp_n(x)$ has degree $n+1$ and $\{p_0,\dots,p_{n+1}\}$ is a basis of the space of real polynomials of degree less than or equal to $n+1$, there are reals c_{nk} such that $xp_n(x)=\sum_{k=0}^{n+1}c_{nk}p_k(x)$. Comparing the coefficients of x^{n+1}, it follows that $c_{n,n+1}$ is the quotient of the leading coefficients of p_n and p_{n+1}. By Proposition 5.3 this yields $c_{n,n+1}=a_n$. Because the basis $\{p_k\}$ is orthonormal, we have

$$c_{nk}=\langle xp_n,p_k\rangle_s=L_s(xp_np_k)=\langle p_n,xp_k\rangle_s,\quad k=0,\dots,n+1.\qquad(5.10)$$

Since xp_k is in the span of $p_0,\dots,p_{k+1}$, (5.10) implies that $c_{nk}=0$ when $k+1<n$. Further, $c_{n,n}=L_s(xp_n^2)=b_n$ by (5.10). Using that $c_{n-1,n}$ is real we derive

$$c_{n,n-1}=\langle p_n,xp_{n-1}\rangle_s=\Big\langle p_n,\sum_{k=0}^n c_{n-1,k}p_k\Big\rangle_s=c_{n-1,n}.$$

Hence $a_{n-1}=c_{n,n-1}$. Putting the preceding together we have proved (5.9).

The formulas for p_1 and p_2 are easily computed from (5.9). □

Corollary 5.7 *The leading term of* $p_n(x)$ *is* $\sqrt{\frac{D_{n-1}}{D_n}}=s_0^{-1/2}\prod_{k=0}^{n-1}a_k^{-1}$ *for* $n\in\mathbb{N}$.

Proof We proceed by induction using the relation $a_k = \sqrt{D_{k-1}D_{k+1}}\, D_k^{-1}$, $k \in \mathbb{N}_0$.

For $n = 1$ we have $\sqrt{\frac{D_0}{D_1}} = D_0^{-1/2}(\sqrt{D_{-1}D_1}\, D_0^{-1})^{-1} = s_0^{-1/2}a_0^{-1}$.

If the assertion holds for $n \in \mathbb{N}$, then

$$\sqrt{\frac{D_n}{D_{n+1}}} = \frac{D_n}{\sqrt{D_{n-1}D_{n+1}}}\sqrt{\frac{D_{n-1}}{D_n}} = a_n^{-1}s_0^{-1/2}\prod_{k=0}^{n-1}a_k^{-1} = s_0^{-1/2}\prod_{k=0}^{n}a_k^{-1}. \qquad \square$$

Equation (5.9) is a *three term recurrence relation* for the polynomials p_n, that is, the polynomial p_{n+1} is determined by

$$p_{n+1}(x) = (x - b_n)a_n^{-1}p_n(x) - a_{n-1}a_n^{-1}p_{n-1}(x), \quad n \in \mathbb{N}_0, \tag{5.11}$$

where $a_{-1} := 1$ and $p_{-1}(x) := 0$. Hence, if $p_{j-1}(x)$ and $p_j(x)$ are given, then Eq. (5.9) (and likewise (5.11)) determines all polynomials $p_n(x)$, $n \geq j+1$, uniquely. This property of a three term reccurance relation will often be used.

The formula $a_n = \sqrt{D_{n-1}D_{n+1}}\, D_n^{-1}$ expresses a_n in terms of determinants involving only the moments. To derive a similar result for the numbers b_n we set

$$\Delta_n = \begin{vmatrix} s_0 & s_1 & s_2 & \cdots & s_{n-1} & s_{n+1} \\ s_1 & s_2 & s_3 & \cdots & s_n & s_{n+2} \\ s_2 & s_3 & s_4 & \cdots & s_{n+1} & s_{n+3} \\ \cdots & & & & & \\ s_n & s_{n+1} & s_{n+2} & \cdots & s_{2n-1} & s_{2n+1} \end{vmatrix}, \quad n \in \mathbb{N}, \text{ and } \Delta_0 = s_1, \tag{5.12}$$

that is, Δ_n is obtained from the Hankel determinant D_n (see (5.1)) by adding 1 to all indices in the last column.

Proposition 5.8 *For any $n, m \in \mathbb{N}_0$, we have*

$$b_{n+1} = \frac{\Delta_{n+1}}{D_{n+1}} - \frac{\Delta_n}{D_n} \quad and \quad \sum_{n=0}^{m}b_n = \frac{\Delta_m}{D_m}. \tag{5.13}$$

Proof First let $m = 0$. Then $(x - b_0)p_0 = a_0p_1$ by (5.6). Since $L_s(p_1) = 0$, we get $p_0s_1 = L_s(p_0x) = L_s(p_0b_0) = p_0b_0s_0$ and hence $b_0 = s_1s_0^{-1} = \Delta_0D_0^{-1}$.

Let $n \in \mathbb{N}$. By developing the determinant in (5.3) after the last row it follows that the coefficients of x^n and x^{n-1} are D_n and $-\Delta_{n-1}$, respectively, so the coefficient of x^n and x^{n-1} in p_n are $\frac{1}{\sqrt{D_{n-1}D_n}}D_n = \sqrt{\frac{D_{n-1}}{D_n}}$, and $-\frac{1}{\sqrt{D_{n-1}D_n}}\Delta_{n-1}$, respectively. Recall that $a_n = \sqrt{D_{n-1}D_{n+1}}\, D_n^{-1}$ by Proposition 5.6. Comparing the coeffients of x^n on both sides of the Eq. (5.9) and inserting $a_n = \sqrt{D_{n-1}D_{n+1}}\, D_n^{-1}$ we obtain

$$-\frac{1}{\sqrt{D_{n-1}D_n}}\Delta_{n-1} = -\frac{\sqrt{D_{n-1}D_{n+1}}}{D_n}\frac{1}{\sqrt{D_nD_{n+1}}}\Delta_n + b_n\sqrt{\frac{D_{n-1}}{D_n}}.$$

From this equation it follows that

$$b_n = \frac{\Delta_n}{D_n} - \frac{\Delta_{n-1}}{D_{n-1}} \quad \text{for} \quad n \in \mathbb{N},$$

which proves the first equality of (5.13). Therefore, since $b_0 = \Delta_0 D_0^{-1}$, we derive

$$\sum_{n=0}^{m} b_n = b_0 + \sum_{n=1}^{m} \left(\frac{\Delta_n}{D_n} - \frac{\Delta_{n-1}}{D_{n-1}} \right) = b_0 + \frac{\Delta_m}{D_m} - \frac{\Delta_0}{D_0} = \frac{\Delta_m}{D_m}. \qquad \square$$

The next proposition contains the three term recurrence relation for the monic OPS. Using the formula $P_n = \sqrt{D_n/D_{n-1}}\, p_n$ (by (5.6)) it is easily derived from the recurrence relation (5.9) for p_n. We omit the details of these simple computations.

Proposition 5.9 *Let $(P_n)_{n \in \mathbb{N}_0}$ be the monic OPS for s, see (5.6). Let a_{n-1} and b_n be as in Proposition 5.6 and $P_{-1} := 0$. Then*

$$P_{n+1}(x) = (x - b_n)P_n(x) - a_{n-1}^2 P_{n-1}(x), \quad n \in \mathbb{N}_0. \tag{5.14}$$

In particular, $P_0(x) = 1$, $P_1(x) = x - b_0$, and $P_2(x) = (x - b_0)(x - b_1) - a_0^2$.

The next result is *Favard's theorem*. It is a converse to Proposition 5.9. Its main direction states that for each set of parameters $\{a_n, b_n : n \in \mathbb{N}_0\}$ with $a_n > 0$ and $b_n \in \mathbb{R}$ the recurrence relation (5.14) defines a monic OPS of some positive definite sequence s and hence of some measure $\mu \in \mathcal{M}_s$.

Theorem 5.10 *Let $(\alpha_n)_{n \in \mathbb{N}_0}$ and $(\beta_n)_{n \in \mathbb{N}_0}$ be complex sequences and set $\alpha_{-1} := 1$. Let $(R_n)_{n \in \mathbb{N}_0}$ denote the sequence of monic polynomials R_n which is uniquely determined by the relations*

$$R_{n+1}(x) = (x - \beta_n)R_n(x) - \alpha_{n-1}R_{n-1}(x), \quad n \in \mathbb{N}_0, \tag{5.15}$$

$$R_{-1}(x) = 0 \quad \text{and} \quad R_0(x) = 1. \tag{5.16}$$

There exists a positive definite real sequence s such that $(R_n)_{n \in \mathbb{N}_0}$ is the monic OPS for s if and only if $\alpha_n > 0$ and $\beta_n \in \mathbb{R}$ for all $n \in \mathbb{N}_0$. If s_0 is a given positive number, then this sequence $s = (s_n)_{n \in \mathbb{N}_0}$ is uniquely determined.

Further, if $\alpha_n > 0$ and $\beta_n \in \mathbb{R}$ for $n \in \mathbb{N}_0$ and $s_0 > 0$ are given, then there exists a measure $\mu \in \mathcal{M}_s$ such that $\mu(\mathbb{R}) = s_0$ and for $j, k \in \mathbb{N}_0, j \neq k$, and $n \in \mathbb{N}$,

$$\int_{\mathbb{R}} R_j(x)R_k(x)\, d\mu(x) = 0, \quad \int R_n^2(x) d\mu(x) = \alpha_{n-1}\alpha_{n-2}\ldots\alpha_0 s_0. \tag{5.17}$$

Proof First note that the sequence (R_n) is indeed uniquely determined by (5.15) and (5.16). Clearly, R_n is monic and has degree n. Hence $(R_n)_{n \in \mathbb{N}_0}$ is a vector space basis of $\mathbb{C}[x]$, so we can define a linear functional L on

$$L(R_0) = L(1) = s_0 \quad \text{and} \quad L(R_n) = 0 \quad \text{for} \quad n \in \mathbb{N}. \tag{5.18}$$

Applying L to both sides of (5.15) we get $L(xR_n) = 0$ for $n \geq 2$. Multiplying (5.15) by x, applying L again by using the latter equality, we obtain $L(x^2 R_n) =$ for $n \geq 3$. Proceeding in this manner we derive

$$L(x^j R_n) = 0 \quad \text{for} \quad j = 0, \ldots, n-1, n \in \mathbb{N}. \tag{5.19}$$

Since $\deg R_m = m$, (5.19) implies that

$$L(R_j R_k) = 0 \quad \text{for} \quad j, k \in \mathbb{N}_0, j \neq k. \tag{5.20}$$

Multiplying (5.15) by x^{n-1} and applying L by using (5.19) again we get $L(x^n R_n) = \alpha_{n-1} L(x^{n-1} R_{n-1})$ and hence $L(x^n R_n) = \alpha_{n-1}\alpha_{n-2} \ldots \alpha_0 s_0$, because $L(R_0) = s_0$. Since R_n is monic, the preceding combined with (5.19) yields

$$L(R_n^2) = \alpha_{n-1}\alpha_{n-2} \ldots \alpha_0 s_0, \ n \in \mathbb{N}. \tag{5.21}$$

After all these technical preparations we are able to prove the assertions.

First suppose that (R_n) is an OPS for some positive definite real sequence s. Then L_s is equal to the functional L defined by (5.18) with $s_0 = L_s(1) > 0$, since $\langle R_n, R_0 \rangle_s = L_s(R_n) = 0$ for $n \in \mathbb{N}$. By the definition of an OPS, $R_n \in \mathbb{R}[x]$, so that $\langle R_n, R_n \rangle_s = L_s(R_n^2) > 0$ for $n \in \mathbb{N}$. Hence, by (5.21), $\alpha_n > 0$ for all $n \in \mathbb{N}$. Since $\alpha_n > 0$ and $R_n \in \mathbb{R}[x]$, it follows from (5.15) that β_n is real.

Conversely, assume that $\alpha_n > 0$ and $\beta_n \in \mathbb{R}$ for $n \in \mathbb{N}_0$. Let $s_0 > 0$ be arbitrary. Then $R_n \in \mathbb{R}[x]$ and $L(R_n^2) > 0$ by (5.21). Combined with (5.20) the latter implies that the linear functional L defined by (5.18) is a positive functional on $\mathbb{C}[x]$ such that $L(p\bar{p}) > 0$ for all $p \in \mathbb{C}[x]$, $p \neq 0$. Therefore, setting $s_n := \tilde{L}(x^n)$, $n \in \mathbb{N}_0$, we get a positive definite sequence $s = (s_n)_{n \in \mathbb{N}_0}$. By (5.20), (R_n) is an OPS for s. Since (R_n) is monic, it is the unique monic OPS associated with s.

Further, by Theorem 3.8, there exists a measure $\mu \in M_+(\mathbb{R})$ such that $s_n = \int x^n d\mu$ for $n \in \mathbb{N}_0$. Then $\mu(\mathbb{R}) = L(1) = s_0$. Since $L = L_s$ and μ is a representing measure for s, the equations (5.17) follow from (5.20) and (5.21). □

Remark 5.11

1. Clearly, if (R_n) is the monic OPS for a positive definite sequence s, so it is for each positive multiple of s. Hence the number $s_0 = L_s(1)$ cannot be determined from the monic OPS (R_n).
2. Favard's theorem for the recurrence relation (5.7) is contained in Theorem 5.14 below, see e.g. Remark 5.15 in the next section. ○

Comparing (5.14) and (5.15) shows that $\alpha_n = a_n^2$ for the monic OPS (P_n), so (5.21) yields the following formula. It also follows from Corollary 5.7.

Corollary 5.12 $\|P_n\|_s^2 = L_s(P_n^2) = a_{n-1}^2 a_{n-2}^2 \ldots a_0^2 s_0$ *for* $n \in \mathbb{N}$.

5.3   The Moment Problem and Jacobi Operators

The three term recurrence relation (5.9) links the moment problem to Jacobi operators. These operators are the basic objects for the *operator-theoretic approach* to the moment problem. This approach will be elaborated in the subsequent chapters.

Let $\mathcal{H}_s$ denote the Hilbert space completion of the unitary space $(\mathbb{C}[x], \langle \cdot, \cdot \rangle_s)$ and X the multiplication operator by the variable x with domain $\mathbb{C}[x]$ on $\mathcal{H}_s$:

$$Xp(x) := xp(x) \text{ for } p \in \mathcal{D}(X) := \mathbb{C}[x].$$

Then X is a *densely defined symmetric operator* with domain $\mathbb{C}[x]$ on the Hilbert space $\mathcal{H}_s$, since

$$\langle Xp, q \rangle_s = L_s(xp\,\overline{q}) = L_s(p\,\overline{xq}) = \langle p, Xq \rangle_s \text{ for } p, q \in \mathbb{C}[x].$$

Let $\{e_n : n \in \mathbb{N}_0\}$ be the standard orthonormal basis of the Hilbert space $l^2(\mathbb{N}_0)$ given by $e_n := (\delta_{kn})_{k \in \mathbb{N}_0}$. Since $\{p_n : n \in \mathbb{N}_0\}$ is an orthonormal basis of $\mathcal{H}_s$, there is a unitary isomorphism U of $\mathcal{H}_s$ onto $l^2(\mathbb{N}_0)$ defined by $Up_n = e_n$. Then, by (5.9), $T := UXU^{-1}$ is a symmetric operator on $l^2(\mathbb{N}_0)$ which acts by

$$Te_n = a_n e_{n+1} + b_n e_n + a_{n-1} e_{n-1}, \quad n \in \mathbb{N}_0, \tag{5.22}$$

where $e_{-1} := 0$. The domain $\mathcal{D}(T) = U(\mathbb{C}[x])$ is the linear span of vectors e_n, that is, $\mathcal{D}(T)$ is the vector space d of finite complex sequences $(\gamma_0, \dots, \gamma_n, 0, \dots)$. For any finite sequence $\gamma = (\gamma_n) \in$ d we obtain

$$T\left(\sum_n \gamma_n e_n\right) = \sum_n \gamma_n(a_n e_{n+1} + b_n e_n + a_{n-1} e_{n-1})$$
$$= \sum_n (\gamma_{n-1} a_{n-1} + \gamma_n b_n + \gamma_{n+1} a_n) e_n$$
$$= \sum_n (a_n \gamma_{n+1} + b_n \gamma_n + a_{n-1} \gamma_{n-1}) e_n \,,$$

where we have set $\gamma_{-1} := 0$. That is,

$$(T\gamma)_0 = a_0 \gamma_1 + b_0 \gamma_0, \ (T\gamma)_n = a_n \gamma_{n+1} + b_n \gamma_n + a_{n-1} \gamma_{n-1}, \ n \in \mathbb{N}. \tag{5.23}$$

Equation (5.23) means that the operator T acts on a sequence $\gamma \in$ d by multiplication with the infinite matrix

$$\mathcal{J} = \begin{pmatrix} b_0 & a_0 & 0 & 0 & 0 & \cdots \\ a_0 & b_1 & a_1 & 0 & 0 & \cdots \\ 0 & a_1 & b_2 & a_2 & 0 & \cdots \\ 0 & 0 & a_2 & b_3 & a_3 & \cdots \\ \cdots\cdots & & \ddots & \ddots & \ddots \end{pmatrix}. \tag{5.24}$$

A matrix J of this form is called a (semi-finite) *Jacobi matrix*. It is symmetric and tridiagonal. The corresponding operator $T \equiv T_J$ (likewise its closure) is called a *Jacobi operator*. Thus we have shown that the multiplication operator X is unitarily equivalent to the Jacobi operator T_J for the matrix (5.24).

The numbers $s_n s_0^{-1}$ can be recovered from the Jacobi operator T by

$$s_n s_0^{-1} = s_0^{-1} \langle x^n 1, 1 \rangle_s = \langle (X)^n p_0, p_0 \rangle_s = \langle T^n e_0, e_0 \rangle, \quad n \in \mathbb{N}_0. \tag{5.25}$$

That is, $s_n s_0^{-1}$ is the entry in the left upper corner of the matrix J^n. Thus, if $s_0 = 1$, then all moments s_n are uniquely determined from the Jacobi matrix. In particular,

$$s_1 s_0^{-1} = \langle T e_0, e_0 \rangle = b_0, \quad s_2 s_0^{-1} = \langle T^2 e_0, e_0 \rangle = b_0^2 + a_0^2,$$

$$s_3 s_0^{-1} = \langle T^3 e_0, e_0 \rangle = b_0^3 + 2a_0^2 b_0 + a_0^2 b_1.$$

Remark 5.13 Let $\tilde{s} = cs$, where $c > 0$, be a multiple of the positive definite sequence s. Then s and $\tilde{s}$ have the same Jacobi matrix J (see Exercise 5.1) and the same Jacobi operator T. Also, the multiplication operators X for $\tilde{s}$ and s are unitarily equivalent and we have $\mu \in \mathcal{M}_s$ if and only if $c\mu \in \mathcal{M}_{\tilde{s}}$. Thus for all self-adjointness problems of X and T and for the determinacy of s we could assume that $s_0 = 1$. ○

Conversely, let a Jacobi matrix (5.24) be given, where $a_n, b_n \in \mathbb{R}$ and $a_n > 0$. We define a linear operator T with domain $\mathcal{D}(T) = d$ on the Hilbert space $l^2(\mathbb{N}_0)$ by (5.23). Clearly, T is a symmetric operator and $T\mathcal{D}(T) \subseteq \mathcal{D}(T)$. Set

$$s = (s_n := \langle T^n e_0, e_0 \rangle)_{n \in \mathbb{N}_0}. \tag{5.26}$$

We prove that the sequence s is a positive definite. For $\xi_0, \ldots, \xi_n \in \mathbb{C}$ and $n \in \mathbb{N}_0$,

$$\sum_{k,l=0}^{n} s_{k+l} \overline{\xi}_k \xi_l = \sum_{k,l=0}^{n} \langle T^{k+l} e_0, e_0 \rangle \overline{\xi}_k \xi_l = \left\| \sum_{l=0}^{n} \xi_l T^l e_0 \right\|^2 \geq 0. \tag{5.27}$$

Hence s is positive semidefinite. By a simple induction argument we show that $T^l e_0 - a_0 \ldots a_{l-1} e_l \in \mathrm{Lin}\, \{e_0, \ldots, e_{l-1}\}$ for $l \in \mathbb{N}$. Therefore, if the expression in (5.27) is zero, we conclude that all ξ_n are zero, since $a_j \neq 0$ for all j. Thus, s is a positive definite real sequence. By (5.26) we have $s_0 = \langle T^0 e_0, e_0 \rangle_s = 1$.

Recall that U is the unitary of $\mathcal{H}_s$ onto $l^2(\mathbb{N}_0)$ defined by $Up_n = e_n, n \in \mathbb{N}_0$. Put $\tilde{p}_n := U^{-1} e_n$ for $n \in \mathbb{N}_0$. Clearly, (5.23) implies (5.22), so the polynomials $\tilde{p}_n$ satisfy the recurrence relation (5.9). Since $a_n > 0$ and $b_n \in \mathbb{R}$ by assumption, it follows from (5.9) that $\tilde{p}_n \in \mathbb{R}[x]$ has degree n and positive leading term. The polynomials $\tilde{p}_n$ are orthonormal in $\mathcal{H}_s$, since the vectors e_n are in $l^2(\mathbb{N}_0)$. Therefore, the $\tilde{p}_n$ are the polynomials of the first kind for s, that is, $p_n = \tilde{p}_n$ for $n \in \mathbb{N}_0$. Hence T is the Jacobi operator associated with s according to the preceding construction.

Summarizing, we have proved the following

Theorem 5.14 *The preceding construction provides a one-to-one correspondence between positive definite sequences s satisfying $s_0 = 1$ and Jacobi matrices J of the form (5.24), where $b_n \in \mathbb{R}$ and $a_n > 0$ for $n \in \mathbb{N}_0$, and a unitary equivalence between the multiplication operator X and the Jacobi operator $T \equiv T_J$ given by a unitary U such that $Up_n = e_n, n \in \mathbb{N}_0$. The Jacobi parameters a_n and b_n in the matrix (5.24) are the numbers occuring in the three term recurrence relation (5.9).*

For notational simplicity we identify the operators X and $T = UXU^{-1}$ in what follows, where U is the unitary of $\mathcal{H}_s$ onto $l^2(\mathbb{N}_0)$ defined by $Up_n = e_n, n \in \mathbb{N}_0$.

Thus, by Theorem 5.14, for any positive definite sequence s the multiplication operator X is unitarily equivalent to the Jacobi operator T_J of the matrix (5.24), where $a_n > 0$ and $b_n \in \mathbb{R}$ are as in Proposition 5.6. Conversely, if J is a Jacobi matrix (5.24) with $a_n > 0$ and $b_n \in \mathbb{R}$, then (5.26) is a positive definite sequence for which the above procedure leads again to the matrix (5.24).

Remark 5.15 Theorem 5.14 can be considered as *Favard's theorem*. Indeed, suppose that numbers $b_n \in \mathbb{R}$ and $a_n > 0$, $n \in \mathbb{N}_0$, are given. Then, by Theorem 5.14, there is a positive definite sequence s with $s_0 = 1$ such that the Jacobi operator $T = T_J$ is unitarily equivalent to the multiplication operator X in $(\mathbb{C}[x], \langle \cdot, \cdot \rangle_s)$ and the polynomials p_n defined by (5.7) form the corresponding orthonormal OPS for s. Further, by Hamburger's Theorem 3.8, there is a Radon measure $\mu \in \mathcal{M}_+(\mathbb{R})$ such that $s_n \equiv \langle T^n e_0, e_0 \rangle = \int x^n \, d\mu$ for $n \in \mathbb{N}_0$.   ∘

5.4 Polynomials of the Second Kind

In this section, we assume that a_n and b_n, $n \in \mathbb{N}_0$, denote the Jacobi parameters of the positive definite sequence s and we set $a_{-1} := 1$. Our aim is to develop another sequence $(q_n)_{n \in \mathbb{N}_0}$ of polynomials associated with s.

For a complex sequence $\gamma = (\gamma_n)_{n \in \mathbb{N}_0}$ we define a complex sequence $\mathcal{T}\gamma$ by

$$(\mathcal{T}\gamma)_n = a_n \gamma_{n+1} + b_n \gamma_n + a_{n-1}\gamma_{n-1} \text{ for } n \in \mathbb{N}_0, \text{ where } \gamma_{-1} := 0. \qquad (5.28)$$

Then $\mathcal{T}$ is a linear mapping of the vector space of all complex sequences. For $\gamma \in$ d, it is obvious that $\mathcal{T}\gamma \in$ d and $\mathcal{T}\gamma = T\gamma$, where $T = T_J$ is the Jacobi operator.

Suppose that $z \in \mathbb{C}$ and consider the *three term recurrence relation*

$$(\mathcal{T}\gamma)_n \equiv a_n \gamma_{n+1} + b_n \gamma_n + a_{n-1}\gamma_{n-1} = z\gamma_n, \qquad (5.29)$$

where $\gamma_{-1} := 0$, for an arbitrary complex sequence $\gamma = (\gamma_n)_{n \in \mathbb{N}_0}$. Clearly, since $a_n > 0$, if we fix two initial data γ_{k-1} and γ_k and assume that the relation (5.29) is satisfied for all $n \geq k$, all terms γ_n, where $n \geq k+1$, are uniquely determined.

We set $\gamma_{-1} = 0$, $\gamma_0 = s_0^{-1/2}$ and assume that (5.29) holds for all $n \in \mathbb{N}_0$. Comparing (5.29) with (5.9) by using that $p_{-1}(z) = 0$ and $p_0(z) = s_0^{-1/2}$ we conclude that γ_n is the value $p_n(z)$ of the polynomial p_n from Proposition 5.1. We abbreviate

$$\mathfrak{p}_z := (p_0(z), p_1(z), p_2(z), \dots), \quad z \in \mathbb{C}. \tag{5.30}$$

Now we set $\gamma_0 = 0$, $\gamma_1 = a_0^{-1} s_0^{1/2}$ and suppose that (5.29) holds for all $n \in \mathbb{N}$. (Note that we do *not* assume (5.29) for $n = 0$.) The numbers γ_n are then uniquely determined and we denote γ_n by $q_n(z)$, $n \in \mathbb{N}_0$. Clearly, the same solution is obtained if we start with the initial data $\gamma_{-1} = -s_0^{1/2}$, $\gamma_0 = 0$ and require (5.29) for all $n \in \mathbb{N}_0$.

Using relation (5.29) it follows easily by induction on n that $q_n(z)$, $n \subset \mathbb{N}$, is a polynomial in z of degree $n-1$. We denote the corresponding sequence by

$$\mathfrak{q}_z := (q_0(z), q_1(z), q_2(z), \dots), \quad z \in \mathbb{C}. \tag{5.31}$$

By definition, $q_0(z) = 0$ and $q_1(z) = a_0^{-1} s_0^{1/2}$. Further, $q_2(z) = (z-b_1)(a_0 a_1)^{-1} s_0^{1/2}$.

It should be emphasized that the numbers a_n, b_n and the polynomials p_n, q_n depend only on the sequence s, but not on any representing measure.

Lemma 5.16 $\mathcal{T}\mathfrak{p}_z = z\mathfrak{p}_z$ and $\mathcal{T}\mathfrak{q}_z = s_0^{1/2} e_0 + z\mathfrak{q}_z$ for all $z \in \mathbb{C}$.

Proof By the recurrence relations we have $(\mathcal{T}\mathfrak{p}_z)_n = zp_n(z) \equiv (z\mathfrak{p}_z)_n$ for $n \in \mathbb{N}_0$ and $(\mathcal{T}\mathfrak{q}_z)_n = zq_n(z) \equiv (z\mathfrak{q}_z)_n$ for $n \in \mathbb{N}$. Using that $\gamma_0=0$ and $\gamma_1=a_0^{-1}s_0^{1/2}$ we compute the zero component $(\mathcal{T}\mathfrak{q}_z)_0$ by

$$(\mathcal{T}\mathfrak{q}_z)_0 = a_0\gamma_1 + b_0\gamma_0 = a_0 a_0^{-1} s_0^{1/2} + 0 = s_0^{1/2} + z\gamma_0 = s_0^{1/2} + zq_0(z). \qquad \square$$

Definition 5.17 The polynomials $q_n(z), n \in \mathbb{N}_0$, are the *orthogonal polynomials of the second kind* associated with the positive definite sequence s.

From the defining relations and intial data for $q_n(z)$ it follows that the polynomials $\widetilde{p}_n(z) := s_0^{-1/2} a_0 q_{n+1}(z)$ satisfy the recurrence relation (5.29) with a_n replaced by $\widetilde{a}_n := a_{n+1}$ and b_n by $\widetilde{b}_n := b_{n+1}$ and the initial data $\widetilde{p}_{-1} = 0, \widetilde{p}_0 = 1$. Therefore, by Theorem 5.14, the polynomials $\widetilde{p}_n(z)$, $n \in \mathbb{N}_0$, are the orthonormal polynomials of first kind with respect to the shifted Jacobi matrix

$$\widetilde{J} = \begin{pmatrix} b_1 & a_1 & 0 & 0 & 0 & \dots \\ a_1 & b_2 & a_2 & 0 & 0 & \dots \\ 0 & a_2 & b_3 & a_3 & 0 & \dots \\ 0 & 0 & a_3 & b_4 & a_4 & \dots \\ \dots \dots \dots & & \ddots & & \ddots & \ddots \end{pmatrix}. \tag{5.32}$$

The corresponding positive definite sequence is $\widetilde{s} = (\widetilde{s}_n)$, where $\widetilde{s}_n = \langle \widetilde{T}^n e_0, e_0 \rangle$ and $\widetilde{T}$ is the Jacobi operator corresponding to the Jacobi matrix $\widetilde{J}$. Hence $q_n, n \in \mathbb{N}$, are orthogonal polynomials (according to Definition 5.4) for the sequence $\widetilde{s}$.

The next proposition contains another useful description of the polynomials q_n.

Let $r(x) = \sum_{k=0}^n \gamma_k x^k \in \mathbb{C}[x], n \in \mathbb{N}$, be a polynomial. For any fixed $z \in \mathbb{C}$,

$$\frac{r(x) - r(z)}{x - z} = \sum_{k=0}^n \gamma_k \frac{x^k - z^k}{x - z} = \sum_{k=1}^n \sum_{l=0}^{k-1} \gamma_k z^{k-l} x^l$$

is also a polynomial in x, so we can apply the functional L_s to it. We shall write $L_{s,x}$ to indicate that x is the corresponding variable.

Proposition 5.18 $q_n(z) = L_{s,x}\left(\frac{p_n(x) - p_n(z)}{x - z}\right)$ for $n \in \mathbb{N}_0$ and $z \in \mathbb{C}$.

Proof Let us denote the polynomial on the right-hand side by $r_n(z)$. From the recurrence relation (5.9) we obtain for $n \in \mathbb{N}$,

$$a_n \frac{p_{n+1}(x) - p_{n+1}(z)}{x - z} + b_n \frac{p_n(x) - p_n(z)}{x - z} + a_{n-1} \frac{p_{n-1}(x) - p_{n-1}(z)}{x - z}$$

$$= \frac{x p_n(x) - z p_n(z)}{x - z} = z \frac{p_n(x) - p_n(z)}{x - z} + p_n(x).$$

Applying the functional $L_{s,x}$ to this identity and using the orthogonality relation $0 = \langle p_n, 1 \rangle_s = L_{s,x}(p_n)$ for $n \geq 1$ we get

$$a_n r_{n+1}(z) + b_n r_n(z) + a_{n-1} r_{n-1}(z) = z r_n(z) \text{ for } n \in \mathbb{N}.$$

Since $p_1(x) = s_0^{-1/2} a_0^{-1}(x - b_0)$ and $p_0 = s_0^{-1/2}$ by Proposition 5.6, we have $r_0(z) = L_{s,x}(0) = 0 = q_0(z)$ and

$$r_1(z) = L_{s,x}\left(\frac{p_1(x) - p_1(z)}{x - z}\right) = s_0^{-1/2} a_0^{-1} L_{s,x}\left(\frac{x - z}{x - z}\right) = s_0^{-1/2} a_0^{-1} s_0 = q_1(z).$$

This shows that the sequence $(r_n(z))$ satisfies the same recurrence relation and initial data as $(q_n(z))$. Therefore, $r_n(z) = q_n(z)$ for $n \in \mathbb{N}_0$. $\qquad\square$

The next corollary expresses the polynomials q_n in terms of the moments.

Corollary 5.19 *Set $q_{n,k}(z) = \sum_{l=0}^{k-1} s_{k-l-1} z^l$ for $k \geq 1$ and $q_{n,0} = 0$. For $n \in \mathbb{N}$,*

$$q_n(z) := \frac{1}{\sqrt{D_{n-1} D_n}} \begin{vmatrix} s_0 & s_1 & s_2 & \dots & s_n \\ s_1 & s_2 & s_3 & \dots & s_{n+1} \\ s_2 & s_3 & s_4 & \dots & s_{n+2} \\ \dots \\ s_{n-1} & s_n & s_{n+1} & \dots & s_{2n-1} \\ q_{n,0}(z) & q_{n,1}(z) & q_{n,2}(z) & \dots & q_{n,n}(z) \end{vmatrix}, z \in \mathbb{C}. \qquad (5.33)$$

Proof From (5.3) it follows that $\frac{p_n(x)-p_n(z)}{x-z}$ is given by the same expression as in (5.3) when the entries x^k of the last row are replaced by $\frac{x^k-z^k}{x-z} = \sum_{l=0}^{k-1} x^{k-l-1} z^l$. Applying the functional $L_{s,x}$ to this determinant gives the determinant in (5.33). By Proposition 5.18, the corresponding polynomial in z is $q_n(z)$.   □

Corollary 5.20 *The monic polynomial associated with* $q_n(z)$ *is*

$$Q_n(z) := s_0^{-1} L_{s,x} \left(\frac{P_n(x) - P_n(z)}{x - z} \right) \quad \text{for} \quad n \in \mathbb{N}_0. \tag{5.34}$$

Further, for $n \in \mathbb{N}$ *and* $z \in \mathbb{C}$ *we have*

$$P_n(z) = s_0^{1/2} a_0 a_1 \ldots a_{n-1} p_n(z), \quad Q_n(z) = s_0^{-1/2} a_0 a_1 \ldots a_{n-1} q_n(z), \tag{5.35}$$

$$\frac{q_n(z)}{p_n(z)} = s_0 \frac{Q_n(z)}{P_n(z)}. \tag{5.36}$$

Proof Let $c_n x^n$ be the leading term of $p_n(x)$. Then $P_n = c_n^{-1} p_n$. Therefore, by (5.34) and Proposition 5.18, $Q_n(z) = s_0^{-1} c_n^{-1} q_n(z)$. Since $\frac{x^n-z^n}{x-z} = \sum_{l=0}^{n-1} x^{n-l-1} z^l$, the leading term of q_n is $z^{n-1} c_n L_{s,x}(1) = c_n s_0 z^{n-1}$. Hence $Q_n(z) = s_0^{-1} c_n^{-1} q_n(z)$ is monic.

By Corollary 5.7, $c_n^{-1} = s_0^{1/2} a_0 a_1 \ldots a_{n-1}$. Since $P_n = c_n^{-1} p_n$ and $Q_n = s_0^{-1} c_n^{-1} q_n$, this implies (5.35) and hence also (5.36).   □

In what follows we will often use the function f_z and the Stieltjes transform I_μ (see Appendix A.2) of a finite Radon measure μ on $\mathbb{R}$. They are defined by

$$f_z(x) := \frac{1}{x - z} \quad \text{and} \quad I_\mu(z) := \int_\mathbb{R} \frac{1}{x - z} d\mu(x), \quad z \in \mathbb{C} \backslash \mathbb{R}. \tag{5.37}$$

Proposition 5.21 *Suppose that* $\mu \in M_s$. *For* $z \in \mathbb{C} \backslash \mathbb{R}$ *and* $n \in \mathbb{N}_0$,

$$\langle f_z, p_n \rangle_{L^2(\mathbb{R},\mu)} = q_n(z) + I_\mu(z) p_n(z), \tag{5.38}$$

$$\| \mathfrak{q}_z + I_\mu(z) \mathfrak{p}_z \|^2 = \sum_{n=0}^\infty |q_n(z) + I_\mu(z) p_n(z)|^2 \leq \frac{\operatorname{Im} I_\mu(z)}{\operatorname{Im} z}. \tag{5.39}$$

In particular, $\mathfrak{q}_z + I_\mu(z) \mathfrak{p}_z \in l^2(\mathbb{N}_0)$. *Moreover, we have equality in the inequality of* (5.39) *if and only if the function* f_z *is in* $\mathcal{H}_s$.

Proof Clearly, the bounded function $f_z(x) = \frac{1}{x-z}$ is in $L^2(\mathbb{R}, \mu)$. We compute

$$\langle f_z, p_n \rangle_{L^2(\mathbb{R},\mu)} = \int_\mathbb{R} \frac{p_n(x) - p_n(z)}{x - z} d\mu(x) + \int_\mathbb{R} \frac{p_n(z)}{x - z} d\mu(x)$$

$$= L_{s,x} \left(\frac{p_n(x) - p_n(z)}{x - z} \right) + p_n(z) I_\mu(z) = q_n(z) + I_\mu(z) p_n(z),$$

which proves (5.38). Here the equality before last holds, since μ is a representing measure for s, and the last equality follows from Proposition 5.18.

The equality in (5.39) is merely the definition of the norm of $l^2(\mathbb{N}_0)$. Since $(\mathbb{C}[x], \langle \cdot, \cdot \rangle_s)$ is a subspace of $L^2(\mathbb{R}, \mu)$, $\{p_n : n \in \mathbb{N}_0\}$ is an orthonormal subset of $L^2(\mathbb{R}, \mu)$. The inequality in (5.39) is just Bessel's inequality for the Fourier coefficients of f_z with respect to this orthonormal set, because

$$\|f_z\|^2_{L^2(\mathbb{R}, \mu)} = \int \frac{1}{|x-z|^2}\, d\mu(x) = \int \frac{1}{z-\bar{z}}\left(\frac{1}{x-z} - \frac{1}{x-\bar{z}}\right) d\mu(x) = \frac{\operatorname{Im} I_\mu(z)}{\operatorname{Im} z}.$$

By an elementary fact from Hilbert space theory, equality in Bessel's inequality holds if and only if f_z belongs to the closed subspace generated by the polynomials p_n, that is, $f_z \in \mathcal{H}_s$. □

Corollary 5.22 $\mathfrak{p}_z \in l^2(\mathbb{N}_0)$ *if and only if* $\mathfrak{q}_z \in l^2(\mathbb{N}_0)$ *for* $z \in \mathbb{C} \backslash \mathbb{R}$.

Proof Let $z \in \mathbb{C} \backslash \mathbb{R}$. Since then $I_\mu(z) \neq 0$ and $\mathfrak{q}_z + I_\mu(z)\mathfrak{p}_z \in l^2(\mathbb{N}_0)$ by Proposition 5.21, it is clear that $\mathfrak{p}_z \in l^2(\mathbb{N}_0)$ is equivalent to $\mathfrak{q}_z \in l^2(\mathbb{N}_0)$. □

5.5 The Wronskian and Some Useful Identities

Suppose that $\gamma = (\gamma_n)_{n \in \mathbb{N}_0}$ and $\beta = (\beta_n)_{n \in \mathbb{N}_0}$ are complex sequences. We define their *Wronskian* as the sequence $W(\gamma, \beta) = (W(\gamma, \beta)_n)_{n \in \mathbb{N}_0}$ with terms

$$W(\gamma, \beta)_n := a_n(\gamma_{n+1}\beta_n - \gamma_n\beta_{n+1}), \quad n \in \mathbb{N}_0, \tag{5.40}$$

Let $\mathcal{T}$ be the linear mapping of complex sequences defined by (5.28), that is,

$$(\mathcal{T}\gamma)_n = a_n\gamma_{n+1} + b_n\gamma_n + a_{n-1}\gamma_{n-1} \text{ for } n \in \mathbb{N}_0, \text{ where } \gamma_{-1} := 0.$$

The following lemma on the Wronskian is the crux for several applications. It will play an essential role in the study of the adjoint of the Jacobi operator in Sect. 6.2.

Lemma 5.23 *Let* $\gamma = (\gamma_n)_{n \in \mathbb{N}_0}$, $\beta = (\beta_n)_{n \in \mathbb{N}_0}$ *be sequences, and* $x, z \in \mathbb{C}$. *Then*

$$\sum_{k=0}^{n} \left((\mathcal{T}\gamma)_k\beta_k - \gamma_k(\mathcal{T}\beta)_k)\right) = W(\gamma, \beta)_n \text{ for } n \in \mathbb{N}_0. \tag{5.41}$$

Let $m, n \in \mathbb{N}_0, n > m$. *If* $(\mathcal{T}\gamma)_k = x\gamma_k$ *and* $(\mathcal{T}\beta)_k = z\beta_k$ *for* $k = m+1, \ldots, n$, *then*

$$(x - z) \sum_{k=m+1}^{n} \gamma_k\beta_k = W(\gamma, \beta)_n - W(\gamma, \beta)_m. \tag{5.42}$$

In particular, if $x = z$ in this case, then

$$W(\gamma, \beta)_n = W(\gamma, \beta)_m. \tag{5.43}$$

Proof We prove the first identity (5.41) by computing

$$\sum_{k=0}^{n} [(\mathcal{T}\gamma)_k \beta_k - \gamma_k (\mathcal{T}\beta)_k)]$$

$$= (a_0\gamma_1 + b_0\gamma_0)\beta_0 - \gamma_0(a_0\beta_1 + b_0\beta_0)$$

$$+ \sum_{k=1}^{n} [(a_k\gamma_{k+1} + b_k\gamma_k + a_{k-1}\gamma_{k-1})\beta_k - \gamma_k(a_k\beta_{k+1} + b_k\beta_k + a_{k-1}\beta_{k-1})]$$

$$= a_0(\gamma_1\beta_0 - \gamma_0\beta_1) + \sum_{k=1}^{n} [(a_k(\gamma_{k+1}\beta_k - \gamma_k\beta_{k+1}) - a_{k-1}(\gamma_k\beta_{k-1} - \gamma_{k-1}\beta_k)]$$

$$= W(\gamma, \beta)_0 + \sum_{k=1}^{n} [W(\gamma, \beta)_k - W(\gamma, \beta)_{k-1}] = W(\gamma, \beta)_n.$$

Equation (5.42) is obtained by applying (5.41) to n and m and taking the difference of both sums. Setting $x = z$ in (5.42) yields (5.43). $\qquad\square$

Now we use Lemma 5.23 to derive some important identities on the polynomials p_n, q_n. Further, we define four polynomials A_n, B_n, C_n, D_n that will be used later in Sect. 7.1. Equation (5.47) is called the *Christoffel–Darboux formula*.

Proposition 5.24 *For $x, z \in \mathbb{C}$ and $n \in \mathbb{N}_0$, we have*

$$A_n(x, z) := (x - z) \sum_{k=0}^{n} q_k(x)q_k(z) = a_n(q_{n+1}(x)q_n(z) - q_n(x)q_{n+1}(z)), \tag{5.44}$$

$$B_n(x, z) := -1 + (x - z) \sum_{k=0}^{n} p_k(x)q_k(z) = a_n(p_{n+1}(x)q_n(z) - p_n(x)q_{n+1}(z)), \tag{5.45}$$

$$C_n(x, z) := 1 + (x - z) \sum_{k=0}^{n} q_k(x)p_k(z) = a_n(q_{n+1}(x)p_n(z) - q_n(x)p_{n+1}(z)), \tag{5.46}$$

$$D_n(x, z) := (x - z) \sum_{k=0}^{n} p_k(x)p_k(z) = a_n(p_{n+1}(x)p_n(z) - p_n(x)p_{n+1}(z)). \tag{5.47}$$

Proof All four identities are derived from Eq. (5.42). As samples we verify (5.45) and (5.47). Recall that the sequences $\mathsf{p}_x = (p_n(x))$, $\mathsf{q}_z = (q_n(z))$, and $\mathsf{q}_z = (q_n(z))$

satisfy the relations

$$(\mathcal{T}\mathfrak{p}_x)_n = xp_n(x), \quad (\mathcal{T}\mathfrak{q}_z)_n = zq_n(z), \quad (\mathcal{T}\mathfrak{p}_z)_n = zp_n(z) \quad \text{for} \quad n \in \mathbb{N}. \tag{5.48}$$

Hence the assumptions of Eq. (5.42) in Lemma 5.23 with $\alpha = \mathfrak{p}_x$, $\beta = \mathfrak{q}_z, \mathfrak{p}_z$, and $m = 0$ are satisfied. Inserting $p_0(x) = s_0^{-1/2}$, $p_1(x) = s_0^{-1/2}a_0^{-1}(x-b_0)$, $q_0(z) = 0$, and $q_1(z) = a_0^{-1}s_0^{1/2}$, we compute

$$W(\mathfrak{p}_x, \mathfrak{q}_z)_0 = a_0(p_1(x)q_0(z) - p_0(x)q_1(z)) = -1, \tag{5.49}$$

$$W(\mathfrak{p}_x, \mathfrak{p}_z)_0 = a_0(p_1(x)p_0(z) - p_0(x)p_1(z)) = (x - z)s_0^{-1}. \tag{5.50}$$

For $n = 0$ the right-hand sides of (5.45) and (5.47) are just $W(\mathfrak{p}_x, \mathfrak{q}_z)_0$ and $W(\mathfrak{p}_x, \mathfrak{p}_z)_0$, respectively, so these equations hold by (5.49) and (5.50). Now suppose that $n \in \mathbb{N}$. Then (5.42) applies with $m = 0$.

From equations (5.42) with $\alpha = \mathfrak{p}_x$, $\beta = \mathfrak{q}_z$ by using (5.49) and (5.40) we derive

$$(x-z)\sum_{k=0}^{n} p_k(x)q_k(z) = (x-z)\sum_{k=1}^{n} p_k(x)q_k(z)$$

$$= W(\mathfrak{p}_x, \mathfrak{q}_z)_n - W(\mathfrak{p}_x, \mathfrak{q}_z)_0 = a_n(p_{n+1}(x)q_n(z) - p_n(x)q_{n+1}(z)) + 1.$$

Applying (5.42) with $\alpha = \mathfrak{p}_x$, $\beta = \mathfrak{p}_z$ combined with (5.50) and (5.40) yields

$$(x-z)\sum_{k=0}^{n} p_k(x)p_k(z) = (x-z)s_0^{-1} + (x-z)\sum_{k=1}^{n} p_k(x)p_k(z)$$

$$= (x - z)s_0^{-1} + W(\mathfrak{p}_x, \mathfrak{p}_z)_n - W(\mathfrak{p}_x, \mathfrak{p}_z)_0 = a_n(p_{n+1}(x)p_n(z) - p_n(x)p_{n+1}(z)).$$

This proves (5.45) and (5.47).   □

Corollary 5.25 *Let $x \in \mathbb{C}$ and $n \in \mathbb{N}_0$. Then*

$$\sum_{k=0}^{n} p_k(x)^2 = a_n[p'_{n+1}(x)p_n(x) - p'_n(x)p_{n+1}(x)], \tag{5.51}$$

$$\frac{1}{a_n} = p_n(x)q_{n+1}(x) - p_{n+1}(x)q_n(x). \tag{5.52}$$

Proof From the identity (5.47) it follows that

$$\sum_{k=0}^{n} p_k(x)p_k(z) = a_n \frac{[p_{n+1}(x) - p_{n+1}(z)]p_n(z) - [p_n(x) - p_n(z)]p_{n+1}(z)}{x - z}.$$

Letting $z \to x$ we obtain (5.51). Equation (5.52) is obtained by setting $x = z$ in (5.46). □

Corollary 5.26 *For $x, z \in \mathbb{C}$ and $n \in \mathbb{N}_0$,*

$$p_{n+1}(x) - p_{n+1}(z) = (x - z) \sum_{k=0}^{n} [p_k(z)q_{n+1}(z) - p_{n+1}(z)q_k(z)]p_k(x), \qquad (5.53)$$

$$q_{n+1}(x) - q_{n+1}(z) = (x - z) \sum_{k=0}^{n} [p_k(z)q_{n+1}(z) - p_{n+1}(z)q_k(z)]q_k(x). \qquad (5.54)$$

Proof First we prove (5.53). We multiply (5.47) by $q_{n+1}(z)$ and (5.45) by $-p_{n+1}(z)$ and add both equations. On the right we get

$$a_n\big(q_{n+1}(z)[p_{n+1}(x)p_n(z) - p_n(x)p_{n+1}(z)] - p_{n+1}(z)[p_{n+1}(x)q_n(z) - p_n(x)q_{n+1}(z)]\big)$$
$$= p_{n+1}(x)\, a_n[q_{n+1}(z)p_n(z) - q_n(z)p_{n+1}(z)]$$
$$= p_{n+1}(x)W(\mathfrak{q}_z, \mathfrak{p}_z)_n = p_{n+1}(z)W(\mathfrak{q}_z, \mathfrak{p}_z)_0 = p_{n+1}(x). \qquad (5.55)$$

Here for the first equality the term $p_n(x)p_{n+1}(z)q_{n+1}(z)$ cancels, while the second is just the definition of $W(\mathfrak{q}_z, \mathfrak{p}_z)_n$. By (5.48) the assumptions of Lemma 5.23 are satisfied, so (5.43) yields $W(\mathfrak{q}_z, \mathfrak{p}_z)_n = W(\mathfrak{q}_z, \mathfrak{p}_z)_0$ for $n \in \mathbb{N}$. For $n = 0$ this is trivial. This gives the third equality. Since $W(\mathfrak{q}_z, \mathfrak{p}_z)_0 = 1$, the last equality in (5.55) holds.

On the left we obtain

$$(x-z) \sum_{k=0}^{n} q_{n+1}(z)p_k(x)p_k(z) + p_{n+1}(z) - (x-z) \sum_{k=0}^{n} p_{n+1}(z)p_k(x)q_k(z)$$

$$= p_{n+1}(z) + (x-z) \sum_{k=0}^{n} [p_k(z)q_{n+1}(z) - p_{n+1}(z)q_k(z)]p_k(x). \qquad (5.56)$$

Comparing (5.55) and (5.56) we obtain (5.53).

The proof of (5.54) is very similar by using (5.46) and (5.44) instead of (5.47) and (5.45). First we multiply (5.46) by $q_{n+1}(z)$ and (5.44) by $-p_{n+1}(z)$. Adding both equations, on the right-hand side we derive

$$a_n\big(q_{n+1}(z)[q_{n+1}(x)p_n(z) - q_n(x)p_{n+1}(z)] - p_{n+1}(z)[q_{n+1}(x)q_n(z) - q_n(x)q_{n+1}(z)]\big)$$
$$= q_{n+1}(x)a_n[q_{n+1}(z)p_n(z) - q_n(z)p_{n+1}(z)]$$
$$= q_{n+1}(x)W(\mathfrak{q}_z, \mathfrak{p}_z)_n = q_{n+1}(x)W(\mathfrak{q}_z, \mathfrak{p}_z)_0 = q_{n+1}(x).$$

The left-hand side gives

$$q_{n+1}(z) + (x-z) \sum_{k=0}^{n} q_{n+1}(z)q_k(x)p_k(z) - (x-z) \sum_{k=0}^{n} p_{n+1}(z)q_k(x)q_k(z)$$

$$= q_{n+1}(z) + (x-z) \sum_{k=0}^{n} [p_k(z)q_{n+1}(z) - p_{n+1}(z)q_k(z)]q_k(x).$$

Now (5.54) follows by comparing the results on both sides.   □

Corollary 5.27 *For any $x, z \in \mathbb{C}$ and $n \in \mathbb{N}_0$, we have*

$$A_n(x,z)D_n(x,z) - B_n(x,z)C_n(x,z) = 1, \tag{5.57}$$

$$D_n(x,0)B_n(z,0) - B_n(x,0)D_n(z,0) = -D_n(x,z). \tag{5.58}$$

Proof Inserting the identities (5.44)–(5.47) from Proposition 5.24 we compute

$$A_n(x,z)D_n(x,z) - B_n(x,z)C_n(x,z)$$

$$= \quad a_n[q_{n+1}(x)q_n(z) - q_n(x)q_{n+1}(z)]\, a_n[(p_{n+1}(x)p_n(z) - p_n(x)p_{n+1}(z)]$$

$$\quad -a_n[p_{n+1}(x)q_n(z) - p_n(x)q_{n+1}(z)]\, a_n[q_{n+1}(x)p_n(z) - q_n(x)p_{n+1}(z)]$$

$$= \quad a_n^2\,[p_{n+1}(x)q_n(x) - p_n(x)q_{n+1}(x)]\,[p_{n+1}(z)q_n(z) - q_{n+1}(z)p_n(z)]$$

$$= \quad B_n(x,x)B_n(z,z) = (-1)(-1) = 1.$$

Likewise, we derive

$$D_n(x,0)B_n(z,0) - B_n(x,0)D_n(z,0)$$

$$= \quad a_n[p_{n+1}(x)p_n(0) - p_n(x)p_{n+1}(0)]\, a_n[p_{n+1}(z)q_n(0) - p_n(z)q_{n+1}(0)]$$

$$\quad -a_n[p_{n+1}(x)q_n(0) - p_n(x)q_{n+1}(0)]\, a_n[p_{n+1}(z)p_n(0) - p_n(z)p_{n+1}(0)]$$

$$= \quad a_n^2\,[p_{n+1}(x)p_n(z) - p_n(x)p_{n+1}(z)](p_{n+1}(0)q_n(0) - p_n(0)q_{n+1}(0)]$$

$$= \quad D_n(x,z)B_n(0,0) = -D_n(x,z). \qquad \qquad □$$

5.6   Zeros of Orthogonal Polynomials

In this section, we derive a number of interesting results about zeros of the orthogonal polynomials .

Proposition 5.28 *Suppose that $p \in \mathbb{R}[x]$ has degree $m \in \mathbb{N}$ and $\langle p(x), x^j \rangle_s = 0$ for $j \in \mathbb{N}_0, j \leq m - 2$. Then the polynomial $p(x)$ has m distinct real zeros.*

More precisely, if μ is a solution of the moment problem for s and J is a closed interval containing supp μ, *then all zeros of p lie in the interior of J.*

Proof First we recall that by Theorem 3.8 there exists a solution μ of the moment problem for s. Let $\lambda_1, \ldots, \lambda_k$ denote the distinct real points in the interior of J, where p changes sign, and put $r(x) = (x - \lambda_1) \ldots (x - \lambda_k)$. If there is no such λ_j, we set $r=1$. Then the polynomial $r(x)p(x)$ does not change sign on J. Hence $q(x) := \tau r(x)p(x) \geq 0$ on J for $\tau=1$ or $\tau= -1$.

We shall prove that $k = m$. Assume to the contrary that $k < m$.

If $k = m - 1$, then p is a real polynomial of degree m which has $m - 1$ distinct real zeros. The latter is only possible if $k = m$, which is a contradiction. Hence $k \leq m - 2$. But then $\langle p, r \rangle_s = 0$ by assumption, so that

$$\int_J q(x)\, d\mu(x) = \tau \int_J p(x)r(x)\, d\mu(x) = \tau \langle p, r \rangle_s = 0.$$

Since $q(x) \geq 0$ on J, this implies that μ has a finite support. But then s is not positive definite, which contradicts our standing assumption.

This proves that $k = m$. Thus p has m distinct real zeros $\lambda_1, \ldots, \lambda_m$. $\square$

In particular, Proposition 5.28 applies orthogonal polynomials and yields

Corollary 5.29 *If $(R_n)_{n \in \mathbb{N}_0}$ is an OPS, then $R_n(x)$ has n distinct real zeros.*

Clearly, Corollary 5.29 holds for the polynomials p_n. As noted in Sect. 5.4, the polynomials $\widetilde{p}_n(x) = s_0^{-1/2} a_0 q_{n+1}(x)$, $n \in \mathbb{N}_0$, are orthonormal polynomials for the positive definite sequence obtained from the shifted Jacobi matrix (5.32). Therefore, Corollary 5.29 applies to $\widetilde{p}_n$, hence also to q_{n+1}, and we have the following

Corollary 5.30 *For $n \in \mathbb{N}$, the polynomials $p_n(x)$ and $q_{n+1}(x)$ have n distinct real zeros.*

The following corollary plays an essential role in the proof of Lemma 7.1 below.

Corollary 5.31 *Let $n \in \mathbb{N}_0$ and $z, z' \in \mathbb{C}$. If $|\text{Im } z| \leq |\text{Im } z'|$, then*

$$|p_n(z)| \leq |p_n(z')| \quad \text{and} \quad |q_n(z)| \leq |q_n(z')|. \tag{5.59}$$

Proof Let r be a polynomial of degree $n \in \mathbb{N}$ which has n distinct real zeros, say $x_1, \ldots, x_n$. Then $r(z) = c(z-x_1) \ldots (z-x_n)$ for some $c \in \mathbb{C}$. Therefore it is easily seen that $|r(z)| \leq |r(z')|$ when $|\text{Im } z| \leq |\text{Im } z'|$. By Corollary 5.30 this applies to p_n and q_{n+1} for $n \in \mathbb{N}$ and gives (5.59). For the constant polynomials p_0, q_0, q_1 the assertion (5.59) is trivial. $\square$

Let us denote the m zeros $\lambda_j^{(m)}$ of the polynomial p_m in increasing order

$$\lambda_1^{(m)} < \lambda_2^{(m)} < \cdots < \lambda_m^{(m)}.$$

Proposition 5.32 *The zeros of $p_n(x)$ and $p_{n+1}(x)$ interlace strictly, that is,*

$$\lambda_1^{(n+1)} < \lambda_1^{(n)} < \lambda_2^{(n+1)} < \cdots < \lambda_n^{(n)} < \lambda_{n+1}^{(n+1)}, \quad n \in \mathbb{N}. \tag{5.60}$$

Proof Since $p_0(x) = 1$, it follows from Eq. (5.51) that

$$p_{n+1}'(x)p_n(x) - p_n'(x)p_{n+1}(x) > 0 \quad \text{for} \quad x \in \mathbb{R}.$$

Setting $x = \lambda_j^{(n+1)}$ therein we get

$$p_{n+1}'(\lambda_j^{(n+1)})p_n(\lambda_j^{(n+1)}) > 0 \quad \text{for} \quad j = 1, \ldots, n+1. \tag{5.61}$$

Since each zero $\lambda_j^{(n+1)}$ is simple and p_{n+1} has a positive leading coefficient, we conclude that $\operatorname{sign} p_{n+1}'(\lambda_j^{(n+1)}) = (-1)^{n+1-j}$. Therefore, $\operatorname{sign} p_n(\lambda_j^{(n+1)}) = (-1)^{n+1-j}$ by (5.61). Thus, by the mean value theorem, $p_n(x)$ has at least one zero in each of the n intervals $(\lambda_j^{(n+1)}, \lambda_{j+1}^{(n+1)})$. Since p_n has only n zeros, we conclude that there is precisely one zero in each of these intervals. □

Corollary 5.33 *The two limits $\alpha_s := \lim_{n\to\infty} \lambda_1^{(n)}$ and $\beta_s := \lim_{n\to\infty} \lambda_n^{(n)}$ exist in $\mathbb{R} \cup \{-\infty\} \cup \{+\infty\}$.*

Proof By (5.60), the sequence $(\lambda_1^{(n)})$ is decreasing and the sequence $(\lambda_n^{(n)})$ is increasing. □

Proposition 5.34 *The zeros of $q_{n+1}(x)$ and $p_{n+1}(x)$ strictly interlace, that is, if $\kappa_1^{(n+1)} < \kappa_2^{(n+1)} < \cdots < \kappa_n^{(n+1)}$ are the zeros of q_{n+1}, then*

$$\lambda_1^{(n+1)} < \kappa_1^{(n+1)} < \lambda_2^{(n+1)} < \cdots < \lambda_n^{(n+1)} < \kappa_n^{(n+1)} < \lambda_{n+1}^{(n+1)}, \quad n \in \mathbb{N}.$$

Proof The proof is similar to the proof of Proposition 5.32. Setting $x = \lambda_j^{(n+1)}$ in formula (5.52) of Corollary 5.25 we obtain

$$p_n(\lambda_j^{(n+1)})q_{n+1}(\lambda_j^{(n+1)}) = a_n^{-1} > 0. \tag{5.62}$$

By (5.60), p_n has precisely one zero in each of the open intervals $(\lambda_j^{(n+1)}, \lambda_{j+1}^{(n+1)})$ and this zero is simple. Hence p_n has different signs at the end points and so does q_{n+1} by (5.62). Therefore, q_{n+1} has at least one zero in $(\lambda_j^{(n+1)}, \lambda_{j+1}^{(n+1)})$ for $j = 1, \ldots, n$. Since q_{n+1} has n zeros, this gives the assertion. □

Since $P_n = \sqrt{D_n/D_{n-1}}\, p_n$, the preceding assertions hold for the monic polynomials P_n as well. In particular, the zeros of P_n and P_{n+1} strictly interlace.

The next proposition says that two arbitrary interlacing finite sequences are zero sequences of some monic orthogonal polynomials.

Proposition 5.35 *Given real numbers* $\lambda_1, \ldots, \lambda_{m+1}, \kappa_1, \ldots, \kappa_m$ *satisfying*

$$\lambda_1 < \kappa_1 < \lambda_2 < \cdots < \kappa_m < \lambda_{m+1}, \tag{5.63}$$

there exists a positive definite sequence $s = (s_n)_{n \in \mathbb{N}_0}$ *such that* $s_0 = 1$ *and a monic OPS* $(P_n)_{n \in \mathbb{N}_0}$ *for* s *such that*

$$P_m(x) = (x - \kappa_1) \cdots (x - \kappa_m) \quad and \quad P_{m+1}(x) = (x - \lambda_1) \cdots (x - \lambda_{m+1}). \tag{5.64}$$

Proof By Favard's Theorem 5.10, it suffices to show that there exist real sequences $(\alpha_n)_{n \in \mathbb{N}_0}$ and $(\beta_n)_{n \in \mathbb{N}_0}$ such that $\alpha_n > 0$ for all n and a sequence $(P_n)_{n \in \mathbb{N}_0}$ of monic real polynomials such that

$$P_{n+1}(x) = (x - \beta_n) P_n(x) - \alpha_n P_{n-1}(x), \quad n \in \mathbb{N}_0, \tag{5.65}$$

where $P_{-1}(x) = 0$, $P_0(x) = 1$, and P_m and P_{m+1} are given by (5.64).

Since $\deg(P_{m+1} - xP_m) \leq m$, there is unique real number β_{m+1} such that

$$R_{m-1}(x) := P_{m+1}(x) - (x - \beta_m) P_m(x) \tag{5.66}$$

has degree at most $m - 1$. From the assumption (5.63) it follows that R_{m-1} changes signs at the zeros of P_m. Hence there is at least one real zero of R_{m-1} between two zeros of P_m. Since $\deg R_{m-1} \leq m - 1$, there is precisely one zero of R_{m-1} between two zeros of P_m. Since $P_{m+1}(\kappa_m) < 0$ by (5.63), we have $R_{m-1}(\kappa_m) < 0$ by (5.66), so the leading coeefficient of R_{m-1} is negative. Hence we can write $R_{m-1} = -\alpha_m P_{m-1}$ with $\alpha_m > 0$ for some unique monic real polynomial P_{m-1}. Then, by construction, (5.65) is satisfied for $n = m$ and the zeros of P_{m-1} and P_m strictly interlace. Therefore, we can continue by induction and construct polynomials $P_{m-2}, \ldots, P_1, P_0 = 1$ such that (5.65) is fulfilled for $n = m - 1, m - 2, \ldots, 0$.

To construct the polynomials P_n for $n \geq m + 2$ it suffices to choose numbers $\alpha_n > 0$ and $\beta_n \in \mathbb{R}$ and define P_n inductively by (5.65). $\quad\square$

Corollary 5.36 *For each monic polynomial* P *of degree* m *with* m *distinct real zeros there exists a monic OPS* $(P_n)_{n \in \mathbb{N}_0}$ *for some positive definite sequence with* $s_0 = 1$ *such that* $P = P_m$.

5.7 Symmetric Moment Problems

As throughout, $s = (s_n)_{n \in \mathbb{N}_0}$ is a real positive definite sequence.

Definition 5.37 We say that s is *symmetric* if $s_{2n+1} = 0$ for all $n \in \mathbb{N}_0$. A measure μ on $\mathbb{R}$ is called *symmetric* if $\mu(M) = \mu(-M)$ for all measurable sets M.

Clearly, s is symmetric if and only if

$$L_s(p(x)) = L_s(p(-x)) \quad \text{for} \quad p \in \mathbb{C}[x]. \tag{5.67}$$

Further, if a measure $\mu \in \mathcal{M}_+(\mathbb{R})$ is symmetric, its odd moments vanish, hence its moment sequence is symmetric. Conversely, each symmetric positive definite sequence s has a symmetric measure $\mu \in \mathcal{M}_s$. (Indeed, if $\nu \in \mathcal{M}_s$, then the measure μ defined by $\mu(M) = \frac{1}{2}(\nu(M) + \nu(-M))$ is symmetric and belongs to $\mathcal{M}_s$.)

Proposition 5.38 *Let s be a symmetric positive definite sequence. Then*

$$p_n(-x) = (-1)^n p_n(x), \quad q_n(x) = (-1)^{n+1} q_n(x), \quad b_n = 0 \quad \text{for} \quad n \in \mathbb{N}_0. \tag{5.68}$$

The operator V defined by $(Vp)(x) = p(-x)$, $p \in \mathbb{C}[x]$, extends to a self-adjoint unitary operator on the Hilbert space $\mathcal{H}_s$ and $VXV^{-1} = -X$.

Proof Let $p, q \in \mathbb{C}[x]$. Using Eq. (5.67) we obtain

$$\langle Vp, q \rangle_s = L_s(p(-x)\overline{q}(x)) = L_s(p(x)\overline{q}(-x)) = \langle p, Vq \rangle_s,$$
$$\langle Vp, Vq \rangle_s = L_s(p(-x)\overline{q}(-x)) = L_s(p(x)\overline{q}(x)) = \langle p, q \rangle_s,$$

so V is a symmetric isometric linear operator on the dense subspace $\mathbb{C}[x]$. Hence V extends by continuity to a self-adjoint unitary on $\mathcal{H}_s$. Obviously, $VXV^{-1} = -X$.

Set $\tilde{p}_n := (-1)^n V(p_n)$. Since p_n has degree n, $\tilde{p}_n$ has a positive leading coefficent. Since V is unitary, $(\tilde{p}_n)_{n\in\mathbb{N}_0}$ is an orthonormal basis of $(\mathbb{C}[x], \langle \cdot, \cdot \rangle_s)$. Therefore, by the uniqueness assertion of Proposition 5.1, $\tilde{p}_n = p_n$, so that $p_n(-x) = (-1)^n p_n(x)$.

Using Corollary 5.18 and Eq. (5.67) we derive for $n \in N_0$,

$$q_n(-x) = L_{s,y}\left(\frac{p_n(y) - p_n(-x)}{y - (-x)}\right) = -(-1)^n L_{s,y}\left(\frac{p_n(-y) - p_n(x)}{-y - x}\right)$$
$$= (-1)^{n+1} L_{s,y}\left(\frac{p_n(y) - p_n(x)}{y - x}\right) = (-1)^{n+1} q_n(x).$$

Finally, we prove that $b_n = 0$. By (5.9) we have the three term recurrence relation

$$x p_n(x) = a_n p_{n+1}(x) + b_n p_n(x) + a_{n-1} p_{n-1}(x).$$

Replacing x by $-x$, using that $p_k(-x) = (-1)^k p_k(x)$, and dividing by $(-1)^n$ we get

$$-x p_n(x) = -a_n p_{n+1}(x) + b_n p_n(x) - a_{n-1} p_{n-1}(x).$$

Adding both equations gives $2b_n p_n = 0$. Hence $b_n = 0$. $\qquad\square$

5.8 Exercises

1. Let s be a positive definite real sequence and set $\tilde{s} := cs$, where $c > 0$. Prove the following for $n \in \mathbb{N}_0$:

   a. $L_{\tilde{s}} = cL_s$ and $D_n(\tilde{s}) = c^{n+1}D_n(s)$.
   b. $\tilde{p}_n = c^{-1/2}p_n$, $\tilde{q}_n = c^{1/2}q_n$, and $\tilde{P}_n = P_n$, $\tilde{Q}_n = Q_n$.
   c. $\tilde{a}_n = a_n$ and $\tilde{b}_n = b_n$.
   d. $\tilde{A}_n(x, z) = cA_n(x, z)$, $\tilde{B}_n(x, z) = B_n(x, z)$, $\tilde{C}_n(x, z) = C_n(x, z)$, $\tilde{D}_n(x, z) = c^{-1}D_n(x, z)$.

   Here all quantities with a tilde refer to the positive definite sequence $\tilde{s}$.
   Hints: Use (5.3) for p_n, Proposition 5.18 for q_n, and Proposition 5.6 for a_n, b_n.

2. Let T be a symmetric linear operator with domain $\mathcal{D}(T)$ on a Hilbert space such that $T\mathcal{D}(T) \subseteq \mathcal{D}(T)$ and let $e \in \mathcal{D}(T)$. Define $s_n = \langle T^n e, e\rangle$ for $n \in \mathbb{N}_0$.

   a. Show that $s = (s_n)_{n \in \mathbb{N}_0}$ is a positive semidefinite sequence.
   b. Show that there exists a positive Radon measure $\mu \in \mathcal{M}(\mathbb{R})$ such that $\mu \in \mathcal{M}_s$.
   c. Show that s is positive definite if and only if the span of vectors $T^n e, n \in \mathbb{N}_0$, is infinite-dimensional.

3. What is the norm of the monic polynomial $P_n(z)$?
4. Let s be positive real sequence. Prove the following:

   a. If $p_n(-x) = (-1)^n p_n(x)$ for all $n \in \mathbb{N}_0$, then s is symmetric.
   b. If $b_n = 0$ for all $n \in \mathbb{N}_0$, then s is symmetric.

In the remaining exercises we develop important examples of orthogonal polynomials. Detailed treatments can be found in standard books such as [Chi1] or [Is].
We begin with the *Chebyshev polynomials T_n*.

5. Show that for each $n \in \mathbb{N}_0$ there is a unique polynomial $T_n \in \mathbb{R}[x]$ such that $T_n(x) = \cos(n \arccos x), x \in \mathbb{R}$, or equivalently, $T_n(\cos\theta) = \cos(n\theta)$, $\theta \in \mathbb{R}$.
6. Show that the polynomials T_n are uniquely defined by the recurrence relation $T_{n+1}(x) = 2xT_n(x) - T_{n-1}(x), n \in \mathbb{N}$, with initial data $T_0(x) = 1, T_1(x) = x$.
7. Show that the leading coefficient of T_n is 2^{n-1} for $n \in \mathbb{N}_0$.
8. Set $s_{2n} = \frac{(2n)!}{2^{2n}(n!)^2}$ and $s_{2n+1} = 0$ for $n \in \mathbb{N}_0$. Let μ be the probability measure on $[-1, 1]$ defined by $d\mu(x) := \pi^{-1}(1 - x^2)^{-1/2}dx$. Show that $s = (s_k)_{k \in \mathbb{N}_0}$ is a moment sequence with representing measure μ.
   Hint: Use [RW, p. 174 and p. 274].
9. Show that $\int_{-1}^1 T_k(x)T_l(x)d\mu(x) = 0$ if $k \neq l$, $\int_{-1}^1 T_k(x)^2 d\mu(x) = \frac{1}{2}$ if $k \in \mathbb{N}$ and $\int_{-1}^1 T_0(x)^2 d\mu = 1$, that is, the orthonormal polynomials p_n of the moment sequence s are $p_0 = T_0$ and $p_n = \sqrt{2}\,T_n$ for $n \in \mathbb{N}$.

Now we turn to the *Hermite polynomials* H_n. Define polynomials H_n by

$$H_{n+1}(x) = 2xH_n(x) - 2nH_{n-1}(x), \ n \in \mathbb{N}, \quad \text{and} \quad H_0(x) = 1, H_1(x) = 2x.$$

10. Let $s_0 = 1, s_{2n+1} = 0, s_{2n} = 2^{-n}(2n-1)!! := 2^{-n} \cdot 1 \cdot 3 \cdots (2n-1)$ for $n \in \mathbb{N}$.
    Let μ be the probability measure on $\mathbb{R}$ given by $d\mu = \frac{1}{\sqrt{\pi}} e^{-x^2} dx$. Show that
    $s = (s_k)_{k \in \mathbb{N}_0}$ is a moment sequence with representing measure μ.
11. Show that μ is the unique representing measure of s.
12. Prove that $\int_{\mathbb{R}} H_k(x)H_l(x)d\mu(x) = 2^k k! \delta_{kl}$ for $k, l \in \mathbb{N}_0$, that is,
    $(\frac{1}{\sqrt{2^n n!}} H_n(X))_{n \in \mathbb{N}_0}$ is the sequence of orthonormal polynomials associated
    with s.
13. Show that $H_n(x) = (-1)^n e^{x^2} \left(\frac{d}{dx}\right)^n e^{-x^2}$ and $\frac{d}{dx} H_{n+1}(x) = 2(n+1)H_n(x)$ for
    $n \in \mathbb{N}_0$.

Next we treat the *Laguerre polynomials* $L_n^\alpha(x)$, where $\alpha > -1$.
Define a sequence of polynomials L_n^α by $L_0^\alpha = 1, L_1^\alpha(x) = -x + 1 + \alpha$, and

$$L_{n+1}^\alpha(x) = \frac{-x + 2n + 1 + \alpha}{n+1} L_n^\alpha(x) - \frac{n+\alpha}{n+1}, \quad n \in \mathbb{N}.$$

14. Let $s_n = \frac{\Gamma(n+1+\alpha)}{\Gamma(\alpha+1)}$ for $\in \mathbb{N}_0$ and let μ be the probability measure on $[0, +\infty)$
    defined by $d\mu = (\Gamma(\alpha + 1))^{-1} x^\alpha e^{-x} dx$. Show that s is a moment sequence
    with representing measure μ.
15. Show that μ is the only representing measure for s.
16. Show that $\int_0^\infty L_k^\alpha(x)L_l^\alpha(x)d\mu = \frac{(\alpha+1)\dots(\alpha+k-1)}{k!} \delta_{kl}$.
17. Show that $L_n^\alpha(x) = \frac{1}{n!} x^{-\alpha} e^x \left(\frac{d}{dx}\right)^n (x^{n+\alpha} e^{-x})$ for $n \in \mathbb{N}_0$.
18. Show that the leading coefficient of $L_n^\alpha(x)$ is $\frac{(-1)^n}{n!}$.

Finally, we develop the *Legendre polynomials* R_n.
Define a sequence of polynomials R_n by

$$(n+1)P_{n+1}(x) = (2n+1)xR_n(x) - nR_{n-1}(x), \ n \in \mathbb{N}, \quad \text{and} \quad R_0(x) = 1, R_1(x) = x.$$

19. Verify that the Lebesgue measure on the interval $[-1, 1]$ has the moment
    sequence $s = (s_k)_{k \in \mathbb{N}_0}$, where $s_{2n} = \frac{2n}{2n+1}$ and $s_{2n+1} = 0$.
20. Show that $\int_{-1}^1 R_k(x)R_l(x)dx = \frac{2}{2k+1} \delta_{kl}$ for $k, l \in \mathbb{N}_0$, that is,
    $p_n(x) = \sqrt{k + \frac{1}{2}} R_n(x), n \in \mathbb{N}$, are the orthonormal polynomials for s.
21. Show that $R_n(x) = \frac{(-1)^n}{2^n n!} \left(\frac{d}{dx}\right)^n ((1 - x^2)^n)$ for $n \in \mathbb{N}_0$.

5.9 Notes

The study of orthogonal polynomials is a large classical subject which is treated in many books such as [Sz], [Chi1], [DX], [Is], [Sim2], [Sim3]. We do not make an attempt to discuss the history of this subject and mention only a few highlights. A number of formulas such as (5.6) and (5.7) go back to E. Heine (1878) [He]. Favard's theorem is in [Fv], see [MA] for a discussion of the history of this result. Proposition 5.35 was proved by B. Wendroff [Wen].

Chapter 6
The Operator-Theoretic Approach to the Hamburger Moment Problem

In this chapter we begin the study of moment problems using self-adjoint operators and self-adjoint extensions on Hilbert spaces. The operator-theoretic approach is a powerful tool and it will be used in the next two chapters as well.

In Sect. 6.1, solutions of the Hamburger moment problem are related to spectral measures of self-adjoint extensions of the multiplication operator X (Theorem 6.1). In Sect. 6.3 we show that the moment problem is determinate if and only if the operator X is essentially self-adjoint (Theorem 6.10) and we characterize von Neumann solutions (Theorem 6.13). The multiplication operator X is unitarily equivalent to the Jacobi operator T. In Sect. 6.2 the adjoint of the operator T is analyzed, while in Sect. 6.4 various determinacy conditions (Theorem 6.16 and Corollary 6.19) are derived. In Sect. 6.5, all self-adjoint extensions of the symmetric operator T on the Hilbert space $l^2(\mathbb{N}_0)$ are described (Theorem 6.23). In Sect. 6.6 we prove Markov's theorem (Theorem 6.29) for determinate moment sequences. Section 6.7 gives a short disgression into continued fractions.

Throughout this chapter, $s = (s_n)_{n \in \mathbb{N}_0}$ is a **positive definite real sequence**.

6.1 Existence of Solutions of the Hamburger Moment Problem

In this section we rederive the existence theorem for the Hamburger moment problem from the spectral theorem for self-adjoint operators. This is not only the shortest, but probably also the most elegant and natural approach to this result.

Theorem 6.1 *Let s be a positive definite real sequence. Then the Hamburger moment problem for s has a solution.*

© Springer International Publishing AG 2017
K. Schmüdgen, *The Moment Problem*, Graduate Texts in Mathematics 277,
DOI 10.1007/978-3-319-64546-9_6

Let A be a self-adjoint operator on a Hilbert space $\mathcal{G}$ such that $\mathcal{H}_s$ is a subspace of $\mathcal{G}$ and $X \subseteq A$. If E_A denotes the spectral measure of A, then

$$\mu_A(\cdot) = \langle E_A(\cdot)1, 1 \rangle_{\mathcal{G}} \tag{6.1}$$

is a representing measure for s. Every representing measure for s is of this form.

Proof Suppose that A is a self-adjoint extension of X on $\mathcal{G}$. By the spectral theorem (see Appendix A.7 or [Sm9]), A has a spectral measure E_A. Since $X \subseteq A$ and hence $(X)^n \subseteq A^n$, the polynomial 1 is in the domain $\mathcal{D}(A^n)$ and we have

$$\int_{\mathbb{R}} x^n \, d\mu_A(x) = \int_{\mathbb{R}} x^n d\langle E_A(x)1, 1 \rangle = \langle A^n 1, 1 \rangle = \langle (X)^n 1, 1 \rangle_s = \langle x^n, 1 \rangle_s = L_s(x^n) = s_n$$

for $n \in \mathbb{N}_0$. This shows that μ_A is a solution of the moment problem for s.

That each solution is of the form (6.1) follows from Proposition 6.2 below.

To prove that the moment problem for s has a solution it therefore suffices to show that the symmetric operator X has a self-adjoint extension. Define $(Jp)(x) = \overline{p(x)}$, $p \in \mathbb{C}[x]$. Then J is a conjugation (see (A.28)) on $\mathbb{C}[x]$ (that is, J is antilinear, $J^2 = I$, and $\langle Jp, Jq \rangle_s = \langle q, p \rangle_s$ for $p, q \in \mathbb{C}[x]$) which commutes with X (that is, $JXp = XJp$ for $p \in \mathbb{C}[x]$). Clearly, J extends by continuity to a conjugation on $\mathcal{H}_s$. Hence, by Proposition A.43, X has a self-adjoint extension on the Hilbert space $\mathcal{H}_s$. □

Proposition 6.2 *Let μ be a representing measure for s. Then*

$$\langle p, q \rangle_s = \langle p, q \rangle_{L^2(\mathbb{R}, \mu)} \text{ for } p, q \in \mathbb{C}[x].$$

The inclusion $\mathbb{C}[x] \subseteq L^2(\mathbb{R}, \mu)$ extends to a unitary operator of $\mathcal{H}_s$ on a closed subspace of $L^2(\mathbb{R}, \mu)$. We identify $\mathcal{H}_s$ with this closed subspace via this unitary mapping. Then the operator A_μ on $L^2(\mathbb{R}, \mu)$ defined by

$$(A_\mu f)(x) = xf(x) \text{ for } f \in \mathcal{D}(A_\mu) := \{f \in L^2(\mathbb{R}, \mu) : xf(x) \in L^2(\mathbb{R}, \mu)\} \tag{6.2}$$

is a self-adjoint extension of the symmetric operator X and

$$\mu(\cdot) = \mu_{A_\mu}(\cdot) \equiv \langle E_{A_\mu}(\cdot)1, 1 \rangle_{L^2(\mathbb{R}, \mu)}.$$

Proof From $\mu \in \mathcal{M}_+(\mathbb{R})$ it follows that $\mathbb{C}[x] \subseteq \mathcal{D}(A_\mu)$. Obviously, $X \subseteq A_\mu$. Since μ is a representing measure for s, we have $L_s(f) = \int f \, d\mu$ for $f \in \mathbb{C}[x]$, so that

$$\langle p, q \rangle_s = L_s(p\bar{q}) = \int p(x)\overline{q(x)} \, d\mu = \langle p, q \rangle_{L^2(\mathbb{R}, \mu)}.$$

Hence the inclusion $\mathbb{C}[x] \subseteq L^2(\mathbb{R}, \mu)$ can be extended by continuity to a unitary embedding of $\mathcal{H}_s$ into $\mathcal{G} := L^2(\mathbb{R}, \mu)$.

From Hilbert space operator theory it is known (see e.g. [Sm9, Example 5.2]) that the operator A_μ is self-adjoint and that the spectral projection $E_{A_\mu}(M)$ acts as the multiplication operator by the characteristic function χ_M of M on $L^2(\mathbb{R}, \mu)$. Hence $\mu(M) = \int \chi_M(x)d\mu = \langle E_{A_\mu}(\cdot)1, 1\rangle_{L^2(\mathbb{R},\mu)}$. The latter proves that $\mu = \mu_{A_\mu}$. □

The following corollary reformulates the second assertion of Theorem 6.1 in terms of the Jacobi operator T associated with s.

Corollary 6.3 *The representing measures of s are precisely the measures of the form*

$$\mu_B(\cdot) = s_0\langle E_B(\cdot)e_0, e_0\rangle_{\mathcal{F}}, \tag{6.3}$$

where B is a self-adjoint extension of T on a possibly larger Hilbert space $\mathcal{F}$ and E_B is the spectral measure of B.

Proof Recall from Sect. 5.3 that $X = U^{-1}TU$, where U is the unitary of $\mathcal{H}_s$ onto $l^2(\mathbb{N}_0)$ defined by $Up_n = e_n, n \in \mathbb{N}_0$. The sets of self-adjoint extensions A of X and B of T are in one-to-one correspondence by a unitary equivalence. It suffices to extend U to a unitary, denoted again by U, of $\mathcal{G}$ onto $\mathcal{F}$ such that $A = U^{-1}BU$. Then $E_A = U^{-1}E_BU$. Since $p_0(x) = s_0^{-1/2}$, we have $U1 = s_0^{1/2}e_0$ and hence

$$\mu_A(\cdot) = \langle E_A(\cdot)1, 1\rangle_{\mathcal{G}} = \langle U^{-1}E_B(\cdot)U1, 1\rangle_{\mathcal{G}}$$
$$= \langle E_B(\cdot)U1, U1\rangle_{\mathcal{F}} = s_0\langle E_B(\cdot)e_0, e_0\rangle_{\mathcal{F}} = \mu_B. \qquad □$$

Note that the measure $\langle E_B(\cdot)e_0, e_0\rangle_{\mathcal{F}}$ in (6.3) is always a probability measure, since e_0 is a unit vector.

By Proposition 6.2, for each representing measure μ the canonical Hilbert space $\mathcal{H}_s$ is a closed subspace of $L^2(\mathbb{R}, \mu)$ and X is a restriction of the self-adjoint operator A_μ on $L^2(\mathbb{R}, \mu)$. We shall see in Sect. 7.4 that for most solutions in the indeterminate case $\mathcal{H}_s$ is a *proper* subspace of $L^2(\mathbb{R}, \mu)$.

The following definition gives a name for those solutions obtained from self-adjoint extensions acting on the *same* Hilbert space $\mathcal{H}_s$.

Definition 6.4 A measure $\mu \in \mathcal{M}_s$ is called a *von Neumann solution* of the moment problem for s if $\mathbb{C}[x]$ is dense in $L^2(\mathbb{R}, \mu)$, or equivalently, if the embedding of $\mathbb{C}[x]$ into $L^2(\mathbb{R}, \mu)$ extends to a unitary operator of $\mathcal{H}_s$ onto $L^2(\mathbb{R}, \mu)$.

6.2 The Adjoint of the Jacobi Operator

Recall from Sect. 5.3 that the Jacobi operator $T \equiv T_J$ is the symmetric linear operator with dense domain d in the Hilbert space $l^2(\mathbb{N}_0)$ given by

$$(T\gamma)_n = a_n\gamma_{n+1} + b_n\gamma_n + a_{n-1}\gamma_{n-1}, \quad n \in \mathbb{N}_0, \tag{6.4}$$

for $(\gamma_n) \in \mathsf{d}$, where $\gamma_{-1} := 0$. The next proposition shows that the adjoint operator T^* is the "maximal operator" on $l^2(\mathbb{N}_0)$ which acts by the same formula (6.4). For this we essentially use the Wronskian defined by (5.40) and Lemma 5.23.

Proposition 6.5 *The adjoint operator T^* is given by*

$$T^*\gamma = \mathcal{T}\gamma \quad for \quad \gamma \in \mathcal{D}(T^*) = \{\gamma \in l^2(\mathbb{N}_0) : \mathcal{T}\gamma \in l^2(\mathbb{N}_0)\}.$$

For $\gamma, \beta \in \mathcal{D}(T^*)$, the limit $W(\gamma, \overline{\beta})_\infty := \lim_{n\to\infty} W(\gamma, \overline{\beta})_n$ exists and

$$\langle T^*\gamma, \beta \rangle - \langle \gamma, T^*\beta \rangle = W(\gamma, \overline{\beta})_\infty . \tag{6.5}$$

Proof Let $\gamma \in l^2(\mathbb{N}_0)$ be such that $\mathcal{T}\gamma \in l^2(\mathbb{N}_0)$. A straightforward computation shows that $\langle T\beta, \gamma \rangle = \langle \beta, \mathcal{T}\gamma \rangle$ for $\beta \in \mathsf{d}$. Therefore, $\gamma \in \mathcal{D}(T^*)$ and $T^*\gamma = \mathcal{T}\gamma$.

Conversely, let $\gamma \in \mathcal{D}(T^*)$ and $n \in \mathbb{N}_0$. Using (6.4) (or (5.22)) we derive

$$\langle e_n, T^*\gamma \rangle = \langle Te_n, \gamma \rangle = a_n\gamma_{n+1} + b_n\gamma_n + a_{n-1}\gamma_{n-1} = (\mathcal{T}\gamma)_n,$$

so that $T^*\gamma = \mathcal{T}\gamma$. This proves the first assertion concerning T^*.

Further, by (5.41) we have

$$\sum_{k=0}^{n} [(\mathcal{T}\gamma)_k\overline{\beta}_k - \gamma_k\overline{(\mathcal{T}\beta)_k)}] = W(\gamma, \overline{\beta})_n.$$

Since $\gamma, \beta, \mathcal{T}\gamma, \mathcal{T}\beta \in l^2(\mathbb{N}_0)$, the limit $n \to \infty$ in the preceding equality exists and we obtain Eq. (6.5).   □

Proposition 6.6 *Suppose that $z \in \mathbb{C}$.*

(i) $\mathcal{N}(T^*-zI) = \{0\}$ *if* $\mathfrak{p}_z \notin l^2(\mathbb{N}_0)$ *and* $\mathcal{N}(T^*-zI) = \mathbb{C}\cdot\mathfrak{p}_z$ *if* $\mathfrak{p}_z \in l^2(\mathbb{N}_0)$.
(ii) *If* $h \in \mathcal{N}(X^*-zI)$ *and* $\langle h, 1 \rangle_s = 0$, *then* $h = 0$.

Proof

(i) From Proposition 6.5 it follows that a sequence γ is in $\mathcal{N}(T^*-zI)$ if and only if $\gamma \in l^2(\mathbb{N}_0)$, $\mathcal{T}\gamma \in l^2(\mathbb{N}_0)$ and the recurrence relation (5.29) holds for $n \in \mathbb{N}_0$, where $\gamma_{-1} = 0$. Since $\gamma_{-1} = 0$, any solution γ of (5.29) is uniquely determined by the number γ_0, so we have $\gamma = \gamma_0\mathfrak{p}_z$. This implies the assertions.

(ii) Passing to the unitarily equivalent operator T the assertion says that $\gamma_0 = 0$ and $\gamma \in \mathcal{N}(T^*-zI)$ imply $\gamma = 0$. Since $\gamma = \gamma_0\mathfrak{p}_z$, this is indeed true.   □

Corollary 6.7 *The symmetric operator T (or equivalently X) has deficiency indices $(0,0)$ or $(1,1)$. The operator T (or X) is essentially self-adjoint if and only if $\mathfrak{p}_z$ is not in $l^2(\mathbb{N}_0)$, or equivalently $\sum_{n=0}^{\infty} |p_n(z)|^2 = \infty$, for one (hence for all) $z \in \mathbb{C}\backslash\mathbb{R}$.*

Proof Since $p_n(x) \in \mathbb{R}[x]$ and hence $\overline{p_n(z)} = p_n(\overline{z})$, we have $\mathfrak{p}_z \in l^2(\mathbb{N}_0)$ if and only if $\mathfrak{p}_{\overline{z}} \in l^2(\mathbb{N}_0)$ for $z \in \mathbb{C}$. Therefore, by Proposition 6.6(i), T has deficiency indices $(0,0)$ or $(1,1)$ and T has deficiency indices $(0,0)$ if and only if $\mathfrak{p}_z \notin l^2(\mathbb{N}_0)$ for some (then for all) $z \in \mathbb{C}\backslash\mathbb{R}$.   □

6.3 Determinacy of the Hamburger Moment Problem

We begin with two technical lemmas which are used in the proofs of Theorems 6.10 and 6.13 below.

Lemma 6.8 *If A and B are different self-adjoint extensions of the multiplication operator X on $\mathcal{H}_s$, then $\langle (A-zI)^{-1}1, 1\rangle_s \neq \langle (B-zI)^{-1}1, 1\rangle_s$ for all $z \in \mathbb{C}\backslash\mathbb{R}$.*

Proof Fix $z \in \mathbb{C}\backslash\mathbb{R}$ and assume to the contrary that

$$\langle (A-zI)^{-1}1, 1\rangle_s = \langle (B-zI)^{-1}1, 1\rangle_s. \tag{6.6}$$

Put $f := (A-zI)^{-1}1 - (B-zI)^{-1}1$. Since $X \subseteq A$ and $X \subseteq B$, we have $A \subseteq X^*$ and $B \subseteq X^*$. Hence $f \in \mathcal{D}(X^*)$ and

$$(X^*-zI)f = (X^*-z)(A-zI)^{-1}1 - (X^*-zI)(B-zI)^{-1}1 = 1 - 1 = 0,$$

so $f \in \mathcal{N}(X^*-zI)$. Since $\langle f, 1\rangle_s = 0$ by (6.6), Proposition 6.6(ii) yields $f = 0$.

Set $g:=(A-zI)^{-1}1$. If g were in $\mathcal{D}(\overline{X})$, then for $h \in \mathcal{N}(X^*-\bar{z}I)$ we would get

$$0 = \langle (X^* - \bar{z}I)h, g\rangle_s = \langle h, (\overline{X}-zI)g\rangle_s = \langle h, (A-zI)(A-zI)^{-1}1\rangle_s = \langle h, 1\rangle_s,$$

so $h = 0$, again by Proposition 6.6(ii). Thus, $\mathcal{N}(X^*-\bar{z}I) = \{0\}$. This is a contradiction, since X has two different self-adjoint extensions and hence its deficiency indices are $(1, 1)$ by Proposition A.42 and Corollary 6.7. Hence g is not in $\mathcal{D}(\overline{X})$.

Let S denote the restriction of A to $\mathcal{D}(\overline{X}) + \mathbb{C}\cdot g$. Then S is symmetric, because A is self-adjoint. Since X, hence $\overline{X}$, has deficiency indices $(1, 1)$ by Corollary 6.7 and $g \notin \mathcal{D}(\overline{X})$, S has deficiency indices $(0, 0)$. Therefore $\overline{S}$ is self-adjoint and hence $\overline{S} = A$. (In fact, the operator S is closed, but this is not needed here.)

Since $f = 0$ and hence $g = (B-zI)^{-1}1$, the same reasoning with B in place of A shows that $\overline{S} = B$. Thus $A = B$, which contradicts our assumption and shows that Eq. (6.6) cannot hold. □

Lemma 6.9 *Suppose that $\mu \in M_+(\mathbb{R})$ is a finite measure. Let $f_z(x)$ denote the function $\frac{1}{x-z}$ from $L^2(\mathbb{R}, \mu)$ for $z \in \mathbb{C}\backslash\mathbb{R}$. Then $E_{z_0} = \mathrm{Lin}\{(f_{z_0})^k, (f_{\bar{z}_0})^k : k \in \mathbb{N}_0\}$ for $z_0 \in \mathbb{C}\backslash\mathbb{R}$ and $E = \mathrm{Lin}\{f_z : z \in \mathbb{C}\backslash\mathbb{R}\}$ are dense linear subspaces of $L^2(\mathbb{R}, \mu)$.*

Proof Let $\varphi \in L^2(\mathbb{R}, \mu)$. Since $\mu(\mathbb{R}) < \infty$, there is a finite complex Radon measure ν_φ on $\mathbb{R}$ given by $d\nu_\varphi = \varphi d\mu$. Its Stieltjes transform $I_{\nu_\varphi}(z) := \int_\mathbb{R} \frac{1}{x-z} d\nu_\varphi(x)$ is holomorphic on $\mathbb{C}\backslash\mathbb{R}$. Using Lebesgue's convergence theorem we obtain

$$(I_{\nu_\varphi})^{(k)}(z) = k! \int_\mathbb{R} \frac{1}{(x - z)^{k+1}} d\nu_\varphi(x) \quad \text{for} \ \ k \in \mathbb{N}_0. \tag{6.7}$$

Let us begin with E_{z_0} and assume that $\varphi \perp E_{z_0}$. Let $k \in \mathbb{N}_0$. Then, by (6.7),

$$0 = k! \, \langle \varphi, (f_{z_0})^{k+1} \rangle_{L^2(\mathbb{R},\mu)} = k! \int \frac{\varphi(x)}{(x-\overline{z}_0)^{k+1}} d\mu(x) = (I_{v_\varphi})^{(k)}(\overline{z}_0).$$

Similarly, $(I_{v_\varphi})^{(k)}(z_0) = 0$ for $k \in \mathbb{N}_0$. Therefore, since I_{v_φ} is holomorphic on $\mathbb{C}\backslash\mathbb{R}$, we conclude that $I_{v_\varphi}(z) = 0$ in the upper half plane and in the lower half-plane. That is, the Stieltjes transform I_{v_φ} is zero on $\mathbb{C}\backslash\mathbb{R}$. From Theorem A.13 it follows that the measure v_φ is zero.

We prove that $\varphi = 0$ μ-a.e. on $\mathbb{R}$. For $\varepsilon > 0$ put $M_\varepsilon := \{x \in \mathbb{R} : |\varphi(x)| \geq \varepsilon\}$. Then, since the measure v_φ is zero and the measure μ is positive, we derive

$$0 = \int_{M_\varepsilon} \overline{\varphi(x)} \, dv_\varphi = \int_{M_\varepsilon} |\varphi(x)|^2 d\mu \geq \int_{M_\varepsilon} \varepsilon^2 d\mu(x) = \varepsilon^2 \mu(M_\varepsilon) \geq 0.$$

Hence $\mu(M_\varepsilon) = 0$. Letting $\varepsilon \to +0$, we get $\mu(\{x \in \mathbb{R} : \varphi(x) \neq 0\}) = 0$, that is, $\varphi = 0$ μ-a.e.. Thus, $\varphi = 0$ in $L^2(\mathbb{R}, \mu)$. This proves that E_{z_0} is dense in $L^2(\mathbb{R}, \mu)$.

Now assume that $\varphi \perp E$. Then $\varphi \perp f_z$ means that $I_{v_\varphi}(\overline{z}) = 0$. Hence the Stieltjes transform $I_{v_\varphi}(z)$ is zero on $\mathbb{C}\backslash\mathbb{R}$, so that $v_\varphi = 0$ by Theorem A.13. As shown in the preceding paragraph this implies $\varphi = 0$. This proves that E is dense in $L^2(\mathbb{R}, \mu)$. $\square$

The following theorem gives an operator-theoretic answer to the *uniqueness problem* for the Hamburger moment problem.

Theorem 6.10 *The moment problem for a positive definite sequence s is determinate if and only if the multiplication operator X (or equivalently, the corresponding Jacobi operator T) is essentially self-adjoint. If this holds and μ is the unique representing measure for s, then $\mathbb{C}[x]$ is dense in $L^2(\mathbb{R}, \mu)$, that is, $\mathcal{H}_s \cong L^2(\mathbb{R}, \mu)$, so μ is a von Neumann solution of the moment problem for s.*

Proof First assume that X is not essentially self-adjoint. Then, by Corollary 6.7, X has deficiency indices $(1, 1)$. Therefore, by Proposition A.42, X has at least two different self-adjoint extensions A and B on $\mathcal{H}_s$. By Theorem 6.1, $\mu_A(\cdot)=\langle E_A(\cdot)1, 1\rangle_s$ and $\mu_B(\cdot)=\langle E_B(\cdot)1, 1\rangle_s$ are representing measures for s. If μ_A were equal to μ_B, then for $z \in \mathbb{C}\backslash\mathbb{R}$ the functional calculus would yield

$$\langle (A-zI)^{-1}1, 1\rangle_s = \int (x-z)^{-1} d\mu_A(x) = \int (x-z)^{-1} d\mu_B(x) = \langle (B-zI)^{-1}1, 1\rangle_s,$$

which contradicts Lemma 6.8. Thus $\mu_A \neq \mu_B$ and s is indeterminate.

Suppose now that X is essentially self-adjoint. Fix $z \in \mathbb{C}\backslash\mathbb{R}$. Since X is essentially self-adjoint, $(X-zI)\mathbb{C}[x]$ is dense in $\mathcal{H}_s$, again by Proposition A.42. Hence there exists a sequence $(r_n(x))$ of polynomials such that $1 = \lim_n (x-z)r_n$ in $\mathcal{H}_s$. (The sequence (r_n) may depend on the number z, but it is crucial that it is independent of any representing measure.) Let μ be an arbitrary representing measure for s. Since μ is finite, the bounded function $\frac{1}{x-z}$ is in $L^2(\mathbb{R}, \mu) \cap L^1(\mathbb{R}, \mu)$.

Using the equations $L_s(p) = \int p\, d\mu$ for $p \in \mathbb{C}[x]$, $\|1\|^2_{L^2(\mathbb{R},\mu)} = s_0^2$, and the Hölder inequality we derive

$$
\begin{aligned}
|I_\mu(z) - L_s(r_n)|^2 &= \left| \int_\mathbb{R} (x - z)^{-1}\, d\mu(x) - \int_\mathbb{R} r_n(x)\, d\mu(x) \right|^2 \\
&\leq \left(\int_\mathbb{R} |(x - z)^{-1} - r_n(x)|\, d\mu(x) \right)^2 \\
&\leq \|1\|^2_{L^2(\mathbb{R},\mu)} \int_\mathbb{R} |(x-z)^{-1} - r_n(x)|^2\, d\mu(x) \qquad (6.8) \\
&= s_0^2 \int_\mathbb{R} |x-z|^{-2} |1 - (x-z)r_n(x)|^2\, d\mu(x) \\
&\leq s_0^2 |\operatorname{Im} z|^{-2} \int_\mathbb{R} |1 - (x-z)r_n(x)|^2\, d\mu(x) \\
&= s_0^2 |\operatorname{Im} z|^{-2} \|1 - (x-z)r_n(x)\|_s^2 \to 0.
\end{aligned}
$$

Therefore $I_\mu(z) = \lim_n L_s(r_n)$ is independent of the representing measure μ. By Theorem A.13 the values of the Stieltjes transform I_μ determine the measure μ. Hence μ is uniquely determined by s, that is, the moment problem for s is determinate.

The preceding inequalities, especially (6.8), imply that for $z \in \mathbb{C}\backslash\mathbb{R}$ the function $f_z(x) = \frac{1}{x-z}$ is in the closure of $\mathbb{C}[x]$ in $L^2(\mathbb{R}, \mu)$. Since the span E of such functions is dense in $L^2(\mathbb{R}, \mu)$ by Lemma 6.9, so is $\mathbb{C}[x]$.   □

Corollary 6.11 *Suppose that $\mu \in \mathcal{M}_+(\mathbb{R})$. Then the moment sequence s of μ is determinate if and only if $\mathbb{C}[x]$ is dense in $L^2(\mathbb{R}, (1 + x^2)d\mu)$.*

Proof Without loss of generality we can assume that s is positive definite. Indeed, otherwise μ has finite support (by Proposition 3.11), hence the image of $\mathbb{C}[x]$ coincides with $L^2(\mathbb{R}, (1 + x^2)d\mu)$, and s is determinate by Corollary 4.2.

By Theorem 6.10, s is determinate if and only if X is essentially self-adjoint, or equivalently by Proposition A.42, $(x + z)\mathbb{C}[x]$ is dense in $\mathcal{H}_s$ for $z = \pm i$. Recall that $\mathcal{H}_s$ is a subspace of $L^2(\mathbb{R}, \mu)$ by Proposition 6.2 and $\mathcal{H}_s \cong L^2(\mathbb{R}, \mu)$ if s is determinate. Therefore, s is determinate if and only if $(x + z)\mathbb{C}[x]$ is dense in $L^2(\mathbb{R}, \mu)$ for $z = \pm i$. It is easily checked that this holds if and only if $\mathbb{C}[x]$ is dense in $L^2(\mathbb{R}, (1 + x^2)d\mu)$. (It suffices to approximate all functions of $C_c(\mathbb{R})$.)   □

By Proposition A.42, the operator X is essentially self-adjoint if $(x \pm i)\mathbb{C}[x]$ is dense in $\mathcal{H}_s$. In the present situation we have the following stronger criterion.

Corollary 6.12 *If there exist a number $z_0 \in \mathbb{C}\backslash\mathbb{R}$ and a sequence $(r_n(x))_{n\in\mathbb{N}}$ of polynomials $r_n \in \mathbb{C}[x]$ such that*

$$
1 = \lim_{n\to\infty} (x - z_0)r_n(x) \quad \text{in } \mathcal{H}_s,
$$

then the moment problem for s is determinate.

Proof Fix $p \in \mathbb{C}[x]$. Since z_0 is a zero of the polynomial $p(x)-p(z_0)$, there is a polynomial $q \in \mathbb{C}[x]$ such that $p(x)-p(z_0) = (x-z_0)q(x)$. Then

$$(x-z_0)(q+p(z_0)r_n) = p-p(z_0) + p(z_0)(x-z_0)r_n \to p - p(z_0) + p(z_0)1 = p.$$

Therefore, because $\mathbb{C}[x]$ is dense in $\mathcal{H}_s$, $(X-z_0I)\mathbb{C}[x]$ is dense in $\mathcal{H}_s$. Since X has equal deficiency indices by Corollary 6.7, X is essentially self-adjoint. Hence s is determinate by Theorem 6.10.   □

Theorem 6.13 *For a measure $\mu \in \mathcal{M}_s$ the following statements are equivalent:*

(i) *μ is von Neumann solution.*
(ii) *$f_z(x) := \frac{1}{x-\bar{z}}$ is in the closure of $\mathbb{C}[x]$ in $L^2(\mathbb{R}, \mu)$ for all $z \in \mathbb{C}\backslash\mathbb{R}$.*
(iii) *$f_{z_0}(x) := \frac{1}{x-z_0}$ is in the closure of $\mathbb{C}[x]$ in $L^2(\mathbb{R}, \mu)$ for one $z_0 \in \mathbb{C}\backslash\mathbb{R}$.*

Proof
(i)→(ii) That μ is a von Neumann solution means that $\mathbb{C}[x]$ is dense in $L^2(\mathbb{R}, \mu)$. Hence the function $f_z \in L^2(\mathbb{R}, \mu)$ is in the closure of $\mathbb{C}[x]$.

(ii)→(iii) is trivial.

(iii)→(i) Set $b := \mathrm{Im}\, z_0$. Let $\mathcal{G}$ denote the closure of $\mathbb{C}[x]$ in $L^2(\mathbb{R}, \mu)$. We first prove by induction on k that $f_{z_0}^k \in \mathcal{G}$ for all $k \in \mathbb{N}$. For $k = 1$ this is true by assumption. Suppose that $f_{z_0}^k \in \mathcal{G}$. Then there is a sequence $(p_n)_{n\in\mathbb{N}}$ of polynomials such that $p_n \to f_{z_0}^k$ in $L^2(\mathbb{R}, \mu)$. We can write $p_n(x) = p_n(z_0) + (x-z_0)q_n(x)$ with $q_n \in \mathbb{C}[x]$. Using that $|f_{z_0}(x)| \leq |b|^{-1}$ on $\mathbb{R}$ we derive

$$\|f_{z_0}^{k+1}(x) - p_n(z_0)(x-z_0)^{-1}-q_n(x)\|_{L^2(\mu)}$$
$$= \|f_{z_0}(x)(f_{z_0}^k(x) - p_n(z_0)-(x - z_0)q_n(x))\|_{L^2(\mu)}$$
$$= \|f_{z_0}(x)(f_{z_0}^k(x) - p_n(x))\|_{L^2(\mu)}$$
$$\leq |b|^{-1}\|f_{z_0}^k(x) - p_n(x)\|_{L^2(\mu)} \to 0 \quad \text{as} \quad n \to \infty.$$

Since $f_{z_0} \in \mathcal{G}$ by assumption and hence $p_n(z_0)(x-z_0)^{-1}-q_n \in \mathcal{G}$, this shows that $f_{z_0}^{k+1} \in \mathcal{G}$, which completes the induction proof.

Clearly, $f_{z_0}^k \in \mathcal{G}$ implies that $f_{\bar{z_0}}^k \in \mathcal{G}$. Hence the vector space E_{z_0} from Lemma 6.9 is contained in $\mathcal{G}$. Since E_{z_0} is dense in $L^2(\mathbb{R}, \mu)$ by Lemma 6.9, $\mathbb{C}[x]$ is dense in $L^2(\mathbb{R}, \mu)$. Hence μ is a von Neumann solution.   □

We close this section by developing an operator-theoretic construction of indeterminate moment sequences.

Example 6.14 For $\alpha \in \mathbb{R}$, let S_α denote the operator $-i\frac{d}{dx}$ with dense domain

$$\mathcal{D}(S_\alpha) = \{f \in AC[0, 2\pi] : f' \in L^2(0, 2\pi), f(2\pi) = e^{i\alpha 2\pi}f(0)\}$$

of $L^2(0, 2\pi)$ with Lebesgue measure. Here $AC[0, 2\pi]$ are the absolutely continuous functions on $[0, 2\pi]$. (Recall that a function f on $[0, 2\pi]$ is called absolutely continuous if there exists a function $h \in L^1(0, 2\pi)$ such that $f(x) = f(a) + \int_a^x f(t)dt$ on $[0, 2\pi]$; in this case $Tf = -ih$.) Then S_α is a self-adjoint operator with spectrum $\sigma(S_\alpha) = \{\alpha + k : k \in \mathbb{Z}\}$, see e.g. [Sm9, p. 16, 34]. Each number $\alpha + k$ is eigenvalue of multiplicity one with normalized eigenfunction $\varphi_{\alpha,k}(x) = \frac{1}{\sqrt{2\pi}} e^{i(\alpha+k)x}$ and the functions $\varphi_{\alpha,k}, k \in \mathbb{Z}$, form an orthonormal basis of the Hilbert space $L^2(0, 2\pi)$. (The latter follows from the spectral theory of self-adjoint operators; it can also be verified directly. That the span of functions $\varphi_{\alpha,k}, k \in \mathbb{Z}$, is dense for $\alpha \in \mathbb{R}$ follows at once from the denseness of functions $\varphi_{0,k}, k \in \mathbb{Z}$.)

We fix a function $f \in C_0^\infty(0, 2\pi), f \neq 0$, and define

$$s_n := \langle (-i)^n f^{(n)}, f \rangle, \ n \in \mathbb{N}_0, \quad \text{and} \quad s - (s_n)_{n \in \mathbb{N}_0}.$$

Since $f^{(n)}(0) = f^{(n)}(2\pi) = 0$ for $n \in \mathbb{N}_0$, f is in the domain of each power $(S_\alpha)^n$ and $(S_\alpha)^n f = (-i)^n f^{(n)}$ for $\alpha \in \mathbb{R}$ and $n \in \mathbb{N}_0$. We develop f with respect to the orthonormal basis $\{\varphi_{\alpha,k} : k \in \mathbb{Z}\}$ and write $f = \sum_k c_{\alpha,k}\varphi_{\alpha,k}$. For $n \in \mathbb{N}_0$ we have $(S_\alpha)^n f = \sum_k c_{\alpha,n}(\alpha + k)^n \varphi_{\alpha,k}$ and hence

$$s_n = \langle (-i)^n f^{(n)}, f \rangle = \langle (S_\alpha)^k, f \rangle = \left\langle \sum_{k \in \mathbb{Z}} c_{\alpha,k}(\alpha + k)^n \varphi_{\alpha,k}, \sum_{l \in \mathbb{Z}} c_{\alpha,l}\varphi_{\alpha,l} \right\rangle$$

$$= \sum_{k \in \mathbb{Z}} c_{\alpha,k}^2 (\alpha + k)^n = \int x^n d\mu_\alpha, \quad \text{where} \quad \mu_\alpha := \sum_{k \in \mathbb{Z}} c_{\alpha,k}^2 \delta_{\alpha+k}.$$

Therefore, s is a moment sequence and $\mu_\alpha, \alpha \in \mathbb{R}$, is a representing measure of s.

Let $\alpha, \beta \in [0, 1), \alpha \neq \beta$. Since $f \neq 0$, there is a $k \in \mathbb{Z}$ such that $c_{\alpha,k}^2 \neq 0$. Hence $\mu_\alpha(\{\alpha + k\}) \neq 0$, while $\mu_\beta(\{\alpha + k\}) = 0$. Thus $\mu_\alpha \neq \mu_\beta$ and s is indeterminate. ∘

6.4   Determinacy Criteria Based on the Jacobi Operator

The following lemma is a main technical ingredient of the proof of Theorem 6.16.

Lemma 6.15 *Suppose that $c = (c_n), \varphi = (\varphi_n), \psi = (\psi_n), \eta = (\eta_n)$, and $\zeta = (\zeta_n)$ are sequences from $l^2(\mathbb{N}_0)$. If $f = (f_n)_{n \in \mathbb{N}_0}$ is a complex sequence satisfying*

$$f_{n+1} = c_n + \varphi_n \sum_{k=0}^{n} \eta_k f_k + \psi_n \sum_{k=0}^{n} \zeta_k f_k \quad \text{for} \ n \in \mathbb{N}_0, \tag{6.9}$$

then $f \in l^2(\mathbb{N}_0)$.

Proof Let $\varepsilon > 0$. Since $c, \varphi, \psi, \eta, \zeta \in l^2(\mathbb{N}_0)$, there exists an $l \in \mathbb{N}$ such that

$$\sum_{n=l}^{m} |c_n|^2 + \|\eta\|^2 \sum_{n=l}^{m} |\varphi_n|^2 + \|\zeta\|^2 \sum_{n=l}^{m} |\psi_n|^2 \le \varepsilon$$

for all $m \ge l$. Suppose that $m \ge l$. Using the inequality

$$\left| \sum_{k=0}^{n} \eta_k f_k \right|^2 \le \left(\sum_{k=0}^{n} |\eta_k|^2 \right) \left(\sum_{k=0}^{n} |f_k|^2 \right) \le \|\eta\|^2 \sum_{j=0}^{m} |f_k|^2, \quad m \ge n,$$

we derive

$$\sum_{n=l}^{m} \left| \varphi_n \sum_{k=0}^{n} \eta_k f_k \right|^2 \le \sum_{n=l}^{m} |\varphi_n|^2 \left| \sum_{k=0}^{n} \eta_k f_k \right|^2 \le \|\eta\|^2 \sum_{j=0}^{m} |f_k|^2 \sum_{n=l}^{m} |\varphi_n|^2 \le \varepsilon \sum_{j=0}^{m} |f_k|^2$$

and similarly

$$\sum_{n=l}^{m} \left| \psi_n \sum_{k=0}^{n} \zeta_k f_k \right|^2 \le \varepsilon \sum_{j=0}^{m} |f_k|^2.$$

From the preceding two inequalities and Eq. (6.9) we therefore obtain

$$\frac{1}{4} \sum_{n=l}^{m} |f_{n+1}|^2 \le \sum_{k=l}^{m} |c_k|^2 + \sum_{n=l}^{m} \left| \varphi_n \sum_{k=0}^{n} \eta_k f_k \right|^2 + \sum_{n=l}^{m} \left| \psi_n \sum_{k=0}^{n} \zeta_k f_k \right|^2 \le \varepsilon + 2\varepsilon \sum_{j=0}^{m} |f_k|^2$$

and hence

$$\left(\frac{1}{4} - 2\varepsilon \right) \sum_{n=l}^{m} |f_{n+1}|^2 \le \varepsilon + 2\varepsilon \sum_{j=0}^{l-1} |f_k|^2 \quad \text{for} \quad m \ge l.$$

Choosing $\varepsilon < \frac{1}{8}$ we conclude that f is in $l^2(\mathbb{N}_0)$. $\square$

The next theorem sharpens Corollary 6.7 and it is the main result of this section.

Theorem 6.16 *For any positive definite sequence s the following are equivalent:*

(i) *The moment problem for s is indeterminate.*
(ii) *The Jacobi operator $T = T_J$ is not essentially self-adjoint.*
(iii) $\mathfrak{p}_z \in l^2(\mathbb{N}_0)$ *for some (equivalently, for all) $z \in \mathbb{C} \backslash \mathbb{R}$.*
(iv) $\mathfrak{q}_z \in l^2(\mathbb{N}_0)$ *for some (equivalently, for all) $z \in \mathbb{C} \backslash \mathbb{R}$.*
(v) $\mathfrak{p}_z \in l^2(\mathbb{N}_0)$ *and $\mathfrak{q}_z \in l^2(\mathbb{N}_0)$ for some (equivalently, for all) $z \in \mathbb{R}$.*
(vi) $\mathfrak{p}_z \in l^2(\mathbb{N}_0)$ *and $\mathfrak{q}_z \in l^2(\mathbb{N}_0)$ for some (equivalently, for all) $z \in \mathbb{C}$.*

Proof First we note that (i)↔(ii) by Theorem 6.10, (ii)↔(iii) by Corollary 6.7, and (iii)↔(iv) by Corollary 5.22. Further, (vi)→(iv) and (vi)→(v) are trivial. The proof is complete once we have shown that (iii) and (iv) together imply (v) and that (v) implies (vi). For this it suffices to prove the following assertion:

If $\mathfrak{p}_z$ and $\mathfrak{q}_z$ are in $l^2(\mathbb{N}_0)$ for some $z \in \mathbb{C}$, then this holds for all $x \in \mathbb{C}$.

Indeed, suppose that $\mathfrak{p}_z$ and $\mathfrak{q}_z$ are in $l^2(\mathbb{N}_0)$. Fix $x \in \mathbb{C}$. Set $f_n = p_n(x)$ and

$$c_n = p_{n+1}(z), \ \varphi_n = (x-z)q_{n+1}(z), \ \psi_n = (z-x)p_{n+1}(z), \ \eta_n = p_n(z), \ \zeta_n = q_n(z)$$

for $n \in \mathbb{N}_0$. Since $\mathfrak{p}_z, \mathfrak{q}_z \in l^2(\mathbb{N}_0)$, the sequences $c = (c_n), \varphi = (\varphi_n), \psi = (\psi_n), \eta = (\eta_n)$, and $\zeta = (\zeta_n)$ are in $l^2(\mathbb{N}_0)$. Further, from the identity (5.53) it follows that

$$p_{n+1}(x) = p_{n+1}(z) + (x-z)q_{n+1}(z) \sum_{k=0}^{n} p_k(z)p_k(x) + (z-x)p_{n+1}(z) \sum_{k=0}^{n} q_k(z)p_k(x).$$

This means that Eq. (6.9) holds. Therefore, all assumptions of Lemma 6.15 are fulfilled, so we obtain $\mathfrak{p}_x = (f_n) \in l^2(\mathbb{N}_0)$. Replacing (5.53) by (5.54) and proceeding verbatim in the same manner we derive that $\mathfrak{q}_x \in l^2(\mathbb{N}_0)$. □

Remark 6.17 The equivalence of conditions (i)–(iv) of Theorem 6.16 was obtained by general operator-theoretic considerations. For the description of self-adjoint extensions in Sect. 6.5 we need that $\mathfrak{p}_0$ and $\mathfrak{q}_0$ are in $l^2(\mathbb{N}_0)$ if s is indeterminate. This is more tricky and follows from the implication (i)→(vi) and also from Lemma 7.1 proved in the next chapter.   ○

Combining Theorem 6.16 (i)→(vi), Proposition 6.5, and Lemma 5.16 we obtain the following important corollary.

Corollary 6.18 *If s is an indeterminate moment sequence, then for all (!) $z \in \mathbb{C}$ the sequences $\mathfrak{p}_z = (p_n(z))_{n \in \mathbb{N}_0}$ and $\mathfrak{q}_z = (q_n(z))_{n \in \mathbb{N}_0}$ are in $\mathcal{D}(T^*)$,*

$$T^*\mathfrak{p}_z = z\mathfrak{p}_z \quad and \quad T^*\mathfrak{q}_z = s_0^{1/2}e_0 + z\mathfrak{q}_z . \tag{6.10}$$

The next corollary contains another sufficient condition for determinacy.

Corollary 6.19 *If $\sum_{n=0}^{\infty} a_n^{-1} = \infty$, then the Jacobi operator T is essentially self-adjoint and the moment problem is determinate.*

Proof Assume to the contrary that T is not essentially self-adjoint. Then, by Theorem 6.16, $\mathfrak{p}_z \in l^2(\mathbb{N}_0)$ and $\mathfrak{q}_z \in l^2(\mathbb{N}_0)$ for $z \in \mathbb{C}\backslash\mathbb{R}$. Since

$$a_n^{-1} = p_n(z)q_{n+1}(z) - p_{n+1}(z)q_n(z),$$

by formula (5.52), we derive

$$\sum_{n=1}^{\infty} a_n^{-1} = \sum_{n=1}^{\infty} [p_n(z)q_{n+1}(z) - p_{n+1}(z)q_n(z)]$$

$$\leq \left(\sum_{n=1}^{\infty} |p_n(z)|^2\right)^{1/2} \left(\sum_{n=1}^{\infty} |q_{n+1}(z)|^2\right)^{1/2} + \left(\sum_{n=1}^{\infty} |p_{n+1}(z)|^2\right)^{1/2} \left(\sum_{n=1}^{\infty} |q_n(z)|^2\right)^{1/2}$$

$$\leq \|\mathfrak{p}_z\|_{l^2(\mathbb{N}_0)} \|\mathfrak{q}_z\|_{l^2(\mathbb{N}_0)} + \|\mathfrak{q}_z\|_{l^2(\mathbb{N}_0)} \|\mathfrak{p}_z\|_{l^2(\mathbb{N}_0)} < \infty,$$

which contradicts the assumption $\sum_{n=0}^{\infty} a_n^{-1} = \infty$. $\qquad\square$

Remark 6.20 Under additional assumptions a converse to Corollary 6.19 holds (see [Bz, Theorem 1.5, p. 507]): *Suppose that the sequence $(b_n)_{n\in\mathbb{N}_0}$ is bounded and $a_{n-1}a_{n+1} \leq a_n^2$ for $n \geq n_0$ and some $n_0 \in \mathbb{N}$. If $\sum_{n=0}^{\infty} a_n^{-1} < \infty$, the Jacobi operator T is not essentially self-adjoint and the moment problem is indeterminate.* ∘

In Sect. 4.2 Carleman's theorem 4.3 was derived from the Denjoy–Carleman theorem 4.4 on quasi-analytic functions. Now we give an operator-theoretic proof of Theorem 4.3 which is based on Corollary 6.19.

Second Proof of Carleman's Theorem 4.3(i) If $s_0 > 0$ and (s_n) satisfies Carleman's condition, then so does $(s_n s_0^{-1})$. Hence we can assume that $s_0 = 1$. Then, by Corollary 5.7, p_n has the leading coefficient $(a_0 \ldots a_{n-1})^{-1}$. Therefore, since $\langle x^k, p_n \rangle_s = 0$ for $0 \leq k < n$, we have

$$\langle (a_0 \ldots a_{n-1})^{-1} x^n, p_n \rangle_s = \langle p_n, p_n \rangle_s = 1.$$

From the Cauchy–Schwarz inequality we obtain

$$1 \leq \|(a_0 \ldots, a_{n-1})^{-1} x^n\|_s^2 \|p_n\|_s^2 = (a_0 \ldots a_{n-1})^{-2} s_{2n},$$

so that

$$s_{2n}^{-\frac{1}{2n}} \leq (a_0 \ldots a_{n-1})^{-\frac{1}{n}}, \quad n \in \mathbb{N}. \tag{6.11}$$

Stirling's formula (see e.g. [RW, p. 45]) yields $(\frac{n}{e})^n \leq n!$ and hence $(\frac{1}{n!})^{1/n} \leq \frac{e}{n}$. Using this fact and the arithmetic-geometric mean inequality we derive

$$\left(\frac{1}{a_0} \cdots \frac{1}{a_{n-1}}\right)^{\frac{1}{n}} = \left(\frac{1}{n!}\right)^{\frac{1}{n}} \left(\frac{1}{a_0} \frac{2}{a_1} \cdots \frac{n}{a_{n-1}}\right)^{\frac{1}{n}} \leq \frac{e}{n} \frac{1}{n} \sum_{k=1}^{n} \frac{k}{a_{k-1}}. \tag{6.12}$$

Let us fix $N \in \mathbb{N}$. For $k \in \mathbb{N}, k < N$, we get

$$\sum_{n=k}^{N} \frac{k}{n^2} \leq \sum_{n=k}^{N} \frac{2k}{n(n+1)} \leq 2k \sum_{n=k}^{N} \left(\frac{1}{n} - \frac{1}{n+1}\right) = 2k \left(\frac{1}{k} - \frac{1}{N+1}\right) < 2. \tag{6.13}$$

Using first (6.11) and (6.12) and finally (6.13) it follows that

$$\sum_{n=1}^{N} s_{2n}^{-\frac{1}{2n}} \le \sum_{n=1}^{N} \frac{e}{n^2} \sum_{k=1}^{n} \frac{k}{a_{k-1}} = \sum_{k=1}^{N} \sum_{n=k}^{N} \frac{e}{n^2} \frac{k}{a_{k-1}} \le \sum_{k=1}^{N} \frac{e}{a_{k-1}} \sum_{n=k}^{N} \frac{k}{n^2} \le 2e \sum_{k=0}^{N-1} \frac{1}{a_k}.$$

Therefore, the assumption $\sum_{n=1}^{\infty} s_{2n}^{-1/2n} = \infty$ implies that $\sum_{n=1}^{\infty} a_n^{-1} = \infty$. Hence s is determinate by Corollary 6.19. □

The case when a Jacobi operator is bounded is clarified by the next proposition. In this case it is obvious that T is essentially self-adjoint, so s is determinate.

Proposition 6.21 *The Jacobi operator T is bounded if and only if both sequences $a = (a_n)_{n \in \mathbb{N}_0}$ and $b = (b_n)_{n \in \mathbb{N}_0}$ are bounded.*

Proof If T is bounded, the sequences a and b are bounded, since

$$|a_{n-1}|^2 + |b_n|^2 + |a_{n+1}|^2 = \|a_n e_{n+1} + b_n e_n + a_{n-1} e_{n-1}\|^2 = \|Te_n\|^2 \le \|T\|^2.$$

Moreover, $\sup_n |a_n| \le \|T\|$ and $\sup_n |b_n| \le \|T\|$.

Conversely, assume that both sequences a and b are bounded, say $|a_n| \le M$ and $|b_n| \le M$ for $n \in \mathbb{N}_0$. Let $\gamma \in d$. Using the triangle inequality in $l^2(\mathbb{N}_0)$ we derive

$$\|T\gamma\| = \left(\sum_n |a_n \gamma_{n-1} + b_n \gamma_n + a_{n-1} \gamma_{n-1}|^2 \right)^{1/2}$$

$$\le \left(\sum_n |a_n \gamma_{n-1}|^2 \right)^{1/2} + \left(\sum_n |b_n \gamma_n|^2 \right)^{1/2} + \left(\sum_n |a_n \gamma_{n-1}|^2 \right)^{1/2} \le 3M\|\gamma\|,$$

that is, T is bounded and $\|T\| \le 3M$. □

6.5 Self-Adjoint Extensions of the Jacobi Operator

If a moment sequence is determinate, the Jacobi operator T is essentially self-adjoint by Theorem 6.10 and hence its closure is the unique self-adjoint extension of T.

Throughout this section, s is an **indeterminate moment sequence**. By Theorem 6.10 and Corollary 6.7, T has deficiency indices $(1, 1)$. Our aim is to describe all self-adjoint extensions of the symmetric operator T on the Hilbert space $\mathcal{H}_s$.

Recall from Corollary 6.18 that for each $z \in \mathbb{C}$ the sequences $\mathfrak{p}_z, \mathfrak{q}_z$ are in $\mathcal{D}(T^*)$ and satisfy $T^* \mathfrak{p}_z = z\mathfrak{p}_z$ and $T^* \mathfrak{q}_z = s_0^{1/2} e_0 + z\mathfrak{q}_z$. In particular,

$$T^* \mathfrak{p}_0 = 0, \quad T^* \mathfrak{q}_0 = s_0^{1/2} e_0, \quad \text{and} \quad p_0(z) = s_0^{-1/2}, \quad q_0(0) = 0.$$

To simplify some computations we set $\mathfrak{p} := s_0^{1/2} \mathfrak{p}_0$ and $\mathfrak{q} := s_0^{-1/2} \mathfrak{q}_0$. Then

$$T^* \mathfrak{p} = 0, \quad T^* \mathfrak{q} = e_0, \quad \text{and} \quad \mathfrak{p} = (1, \dots), \quad \mathfrak{q} = (0, \dots). \tag{6.14}$$

These properties of $\mathfrak{p}$ and $\mathfrak{q}$ will play an essential role in what follows.

Let $\dotplus$ denote the direct sum. Recall that $\overline{T}$ is the closure of the operator T.

Lemma 6.22 $\mathcal{D}(T^*) = \mathcal{D}(\overline{T}) \dotplus \mathbb{C} \cdot \mathfrak{q}_0 \dotplus \mathbb{C} \cdot \mathfrak{p}_0.$

Proof Obviously, $\mathcal{D}(\overline{T}) + \mathbb{C} \cdot \mathfrak{q}_0 + \mathbb{C} \cdot \mathfrak{p}_0 = \mathcal{D}(\overline{T}) + \mathbb{C} \cdot \mathfrak{q} + \mathbb{C} \cdot \mathfrak{p} \subseteq \mathcal{D}(T^*)$. It suffices to prove the converse inclusion of the latter.

Using (6.14) and the fact that the operator $\overline{T}$ is symmetric we compute

$$\langle T^*(\varphi + c_0\mathfrak{q} + c_1\mathfrak{p}), \psi + d_0\mathfrak{q} + d_1\mathfrak{p} \rangle - \langle \varphi + c_0\mathfrak{q} + c_1\mathfrak{p}, T^*(\psi + d_0\mathfrak{q} + d_1\mathfrak{p}) \rangle$$

$$= \langle \overline{T}\varphi + c_0 e_0, \psi + d_0\mathfrak{q} + d_1\mathfrak{p} \rangle - \langle \varphi + c_0\mathfrak{q} + c_1\mathfrak{p}, \overline{T}\psi + d_0 e_0 \rangle$$

$$= \langle \varphi, d_0 e_0 \rangle + \langle c_0 e_0, \psi + d_1\mathfrak{p} \rangle - \langle c_0 e_0, \psi \rangle - \langle \varphi + c_1\mathfrak{p}, d_0 e_0 \rangle$$

$$= \varphi_0 \overline{d_0} + c_0 \left(\overline{\psi_0} + \overline{d_1} \right) - c_0 \overline{\psi_0} - (\varphi_0 + c_1) \overline{d_0} = c_0 \overline{d_1} - c_1 \overline{d_0} \qquad (6.15)$$

for arbitrary $\varphi, \psi \in \mathcal{D}(\overline{T})$ and $c_0, c_1, d_0, d_1 \in \mathbb{C}$.

Since T has deficiency indices $(1, 1)$, we have $\dim \mathcal{D}(T^*)/\mathcal{D}(\overline{T}) = 2$ by formula (A.27) in Appendix A.7. Therefore, to prove that $\mathcal{D}(T^*) \subseteq \mathcal{D}(\overline{T}) + \mathbb{C} \cdot \mathfrak{q} + \mathbb{C} \cdot \mathfrak{p}$ it suffices to show the vectors $\mathfrak{q}$ and $\mathfrak{p}$ are linearly independent modulo $\mathcal{D}(\overline{T})$. Indeed, if $c_0\mathfrak{q} + c_1\mathfrak{p} \in \mathcal{D}(\overline{T})$, then

$$\langle T^*(c_0\mathfrak{q} + c_1\mathfrak{p}), d_0\mathfrak{q} + d_1\mathfrak{p} \rangle = \langle c_0\mathfrak{q} + c_1\mathfrak{p}, T^*(d_0\mathfrak{q} + d_1\mathfrak{p}) \rangle$$

and hence $c_0 \overline{d_1} - c_1 \overline{d_0} = 0$ by (6.15) for arbitrary $d_0, d_1 \in \mathbb{C}$. Therefore, $c_0 = c_1 = 0$, so $\mathfrak{q}$ and $\mathfrak{p}$ are linearly independent modulo $\mathcal{D}(\overline{T})$. $\qquad\square$

Theorem 6.23 *The self-adjoint extensions of the Jacobi operator T on $\mathcal{H}_s \cong l^2(\mathbb{N}_0)$ are precisely the operators $T_t = T^* \lceil \mathcal{D}(T_t)$, $t \in \mathbb{R} \cup \{\infty\}$, where*

$$\mathcal{D}(T_t) = \mathcal{D}(\overline{T}) \dotplus \mathbb{C} \cdot (\mathfrak{q}_0 + t\mathfrak{p}_0) \text{ for } t \in \mathbb{R}, \ \mathcal{D}(T_\infty) = \mathcal{D}(\overline{T}) \dotplus \mathbb{C} \cdot \mathfrak{p}_0. \qquad (6.16)$$

Further, if the symmetric operator T is positive, so is the self-adjoint operator T_∞.

Proof Let A be a self-adjoint extension of T on the Hilbert space $\mathcal{H}_s$. Because $T \subseteq A$ and hence $T^* \supseteq A^* = A$, the operator A is completely described by its domain $\mathcal{D}(A)$. Since T has deficiency indices $(1, 1)$ and hence $\dim \mathcal{D}(T^*)/\mathcal{D}(\overline{T}) = 2$, we conclude that $\dim \mathcal{D}(A)/\mathcal{D}(\overline{T}) = 1$. Thus, up to complex multiples, there exists a unique $\eta \in \mathcal{D}(A)$ which is not in $\mathcal{D}(\overline{T})$. By Lemma 6.22 the vector η can be chosen to be of the form $\eta = c_0\mathfrak{q}_0 + c_1\mathfrak{p}_0 \in \mathcal{D}(A)$. Further, upon scaling we can write $\eta = s_0\mathfrak{q} + t\mathfrak{p}$ with $t \in \mathbb{C}$ or $\eta = \mathfrak{p}$.

First we treat the case $\eta = s_0\mathfrak{q} + t\mathfrak{p}$. Let $c_0, d_0, d_1 \in \mathbb{C}$ and $\psi \in \mathcal{D}(\overline{T})$. Since $\varphi + c_0\eta = \varphi + c_0 s_0\mathfrak{q} + c_0 t\mathfrak{p} \in \mathcal{D}(A)$ and $A \subseteq T^*$, it follows from (6.15) that

$$\langle A(\varphi + c_0\eta), \psi + d_0\mathfrak{q} + d_1\mathfrak{p} \rangle - \langle \varphi + c_0\eta, T^*(\psi + d_0\mathfrak{q} + d_1\mathfrak{p}) \rangle$$

$$= c_0(s_0 \overline{d_1} - t \overline{d_0}). \qquad (6.17)$$

Let $d_0\mathfrak{q} + d_1\mathfrak{p}$ be a nonzero vector in $\mathcal{D}(A)$. Then $d_0\mathfrak{q} + d_1\mathfrak{p}$ is a multiple of η, so that $s_0 d_1 = d_0 t$ and $d_0 \neq 0$. Hence $c_0(s_0 \overline{d_1} - t\,\overline{d_0}) = c_0 s_0 \overline{d_0}(\overline{t} - t)$. Therefore it follows from (6.17) that the operator A is symmetric if and only if (6.17) vanishes for all $c_0 \in \mathbb{C}$, or equivalently, if t is real. Since $\eta = s_0\mathfrak{q} + t\mathfrak{p} = s_0^{1/2}(\mathfrak{q}_0 + t\mathfrak{p}_0) \in \mathcal{D}(A)$, then we have $A = T_t$.

Now let $\eta = \mathfrak{p}$. Then $A = T_\infty$. If $c_1, d_0, d_1 \in \mathbb{C}$ and $\psi \in \mathcal{D}(\overline{T})$, then by (6.15),

$$\langle A(\varphi + c_1\eta), \psi + d_0\mathfrak{q} + d_1\mathfrak{p}\rangle - \langle \varphi + c_1\eta, T^*(\psi + d_0\mathfrak{q} + d_1\mathfrak{p})\rangle = c_1\,\overline{d_0}. \qquad (6.18)$$

If $d_0\mathfrak{q} + d_1\mathfrak{p} \in \mathcal{D}(T_\infty)$, then $d_0 = 0$ and the right-hand side of (6.18) vanishes. This proves that $A = T_\infty$ is symmetric.

We have shown that all operators T_t, $t \in \mathbb{R} \cup \{\infty\}$, are symmetric. It remains to prove that they are self-adjoint. Assume to the contrary that $A = T_t$ is not self-adjoint. Then $A \neq A^*$. Since $T \subseteq A \subseteq A^* \subseteq T^*$ and $\dim \mathcal{D}(T^*)/\mathcal{D}(A) = 1$ (by $\dim \mathcal{D}(T^*)/\mathcal{D}(\overline{T}) = 2$), we conclude that $A^* = T^*$. Hence the right-hand side of Eq. (6.17) resp. (6.18) has to vanish. Since $c_0, d_0, d_1 \in \mathbb{C}$ resp. $c_1, d_0 \in \mathbb{C}$ are arbitrary, this leads to a contradiction.

To prove the last assertion let $\varphi \in \mathcal{D}(\overline{T})$ and $c \in \mathbb{C}$. Since the symmetric operator T is positive, so is its closure $\overline{T}$. Using that $T_\infty\mathfrak{p} = T^*\mathfrak{p} = 0$ we obtain

$$\langle T_\infty(\varphi + c\mathfrak{p}), \varphi + c\mathfrak{p}\rangle = \langle \overline{T}\varphi, \varphi + c\mathfrak{p}\rangle = \langle \overline{T}\varphi, \varphi\rangle + \overline{c}\langle \varphi, T^*\mathfrak{p}\rangle = \langle \overline{T}\varphi, \varphi\rangle \geq 0.$$

This shows that T_∞ is positive.   □

For $t \in \mathbb{R} \cup \{\infty\}$ we set $\mu_t(\cdot) := s_0 \langle E_t(\cdot)e_0, e_0\rangle$, where E_t denotes the spectral measure of the self-adjoint operator T_t. The next lemma plays an essential role in the proof of Theorem 7.6 below. It shows how the parameter $t \subset \mathbb{R}$ can be recovered from the measure μ_t and from the operator T_t.

Lemma 6.24 *For $t \in \mathbb{R}$, the operator T_t is invertible and*

$$\lim_{y\in\mathbb{R}, y\to 0} I_{\mu_t}(yi) = s_0 \lim_{y\in\mathbb{R}, y\to 0} \langle (T_t - yiI)^{-1}e_0, e_0\rangle = s_0\langle T_t^{-1}e_0, e_0\rangle = t. \qquad (6.19)$$

Further, $\lim_{y\in\mathbb{R}, y\to 0} |I_{\mu_\infty}(yi)| = +\infty$.

Proof Let $t \in \mathbb{R}$. Since $\mathcal{N}(T_t) \subseteq \mathcal{N}(T^*) = \mathbb{C} \cdot \mathfrak{p}$ by Proposition 6.6(i) and $\mathfrak{p} \notin \mathcal{D}(T_t)$, we have $\mathcal{N}(T_t) = \{0\}$, so the operator T_t is invertible. Recall that $\mathfrak{q} + s_0^{-1}t\mathfrak{p} = s_0^{-1/2}(\mathfrak{q}_0 + t\mathfrak{p}_0) \in \mathcal{D}(T_t)$. From $T^*\mathfrak{p} = 0$ and $T^*\mathfrak{q} = e_0$ by (6.14) it follows that $T_t(\mathfrak{q} + s_0^{-1}t\mathfrak{p}) = T^*(\mathfrak{q} + s_0^{-1}t\mathfrak{p}) = e_0$. Thus $e_0 \in \mathcal{D}(T_t^{-1})$ and $T_t^{-1}e_0 = \mathfrak{q} + s_0^{-1}t\mathfrak{p}$, so that

$$s_0\langle T_t^{-1}e_0, e_0\rangle = \langle s_0\mathfrak{q} + t\mathfrak{p}, e_0\rangle = t. \qquad (6.20)$$

Since $e_0 \in \mathcal{D}(T_t^{-1})$, the function $h(x) = x^{-2}$ is μ_t-integrable. In particular, $\mu_t(\{0\}) = 0$. We set $h_y(x) = |(x - yi)^{-1} - x^{-1}|^2$ for $y \in (-1, 1)$. Then $h_y(x) \to 0$

μ_t–a.e. on $\mathbb{R}$ as $y \to 0$ and $|h_y(x)| \leq h(x)$ for $y \in (-1, 1)$. Therefore, Lebesgue's dominated convergence theorem applies and using the functional calculus for the self-adjoint operator T_t we derive

$$\|(T_t - yiI)^{-1}e_0 - T_t^{-1}e_0\|^2 = \int_{\mathbb{R}} |(x - yi)^{-1} - x^{-1}|^2 \, d\langle E_t(x)e_0, e_0 \rangle$$

$$= s_0^{-1} \int_{\mathbb{R}} |(x - yi)^{-1} - x^{-1}|^2 \, d\mu_t(x) \to 0 \text{ as } y \to 0.$$

Therefore, $\lim_{y \to 0} (T_t - yiI)^{-1}e_0 = T_t^{-1}e_0$. Combining this with (6.20) and using the equality $I_{\mu_t}(yi) = s_0 \langle (T_t - yiI)^{-1}e_0, e_0 \rangle$, by the functional calculus, we get (6.19).

Since $T_\infty e_0 = 0$, $\mu_\infty(\{0\}) > 0$. Hence it follows from $|I_{\mu_\infty}(yi)| \geq \mu_\infty(\{0\})|y|^{-1}$ that $|I_{\mu_\infty}(yi)| \to +\infty$ as $\mathbb{R} \ni y \to 0$.   □

In the remaining part of this section we consider *symmetric* moment sequences.

Proposition 6.25 *Suppose that the indeterminate moment sequence s is symmetric, that is, $s_{2n+1} = 0$ for $n \in \mathbb{N}_0$. Let V be the self-adjoint unitary operator (see Proposition 5.38) of the Hilbert space $\mathcal{H}_s$ defined by $V(p)(x) = p(-x)$, $p \in \mathbb{C}[x]$. Then $VT_0V^{-1} = -T_0$ and $VT_\infty V^{-1} = -T_\infty$. Further, if $t \in \mathbb{R} \cup \{\infty\}$ and $V\mathcal{D}(T_t) \subseteq \mathcal{D}(T_t)$, then $t = 0$ or $t = \infty$.*

Proof Since s is symmetric, $p_n(-x) = (-1)^n p_n(x)$ and $q_n(x) = (-1)^{n+1}q_n(x)$ by (5.68). Therefore, $p_{2k+1}(0) = q_{2k}(0) = 0$ for all $k \in \mathbb{N}_0$. Hence the Fourier series of $\mathfrak{p}_0$ and $\mathfrak{q}_0$ with respect to the orthonormal basis $(p_n(x))_{n \in \mathbb{N}_0}$ of the Hilbert space $l^2(\mathbb{N}_0) \cong \mathcal{H}_s$ have the form

$$\mathfrak{p} \cong \sum_{k=0}^{\infty} p_{2k}(0)p_{2k}(x) \quad \text{and} \quad \mathfrak{q} \cong \sum_{k=0}^{\infty} q_{2k+1}(0)p_{2k+1}(x).$$

Hence $V\mathfrak{p}_0 = \mathfrak{p}_0$ and $V\mathfrak{q}_0 = -\mathfrak{q}_0$, so that $V(\mathfrak{q}_0 + t\mathfrak{p}_0) = -\mathfrak{q}_0 + t\mathfrak{p}_0$. Thus, by (6.16), the relation $V\mathcal{D}(T_t) \subseteq \mathcal{D}(T_t)$ holds if and only if the parameter t is 0 or ∞.

Recall that $VTV^{-1} = -T$ by Proposition 5.38. (As throughout, we identify X and T via the canonical unitary isomorphism of $\mathcal{H}_s$ and $l^2(\mathbb{N}_0)$.) Hence $V\mathcal{D}(T) = \mathcal{D}(T)$ and $VT^*V^{-1} = -T^*$. As noted in the preceding paragraph, $V\mathcal{D}(T_t) = \mathcal{D}(T_t)$ for $t = 0, \infty$. Therefore, since $T_t \subseteq T^*$, we conclude that $VT_tV^{-1} = -T_t$ for $t = 0, \infty$.   □

Corollary 6.26 *If s is a symmetric indeterminate moment sequence, then μ_0 and μ_∞ are different symmetric (!) representing measures for s.*

Proof Let $t = 0$ or $t = \infty$. Then $VT_tV^{-1} = -T_t$ by Proposition 6.25. We define a positive Radon measure $\tilde{\mu}_t$ on $\mathbb{R}$ by $d\tilde{\mu}_t(-x) = d\mu_t(x)$, $x \in \mathbb{R}$. Clearly, it follows from that equality $-T_t = VT_tV^{-1}$ that $(-T_t-zI)^{-1} = V(T_t-zI)^{-1}V^{-1}$ for $z \in \mathbb{C}\backslash\mathbb{R}$. Using the relations $V^{-1}e_0 = e_0$ and $V = V^*$ we derive

$$I_{\tilde{\mu}_t}(z) = s_0 \langle (-T_t - zI)^{-1}e_0, e_0 \rangle = s_0 \langle V(T_t - zI)^{-1}V^{-1}e_0, e_0 \rangle$$
$$= s_0 \langle (T_t - zI)^{-1}e_0, e_0 \rangle = I_{\mu_t}(z) \quad \text{for } z \in \mathbb{C}\backslash\mathbb{R},$$

that is, the Stieltjes transforms of $\tilde{\mu}_t$ and μ_t coincide. Hence $\tilde{\mu}_t = \mu_t$ by Theorem A.13. This means that μ_t is symmetric.

To prove that $\mu_0 \neq \mu_\infty$ we assume to the contrary that $\mu_0 = \mu_\infty$. Then we have $\langle (T_0-zI)^{-1}e_0, e_0 \rangle = \langle (T_\infty-zI)^{-1}e_0, e_0 \rangle$. This contradicts Lemma 6.8, since T_0 and T_∞ are different self-adjoint extensions of T. (Note that Lemma 6.8 was formulated for the operator X, but X is unitarily equivalent to T.) $\qquad\square$

6.6 Markov's Theorem

In this section $s = (s_n)_{n \in \mathbb{N}_0}$ is a positive definite sequence and $a = (a_n)_{n \in \mathbb{N}_0}$ and $b = (b_n)_{n \in \mathbb{N}_0}$ are the sequences in the corresponding Jacobi matrix J.

Let J_n be the operator on $\mathbb{C}^n$ defined by the truncated Jacobi matrix

$$J_n = \begin{pmatrix} b_0 & a_0 & 0 & \dots 0 & 0 & 0 \\ a_0 & b_1 & a_1 & \dots 0 & 0 & 0 \\ 0 & a_1 & b_2 & \dots 0 & 0 & 0 \\ \dots & \dots & \dots & \dots & \dots & \dots \\ 0 & 0 & 0 & \dots a_{n-3} & b_{n-2} & a_{n-2} \\ 0 & 0 & 0 & \dots 0 & a_{n-2} & b_{n-1} \end{pmatrix}, \quad n \in \mathbb{N}. \tag{6.21}$$

This matrix and Lemmas 6.27 and 6.28 will be used later in Sect. 8.4 as well.

The next lemma gives another description of the monic polynomial P_n and it shows that eigenvalues and eigenvectors of the self-adjoint operator J_n can be nicely expressed in terms of zeros of the orthogonal polynomials p_k.

Lemma 6.27

(i) $P_n(z) = \det(zI_n - J_n)$ for $n \in \mathbb{N}, z \in \mathbb{C}$.
(ii) Let $\lambda_1, \dots, \lambda_n$ be the zeros of p_n and $y^{(j)} = (p_0(\lambda_j), \dots, p_{n-1}(\lambda_j)) \in \mathbb{C}^n$. Then we have $J_n y^{(j)} = \lambda_j y^{(j)}$ for $j = 0, \dots, n-1$ and $n \in \mathbb{N}$.

Proof

(i) By developing $\det(zI_{n+1}-J_{n+1})$ after the last row we obtain

$$\det(zI_{n+1}-J_{n+1}) = (z - b_n)\det(zI_n-J_n) - a_{n-1}^2 \det(zI_{n-1}-J_{n-1}) \tag{6.22}$$

for $n \in \mathbb{N}$. Note that (6.22) is also valid for $n = 1$ by setting $\det(zI_0 - J_0) := 1$. From (6.22) and Proposition 5.9 it follows that the sequence of polynomials $\det(zI_n - J_n)$ satisfy the same reccurence relation (5.14) and intial data $\det(zI_0 - J_0) = P_0(x)$ and $\det(zI_1 - J_1) = z - b_0 = P_1(z)$ as P_n. Hence $P_n(z) = \det(zI_n - J_n)$ for $n \in \mathbb{N}$.

(ii) For $n = 1$ the assertion is easily checked, so we can assume that $n \geq 2$. Consider an equation $J_n y = zy$, where $y = (y_0, \ldots, y_{n-1}) \in \mathbb{C}^n$, $y \neq 0$, and $z \in \mathbb{C}$. For the first $n-1$ components this equation means that $y_0, \ldots, y_{n-1}$ satisfy the same recurrence relations (5.29) as the polynomials $p_0(z), \ldots, p_{n-1}(z)$ do with $y_{-1} = p_{-1}(z) := 0$. Hence $y_k = c p_k(z)$ for $k = 0, \ldots, n-1$ and some nonzero $c \in \mathbb{C}$. Inserting this into the n-th component of $J_n y = zy$ yields

$$z p_{n-1}(z) = a_{n-2} p_{n-2}(z) + b_{n-1} p_{n-1}(z).$$

On the other hand, by the relation (5.29) we have

$$z p_{n-1}(z) = a_{n-1} p_n(z) + + b_{n-1} p_{n-1}(z) + a_{n-2} p_{n-2}(z).$$

Since $a_{n-1} \neq 0$, it follows that $J_n y = zy$ holds if and only if $p_n(z) = 0$. Hence the eigenvalues of J_n are precisely the zeros λ_j of p_n and the corresponding eigenvectors are the vectors $y^{(j)}$.   □

Lemma 6.28 *Let K_n be the matrix which is obtained from J_n by removing the first row and the first column. Then*

$$\det (zI - K_n) = Q_n(z) \ \text{ for } \ n \in \mathbb{N}, n \geq 2, \ z \in \mathbb{C}, \tag{6.23}$$

$$\langle (J_n - zI)^{-1} e_0, e_0 \rangle = -\frac{Q_n(z)}{P_n(z)} \ \text{ for } \ z \in \rho(J_n). \tag{6.24}$$

Proof It suffices to prove the assertion for $n \geq 3$. Consider an equation $K_n y = zy$ for $y = (y_1, \ldots, y_{n-1}) \in \mathbb{C}^{n-1}$, $y \neq 0$, and $z \in \mathbb{C}$. Recall that K_n is the $(n-1) \times (n-1)$ matrix in left upper corner of the shifted Jacobi matrix $\tilde{J}$ from (5.32). Therefore, setting $y_0 := 0$ and using that $q_0 = 0$, the first $n-2$ components of the equation $K_n y = zy$ coincide with the recurrence relation for the polynomials $q_1(z), \ldots, q_{n-1}(z)$, see Sect. 5.4. Hence $y_j = c q_j(z)$ for $j = 1, \ldots, n-1$ and some $c \in \mathbb{C}, c \neq 0$. For the $(n-1)$-th component it follows in a similar manner as in the proof of Lemma 6.27 that z is an eigenvalue of K_n if and only if $q_n(z) = 0$, or equivalently, $Q_n(z) = 0$. Thus, $\det (zI - K_n)$ and $Q_n(z)$ are monic polynomials of degree $n - 1$ having the same zeros. Hence these polynomials coincide. This proves (6.23).

Finally, we prove (6.24). Let $z \in \rho(J_n)$. Setting $f = (f_0, \cdots, f_n) := (J_n - zI)^{-1} e_0$, to compute f_0 we apply Cramer's rule and get

$$\langle (J_n - zI)^{-1} e_0, e_0 \rangle = f_0 = \frac{\begin{vmatrix} 1 & a_0 & 0 & \dots 0 & & 0 \\ 0 & b_1 - z & a_1 & \dots 0 & & 0 \\ \dots\dots & & \dots\dots\dots & & \dots \\ 0 & 0 & 0 & \dots & b_{n-1} - z & a_{n-1} \\ 0 & 0 & 0 & \dots & a_{n-1} & b_n - z \end{vmatrix}}{\det (J_n - zI)}$$

$$= \frac{\det (K_n - zI)}{\det (J_n - zI)} = -\frac{\det (zI - K_n)}{\det (zI - J_n)} = -\frac{Q_n(z)}{P_n(z)}.$$

Here for the last equality we used (6.23) and Lemma 6.27(i). $\square$

Equation (6.25) in the following result is called *Markov's theorem*.

In particular, if supp μ is bounded, or equivalently, if the operator $T \cong X$ is bounded, then s is determinate and the equality (6.25) holds.

Theorem 6.29 *Suppose that s is a positive definite determinate moment sequence. If μ is the representation measure for s and R is an interval containing* supp μ, *then*

$$-\lim_{n \to \infty} \frac{q_n(z)}{p_n(z)} = -s_0 \lim_{n \to \infty} \frac{Q_n(z)}{P_n(z)} = \int_{\mathbb{R}} \frac{d\mu(x)}{x - z} \quad for \quad z \in \mathbb{C} \backslash R. \tag{6.25}$$

Proof The first equality in (6.25) follows at once from (5.36). To prove the second equality we extend J_n to a finite rank operator $\tilde{J}_n$ on $l^2(\mathbb{N}_0)$ by setting

$$\tilde{J}_n := \begin{pmatrix} J_n & 0 \\ 0 & 0 \end{pmatrix} \quad \text{on} \quad l^2(\mathbb{N}_0) = \mathbb{C}^n \oplus l^2(\mathbb{N}_n),$$

where $\mathbb{N}_n = \{k \in \mathbb{N} : k \geq n\}$. Fix a number $z \in \mathbb{C} \backslash R$.

Because s is determinate, $T \cong X$ is essentially self-adjoint by Theorem 6.10 and $\mathcal{H}_s \cong L^2(\mathbb{R}, \mu)$. Since the multiplication operator A_μ by the variable x in $L^2(\mathbb{R}, \mu)$ (see Proposition 6.2) is a self-adjoint extension of X and so of $\overline{M}_x$, we have $A_\mu = \overline{M}_x$. Therefore, the spectrum of the operator $\overline{T} \cong \overline{M}_x = A_\mu$ is the support of μ and so a subset of $\mathbb{R}$. Hence, since $z \in \mathbb{C} \backslash R$, $(T - zI)$d is dense in $l^2(\mathbb{N}_0)$. Further, since $P_n(z) = \det (zI - J_n)$ by Lemma 6.27(i), the spectrum of J_n, hence of $\tilde{J}_n$, consists of the zeros of P_n, so it is also contained in R by Corollary 5.28. Thus, z is in the resolvent sets of $\tilde{J}_n$ and $\overline{T}$.

Let $\varphi = (\varphi_0, \dots, \varphi_k, 0, \dots) \in$ d. If $n \geq k$, then we have $(T - zI)\varphi \in \mathbb{C}^{n+1}$ and $(\tilde{J}_n - zI)\varphi = (T - zI)\varphi$, so that

$$\left((\tilde{J}_n - zI)^{-1} - (\overline{T} - zI)^{-1} \right)(T - zI)\varphi$$

$$= (\tilde{J}_n - zI)^{-1}(\tilde{J}_n - zI)\varphi - (\overline{T} - zI)^{-1}(T - zI)\varphi = \varphi - \varphi = 0.$$

Hence, since $\|(\tilde{J}_n - zI)^{-1}\| \leq (\text{dist}(z, R))^{-1}$ and $(T - zI)$d is dense, it follows that

$$\lim_{n\to\infty} (\widetilde{J}_n - zI)^{-1}\psi = (\overline{T} - zI)^{-1}\psi \quad \text{for} \quad \psi \in l^2(\mathbb{N}_0). \tag{6.26}$$

Since $\overline{T}$ is self-adjoint, we have $\mu(\cdot) = s_0\langle E_{\overline{T}}(\cdot)e_0, e_0\rangle$ by formula (6.3) in Corollary 6.3. Therefore, applying (6.24) and (6.26) with $\psi = e_0$, we derive

$$-s_0 \lim_{n\to\infty} \frac{Q_n(z)}{P_n(z)} = s_0 \lim_{n\to\infty} \langle (J_n - zI)^{-1}e_0, e_0\rangle = s_0 \lim_{n\to\infty} \langle (\widetilde{J}_n - zI)^{-1}e_0, e_0\rangle$$

$$= s_0 \langle (\overline{T} - zI)^{-1}e_0, e_0\rangle = s_0 \int_{\mathbb{R}} (x-z)^{-1}d\langle E_{\overline{T}}(x)e_0, e_0\rangle = \int_{\mathbb{R}} (x-z)^{-1}d\mu(x),$$

which proves the second and main equality of (6.25). □

6.7　Continued Fractions

We begin with general continued fractions. Let $(\alpha_n)_{n\in\mathbb{N}}$ and $(\beta_n)_{n\in\mathbb{N}_0}$ be complex sequences. An (infinite) *continued fraction* is a formal expression

$$\beta_0 + \cfrac{\alpha_1}{\beta_1 + \cfrac{\alpha_2}{\beta_2 + \cfrac{\ddots + \cfrac{\alpha_n}{\beta_n + \ddots}}{}}} . \tag{6.27}$$

Just as in case of infinite series we want to associate a number to this expression. For this reason we set

$$C_n = \beta_0 + \cfrac{\alpha_1}{\beta_1 + \cfrac{\alpha_2}{\beta_2 + \cfrac{\ddots + \frac{\alpha_n}{\beta_n}}{}}} . \tag{6.28}$$

Note that it may happen that C_n is not defined if some denominator is zero. To save space the expressions (6.27) and (6.28) are written as

$$\beta_0 + \frac{\alpha_1|}{|\beta_1} + \frac{\alpha_2|}{|\beta_2} + \cdots + \frac{\alpha_n|}{|\beta_n} + \cdots,$$

$$C_n = \beta_0 + \frac{\alpha_1|}{|\beta_1} + \frac{\alpha_2|}{|\beta_2} + \cdots + \frac{\alpha_n|}{|\beta_n}.$$

Definition 6.30 The continued fraction (6.27) *converges* to $C \in \mathbb{C}$ if the numbers C_n are defined up to a finite subset of $\mathbb{N}_0$ and $\lim_{n\to\infty} C_n = C$. In this case we write

$$C = \beta_0 + \frac{\alpha_1|}{|\beta_1} + \frac{\alpha_2|}{|\beta_2} + \cdots + \frac{\alpha_n|}{|\beta_n} + \cdots.$$

Proposition 6.31 *For the numbers C_n we have $C_n = \frac{A_n}{B_n}$ for $n \in \mathbb{N}_0$, where*

$$A_n = \beta_n A_{n-1} + \alpha_n A_{n-2}, \quad A_0 = \beta_0, \quad A_{-1} = 1, \tag{6.29}$$

$$B_n = \beta_n B_{n-1} + \alpha_n B_{n-2}, \quad B_0 = 1, \quad B_{-1} = 0. \tag{6.30}$$

If $B_n \neq 0$ for $n = 1, \ldots, m$, then

$$C_m = \frac{A_m}{B_m} = \beta_0 + \sum_{n=1}^{m} \frac{(-1)^{n+1}\alpha_1 \ldots \alpha_n}{B_{n-1}B_n}. \tag{6.31}$$

Proof The first assertion will be proved by induction on n. We define A_n and B_n recursively by (6.29) and (6.30). Then we have to show that $C_n = \frac{A_n}{B_n}$.

For $n = 0$ this holds. Suppose that $C_n = \frac{A_n}{B_n}$. Recall that C_{n+1} is obtained if we replace β_n by $\beta_n + \frac{\alpha_{n+1}}{\beta_{n+1}}$ in the formula for C_n. Then, using the induction hypothesis and formulas (6.29) and (6.30), we derive

$$\begin{aligned}
C_{n+1} &= \frac{(\beta_n + \frac{\alpha_{n+1}}{\beta_{n+1}})A_{n-1} + \alpha_n A_{n-2}}{(\beta_n + \frac{\alpha_{n+1}}{\beta_{n+1}})B_{n-1} + \alpha_n B_{n-2}} \\
&= \frac{\beta_{n+1}(\beta_n A_{n-1} + \alpha_n A_{n-2}) + \alpha_{n+1}A_{n-1}}{\beta_{n+1}(\beta_n B_{n-1} + \alpha_n B_{n-2}) + \alpha_{n+1}B_{n-1}} \\
&= \frac{\beta_{n+1}A_n + \alpha_{n+1}A_{n-1}}{\beta_{n+1}B_n + \alpha_{n+1}B_{n-1}} = \frac{A_{n+1}}{B_{n+1}},
\end{aligned}$$

which completes the induction proof.

Next we prove (6.31). Subtracting (6.29) multiplied by B_{n-1} and (6.29) multiplied by A_{n-1} we obtain

$$A_n B_{n-1} - B_n A_{n-1} = -\alpha_n(A_{n-1}B_{n-2} - B_{n-1}A_{n-2}).$$

Repeated application of this relation yields

$$A_n B_{n-1} - B_n A_{n-1} = (-1)^{n+1}\alpha_1 \ldots \alpha_n,$$

which can be written as

$$\frac{A_n}{B_n} - \frac{A_{n-1}}{B_{n-1}} = \frac{(-1)^{n+1}\alpha_1 \ldots \alpha_n}{B_{n-1}B_n}. \tag{6.32}$$

Now (6.31) follows by summing over $n = 1, \ldots, m$ and using that $\frac{A_0}{B_0} = \beta_0$. $\quad\square$

Formulas (6.29) and (6.30) are three term recurrence relations that resemble the relations for monic orthogonal polynomials. Equation (6.31) implies that the continued fraction converges if and only if the corresponding series in (6.31) does.

Now we will use the (positive definite) moment sequence $s = (s_n)_{n \in \mathbb{N}_0}$. If a_n, b_n denote the corresponding Jacobi parameters a_n, b_n, we set

$$\beta_0 = 0, \quad \beta_n = z - b_{n-1}, \quad \alpha_1 = s_0, \quad \alpha_{n+1} = -a_{n-1}^2 \quad \text{for } n \in \mathbb{N}.$$

Replacing n by $n + 1$ in (6.30) we obtain for the denominators

$$B_{n+1} = (z - b_n)B_n - a_{n-1}^2 B_{n-1}, \quad B_0 = 1, B_{-1} = 0, \ n \in \mathbb{N},$$

and $B_1 = (z - b_0)B_0 + s_0 \cdot B_{-1} = z - b_0$. These are precisely the recurrence relation (5.14) and the intial data for the monic orthogonal polynomials $P_n(z)$. Therefore, we have $B_n = P_n(z)$ for $n \in \mathbb{N}_0$. In particular, $B_2 = (z - b_1)(z - b_0) - a_0^2 = P_2(z)$.

For the numerators A_n we obtain in a similar manner

$$A_{n+1} = (z - b_n)A_n - a_{n-1}^2 A_{n-1}, \quad A_0 = 0, A_{-1} = 1, \ n \in \mathbb{N},$$

and $A_1 = (z - b_0)A_0 + s_0 \cdot A_{-1} = s_0$. From Corollary 5.20 it follows that $s_0^{-1} A_n$ satisfies the same recurrence relation and initial data as the corresponding polynomials $Q_n(z)$. Hence $A_n = s_0 Q_n(z)$ for $n \in \mathbb{N}_0$. In particular, $A_2 = (z - b_1)s_0 = s_0 Q_2(z)$.

Summarizing the preceding, we obtain

$$C_{n+1} = \frac{A_{n+1}}{B_{n+1}} = \frac{s_0 Q_{n+1}(z)}{P_{n+1}(z)} = \frac{s_0|}{|z - b_0} + \frac{-a_0^2|}{|z - b_1} + \cdots + \frac{-a_{n-1}^2|}{|z - b_n}, \quad n \in \mathbb{N}. \tag{6.33}$$

Now we assume that the moment sequence s is *determinate*. Let μ be its representing measure. Then, by Markov's theorem 6.29,

$$\int_{\mathbb{R}} \frac{d\mu(x)}{z - x} = \lim_{n \to \infty} \frac{s_0 Q_{n+1}(z)}{P_{n+1}(z)}, \quad z \in \mathbb{C} \backslash \mathbb{R}.$$

Passing to the limit in (6.33) and inserting the latter equality yields

$$-I_\mu(z) = \int_{\mathbb{R}} \frac{d\mu(x)}{z - x} = \frac{s_0|}{|z - b_0} + \frac{-a_0^2|}{|z - b_1} + \cdots + \frac{-a_{n-1}^2|}{|z - b_n} + \ldots, \quad z \in \mathbb{C} \backslash \mathbb{R}. \tag{6.34}$$

Formula (6.34) provides an expansion of the negative Stieltjes transform $-I_\mu(z)$ of μ as a continued fraction. It connects continued fractions and moment problems. We will not follow this path in this book and prefer to use other methods.

Further, suppose that μ is supported by a bounded interval $[a, b]$. Then

$$-I_\mu(z) = \int \frac{d\mu(x)}{z - x} = \frac{s_0}{z} + \frac{s_1}{z^2} + \cdots + \frac{s_n}{z^{n+1}} + \cdots \qquad (6.35)$$

for $|z| > \max(|a|, |b|)$. (Indeed, then $\frac{1}{z-x} = \sum_{n=0}^{\infty} \frac{x^n}{z^{n+1}}$ converges uniformly on $[a, b]$, so we can interchange summation and integration.) For general representing measures there is an asymptotic expansion of the form (6.35), see Proposition 7.12 below.

The interplay between the two expansions (6.34) and (6.35) is one of the interesting features of the classical theory of one-dimensional moment problems.

6.8 Exercises

1. Let s be a Hamburger moment sequence and μ a representing measure for s. Prove that the following statements are equivalent:

    (i) The moment problem for s is determinate.
    (ii) There exists a number $z \in \mathbb{C}\backslash\mathbb{R}$ such that $\mathsf{p}_z \notin l^2(\mathbb{N}_0)$ and $\mathsf{q}_z \notin l^2(\mathbb{N}_0)$.
    (iii) There is a number $z_0 \in \mathbb{C}\backslash\mathbb{R}$ such that $(x - z_0)\mathbb{C}[x]$ is dense in $L^2(\mathbb{R}, \mu)$.
    (iv) There exist a number $z_0 \in \mathbb{C}\backslash\mathbb{R}$ and a sequence $(r_n)_{n\in\mathbb{N}}$ of polynomials $r_n \in \mathbb{C}[x]$ such that $\lim_{n\to\infty}(x - z_0)r_n(x) = 1$ in $L^2(\mathbb{R}, \mu)$.

2. Let $\mu, \nu \in \mathcal{M}_+(\mathbb{R})$. Suppose that there exists a $c > 0$ such that $\mu(M) \leq c\nu(M)$ for each Borel subset M of $\mathbb{R}$. Prove that if ν is determinate, so is μ.

3. Let $s = (s_n)_{n\in\mathbb{N}_0}$ and $t = (t_n)_{n\in\mathbb{N}_0}$ be moment sequences such that the moment sequence $s + t := (s_n + t_n)_{n\in\mathbb{N}_0}$ is determinate. Prove that s and t are determinate.

4. Let $\mu \in \mathcal{M}_+(\mathbb{R})$ and let $a, b, c \in \mathbb{R}, a < c < b$ Suppose that $\operatorname{supp}\mu \subseteq \mathbb{R}\backslash[a, b]$. Prove that μ is determinate if and only if $(x - c)\mathbb{C}[x]$ is dense in $L^2(\mathbb{R}, \mu)$.

5. Prove the "Plücker identity" for the Wronskian defined by (5.40):

$$W(\alpha, \beta)_n W(\gamma, \delta)_n - W(\alpha, \gamma)_n W(\beta, \delta)_n + W(\alpha, \delta)_n W(\beta, \gamma)_n = 0, n \in \mathbb{N}_0,$$

where $\alpha, \beta, \gamma, \delta$ are arbitrary complex sequences.

6. (*Translation of Hamburger moment sequences*)
Let $s = (s_n)_{n\in\mathbb{N}_0}$ be a moment sequence and $\gamma \in \mathbb{R}$. Define $s(\gamma)_n = \sum_{k=0}^{n} \binom{n}{k} \gamma^k s_{n-k}$ for $n \in \mathbb{N}_0$.

    a. Show that $s(\gamma) = (s(\gamma)_n)_{n\in\mathbb{N}_0}$ is also a moment sequence.
    b. Show that $\mu \in \mathcal{M}_s$ if and only if $\mu_\gamma \in \mathcal{M}_{s(\gamma)}$, where $d\mu_\gamma(x) := d\mu(x - \gamma)$.
    c. Show that s is determinate if and only if $s(\gamma)$ is determinate.

7. Let $\mu \in \mathcal{M}_+(\mathbb{R})$ be symmetric. Assume that $(-c, c) \cap \operatorname{supp}\mu = \emptyset$ for some $c > 0$. Show that the sequence $\left(\frac{q_n(0)}{p_n(0)}\right)_{n\in\mathbb{N}}$ does not converge as $n \to \infty$.

6.9   Notes

The Hamburger moment problem was first studied extensively by H. Hamburger [Hm]. Important classical result were obtained by M. Riesz [Rz2]. Among other things he discovered the operator-theoretic characterization of determinacy (Corollary 6.11). One-dimensional moment problems in the context of the extension theory of symmetric operators were treated in [DM]. The terminology of a von Neumann solution is from [Sim1]. Our approach partly follows [Sm9, Chapter 16] and [Sim1].

Markov's theorem 6.29 goes back to A.A. Markov [Mv1]. It is proved in [Chi1] for measures with bounded support and in [VA] and [Be] for determinate measures. Our approach to Theorem 6.29 and Exercise 6.7 are taken from C. Berg [Be].

Continued fractions are treated in [Wl] and [JT].

Chapter 7
The Indeterminate Hamburger Moment Problem

In this chapter we assume that s is an **indeterminate Hamburger moment sequence**. Our aim is to analyze the structure of the set $\mathcal{M}_s$ of all solutions of the moment problem for s. The central result in this respect is Nevanlinna's theorem (Theorem 7.13) on the parametrization of $\mathcal{M}_s$ in terms of the set $\mathfrak{P} \cup \{\infty\}$, where $\mathfrak{P}$ are the holomorphic functions on the upper half plane with nonnegative imaginary parts. It provides a one-to-one correspondence between Stieltjes transforms of measures $\mu \in \mathcal{M}_s$ and elements $\phi \in \mathfrak{P} \cup \{\infty\}$ given by a fractional linear transformation (7.16) with respect to four distinguished entire functions A, B, C, D. Note that in contrast to the moments the Stieltjes transform determines a finite Radon measure uniquely (by Theorem A.13)!

In Sect. 7.1 these four Nevanlinna functions A, B, C, D are defined and investigated. Sections 7.2 and 7.5 contain fundamental results on von Neumann solutions (Theorems 7.6, 7.7, and 7.15). The family of Weyl circles is introduced in Sect. 7.3. In Sect. 7.4 the celebrated Nevanlinna theorem is proved. In Sect. 7.6 we give a short excursion into Nevanlinna–Pick interpolation and derive basic results on the existence of a solution and on rational Nevanlinna functions (Theorems 7.20 and 7.22). In Sect. 7.7 solutions of finite order are studied and a number of characterizations of these solutions is given (Theorem 7.27 and 7.33).

7.1 The Nevanlinna Functions $A(z), B(z), C(z), D(z)$

The crucial technical step for the definition of the Nevanlinna functions is contained in the following lemma.

Lemma 7.1 *For any $z \in \mathbb{C}$ the series $\sum_{n=0}^{\infty} |p_n(z)|^2$ and $\sum_{n=0}^{\infty} |q_n(z)|^2$ converge. The sums are uniformly bounded on compact subsets of the complex plane.*

Proof Because the moment problem for s is indeterminate, it has at least two different solutions μ and ν. The corresponding Stieltjes transforms I_μ and I_ν

© Springer International Publishing AG 2017

K. Schmüdgen, *The Moment Problem*, Graduate Texts in Mathematics 277,

DOI 10.1007/978-3-319-64546-9_7

are holomorphic functions on $\mathbb{C}\backslash\mathbb{R}$. Since $\mu \neq \nu$, they do not coincide by Theorem A.13. Hence the set $\mathcal{Z}:=\{z \in \mathbb{C}\backslash\mathbb{R} : I_\mu(z)=I_\nu(z)\}$ has no accumulation point in $\mathbb{C}\backslash\mathbb{R}$.

Let M be a compact subset of $\mathbb{C}$. We choose $b > 0$ such that $|z| \leq b$ for all $z \in M$ and the line segment $L:=\{z \in \mathbb{C} : \text{Im } z = b, |\text{Re } z| \leq b\}$ does not intersect $\mathcal{Z}$. By the first condition and Corollary 5.31, the suprema of $|p_n(z)|$ and $|q_n(z)|$ over M are less than or equal to the corresponding suprema over L. Hence it suffices to prove the uniform boundedness of both sums on the set L.

Suppose that $z \in L$. Then we derive

$$|I_\mu(z) - I_\nu(z)|^2 \sum_n |p_n(z)|^2$$
$$= \sum_n |(q_n(z) + I_\mu(z)p_n(z)) - (q_n(z) + I_\nu(z)p_n(z))|^2$$
$$\leq 2 \sum_n |q_n(z) + I_\mu(z)p_n(z)|^2 + 2 \sum_n |q_n(z) + I_\nu(z)p_n(z)|^2$$
$$\leq 2b^{-1}(\text{Im } I_\mu(z) + \text{Im } I_\nu(z)) \leq 2b^{-1}(|I_\mu(z)| + |I_\nu(z)|),$$

where the inequality before last follows from inequality (5.39) in Proposition 5.37. Since the function $|I_\mu(z)-I_\nu(z)|$ has a positive infimum on L (because L has a positive distance from the set $\mathcal{Z}$) and $I_\mu(z)$ and $I_\nu(z)$ are bounded on L, the preceding inequality implies that the sum $\sum_n |p_n(z)|^2$ is finite and uniformly bounded on L.

Using once more (5.39) and proceeding in a similar manner we derive

$$\sum_n |q_n(z)|^2 \leq 2 \sum_n |q_n(z) + I_\mu(z)p_n(z)|^2 + 2|I_\mu(z)|^2 \sum_n |p_n(z)|^2$$
$$\leq 2b^{-1}|I_\mu(z)| + 2|I_\mu(z)|^2 \sum_n |p_n(z)|^2$$

for $z \in L$. Hence the boundedness of the sum $\sum_n |p_n(z)|^2$ on L implies the boundedness of $\sum_n |q_n(z)|^2$ on L. $\square$

Lemma 7.2 *For any sequence $c = (c_n) \in l^2(\mathbb{N}_0)$ the equations*

$$f(z) = \sum_{n=0}^{\infty} c_n p_n(z) \quad and \quad g(z) = \sum_{n=0}^{\infty} c_n q_n(z) \tag{7.1}$$

define entire functions $f(z)$ and $g(z)$ on the complex plane.

Proof We carry out the proof for $f(z)$. Since $(p_n(z)) \in l^2(\mathbb{N}_0)$ by Lemma 7.1 and $(c_n) \in l^2(\mathbb{N}_0)$, the series $f(z)$ converges for all $z \in \mathbb{C}$. We have

$$\left|f(z) - \sum_{n=0}^{k} c_n p_n(z)\right|^2 = \left|\sum_{n=k+1}^{\infty} c_n p_n(z)\right|^2 \leq \left(\sum_{n=k+1}^{\infty} |c_n|^2\right)\left(\sum_{n=0}^{\infty} |p_n(z)|^2\right)$$

for $k \in \mathbb{N}$ and $z \in \mathbb{C}$. Therefore, since $(c_n) \in l^2(\mathbb{N}_0)$ and the sum $\sum_n |p_n(z)|^2$ is bounded on compact sets (by Lemma 7.1), it follows that $\sum_{n=0}^{k} c_n p_n(z) \to f(z)$ as $k \to \infty$ uniformly on compact subsets of $\mathbb{C}$. Hence $f(z)$ is holomorphic on the whole complex plane. $\qquad\square$

From Lemmas 7.1 and 7.2 we conclude that

$$A(z, w) := (z-w) \sum_{n=0}^{\infty} q_n(z)q_n(w), \qquad B(z, w) := -1 + (z-w) \sum_{n=0}^{\infty} p_n(z)q_n(w),$$

$$C(z, w) := 1 + (z-w) \sum_{n=0}^{\infty} q_n(z)p_n(w), \quad D(z, w) := (z-w) \sum_{n=0}^{\infty} p_n(z)p_n(w)$$

are entire functions in each of the complex variables z and w. They are the limits of the polynomials $A_k(z, w), B_k(z, w), C_k(z, w), D_k(z, w)$, respectively, which have been defined in Proposition 5.24. By passing to the limit $n \to \infty$ in formula (5.57) of Corollary 5.27 we obtain the important identity

$$A(z, w)D(z, w) - B(z, w)C(z, w) = 1. \qquad (7.2)$$

From the above formulas it is obvious that

$$A(z, w) = -A(w, z), \; B(z, w) = -C(w, z), \; D(z, w) = -D(w, z), \quad z, w \in \mathbb{C}. \qquad (7.3)$$

Definition 7.3 The four functions $A(z, w), B(z, w), C(z, w), D(z, w)$ are called the *Nevanlinna functions* associated with the indeterminate moment sequence s.

These four functions are a fundamental tool in the study of the indeterminate moment problem. It should be emphasized that they depend only on the indeterminate moment sequence s.

We shall see by Theorems 7.6 and 7.13 below that the entire functions

$$A(z) := A(z, 0), B(z) := B(z, 0), C(z) := C(z, 0), D(z) := D(z, 0)$$

will enter in the parametrization of solutions. Often these four entire functions $A(z)$, $B(z)$, $C(z)$, $D(z)$ are called the *Nevanlinna functions* associated with s.

A number of facts on these functions are collected in the next proposition. Scalar products and norms refer always to the Hilbert space $l^2(\mathbb{N}_0)$.

Proposition 7.4 *Suppose that $z, w \in \mathbb{C}$. Then we have:*

(i) $D(z, 0)B(w, 0) - B(z, 0)D(w, 0) = -D(z, w)$.
(ii) $A(z, w) = (z - w)\langle \mathfrak{q}_z, \mathfrak{q}_{\overline{w}} \rangle$, $D(z, w) = (z - w)\langle \mathfrak{p}_z, \mathfrak{p}_{\overline{w}} \rangle$.
$B(z, w) + 1 = (z - w)\langle \mathfrak{p}_z, \mathfrak{q}_{\overline{w}} \rangle$, $C(z, w) - 1 = (z - w)\langle \mathfrak{q}_z, \mathfrak{p}_{\overline{w}} \rangle$.
(iii) $\operatorname{Im}(B(z)\overline{D(z)}) = \operatorname{Im} z \, \|\mathfrak{p}_z\|^2$.
(iv) $D(z) \neq 0$ and $D(z)t + B(z) \neq 0$ for $z \in \mathbb{C} \backslash \mathbb{R}$ and $t \in \mathbb{R}$.
(v) $D(z)\zeta + B(z) \neq 0$ for all $z, \zeta \in \mathbb{C}$, $\operatorname{Im} z > 0$ and $\operatorname{Im} \zeta \geq 0$.

Proof

(i) follows from the formula (5.58) by passing to the limit $n \to \infty$.

(ii) Since p_n and q_n have real coefficients, $\overline{p_n(w)} = p_n(\overline{w})$ and $\overline{q_n(w)} = q_n(\overline{w})$. Hence the formulas follow at once from the definitions of A, B, C, D and $\mathfrak{p}_z, \mathfrak{q}_w$.

(iii) Using (i) and the second equality from (ii) we compute

$$B(z)\overline{D(z)} - \overline{B(z)}D(z) = D(z, \overline{z}) = (z - \overline{z})\|\mathfrak{p}_z\|^2.$$

(iv) Let $z \in \mathbb{C}\backslash\mathbb{R}$. Since $p_0(z) \neq 0$, Im $z \|\mathfrak{p}_z\|_s^2 \neq 0$ and hence $D(z) \neq 0$ by (iii).
Assume to the contrary that $D(z)t + B(z) = 0$ for some $z \in \mathbb{C}\backslash\mathbb{R}$ and $t \in \mathbb{R}$. Then we have $-t = B(z)D(z)^{-1}$ and hence

$$0 = \mathrm{Im}\,(B(z)D(z)^{-1}|D(z)|^2) = \mathrm{Im}\,(B(z)\overline{D(z)}) = \mathrm{Im}\,z\,\|\mathfrak{p}_z\|^2$$

by (iii), which is a contradiction.

(v) follows in a similar manner as the last assertion of (iv). $\qquad\square$

Proposition 7.5 *If $\mu \in \mathcal{M}_s$ is a von Neumann solution, then*

$$I_\mu(z) = -\frac{A(z,w) + I_\mu(w)C(z,w)}{B(z,w) + I_\mu(w)D(z,w)} \quad \textit{for } z, w \in \mathbb{C}\backslash\mathbb{R}. \tag{7.4}$$

Formula (7.4) determines *all* values of $I_\mu(z)$ on $\mathbb{C}\backslash\mathbb{R}$ provided *one* fixed value $I_\mu(w)$ is given.

Proof Since μ is a von Neumann solution, $\mathcal{H}_s \cong L^2(\mathbb{R}, \mu)$. Hence $\{p_n : n \in \mathbb{N}_0\}$ is an orthonormal basis of $L^2(\mathbb{R}, \mu)$, so by (5.38) for all $z, w \in \mathbb{C}\backslash\mathbb{R}$ we have

$$f_z = \sum_{n=0}^\infty (q_n(z) + I_\mu(z)p_n(z))p_n, \quad f_{\overline{w}} = \sum_{n=0}^\infty (q_n(\overline{w}) + I_\mu(\overline{w})p_n(\overline{w}))p_n.$$

Using these formulas, the Parseval identity and Lemma 5.24 we derive

$$I_\mu(z) - I_\mu(w) = (z - w)\int \frac{1}{x-z}\frac{1}{x-w}\,d\mu(x) = (z-w)\,\langle f_z, f_{\overline{w}}\rangle_\mu$$

$$= (z-w)\sum_{n=0}^\infty (q_n(z) + I_\mu(z)p_n(z))\overline{(q_n(\overline{w}) + I_\mu(\overline{w})p_n(\overline{w}))}$$

$$= (z-w)\sum_{n=0}^\infty (q_n(z) + I_\mu(z)p_n(z))(q_n(w) + I_\mu(w)p_n(w))$$

$$= A(z,w) + I_\mu(z)(B(z,w)+1) + I_\mu(w)(C(z,w)-1) + I_\mu(z)I_\mu(w)D(z,w).$$

Eliminating $I_\mu(z)$ in the last equation we obtain (7.4). $\qquad\square$

7.2 Von Neumann Solutions

Recall that the self-adjoint extensions of the Jacobi operator T on the Hilbert space $\mathcal{H}_s$ are the operators T_t, $t \in \mathbb{R} \cup \{\infty\}$, from Theorem 6.23. If E_t denotes the spectral measure of T_t, we set

$$\mu_t(\cdot) := s_0 \, \langle E_t(\cdot)e_0, e_0 \rangle. \tag{7.5}$$

Theorem 7.6 *The measures μ_t, where $t \in \mathbb{R} \cup \{\infty\}$, are precisely the von Neumann solutions for the indeterminate moment sequence s. For $z \in \mathbb{C} \backslash \mathbb{R}$, we have*

$$s_0 \langle (T_t - zI)^{-1} e_0, e_0 \rangle = I_{\mu_t}(z) \equiv \int_{\mathbb{R}} \frac{1}{x - z} \, d\mu_t(x) = -\frac{A(z) + tC(z)}{B(z) + tD(z)}, \tag{7.6}$$

where for $t = \infty$ the fraction on the right-hand side has to be set equal to $-\frac{C(z)}{D(z)}$.

Proof The measures μ_t are indeed the von Neumann solutions, since the operators T_t exhaust the set of all self-adjoint extensions of T on $\mathcal{H}_s$ by Theorem 6.23.

The first equality of (7.6) follows at once from the definition of μ_t and the functional calculus for the resolvent of the self-adjoint operator T_t.

The main assertion of Theorem 7.6 is the last equality of (7.6). For this we apply formula (7.4) to $\mu = \mu_t$, $w = yi$ and pass to the limit $y \to 0$. Then the holomorphic function $A(z, w)$ in w tends to $A(z, 0) = A(z)$. Similarly the limits of $B(z, w), C(z, w), D(z, w)$ are $B(z), C(z), D(z)$, respectively.

Let $t \in \mathbb{R}$. Then $\lim_{y \to 0} I_{\mu_t}(yi) = t$ by Lemma 6.24 and $D(z)t + B(z) \neq 0$ on $\mathbb{C} \backslash \mathbb{R}$ by Proposition 7.4(iv), so the right-hand side of (7.4) tends to $-\frac{A(z) + tC(z)}{B(z) + tD(z)}$.

Now let $t = \infty$. Since $\lim_{y \to 0} |I_{\mu_\infty}(yi)| = +\infty$ by Lemma 6.24 and $D(z) \neq 0$ for $z \in \mathbb{C} \backslash \mathbb{R}$ by Proposition 7.4(iv), in this case the limit of (7.4) is $-\frac{C(z)}{D(z)}$. □

Since $A(z), B(z), C(z), D(z)$ are entire functions, it follows from Eq. (7.6) that the Stieltjes transform $I_{\mu_t}(z)$, $t \in \mathbb{R} \cup \{\infty\}$, is a meromorphic function.

Further, the numerator and denominator in (7.6) have no common zero. Indeed, if $t \in \mathbb{R}$ and z were a zero of $A + tC$ and $B + tD$, then

$$A(z)D(z) - B(z)C(z) = (-tC(z))D(z) - (-tD(z))C(z) = 0,$$

which contradicts (7.2). Similarly, for $t = \infty$, C and D have no common zero.

The following theorem describes the structure of von Neumann solutions.

Theorem 7.7 *Suppose that s is an indeterminate Hamburger moment sequence.*

(i) *Each von Neumann solution μ_t of s has a discrete unbounded support. The numbers in supp μ_t are precisely the zeros of the entire function $B(z) + tD(z)$ for $t \in \mathbb{R}$ resp. $D(z)$ for $t = \infty$. The set supp μ_t is the spectrum of the self-adjoint operator T_t.*   •

(ii) *For each number $x \in \mathbb{R}$, there is a unique $t_x \in \mathbb{R} \cup \{\infty\}$ such that $x \in \operatorname{supp} \mu_{t_x}$.*
If $t, \tilde{t} \in \mathbb{R} \cup \{\infty\}$ and $t \neq \tilde{t}$, then the supports of μ_t and $\mu_{\tilde{t}}$ are disjoint.

(iii) *If $x \in \operatorname{supp} \mu_t$, then x is a simple eigenvalue of the operator T_t and*

$$\mu_t(\{x\}) = \|\mathfrak{p}_x\|^{-2} = \left(\sum_{n=0}^{\infty} |p_n(x)|^2 \right)^{-1}. \tag{7.7}$$

Proof

(i) By Proposition A.15, a closed subset K of $\mathbb{R}$ belongs to $\operatorname{supp} \mu_t$ if and only if the Stieltjes transform $I_{\mu_t}(z)$ has a holomorphic extension to $\mathbb{C} \backslash K$. Hence $\operatorname{supp} \mu_t$ is the set of poles of the meromorphic function I_{μ_t}. Since the numerator and denominator in (7.6) have no common zero, these are precisely the zeros of the denominator. Being the zero set of an entire function, $\operatorname{supp} \mu_t$ is discrete.

Since μ_t is a von Neumann solution, e_0 is a cyclic vector for T_t. Hence T_t acts as the multiplication operator by the variable x in $L^2(\mathbb{R}, \mu_t)$ and $\operatorname{supp} \mu_t$ is the spectrum of T_t, see [Sm9, Section 5.4]. The operator T_t is unbounded and so is its spectrum $\operatorname{supp} \mu_t$.

(ii) If $D(x) \neq 0$, then x is a pole of I_{μ_t} for $t = -B(x)D(x)^{-1}$, so that $x \in \operatorname{supp} \mu_t$ by (i). Similarly, if $D(x) = 0$, then x is a pole of I_{μ_∞} and hence $x \in \operatorname{supp} \mu_\infty$ by (i).

To prove the uniqueness assertion, assume that $x \in \operatorname{supp} \mu_t$ and $x \in \operatorname{supp} \mu_{\tilde{t}}$ for $t, \tilde{t} \in \mathbb{R} \cup \{\infty\}$. If $t \in \mathbb{R}$ and $\tilde{t} \in \mathbb{R}$, then $B(x) + tD(x) = B(x) + \tilde{t}D(x) = 0$. Hence $D(x) \neq 0$ (otherwise $AD - BC = 0$, which contradicts (7.2)) and therefore $t = \tilde{t}$. If $t \in \mathbb{R}$ and $\tilde{t} = \infty$, then $B(x) + tD(x) = 0$ and $D(x) = 0$, so that $B(x) = 0$. Again this contradicts (7.2). Thus, $t = \tilde{t}$.

(iii) Let $x \in \operatorname{supp} \mu_t$. Then x is in the spectrum of T_t by (i). Because this set is discrete, x is an eigenvalue of T_t. Since $T_t \subseteq T^*$, it follows from Proposition 6.6(i) that all corresponding eigenvectors are multiples of $\mathfrak{p}_x = (p_n(x))_{n \in \mathbb{N}_0}$. Thus $\|\mathfrak{p}_x\|^{-1}\mathfrak{p}_x$ is a normalized simple eigenvector and the spectral projection $E_t(\{x\})$ of T_t is the rank one projection $E_t(\{x\}) = \|\mathfrak{p}_x\|^{-2}\langle \cdot, \mathfrak{p}_x \rangle \mathfrak{p}_x$. Since $\langle e_0, \mathfrak{p}_x \rangle = p_0(x) = s_0^{-1/2}$, we obtain

$$\mu_t(\{x\}) = s_0 \langle E_t(\{x\})e_0, e_0 \rangle = s_0 \|\mathfrak{p}_x\|^{-2}\langle e_0, \mathfrak{p}_x \rangle \langle \mathfrak{p}_x, e_0 \rangle = \|\mathfrak{p}_x\|^{-2},$$

which proves (7.7). ∎

Definition 7.8 A von Neumann solution of an indeterminate Hamburger moment sequence is called *Nevanlinna extremal*, or briefly, *N-extremal*.

By Theorem 7.7 (i) and (iii), each N-extremal solution μ is the form

$$\mu = \sum_{k=1}^{\infty} m_k \delta_{x_k},$$

where $m_k > 0$, the points $x_k \in \mathbb{R}$ are pairwise distinct, and $\lim_{k \to \infty} |x_k| = \infty$. Further, $\{x_k : k \in \mathbb{N}\}$ is the zero set of an entire function specified in Theorem 7.7(i).

7.3 Weyl Circles

An instructive tool in the theory of indeterminate moment problems is provided by the Weyl circles.

Definition 7.9 For $z \in \mathbb{C} \backslash \mathbb{R}$ the *Weyl circle* K_z is the closed circle in the complex plane with radius ρ_z and center C_z given by

$$\rho_z := \frac{1}{|z-\bar{z}| \, \|\mathfrak{p}_z\|_s^2} = \frac{z-\bar{z}}{|z-\bar{z}|} \frac{1}{D(z,\bar{z})} \, , \quad C_z := -\frac{(z-\bar{z})^{-1} + \langle \mathfrak{q}_z, \mathfrak{p}_z \rangle_s}{\|\mathfrak{p}_z\|_s^2} = -\frac{C(z,\bar{z})}{D(z,\bar{z})} \, .$$

The two equalities in this definition follow easily from Proposition 7.4(iii).

The proof of the next proposition shows that in the indeterminate case the inequality (5.39) means that the number $I_\mu(z)$ belongs to the Weyl circle K_z.

Proposition 7.10 *Suppose that* $\mu \in \mathcal{M}_s$. *Then the number* $I_\mu(z)$ *lies in the Weyl circle* K_z *for each* $z \in \mathbb{C} \backslash \mathbb{R}$. *The measure* μ *is a von Neumann solution if and only if* $I_\mu(z)$ *belongs to the boundary* ∂K_z *for one (hence for all)* $z \in \mathbb{C} \backslash \mathbb{R}$.

Proof We fix $z \in \mathbb{C} \backslash \mathbb{R}$ and abbreviate $\zeta := I_\mu(z)$. The inequality (5.39) says that

$$\|\mathfrak{q}_z\|_s^2 + \zeta \, \overline{\langle \mathfrak{q}_z, \mathfrak{p}_z \rangle_s} + \bar{\zeta} \, \langle \mathfrak{q}_z, \mathfrak{p}_z \rangle_s + |\zeta|^2 \|\mathfrak{p}_z\|_s^2 \equiv \|\mathfrak{q}_z + \zeta \mathfrak{p}_z\|_s^2 \le \frac{\zeta - \bar{\zeta}}{z - \bar{z}} \, . \tag{7.8}$$

The inequality in (7.8) can be rewritten as

$$|\zeta|^2 \|\mathfrak{p}_z\|_s^2 + \zeta \left(\overline{\langle \mathfrak{q}_z, \mathfrak{p}_z \rangle_s} + (z-\bar{z})^{-1} \right) + \bar{\zeta} \left(\langle \mathfrak{q}_z, \mathfrak{p}_z \rangle_s + (z-\bar{z})^{-1} \right) + \|\mathfrak{q}_z\|_s^2 \le 0 \, .$$

The latter inequality is equivalent to

$$\|\mathfrak{p}_z\|_s^2 \left| \zeta + \|\mathfrak{p}_z\|_s^{-2} (\langle \mathfrak{q}_z, \mathfrak{p}_z \rangle_s + (z-\bar{z})^{-1}) \right|^2 \le \|\mathfrak{p}_z\|_s^{-2} |\langle \mathfrak{q}_z, \mathfrak{p}_z \rangle_s + (z-\bar{z})^{-1}|^2 - \|\mathfrak{q}_z\|_s^2$$

and hence to

$$\left| \zeta + \|\mathfrak{p}_z\|_s^{-2} (\langle \mathfrak{q}_z, \mathfrak{p}_z \rangle_s + (z-\bar{z})^{-1}) \right|^2 \le \|\mathfrak{p}_z\|_s^{-4} \left(|\langle \mathfrak{q}_z, \mathfrak{p}_z \rangle_s + (z-\bar{z})^{-1}|^2 - \|\mathfrak{p}_z\|_s^2 \|\mathfrak{q}_z\|_s^2 \right).$$

This shows that ζ lies in a circle with center C_z given above and radius

$$\tilde{\rho}_z = \|\mathfrak{p}_z\|_s^{-2} \sqrt{|\langle \mathfrak{q}_z, \mathfrak{p}_z \rangle_s + (z-\bar{z})^{-1}|^2 - \|\mathfrak{p}_z\|_s^2 \|\mathfrak{q}_z\|_s^2} \tag{7.9}$$

provided the expression under the square root is nonnegative. Using Proposition 7.4(iii), the relation $\overline{C(z,\bar{z})} = -B(z,\bar{z})$, and (7.2) we compute

$$
\begin{aligned}
\tilde{\rho}_z &= \frac{z - \bar{z}}{D(z,\bar{z})} \sqrt{\frac{|C(z,\bar{z})|^2}{|z - \bar{z}|^2} - \frac{A(z,\bar{z})D(z,\bar{z})}{(z - \bar{z})^2}} \\
&= \frac{z - \bar{z}}{D(z,\bar{z})} \sqrt{\frac{-C(z,\bar{z})B(z,\bar{z})}{|z - \bar{z}|^2} + \frac{A(z,\bar{z})D(z,\bar{z})}{|z - \bar{z}|^2}} \\
&= \frac{z - \bar{z}}{D(z,\bar{z})} \frac{1}{|z - \bar{z}|} = \frac{1}{\|\mathfrak{p}_z\|_s^2 \, |z - \bar{z}|} .
\end{aligned}
$$

Thus $\tilde{\rho}_z$ is equal to the radius ρ_z of the Weyl circle and we have proved that $\zeta \in K_z$.

The preceding proof shows that $I_\mu(z) = \zeta \in \partial K_z$ if and only if we have equality in the inequality (7.8) and hence in (5.39). The latter is equivalent to the relation $f_z \in \mathcal{H}_s$ by Proposition 5.21 and so to the fact that μ is a von Neumann solution by Proposition 6.13. This holds for fixed and hence for all $z \in \mathbb{C} \backslash \mathbb{R}$.   □

Let $z, w \in \mathbb{C}$. Since $A(z,w)D(z,w) - B(z,w)C(z,w) = 1$ by (7.2), the fractional linear transformation $H_{z,w}$ defined by

$$
\xi = H_{z,w}(\zeta) := -\frac{A(z,w) + \zeta C(z,w)}{B(z,w) + \zeta D(z,w)} \tag{7.10}
$$

is a bijection of the extended complex plane $\overline{\mathbb{C}} = \mathbb{C} \cup \{\infty\}$ with inverse given by

$$
\zeta = H_{z,w}^{-1}(\xi) = -\frac{A(z,w) + \xi B(z,w)}{C(z,w) + \xi D(z,w)} . \tag{7.11}
$$

Some properties of these transformations $H_{z,w}$ can be found in Exercises 7.5 and 7.6. Here we will use only the transformations $H_z := H_{z,0}$. Set $\overline{\mathbb{R}} := \mathbb{R} \cup \{\infty\}$ and recall that $\mathbb{C}_+ = \{z \in \mathbb{C} : \text{Im } z > 0\}$. The next lemma is illustrated in Fig. 7.1.

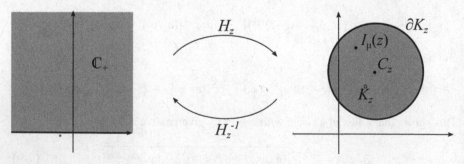

Fig. 7.1 The transformation H_z and the Weyl circle K_z

Lemma 7.11

(i) H_z is a bijection of $\overline{\mathbb{R}}$ onto the boundary $\partial K_z = \{I_{\mu_t}(z) : t \in \overline{\mathbb{R}}\}$ of the Weyl circle K_z for $z \in \mathbb{C} \setminus \mathbb{R}$.

(ii) H_z is a bijection of $\mathbb{C}_+$ onto the interior $\overset{\circ}{K}_z$ of the Weyl circle K_z for $z \in \mathbb{C}_+$.

(iii) $K_z \subseteq \mathbb{C}_+$ for $z \in \mathbb{C}_+$.

Proof

(i) By Theorem 7.6, H_z maps $\overline{\mathbb{R}}$ on the set $\{I_{\mu_t}(z) : t \in \overline{\mathbb{R}}\}$. Since $I_{\mu_t}(z) \in \partial K_z$ by Proposition 7.10, H_z maps $\overline{\mathbb{R}}$ into ∂K_z. But fractional linear transformations map generalized circles bijectively onto generalized circles. Hence H_z maps $\overline{\mathbb{R}}$ onto ∂K_z.

(ii) From (i) it follows that H_z is a bijection of either the upper half-plane or the lower half-plane on the interior of K_z. It therefore suffices to find one point $\xi \in \overset{\circ}{K}_z$ for which $H_z^{-1}(\xi) \in \mathbb{C}_+$. Since $I_{\mu_0}(z), I_{\mu_\infty}(z) \in \partial K_z$ by (i),

$$\xi := (I_{\mu_0}(z) + I_{\mu_\infty}(z))/2 = (-A(z)B(z)^{-1} - C(z)D(z)^{-1})/2 \in \overset{\circ}{K}_z .$$

Here the second equality follows from Theorem 7.6. Inserting this expression into (7.11) we easily compute

$$H_z^{-1}(\xi) = B(z)D(z)^{-1} = |D(z)|^{-2}B(z)\overline{D(z)} .$$

Hence $H_z^{-1}(\xi) \in \mathbb{C}_+$ by Proposition 7.4(iii), since $z \in \mathbb{C}_+$.

(iii) Since $z \in \mathbb{C}_+$, we have $I_{\mu_t}(z) \in \mathbb{C}_+ \cap \partial K_z$ by (i). Hence $K_z \subseteq \mathbb{C}_+$. $\qquad\square$

7.4 Nevanlinna Parametrization

First we prove a classical result due to Hamburger and Nevanlinna which is of interest in itself. It characterizes solutions of the moment problem in terms of the asympotic behaviour of their Stieltjes transforms.

Proposition 7.12

(i) If $s = (s_n)_{n \in \mathbb{N}_0}$ is an arbitrary (!) Hamburger moment sequence and $\mu \in \mathcal{M}_s$, then for each $n \in \mathbb{N}_0$,

$$\lim_{y \in \mathbb{R}, y \to \infty} y^{n+1} \left(I_\mu(iy) + \sum_{k=0}^{n} \frac{s_k}{(iy)^{k+1}} \right) = 0, \qquad (7.12)$$

where for fixed n the convergence is uniform on the set $\mathcal{M}_s$.

(ii) *Let $s = (s_n)_{n \in \mathbb{N}_0}$ be a real sequence and $\mathcal{I}(z)$ a Pick function. If (7.12) is satisfied (with I_μ replaced by $\mathcal{I}$) for all $n \in \mathbb{N}_0$, then s is Hamburger moment sequence and there exists a unique measure $\mu \in \mathcal{M}_s$ such that $\mathcal{I}(z) = I_\mu(z)$ for $z \in \mathbb{C} \backslash \mathbb{R}$.*

Proof

(i) First we rewrite the sum in Eq. (7.12) as

$$\sum_{k=0}^{n} \frac{s_k}{(iy)^{k+1}} = \sum_{k=0}^{n} \frac{(-i)^{k+1}}{y^{k+1}} \int_{\mathbb{R}} x^k d\mu(x) = -i\frac{1}{y^{n+1}} \int_{\mathbb{R}} \sum_{k=0}^{n} (-ix)^k y^{n-k} \, d\mu(x)$$

$$= -i\frac{1}{y^{n+1}} \int_{\mathbb{R}} \frac{(-ix)^{n+1} - y^{n+1}}{-ix - y} \, d\mu(x) = \frac{1}{y^{n+1}} \int_{\mathbb{R}} \frac{(-ix)^{n+1} - y^{n+1}}{x - iy} \, d\mu(x).$$

$$(7.13)$$

Therefore, since $|x - iy|^{-1} \leq |y|^{-1}$ for $x, y \in \mathbb{R}$, $y \neq 0$, we obtain

$$\left| y^{n+1} \left(I_\mu(iy) + \sum_{k=0}^{n} \frac{s_k}{(iy)^{k+1}} \right) \right|$$

$$= \left| \int_{\mathbb{R}} \frac{y^{n+1}}{x - iy} \, d\mu(x) + \int_{\mathbb{R}} \frac{(-ix)^{n+1} - y^{n+1}}{x - iy} \, d\mu(x) \right|$$

$$= \left| \int_{\mathbb{R}} \frac{(-ix)^{n+1}}{x - iy} \, d\mu(x) \right| \leq |y|^{-1} \int_{\mathbb{R}} |x|^{n+1} \, d\mu(x) \leq |y|^{-1} c_n,$$

where $c_n := s_{n+1}$ if n is odd and $c_n := s_n + s_{n+2}$ if n is even. Since c_n does not depend on the measure μ, we conclude that (7.12) holds uniformly on the set $\mathcal{M}_s$.

(ii) Condition (7.12) for $n = 0$ implies that $\lim_{y \to \infty} y\mathcal{I}(iy) = is_0$. Therefore, by Theorem A.14, the Pick function $\mathcal{I}$ is the Stieltjes transform I_μ of a finite positive Radon measure μ on $\mathbb{R}$ and μ is uniquely determined. Since $\mu(\mathbb{R}) < \infty$, Lebesgue's dominated convergence theorem applies and yields

$$s_0 = \lim_{y \to \infty} -iy\mathcal{I}(iy) = \lim_{y \to \infty} -iyI_\mu(iy) = \lim_{y \to \infty} \int_{\mathbb{R}} \frac{-iy}{x - iy} \, d\mu(x) = \int_{\mathbb{R}} d\mu(x).$$

The main part of the proof is to show that the n-th moment of μ exists and is equal to s_n for all $n \in \mathbb{N}_0$. We proceed by induction on n. For $n = 0$, this was just proved. Let $n \in \mathbb{N}$ and assume that μ has the moments $s_0, \ldots, s_{2n-2}$. Then, by the preceding proof of (i), formula (7.13) is valid with n replaced by $2n-2$. We use this formula in the case $2n-2$ to derive the second equality below and compute

$$(iy)^{2n+1}\left(\sum_{k=0}^{2n}\frac{s_k}{(iy)^{k+1}}+I_\mu(iy)\right)$$

$$=s_{2n}+iys_{2n-1}+i^{2n+1}y^2\left(y^{2n-1}\sum_{k=0}^{2n-2}\frac{s_k}{(iy)^{k+1}}+y^{2n-1}I_\mu(iy)\right)$$

$$=s_{2n}+iys_{2n-1}+i^{2n+1}y^2\left(\int_{\mathbb{R}}\frac{(-ix)^{2n-1}-y^{2n-1}}{x-iy}\,d\mu(x)+\int_{\mathbb{R}}\frac{y^{2n-1}}{x-iy}\,d\mu(x)\right)$$

$$=s_{2n}+iys_{2n-1}-y^2\int_{\mathbb{R}}\frac{x^{2n-1}}{x-iy}\,d\mu(x)$$

$$=s_{2n}-\int_{\mathbb{R}}\frac{x^{2n}}{(x/y)^2+1}\,d\mu(x)+iy\left(s_{2n-1}-\int_{\mathbb{R}}\frac{x^{2n-1}}{(x/y)^2+1}\,d\mu(x)\right).$$

By assumption (7.12) the term in the first line converges to zero as $y\to\infty$. Considering the real part and using Lebesgue's monotone convergence theorem we get

$$s_{2n}=\lim_{y\to\infty}\int_{\mathbb{R}}\frac{x^{2n}}{(x/y)^2+1}\,d\mu(x)=\int_{\mathbb{R}}x^{2n}d\mu(x)<\infty. \qquad (7.14)$$

The imaginary part, hence the imaginary part divided by y, also converges to zero as $y\to\infty$. Since

$$\frac{|x|^{2n-1}}{(x/y)^2+1}\le|x|^{2n-1}\le1+x^{2n}$$

and $1+x^{2n}$ is μ-integrable by (7.14), the dominated convergence theorem yields

$$s_{2n-1}=\lim_{y\to\infty}\int_{\mathbb{R}}\frac{x^{2n-1}}{(x/y)^2+1}\,d\mu(x)=\int_{\mathbb{R}}x^{2n-1}d\mu(x). \qquad (7.15)$$

By (7.14) and (7.15) the induction proof is complete. $\qquad\square$

Let $\mathfrak{P}$ denote the Pick functions (see Appendix A.2). We identify $t\in\mathbb{R}$ with the constant function t; then $\mathbb{R}$ becomes a subset of $\mathfrak{P}$. Set $\overline{\mathfrak{P}}:=\mathfrak{P}\cup\{\infty\}$.

The main result in this section is the following theorem of R. Nevanlinna. It expresses the Stieltjes transforms of representing measures of s by a fractional transformation of functions from the parameter space $\overline{\mathfrak{P}}$.

Theorem 7.13 *Suppose s is an indeterminate moment sequence. There is a one-to-one correspondence between functions $\phi \in \overline{\mathfrak{P}}$ and measures $\mu \in \mathcal{M}_s$ given by*

$$I_{\mu_\phi}(z) \equiv \int_{\mathbb{R}} \frac{1}{x-z} \, d\mu_\phi(x) = -\frac{A(z) + \phi(z)C(z)}{B(z) + \phi(z)D(z)} \equiv H_z(\phi(z)), \ z \in \mathbb{C}_+. \qquad (7.16)$$

Proof Suppose that $\mu \in \mathcal{M}_s$. If μ is a von Neumann solution, then by Theorem 7.6 there exists a $t \in \overline{\mathbb{R}}$ such that $I_\mu(z) = H_z(t)$ for all $z \in \mathbb{C}_+$.

Assume that μ is not a von Neumann solution and define $\phi(z) := H_z^{-1}(I_\mu(z))$. Let $z \in \mathbb{C}_+$. Then $I_\mu(z) \in \overset{\circ}{K}_z$ by Proposition 7.10 and hence $\phi(z) = H_z^{-1}(I_\mu(z)) \in \mathbb{C}_+$ by Lemma 7.11(ii). That is, $\phi(\mathbb{C}_+) \subseteq \mathbb{C}_+$. We show that $C(z)+I_\mu(z)D(z) \neq 0$. Indeed, otherwise $I_\mu(z) = -C(z)D(z)^{-1} = H_z(\infty) \in \partial K_z$ by Lemma 7.11(i) which contradicts the fact that $I_\mu(z) \in \overset{\circ}{K}_z$. Thus, $\phi(z)$ is the quotient of two holomorphic functions on $\mathbb{C}_+$ with nonvanishing denominator function. Therefore, ϕ is holomorphic on $\mathbb{C}_+$. This proves that $\phi \in \mathfrak{P}$. By the definition of ϕ we have $H_z(\phi(z)) = I_\mu(z)$ on $\mathbb{C}_+$, that is, (7.16) holds.

Conversely, suppose that $\phi \in \overline{\mathfrak{P}}$. If $\phi = t \in \overline{\mathbb{R}}$, then by Theorem 7.6 there is a von Neumann solution $\mu_t \in \mathcal{M}_s$ such that $I_{\mu_t}(z) = H_z(t)$.

Suppose now that ϕ is not in $\overline{\mathbb{R}}$. Let $z \in \mathbb{C}_+$ and define $\mathcal{I}(z) = H_z(\phi(z))$. Then $\phi(z) \in \mathbb{C}_+$ and hence $\mathcal{I}(z) = H_z(\phi(z)) \in \overset{\circ}{K}_z \subseteq \mathbb{C}_+$ by Lemma 7.11 (ii) and (iii). From Proposition 7.4(v) it follows that $B(z) + \phi(z)D(z) \neq 0$. Therefore, $\mathcal{I}$ is a holomorphic function on $\mathbb{C}_+$ with values in $\mathbb{C}_+$, that is, $\mathcal{I} \in \mathfrak{P}$.

To prove that $\mathcal{I} = I_\mu$ for some $\mu \in \mathcal{M}_s$ we want to apply Proposition 7.12(ii). For this we have to check that condition (7.12) is fulfilled. Indeed, by Proposition 7.12(i), given $\varepsilon > 0$ there exists a $Y_\varepsilon > 0$ such that

$$\left| y^{n+1} \left(I_\mu(iy) + \sum_{k=0}^{n} \frac{s_k}{(iy)^{k+1}} \right) \right| < \varepsilon \ \text{ for all } \ y \geq Y_\varepsilon \qquad (7.17)$$

and for all $\mu \in \mathcal{M}_s$. (Here it is crucial that Y_ε does not depend on μ and that (7.17) is valid for all measures $\mu \in \mathcal{M}_s$!) Fix a $y \geq Y_\varepsilon$. Since $\mathcal{I}(iy) = H_{iy}(\phi(iy))$ is in the interior of the Weyl circle K_{iy} by Lemma 7.11(ii), $\mathcal{I}(iy)$ is a convex combination of two points from the boundary ∂K_{iy}. By Lemma 7.11(i), all points of ∂K_{iy} are of the form $I_{\mu_t}(iy)$ for some $t \in \overline{\mathbb{R}}$. Since (7.17) holds for all $I_{\mu_t}(iy)$ and $\mathcal{I}(iy)$ is a convex combination of values $I_{\mu_t}(iy)$, (7.17) remains valid if $I_\mu(iy)$ is replaced by $\mathcal{I}(iy)$. This shows that $\mathcal{I}(z)$ fulfills the assumptions of Proposition 7.12(ii), so that $\mathcal{I} = I_\mu$ for some measure $\mu \in \mathcal{M}_s$.

By Theorem A.13, the positive measure μ_ϕ is uniquely determined by the values of its Stieltjes transform I_{μ_ϕ} on $\mathbb{C}_+$. Therefore, since I_{μ_ϕ} and $\phi \in \overline{\mathfrak{P}}$ correspond to each other uniquely by the relation $I_\mu(z) = H_z(\phi(z))$ on $\mathbb{C}_+$, (7.16) gives a one-to-one correspondence between $\mu \in \mathcal{M}_s$ and $\phi \in \overline{\mathfrak{P}}$. $\qquad \square$

Let us briefly discuss and summarize some of the results obtained so far.

Theorem 7.13 provides a complete parametrization of the solution set $\mathcal{M}_s$ in the indeterminate case in terms of the set $\overline{\mathfrak{P}}$. However, the one-to-one correspondence $\phi \leftrightarrow \mu_\phi$ between $\overline{\mathfrak{P}}$ and $\mathcal{M}_s$ given by (7.16) is highly nonlinear and very implicit. From Eq. (7.16) we derive that $\phi(z)$ is obtained from $I_{\mu_\phi}(z)$ by

$$\phi(z) = -\frac{A(z) + I_{\mu_\phi}(z)B(z)}{C(z) + I_{\mu_\phi}(z)D(z)}, \quad z \in \mathbb{C}_+.$$

The subset $\mathbb{R} \cup \{\infty\}$ of $\overline{\mathfrak{P}}$ is in one-to-one correspondence to the von Neumann solutions, or equivalently, to the self-adjoint extensions of T on the Hilbert space $\mathcal{H}_s \cong l^2(\mathbb{N}_0)$. The nonconstant Pick functions correspond to self-adjoint extension on a strictly larger Hilbert space (see Theorem 6.1).

Fix $z \in \mathbb{C}_+$. The values $I_\mu(z)$ for all von Neumann solutions $\mu \in \mathcal{M}_s$ fill the boundary ∂K_z of the Weyl circle, while the numbers $I_\mu(z)$ for all other solutions $\mu \in \mathcal{M}_s$ lie in the interior $\overset{\circ}{K}_z$. By taking convex combinations of von Neumann solutions it follows that each number of $\overset{\circ}{K}_z$ is of the form $I_\mu(z)$ for some $\mu \in \mathcal{M}_s$.

By Theorem 1.19, the solution set $\mathcal{M}_s$ is compact in the vague topology. For each indeterminate moment sequence the set $\mathcal{M}_s$ is "very large". We illustrate this by stating two results without proofs from [BC1, Theorem 1]:

The subset of measures $\mu \in \mathcal{M}_s$ of the form $d\mu = f(x)dx$ for some nonnegative function $f \in C^\infty(\mathbb{R})$ is dense in $\mathcal{M}_s$ with respect to the vague topology. The set of measures of finite order (as defined in Sect. 7.7 below) is also dense in $\mathcal{M}_s$.

All solutions of finite order are extreme points of $\mathcal{M}_s$ (see Exercise 7.9). Hence $\mathcal{M}_s$ is a convex compact set (in the vague topology) with dense set of extreme points! Recall from Theorem 1.21 that a measure $\mu \in \mathcal{M}_s$ is an extreme point of $\mathcal{M}_s$ if and only if $\mathbb{C}[x]$ is dense in $L^1(\mathbb{R}, \mu)$.

Remark 7.14 It is easily seen that the map $\psi \mapsto \phi := -\psi^{-1}$ is a bijection of the set $\overline{\mathfrak{P}}$. Inserting this into (7.16) we obtain

$$I_{\mu_\phi}(z) = -\frac{A(z) - \psi(z)^{-1}C(z)}{B(z) - \psi(z)^{-1}D(z)} = -\frac{A(z)\psi(z) - C(z)}{B(z)\psi(z) - D(z)}, \quad \psi \in \overline{\mathfrak{P}}. \tag{7.18}$$

The fraction on the right-hand side of (7.18) is another equivalent form of parametrization of solutions which often occurs in the literature (for instance, in Akhiezer's book [Ak]). Our convention (7.16) follows [Sim1]. $\circ$

7.5   Maximal Point Masses

The following theorem and its subsequent corollary contain a remarkable property of von Neumann solutions concerning maximal point masses.

Theorem 7.15 *Let μ be a representing measure of the indeterminate moment sequence s. Suppose that μ is not a von Neumann solution. Then, for any $x \in \mathbb{R}$ there is a von Neumann solution μ_t of s such that $\mu_t(\{x\}) > \mu(\{x\})$.*

The following lemma is used in the proof of Theorem 7.15.

Lemma 7.16 *Suppose that $\phi \in \mathfrak{P}$ and $\phi \notin \mathbb{R}$. Let $x \in \mathbb{R}$ and $t \in \mathbb{R}$. If the limit $L(x, t) := \lim_{\varepsilon \to +0} \frac{\phi(x+i\varepsilon)-t}{i\varepsilon}$ exists and is a real number, then $L(x, t) > 0$.*

Proof We use the canonical representation (A.4) of the Pick function

$$\phi(z) = a + bz + \int_{\mathbb{R}} \frac{1+zt}{t-z}\, d\nu(t),$$

where $a, b \in \mathbb{R}$, $b \geq 0$, and ν is a finite positive Radon measure on $\mathbb{R}$. Since $L(x, t)$ is real by assumption, we derive

$$L(x, t) = \lim_{\varepsilon \to +0} \frac{\phi(x+i\varepsilon)-t}{i\varepsilon} = \lim_{\varepsilon \to +0} \mathrm{Re}\, \frac{\phi(x+i\varepsilon)-t}{i\varepsilon} = \lim_{\varepsilon \to +0} \frac{\mathrm{Im}\, \phi(x+i\varepsilon)}{\varepsilon}$$

$$= \lim_{\varepsilon \to +0} \left(b + \int_{\mathbb{R}} \frac{1+t^2}{(t-x)^2 + \varepsilon^2}\, d\nu(t) \right) = b + \int_{\mathbb{R}} \frac{1+t^2}{(t-x)^2}\, d\nu(t),$$

where the last equality holds by Lebesgue's monotone convergence theorem. The right-hand side is obviously non-negative. If it were zero, then we would have $b = 0$ and $\nu \equiv 0$, so that $\phi \in \mathbb{R}$, which contradicts the assumption. Thus $L(x, t) > 0$. □

Proof of Theorem 7.15 First assume that $D(x) \neq 0$. Put $t = -B(x)D(x)^{-1}$. Then we have $x \in \mathrm{supp}\,\mu_t$ and $\mu_t(\{x\}) > 0$ by Theorem 7.7. Therefore the assertion is trivial if $\mu(\{x\}) = 0$, so we can assume that $\mu(\{x\}) > 0$.

Since $A(x)D(x) - B(x)C(x) = 1$, we have

$$D(x)^{-1} = D(x)^{-1}(A(x)D(x) - B(x)C(x)) = A(x) + tC(x). \tag{7.19}$$

By Theorem 7.13 there is a unique $\phi \in \overline{\mathfrak{P}}$ such that $\mu = \mu_\phi$. Since μ is not a von Neumann solution, $\phi \notin \mathbb{R} \cup \{\infty\}$. From the Stieltjes–Perron formula (A.8) and Eq. (7.16) we obtain

$$\mu(\{x\}) = \lim_{\varepsilon \to +0} (-i\varepsilon) I_\mu(x + i\varepsilon) = \lim_{\varepsilon \to +0} i\varepsilon\, \frac{A(x + i\varepsilon) + \phi(x + i\varepsilon)C(x + i\varepsilon)}{B(x + i\varepsilon) + \phi(x + i\varepsilon)D(x + i\varepsilon)}. \tag{7.20}$$

Therefore, since $\mu(\{x\}) > 0$, we have $\lim_{\varepsilon \to +0} |I_\mu(x + i\varepsilon)| = +\infty$ and hence

$$\lim_{\varepsilon \to +0} \phi(x + i\varepsilon) = \lim_{\varepsilon \to +0} H_{x+i\varepsilon}^{-1}(I_\mu(x + i\varepsilon))$$

$$= \lim_{\varepsilon \to +0} -\frac{A(x + i\varepsilon) + I_\mu(x + i\varepsilon)B(x + i\varepsilon)}{C(x + i\varepsilon) + I_\mu(x + i\varepsilon)D(x + i\varepsilon)} = -\frac{B(x)}{D(x)} = t. \tag{7.21}$$

Now we compute the limits of the numerator and denominator of the fraction in the right-hand side of Eq. (7.20), where the factor $i\varepsilon$ is included into the denominator. By (7.19) and (7.21) we obtain for the numerator

$$\lim_{\varepsilon \to +0} \left(A(x + i\varepsilon) + \phi(x + i\varepsilon)C(x + i\varepsilon) \right) = A(x) + tC(x) = D(x)^{-1}. \qquad (7.22)$$

For the denominator we derive

$$\lim_{\varepsilon \to +0} \frac{B(x + i\varepsilon) + \phi(x + i\varepsilon)D(x + i\varepsilon)}{i\varepsilon}$$

$$= \lim_{\varepsilon \to +0} \frac{B(x + i\varepsilon) - B(x) + t(D(x + i\varepsilon) - D(x)) + D(x + i\varepsilon)(\phi(x + i\varepsilon) - t)}{i\varepsilon}$$

$$= B'(x) + tD'(x) + D(x) \lim_{\varepsilon \to +0} \frac{\phi(x + i\varepsilon) - t}{i\varepsilon}. \qquad (7.23)$$

By (7.22), the limit of the numerator in the right-hand side of (7.20) exists and is real. Therefore, since $\mu(\{x\}) > 0$, the limit of the denominator exists as well. Since $D(x) \neq 0$, it follows from (7.23) that the limit $L(x, t) := \lim_{\varepsilon \to +0} \frac{\phi(x+i\varepsilon)-t}{i\varepsilon}$ exists. Since the limits in (7.20) and (7.22) are real, $L(x, t)$ is real. Hence $L(x, t) > 0$ by Lemma 7.16. Inserting the numerator and denominator limits into (7.20) we get

$$\mu(\{x\}) = \frac{1}{D(x)(B'(x) + tD'(x)) + D(x)^2 L(x, t)}. \qquad (7.24)$$

On the other hand, we compute the mass $\mu_t(\{x\})$ by applying again formula (7.20) with μ replaced by μ_t and ϕ by t. Then, by (7.19), we obtain

$$\mu_t(\{x\}) = \lim_{\varepsilon \to +0} (-i\varepsilon)I_{\mu_t}(x + i\varepsilon) = \lim_{\varepsilon \to +0} i\varepsilon \frac{A(x + i\varepsilon) + tC(x + i\varepsilon)}{B(x + i\varepsilon) + tD(x + i\varepsilon)}$$

$$= (A(x) + tC(x)) \lim_{\varepsilon \to +0} \frac{i\varepsilon}{B(x + i\varepsilon) - B(x) + t(D(x + i\varepsilon) - D(x))}$$

$$= \frac{A(x) + tC(x)}{B'(x) + tD'(x)} = \frac{1}{D(x)(B'(x) + tD'(x))}. \qquad (7.25)$$

Recall that $D(x) \neq 0$ and $L(x, t) > 0$. Therefore, comparing (7.24) and (7.25) it follows that $\mu_t(\{x\}) > \mu(\{x\})$. This proves the assertion in the case when $D(x) \neq 0$.

If $D(x) = 0$, then $B(x) \neq 0$ and the same proof goes through verbatim by using the second parametrization (7.18) of solutions. $\qquad \square$

The following corollary combines some assertions of Theorems 7.7 and 7.15.

Corollary 7.17 *Let s be an indeterminate Hamburger moment sequence. For each $x \in \mathbb{R}$ there exists a unique von Neumann solution μ_t of s such that $\mu_t(\{x\}) > 0$. For any solution $\mu \neq \mu_t$ of the moment problem for s we have $\mu(\{x\}) < \mu_t(\{x\})$.*

Proof The existence assertion on μ_t is contained in Theorem 7.7. Let $\mu \neq \mu_t$ be another solution. If μ is a von Neumann solution, then $\mu(\{x\}) = 0$ by Theorem 7.7(ii). If μ is not a von Neumann solution, then $\mu(\{x\}) < \mu_t(\{x\})$ by Theorem 7.15. $\square$

Corollary 7.18 *Let* $\mu = \sum_{k=1}^{\infty} m_k \delta_{x_k}$ *be an N-extremal solution of an indeterminate moment sequence, where all* $m_k > 0$ *and the points* $x_k \in \mathbb{R}$ *are pairwise distinct. Then, the measure* $\mu_k := \mu - m_k \delta_{x_k}$ *is determinate for* $k \in \mathbb{N}$*. In particular,* μ *is the sum of the two determinate measures* μ_k *and* $m_k \delta_{x_k}$*.*

Proof Assume to the contrary that μ_k is indeterminate. By Corollary 7.17, there is a von Neumann solution ν_k of the moment sequence of μ_k such that $\nu_k(\{x_k\}) > 0$. Clearly, $\nu := \nu_k + m_k \delta_{x_k}$ has the same moment sequence as μ and $\nu(\{x_k\}) > m_k = \mu(\{x_k\})$. This is impossible by the last assertion of Corollary 7.17. $\square$

Remark 7.19 Retain the assumptions of Corollary 7.18 and assume (upon translation) that $x_1 = 0$. Then $s_n(\mu) = \int x^n d\mu = \int x^n d\mu_1 = s_n(\mu_1)$ for $n \in \mathbb{N}$. Thus, except for the first moment, the *indeterminate* measure μ and the *determinate* measure μ_1 have the same moments and so the same growth of moment sequences! That is, there is no characterization of determinacy by growth conditions of the moment sequence. The determinate moment sequence of μ_1 cannot satisfy Carleman's condition (4.2), since otherwise μ would be determinate by Carleman's theorem 4.3.   ○

For an indeterminate moment sequence s we define a function

$$\rho^s(z) := \|\mathfrak{p}_z\|_s^{-2} = \left(\sum_{n=0}^{\infty} |p_n(z)|^2 \right)^{-1}, \quad z \in \mathbb{C}.$$

This function plays an important role in the study of the moment problem for s.

By Definition 7.9, $\rho_z = |z - \bar{z}|^{-1} \rho^s(z)$ is the radius of the Weyl circle K_z for $z \in \mathbb{C} \backslash \mathbb{R}$. By Corollary 7.17 and Theorem 7.7(iii), for $x \in \mathbb{R}$ the number $\rho^s(x)$ is the *maximal mass* of the one point set $\{x\}$ among all solutions of the moment problem for s. This maximum is attained at a unique solution: the von Neumann solution μ_t for which $x \in \operatorname{supp} \mu_t$, that is, $t = -B(x)D(x)^{-1}$ if $D(x) \neq 0$ and $t = \infty$ if $D(x) = 0$.

7.6 Nevanlinna–Pick Interpolation

Recall that each Pick function $\Phi \in \mathfrak{P}$ has a representation

$$\Phi(z) = a + bz + \int_{\mathbb{R}} \left(\frac{1}{t-z} - \frac{t}{1+t^2} \right) d\nu(t), \quad z \in \mathbb{C} \backslash \mathbb{R}, \tag{7.26}$$

where $a, b \in \mathbb{R}$, $b \geq 0$, and ν is a positive measure such that $\int (1+t^2)^{-1} d\nu(t) < \infty$.

Given a function $\eta : \mathcal{Z} \to \mathbb{C}_+$ defined on a subset $\mathcal{Z}$ of the upper half-plane $\mathbb{C}_+ = \{z \in \mathbb{C} : \Im z > 0\}$, the *Nevanlinna–Pick interpolation problem* asks:

When does there exist a function $\Phi \in \mathfrak{P}$ such that $\Phi(z) = \eta(z)$ for all $z \in \mathcal{Z}$?

As in the case of moment problems, some appropriate positivity condition is necessary and sufficient for the existence of a solution.

Theorem 7.20 *Let $\mathcal{Z}$ be a subset of $\mathbb{C}_+$ and $\eta : \mathcal{Z} \to \mathbb{C}_+$ a function on $\mathcal{Z}$. There exists a Pick function Φ such that $\Phi(z) = \eta(z)$ for all $z \in \mathcal{Z}$ if and only if for each finite set Z of pairwise distinct elements $z_0, \ldots, z_n \in \mathcal{Z}$ the matrix*

$$\mathcal{K}(\mathcal{Z}) := \left(\frac{\eta(z_i) - \overline{\eta(z_j)}}{z_i - \bar{z}_j} \right)^n_{i,j=0} \tag{7.27}$$

is positive semidefinite. If the set $\mathcal{Z}$ is finite, Φ can be chosen rational.

Proof Assume first that there is a function $\Phi \in \mathfrak{P}$ such that $\Phi(z) = \eta(z)$ for $z \in \mathcal{Z}$. Then using the canonical representation (7.26) we derive

$$\frac{\eta(z_i) - \overline{\eta(z_j)}}{z_i - \bar{z}_j} = \frac{\Phi(z_i) - \overline{\Phi(z_j)}}{z_i - \bar{z}_j} = b + \int_{\mathbb{R}} \frac{d\nu(x)}{(x - z_i)(x - \bar{z}_j)}. \tag{7.28}$$

Hence for $\xi_0, \ldots, \xi_n \in \mathbb{C}$ we obtain

$$\sum_{i,j=0}^n \frac{\eta(z_i) - \overline{\eta(z_j)}}{z_i - \bar{z}_j} \xi_i \bar{\xi}_j = b \left| \sum_{i=0}^n \xi_i \right|^2 + \int_{\mathbb{R}} \left| \sum_{i=0}^n \frac{\xi_i}{x - z_i} \right|^2 d\nu(x) \geq 0, \tag{7.29}$$

which proves that the matrix $\mathcal{K}(\mathcal{Z})$ is positive semidefinite.

Now we prove the converse implication. Upon a linear transformation $z \mapsto az+b$ and a shift $\eta \mapsto \eta - c$ with real a, b, c we can assume that $i \in \mathcal{Z}$ and $\Re \eta(i) = 0$. For $z \in \mathbb{C} \backslash \mathbb{R}$, let φ_z denote the function

$$\varphi_z(x) = \frac{1 + xz}{x - z}, \quad x \in \mathbb{R}.$$

Next we verify that the functions $\{\varphi_z : z \in \mathbb{C} \backslash \mathbb{R}, z \neq -i\}$ are linearly independent over $\mathbb{C}$. Indeed, suppose that $\lambda_0 \varphi_i + \sum_{j=1}^k \lambda_j \varphi_{z_j} = 0$, where $z_1, \ldots, z_k \in \mathbb{C} \backslash \mathbb{R}$ are distinct, $z_j \neq -i, i$, and $\lambda_j \in \mathbb{C}$. Note that $\varphi_i(x) = i$ for $x \in \mathbb{R}$. Then we obtain

$$\sum_{j=1}^k \lambda_j (1 + z_j^2)(x - z_j)^{-1} + \sum_{j=1}^k \lambda_j z_j + \lambda_0 i = 0, \quad x \in \mathbb{R}.$$

Since $z_j \neq -i, i$ and hence $1 - z_j^2 \neq 0$ for $j \neq 0$, the preceding equality implies that $\lambda_j = 0$ for $j = 1, \ldots, k$ and hence also $\lambda_0 = 0$.

Since $\lim_{x\to\pm\infty} \varphi_z(x) = z$, φ_z is a continuous function on the one point compactification $\overline{\mathbb{R}} = \mathbb{R} \cup \{\infty\}$ of $\mathbb{R}$ by setting $\varphi(\infty) = z$. Let E be the set of functions

$$f(x) = \mathrm{i}\lambda_0\varphi_\mathrm{i} + \sum_{j=1}^{n} \left(\lambda_j\varphi_{z_j} + \overline{\lambda}_j\varphi_{\overline{z}_j}\right), \tag{7.30}$$

where $\lambda_0 \in \mathbb{R}, \lambda_1,\dots,\lambda_n \in \mathbb{C}$ and $z_1,\dots,z_n \in \mathcal{Z}\backslash\mathrm{i}$ are distinct. Since $\overline{\varphi_z} = \varphi_{\overline{z}}$, each function f is real-valued and E is a real subspace $C(\overline{\mathbb{R}}; \mathbb{R})$. Since $\mathcal{Z} \subseteq \mathbb{C}_+$, it follows from the linear independence of the functions φ_z shown above that the numbers λ_j in (7.30) are uniquely determined by f. Hence there is a well-defined (!) linear functional $L : E \to \mathbb{R}$ defined by

$$L(f) = \lambda_0\mathrm{i}\,\eta(\mathrm{i}) + \sum_{j=1}^{n} \left(\lambda_j\eta(z_j) + \overline{\lambda}_j\,\overline{\eta(z_j)}\right). \tag{7.31}$$

Note that $\eta(\mathrm{i})$ is defined, because $\mathrm{i} \in \mathcal{Z}$, and $\mathrm{i}\eta(\mathrm{i})$ is real, since $\Re\,\eta(\mathrm{i}) = 0$. Further, since $\eta(\mathrm{i}) \in \mathbb{C}_+$, the constant function $g(x) := -\mathrm{i}\eta(\mathrm{i}) = \Im\eta(\mathrm{i}) > 0$ is in E.

Let $L_{\mathbb{C}} : E_{\mathbb{C}} \to \mathbb{C}$ be the extension of L to a linear functional on the complex vector space $E_{\mathbb{C}} = E + \mathrm{i}E$. Clearly, $\varphi_z, \varphi_{\overline{z}} \in E_{\mathbb{C}}$ for $z \in \mathcal{Z}$ and (7.31) implies that

$$L_{\mathbb{C}}(\varphi_z) = \eta(z) \quad \text{and} \quad L_{\mathbb{C}}(\varphi_{\overline{z}}) = \overline{\eta(z)} \quad \text{for } z \in \mathcal{Z}. \tag{7.32}$$

The crucial step of this proof is to show that $L(f) \geq 0$ for $f \in E_+$. This is where the positivity assumption comes in. Suppose that the function f from (7.30) is in E_+. Recall that $\varphi_\mathrm{i} = \mathrm{i}$. Hence we can write f as

$$f(x) = |x - z_1|^{-2} \dots |x - z_n|^{-2} p(x)$$

for some polynomial p. Since $f(x) \geq 0$ on $\overline{\mathbb{R}}$, $p(x) \geq 0$ on $\mathbb{R}$. Therefore, by Proposition 3.1, $p = q_1^2 + q_2^2$ for $q_1, q_2 \in \mathbb{R}[x]$. Setting $q := q_1 + \mathrm{i}q_2 \in \mathbb{C}[x]$, we have $p(x) = \overline{q(x)}q(x)$. Put

$$h(x) := (x - z_1)^{-1} \dots (x - z_n)^{-1} q(x).$$

Since f is bounded on $\mathbb{R}$, $\deg(p) \leq 2n$, so that $\deg(q) \leq n$. Therefore, since the numbers z_j are distinct, $h(x)$ is a linear combination of some constant and partial fractions $(x - z_j)^{-1}$. Setting $z_0 = \mathrm{i}$ and using that $(x - z_j)^{-1} = (z_j - \mathrm{i})^{-1}(\frac{x-\mathrm{i}}{x-z_j} - 1)$ for $j = 1,\dots,k$, it follows that $h(x)$ can be written as

$$h(x) = \sum_{j=0}^{n} \frac{x - \mathrm{i}}{x - z_j}\,\xi_j \tag{7.33}$$

with $\xi_j \in \mathbb{C}$. Then, inserting the corresponding functions, we compute

$$f(x) = |h(x)|^2 = \sum_{i,j=0}^{n} \frac{1+x^2}{(x-z_i)(x-\bar{z}_j)} \xi_i \bar{\xi}_j = \sum_{i,j=0}^{n} \frac{\varphi_{z_i}(x) - \varphi_{\bar{z}_j}(x)}{z_i - \bar{z}_j} \xi_i \bar{\xi}_j. \quad (7.34)$$

Therefore, by (7.32),

$$L(f) = L_{\mathbb{C}}(f) = \sum_{i,j=0}^{n} \frac{\eta(z_i) - \overline{\eta(z_j)}}{z_i - \bar{z}_j} \xi_i \bar{\xi}_j. \quad (7.35)$$

Since the matrix $\mathcal{K}(\mathcal{Z})$ is positive semidefinite by (7.27), $L(f) \geq 0$ by (7.35).

Thus L and E satisfy the assumptions of Proposition 1.9, so there exists a (finite) measure $\tilde{\nu} \in M_+(\overline{\mathbb{R}})$ such that $L(f) = \int f d\tilde{\nu}$ for $f \in E$. Define $\nu(M) = \tilde{\nu}(M)$ for a Borel subset M of $\mathbb{R}$ and $b = \tilde{\nu}(\{\infty\})$. Then ν is a finite measure of $M_+(\mathbb{R})$ and

$$L(f) = bf(\infty) + \int_{\mathbb{R}} f(x) \, d\nu(x), \quad f \in E. \quad (7.36)$$

Clearly, (7.36) extends to $L_{\mathbb{C}}(f)$ and $f \in E_{\mathbb{C}}$. By (7.32), $L_{\mathbb{C}}(\varphi_z) = \eta(z)$ and hence

$$\eta(z) = L_{\mathbb{C}}(\varphi_z) = bz + \int_{\mathbb{R}} \frac{1+xz}{x-z} \, d\nu(x), \quad \text{for } z \in \mathcal{Z}. \quad (7.37)$$

The right-hand side of (7.37) defines a function $\Phi \in \mathfrak{P}$. This completes the proof of the converse implication.

If the set $\mathcal{Z}$ is finite, the vector space E is finite-dimensional. Then Proposition 1.26 applies instead of Proposition 1.9 and yields a finitely atomic measure $\tilde{\nu}$. Then (7.37) gives a rational function $\Phi \in \mathfrak{P}$. $\qquad \square$

Corollary 7.21 *Let $z_0, \ldots, z_n \in \mathbb{C}_+$ be pairwise distinct and $w_0, \ldots, w_n \in \mathbb{C}_+$. There exists a function $\Phi \in \mathfrak{P}$ such that $\Phi(z_j) = w_j$ for $j = 0, \ldots, n$ if and only if the matrix*

$$\mathcal{K} = (K_{ij})_{i,j=0}^{n}, \quad \text{where} \quad K_{ij} = \frac{w_i - \bar{w}_j}{z_i - \bar{z}_j}, \quad (7.38)$$

is positive semidefinite. In this case the function Φ can be chosen rational.

Proof We apply Theorem 7.20 with $\mathcal{Z} := \{z_0, \ldots, z_n\}$ and $\eta(z_j) := w_j, j = 0, \ldots, n$. Then the necessity of the positive semidefiniteness of $\mathcal{K}$ is stated in Theorem 7.20 and its sufficiency follows from the preceding proof. Indeed, by the definition of $\mathcal{Z}$ all functions f of E are of the form (7.30), so by the above proof it suffices that the single matrix $\mathcal{K}$ is positive semidefinite. $\qquad \square$

The degree of a rational function $\varphi(z) = \frac{p(z)}{q(z)}$, where $p, q \in \mathbb{C}[z]$ have no common zero, is defined by

$$\deg(\varphi) := \max(\deg(p), \deg(q)). \qquad (7.39)$$

Theorem 7.22 *Let $z_0, \dots, z_n \in \mathbb{C}_+$ be pairwise distinct and $w_0, \dots, w_n \in \mathbb{C}_+$. Suppose that there exists a $\Phi \in \mathfrak{P}$ such that $\Phi(z_j) = w_j$ for $j = 0, \dots, n$. Then the following are equivalent:*

(i) *The matrix (7.38) has the eigenvalue 0.*
(ii) *Φ is real rational function with degree at most n.*
(iii) *The interpolating function $\Phi \in \mathfrak{P}$ is uniquely determined.*

Proof For $\xi_0, \dots, \xi_n \in \mathbb{C}$ we compute (see (7.29))

$$\sum_{i,j=0}^{n} K_{ij}\, \xi_i \bar{\xi}_j = \sum_{i,j=0}^{n} \frac{\Phi(z_i) - \overline{\Phi(z_j)}}{z_i - \bar{z}_j}\, \xi_i \bar{\xi}_j = b \left| \sum_{i=0}^{n} \xi_i \right|^2 + \int_{\mathbb{R}} \left| \sum_{i=0}^{n} \frac{\xi_i}{x - z_i} \right|^2 d\nu(x).$$

$$(7.40)$$

(i)→(ii) Let $(\xi_0, \dots, \xi_n)$ be an eigenvector of $\mathcal{K}$ for the eigenvalue 0. Then the expression in (7.40) is zero. Set $p(x) = \prod_{i=0}^{n}(x - z_i)$. Then there is a polynomial q of degree at most n such that $\sum_{i=0}^{n} \frac{\xi_i}{x - z_i} = \frac{q(x)}{p(x)}$. Since (7.40) vanishes, we obtain $\int \left| \frac{q(x)}{p(x)} \right|^2 d\nu(x) = 0$. Therefore, ν is supported on the zero set $\mathcal{Z}(q)$ by Proposition 1.23. This has at most n points. If $b \neq 0$, we have in addition $\sum_{i=0}^{n} \xi_i = 0$ by (7.40). This implies that $\deg(q) \leq n - 1$, so that ν is supported at at most $n - 1$ points. In both cases $b = 0$ and $b \neq 0$ it follows from the canonical representation (7.26) that Φ is a real rational function of degree at most n.

(ii)→(i) Suppose that Φ is rational and $\deg(\Phi) \leq n$. Since Φ is holomorphic outside the support of ν, it follows at once from (7.26) that ν is supported at k points $x_1, \dots, x_k$, where $k \leq n$ if $b = 0$ and $k \leq n - 1$ if $b \neq 0$. Put $m = k$ if $b = 0$ and $m = k + 1$ if $b \neq 0$. In either case $m \leq n$. For $\xi = (\xi_0, \dots, \xi_n) \in \mathbb{C}^{n+1}$ we define $h(\xi) = (h_1(\xi), \dots, h_m(\xi)) \in \mathbb{C}^m$, where

$$h_j(\xi) = \sum_{i=0}^{n} \frac{\xi_i}{x_j - z_i}, \quad j = 1, \dots, k,$$

and $h_m(\xi) = \sum_{i=0}^{n} \xi_i$ if $b \neq 0$. Since $m \leq n$, the mapping $h : \mathbb{C}^{n+1} \to \mathbb{C}^m$ has a nontrivial kernel. If $\xi \neq 0$ is in this kernel, then $\sum_{i,j=0}^{n} K_{ij} \xi_i \bar{\xi}_j = 0$ by (7.40). Since the matrix (7.38) is positive semidefinite by Corollary 7.21, we conclude that ξ is an eigenvector for the eigenvalue 0.

(i)→(iii) Let $\tilde{\Phi}$ be a solution of the interpolation problem. Then $\tilde{\Phi}$ is a real rational function of degree at most n by (i)→(ii). Since $\tilde{\Phi}$ is real on $\mathbb{R}$, we have $\tilde{\Phi}(\bar{z}) = \overline{\tilde{\Phi}(z)}$ for $z \in \mathbb{C} \backslash \mathbb{R}$ by Schwarz' reflection principle. Hence $\tilde{\Phi}(z_j) = w_j$ and

$\tilde{\Phi}(\bar{z}_j) = \bar{w}_j$ for $j = 0, \dots, n$. Hence $\tilde{\Phi}$ is uniquely determined, because a rational function of degree at most n is determined by $2n + 1$ distinct points. (Indeed, if $\frac{q_1}{p_1}(v_j) = \frac{q_2}{p_2}(v_j)$ for distinct $v_1 \dots, v_{2n+1}$, then $q_1(v_j)p_2(v_j) - q_2(v_j)p_1(v_j) = 0$. Since $\deg(q_1p_2 - q_2p_1) \leq 2n$, this implies $q_1p_2 - q_2p_1 = 0$, so that $\frac{q_1}{p_1} = \frac{q_2}{p_2}$.)

(iii)$\rightarrow$(i) Assume to the contrary that 0 is not an eigenvalue of the matrix (7.38). We use the setup of the proof of Theorem 7.20. Let $E \subseteq C(\overline{\mathbb{R}}; \mathbb{R})$ be the real vector space of functions (7.30) with n fixed. Suppose that $f \in E_+$ and $L(f) = 0$. Since 0 is not an eigenvalue, the matrix (7.38) is positive definite. Therefore, by (7.35), $L(f) = 0$ implies that all numbers ξ_j in (7.33) are zero, so that $h = 0$ and hence $f = 0$. Thus, L is strictly E_+-positive. Hence, by Theorem 1.30(ii), L has different representing measures $\tilde{\nu}$ and these measures give different interpolating functions Φ by the right-hand side of (7.37). This contradicts (iii) and completes the proof. $\square$

There is an alternative proof of the last implication (iii)$\rightarrow$(i). One can replace Theorem 1.30(ii) by Theorem 9.7 (proved in Sect. 9.1 below) on the truncated moment problem. Let us sketch the necessary modifications. First we note that $\dim E = 2n+1$. (To see this it suffices to recall that the complex space $E_{\mathbb{C}}$ is spanned by the linearly independent functions $1, \varphi_{z_j}, \varphi_{\bar{z}_j}, j = 1, \dots, n$, so it has dimension $2n+1$.) For each $f \in E$ there exists a unique polynomial $p_f \in \mathbb{R}[x]_{2n}$ such that

$$f(x) = |x - z_1|^{-2} \dots |x - z_n|^{-2} p_f(x).$$

Since $\dim \mathbb{R}[x]_{2n} = \dim E = 2n+1$, the map $f \mapsto p_f$ is a linear bijection of E onto $\mathbb{R}[x]_{2n}$. Hence there exists a linear functional on $\mathbb{R}[x]_{2n}$ defined by $\tilde{L}(p_f) = L(f)$, $f \in E$. Let $q \in \mathbb{R}[x]_n$, $q \neq 0$. Then $q^2 = p_f$ for some $f \in E$. Since $q \neq 0$, we have $f \neq 0$ and $f \in E_+$, so that $L(f) > 0$ as shown in the preceding proof of (iii)$\rightarrow$(i). Thus, $\tilde{L}(q^2) = \tilde{L}(p_f) = L(f) > 0$. Therefore, by Theorem 9.7, there exists a one-parameter family μ_t of finitely atomic representing measures for the functional $\tilde{L}$. Hence, setting $d\nu_t = |x - z_1|^2 \dots |x - z_n|^2 d\mu_t$, we obtain a one-parameter family of representing measures for L and so of interpolating functions Φ by (7.37).

7.7 Solutions of Finite Order

Throughout this section we assume that μ is a measure in $\mathcal{M}_+(\mathbb{R})$. This means that all moments of μ are finite. Let s denote its moment sequence. Recall from Proposition 6.2 that the canonical Hilbert space $\mathcal{H}_s$ is a closed subspace of $L^2(\mathbb{R}, \mu)$.

Definition 7.23 For a measure $\mu \in \mathcal{M}_+(\mathbb{R})$ the *order* of μ is defined by

$$\text{ord}(\mu) = \dim (L^2(\mathbb{R}, \mu) \ominus \mathcal{H}_s). \tag{7.41}$$

Here dim means the cardinality of an orthonormal basis of the Hilbert space.

Comparing Definitions 6.4 and 7.23 we see that the von Neumann solutions are precisely the solutions of order 0. Thus, $\mu \in \mathcal{M}_+(\mathbb{R})$ has order 0 if and only if either its moment sequence is determinate (by Theorem 6.10) or it is an N-extremal solution (Definition 7.8) of an indeterminate moment sequence. Therefore, for each measure of *nonzero* order the corresponding moment sequence is *indeterminate*.

Measures of finite order are, after N-extremal measures, the simplest solutions of indeterminate moment problems. In this section we derive a number of characterizations of such measures.

We begin with some preliminaries. A crucial role is played by the functions

$$f_z(x) = \frac{1}{x-z}, \quad \text{where} \quad z \in \mathbb{C}\backslash\mathbb{R},\ x \in \mathbb{R}.$$

For $n_1,\ldots,n_k \in \mathbb{N}_0$ and pairwise distinct $z_1,\ldots,z_k \in \mathbb{C}\backslash\mathbb{R}$ we define a closed linear subspace of $L^2(\mathbb{R},\mu)$ by

$$\mathcal{H}(z_1,\ldots,z_k;n_1,\ldots,n_k) := \mathcal{H}_s + \text{Lin}\{f_{z_l}^j : j=1,\ldots,n_l,\ l=1,\ldots,k\}. \quad (7.42)$$

(If $n_l = 0$ for some l, the corresponding term in (7.42) will be set zero.) Further, we shall use the bounded operator $V(z,w)$ of $L^2(\mathbb{R},\mu)$ with bounded inverse defined by

$$V(z,w) := (A_\mu - zI)(A_\mu - wI)^{-1} = I + (w-z)(A_\mu - wI)^{-1},\quad z,w \in \mathbb{C}\backslash\mathbb{R},$$

where A_μ is the multiplication operator by the variable x on $L^2(\mathbb{R},\mu)$, see (6.2).

Lemma 7.24

(i) $(A_\mu - zI)^{-1}\mathcal{H}_s \subseteq \mathcal{H}_s + \mathbb{C}\cdot f_z$ for $z \in \mathbb{C}\backslash\mathbb{R}$.
(ii) $f_z^k f_w^m \in \text{Lin}\{f_z^j, f_w^l : j=1,\ldots,k, l=1,\ldots,m\}$ for $z,w \in \mathbb{C}\backslash\mathbb{R}, z \neq w, k,n \in \mathbb{N}$.
(iii) *Suppose that* $z_1,\ldots,z_k \in \mathbb{C}\backslash\mathbb{R}$ *are pairwise distinct and* $n_1,\ldots,n_k \in \mathbb{N}_0$. *If* $f_{z_j}^{n_j+1} \in \mathcal{H}(z_1,\ldots,z_k;n_1,\ldots,n_k)$, *then* $f_{z_j}^k \in \mathcal{H}(z_1,\ldots,z_k;n_1,\ldots,n_k)$ *for* $k \in \mathbb{N}$.

Proof

(i) Let $p \in \mathbb{C}[x]$. Then $q_z(x) := \frac{p(x)-p(z)}{x-z}$ is a polynomial in x and

$$((A_\mu - zI)^{-1}p)(x) = (x-z)^{-1}p(x) = \frac{p(x)-p(z)}{x-z} + p(z)(x-z)^{-1}$$

$$= q_z(x) + p(z)f_z(x) \in \mathcal{H}_s + \mathbb{C}\cdot f_z.$$

Thus $(A_\mu - zI)^{-1}\mathbb{C}[x] \subseteq \mathcal{H}_s + \mathbb{C}\cdot f_z$. Since $\mathcal{H}_s + \mathbb{C}\cdot f_z$ is closed in $L^2(\mathbb{R},\mu)$, $\mathbb{C}[x]$ is dense in $\mathcal{H}_s$ and $(A_\mu - zI)^{-1}$ is bounded, it follows that $(A_\mu - zI)^{-1}\mathcal{H}_s \subseteq \mathcal{H}_s + \mathbb{C}\cdot f_z$.

(ii) The assertion follows by induction on $j+l$ easily from the identity

$$f_z^j f_w^l = (f_z^j f_w^{l-1} - f_z^{j-1}f_w^l)(z-w)^{-1}, \quad j,l \in \mathbb{N}.$$

(iii) For notational simplicity let $j = 1$ and write $z = z_1, n = n_1$.

First let $n \geq 1$. By the assumption we can write $f_z^{n+1} = \sum_{l=1}^{n} \gamma_l f_z^l + g$, where $\gamma_l \in \mathbb{C}$ and $g \in \mathcal{H}(z_2, \ldots, z_k; n_2, \ldots, n_k)$. Applying $(A_\mu - zI)^{-1}$ we obtain

$$f_z^{n+2} = \sum_{l=1}^{n} \gamma_l f_z^{l+1} + (A_\mu - zI)^{-1} g. \tag{7.43}$$

From (i) and (ii) it follows that $(A_\mu - zI)^{-1} g \in \mathcal{H}(z, z_2, \ldots, z_k; 1, n_2, \ldots, n_k)$. Since $f_z^{n+1} \in \mathcal{H}(z, \ldots, z_k; n, \ldots, n_k)$ by assumption, $\sum_{l=1}^{n} \gamma_l f_z^{l+1} \in \mathcal{H}(z, \ldots, z_k; n, \ldots, n_k)$. Thus, if $n \geq 1$, both summands in (7.43) are in $\mathcal{H}(z, \ldots, z_k; n, \ldots, n_k)$ and so is f_z^{n+2}.

Let $n = 0$. Then $f_z \in \mathcal{H}(z_2, \ldots, z_k; n_2, \ldots, n_k)$ by assumption, so by (i) and (ii),

$$f_z^2 \in \mathbb{C} \cdot f_z + \mathcal{H}(z_2, \ldots, z_k; n_2, \ldots, n_k) \subseteq \mathcal{H}(z, z_2, \ldots, z_k; 0, n_2, \ldots, n_k).$$

This completes the proof of the assertion for $k = n + 1$. Proceeding in a similar manner by induction it follows that $f_z^k \in \mathcal{H}(z, \ldots, z_k; n, \ldots, n_k)$ for all $k \in \mathbb{N}$. □

Lemma 7.25 *Let* $w, z_1, \ldots, z_k \in \mathbb{C} \backslash \mathbb{R}$ *be pairwise distinct and* $n_1, \ldots, n_k \in \mathbb{N}_0$. *Then, for* $l = 1, \ldots, k$, *we have*

(i) $V(z_l, w) \mathcal{H}(z_1, \ldots, z_k; n_1, \ldots, n_l + 1, \ldots, n_k) = \mathcal{H}(z_1, \ldots, z_k, w; n_1, \ldots, n_k, 1)$.

(ii) $V(w, z_l) \mathcal{H}(z_1, \ldots, z_k, w; n_1, \ldots, n_k, 1) = \mathcal{H}(z_1, \ldots, z_k; n_1, \ldots, n_l + 1, \ldots, n_k)$.

Proof From Lemma 7.24(ii) we conclude that $V(w, z_l)$ and $V(z_l, w)$ map the corresponding subspaces *into* the spaces on the right. Therefore, the composition $I = V(z_l, w) V(w, z_l)$ maps $\mathcal{H}(z_1, \ldots, z_k, w; n_1, \ldots, n_k, 1)$ into, hence onto, itself. This implies that we have equality in (ii) and similarly in (i). □

Proposition 7.26 *Suppose* $\mu \in \mathcal{M}_+(\mathbb{R})$ *and* $\mathrm{ord}(\mu) = n \in \mathbb{N}$. *Let* $z_1, \ldots, z_k \in \mathbb{C} \backslash \mathbb{R}$ *be pairwise distinct and* $n_1, \ldots, n_k \in \mathbb{N}$ *such that* $n = n_1 + \cdots + n_k$. *Then*

$$L^2(\mathbb{R}, \mu) = \mathcal{H}_s + \mathrm{Lin}\{f_{z_l}^j : j = 1, \ldots, n_l, l = 1, \ldots, k\}.$$

Proof Let $\mathcal{A} = \mathrm{Lin}\{f_z^k : z \in \mathbb{C} \backslash \mathbb{R}, k \in \mathbb{N}\}$. By Lemma 7.24(ii), $\mathcal{A}$ is closed under multiplication, so $\mathcal{A}$ is a $*$-subalgebra of the C^*-algebra $C_0(\mathbb{R})$ of continuous functions on $\mathbb{R}$ vanishing at infinity. Obviously, $\mathcal{A}$ separates the points of $\mathbb{R}$. Hence, by the Stone–Weierstrass theorem [Cw, Corollary 8.3], $\mathcal{A}$ is norm dense in $C_0(\mathbb{R})$. Since the measure μ is finite, this implies that $\mathcal{A}$ is dense in $L^2(\mathbb{R}, \mu)$. Since $\mathrm{ord}(\mu) = n$, there exists a finite-dimensional subspace $\mathcal{B}$ of $\mathcal{A}$ such that $L^2(\mathbb{R}, \mu) = \mathcal{H}_s + \mathcal{B}$. The latter implies that $L^2(\mathbb{R}, \mu) = \mathcal{H}_s + \mathcal{A}$.

Let $\{v_1, \ldots, v_r\}$ be a maximal subset of $\mathbb{C}\backslash\mathbb{R}$ such that $F_1 := \{f_{v_1}, \ldots, f_{v_r}\}$ is linearly independent modulo $\mathcal{H}_s$, where maximality means that $f_v \in \mathcal{H}_s + \mathrm{Lin}\, F_1$ for any $v \in \mathbb{C}\backslash\mathbb{R}$ distinct from all v_j. It is clear that such a set always exists, since $\mathrm{ord}(\mu) = n$ and $L^2(\mathbb{R}, \mu) = \mathcal{H}_s + \mathcal{A}$. Further, there exist numbers $m_1, \ldots, m_r \in \mathbb{N}$ such that $F = \{f_{v_l}^j : j = 1, \ldots, m_l, l = 1, \ldots, r\}$ is a maximal set which is linearly independent modulo $\mathcal{H}_s$. Here the maximality means that $f_{v_l}^{m_l+1} \in \mathcal{H}_s + \mathrm{Lin}\, F$ for all $l = 1, \ldots, r$. Then, it follows from the definition (7.42) that

$$\mathcal{G} := \mathcal{H}(v_1, \ldots, v_r; m_1, \ldots, m_r) = \mathcal{H}_s + \mathrm{Lin}\, F.$$

If $z \in \mathbb{C}\backslash\mathbb{R}$, $z \neq v_j$ for all j, then $f_z \in \mathcal{G}$ and hence $f_z^k \in \mathcal{G}$ for all $k \in \mathbb{N}$ by Lemma 7.24(iii). Likewise, $f_{v_j}^{m_j+1} \in \mathcal{G}$ implies $f_{v_j}^k \in \mathcal{G}$ for $k \in \mathbb{N}$, again by Lemma 7.24(iii). This proves that $\mathcal{A} \subseteq \mathcal{G}$. Since $L^2(\mathbb{R}, \mu) = \mathcal{H}_s + \mathcal{A}$ as shown in the preceding paragraph, we get $L^2(\mathbb{R}, \mu) = \mathcal{G}$. Because F is linearly independent modulo $\mathcal{H}_s$,

$$n = \mathrm{ord}(\mu) = \dim L^2(\mathbb{R}, \mu)/\mathcal{H}_s = \dim \mathrm{Lin}\, F = m_1 + \cdots + m_r.$$

Thus, by the preceding we have shown that

$$L^2(\mathbb{R}, \mu) = \mathcal{H}(v_1, \ldots, v_r; m_1, \ldots, m_r) = \mathcal{H}_s + \mathrm{Lin}\, \{f_{v_l}^j : j = 1, \ldots, m_l, l = 1, \ldots, r\}.$$

This equality is of the required form except for the fact that we have to take our *given* functions f_{z_l} and numbers n_l instead of f_{v_l} and m_l, respectively. To remedy this we now use Lemma 7.25.

First let us choose $w_1, \ldots, w_n \in \mathbb{C}\backslash\mathbb{R}$ such that both $w_1, \ldots, w_n, v_1 \ldots, v_r$ and $w_1, \ldots, w_n, z_1, \ldots, z_k$ are pairwise distinct. Then, by Lemma 7.25(i), we can find a products of operators $V(v_l, w_i)$ that maps $L^2(\mathbb{R}, \mu) = \mathcal{H}(v_1, \ldots, v_r; m_1, \ldots, m_r)$ onto $\mathcal{H}(w_1, \ldots, w_n; 1, \ldots, 1)$. Further, by Lemma 7.25(ii), there is a product of operators $V(w_i, z_j)$ which maps $\mathcal{H}(w_1, \ldots, w_n; 1, \ldots, 1)$ onto $\mathcal{H}(z_1, \ldots, z_k; n_1, \ldots, n_k)$. Since all operators $V(v_l, w_i)$ and $V(w_i, z_j)$ are isomorphisms of $L^2(\mathbb{R}, \mu)$, we obtain

$$L^2(\mathbb{R}, \mu) = \mathcal{H}(z_1, \ldots, z_k; n_1, \ldots, n_k) = \mathcal{H}_s + \mathrm{Lin}\{f_{z_l}^j : j=1, \ldots, n_l, l=1, \ldots, k\}. \quad \square$$

Proposition 7.26 says that $\{f_{z_l}^j : j = 1, \ldots, n_l, l = 1, \ldots, k\}$ forms a basis of the quotient space $L^2(\mathbb{R}, \mu)/\mathcal{H}_s$ if $\mathrm{ord}(\mu) = n_1 + \ldots + n_k \in \mathbb{N}$ and $z_1, \ldots, z_k \in \mathbb{C}\backslash\mathbb{R}$ are pairwise distinct. This is a crucial step for the following theorem, which characterizes measures of finite order in terms of density and determinacy conditions.

Theorem 7.27 *Suppose that $\mu \in \mathcal{M}_+(\mathbb{R})$. Let $z_1, \ldots, z_k \in \mathbb{C}\backslash\mathbb{R}$ be pairwise distinct numbers and $z \in \mathbb{C}\backslash\mathbb{R}$. Let $n_1, \ldots, n_k \in \mathbb{N}$ and set $n = n_1 + \cdots + n_k$. Define measures μ_n and μ_{n+1} of $\mathcal{M}_+(\mathbb{R})$ by*

$$d\mu_n(x) = \prod_{l=1}^{k} |x - z_l|^{-2n_l} d\mu(x), \quad d\mu_{n+1}(x) = |x - z|^{-2} d\mu_n(x).$$

Then the following statements are equivalent:

  (i) $\operatorname{ord}(\mu) \leq n$.
 (ii) $L^2(\mathbb{R}, \mu) = \mathcal{H}_s + \operatorname{Lin}\{f^j_{z_l} : j = 1, \ldots, n_l, l = 1, \ldots, k\}$.
(iii) $\mathbb{C}[x]$ *is dense in* $L^2(\mathbb{R}, \mu_n)$.
 (iv) μ_{n+1} *is determinate.*

Proof In this proof we abbreviate $\mathcal{F} = \operatorname{Lin}\{f^j_{z_l} : j = 1, \ldots, n_l, l = 1, \ldots, k\}$.

(i)$\rightarrow$(ii) Assume that $m := \operatorname{ord}(\mu) \leq n$. If $m = 0$, then $L^2(\mathbb{R}, \mu) = \mathcal{H}_s$ and the assertion trivially holds. Now let $m \in \mathbb{N}$. We choose natural numbers $i \leq k$ and $m_j \leq n_j$ for $j = 1, \ldots, i$ such that $m = m_1 + \cdots + m_i$. Then, by Proposition 7.26, $L^2(\mathbb{R}, \mu) = \mathcal{H}_s + \operatorname{Lin}\{f^j_{z_l} : j = 1, \ldots, m_l, l = 1, \ldots, i\}$. By adding further powers $f^j_{z_l}$ if $m < n$ this obviously implies $L^2(\mathbb{R}, \mu) = \mathcal{H}_s + \mathcal{F}$, which proves (ii).

(ii)$\rightarrow$(i) is trivial, since $\dim \mathcal{F} \leq n_1 + \cdots + n_k = n$.

(ii)$\leftrightarrow$(iii) From the definition of the measure μ_n we see that the mapping U defined by $(Uf)(x) = \prod_{l=1}^{k}(x - z_l)^{n_l} f(x)$ is a unitary operator of $L^2(\mathbb{R}, \mu)$ onto $L^2(\mathbb{R}, \mu_n)$. Decomposition into partial fractions yields an identity

$$\prod_{l=1}^{k}(x - z_l)^{-n_l} = \sum_{l=1}^{k} \sum_{j=1}^{n_l} \frac{a_{lj}}{(x - z_l)^j}, \qquad (7.44)$$

where $a_{lj} \in \mathbb{C}$. Equation (7.44) implies that U maps $\mathbb{C}[x] + \mathcal{F}$ into $\mathbb{C}[x]$. From Lemma 7.24(ii) it follows that U maps $\mathbb{C}[x] + \mathcal{F}$ onto $\mathbb{C}[x]$. Since $\mathbb{C}[x]$ is dense in $\mathcal{H}_s$, (ii) is equivalent to the density of $\mathbb{C}[x] + \mathcal{F}$ in $L^2(\mathbb{R}, \mu)$. Hence $\mathbb{C}[x] + \mathcal{F}$ is dense in $L^2(\mathbb{R}, \mu)$ if and only if $U(\mathbb{C}[x] + \mathcal{F}) = \mathbb{C}[x]$ is in $L^2(\mathbb{R}, \mu_n)$. Thus, (ii)$\leftrightarrow$(iii).

(iii)$\leftrightarrow$(iv) By Theorem 6.13, $\mathbb{C}[x]$ is dense in $L^2(\mathbb{R}, \mu_n)$ if and only if the function $f_z = (x - z)^{-1}$ is in the closure of $\mathbb{C}[x]$ in $L^2(\mathbb{R}, \mu_n)$. From Corollary 6.12 (or from Exercise 6.1) it follows that μ_{n+1} is determinate if and only if 1 is in the closure of $(x - z)\mathbb{C}[x]$ in $L^2(\mathbb{R}, \mu_{n+1})$. But both conditions are equivalent, since

$$\int |(x-z)^{-1} - p(x)|^2 d\mu_n(x) = \int |1 - (x-z)p(x)|^2 |x-z|^{-2} d\mu_n(x)$$

$$= \int |1 - (x-z)p(x)|^2 d\mu_{n+1}(x) \quad \text{for} \quad p \in \mathbb{C}[x]. \qquad \square$$

Since statement (i) of Theorem 7.27 does not depend on the choice of z_j, n_j, this holds for the assertions (ii)–(iv) as well. We elaborate on this in some corollaries.

Corollary 7.28 *Let* $\mu \in \mathcal{M}_+(\mathbb{R})$, $w_j \in \mathbb{C} \backslash \mathbb{R}$ *for* $j \in \mathbb{N}$. *Define* $\mu_k \in \mathcal{M}_+(\mathbb{R})$ *by*

$$d\mu_k := \prod_{j=1}^{k} |x - w_j|^{-2} d\mu, \quad k \in \mathbb{N}, \quad \mu_0 := \mu. \qquad (7.45)$$

Then the following statements are equivalent:

(i) ord(μ) *is finite.*
(ii) *There exists a $k \in \mathbb{N}$ such that $\mathbb{C}[x]$ is dense in $L^2(\mathbb{R}, \mu_k)$ for some, equivalently for arbitrary, numbers $w_1, \ldots, w_k \in \mathbb{C} \backslash \mathbb{R}$.*
(iii) *There exists an $m \in \mathbb{N}$ such that μ_m is determinate for some, equivalently for arbitrary, numbers $w_1, \ldots, w_m \in \mathbb{C} \backslash \mathbb{R}$.*

Further, ord$(\mu) \leq k$ *if (ii) holds and* ord$(\mu) \leq m - 1$ *if (iii) is satisfied.*

Proof Let $w_1, \ldots, w_n \in \mathbb{C} \backslash \mathbb{R}$ be arbitrary. We denote by $z_1, \ldots, z_k$ the pairwise distinct ones among them and by n_j the multiplicity of w_j in the sequence $\{w_1, \ldots, w_n\}$. Then we are in the setup of Theorem 7.27 and the equivalence of statements (i), (iii), (iv) therein yields the assertions. $\qquad \square$

The next corollary follows at once from the last statement in Corollary 7.28.

Corollary 7.29 *Retain the assumptions and the notation of Corollary 7.28, that is, $w_j \in \mathbb{C} \backslash \mathbb{R}$ are arbitrary, and μ_k is defined by (7.45). Let $n \in \mathbb{N}$. Then:*

(i) ord$(\mu) = n$ *if and only if $\mathbb{C}[x]$ is dense in $L^2(\mu_n)$, but not in $L^2(\mu_{n-1})$.*
(ii) ord$(\mu) = n$ *if and only if μ_{n+1} is determinate, but μ_n is not determinate.*

Corollary 7.30 *For a measure $\mu \in \mathcal{M}_+(\mathbb{R})$ the following are equivalent:*

(i) ord(μ) *is finite.*
(ii) $L^2(\mathbb{R}, \mu) = \mathcal{H}_s + \mathrm{Lin}\{f_{w_1}, \ldots, f_{w_n}\}$ *for some $n \in \mathbb{N}$ and some, equivalently for arbitrary, pairwise distinct numbers $w_1, \ldots, w_n \in \mathbb{C} \backslash \mathbb{R}$.*
(iii) $L^2(\mathbb{R}, \mu) = \mathcal{H}_s + \mathrm{Lin}\{f_w, f_w^2, \ldots, f_w^n\}$ *for some $n \in \mathbb{N}$ and some, equivalently for arbitrary, $w \in \mathbb{C} \backslash \mathbb{R}$.*

Proof We regroup $\{w_1, \ldots, w_n\}$ as in the proof of Corollary 7.28 and apply Theorem 7.27 (i)$\leftrightarrow$(ii). $\qquad \square$

Corollary 7.31 *If $\mu \in \mathcal{M}_+(\mathbb{R})$ has order $n \in \mathbb{N}$, then the support of μ is a discrete unbounded set.*

Proof We retain the notation (7.45). By Corollary 7.29 (i) and (ii), $\mathbb{C}[x]$ is dense in $L^2(\mathbb{R}, \mu_n)$ and μ_n is not determinate. Thus μ_n is a von Neumann solution of an indeterminate moment problem. Hence the support of μ_n is discrete and unbounded by Theorem 7.7. Since $d\mu = \prod_{j=1}^{n} |x - w_j|^2 d\mu_n(x)$, so is the support of μ. $\qquad \square$

The next proposition relates measures of finite order to moment problems with constraints on their Stieltjes transforms.

Let $z_1, \ldots, z_n \in \mathbb{C}_+$ be pairwise distinct and $w_1, \ldots, w_n \in \mathbb{C}_+$. We abbreviate

$$r(x) = |x - z_1|^2 \cdots |x - z_n|^2.$$

For $p \in \mathbb{C}[x]$ there is a decomposition of the rational function $\frac{p}{r}$ as a sum of partial fractions

$$\frac{p(x)}{r(x)} = q_p(x) + \sum_{j=1}^{n} \left(\frac{a_{j,p}}{x - z_j} + \frac{b_{j,p}}{x - \bar{z}_j} \right), \tag{7.46}$$

where $a_{j,p}, b_{j,p} \in \mathbb{C}$ and $q_p \in \mathbb{C}[x]$. The constants $a_{j,p}, b_{j,p}$ and the polynomial q_p are uniquely determined by p and $z_1 \dots, z_n$. Note that $\overline{a_{j,p}} = b_{j,\bar{p}}$ and $\overline{q_p} = q_{\bar{p}}$.

For a linear functional L on $\mathbb{C}[x]$, we define a linear functional on $\mathbb{C}[x]$ by

$$\Lambda(L)_{(z_1,\dots,z_n;w_1,\dots,w_n)}(p) = L(q_p) + \sum_{j=1}^{n} (a_{j,p} w_j + b_{j,p} \bar{w}_j), \; k \in \mathbb{N}_0. \tag{7.47}$$

Because of (7.46) the functional L and the numbers w_j can be recovered from the functional $\Lambda(L)_{(z_1,\dots,z_n;w_1,\dots,w_n)}$ by the formulas

$$L(p(x)) = \Lambda(L)_{(z_1,\dots,z_n;w_1,\dots,w_n)}(p(x)r(x)), \tag{7.48}$$

$$w_j = \Lambda(L)_{(z_1,\dots,z_n;w_1,\dots,w_n)}(r(x)(x - z_j)^{-1}). \tag{7.49}$$

(The latter expression is well-defined, since $r(x)(x - z_j)^{-1}$ is a polynomial.)

Proposition 7.32 *Suppose that s is a moment sequence. Let $z_1,\dots,z_n \in \mathbb{C}_+$ be pairwise distinct and $w_1,\dots,w_n \in \mathbb{C}_+$. There is a bijection between all solutions $\mu \in \mathcal{M}_s$ satisfying $I_\mu(z_j) = w_j$ for $j = 1,\dots,n$ and solutions $\mu_n \in \mathcal{M}_{\tilde{s}}$, where $\tilde{s} = (\tilde{s}_k)_{k \in \mathbb{N}_0}$ and $\tilde{s}_k := \Lambda(L_s)_{(z_1,\dots,z_n;w_1,\dots,w_n)}(x^k), \; k \in \mathbb{N}_0$, is defined by (7.47) for $p = x^k$. (Note that both sets of solutions may be empty.) This bijection is given by*

$$d\mu \;\leftrightarrow\; d\mu_n := \prod_{j=1}^{n} |x - z_j|^{-2} d\mu \equiv r(x)^{-1} d\mu. \tag{7.50}$$

Proof First, let $\mu \in \mathcal{M}_s$ be such that $I_\mu(z_j) = w_j$ for $j = 1,\dots,n$. Let $k \in \mathbb{N}_0$. Then, using (7.50), (7.46), and finally (7.47), we derive

$$\int x^k d\mu_n(x) = \int x^k r(x)^{-1} d\mu(x)$$

$$= \int q_{x^k}(x) d\mu(x) + \sum_{j=1}^{n} \left(a_{j,x^k} \int \frac{d\mu(x)}{x - z_j} + b_{j,x^k} \int \frac{d\mu(x)}{x - \bar{z}_j} \right)$$

$$= \Lambda(L_s)_{(z_1,\dots,z_n;I_\mu(z_1),\dots,I_\mu(z_n))}(x^k) = \Lambda(L_s)_{(z_1,\dots,z_n;w_1,\dots,w_n)}(x^k) = \tilde{s}_k,$$

that is, $\mu_n \in \mathcal{M}_{\tilde{s}}$.

Now suppose that $\mu_n \in \mathcal{M}_{\tilde{s}}$. Using the formulas (7.48) and (7.49) we obtain

$$s_k = L_s(x^k) = \Lambda(L_s)_{(z_1,\dots,z_n;w_1,\dots,w_n)}(x^k r(x)) = \int x^k r(x) d\mu_n(x) = \int x^k d\mu(x),$$

$$w_j = \Lambda(L_s)_{(z_1,\ldots,z_n;w_1,\ldots,w_n)}\big(r(x)(x-z_j)^{-1}\big)$$

$$= \int r(x)(x-z_j)^{-1}\,d\mu_n(x) = \int (x-z_j)^{-1}d\mu(x) = I_\mu(z_j)$$

for $k \in \mathbb{N}_0$ and $j = 1,\ldots,n$. Thus, $\mu \in \mathcal{M}_s$ and $I_\mu(z_j) = w_j, j = 1,\ldots,n$.   □

The next theorem states that solutions of finite order are precisely those corresponding to *rational* functions $\Phi \in \mathfrak{P}$ in the Nevanlinna parametrization of Theorem 7.13. In the proof we use Theorem 7.22 on Nevanlinna–Pick interpolation.

Theorem 7.33 *Suppose that s is an indeterminate Hamburger moment sequence. A representing measure $\mu_\Phi \in \mathcal{M}_s$ (given by (7.16) with $\Phi \in \mathfrak{P}$) has a finite order if and only if Φ is a rational function. In this case, $\mathrm{ord}(\mu_\Phi) = \deg(\Phi)$, where $\deg(\Phi)$ is defined by (7.39).*

Proof We fix $z_1,\ldots,z_{n+1} \in \mathbb{C}_+$ pairwise distinct and abbreviate $\mu := \mu_\Phi$.
As in Theorem 7.27 and Proposition 7.32 we define $\mu_{n+1} \in \mathcal{M}_+(\mathbb{R})$ by

$$d\mu_{n+1}(x) = \prod_{j=1}^{n+1} |x-z_j|^{-2}d\mu(x).$$

Then $v_j := I_\mu(z_j)$ and $w_j := H_{z_j}^{-1}(v_j)$ are in $\mathbb{C}_+$ for $j = 1,\ldots,n+1$. By (7.16) we have $H_{z_j}(\Phi(z_j)) = I_\mu(z_j) = v_j$ and hence $\Phi(z_j) = H_{z_j}^{-1}(v_j) = w_j$. That is, Φ is a solution of the Nevanlinna–Pick interpolation problem

$$\Psi(z_j) = w_j, j = 1,\ldots,n+1, \quad \text{for} \quad \Psi \in \mathfrak{P}. \tag{7.51}$$

Suppose μ has order $n \in \mathbb{N}$. Let $\tilde{\Phi} \in \mathfrak{P}$ be another solution of the interpolation problem (7.51). Then $\tilde{\mu} := \mu_{\tilde{\Phi}} \in \mathcal{M}_s$ (by Theorem 7.13) and $\tilde{\Phi}(z_j) = w_j$, so that $v_j = H_{z_j}(w_j) = H_{z_j}(\tilde{\Phi}(z_j)) = I_{\tilde{\mu}}(z_j)$ by (7.16) for $j = 1,\ldots,n+1$. Hence, by Proposition 7.32, $(\tilde{\mu})_{n+1}$ and μ_{n+1} have the same moment sequence. But, since $\mathrm{ord}(\mu) = n$, the measure μ_{n+1} is determinate by Corollary 7.29(ii). Therefore, $(\tilde{\mu})_{n+1} = \mu_{n+1}$ which in turn implies that $\mu_{\tilde{\Phi}} \equiv \tilde{\mu} = \mu \equiv \mu_\Phi$ and hence $\tilde{\Phi} = \Phi$ by Theorem 7.13. This shows that the interpolation problem (7.51) has a *unique* solution. It follows from Theorem 7.22 (iii)→(ii) that Φ is rational and $\deg(\Phi) \leq n$.

Now suppose that Φ is rational and $\deg(\Phi) \leq n$. We proceed as in the preceding paragraph but in reverse order. Let ν be a solution of the moment problem for μ_{n+1} and define $\tilde{\nu} \in \mathcal{M}_+(\mathbb{R})$ by $d\tilde{\nu} = \prod_{j=1}^{n+1} |x-z_j|^2 d\nu$. Then $(\tilde{\nu})_{n+1} = \nu$. From Proposition 7.32, applied in the converse direction, it follows that $I_{\tilde{\nu}}(z_j) = I_\mu(z_j) = v_j$, $j = 1,\ldots,n+1$, and $\tilde{\nu} \in \mathcal{M}_s$. Therefore, $\tilde{\nu} = \mu_{\tilde{\Phi}}$ for some $\tilde{\Phi} \in \mathfrak{P}$ by Theorem 7.13. Then we have $v_j = I_{\tilde{\nu}}(z_j) = H_{z_j}(\tilde{\Phi}(z_j))$ by (7.16), so that $\tilde{\Phi}(z_j) = H_{z_j}^{-1}(v_j) = w_j$. Hence $\tilde{\Phi}$ solves the interpolation problem (7.51) as well. Since Φ is rational and $\deg(\Phi) \leq n$, the interpolation problem (7.51) has a unique solution by

Theorem 7.22 (ii)→(iii). Thus, $\tilde{\Phi} = \Phi$ which implies $\tilde{\nu} = \mu$ and hence $\nu = \mu_{n+1}$. This proves that μ_{n+1} is determinate. Therefore, $\mathrm{ord}(\mu) \leq n$ by Corollary 7.27.

In the paragraph before last it was shown that $\mathrm{ord}(\mu_\Phi) = n$ implies $\deg(\Phi) \leq n$. If we had $\deg(\Phi) \leq n-1$, then $\mathrm{ord}(\mu_\Phi) \leq n-1$ by the preceding paragraph, which is a contradiction. Thus we have proved that $\mathrm{ord}(\mu_\phi) = \deg(\Phi)$. $\qquad\qquad\square$

7.8 Exercises

In this section, we assume that s is an **indeterminate** Hamburger moment sequence.

1. Let M be a compact subset of $\mathbb{C}$. Then $c_M := \sup_{z \in M} \left(\sum_{n=0}^{\infty} |p_n(z)|^2 \right)^{1/2} < \infty$ by Lemma 7.1. Show that $|p(z)| \leq c_M \|p\|_s$ for $p \in \mathbb{C}[x]$ and $z \in M$.
2. Let $z_1, z_2, z_3, z_4 \in \mathbb{C}$. Prove the following identities:

$$A(z_1, z_2)D(z_3, z_4) - B(z_3, z_2)C(z_1, z_4) + B(z_3, z_1)C(z_2, z_4) = 0,$$

$$A(z_1, z_2)C(z_3, z_4) + A(z_3, z_1)C(z_2, z_4) + A(z_2, z_3)C(z_1, z_4) = 0,$$

$$D(z_1, z_2)B(z_3, z_4) + D(z_3, z_1)B(z_2, z_4) + D(z_2, z_3)B(z_1, z_4) = 0.$$

   Hint: Verify the corresponding identities for A_k, B_k, C_k, D_k. Use Lemma 5.24.
3. Let $z, w \in \mathbb{C}$. Show that

$$A(z, w) = A(z)C(w) - C(z)A(w), \quad B(z, w) = A(z)D(w) - C(z)B(w),$$

$$C(z, w) = B(z)C(w) - D(z)A(w), \quad D(z, w) = B(z)D(w) - D(z)B(w).$$

4. (*Reproducing kernel*) Recall that $p_k, k \in \mathbb{N}_0$, are the orthonormal polynomials.

   a. Show that the series

$$K(z, w) = \sum_{k=0}^{\infty} p_k(z)p_k(w), \quad (z, w) \in \mathbb{C}^2,$$

   converges uniformly on compact subsets of $\mathbb{C}^2$ to a holomorphic function K, called the *reproducing kernel* for s, such that $D(z, w) = (z - w)K(z, w)$.

   b. Show that for each representing measure $\mu \in \mathcal{M}_s$ and polynomial $f \in \mathbb{C}[x]$,

$$\int_{\mathbb{R}} K(z, x)f(x)d\mu(x) = f(z), \quad z \in \mathbb{C}.$$

   c. Show that the preceding equality remains valid for each holomorphic function $f(z) = \sum_{k=0}^{\infty} c_k p_k(z)$, where $(c_n) \in l^2(\mathbb{N}_0)$.

5. Prove that the fractional linear transformation $H_{z,w}$ defined by (7.10) maps

    a. ∂K_w onto ∂K_z if $\operatorname{Im} z \neq 0$ and $\operatorname{Im} w \neq 0$,

    b. ∂K_w onto $\overline{\mathbb{R}}$ if $\operatorname{Im} z = 0$ and $\operatorname{Im} w \neq 0$,

    c. $\overline{\mathbb{R}}$ onto ∂K_z if $\operatorname{Im} z \neq 0$ and $\operatorname{Im} w = 0$,

    d. $\overline{\mathbb{R}}$ onto $\overline{\mathbb{R}}$ if $\operatorname{Im} z = \operatorname{Im} w = 0$.

6. Let $H_{z,w}$ be the transformation defined by (7.10) and $H_z := H_{z,0}$. Prove that

$$H_{z,w}H_{w,v} = H_{z,v}, \ H(z,z) = I, \ (H_{z,w})^{-1} = H_{w,z}, \ H_{z,w} = H_z(H_w)^{-1}, \quad z,w,v \in \mathbb{C}.$$

    Hint: Use (7.4) and (7.3).

7. Show that for each $n \in \mathbb{N}$ there is a continuum of measures $\mu \in \mathcal{M}_s$ of order n.

8. Suppose that $\mu \in \mathcal{M}_s$ has order $n \in \mathbb{N}$. Let $p \in \mathbb{R}[x]$ and define $v \in \mathcal{M}_+(\mathbb{R})$ by $dv = (1 + p(x)^2)^{-1}d\mu$. What is the order of v?

9. Show that each measure in $\mathcal{M}_s$ of finite order is an extreme point of the set $\mathcal{M}_s$. Hint: Use Proposition 1.21 and (for instance) Theorem 7.27 (iii).

10. Consider the Nevanlinna–Pick interpolation problem in Theorem 7.22 and sharpen the equivalence of (i) and (ii) therein: Show that 0 is an eigenvalue of multiplicity k if and only if the rational function Φ has degree $n + 1 - k$.

11. Let $\Phi = \frac{p}{q} \in \mathfrak{P}$, where $p,q \in \mathbb{R}[x]$ have no common zeros. Suppose that the support of the measure v in the representation (7.26) consists of n points.

    a. Show that $\deg(\Phi) = n$ and discuss the possible degrees of p and q.

    b. Express the polynomials p and q in terms of atoms and masses of v and of the constant b in (7.26).

12. Collect characterizations of N-extremal solutions among all solutions (denseness of $\mathbb{C}[x]$ in $L^2(\mathbb{R}, \mu)$, orthonormal basis $\{p_k : k \in \mathbb{N}_0\}$ of $L^2(\mathbb{R}, \mu)$, values of the Stieltjes transform $I_\mu(z)$ for $z \in \mathbb{C}_+$, Nevanlinna parametrization, order).

13. Let $\mu = \sum_{k=1}^{\infty} m_k \delta_{x_k}$ be an N-extremal solution of s, with $m_k > 0$ and $x_k \in \mathbb{R}$ pairwise distinct. (Proofs of the following results can be found in [BC1].)

    a. Show that the measure $v := \mu - \sum_{k=1}^{r} m_k \delta_{x_k}$ is determinate for each $r \in \mathbb{N}$.

    b. Let $m_0 > 0$ and $x_0 \in \mathbb{R}$. Suppose that $x_0 \neq x_k$ for all $k \in \mathbb{N}$. Show that the measure $v = m_0 \delta_{x_0} + \sum_{k=2}^{\infty} m_k \delta_{x_k}$ is an N-extremal solution of an indeterminate moment sequence.

7.9 Notes

Theorem 7.13 was proved in 1924 by R. Nevanlinna [Nv1]. Proposition 7.12 is due to H. Hamburger [Hm] and R. Nevanlinna [Nv1]. The two-parameter Nevanlinna functions and fractional transformations appeared in [BCa2]. It is difficult to determine explicit examples of Nevanlinna functions A, B, C, D. The first such examples were calculated in [ChiI], [IM], [BV], [CI].

The existence Theorem 7.20 on Nevanlinna–Pick interpolation is due to G. Pick [Pi] for finite sets $\mathcal{Z}$ and due to R. Nevanlinna [Nv2] for countable sets, see also [Ak, Theorem 3.3.3]. Operator-theoretic approaches to Nevanlinna–Pick interpolation are given in [SK] and [AM]. Characterizations of solutions of finite order such as in Theorem 7.27 were first obtained by H. Buchwalther and G. Cassier [BCa1]; further results are in [Sim1] and [Ge].

There are important results about growth properties of Nevanlinna functions. Already M. Riesz [Rz2] had shown (see e.g. [Ak, p. 101]) that the entire functions $f = A, B, C, D$ are of minimal exponential type, that is, for each $\varepsilon > 0$ there exists a $K_\varepsilon > 0$ such that $|f(z)| \leq K_\varepsilon e^{\varepsilon|z|}$ for $z \in \mathbb{C}$. C. Berg and H.L. Petersen [BP] proved that the four functions have the same order and type, called the order and type of the indeterminate moment sequence s. Further results are given in [BS2].

Chapter 8
The Operator-Theoretic Approach to the Stieltjes Moment Problem

This chapter is devoted to a detailed study of Stieltjes moment problems by using positive self-adjoint extensions of positive symmetric operators on Hilbert spaces.

In Sect. 8.2 we rederive the existence theorem for the Stieltjes moment problem by operator-theoretic methods (Theorem 8.2). Since the Jacobi operator T for a Stieltjes moment sequence s is positive, it has a largest positive self-adjoint extension on $\mathcal{H}_s$, the Friedrichs extension, and a smallest positive self-adjoint extension, the Krein extension. By the corresponding spectral measures this leads to two distinguished solutions μ_F and μ_K of the Stieltjes moment problem for s. In Sect. 8.3 we give an operator-theoretic characterization of Stieltjes determinacy by showing that the Stieltjes moment problem is determinate if and only if the Jacobi operator T has a unique *positive* self-adjoint extension on $\mathcal{H}_s$ (Theorem 8.7). The relationship between Hamburger determinacy and Stieltjes determinacy is discussed. In Sect. 8.4 we prove that for any other solution μ of the Stieltjes moment problem the Stieltjes transforms satisfy $I_{\mu_F}(x) \leq I_\mu(x) \leq I_{\mu_K}(x)$ for $x < 0$ (Theorem 8.18). Further, an approximation theorem for the Stieltjes transforms $I_{\mu_F}(x)$ and $I_{\mu_K}(x)$ is obtained (Theorem 8.16). Sections 8.5 and 8.6 develop the Nevanlinna parametrization of solutions (Theorem 8.24) and the Weyl circle description, respectively, for an indeterminate Stieltjes moment sequence.

8.1 Preliminaries on Quadratic Forms on Hilbert Spaces

In this short section we collect some facts on forms and positive self-adjoint operators that will be used in this chapter; all of them can be found in the book [Sm9].

Suppose that $\mathcal{H}$ is a Hilbert space. A *positive quadratic form* $\mathbf{s}$ on a linear subspace $\mathcal{D}[\mathbf{s}]$ of $\mathcal{H}$ is a mapping $\mathbf{s}[\cdot,\cdot] : \mathcal{D}[\mathbf{s}] \times \mathcal{D}[\mathbf{s}] \to \mathbb{C}$ which is linear in the first variable, antilinear in the second and satisfies $\mathbf{s}[\varphi, \varphi] \geq 0$ for $\varphi \in \mathcal{D}[\mathbf{s}]$.

© Springer International Publishing AG 2017   177
K. Schmüdgen, *The Moment Problem*, Graduate Texts in Mathematics 277,
DOI 10.1007/978-3-319-64546-9_8

Such a form s is called *closed* if for each sequence $(\varphi_n)_{n \in \mathbb{N}}$ from $\mathcal{D}[\mathbf{s}]$ such that $\lim_{n,k \to \infty} \mathbf{s}[\varphi_n - \varphi_k, \varphi_n - \varphi_k] = 0$ and $\lim_{n \to \infty} \varphi_n = \varphi$ in $\mathcal{H}$ for some $\varphi \in \mathcal{H}$ we have $\varphi \in \mathcal{D}[\mathbf{s}]$ and $\lim_{n \to \infty} \mathbf{s}[\varphi_n - \varphi, \varphi_n - \varphi] = 0$.

A symmetric operator T on $\mathcal{H}$ is said to be *positive* if $\langle T\varphi, \varphi \rangle \geq 0$ for $\varphi \in \mathcal{D}(T)$. For a positive symmetric operator T its *greatest lower bound* is the number

$$m(T) := \sup \{\lambda \in \mathbb{R} : \langle T\varphi, \varphi \rangle \geq \lambda \langle \varphi, \varphi \rangle \text{ for } \varphi \in \mathcal{D}(T)\}. \tag{8.1}$$

Let A be a positive self-adjoint operator. Then the spectral measure E_A is supported on $[m(A), +\infty) \subseteq \mathbb{R}_+$ and A has a unique positive square root $A^{1/2}$ given by $A^{1/2} = \int_0^\infty \lambda^{1/2} \, dE_A(\lambda)$. There exists a unique closed positive quadratic form $\mathbf{s}_A$:

$$\mathcal{D}[\mathbf{s}_A] = \mathcal{D}(A^{1/2}) \quad \text{and} \quad \mathbf{s}_A[\varphi, \psi] = \langle A^{1/2}\varphi, A^{1/2}\psi \rangle \text{ for } \varphi, \psi \in \mathcal{D}[\mathbf{s}_A].$$

Conversely, for each densely defined closed positive quadratic form $\mathbf{s}$ there exists a unique positive self-adjoint operator A such that $\mathbf{s} = \mathbf{s}_A$, see [Sm9, Theorem 10.17].

Let $\mathbf{s}_1, \mathbf{s}_2$ be positive quadratic forms on $\mathcal{H}$. We define $\mathbf{s}_1 \leq \mathbf{s}_2$ if $\mathcal{D}[\mathbf{s}_2] \subseteq \mathcal{D}[\mathbf{s}_1]$ and $\mathbf{s}_1[\varphi, \varphi] \leq \mathbf{s}_2[\varphi, \varphi]$ for all $\varphi \in \mathcal{D}[\mathbf{s}_2]$.

Let $\mathcal{G}_1$ and $\mathcal{G}_2$ be closed linear subspaces of $\mathcal{H}$ and let A_1 and A_2 be positive self-adjoint operators on $\mathcal{G}_1$ and $\mathcal{G}_2$, respectively. We write $A_1 \leq A_2$ if $\mathbf{s}_{A_1} \leq \mathbf{s}_{A_2}$, or equivalently, $\mathcal{D}(A_2^{1/2}) \subseteq \mathcal{D}(A_1^{1/2})$ and $\|A_1^{1/2}\varphi\| \leq \|A_2^{1/2}\varphi\|$ for $\varphi \in \mathcal{D}(A_2^{1/2})$.

Proposition 8.1 *Let A_1 and A_2 be as above. Then $A_1 \leq A_2$ if and only if*

$$(A_2 - \lambda I)^{-1} \leq (A_1 - \lambda I)^{-1}$$

for one (then for all) $\lambda < 0$. Here $(A_j - \lambda I)^{-1}$ denotes the operator of $\mathbf{B}(\mathcal{H})$ which is the inverse $(A_j - \lambda I)^{-1}$ on $\mathcal{G}_j$ and 0 on $\mathcal{G}_j^\perp$, $j = 1, 2$.

Proof [Sm9, Corollary 10.13] in the case $\mathcal{G}_1 = \mathcal{G}_2 = \mathcal{H}$. The general case is easily obtained by minor modifications.   □

Now suppose that T is a densely defined *positive* symmetric operator on $\mathcal{H}$. Then T always has a positive self-ajoint extension on $\mathcal{H}$. There is a largest positive self-adjoint extension, called the *Friedrichs extension* and denoted by T_F, and a smallest positive self-adjoint extension, called the *Krein extension* and denoted by T_K, with respect to the order relation "$\leq$" [Sm9, Corollary 13.15]. That is, if A is an arbitrary positive self-adjoint extension of T on $\mathcal{H}$, then we have

$$(T_F + \lambda I)^{-1} \leq (A + \lambda I)^{-1} \leq (T_K + \lambda I)^{-1} \quad \text{for } \lambda > 0.$$

The Friedrichs extension is defined as follows. It can be shown that the positive quadratic form $\mathbf{s}$ defined by $\mathbf{s}[\varphi, \psi] = \langle T\varphi, \psi \rangle$, $\varphi, \psi \in \mathcal{D}[\mathbf{s}] := \mathcal{D}(T)$, has a smallest closed extension $\bar{\mathbf{s}}$. Since $\bar{\mathbf{s}}$ is densely defined, closed, and positive, it is the quadratic form of a unique positive self-adjoint operator. This operator is the

Friedrichs extension T_F, see [Sm9, Theorem 10.17]. From this construction of T_F it follows that T and T_F have the same lower bounds, that is,

$$m_s := m(T) = m(T_F). \tag{8.2}$$

The Krein extension T_K can be nicely described in the case when $m(T) > 0$. Then we have (see [Sm9, formulas (14.67–68)])

$$\mathcal{D}(T_K) = \mathcal{D}(\overline{T}) \dotplus \mathcal{N}(T^*); \ T_K(\varphi + \zeta) = \overline{T}\varphi \ \text{for} \ \varphi \in \mathcal{D}(\overline{T}), \zeta \in \mathcal{N}(T^*). \tag{8.3}$$

If $\mathcal{N}(T^*) \neq \{0\}$, then 0 is an eigenvector of T_K. Note that T_F is invariant under translation, that is, $(T - \lambda I)_F = T_F - \lambda I$, but T_K is not in general.

8.2 Existence of Solutions of the Stieltjes Moment Problem

In this short section we apply Hilbert space operator theory to solve the Stieltjes moment problem. More precisely, we use the Friedrichs extension of a positive symmetric operator and the spectral theorem for self-adjoint operators [Sm9]. The following theorem is the counterpart of Theorem 6.1 for the Stieltjes moment problem.

Theorem 8.2 *Suppose that* $s = (s_n)_{n \in \mathbb{N}_0}$ *is a positive definite real sequence such that the sequence* $Es = (s_{n+1})_{n \in \mathbb{N}_0}$ *is positive semidefinite. Then the Stieltjes moment problem for s is solvable.*

If A is a positive self-adjoint extension of the symmetric operator X on a possibly larger Hilbert space $\mathcal{G}$ *(that is,* $\mathcal{H}_s \subseteq \mathcal{G}$ *and* $X \subseteq A$*) and* E_A *is the spectral measure of A, then* $\mu_A(\cdot) = \langle E_A(\cdot)1, 1 \rangle_{\mathcal{G}}$ *is a solution of the Stieltjes moment problem for s. Each solution of the Stieltjes moment problem for s is of this form.*

Proof The proof follows the lines of the proof of Theorem 6.1 and we explain only the necessary modifications.

Let $p(x) = \sum_{j=0}^{n} c_j x^j \in \mathbb{C}[x]$. Then $xp(x)\overline{p}(x) = \sum_{j,k=0}^{n} c_j \overline{c_k} x^{j+k+1}$. Therefore, since the sequence Es is positive semidefinite, we obtain

$$\langle Xp, p \rangle_s = L(xp\overline{p}) = \sum_{j,k=0}^{n} c_j \overline{c_k} s_{j+k+1} \geq 0. \tag{8.4}$$

This shows that the symmetric operator X is positive. The Friedrichs extension of the densely defined positive operator X is a positive self-adjoint extension. Hence X has at least one positive self-adjoint extension on $\mathcal{H}_s$.

For any positive self-adjoint extension A of X, the spectral measure E_A is supported on $[0, +\infty)$, so μ_A is a solution of the Stieltjes moment problem for s.

Conversely, if μ is a solution of the Stieltjes moment problem for s, the self-adjoint operator A_μ from Proposition 6.2 is a *positive* self-adjoint extension of X acting on the (possibly larger) Hilbert space $L^2([0, +\infty), \mu)$. $\qquad\square$

Recall that the Jacobi operator T on $l^2(\mathbb{N}_0)$ is unitarily equivalent to X. Thus Theorem 8.2 yields at once the following counterpart of Corollary 6.3.

Corollary 8.3 *If s is as in Theorem 8.2, then the solutions of the Stieltjes moment problem for s are precisely the measures of the form $\mu_B(\cdot) = s_0 \langle E_B(\cdot)e_0, e_0 \rangle_\mathcal{F}$, where B is a positive (!) self-adjoint extension of T on a possibly larger Hilbert space $\mathcal{F}$.*

Suppose that s is a positive definite Stieltjes moment sequence. Then, by (8.4), the symmetric operator $X \cong T$ on $\mathcal{H}_s \cong l^2(\mathbb{N}_0)$ is positive. Let m_s denote the greatest lower bound $m(T)$ of the operator T, see (8.1). From the extension theory of positive symmetric operators (see Sect. 8.1) it is known that T has a largest positive self-adjoint extension on $\mathcal{H}_s$, the *Friedrichs extension T_F*, and a smallest positive self-adjoint extension on $\mathcal{H}_s$, the *Krein extension T_K*. By (8.2), we have

$$m_s = m(T) = m(T_F) \geq 0. \tag{8.5}$$

By Corollary 8.3, the spectral measures E_{T_F} and E_{T_K} give rise to solutions μ_F and μ_K, respectively, of the Stieltjes moment problem for s.

Definition 8.4 $\mu_F(\cdot) := s_0 \langle E_{T_F}(\cdot)e_0, e_0 \rangle$ is the *Friedrichs solution* and $\mu_K(\cdot) := s_0 \langle E_{T_K}(\cdot)e_0, e_0 \rangle$ is the *Krein solution* of the Stieltjes moment problem for s.

These two distinguished solutions μ_F and μ_K will play a crucial role in this chapter. Both solutions come from self-adjoint extensions of T on the Hilbert space $\mathcal{H}_s \cong l^2(\mathbb{N}_0)$, so they are von Neumann solutions according to Definition 6.4.

8.3 Determinacy of the Stieltjes Moment Problem

Suppose that s is a Stieltjes moment sequence and μ is a solution of the Stieltjes moment problem for s. If μ is the only representing measure of s supported on $[0, +\infty)$, then we say that s, and likewise μ, is *determinate* or *Stieltjes determinate* if confusion can arise. But s may be indeterminate as a Hamburger moment sequence, that is, s may have different representing measures on $\mathbb{R}$ (see Example 8.11 below). Then the Stieltjes moment sequence s is called *Hamburger indeterminate*. In order to distinguish these cases unambiguously we will speak about Stieltjes determinacy and Hamburger determinacy of Stieltjes moment sequences in what follows. Obviously, if s is Hamburger determinate, it is also Stieltjes determinate.

If s is Hamburger indeterminate, then μ_F and μ_K are N-extremal solutions of the Hamburger moment problem for s according to Definition 7.8.

Let us begin with Hamburger indeterminate Stieltjes moment sequences.

Proposition 8.5 *Suppose that s is a Stieltjes moment sequence which is Hamburger indeterminate. Then the following are equivalent:*

(i) $X \cong T$ *has a unique positive self-adjoint extension on* $\mathcal{H}_s \cong l^2(\mathbb{N}_0)$.

(ii) 0 *is an eigenvalue of the Friedrichs extension T_F of T.*

(iii) $m_s = 0$.

Proof

(i)$\rightarrow$(ii) The operator T_∞ from Theorem 6.23 is a positive self-adjoint extension of T satisfying $T_\infty \mathfrak{p}_0 = T^* \mathfrak{p}_0 = 0$, that is, 0 is an eigenvalue of T_∞. Since T has a unique positive self-adjoint extension by (i), we have $T_\infty = T_F$. This proves (ii).

(ii)$\rightarrow$(iii) is trivial.

(iii)$\rightarrow$(i) Let A be an arbitrary positive self-adjoint extension of T on $l^2(\mathbb{N}_0)$. The Friedrichs extension T_F is the largest positive self-adjoint extension of T, so that $A \leq T_F$. Hence, by Proposition 8.1,

$$(T_F + I)^{-1} \leq (A + I)^{-1} \leq 1, \tag{8.6}$$

where the second inequality holds because A is positive. By (iii) and (8.5) we have $m_s = m(T_F) = 0$. Therefore, $\|(T_F + I)^{-1}\| = 1$ and hence $\|(A + I)^{-1}\| = 1$ by (8.6). Since $A \geq 0$ and A has a discrete spectrum by Theorem 7.7(i), it follows from the equality $\|(A + I)^{-1}\| = 1$ that 0 is an eigenvalue of A. But 0 is also in the spectrum of T_F, because $m(T_F) = 0$. Therefore, from Theorem 7.7(ii) it follows that $A = T_F$. This shows that T_F is the unique self-adjoint extension of T on $l^2(\mathbb{N}_0)$. $\qquad\square$

Corollary 8.6 *Let s be a Stieltjes moment sequence which is Hamburger indeterminate. Then the operator T_∞ from Theorem 6.23 is the Krein extension T_K of T and*

$$\mathcal{D}(T_K) = \mathcal{D}(T_\infty) = \mathcal{D}(\overline{T}) \dotplus \mathbb{C} \cdot \mathfrak{p}_0 = \mathcal{D}(\overline{T}) \dotplus \mathcal{N}(T^*), \tag{8.7}$$

$$T_K(\varphi + \lambda \mathfrak{p}_0) = \overline{T}\varphi \quad for \quad \varphi \in \mathcal{D}(\overline{T}), \; \lambda \in \mathbb{C}. \tag{8.8}$$

Proof Because s is Hamburger indeterminate, we have $\mathfrak{p}_0 \in l^2(\mathbb{N}_0)$ by Theorem 6.16 and hence $\mathcal{N}(T^*) = \mathbb{C} \cdot \mathfrak{p}_0$ by Proposition 6.6(i).

First let $m_s = 0$. Then, by Proposition 8.5, T has only one positive self-adjoint extension on $l^2(\mathbb{N}_0)$. Since T_K (by definition) and T_∞ (by Theorem 6.23) are such extensions, $T_\infty = T_K$ and (6.16) implies (8.7) and (8.8).

Now suppose that $m_s \neq 0$. Then $m_s > 0$ and hence $\mathcal{D}(T_K) = \mathcal{D}(\overline{T}) \dotplus \mathcal{N}(T^*)$ by (8.3). Therefore, since $\mathcal{N}(T^*) = \mathbb{C} \cdot \mathfrak{p}_0$, we obtain $\mathcal{D}(T_\infty) = \mathcal{D}(T_K)$. But T_∞ and T_K are restrictions of T^*, so that $T_\infty = T_K$ and (6.16) yields (8.7) and (8.8). $\qquad\square$

The main result in this section is the following operator-theoretic characterization of Stieltjes determinacy. It is the counterpart of the corresponding result (Theorem 6.10) for the Hamburger moment problem.

Theorem 8.7 *Suppose that s is a positive definite Stieltjes moment sequence. Then s is Stieltjes determinate if and only if the symmetric operator X, or equivalently the Jacobi operator T, has a unique positive self-adjoint extension on the Hilbert space $\mathcal{H}_s \cong l^2(\mathbb{N}_0)$.*

Proof First we assume that X has two different positive self-adjoint extensions, say A and B, on $\mathcal{H}_s$. We repeat the reasoning from the proof of Theorem 6.10. Then $\mu_A(\cdot) = \langle E_A(\cdot)1, 1\rangle$ and $\mu_B(\cdot) = \langle E_B(\cdot)1, 1\rangle$ are representing measures for s. They are supported on $[0, +\infty)$, because A and B are positive. If μ_A were equal to μ_B, then we would have $\langle (A-zI)^{-1}1, 1\rangle = \langle (B-zI)^{-1}1, 1\rangle$ for $z \in \mathbb{C}\backslash\mathbb{R}$ by the functional calculus of self-adjoint operators. This contradicts Lemma 6.8. Hence $\mu_A \neq \mu_B$, so s is Stieltjes indeterminate.

Now we assume that X has a unique positive self-adjoint extension on $\mathcal{H}_s$. If s is Hamburger determinate, it is Stieltjes determinate and we are finished. Suppose now that s is Hamburger indeterminate. Then μ_F is N-extremal and 0 is an eigenvalue of T_F by Proposition 8.5. Since the multiplication operator X on $\mathcal{H}_s$ and the Jacobi operator T on $l^2(\mathbb{N}_0)$ are unitarily equivalent, so are their Friedrichs extensions X_F and T_F, and we have $\mu_F(\cdot) = \langle E_{X_F}(\cdot)1, 1\rangle_s$. Then 0 is an eigenvalue of X_F. Let $f \in \mathcal{H}_s$ be a corresponding unit eigenvector. From the definition of the Friedrichs extension it follows that there exists a sequence $(f_n)_{n\in\mathbb{N}}$ from $\mathcal{D}(X) = \mathbb{C}[x]$ such that $\lim_n f_n = f$ in $\mathcal{H}_s \cong L^2(\mathbb{R}_+, \mu_F)$ and $\lim_n \langle Xf_n, f_n\rangle_s = \langle X_F f, f\rangle_s = 0$.

Let μ be an arbitrary solution of the Stieltjes moment problem for s. Since we have $f_n \to f$ in $\mathcal{H}_s$, $(f_n)_{n\in\mathbb{N}}$ is a Cauchy sequence in $(\mathbb{C}[x], \|\cdot\|_s)$ and so in $L^2(\mathbb{R}_+, \mu)$ by Proposition 6.2. Hence $f_n \to g$ in $L^2(\mathbb{R}_+, \mu)$ for some $g \in L^2(\mathbb{R}_+, \mu)$. Clearly, $\|g\|_{L^2(\mathbb{R}_+,\mu)} = 1$, since $\|f_n\|_{L^2(\mathbb{R}_+,\mu)} = \|f_n\|_s \to \|f\|_s = 1$. Then

$$\int_0^\infty |\sqrt{x}f_n|^2 \, d\mu = \int_0^\infty xf_n\overline{f_n} \, d\mu = L_s(xf_n\overline{f_n}) = \langle Xf_n, f_n\rangle_s \to \langle X_F f, f\rangle_s = 0.$$

Therefore, for each function $\varphi \in C_c(\mathbb{R}_+; \mathbb{R})$ we obtain

$$\int_0^\infty \sqrt{x}f_n \, \varphi \, d\mu = \int_0^\infty f_n \sqrt{x} \, \varphi \, d\mu \to 0 = \int_0^\infty g \sqrt{x} \, \varphi \, d\mu.$$

This implies that $g(x) = 0$ μ-a.e. on $(0, +\infty)$. Since $f_n \to f$ in $L^2(\mathbb{R}_+, \mu_F)$, we have in particular $f(x) = 0$ μ_F-a.e. on $(0, +\infty)$. (This also follows from the fact that $X_F f = 0$.) Thus, since $g \in L^2(\mathbb{R}_+, \mu)$ and $f \in L^2(\mathbb{R}_+, \mu_F)$ are unit vectors, we get

$$\mu(\{0\})|g(0)|^2 = \mu_F(\{0\})|f(0)|^2 = 1. \tag{8.9}$$

Further, we have

$$\int_0^\infty f_n \, d\mu = L_s(f_n) = \int_0^\infty f_n \, d\mu_F \to \int_0^\infty g \, d\mu = \int_0^\infty f \, d\mu_F.$$

Since $g(x) = 0$ μ-a.e. and $f(x) = 0$ μ_F-a.e. on $(0, +\infty)$, the latter equality yields

$$\mu(\{0\})g(0) = \mu_F(\{0\})f(0). \tag{8.10}$$

Combining (8.9) and (8.10) we obtain $\mu(\{0\}) = \mu_F(\{0\}) > 0$. Because μ_F is a von Neumann solution of the indeterminate Hamburger moment sequence s, it follows from Corollary 7.17 that $\mu = \mu_F$. This proves that s is Stieltjes determinate. $\square$

We close this section by deriving three useful corollaries. An immediate consequence of Theorem 8.7 and Proposition 8.5 is the following.

Corollary 8.8 *Let s be a Stieltjes moment sequence which is Hamburger indeterminate. The following are equivalent:*

(i) *s is Stieltjes determinate.*
(ii) *0 is an eigenvalue of the Friedrichs extension T_F of T.*
(iii) *$m_s = 0$.*

Corollary 8.9 *Let s be a determinate Stieltjes moment sequence with representing measure μ. If $\mu(\{0\}) = 0$, then s is Hamburger determinate.*

Proof Since s is Stieltjes determinate, $\mu = \mu_F$. The multiplication operator A_μ by the variable x on $L^2([0, +\infty), \mu)$ and the Friedrichs extension T_F are positive self-adjoint extensions of $X \cong T$ on $\mathcal{H}_s \cong L^2([0, +\infty), \mu) = L^2([0, +\infty), \mu_F)$. Hence $A_\mu = T_F$ by Theorem 8.7. Since $\mu(\{0\}) = 0$, 0 is not an eigenvalue of $A_\mu = T_F$. Hence s cannot be Hamburger indeterminate by Corollary 8.8. $\square$

Corollary 8.10 *Suppose that s is an indeterminate Stieltjes moment sequence. Then $m(T_F) > 0$, $\operatorname{supp}\mu_F \subseteq [m(T_F), +\infty)$, and 0 is in the resolvent set of the Friedrichs extension T_F, that is, $(T_F)^{-1} \in B(\mathcal{H}_s)$.*

Proof Since s is Stieltjes indeterminate, it is Hamburger indeterminate and T has at least two different positive self-adjoint extensions on $l^2(\mathbb{N}_0)$ by Theorem 8.7. Therefore, $m_s = m(T_F) > 0$ by Proposition 8.5 and (8.5). From the theory of self-adjoint operators it follows that the spectrum of T_F, hence the support of the measure μ_F, is contained in $[m(T_F), +\infty)$, so that 0 is in the resolvent set of T_F. $\square$

Example 8.11 (A determinate Stieltjes moment sequence that is Hamburger indeterminate) Let s be an indeterminate Stieltjes moment sequence. Then $m_s = m(T_F) > 0$ by Corollary 8.10. Let $\tilde{s} = (\tilde{s}_n)_{n \in \mathbb{N}_0}$ denote the shifted sequence of s by $-m_s$, that is, $\tilde{s}_n = \sum_{k=0}^n \binom{n}{k}(-m_s)^k s_{n-k}$ for $n \in \mathbb{N}_0$, see Exercise 6.5. Then $\tilde{s}$ is Hamburger indeterminate (because s is Hamburger indeterminate) and $m_{\tilde{s}} = 0$. By Corollary 8.8, $m_{\tilde{s}} = 0$ implies that $\tilde{s}$ is Stieltjes determinate. $\circ$

8.4　Friedrichs and Krein Approximants

In this section we suppose that s is a **positive definite Stieltjes moment sequence**. We develop two sequences of matrices and two related sequences of quotients of polynomials as approximants for the Friedrichs and Krein extensions and as useful tools in the proofs of our main results (Theorems 8.16 and 8.18).

The truncated Jacobi matrix (6.21) is called the *Friedrichs approximant* and denoted by $A_F^{[n]}$, that is, we set

$$A_F^{[n]} := J_n = \begin{pmatrix} b_0 & a_0 & 0 & \dots 0 & 0 & 0 \\ a_0 & b_1 & a_1 & \dots 0 & 0 & 0 \\ 0 & a_1 & b_2 & \dots 0 & 0 & 0 \\ \dots\dots\dots\dots & \dots\dots & \dots \\ 0 & 0 & 0 & \dots a_{n-3} & b_{n-2} & a_{n-2} \\ 0 & 0 & 0 & \dots 0 & a_{n-2} & b_{n-1} \end{pmatrix}, \quad n \in \mathbb{N}. \tag{8.11}$$

The *Krein approximant* $A_K^{[n]}$ is defined by

$$A_K^{[n]} = \begin{pmatrix} b_0 & a_0 & 0 & \dots 0 & 0 \\ a_0 & b_1 & a_1 & \dots 0 & 0 \\ 0 & a_1 & b_2 & \dots 0 & 0 \\ \dots\dots\dots\dots\dots \\ 0 & 0 & 0 & \dots b_{n-2} & a_{n-2} \\ 0 & 0 & 0 & \dots a_{n-2} & b_{n-1} - \alpha_{n-1} \end{pmatrix}, \quad n \in \mathbb{N}, \tag{8.12}$$

where α_{n-1} is chosen according to the following lemma.

Lemma 8.12 *There is a unique positive number α_{n-1} such that $A_K^{[n]}$ has the eigenvalue zero. A corresponding eigenvector is $(p_0(0), \dots, p_{n-1}(0))$. The matrix $A_K^{[n]}$ is positive semidefinite, $p_{n-1}(0) \neq 0$, and we have*

$$(b_{n-1} - \alpha_{n-1})p_{n-1}(0) + a_{n-2}p_{n-2}(0) = 0, \tag{8.13}$$

$$\alpha_{n-1} = -a_{n-1}\frac{p_n(0)}{p_{n-1}(0)}, \tag{8.14}$$

$$(b_n - \alpha_n)\alpha_{n-1} - a_{n-1}^2 = 0. \tag{8.15}$$

Proof The assertion is obvious for $n = 1$, so we assume that $n \geq 2$. Fix α_{n-1} and $y = (y_0, \dots, y_{n-1}) \in \mathbb{C}^n$ with $y_0 := p_0(0)$. Let us consider the equation $A_K^{[n]}y = 0$. For the first $n-1$ components of y this is equivalent to the recurrence relations (5.9) for $x = 0$ with the same intial data. Therefore, $y_k = p_k(0)$ for $k = 0, \dots, n-1$. For the n-th component this is precisely equation (8.13).

The sequence s is positive definite, hence is $A_F^{[n]}$. Therefore, $p_{n-1}(0) \neq 0$, because otherwise $A_F^{[n]}$ would have the eigenvalue zero. Since $p_{n-1}(0) \neq 0$, (8.13) has a unique solution α_{n-1}. Let C_n denote the matrix with entry one at the right lower corner and zero otherwise. Then $A_F^{[n]} = A_K^{[n]} + \alpha_{n-1} C_n$ by definition. If $\alpha_{n-1} \leq 0$, then $0 \leq A_F^{[n]} \leq A_K^{[n]}$ and hence $A_F^{[n]}$ would not be positive definite. Hence $\alpha_{n-1} > 0$.

The recurrence relation (5.9) gives $a_{n-1}p_n(0) + b_{n-1}p_{n-1}(0) + a_{n-2}p_{n-2}(0) = 0$. Combined with (8.13) this yields (8.14). Replacing now $n - 1$ by n in Eq. (8.13) and comparing this with (8.14) we obtain (8.15).   □

Lemma 8.13 *For $n \in \mathbb{N}_0$ we define*

$$M_n(z) = P_n(z) - \frac{P_n(0)}{P_{n-1}(0)} P_{n-1}(z), \ N_n(z) = Q_n(z) - \frac{P_n(0)}{P_{n-1}(0)} Q_{n-1}(z).$$

Then

$$\langle (A_K^{[n]} - zI)^{-1} e_0, e_0 \rangle = -\frac{N_n(z)}{M_n(z)}, \quad z \in \rho(A_K^{[n]}). \tag{8.16}$$

Proof In this proof we use Lemmas 6.27 and 6.28. Let $B_K^{[n]}$ be the matrix obtained from $A_K^{[n]}$ by removing the first row and the first column. By developing the determinant and using Lemma 6.27(i) we get

$$\det(zI - A_K^{[n]}) = \det(zI - A_F^{[n]}) - \alpha_{n-1} \det(zI - A_F^{[n-1]}) = P_n(z) - \alpha_{n-1}P_{n-1}(z).$$

Since the matrix $A_K^{[n]}$ has the eigenvalue 0 by Lemma 8.12, we have $\det A_K^{[n]} = 0$. Hence $P_n(0) = \alpha_{n-1}P_{n-1}(0)$ and

$$\det(zI - A_K^{[n]}) = P_n(z) - P_n(0)P_{n-1}(0)^{-1}P_{n-1}(z) = M_n(z). \tag{8.17}$$

Similarly, applying Lemma 6.28(i) we derive

$$\det(zI - B_K^{[n]}) = \det(zI - B_F^{[n]}) - \alpha_{n-1} \det(zI - B_F^{[n-1]})$$
$$= Q_n(z) - \alpha_{n-1}Q_{n-1}(z) = Q_{n+1}(z) - P_n(0)P_{n-1}(0)^{-1} Q_{n-1}(z) = N_n(z).$$

As in the proof of Lemma 6.28 we use Cramer's rule and obtain

$$\langle (A_K^{[n]} - zI)^{-1} e_0, e_0 \rangle = \frac{\det(B_K^{[n]} - zI)}{\det(A_K^{[n]} - zI)} = -\frac{\det(zI - B_K^{[n]})}{\det(zI - A_K^{[n]})} = -\frac{N_n(z)}{M_n(z)}$$

for z in the resolvent set $\rho(A_K^{[n]})$. This completes the proof of Lemma 8.13.   □

Lemma 8.14 *For $x < m_s$ and $n \in \mathbb{N}_0$, we have*

$$\langle (A_F^{[n]} - xI)^{-1} e_0, e_0 \rangle \leq \langle (A_F^{[n+1]} - xI)^{-1} e_0, e_0 \rangle, \tag{8.18}$$

$$\lim_{n \to +\infty} \langle (A_F^{[n]} - xI)^{-1} e_0, e_0 \rangle = \langle (T_F - xI)^{-1} e_0, e_0 \rangle. \tag{8.19}$$

Proof In this proof we use some facts on forms and self-adjoint operators, see Sect. 8.1. Let s_n denote the positive quadratic form defined by

$$s_n(f,g) = \langle A_F^{[n]} f, g \rangle, \quad f,g \in \mathcal{D}[s_n] := \{(f_0, \ldots, f_{n-1}, 0, 0, \ldots) : f_j \in \mathbb{C}\}.$$

Then $\mathcal{D}[s_n] \subseteq \mathcal{D}[s_{n+1}]$ and $s_n(f,f) = s_{n+1}(f,f)$ for $f \in \mathcal{D}[s_n]$. By the definition of the order relation of forms this means that $s_{n+1} \leq s_n$. Therefore, by Proposition 8.1,

$$(A_F^{[n]} - xI)^{-1} \leq (A_F^{[n+1]} - xI)^{-1}. \tag{8.20}$$

(By the convention in Proposition 8.1 the resolvents are defined to be 0 on the orthogonal complements of $\mathcal{D}[s_n]$ and $\mathcal{D}[s_{n+1}]$ in $l^2(\mathbb{N}_0)$.) Clearly, (8.20) implies (8.18).

Since $m_s > x$, the sequence $((A_F^{[n]} - xI)^{-1})_{n \in \mathbb{N}}$ of bounded positive self-adjoint operators on $l^2(\mathbb{N}_0)$ is monotonically increasing by (8.20) and bounded from above by $(m_s - x)^{-1} I$ (since $A_F^{[n]} \geq m_s I$). Hence it converges strongly to a bounded positive self-adjoint operator S such that $S \leq (m_s - x)^{-1} I$.

Let $f \in \mathcal{N}(S)$. Then $0 = \langle Sf, f \rangle \geq \langle (A_F^{[n]} - xI)^{-1} f, f \rangle \geq 0$. Since $A_F^{[n]} \geq m_s I$, this implies $(A_F^{[n]} - xI)^{-1} f = 0$ for all $n \in \mathbb{N}$. Hence $f \in \cap_n \mathcal{D}[s_n]^\perp = \{0\}$, so that $f = 0$. Thus, S has a trivial kernel. Therefore, since $S \leq (m_s - x)^{-1} I$, it follows that $A := S^{-1} + xI$ is a positive self-adjoint operator on $l^2(\mathbb{N}_0)$ and

$$(A_F^{[n]} - xI)^{-1} \leq S = (A - xI)^{-1}, \quad n \in \mathbb{N}. \tag{8.21}$$

We prove that $A = T_F$. Let s_A denote the positive quadratic form associated with A. By definition the Friedrichs extension T_F of T is the positive self-adjoint operator associated with the closure $\bar{s}_\infty$ of the quadratic form defined by

$$s_\infty(f,g) = \langle Tf, g \rangle, \quad f, g \in \mathcal{D}[s_\infty] = \text{d}.$$

By (8.21) and Proposition 8.1, $s_A \leq s_n$. Therefore, $\mathcal{D}[s_n] \subseteq \mathcal{D}[s_A]$ for all $n \in \mathbb{N}$ and hence $\mathcal{D}[s_\infty] = \cup_n \mathcal{D}[s_n] \subseteq \mathcal{D}[s_A]$. Further, $s_n(f,f) = s_\infty(f,f)$ for $f \in \mathcal{D}[s_n]$ and

$$s_A(f,f) \leq \lim_n s_n(f,f) = s_\infty(f,f) \quad \text{for } f \in \mathcal{D}[s_\infty].$$

The preceding facts show that $s_A \leq s_\infty \leq s_n$. Since s_n is closable, $s_A \leq \bar{s}_\infty \leq s_n$. Applying again Proposition 8.1 we conclude that

$$(A_F^{[n]} - xI)^{-1} \leq (T_F - xI)^{-1} \leq (A - xI)^{-1} = S.$$

But S is the strong limit of the sequence $((A_F^{[n]} - xI)^{-1})_{n \in \mathbb{N}}$. Hence the latter implies that $(T_F - xI)^{-1} = (A - xI)^{-1} = S$. Therefore, $T_F = A$. Thus, for $f \in l^2(\mathbb{N}_0)$,

$$(T_F - xI)^{-1}f = (A - xI)^{-1}f = Sf = \lim_n (A_F^{[n]} - xI)^{-1}f,$$

which in turn yields (8.19). □

Now we treat the Krein approximants. We extend $A_K^{[n]}$ to a positive self-adjoint finite rank operator $T_K^{[n]}$ on $l^2(\mathbb{N}_0)$ by filling up the matrix with zeros, that is,

$$T_K^{[n]} := \begin{pmatrix} A_K^{[n]} & 0 \\ 0 & 0 \end{pmatrix}.$$

Then it is obvious that

$$\langle (T_K^{[n]} - zI)^{-1}e_0, e_0 \rangle = \langle (A_K^{[n]} - zI)^{-1}e_0, e_0 \rangle, \quad z \in \rho(T_K^{[n]}) = \rho(A_K^{[n]}). \tag{8.22}$$

The following Eq. (8.24) says that the self-adjoint operator T_K is the strong resolvent limit of the sequence $(T_K^{[n]})_{n \in \mathbb{N}}$.

Lemma 8.15 For $x < 0$, $n \in \mathbb{N}_0$, and $f \in l^2(\mathbb{N}_0)$,

$$\langle (T_K^{[n+1]} - xI)^{-1}e_0, e_0 \rangle \leq \langle (T_K^{[n]} - xI)^{-1}e_0, e_0 \rangle, \tag{8.23}$$

$$\lim_{n \to +\infty} (T_K^{[n]} - xI)^{-1}f = (T_K - xI)^{-1}f. \tag{8.24}$$

Proof The nonzero part of the matrix $T_K^{[n+1]} - T_K^{[n]}$ is the block matrix

$$\mathcal{D}_n = \begin{pmatrix} \alpha_{n-1} & a_{n-1} \\ a_{n-1} & b_n - \alpha_n \end{pmatrix}.$$

By (8.15), $\det \mathcal{D}_n = \alpha_{n-1}(b_n - \alpha_n) - a_{n-1}^2 = 0$. In particular, $\alpha_{n-1}(b_n - \alpha_n) \geq 0$. Hence $b_n - \alpha_n \geq 0$, since $\alpha_{n-1} > 0$ by Lemma 8.12, and $\text{Tr}\,\mathcal{D}_n = b_n - \alpha_n + a_{n-1} \geq 0$, since $a_{n-1} > 0$. Since $\det \mathcal{D}_n = 0$ and $\text{Tr}\,\mathcal{D}_n \geq 0$, it follows that $\mathcal{D}_n \geq 0$ and hence $T_K^{[n+1]} - T_K^{[n]} \geq 0$. Therefore $(T_K^{[n+1]} - xI)^{-1} \leq (T_K^{[n]} - xI)^{-1}$, which implies (8.23).

Now we prove (8.24). Since $\|(T_K^{[n]} - xI)^{-1}\| \leq |x|^{-1}$ for all $n \in \mathbb{N}$, it is easily shown that the set of $f \in l^2(\mathbb{N}_0)$ for which (8.24) holds is closed. Further, (8.24) is valid for $f \in (T - xI)\mathcal{D}(T)$. Indeed, then $T_K^{[n]}f = Tf = T_Kf$ for some $n \in \mathbb{N}$ and hence $(T_K^{[n]} - xI)^{-1}f = (T_K - xI)^{-1}f$.

First assume that s is Hamburger determinate. Then T is essentially self-adjoint by Theorem 6.10. Hence $\overline{T} = T_K$. Since $T \geq 0$ and $x < 0$, $(T - xI)\mathcal{D}(T)$ is dense in $l^2(\mathbb{N}_0)$ by Proposition A.42(iv). Thus (8.24) holds on $l^2(\mathbb{N}_0)$ as noted above.

Now suppose s is Hamburger indeterminate. Then we have $\mathcal{D}(T_K) = \mathcal{D}(\overline{T}) + \mathbb{C} \cdot \mathfrak{p}_0$ and $T_K \mathfrak{p}_0 = 0$ by (8.7). From $(T_K - xI)\mathfrak{p}_0 = -x\mathfrak{p}_0$ we get $(T_K - xI)^{-1}\mathfrak{p}_0 = -x^{-1}\mathfrak{p}_0$. Put $\mathfrak{p}_0^{[n]} := (p_0(0), \ldots, p_n(0), 0, \ldots)$. Since $T_K^{[n]}\mathfrak{p}_0^{[n]} = 0$ by Lemma 8.12, we have $(T_K^{[n]} - xI)^{-1}\mathfrak{p}_0^{[n]} = -x^{-1}\mathfrak{p}_0^{[n]}$. Then, as $n \to \infty$,

$$\|(T_K^{[n]} - xI)^{-1}\mathfrak{p}_0 - (T_K - xI)^{-1}\mathfrak{p}_0\| = \|(T_K^{[n]} - xI)^{-1}\mathfrak{p}_0 - x^{-1}\mathfrak{p}_0\|$$

$$= \|(T_K^{[n]} - xI)^{-1}(\mathfrak{p}_0 - \mathfrak{p}_0^{[n]}) + x^{-1}(\mathfrak{p}_0^{[n]} - \mathfrak{p}_0)\| \leq 2|x|^{-1}\|(\mathfrak{p}_0^{[n]} - \mathfrak{p}_0)\| \to 0.$$

This proves (8.24) for $f = \mathfrak{p}_0$. Because $T_K \geq 0$ is self-adjoint and $x < 0$, it follows from Proposition A.42(iv) and (8.7) that

$$(T_K - xI)\mathcal{D}(T_K) = (\overline{T} - xI)\mathcal{D}(\overline{T}) + \mathbb{C} \cdot \mathfrak{p}_0 = \overline{(T - xI)\mathcal{D}(T)} + \mathbb{C} \cdot \mathfrak{p}_0 = l^2(\mathbb{N}_0).$$

Therefore, since (8.24) is valid for $f \in (T - xI)\mathcal{D}(T)$ and $f = \mathfrak{p}_0$ as shown above, (8.24) holds for all $f \in l^2(\mathbb{N}_0)$. $\square$

The zeros of $P_n(x)$ are contained in $[m_s, +\infty)$ by Proposition 5.28 (or by Lemma 6.27(i)). Hence $P_n(x) \neq 0$ for $x < m_s$. Moreover, $M_n(x) \neq 0$ for $x < 0$ by (8.17).

Putting the preceding together we obtain our main approximation result.

Theorem 8.16 *Suppose that s is a positive definite Stieltjes moment sequence. For any $x < m_s$ the sequence $\left(-\frac{Q_n(x)}{P_n(x)}\right)_{n \in \mathbb{N}_0}$ is bounded increasing and*

$$-s_0 \lim_{n \to \infty} \frac{Q_n(x)}{P_n(x)} = \int_0^\infty \frac{d\mu_F(y)}{y - x}. \tag{8.25}$$

For $x < 0$ the sequence $\left(-\frac{N_n(x)}{M_n(x)}\right)_{n \in \mathbb{N}_0}$ is bounded decreasing and

$$-s_0 \lim_{n \to \infty} \frac{N_n(x)}{M_n(x)} = \int_0^\infty \frac{d\mu_K(y)}{y - x}. \tag{8.26}$$

Proof Recall that $A_F^{[n]}$ acting as an operator on $\mathbb{C}^n$ is just the operator J_n in (6.24). Combining (6.24), (8.18), and (8.19) it follows that $\left(-\frac{Q_n(x)}{P_n(x)}\right)_{n \in \mathbb{N}_0}$ is a bounded increasing sequence converging to $\langle (T_F - xI)^{-1}e_0, e_0 \rangle$. Since $\mu_F(\cdot) = s_0 \langle E_{T_F}(\cdot)e_0, e_0 \rangle$ by Definition 8.4, we have $s_0 \langle (T_F - xI)^{-1}e_0, e_0 \rangle = \int_0^\infty \frac{d\mu_F(y)}{y - x}$ by the functional calculus, whence (8.25) follows.

Similarly, we conclude from (8.16), (8.22), (8.23), and (8.24) that the sequence $\left(-\frac{N_n(x)}{M_n(x)}\right)$ is bounded, decreasing, and that it converges to $\langle (T_K - xI)^{-1}e_0, e_0 \rangle$. Combined with $s_0 \langle (T_K - xI)^{-1}e_0, e_0 \rangle = \int_0^\infty \frac{d\mu_K(y)}{y - x}$ this yields (8.26). $\square$

Remark 8.17 The assertion of Theorem 8.16 also holds for a *determinate* positive definite Stieltjes moment sequence s. In this case $\mu := \mu_K = \mu_F$ is the unique solution of the Stieltjes moment problem for s and it follows from Theorem 8.16 that $\left(-s_0 \frac{Q_n(x)}{P_n(x)}\right)_{n\in\mathbb{N}_0}$ and $\left(-s_0 \frac{N_n(x)}{M_n(x)}\right)_{n\in\mathbb{N}_0}$ are monotone sequences converging from below resp. from above to the Stieltjes transform $I_\mu(x) = \int_0^\infty (y-x)^{-1} d\mu(y)$ of the representing measure μ for $x < 0$.   ○

Our second main result in this section is the following theorem.

Theorem 8.18 *Suppose that s is an indeterminate Stieltjes moment sequence. If μ is an arbitrary solution of the Stieltjes moment problem for s, then*

$$\int_0^\infty \frac{d\mu_F(y)}{y-x} \leq \int_0^\infty \frac{d\mu(y)}{y-x} \leq \int_0^\infty \frac{d\mu_K(y)}{y-x} \quad \text{for } x < 0, \tag{8.27}$$

$$\int_0^\infty \frac{d\mu(y)}{y-x} < \int_0^\infty \frac{d\mu_K(y)}{y-x} \quad \text{if } \mu \neq \mu_K, \, x < 0, \tag{8.28}$$

$$\int_0^\infty \frac{d\mu_F(y)}{y-x} < \int_0^\infty \frac{d\mu(y)}{y-x} \quad \text{if } \mu \neq \mu_F, \, x \leq 0. \tag{8.29}$$

Remark 8.19 It should be emphasized that both inequalities (8.28) and (8.29) are strict and that (8.29) also holds for $x = 0$. In this case the integral $\int_0^\infty y^{-1} d\mu(y)$ in (8.29) can be infinite, while $\int_0^\infty y^{-1} d\mu_F(y)$ is always finite, since $\operatorname{supp} \mu_F \subseteq [m(T_F), +\infty)$ and $m(T_F) > 0$ by Corollary 8.10.   ○

Proof The proofs of the two strict inequalities (8.28) and (8.29) will be given at the end of the proof of Theorem 8.24 below. Here we only prove the inequalities (8.27).

Since an indeterminate Stieltjes moment sequence is obviously positive definite, Theorem 8.16 applies and by (8.25) and (8.26) it suffices to show that

$$-s_0 \frac{Q_n(x)}{P_n(x)} \leq \int_0^\infty \frac{d\mu(y)}{y-x} \leq -s_0 \frac{N_n(x)}{M_n(x)} \quad \text{for } x < 0. \tag{8.30}$$

Recall that $\mathcal{H}_s$ is a subspace of $L^2(\mathbb{R}, \mu)$ by Proposition 6.2. Now we derive

$$0 \leq \int_0^\infty \frac{P_n(y)^2}{y-x} d\mu(y)$$

$$= \int_0^\infty P_n(y) \frac{P_n(y)-P_n(x)}{y-x} d\mu(y) + P_n(x) \int_0^\infty \frac{P_n(y)}{y-x} d\mu(y)$$

$$\overset{(1)}{=} P_n(x) \int_0^\infty \frac{P_n(y)}{y-x} d\mu(y)$$

$$= P_n(x) \int_0^\infty \frac{P_n(y)-P_n(x)}{y-x} d\mu(y) + P_n(x)^2 \int_0^\infty \frac{d\mu(y)}{y-x}$$

$$\overset{(2)}{=} P_n(x) L_{s,y}\left(\frac{P_n(y) - P_n(x)}{y - x}\right) + P_n(x)^2 I_\mu(x)$$

$$\overset{(3)}{=} P_n(x) s_0 Q_n(x) + P_n(x)^2 I_\mu(x).$$

Dividing this inequality by $P_n(x)^2$ leads to the left inequality of (8.30).

Let us explain why the three equalities (1)–(3) of the preceding derivation hold. Since $(P_n(y) - P_n(x))/(y - x)$ is a polynomial in y of degree less than $\deg P_n = n$, it is orthogonal to $P_n(y)$ in $\mathcal{H}_s$ and so in $L^2(\mathbb{R}_+, \mu)$. Thus,

$$\int_0^\infty P_n(y) \frac{P_n(y) - P_n(x)}{y - x} \, d\mu(y) = 0.$$

This has been used in (1). Because μ is a solution of the Stieltjes moment problem for s, equality (2) holds. Finally, (3) follows from (5.34).

Now we turn to the right inequality of (8.30). Since $M_n(0) = 0$ by the definition of M_n, $M_n(y)y^{-1}$ is a polynomial of degree $n - 1$ in y. Hence

$$\frac{M_n(y)y^{-1} - M_n(x)x^{-1}}{y - x}$$

is a polynomial of degree at most $n - 2$ in y. Therefore, it is orthogonal to $P_{n-1}(y)$ and $P_n(y)$ and hence to $M_n(y)$ in $\mathcal{H}_s$ and so in $L^2(\mathbb{R}, \mu)$. This gives the equality (4) below. Further, from the definitions of N_n and M_n (see Lemma 8.13) and the equation $s_0 Q_n(x) = L_{s,y}\left(\frac{P_n(y) - P_n(x)}{y - x}\right)$ by (5.34) it follows that

$$\int_0^\infty \frac{M_n(y) - M_n(x)}{y - x} \, d\mu(y) = L_{s,y}\left(\frac{M_n(y) - M_n(x)}{y - x}\right) = s_0 N_n(x).$$

This relation is inserted in equality (5) below. Using the preceding facts we derive

$$0 \le \int_0^\infty \frac{M_n(y)^2}{(y - x)y} \, d\mu(y)$$

$$= \int_0^\infty M_n(y) \frac{M_n(y)y^{-1} - M_n(x)x^{-1}}{y - x} \, d\mu(y) + M_n(x)x^{-1} \int_0^\infty \frac{M_n(y)}{y - x} \, d\mu(y)$$

$$\overset{(4)}{=} M_n(x)x^{-1} \int_0^\infty \frac{M_n(y)}{y - x} \, d\mu(y)$$

$$= M_n(x)x^{-1} \int_0^\infty \frac{M_n(y) - M_n(x)}{y - x} \, d\mu(y) + M_n(x)^2 x^{-1} \int_0^\infty \frac{d\mu(y)}{y - x}$$

$$\overset{(5)}{=} M_n(x)x^{-1} s_0 N_n(x) + M_n(x)^2 x^{-1} I_\mu(y).$$

Dividing now by $M_n(x)^2 x^{-1} < 0$ yields the right inequality of (8.30).   □

By Theorem 8.16, the sequence $\left(-\frac{Q_n(x)}{P_n(x)}\right)_{n\in\mathbb{N}_0}$ is monotonically increasing and the sequence $\left(x\frac{N_n(x)}{M_n(x)}\right)_{n\in\mathbb{N}_0}$ is monotonically decreasing for $x < 0$. Letting $x \to -0$ it follows that $\left(-\frac{Q_n(0)}{P_n(0)}\right)$ is also monotonically increasing and $\left(\frac{N_n(0)}{M_n'(0)}\right)$ is monotonically decreasing. (Note that $P_n(0) \neq 0$ by Lemma 8.12 and $\lim_{x\to-0}\frac{xN_n(x)}{M_n(x)} = \frac{N_n(0)}{M_n'(0)}$, since $M_n(0) = 0$.) Hence the limits γ_s and β_s of these sequences exist. They enter into the following corollary.

Corollary 8.20 *For each positive definite Stieltjes moment sequence s we have*

$$\gamma_s := -s_0 \lim_{n\to\infty} \frac{Q_n(0)}{P_n(0)} = \int_0^\infty \frac{d\mu_F(y)}{y}, \quad \beta_s := s_0 \lim_{n\to\infty} \frac{N_n(0)}{M_n'(0)} = \mu_K(\{0\}).$$

$$(8.31)$$

Proof From Theorem 8.16 we obtain

$$-s_0 \frac{Q_n(x)}{P_n(x)} \le \int_0^\infty \frac{d\mu_F(y)}{y-x} \le \int_0^\infty \frac{d\mu_F(y)}{y}$$

for $x < 0$. Letting $x \to -0$ and then $n \to \infty$ we get $\gamma_s \le \int y^{-1} d\mu_F(y)$.

We prove the converse inequality. Recall that $J_n = A_F^{[n]}$ by (8.11). Hence, by Lemma 6.28 we have $\langle (A_F^{[n]} - xI)^{-1} e_0, e_0 \rangle = -\frac{Q_n(x)}{P_n(x)}$ for $x < 0$. Since $A_F^{[n]} \ge 0$, the left-hand side increases as $x \to -0$ and we obtain

$$-s_0 \frac{Q_n(x)}{P_n(x)} \le -s_0 \frac{Q_n(0)}{P_n(0)} \le \gamma_s, \quad x < 0.$$

Letting $n \to \infty$ and using (8.25) this yields $\int (y-x)^{-1} d\mu_F(y) \le \gamma_s$. Passing to the limit $x \to -0$ by using Lebesgue's monotone convergence theorem we conclude that $\int y^{-1} d\mu_F(y) \le \gamma_s$. This completes the proof of the first equality in (8.31).

By Theorem 8.16, the sequence $(-s_0 \frac{N_n(x)}{M_n(x)})$ converges to $\int (y-x)^{-1} d\mu_K(y)$ from above for $x < 0$. Therefore, multiplying by $-x > 0$, we obtain

$$\mu_K(\{0\}) \le \int_0^\infty \frac{(-x)d\mu_K(y)}{y-x} \le s_0 \frac{xN_n(x)}{M_n(x)}.$$

Passing to the limits $x \to -0$ and then $n \to \infty$ yields $\mu_K(\{0\}) \le \beta_s$.

Conversely, the expression $\langle (-x)(A_K^{[n]} - xI)^{-1} e_0, e_0 \rangle = \frac{xN_n(x)}{M_n(x)}$ (by Lemma 8.13) decreases as $x \to -0$, since $A_K^{[n]} \ge 0$. Therefore, for $x < 0$,

$$s_0 \frac{xN_n(x)}{M_n(x)} \ge s_0 \frac{N_n(0)}{M_n'(0)}.$$

Now we take the limit $n \to \infty$. Because of (8.26) we then obtain

$$\int_0^\infty \frac{(-x)d\mu_K(y)}{y-x} \geq \beta_s, \quad x < 0. \tag{8.32}$$

Since $\left|\frac{-x}{y-x}\right| \leq 1$ for $y \geq 0$ and $\lim_{x\to-0}(\frac{-x}{y-x})$ is the characteristic function of the point 0, Lebesgue's dominated convergence theorem applies for the limit $x \to -0$ in (8.32) and yields $\mu_K(\{0\}) \geq \beta_s$. This proves the second equality in (8.31). $\square$

Note that γ_s can be infinite if the Stieltjes moment sequence s is determinate. However, if s is indeterminate, then γ_s is finite (by Corollary 8.21). In this case the number γ_s enters into the Nevanlinna parametrization given in Theorem 8.24 below.

Corollary 8.21 *Let s be a positive definite Stieltjes moment sequence. Then s is Stieltjes indeterminate if and only if $\gamma_s < \infty$ and $\beta_s \neq 0$.*

Proof Let s be Stieltjes indeterminate. Then $m(T_F) > 0$ by Corollary 8.10 and $\mathrm{supp}\,\mu_F \subseteq [m(T_F), +\infty)$, so that $\gamma_s = \int_0^\infty y^{-1}d\mu_F(y) < \infty$. Obviously, s is also Hamburger indeterminate. Hence $\mathfrak{p}_0 \in \mathcal{D}(T_K)$ by Corollary 8.6. Since $T_K\mathfrak{p}_0 = T^*\mathfrak{p}_0 = 0$, 0 is an eigenvalue of T_K and hence $\beta_s = \mu_K(\{0\}) = \|\mathfrak{p}_0\|^{-2} \neq 0$ by formula (7.7).

Conversely, assume that $\gamma_s < \infty$ and $\beta_s \neq 0$. Since then $\int_0^\infty y^{-1}d\mu_F(y) < \infty$ and $\mu_K(\{0\}) \neq 0$, we have $\mu_F \neq \mu_K$. Hence s is Stieltjes indeterminate. $\square$

The next corollary uses the inquality (8.29) which will be proved only in the next section. It can be used to construct determinate moment sequences of "fast growth".

Corollary 8.22 *Suppose that $s = (s_n)_{n\in\mathbb{N}_0}$ is an indeterminate Stieltjes moment sequence. Set $c := \int_0^\infty y^{-1}d\mu_F(y)$, $\tilde{s}_0 = 1$, and $\tilde{s}_n = c^{-1}s_{n-1}$ for $n \in \mathbb{N}$. Then $\tilde{s} = (\tilde{s}_n)_{n\in\mathbb{N}_0}$ is a Stieltjes moment sequence which is Hamburger determinate.*

Proof First note that $c \in (0, +\infty)$, since $\mathrm{supp}\,\mu_F \subseteq [m(T_F), +\infty)$ and $m(T_F) > 0$ by Corollary 8.10. Clearly, the measure ν_0 given by $d\nu_0 = c^{-1}y^{-1}d\mu_F$ is supported on $\mathbb{R}_+$ and has the moments $\tilde{s}_n$. Hence $\tilde{s}$ is a Stieltjes moment sequence.

Let ν be a solution of the Stieltjes moment problem for $\tilde{s}$. Then the measure μ given by $d\mu(y) := cyd\nu(y)$ has the moment sequence s, is supported on $\mathbb{R}_+$, and

$$\int_0^\infty y^{-1}d\mu(y) = c\int_0^\infty d\nu(y) = c\tilde{s}_0 = c = \int_0^\infty y^{-1}d\mu_F(y).$$

Therefore, it follows from statement (8.29) in Theorem 8.18, applied with $x = 0$, that $\mu = \mu_F$. This implies $\nu = \nu_0$. Hence $\tilde{s}$ is a determinate Stieltjes moment sequence with representing measure ν_0. Since $\nu_0(\{0\}) = 0$, $\tilde{s}$ is also Hamburger determinate by Corollary 8.9. $\square$

8.5 Nevanlinna Parametrization for the Indeterminate Stieltjes Moment Problem

For $\gamma \in \mathbb{R}$ let $\mathfrak{P}_\gamma$ denote the set of Pick functions $\Phi \in \mathfrak{P}$ which are holomorphic on $\mathbb{C}\backslash\mathbb{R}_+$ and map $(-\infty, 0)$ into $[\gamma, \infty)$. Note that all constant functions equal to a number $t \in [\gamma, \infty)$ are contained in $\mathfrak{P}_\gamma$. Set $\overline{\mathfrak{P}}_\gamma := \mathfrak{P}_\gamma \cup \{\infty\}$.

Proposition 8.23 *A function Φ belongs to the class $\mathfrak{P}_\gamma$ if and only if*

$$\Phi(z) = \alpha + \int_0^\infty \frac{d\tau(x)}{x - z}, \quad z \in \mathbb{C}\backslash[\gamma, \infty), \tag{8.33}$$

where $\alpha \geq \gamma$ and τ is a positive Borel measure on $\mathbb{R}_+$ satisfying $\int_0^\infty \frac{d\tau(x)}{x+1} < \infty$.

Proof Clearly, any function Φ of the form (8.33) is holomorphic on $\mathbb{C}\backslash\mathbb{R}_+$. Since $a \geq \gamma$ and $\int_0^\infty (x - z)d\tau(x) \geq 0$ for $z < 0$, Φ maps $(-\infty, 0)$ into $[\gamma, \infty)$, so that $\Phi \in \mathfrak{P}_\gamma$.

Conversely, let $\Phi \in \mathfrak{P}_\gamma$. Then $\Phi \in \mathfrak{P}$, so by formula (A.5) it has a representation

$$\Phi(z) = a + bz + \int_{\mathbb{R}} \left(\frac{1}{x - z} - \frac{x}{1 + x^2} \right) d\tau(x), \quad z \in \mathbb{C}\backslash\mathbb{R}, \tag{8.34}$$

where $a, b \in \mathbb{R}$, $b \geq 0$, and τ is a positive measure such that $\int (1+x^2)^{-1} d\tau(x) < \infty$. Since Φ is holomorphic on $\mathbb{C}\backslash\mathbb{R}_+$, Proposition A.15 implies that $\operatorname{supp} \tau \subseteq \mathbb{R}_+$.

We consider the limit $z \to -\infty$ in (8.34). The integrand converges monotonically decreasing on $\mathbb{R}_+$ to the function $-\frac{x}{1+x^2}$. If $b > 0$, it follows from Lebesgue's monotone convergence theorem that $\Phi(z) \to -\infty$ as $z \to -\infty$. This contradicts $\Phi \in \mathfrak{P}_\gamma$. Thus $b = 0$. Applying the limit $z \to -\infty$ in (8.34) once more, $\Phi \subset \mathfrak{P}_\gamma$ implies that $c := \int_0^\infty x(1 + x^2)^{-1} d\tau(x) < \infty$ and $\Phi(z) \to a - c$. This yields $\int_0^\infty (1 + x)^{-1} d\tau(x) < \infty$ and $\alpha := a - c \geq \gamma$, since $\Phi \in \mathfrak{P}_\gamma$. Then (8.34) gives (8.33). $\qquad\square$

Suppose that s is an indeterminate Stieltjes moment sequence. Let γ_s denote the positive real number (see Corollary 8.21) defined by (8.31). Recall that $m(T_F) > 0$, $\operatorname{supp} \mu_F \subseteq [m(T_F), +\infty)$, and $(T_F)^{-1} \in \mathbf{B}(\mathcal{H}_s)$ by Corollary 8.10. If E_{T_F} denotes the spectral measure of the self-adjoint operator T_F, the functional calculus yields

$$\gamma_s \equiv \int_0^\infty y^{-1} d\mu_F(y) = s_0 \int_0^\infty y^{-1} d\langle E_{T_F}(y)e_0, e_0\rangle = s_0 \langle (T_F)^{-1}e_0, e_0\rangle. \tag{8.35}$$

The positive number γ_s is called the *Friedrichs parameter* of s.

The following theorem is the counterpart of Nevanlinna's Theorem 7.13 for the Stieltjes moment problem.

Theorem 8.24 *Suppose that s is an indeterminate Stieltjes moment sequence. Then there is a one-to-one correspondence between functions* $\phi \in \overline{\mathfrak{P}}_{\gamma_s}$ *and solutions* μ *of the Stieltjes moment problem for s given by*

$$I_{\mu_\phi}(z) \equiv \int_0^\infty \frac{1}{x-z}\, d\mu_\phi(x) = -\frac{A(z) + \phi(z)C(z)}{B(z) + \phi(z)D(z)} \equiv H_z(\phi(z)), \ z \in \mathbb{C}_+.$$

(8.36)

Proof Since s is Stieltjes indeterminate, it is Hamburger indeterminate. Thus Theorem 7.13 applies and gives a one-to-one correspondence $\mu_\Phi \leftrightarrow \Phi$ between solutions μ_Φ of the Hamburger moment problem for s and functions $\Phi \in \overline{\mathfrak{P}}$. Therefore, it suffices to show that supp $\mu_\Phi \subseteq \mathbb{R}_+$ if and only if $\Phi \in \overline{\mathfrak{P}}_{\gamma_s}$.

Recall that the solutions μ_t for $t \in \mathbb{R} \subseteq \overline{\mathfrak{P}}$ correspond to the self-adjoint operators T_t from Theorem 6.23. Since $T_K = T_\infty$ by Corollary 8.6, the assertion holds for $\mu = \mu_K$ and $\Phi = \infty$. Further, $s_0 \langle T_t^{-1} e_0, e_0 \rangle = t$ for $t \in \mathbb{R}$ by Lemma 6.24. Hence Eq. (8.35) implies that $T_F = T_{\gamma_s}$, so that the assertion is also valid for $\Phi = \gamma_s$.

Let us fix $x \in \mathbb{R}$ and consider the fractional linear transformation (see (7.10))

$$H_x(t) := -\frac{A(x) + tC(x)}{B(x) + tD(x)}.$$

Since $A(x), B(x), C(x), D(x) \in \mathbb{R}$ and $A(x)D(x) - B(x)C(x) = 1$, H_x is a bijection of $\overline{\mathbb{R}}$ on $\overline{\mathbb{R}}$, where $\overline{\mathbb{R}} := \mathbb{R} \cup \{\infty\}$. From the relation $H_x'(t) = (B(x) + tD(x))^{-2} > 0$ it follows that H_x is strictly increasing on $\mathbb{R}$ outside the pole $t = -\frac{B(x)}{D(x)}$. Therefore, since $H_x(\gamma_s) = I_{\mu_F}(x)$ and $H_x(\infty) = I_{\mu_K}(x)$ as shown in the preceding paragraph, H_x is a bijection of $[\gamma_s, \infty]$ on the interval $[I_{T_F}(x), I_{T_K}(x)]$.

Now suppose that $\Phi \in \mathfrak{P}_{\gamma_s}$. Then $I_{\mu_\Phi}(z) = H_z(\Phi(z))$ is holomorphic on $\mathbb{C}\backslash\mathbb{R}$. Let $x < 0$. Then $\Phi(x) \in [\gamma_s, \infty)$ and hence $H_x(\phi(x)) \in [I_{T_F}(x), I_{T_K}(x)]$. Hence the denominator of $H_x(\Phi(x))$ does not vanish, because otherwise $H_x(\Phi(x)) = \infty$. Therefore, since the functions A, B, C, D are entire and $\Phi(z)$ is holomorphic on $\mathbb{C}\backslash\mathbb{R}_+$, $I_{\mu_\Phi}(z) = H_z(\Phi(z))$ is holomorphic on $\mathbb{C}\backslash\mathbb{R}_+$. Hence supp $\mu_\Phi \subseteq \mathbb{R}_+$ by Proposition A.15. This proves one direction of Theorem 8.24.

To prove the converse direction we assume that $\mu \neq \mu_K$ is a solution of the Stieltjes moment problem for s. Then, by formula (8.27) in Theorem 8.18,

$$I_{\mu_F}(x) \le I_\mu(x) \le I_{\mu_K}(x) \quad \text{for} \quad x < 0.$$

(8.37)

Since $\mu_K \cong \mu_\infty$, by Theorem 7.13 there is a unique $\Phi \in \mathfrak{P}$ such that $\mu = \mu_\Phi$. We have to show that $\Phi \in \mathfrak{P}_{\gamma_s}$.

First we verify that $\Phi(z) \neq 0$ for $z \in \mathbb{C}_+ \equiv \{z \in \mathbb{C} : \text{Im}\, z > 0\}$. Indeed, assume to the contrary that $\Phi(z_0) = 0$ for some $z_0 \in \mathbb{C}_+$. Then, since $\Phi \in \mathfrak{P}_{\gamma_s}$, $\Phi(\mathbb{C}_+)$ is not open. Therefore Φ is constant, so that $\Phi = 0 \in \mathbb{R}$. Lemma 6.24 implies that

$\langle T_0^{-1} e_0, e_0 \rangle = 0$. From the functional calculus of self-adjoint operators we obtain

$$0 = s_0 \langle T_0^{-1} e_0, e_0 \rangle = s_0 \int_{\mathbb{R}} y^{-1} d\langle E_{T_0}(y) e_0, e_0 \rangle = \int_0^\infty y^{-1} d\mu_\Phi(y) < \infty.$$

Hence $\mu_\Phi = 0$, which is a contradiction, since μ_Φ solves an indeterminate Stieltjes moment problem. This proves that $\Phi(z) \neq 0$ for $z \in \mathbb{C}_+$.

Hence $\Psi := 1/\Phi$ is a holomorphic function on $\mathbb{C}_+$ and we have

$$\Psi(z) = \frac{1}{\Phi(z)} = \frac{1}{H_z^{-1}(I_\mu(z))} = -\frac{C(z) + I_\mu(z)D(z)}{A(z) + I_\mu(z)B(z)}. \tag{8.38}$$

Assume that $A(x_0) + I_\mu(x_0)B(x_0) = 0$ for some $x_0 < 0$. Then, since the numerator and denominator of the fractional linear transformation do not vanish simultaneously, $C(x_0) + I_\mu(x_0)D(x_0) \neq 0$ and hence

$$\Phi(x_0) := \lim_{z \in \mathbb{C}_+, z \to x_0} \Phi(z) = - \lim_{z \in \mathbb{C}_+, z \to x_0} \frac{A(z) + I_\mu(z)B(z)}{C(z) + I_\mu(z)D(z)} = 0. \tag{8.39}$$

Since $\Phi(z) = H_z^{-1}(I_\mu(z))$, this implies that $\Phi(x_0) = H_{x_0}^{-1}(I_\mu(x_0))$. By (8.37) we have $H_{x_0}^{-1}(I_\mu(x_0)) \in [\gamma_s, \infty]$. Since $\Phi(x_0) = 0$ and $\gamma_s > 0$, this is a contradiction. Thus we have proved that $A(x) + I_\mu(x)B(x) \neq 0$ for all $x < 0$.

Therefore, since μ solves the Stieltjes moment problem and hence I_μ is continuous on $\mathbb{C}\backslash\mathbb{R}_+$, it follows from (8.38) that Ψ has a continuous extension to $\mathbb{C}_+ \cup (-\infty, 0)$ with real values on $(-\infty, 0)$. Hence, by Schwarz' reflection principle, Ψ has a holomorphic extension to $\mathbb{C}\backslash\mathbb{R}_+$. From (8.37) we conclude that $\Phi(x) = H_x^{-1}(I_\mu(x)) \in [\gamma_s, \infty]$ and therefore $\Psi(x) \in [0, \gamma_s^{-1}]$ for $x < 0$.

Next we show that $\Psi(x) \neq 0$ on $(-\infty, 0)$. Assume to the contrary that $\Psi(x_0) = 0$ for some $x_0 < 0$. Since $\Psi(z) = 1/\Phi(z) \neq 0$ on $\mathbb{C}\backslash\mathbb{R}_+$ and $\Psi(x) \in [0, \gamma_s^{-1}]$ for all $x < 0$, we have $(-\infty, 0) \cap \Psi(\mathbb{C}\backslash\mathbb{R}_+) = \emptyset$. Hence $\Psi(\mathbb{C}\backslash\mathbb{R}_+)$ is not open, so Ψ is constant. Since $\Psi(x_0) = 0$, $\Psi(z) \equiv 0$. But $\Psi = 1/\Phi$ with $\Phi \in \mathfrak{P}$, so we have a contradiction.

Putting the preceding together we have shown that $\Phi = 1/\Psi$ has a holomorphic extension to $\mathbb{C}\backslash\mathbb{R}_+$ and $\Phi(x) \in [\gamma_s, \infty)$ for $x \in (-\infty, 0)$. That is, $\Phi \in \mathfrak{P}_{\gamma_s}$. This proves the converse direction and completes the proof of Theorem 8.24.

Finally, we prove the two inequalities (8.28) and (8.29) from Theorem 8.18.

Since $\Psi(x) \neq 0$ as shown in the paragraph before last, $\Phi(x) \neq \infty$ and therefore $I_\mu(x) = H_x(\Phi(x)) \neq H_x(\infty) = I_{\mu_K}(x)$. Since $I_\mu(x) \leq I_{\mu_K}(x)$ by (8.37), this yields $I_\mu(x) < I_{\mu_K}(x)$. This proves the first inequality (8.28).

Now we turn to the proof of (8.29). Let $\mu \neq \mu_F$ be a solution of the Stieltjes moment problem for s. Then $\mu = \mu_\Phi$, where $\Phi(z) := H_z^{-1}(I_\mu(z))$ for $z \in \mathbb{C}_+$.

First we assume to the contrary that $I_\mu(x) = I_{\mu_F}(x)$ for some $x < 0$. Then we have $\Phi(x) = H_x^{-1}(I_\mu(x)) = H_x^{-1}(I_{\mu_F}(x)) = \gamma_s$, so $\Psi(x) = \gamma_s^{-1}$ is the right end point of the interval $[0, \gamma_s^{-1}]$. Arguing as above, $\Psi(\mathbb{C}\backslash\mathbb{R}_+)$ is not open, so Ψ and hence Φ

are constant. Thus $\Phi(z) = \Phi(x) = \gamma_s$ and therefore $I_\mu(z) = H_z(\Phi(z)) = H_z(\gamma_s) = I_{\mu_F}(z)$ for all $z \in \mathbb{C}\backslash\mathbb{R}_+$. Hence $\mu = \mu_F$, which is the desired contradiction.

Now let $x = 0$. Since μ solves the Stieltjes moment problem, $\Phi \in \mathfrak{P}_{\gamma_s}$, so Φ is of the form (8.33). This implies that $\Phi(z) \in [\gamma_s, \infty)$ and $\Phi(z)$ is monotone increasing on the interval $(-\infty, 0)$. Hence the limit $\phi_0 := \lim_{z\to -0} \Phi(z) \in [\gamma_s, +\infty]$ exists. We have $\phi_0 \neq \gamma_s$. (Indeed, otherwise $\Phi(z) \geq \gamma_s$ and the mononiticity of $\Phi(z)$ on $(-\infty, 0)$ imply that $\alpha = \gamma_s$ and $\tau = 0$; then $\Phi = \gamma_s$ and hence $\mu = \mu_\Phi = \mu_F$, which is a contradiction.) Since $A(0) = D(0) = 0$ and $C(0) = -B(0) = 1$, we have $H_0(t) = t$ for $t \in \mathbb{R}$. Clearly, $I_\mu(z) \leq \int_0^\infty y^{-1} d\mu$ for $z < 0$. Then, by the preceding,

$$\int_0^\infty y^{-1} d\mu_F(y) = I_{\mu_F}(0) = H_0(\gamma_s) = \gamma_s < \phi_0$$

$$= \lim_{z\to -0} H_z(\Phi(z)) = \lim_{z\to -0} I_\mu(z) \leq \int_0^\infty y^{-1} d\mu(y).$$

This proves (8.29) for $x = 0$ and it completes the proofs of (8.28) and (8.29).   □

We briefly repeat some facts from the preceding proof. In the parametrization (8.36) the Friedrichs solution μ_F corresponds to the Friedrichs parameter $\Phi = \gamma_s$, while the Krein solution μ_K is obtained for $\Phi = \infty$. The von Neumann solutions of the Stieltjes moment problem for s (that is, the solutions μ for which $\mathbb{C}[x]$ is dense in $L^2(\mathbb{R}, \mu)$) are precisely the measures μ_t for constants $\Phi = t$ with $t \in [\gamma_s, +\infty]$.

Remark 8.25 The parametrization (7.18) is related to our Nevanlinna parametrization (7.16) by taking $-\Phi^{-1}$ instead of Φ. Hence in this parametrization (7.18) Friedrichs and Krein solutions correspond to the parameters $-\gamma_s^{-1}$ and 0, respectively.   ○

8.6   Weyl Circles for the Indeterminate Stieltjes Moment Problem

In this section we suppose that s is an **indeterminate Stieltjes moment sequence** and $z \in \mathbb{C}_+$. For $t \in \mathbb{R}$, let $W(z, t)$ denote the cone in the complex plane given by

$$W(z, t) := \{w \in \mathbb{C} : 0 \leq \arg(w - t) \leq \pi - \arg(z)\}.$$

One easily verifies that

$$w \in W(z, t) \iff \operatorname{Im}(w) \geq 0 \quad \text{and} \quad \operatorname{Im}(z(w - t)) \geq 0. \tag{8.40}$$

Note that $W(z, t) = t + W(z, 0)$ and $W(z, t) \subseteq W(z, t')$ if $t' \leq t$.

Proposition 8.26

(i) *If $\Phi \in \mathfrak{P}_{\gamma_s}$, then $\Phi(z) \in W(z, \gamma_s)$.*

(ii) *For each $w \in W(z, \gamma_s)$ there exists a $\Phi \in \mathfrak{P}_{\gamma_s}$ such that $w = \Phi(z)$.*

(iii) *Suppose that $w \in W(z, \gamma_s)$ and $\mathrm{Im}(z(w - \gamma_s)) = 0$. Then there is a unique function $\Phi \in \mathfrak{P}_{\gamma_s}$ such that $\Phi(z) = w$. This function is*

$$\Phi(z') = \gamma_s - \frac{z(\gamma_s - w)}{z'}, \quad z' \in \mathbb{C}_+.$$

Proof

(i) By Proposition 8.23, $\Phi \in \mathfrak{P}_{\gamma_s}$ is of the form (8.33) with $\alpha \geq \gamma_s$. Since

$$\mathrm{Im}\left(z \int_0^\infty \frac{d\tau(x)}{x - z}\right) = \int_0^\infty \mathrm{Im}\left(\frac{z}{x - z}\right) d\tau(x) \geq 0$$

and $\mathrm{Im}\left(\int_0^\infty (x - z)^{-1} d\tau(x)\right) \geq 0$, (8.40) implies that $\int_0^\infty (x - z)^{-1} d\tau(x) \in W(z, 0)$. Therefore, since $\alpha \geq \gamma_s$, we deduce that

$$\Phi(z) = \alpha + \int_0^\infty \frac{d\tau(x)}{x - z} \in \alpha + W(z, 0) = W(z, \alpha) \subseteq W(z, \gamma_s).$$

(ii) Let $z = a + ib$ and $w = u + iv \in W(z, \gamma_s)$, where $a, b, u, v \in \mathbb{R}$. Since $z \in \mathbb{C}_+$, $b > 0$. Further, since $w \in W(z, \gamma_s)$, it follows from (8.40) that $v \geq 0$ and

$$\mathrm{Im}(z(w - \gamma_s)) = \mathrm{Im}((a + ib)(u - \gamma_s + iv)) = b(u - \gamma_s) + av \geq 0. \tag{8.41}$$

First let $v = 0$. Then $u \geq \gamma_s$. Hence $\Phi := u \in \mathfrak{P}_{\gamma_s}$ and $\Phi(z) = w$. Now suppose $v > 0$. Then $x_0 := v^{-1}(b(u - \gamma_s) + av) \geq 0$ by (8.41). Define a measure $\tau = \frac{v}{b}|x_0 - z|^2 \delta_{x_0}$. By Proposition 8.23, $\Phi(z') := \gamma_s + \int_0^\infty (x - z')^{-1} d\tau(x) \in \mathfrak{P}_{\gamma_s}$. We compute

$$\Phi(z) = \gamma_s + \frac{v}{b}|x_0 - z|^2 (x_0 - z)^{-1} = \gamma_s + \frac{v}{b}(x_0 - a + ib) = u + iv = w.$$

(iii) Suppose that $\Phi \in \mathfrak{P}_{\gamma_s}$ and $\Phi(z) = w$. By Proposition 8.23, Φ is of the form (8.33). Inserting $\Phi(z) = w$ therein we derive

$$0 = \mathrm{Im}(z(w - \gamma_s)) = \mathrm{Im}(z(\alpha - \gamma_s)) + \int_0^\infty \mathrm{Im}\left(\frac{z}{x - z}\right) d\tau(x)$$

$$= \mathrm{Im}(z)(\alpha - \gamma_s) + \mathrm{Im}(z) \int_0^\infty \frac{x}{|x - z|^2} d\tau(x).$$

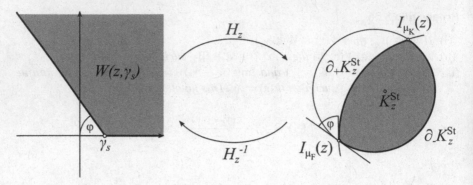

Fig. 8.1 The transformation H_z and the set K_z^{St}

Since $\mathrm{Im}(z) > 0$ and $\alpha \geq \gamma_s$, we deduce that $\alpha = \gamma_s$ and $\tau = c\delta_0, c \geq 0$. But then $w = \Phi(z) = \gamma_s - cz^{-1}$, so that $c = z(\gamma_s - w)$. Hence $\Phi(z') = \gamma_s - \frac{z(\gamma_s - w)}{z'}$.

Let Φ be this function. Clearly, $\Phi(z) = w$. Since $\mathrm{Im}(z(w - \gamma_s)) = 0$, we derive from (8.41) that $c = z(\gamma_s - w) = v(a^2 b^{-1} + b) \geq 0$. This implies that $\Phi \in \mathfrak{P}_{\gamma_s}$. $\qquad\qquad\qquad\qquad\qquad\qquad\qquad\qquad\qquad\qquad\qquad\qquad\qquad\square$

By Theorem 8.24, the values of Stieltjes transforms of all solutions of the Stieltjes moment problem for s are $I_{\mu_\phi}(z) = H_z(\phi(z))$, where $\phi \in \overline{\mathfrak{P}}_{\gamma_s}$. From Proposition 8.26 we deduce that these are precisely the numbers of the set

$$K_z^{St} = H_z(W(z, \gamma_s) \cup \{\infty\})$$

which is the right gray shaded area with boundary in Fig. 8.1.

Let us consider the boundary of the set K_z^{St}. Recall that K_z denotes the Weyl circle defined in Sect. 7.3. Then the boundary ∂K_z^{St} is the disjoint union of the sets

$$\partial_- K_z^{St} := \partial K_z^{St} \cap \partial K_z = H_z([\gamma_s, +\infty) \cup \{\infty\}),$$

$$\partial_+ K_z^{St} := \partial K_z^{St} \backslash \partial K_z = \{H_z(w) : w \in \partial W(z, \gamma_s)), \mathrm{Im}(w) > 0\}.$$

From the discussion after Theorem 7.13 we know that for each $w \in \partial_- K_z^{St} \subseteq \partial K_z$ there exists a unique solution μ of the Hamburger moment problem for s satisfying $I_\mu(z) = w$. Since $w \in \partial K_z^{St}$, μ is also the *unique* Stieltjes solution such that $I_\mu(z) = w$. Further, μ is a von Neumann solution, that is, $\mathrm{ord}(\mu) = 0$.

Now let $w \in \partial_+ K_z^{St}$. One easily verifies (see (8.40)) that $\mathrm{Im}(z(w - \gamma_s)) = 0$. Therefore, by Proposition 8.26(iii), there is a unique function $\Phi \in \mathfrak{P}_{\gamma_s}$ such that $\Phi(z) = w$. Hence it follows from Theorem 8.24 that there exists a *unique* solution μ of the Stieltjes moment problem for s satisfying $I_\mu(z) = w$. In this case, $\mathrm{ord}(\mu) = 1$.

8.7 Exercises

1. Show $s = (s_j)_{j \in \mathbb{N}_0}$ is a Stieltjes moment sequence if and only if the sequence $(s_0, 0, s_1, 0, s_2, \dots)$ is positive semidefinite.

2. Let $s = (s_j)_{j \in \mathbb{N}_0}$ be a Hamburger moment sequence. Show that $(s_{2j})_{j \in \mathbb{N}_0}$ is a Stieltjes moment sequence.

3. Let $s = (s_j)_{j \in \mathbb{N}_0}$ be a Stieltjes moment sequence. Show that $s_{m+n}^2 \leq s_k s_l$ for all $m, n, k, l \in \mathbb{N}_0$ such that $k + l = 2(m + n)$.

4. Let s be a Stieltjes moment sequence. Show that $m_s = \inf\{\lambda : \lambda \in \operatorname{supp} \mu_F\}$.

5. (*A determinate sequence t growing faster than an indeterminate sequence s.*) Find Stieltjes moment sequences $s = (s_n)_{n \in \mathbb{N}_0}$ and $t = (t_n)_{n \in \mathbb{N}_0}$ such that s is Stieltjes indeterminate, t is Hamburger determinate, and $\lim_{n \to \infty} \frac{s_n}{t_n} = 0$.

   Hint: Use Corollary 8.22 and Examples 4.18 for $\alpha = \frac{1}{4}$ and 4.23.

6. Let $(\alpha_n)_{n \in \mathbb{N}_0}$ and $(\beta_n)_{n \in \mathbb{N}_0}$ be positive sequences. Prove that there exists a Stieltjes moment sequence $s = (s_n)_{n \in \mathbb{N}_0}$ such that $s_n \geq \alpha_n$ and $s_{n+1} \geq \beta_n s_n$ for $n \in \mathbb{N}_0$.

7. Let $z \in \mathbb{C}_+$, $t \in \mathbb{R}$, and $w \in \mathbb{C}$.

   a. Show that $w \in W(z, t)$ if and only if $\operatorname{Im}(w) \geq 0$ and $\frac{\operatorname{Re}(z)}{\operatorname{Im}(z)} \operatorname{Im}(w) + \operatorname{Re}(w) \geq t$.

   b. Show that if w is an interior point of $W(z, t)$, then $\frac{\operatorname{Re}(z)}{\operatorname{Im}(z)} < \frac{\operatorname{Re}(w-t)}{\operatorname{Im}(w-t)}$.

8. Suppose that $z \in \mathbb{C}_+$ and $u \in \mathbb{C}_+$. Let $n \in \mathbb{N}$ and $y, \tilde{y} \in \mathbb{R}, y \neq 0, \tilde{y} \neq 0$.

   a. Show that there are numbers $c_j > 0$ and pairwise distinct points $x_j \in \mathbb{R}$ such that $c_j(x_j - z) = \frac{1}{n}(u - 2yj)$ for $j = 1, \dots, n$.

   b. Define $\Phi_y(z') := -y(n + 1) + \sum_{j=1}^{n} \frac{c_j}{x_j - z'}$, $z' \in \mathbb{C}_+$. Show that $\Phi_y \in \mathfrak{P}$ is a rational function of degree n satisfying $\Phi_y(z) = u$.

   c. Show that $\Phi_y \neq \Phi_{\tilde{y}}$ for $y \neq \tilde{y}$.

9. Suppose that s is an indeterminate Stieltjes moment sequence and $z \in \mathbb{C}_+$. Let v be an interior point of $W(z, \gamma_s)$, w an interior point of K_z^{St}, and $n \in \mathbb{N}$.

   a. Show that there are infinitely many rational functions $\Phi \in \mathfrak{P}_{\gamma_s}$ of degree n such that $\Phi(z) = v$.

   b. Show that there are infinitely many solutions μ of the Stieltjes moment problem for s such that $\operatorname{ord}(\mu) = n$ and $I_\mu(z) = w$.

   Hint for b: Use the construction sketched in Exercise 8. Show that the numbers x_j can be chosen positive for small $y > 0$; details can be found in [Ge].

8.8 Notes

The theory of the Stieltjes moment problem goes back to T. Stieltjes' famous memoir [Stj], which contains many basic results.

The Friedrichs parameter has been identified in [Pd1]. Since these papers use the parametrization in the form (7.18), their number is our $-\gamma_s^{-1}$. The Nevanlinna parametrization for indeterminate Stieltjes moment sequences was first given by H.L. Pedersen [Pd1]; it is also contained in [Sim1].

The use of the operators T_F and T_K and the corresponding approximants is due to B. Simon [Sim1]. Our operator-theoretic approach is based on [Sim1] and [Ge].

Part II
The One-Dimensional Truncated Moment Problem

Chapter 9
The One-Dimensional Truncated Hamburger and Stieltjes Moment Problems

In this chapter we are concerned with the following problem:

Let $s = (s_j)_{j=0}^m$ be a real m-sequence, where $m \in \mathbb{N}_0$. When does there exist a Radon measure μ on $\mathbb{R}$ such that $s_j = \int_{\mathbb{R}} x^j \, d\mu(x)$ for all $j = 0, \ldots, m$?

This is the *truncated Hamburger moment problem* and in the affirmative case s is called a *truncated Hamburger moment sequence*. If we require in addition that the measure μ is supported on $\mathbb{R}_+$, we get the *truncated Stieltjes moment problem*.

In Sects. 9.1 and 9.3 the special case of a positive definite $2n$-sequence is studied in detail. Using quasi-orthogonal polynomials Gauss' quadrature formulas are derived (Theorems 9.4 and 9.6) and a one-parameter family of $(n + 1)$-atomic solutions is constructed (Theorem 9.7). The associated reproducing kernel space and the Christoffel function are investigated in Sect. 9.3. In the short Sect. 9.2 we apply some result from Sect. 9.1 to reprove Hamburger's and Markov's theorem in the positive definite case.

The remaining part of the chapter deals with positive semidefinite finite sequences. In Sect. 9.4 such sequences are characterized by integral representations (Theorems 9.15 and 9.19). In Sect. 9.5 the Hankel rank of a positive semidefinite $2n$-sequence is introduced. The integral representation and the Hankel rank enter into the treatment of truncated moment problems in Sect. 9.6. Here basic existence theorems for the truncated Hamburger and Stieltjes moment problems in the even case $m = 2n$ (Theorems 9.27 and 9.36) and in the odd case $m = 2n + 1$ (Theorems 9.32 and 9.35) are obtained. Further, neccesary and sufficient conditions for the uniqueness of the representing measures are given.

Let us recall some standard notations. The real polynomials of degree at most n are denoted by $\mathbb{R}[x]_n$. For a sequence $s = (s_j)_{j=0}^m$ and $2n \leq m$, the Hankel matrix $H_n(s)$ is defined by $H_n(s) = (s_{i+j})_{i,j=0}^n$, the corresponding Hankel determinant is $D_n(s) := \det H_n(s)$, and L_s is the Riesz functional on $\mathbb{R}[x]_m$ defined by $L_s(x^j) = s_j$.

© Springer International Publishing AG 2017
K. Schmüdgen, *The Moment Problem*, Graduate Texts in Mathematics 277,
DOI 10.1007/978-3-319-64546-9_9

9.1 Quadrature Formulas and the Truncated Moment Problem for Positive Definite $2n$-Sequences

Throughout this section, we assume that $n \in \mathbb{N}$ and $s = (s_j)_{j=0}^{2n}$ is a real **positive definite** $2n$-sequence. That s is positive definite means that

$$\sum_{k,l=0}^{n} s_{k+l}\xi_k\xi_l > 0 \quad \text{for all} \quad (\xi_0,\dots,\xi_n)^T \in \mathbb{R}^{n+1}, \ (\xi_0,\dots,\xi_n)^T \neq 0. \tag{9.1}$$

In terms of the Riesz functional L_s on $\mathbb{R}[x]_{2n}$, condition (9.1) is equivalent to the requirement $L_s(p^2) > 0$ for all $p \in \mathbb{R}[x]_n, p \neq 0$.

Lemma 9.1 *Let $s_{2j+1}, j \geq n$, be given real numbers. There exist real numbers s_{2i} for $i \in \mathbb{N}, i \geq n + 1$, such that $s = (s_k)_{k\in\mathbb{N}_0}$ is a positive definite sequence.*

Proof Since $(s_j)_{j=0}^{2n}$ is positive definite, the Hankel matrix $H_n(s)$ is positive definite. Hence the Hankel determinant $D_n(s)$ of s is positive. Let s_{2n+2} be a real number. The Hankel determinant D_{n+1} for the sequence $(s_j)_{j=0}^{2n+2}$ is of the form

$$D_{n+1} = s_{2n+2}D_n(s) + p(s_0,\dots,s_{2n},s_{2n+1}).$$

Here p is a real polynomial p in $s_0,\dots,s_{2n},s_{2n+1}$ that does not depend on s_{2n+2}. Since $D_n(s) > 0$, we have $D_{n+1} > 0$ for sufficiently large s_{2n+2}. Proceeding by induction this extension procedure leads to a positive definite sequence s. $\square$

Combining Lemma 9.1 with Hamburger's Theorem 3.8 we obtain the following

Corollary 9.2 *The truncated Hamburger moment problem for each positive definite real $2n$-sequence is solvable.*

As in Chap. 5 we define a scalar product $\langle \cdot, \cdot \rangle_s$ on $\mathbb{R}[x]_{n+1}$ by

$$\langle p, q \rangle_s = L_s(pq), \quad p, q \in \mathbb{R}[x]_{n+1}. \tag{9.2}$$

For this scalar product on $\mathbb{R}[x]_{n+1}$ and at a few other places we need an extension of s to a positive definite $(2n + 2)$-sequence. Such an extension exists by Lemma 9.1. We fix this extension. Note that (9.2) also depends on the numbers s_{2n+1}, s_{2n+2}.

Definition 9.3 A polynomial $P \in \mathbb{R}[x]_{n+1}, P \neq 0$, is called *quasi-orthogonal* of rank $n + 1$ if

$$L_s(Px^j) = 0 \quad \text{for} \quad j = 0,\dots,n-1. \tag{9.3}$$

(Since $\deg(Px^j) \leq 2n$, (9.3) does not require an extension of the $2n$-sequence s.)

Proceeding as in Sect. 5.1 we obtain a unique sequence $p_0,\dots,p_{n+1}$ of orthonormal polynomials of the unitary space $(\mathbb{R}_{n+1}[x], \langle \cdot, \cdot \rangle_s)$ such that $\deg p_k = k$ and the leading coefficient of p_k is positive for $k = 0, \ldots, n+1$. Since then $L_s(p_k x^j) = 0$ for $j < k$, p_n and p_{n+1} are quasi-orthogonal of rank $n+1$. Therefore, for any $t \in \mathbb{R}$,

$$P_{(t)}(x) = p_{n+1}(x) + t p_n(x) \tag{9.4}$$

is quasi-orthogonal of rank $n+1$ and degree $n+1$. Up to constant multiples all quasi-orthogonal polynomials of rank $n+1$ and degree $n+1$ are of this form.

In what follows we fix a quasi-orthogonal polynomial P of rank $n+1$ and degree $n+1$. By Proposition 5.28, P has $n+1$ real zeros $\lambda_1 < \lambda_2 < \cdots < \lambda_{n+1}$. Let us introduce the quantities

$$\pi_j(x) := \frac{P(x)}{(x - \lambda_j) P'(\lambda_j)} \quad \text{and} \quad m_j := L_s(\pi_j^2) > 0, \quad j = 1, \ldots, n+1, \tag{9.5}$$

$$Q(z) := L_{s,x}\left(\frac{P(x) - P(z)}{x - z} \right) \tag{9.6}$$

and define an $(n+1)$-atomic measure μ_P by

$$\mu_P = \sum_{j=1}^{n+1} m_j \delta_{\lambda_j}. \tag{9.7}$$

Since $\frac{P(x) - P(z)}{x - z}$ is a polynomial in x of degree at most n, the functional $L_{s,x}$ applies to this polynomial and $Q(z)$ is a polynomial of degree at most n. Clearly,

$$\pi_j(x) = \frac{(x - \lambda_1) \ldots (x - \lambda_{j-1})(x - \lambda_{j+1}) \ldots (x - \lambda_{n+1})}{(\lambda_j - \lambda_1) \ldots (\lambda_j - \lambda_{j-1})(\lambda_j - \lambda_{j+1}) \ldots (\lambda_j - \lambda_{n+1})},$$

$$\pi_j(\lambda_i) = \delta_{ij} \quad \text{for } i, j = 1, \ldots, n+1. \tag{9.8}$$

Now can can state and prove the first main theorem of this section.

Theorem 9.4 *For each polynomial $f \in \mathbb{R}[x]_{2n}$ we have*

$$L_s(f) = \sum_{j=1}^{n+1} m_j f(\lambda_j) \equiv \int f(x) \, d\mu_P(x). \tag{9.9}$$

Further, for $j = 1, \ldots, n+1$ and $z \in \mathbb{C} \backslash \{\lambda_1, \ldots, \lambda_{n+1}\}$,

$$m_k = \frac{Q(\lambda_k)}{P'(\lambda_k)} = L_s(\pi_k) \quad \text{and} \quad -\frac{Q(z)}{P(z)} = \sum_{j=1}^{n+1} \frac{m_j}{\lambda_j - z} \equiv \int \frac{d\mu_P(x)}{x - z}. \tag{9.10}$$

Proof For each $p \in \mathbb{R}[x]_n$ we have the Lagrange interpolation formula

$$p(x) = \sum_{j=1}^{n+1} \pi_j(x)\, p(\lambda_j). \tag{9.11}$$

(To prove (9.11) it suffices to note the the difference of both sides is a polynomial of degree at most n that vanishes at $n + 1$ points λ_j, so it is identically zero.)

Now fix $f \in \mathbb{R}[x]_{2n}$. Since $\deg(P) = n + 1$, there are polynomials $q_f \in \mathbb{R}[x]_{n-1}$ and $p_f \in \mathbb{R}[x]_n$ such that

$$f(x) = P(x)q_f(x) + p_f(x). \tag{9.12}$$

The defining relation (9.3) for the quasi-orthogonal polynomial P yields $L_s(Pq_f) = 0$, so that $L_s(f) = L_s(p_f)$. Further, since $P(\lambda_j) = 0$ and hence $p_f(\lambda_j) = f(\lambda_j)$ by (9.12), combining the relations $L_s(f) = L_s(p_f)$ and (9.11) we obtain

$$L_s(f) = \sum_{j=1}^{n+1} L_s(\pi_j) f(\lambda_j). \tag{9.13}$$

Applying this formula to $f = \pi_k^2$ by using (9.5) and $\pi_j(\lambda_k) = \delta_{jk}$ by (9.8) we get

$$m_k = L_s(\pi_k^2) = \sum_{j=1}^{n+1} L_s(\pi_j)\pi_k(\lambda_j)^2 = L_s(\pi_k)\,, \quad k = 1,\ldots,n+1. \tag{9.14}$$

Inserting (9.14) into (9.13) we obtain (9.9).

Combining (9.14) with (9.5) and (9.6) we get

$$m_k = L_s(\pi_k) = L_s\left(\frac{P(x)}{(x-\lambda_k)P'(\lambda_k)}\right) = \frac{1}{P'(\lambda_k)} L_s\left(\frac{P(x)-P(\lambda_k)}{x-\lambda_k}\right) = \frac{Q(\lambda_k)}{P'(\lambda_k)}.$$

This yields the first half of (9.10).

Finally, we prove the second half of (9.10). Since $\deg P = n + 1$, P has $n + 1$ simple real zeros, and $\deg Q \le n$, there exists a partial fraction decomposition

$$-\frac{Q(z)}{P(z)} = \sum_{j=1}^{n+1} \frac{\mu_j}{\lambda_j - z} \tag{9.15}$$

with $\mu_j \in \mathbb{R}$. Here the coefficients μ_j are

$$\mu_j = \lim_{z\to\lambda_j} \frac{(z-\lambda_j)Q(z)}{P(z)} = \frac{Q(\lambda_j)}{P'(\lambda_j)} = m_j$$

by the first part of (9.10). Inserting $\mu_j = m_j$ into (9.15) gives the second half of (9.10). $\qquad\square$

Formula (9.9) is usually called a *Gaussian quadrature formula*. If the functional L_s is of the form $L_s(f) = \int f(x)d\mu(x)$ for some measure μ, then (9.9) reads as

$$\int f(x)\,d\mu(x) = \sum_{j=1}^{n+1} m_j f(\lambda_j). \qquad (9.16)$$

The numbers λ_j are called *nodes* and the numbers m_j *weights* of the quadrature formula. For general functions f the sum on the right of (9.16) can be considered as an approximation of the integral on the left.

Theorem 9.4 says that the identity (9.16) holds for all polynomials $f \in \mathbb{R}[x]_{2n}$. But (9.16) does not hold for $f \in \mathbb{R}[x]_{2n+2}$. For instance, if $f(x) = \prod_{j=1}^{n+1}(x - \lambda_j)^2$, then the right-hand side of (9.16) vanishes, but $L_s(f) = \int f\,d\mu > 0$, because s is positive definite.

Before we continue we derive a nice application of formula (9.9).

Corollary 9.5 *Let P and $\tilde{P}$ be two quasi-orthogonal polynomials of rank $n + 1$ and degree $n + 1$ with zeros $\lambda_1 < \cdots < \lambda_{n+1}$ and $\tilde{\lambda}_1 < \cdots < \tilde{\lambda}_{n+1}$, respectively. Then either $\tilde{P}$ is a multiple of P or the sequences $(\lambda_1, \cdots, \lambda_{n+1})$ and $(\tilde{\lambda}_1, \cdots, \tilde{\lambda}_{n+1})$ are stricly interlacing, that is, for any k there is a j such that $\lambda_k < \tilde{\lambda}_j < \lambda_{k+1}$. In particular, if $t, \tilde{t} \in \mathbb{R}$ and $t \neq \tilde{t}$, then the zeros of the quasi-orthogonal polynomials $P_{(t)}$ and $P_{(\tilde{t})}$ defined by (9.4) are strictly interlacing.*

Proof Fix $k \in \{1, \ldots, n\}$ and put

$$f_k(x) = \frac{1}{(x - \lambda_k)(x - \lambda_{k+1})} \prod_{j=1}^{n+1}(x - \lambda_j)^2.$$

Then $f_k \in \mathbb{R}[x]_{2n}$, so (9.9) applies and yields $L_s(f_k) = 0$, since $f_k(\lambda_j) = 0$ for $j = 1, \ldots, n + 1$. Applying the same formula with P replaced by $\tilde{P}$ we get

$$L_s(f_k) = \sum_{j=1}^{n+1} \widetilde{m}_j f_k(\tilde{\lambda}_j) = 0. \qquad (9.17)$$

The zeros of f_k are precisely the numbers λ_l and we have $f_k(x) < 0$ for $x \in (\lambda_k, \lambda_{k+1})$ and $f_k(x) \geq 0$ for $x \leq \lambda_k$ and $x \geq \lambda_{k+1}$. Therefore, if there were no j such that $\lambda_k < \tilde{\lambda}_j < \lambda_{k+1}$, then all (!) numbers $\tilde{\lambda}_j$ would have to be contained in the zero set of f_k. This implies that $\lambda_k = \tilde{\lambda}_k$ for all k, so P is a constant multiple of $\tilde{P}$. $\qquad\square$

As in Sect. 5.2 the orthonormal polynomials $p_k, k \leq n$, satisfy the recurrence relation (5.9); let a_n be the corresponding Jacobi coefficient from (5.9).

In the case $P = p_{n+1}$ we have the following stronger result than Theorem 9.4.

Theorem 9.6 *If $P = p_{n+1}$, then formula (9.9) holds for all $f \in \mathbb{R}[x]_{2n+1}$ and*

$$m_j^{-1} = a_n p_n(\lambda_j) p_{n+1}'(\lambda_j) = \sum_{k=0}^n p_k(\lambda_j)^2, \ j = 1, \dots, n+1. \qquad (9.18)$$

Conversely, if there are real numbers $\tilde{\lambda}_j$ and positive numbers $\tilde{m}_j, j = 1 \dots, n+1$, such that $\tilde{\lambda}_1 < \cdots < \tilde{\lambda}_{n+1}$ and

$$L_s(f) = \sum_{k=1}^{n+1} \tilde{m}_j f(\tilde{\lambda}_k) \quad \text{for all} \quad f \in \mathbb{R}[x]_{2n+1}, \qquad (9.19)$$

then $\tilde{\lambda}_j = \lambda_j$ and $\tilde{m}_j = m_j$ for $j=1, \dots, n + 1$ and formula (9.19) coincides with (9.9).

Proof Let $f \in \mathbb{R}[x]_{2n+1}$. Then the polynomial q_f in (9.12) belongs to $\mathbb{R}[x]_n$. Therefore, since $P = p_{n+1}$, we have $L_s(Pq_f) = 0$. Proceeding now as in the proof of Theorem 9.4 it follows that formula (9.9) holds for f as well.

Next we prove (9.18). Since $p_0, \dots, p_{n+1}$ are orthonormal polynomials for the extended sequence (by Lemma 9.1), the formulas (5.47) and (5.51) proved in Sect. 5.5 are valid. We set $z = \lambda_j$ in the Christoffel formula (5.47) and remember that $p_{n+1}(\lambda_j) = 0$. Inserting the definition (9.5) of $\pi_j(x)$ for $P = p_{n+1}$ we derive

$$\sum_{k=0}^n p_k(x) p_k(\lambda_j) = a_n \frac{p_{n+1}(x) p_n(\lambda_j)}{x - \lambda_j} = a_n \pi_j(x) p_{n+1}'(\lambda_j) p_n(\lambda_j).$$

Applying the functional L_s by using that $p_0 = s_0^{-1/2}$ and $m_j = L_s(\pi_j)$ we get

$$1 = s_0^{-1} L_s(1) = L_s(p_0) p_0(\lambda_j) = \sum_{k=0}^n L_s(p_k) p_k(\lambda_j) = a_n p_n(\lambda_j) p_{n+1}'(\lambda_j) m_j$$

which yields the first equality of (9.18). The second equality follows from the first by setting $x = \lambda_j$ in formula (5.51) and using once again that $p_{n+1}(\lambda_j) = 0$.

Finally, we prove the uniqueness assertion. We set

$$\tilde{P}(x) = (x - \tilde{\lambda}_1) \dots (x - \tilde{\lambda}_{n+1})$$

and define $\tilde{\pi}_j$ and $\tilde{m}_j$ by (9.8) with P replaced by $\tilde{P}$ and λ_j by $\tilde{\lambda}_j$.

By $\deg \tilde{\pi}_j = n$ and $P = p_{n+1}$, we have $L_s(P \tilde{\pi}_j) = 0$. Since $\deg(P \tilde{\pi}_j) \leq 2n + 1$, formula (9.19) applies to $P \tilde{\pi}_j$ as well. Using the relation $\tilde{\pi}_j(\tilde{\lambda}_k) = \delta_{kj}$ (by 9.8)) we obtain $0 = L_s(P \tilde{\pi}_j) = \tilde{m}_j P(\tilde{\lambda}_j)$. Thus, $P(\tilde{\lambda}_j) = 0$, since $\tilde{m}_j > 0$. That is, the numbers $\tilde{\lambda}_j$ are the zeros of P. Therefore, $\tilde{\lambda}_j = \lambda_j$ for $j = 1, \dots, n + 1$ and $\tilde{P}$ is a constant multiple of $P = p_{n+1}$. Hence $\tilde{\pi}_j = \pi_j$, so that $m_j = L_s(\pi_j) = L_s(\tilde{\pi}_j) = \tilde{m}_j$ by (9.10) for $j = 1, \ldots, n+1$. $\qquad\square$

The next theorem summarizes and restates some of the preceding results on the quadrature formula (9.9) in terms of the truncated Hamburger moment problem.

Theorem 9.7 *Suppose that $s = (s_j)_{j=0}^{2n}$ is a real positive definite $2n$-sequence. Let μ_t denote the $(n+1)$-atomic measure $\mu_{P_{(t)}}$ for the quasi-orthogonal polynomial $P_{(t)}$ defined by (9.4) and (9.7), respectively.*

  (i) *Each measure μ_t, $t \in \mathbb{R}$, is a solution of the truncated Hamburger moment problem for s, that is, we have $s_j = \int x^j \, d\mu_t(x)$ for $j = 0, \ldots, 2n$.*

  (ii) *For $t, \tilde{t} \in \mathbb{R}$, $t \neq \tilde{t}$, the atoms of μ_t and $\mu_{\tilde{t}}$ are strictly interlacing, so μ_t and $\mu_{\tilde{t}}$ have disjoint supports.*

  (iii) *For any $\zeta \in \mathbb{R}$ that is not a zero of $p_n(x)$, there exists a unique $t \in \mathbb{R}$ such that ζ is an atom of μ_t.*

  (iv) *For each real number s_{2n+1} there is a unique $(n+1)$-atomic measure ν such that $s_0, \ldots, s_{2n+1}$ are the moment of ν. In fact, ν is the measure μ_0 when the sequence $(s_j)_{j=0}^{2n+1}$ is extended to a positive definite $(2n+2)$-sequence $(s_j)_{j=0}^{2n+2}$.*

Proof (i) and (ii) follow from formula (9.9) and Corollary 9.5, respectively.

  (iii) Setting $t := -p_n(\zeta)^{-1} p_{n+1}(\zeta)$, the number ζ is a zero of the quasi-orthogonal polynomial

$$P_{(t)}(x) = p_{n+1}(x) - p_n(\zeta)^{-1} p_{n+1}(\zeta) p_n(x)$$

and hence an atom of μ_t. (ii) yields the uniqueness assertion.

  (iv) We extend $(s_j)_{j=0}^{2n+1}$ to a positive definite real sequence $(s_n)_{n \in \mathbb{N}_0}$ (by Lemma 9.1) and put $P := P_{(0)} \equiv p_{n+1}$. Then, by Theorem 9.6, (9.9) holds for the polynomials $x^0, \ldots, x^{2n+1}$ which means that the measure μ_0 has the moments $s_0, \ldots, s_{2n+1}$. The uniqueness of $\nu = \mu_0$ follows from the second part of Theorem 9.6. $\qquad\square$

We restate assertion (iv) of the preceding theorem separately as

Corollary 9.8 *For each positive definite real sequence $(s_j)_{j=0}^{2n}$ and each real number ξ there exists a unique $(n+1)$-atomic measure ν_ξ such that*

$$s_j = \int x^j \, d\nu_\xi(x) \ \text{for } j = 0, \ldots, 2n \ \text{ and } \ \xi = \int x^{2n+1} \, d\nu_\xi(x).$$

9.2 Hamburger's Theorem and Markov's Theorem Revisited

In this section, we assume that $s = (s_n)_{n \in \mathbb{N}_0}$ is a real **positive definite sequence**. Recall that p_n and q_n are the orthogonal polynomials of the first and the second kind, respectively, associated with s.

We apply Theorem 9.4 with $P = p_{n+1}, k = n + 1$, and set $\mu_k := \mu_P$. Then $q_k = Q$ by (9.6). Let $\lambda_1^{(n)} < \cdots < \lambda_n^{(n)}$ denote the zeros of p_k and $m_j^{(k)}$ the corresponding weights. From (9.9), applied with $f = x^j$, and (9.10) we obtain

$$s_j = \int_{\mathbb{R}} x^j \, d\mu_k(x), \quad j = 0, \ldots, k - 1, \tag{9.20}$$

$$-\frac{q_k(z)}{p_k(z)} = \sum_{j=1}^{k} \frac{m_j^{(k)}}{\lambda_j^{(k)} - z} = \int_{\mathbb{R}} \frac{d\mu_k(x)}{x - z}, \quad z \in \mathbb{C} \setminus \{\lambda_1^{(k)}, \ldots, \lambda_k^{(k)}\}. \tag{9.21}$$

Since $\mu_k(\mathbb{R}) = s_0$ for all k, $(s_0^{-1} \mu_k)_{k \in \mathbb{N}_0}$ is a sequence of probability measures. Hence it follows from Theorem A.9 that there exists a subsequence $(\mu_{k_n})_{n \in \mathbb{N}_0}$ which converges vaguely to some measure $\mu \in M_+(\mathbb{R})$.

Let L^{μ_k} be the linear functional on $\mathbb{R}[x]_{k-1}$ defined by $L^{\mu_k}(f) = \int f \, d\mu_k$. Then, by (9.20), $(L^{\mu_k})_{k \in \mathbb{N}}$ is a directed sequence of linear functionals on $\mathbb{R}[x]$ according to Definition 1.17. Hence $\lim_{n \to \infty} L^{\mu_{k_n}} = L^{\mu}$ by Lemma 1.18 (or Theorem 1.20) and

$$L^{\mu}(x^j) = L^{\mu_{k_n}}(x^j) = \int_{\mathbb{R}} x^j \, d\mu_{k_n} = s_j \text{ for } k_n > j$$

by (9.20). Therefore, μ *is a representing measure for* s, that is, μ solves the Hamburger moment problem for s. Thus we have given a *third proof* of the main implication (ii)→(i) of Hamburger's theorem 3.8 for positive definite sequences.

By formula (5.60) the sequence $(\lambda_1^{(k)})_{k \in \mathbb{N}}$ is decreasing and the sequence $(\lambda_k^{(k)})_{k \in \mathbb{N}}$ is increasing. Therefore, the limits

$$\alpha_s := \lim_{k \to \infty} \lambda_1^{(k)} \in \{-\infty\} \cup \mathbb{R} \quad \text{and} \quad \beta_s := \lim_{k \to \infty} \lambda_k^{(k)} \in \mathbb{R} \cup \{+\infty\}$$

exist, see Corollary 5.33. Let $\mathcal{J}_s$ denote the smallest closed interval which contains (α_s, β_s). That is, $\mathcal{J}_s$ is the smallest closed interval which contains *all zeros* of the orthonormal polynomials $p_k, k \in \mathbb{N}$. Note that α_s, β_s, and $\mathcal{J}_s$ depend only on the sequence s, but not on any representing measure. Since $\operatorname{supp} \mu_k \subseteq [\lambda_1^{(k)}, \lambda_k^{(k)}] \subseteq \mathcal{J}_s$ for $k \in \mathbb{N}$ by (9.21), it follows that $\operatorname{supp} \mu \subseteq \mathcal{J}_s$. In the literature [Chi1] a solution of the moment problem with support contained in $\mathcal{J}_s$ is called a *natural solution*.

Now we use formula (9.21) to develop a second approach to Markov's theorem 6.29.

Theorem 9.9 *Suppose that s is a determinate positive definite sequence s. If μ and $\mathcal{J}_s$ are as above, then*

$$-\lim_{k \to \infty} \frac{q_k(z)}{p_k(z)} = \int_{\mathcal{J}_s} \frac{d\mu(x)}{x - z} \quad \text{for} \quad z \in \mathbb{C} \setminus \mathcal{J}_s, \tag{9.22}$$

where the convergence is uniform on compact subsets of $\mathbb{C} \setminus \mathcal{J}_s$.

Proof By Theorem 1.20, $(\mu_k)_{k\in\mathbb{N}}$ converges vaguely to μ. Let $z \in \mathbb{C}\backslash\mathcal{J}_s$. Then the function $f_z(x) = \frac{1}{x-z}$ is in $C_0(\mathcal{J}_s)$. Therefore, since $\operatorname{supp}\mu_k \subseteq \mathcal{J}_s$ and $\operatorname{supp}\mu \subseteq \mathcal{J}_s$, Proposition A.9 implies that

$$\lim_{k\to\infty} \int_{\mathcal{J}_s} \frac{d\mu_n(x)}{x-z} = \int_{\mathcal{J}_s} \frac{d\mu(x)}{x-z}. \tag{9.23}$$

Then (9.22) is obtained by inserting (9.21) on the left-hand side of (9.23).

Now we prove the assertion concerning the uniform convergence. Let K be a compact subset of $\mathbb{C}\backslash\mathcal{J}_s$. Then $c := \operatorname{dist}(K, \mathcal{J}_s) > 0$. By (9.21), the convergence in (9.22) is uniform on K provided this holds for the convergence in (9.23). Let $\varepsilon > 0$ be given. Since K is compact, there are finitely many numbers $z_1, \ldots, z_k$ such that the open discs centered at z_j with radius ε cover K. Given $z \in K$ we choose j such that $|z - z_j| < \varepsilon$. Therefore, we have

$$\left| \frac{1}{x-z} - \frac{1}{x-z_j} \right| = \frac{|z-z_j|}{|(x-z)(x-z_j)|} \leq \frac{\varepsilon}{c^2} \quad \text{for} \quad x \in \mathcal{J}_s.$$

Let $k \in \mathbb{N}$. Then, using that $\operatorname{supp}\mu_k \subseteq \mathcal{J}_s$, $\operatorname{supp}\mu \subseteq \mathcal{J}_s$, and $\mu_k(\mathbb{R}) = \mu(\mathbb{R}) = s_0$, we derive

$$\left| \int \frac{d\mu_k(x)}{x-z} - \int \frac{d\mu(x)}{x-z} \right| \tag{9.24}$$

$$\leq \left| \int \frac{d\mu_k(x)}{x-z} - \int \frac{d\mu_k(x)}{x-z_j} \right| + \left| \int \frac{d\mu_k(x)}{x-z_j} - \int \frac{d\mu(x)}{x-z_j} \right| + \left| \int \frac{d\mu(x)}{x-z_j} - \int \frac{d\mu(x)}{x-z} \right|$$

$$\leq s_0 \frac{\varepsilon}{c^2} + \left| \int \frac{d\mu(x)}{x-z_j} - \int \frac{d\mu_k(x)}{x-z_j} \right| + s_0 \frac{\varepsilon}{c^2}. \tag{9.25}$$

For each $j = 1, \ldots, k$, the middle term in (9.25) converges to zero by (9.23). Hence for sufficiently large k the expression in (9.24) is less than $3\varepsilon s_0 c^{-2}$. This proves that the convergence in (9.23) is uniform on K. $\qquad\square$

9.3 The Reproducing Kernel and the Christoffel Function

Throughout this section, $s = (s_j)_{j=0}^{2n}$ is again a real **positive definite $2n$-sequence**.

Let $(\mathbb{C}[x]_n, \langle\cdot,\cdot\rangle_s)$ be the complex Hilbert space with scalar product defined by $\langle p, q\rangle_s = L_s(p\,\overline{q})$, $p, q \in \mathbb{C}[x]_n$, where $\overline{q}(x) := \sum_j \overline{a}_j x^j$ for $q(x) = \sum_j a_j x^j \in \mathbb{C}[x]_n$.

As above, $p_0, \ldots, p_n$ denote the corresponding orthonormal polynomials. Recall that $p_j \in \mathbb{R}[x]$ and hence $\overline{p_j(y)} = p_j(\overline{y})$ for $y \in \mathbb{C}$. Then

$$K_n(x, y) := \sum_{k=0}^{n} p_k(x)\overline{p_k(y)}, \quad x, y \in \mathbb{C}, \tag{9.26}$$

is a *reproducing kernel* on the Hilbert space $(\mathbb{C}[x]_n, \langle \cdot, \cdot \rangle_s)$, that is, we have

$$p(y) = \langle p(x), K_n(x, y) \rangle_s \quad \text{and} \quad \langle K_n(x, y), p(x) \rangle_s = \overline{p(y)}, \quad p \in \mathbb{C}[x]_n, \ y \in \mathbb{C}.$$
(9.27)

As a sample we prove the second equality; the proof of the first equality is similar. We write $p = \sum_{j=0}^n c_j p_j(x)$ and derive

$$\langle K_n(x, y), p(x) \rangle_{s,x} = \sum_{j,k=0}^n \overline{p_k(y)} \, \overline{c_j} \, \langle p_k(x), p_j(x) \rangle_s = \sum_{j,k=0}^n \overline{c_j p_k(y)} \, \delta_{jk} = \overline{p(y)}.$$

Note that $K_n(z, z) = \sum_{k=0}^n |p_k(z)|^2 > 0$ for all $z \in C$, since $p_0(z) > 0$.

Definition 9.10 The n-th *Christoffel function* is

$$\rho_n(z) := K_n(z, z)^{-1} = \left(\sum_{k=0}^n |p_k(z)|^2 \right)^{-1}, \quad z \in \mathbb{C}.$$
(9.28)

The following proposition expresses the kernel $K_n(x, y)$ and the Christoffel function $\rho_n(z) = K_n(z, z)^{-1}$ in terms of determinants involving the moments.

Proposition 9.11 *For $x, y \in \mathbb{C}$ and $n \in \mathbb{N}_0$ we have*

$$K_n(x, \overline{y}) = -D_n^{-1} \begin{vmatrix} 0 & 1 & x & \dots & x^n \\ 1 & s_0 & s_1 & \dots & s_n \\ y & s_1 & s_2 & \dots & s_{n+1} \\ y^2 & s_2 & s_3 & \dots & s_{n+2} \\ \dots\dots\dots & & \dots\dots \\ y^n & s_n & s_{n+1} & \dots & s_{2n} \end{vmatrix}.$$
(9.29)

Proof Let $G(x, y)$ denote the determinant on the right. Fix $y \in \mathbb{C}$ and $j \in \{0, \dots, n\}$. We compute $L_{s,x}(x^j G(x, y))$. That is, we multiply the first row in the determinant by x^j and then apply the functional $L_{s,x}$ by using the multilinearity of determinants. As a result, the first row of $G(x, y)$ is replaced by $(0, s_j, s_{j+1}, \dots, s_{n+j})$. Then we subtract the $(j + 2)$-th row $(y^j, s_j, s_{j+1}, \dots, s_{n+j})$. This does not change the determinant and yields the first row $(-y^j, 0, \dots, 0)$. Finally, we develop the resulting determinant after the first row and get $L_{s,x}(G(x, y)x^j) = -y^j D_n$.

Set $H(x, \overline{y}) := -D_n^{-1} G(x, y)$. Since $G(x, y) \in \mathbb{R}[x, y]$, we have

$$\langle x^j, H(x, y) \rangle_{s,x} = -D_n^{-1} L_{s,x}(x^j G(x, y)) = -D_n^{-1}(-y^j D_n) = y^j.$$

This implies that $\langle p(x), H(x, y) \rangle_{s,x} = p(y)$ for all $p \in \mathbb{C}[x]_n$. In particular,

$$\langle K_n(x, z), H(x, y) \rangle_{s,x} = K_n(y, z) \quad \text{for } y, z \in \mathbb{C}.$$
(9.30)

On the other hand, it follows from the definition of G that $\overline{G(z,\bar{y})} = G(y,\bar{z})$. This implies that $\overline{H(z,y)} = H(y,z)$. Therefore, by the second equality of (9.27),

$$\langle K_n(x,z), H(x,y)\rangle_{s,x} = \overline{H(z,y)} = H(y,z) \quad \text{for } y, z \in \mathbb{C}. \tag{9.31}$$

Comparing (9.30) and (9.31) we get $K_n(y,z) = H(y,z)$, which is the assertion. $\square$

The next proposition characterizes the function ρ_n by an extremal property.

Proposition 9.12 *For any $z \in \mathbb{C}$ we have*

$$\min\{L_s(p\,\bar{p}) : p \in \mathbb{C}[x]_n, \ p(z) = 1\} = K_n(z,z)^{-1} = \rho_n(z) \tag{9.32}$$

and the minimum is attained at p if and only if $p(x) = K_n(x,z)\rho_n(z)$.

The proof will be derived below from Proposition 9.14. Since the latter result is used in Sect. 10.7 in the real setting, we formulate it in the real and complex cases. It is based on the following fact from elementary Hilbert space theory.

Lemma 9.13 *Let $f \neq 0$ be an element of a real or complex unitary space $\mathcal{H}$. Then*

$$\min\{\|g\|^2 : g \in \mathcal{H}, \ \langle g,f\rangle = 1\} = \|f\|^{-2} \tag{9.33}$$

and the minimum (9.33) is attained at g if and only if $g = f\|f\|^{-2}$.

Proof Put $g_0 = f\|f\|^{-2}$. If $\langle g,f\rangle = 1$, then $\langle g - g_0, f\rangle = \langle g,f\rangle - \|f\|^{-2}\langle f,f\rangle = 0$. Thus $g - g_0 \perp g_0$. Therefore, by Pythagoras' theorem,

$$\|g\|^2 = \|g_0 + g - g_0\|^2 = \|g_0\|^2 + \|g - g_0\|^2.$$

Hence the minimum is attained at g if and only if $g = g_0$. $\square$

Proposition 9.14 *Let $\mathbb{K} = \mathbb{C}$ or $\mathbb{K} = \mathbb{R}$. Suppose that $s = (s_j)_{j=0}^{2n}$ is a positive definite real sequence and let $p_0, \ldots, p_n$ be the corresponding orthonormal polynomials. Then, for any $z \in \mathbb{K}$ we have*

$$\min\{L_s(p\,\bar{p}) : p \in \mathbb{K}[x]_n, \ p(z) = 1\} = \left(\sum_{j=0}^{n} |p_j(z)|^2\right)^{-1} \tag{9.34}$$

and the minimum is attained at p if and only if

$$p(x) = \sum_{j=0}^{n} p_j(x)\overline{p_j(z)}\left(\sum_{i=0}^{n} |p_i(z)|^2\right)^{-1}. \tag{9.35}$$

Proof Let $p \in \mathbb{K}[x]_n$. We develop p with respect to the orthonormal basis $\{p_j\}$ of the unitary space $(\mathbb{K}[x]_n, \langle \cdot, \cdot \rangle_s)$ and obtain

$$p = \sum_{j=0}^{n} c_j p_j, \text{ where } c_j = \langle p, p_j \rangle_s, \text{ and } L_s(p\overline{p}) = \|p\|_s^2 = \sum_{j=0}^{n} |c_j|^2.$$

Then $p(z) = \sum_j c_j p_j(z)$. Hence, the problem in (9.34) is to minimize $\sum_j |c_j|^2$ for $g := (c_0, \dots, c_n)^T \in \mathbb{K}^{n+1}$ under the constraint $\sum_j c_j p_j(z) = 1$. This is solved by Lemma 9.13, applied to the unitary space $\mathbb{K}^{n+1}$ and $f := (\overline{p_0(z)}, \dots, \overline{p_n(z)})^T$. The minimum is $\|f\|^{-2}$ and it is attained if and only if $c_j = \overline{p_j(z)} \|f\|^{-2}$ for $j = 0, \dots, n$. Inserting these facts yields the asserted formulas (9.34) and (9.35). $\qquad\square$

Proof of Proposition 9.12 We apply Proposition 9.14 in the case $\mathbb{K} = \mathbb{C}$. By (9.28) the minimum in (9.34) is $\rho_n(z)$. Comparing (9.35) and (9.26) we conclude that the minimum is attained if and only if $p(x) = K_n(x, z)\rho_n(z)$. $\qquad\square$

9.4   Positive Semidefinite $2n$-Sequences

In the rest of this chapter we investigate positive **semidefinite** finite sequences.

The following theorem characterizes the positive semidefiniteness of a $2n$-sequence in terms of an integral representation (9.36).

Theorem 9.15 *For any real sequence $s = (s_j)_{j=0}^{2n}$ the following five statements are equivalent:*

(i) *s is positive semidefinite, that is,*

$$\sum_{k,l=0}^{n} s_{k+l} c_k c_l \geq 0 \quad \text{for } (c_0, \dots, c_n)^T \in \mathbb{R}^{n+1}.$$

(ii) $H_n(s) \succeq 0.$
(iii) $L_s(p^2) \geq 0$ *for all $p \in \mathbb{R}[x]_n$.*
(iv) *There exists a Radon measure μ on $\mathbb{R}$ and a real number $a \geq 0$ such that $\mathbb{R}[x]_{2n} \subseteq \mathcal{L}^1(\mathbb{R}, \mu)$,*

$$s_j = \int_{\mathbb{R}} x^j \, d\mu(x) \text{ for } j = 0, \dots, 2n-1, \text{ and } s_{2n} = a + \int_{\mathbb{R}} x^{2n} \, d\mu(x).$$
$$\tag{9.36}$$

(v) *There exists a k-atomic measure μ on $\mathbb{R}$, where $k \leq 2n + 1$, and a constant $a \geq 0$ such that (9.36) holds.*

Proof The equivalence of (i), (ii), and (iii) is easily verified by the same reasoning used in the proof of Hamburger' theorem 3.8; we do not repeat it here.

(iii)→(v) The proof is based on Proposition 1.26. Let $\mathcal{X} := \mathbb{R} \cup \{\infty\}$ denote the one point compactification of $\mathbb{R}$. Then the functions

$$u_j(x) := x^j(1 + x^{2n})^{-1}, \quad j = 0, \ldots, 2n,$$

extend to continuous functions on $\mathcal{X}$ by setting $u_j(\infty) = 0$ for $j = 0, \ldots, 2n - 1$ and $u_{2n}(\infty) = 1$. Let E be the span of these functions in $C(\mathcal{X}; \mathbb{R})$.

We define a linear functional $\tilde{L}$ on E by $\tilde{L}(u_j) = L_s(x^j), j = 0, \ldots, 2n$. Let $f \in E_+$. Then $g := (1 + x^{2n})f \in \mathbb{R}[x]_{2n}$ is nonnegative on $\mathbb{R}$. Therefore, by Proposition 3.1, we have $g = p^2 + q^2$ with polynomials $p, q \in \mathbb{R}[x]_n$. Thus we obtain $f = (p^2 + q^2)(1 + x^{2n})^{-1}$ and hence $\tilde{L}(f) = L_s(p^2 + q^2)$ by the definition of $\tilde{L}$. Since $L_s(p^2) \geq 0$ and $L_s(q^2) \geq 0$ by (iii), $\tilde{L}(f) \geq 0$. Thus, Proposition 1.26 applies with $g = u_{2n}$ and there is a k-atomic positive measure $\tilde{\mu}$ on $\mathcal{X}, k \leq 2n + 1$, such that $\tilde{L}(f) = \int f \, d\tilde{\mu}$ for $f \in E$.

Set $a = \tilde{\mu}(\{\infty\})$. We define atomic measures $\hat{\mu}$ and μ on $\mathbb{R}$ by $\hat{\mu}(M) := \tilde{\mu}(M)$ for $M \subseteq \mathbb{R}$ and $d\mu := (1 + x^{2n})^{-1} d\hat{\mu}$. For $j = 0, \ldots, 2n$ we compute

$$s_j = L_s(x^j) = \tilde{L}(u_j) = \int_{\mathcal{X}} u_j(x) \, d\tilde{\mu} = au_j(\infty)$$

$$+ \int_{\mathbb{R}} \frac{x^j}{1 + x^{2n}} \, d\hat{\mu} = a\delta_{j,2n} + \int_{\mathbb{R}} x^j \, d\mu,$$

which proves (v).

(v)→(iv) is trivial.

(iv)→(ii) Let $(c_0, \ldots, c_n)^T \in \mathbb{R}^{n+1}$. Using (9.36) we derive

$$\sum_{k,l=0}^{n} s_{k+l} c_k c_l = ac_n^2 + \sum_{k,l=0}^{n} \int_{\mathbb{R}} c_k c_l x^{k+l} d\mu = ac_n^2 + \int_{\mathbb{R}} \left(\sum_{k=0}^{n} c_k x^k \right)^2 d\mu \geq 0,$$

$$(9.37)$$

since $a \geq 0$. This proves (ii). □

The positive semidefiniteness of s is *not sufficient* for representing *all* numbers s_j as moments $s_j = \int x^j d\mu$, see Example 9.17 below. It implies only a representation (9.36) with $s_{2n} \geq \int x^{2n} d\mu$. Some authors consider (9.36) as the "right version" of the Hamburger truncated moment problem. We require that s_{2n} is the $(2n)$-th moment of μ as well, that is, $s_{2n} = \int x^{2n} d\mu$. To obtain the latter equality additional conditions are needed; they depend on whether or not we are dealing with the Hamburger or Stieltjes truncated moment problem and in the even or odd case.

Our main existence results on the Hamburger truncated moment problem are derived from the following corollary.

Corollary 9.16 *Let $s = (s_j)_{j=0}^{2n}$ be a real sequence and suppose that $H_n(s) \succeq 0$.*

(i) *There is a k-atomic positive measure μ on $\mathbb{R}$, $k \le 2n + 1$, such that*

$$L_s(p) = \int_{\mathbb{R}} p(x)\, d\mu \quad for \quad p \in \mathbb{R}[x]_{2n-1}. \tag{9.38}$$

If $L_s(f^2) = 0$ for some $f \in \mathbb{R}[x]_n$, then $\operatorname{supp}\mu \subseteq \mathcal{Z}(f)$.

(ii) *Suppose that there exists a polynomial $f \in \mathbb{R}[x]$ of degree n such that $L_s(f^2) = 0$. Then s is a truncated Hamburger moment sequence and $\operatorname{supp}\mu \subseteq \mathcal{Z}(f)$ for each representing measure μ of s.*

Proof

(ii) Let $f(x) = \sum_{k=0}^{n} c_k x^n$. Using (9.36) and (9.37) we get

$$0 = L_s(f^2) = \sum_{k,l=0}^{n} s_{k+l} c_k c_l = a c_n^2 + \int_{\mathbb{R}} f(x)^2\, d\mu. \tag{9.39}$$

Since $\deg(f) = n$, $c_n \neq 0$. Hence $a = 0$, so that s is truncated moment sequence by (9.36). Inserting $a = 0$ into (9.39), Proposition 1.23 implies that $\operatorname{supp}\mu \subseteq \mathcal{Z}(f)$.

(i) The first assertion is only a reformulation of Theorem 9.15 (i)→(v).

If $\deg(f) = n$, the second assertion follows from (ii). If $\deg(f) < n$, then $L_s(f^2) = \int f^2\, d\mu = 0$ by (9.38), so that $\operatorname{supp}\mu \subseteq \mathcal{Z}(f)$ again by Proposition 1.23. $\qquad\square$

Example 9.17 Let $s = (0, 0, 1)$. Then $H_1(s) \succeq 0$. Since $s_0 = 0$, s cannot be given by a Radon measure on $\mathbb{R}$, but s has a representation (9.36) with $\mu = 0$ and $a = 1$. $\circ$

Example 9.18 Let $s = (16, 0, 4, 0, 4)$. Then the sequence s has a representation (9.36) with $\mu = 8\delta_{-\frac{1}{2}} + 8\delta_{\frac{1}{2}}$ and $a = 3$. Since $\det H_2(s) > 0$, s is *positive definite*, so it also has representing measures on $\mathbb{R}$ by Corollary 9.2; one such measure is $\nu := 2\delta_{-1} + 12\delta_0 + 2\delta_1$. $\circ$

The next theorem is the counterpart of Theorem 9.15 for the positive half-line.

Theorem 9.19 *Let $s = (s_j)_{j=0}^{m}$ be a real sequence and set $Es := (s_1, \dots, s_m)$. The following are equivalent:*

(i) *$m = 2n$: $H_n(s) \succeq 0$ and $H_{n-1}(Es) \succeq 0$, that is,*

$$\sum_{k,l=0}^{n} s_{k+l} c_k c_l \ge 0 \quad and \quad \sum_{k,l=0}^{n-1} s_{k+l+1} c_k c_l \ge 0 \quad for \ (c_0, \dots, c_n)^T \in \mathbb{R}^{n+1}.$$

$m = 2n + 1$: $H_n(s) \succeq 0$ and $H_n(Es) \succeq 0$, that is,

$$\sum_{k,l=0}^{n} s_{k+l} c_k c_l \geq 0 \quad and \quad \sum_{k,l=0}^{n} s_{k+l+1} c_k c_l \geq 0 \quad for \ (c_0, \ldots, c_n)^T \in \mathbb{R}^{n+1}.$$

(ii) $m = 2n$: $L_s(p^2) \geq 0$ and $L_s(xq^2) \geq 0$ for $p \in \mathbb{R}[x]_n, q \in \mathbb{R}[x]_{n-1}$.
   $m = 2n + 1$: $L_s(p^2) \geq 0$ and $L_s(xq^2) \geq 0$ for $p, q \in \mathbb{R}[x]_n$.
(iii) *There exists a Radon measure μ on $\mathbb{R}_+$ and a real number $a \geq 0$ such that* $\mathbb{R}[x]_m \subseteq \mathcal{L}^1(\mathbb{R}_+, \mu)$,

$$s_j = \int_0^{\infty} x^j \, d\mu(x) \ for \ j = 0, \ldots, m-1, \quad and \quad s_m = a + \int_0^{\infty} x^m \, d\mu(x).$$
(9.40)

(iv) *There exists a k-atomic measure μ on $\mathbb{R}_+$, where $k \leq m + 1$, and a constant $a \geq 0$ such that (9.40) holds.*

Proof The proof follows the same reasoning as the proof of Theorem 9.15. We sketch only the proof of the main implication (ii)→(iv). Let $\mathcal{X} = \mathbb{R}_+ \cup \{+\infty\}$ denote the one point compactification of $\mathbb{R}_+$ and E the span of continuous functions

$$u_j(x) := x^j (1 + x^m)^{-1}, \quad j = 0, \ldots, m,$$

on $\mathcal{X}$, where $u_j(+\infty) := 0$ if $j = 0, \ldots, m-1$ and $u_m(+\infty) := 1$.

We define a linear functional $\tilde{L}$ on E by $\tilde{L}(u_j) := L_s(x_j)$. Let $f \in E_+$. Then $g = (1 + x^m)f \in \mathbb{R}[x]_m$ is nonnegative on $\mathbb{R}_+$. Therefore, by Proposition 3.2, the polynomial g is of the form $g = p + xq$, where $p \in \sum_n^2, q \in \sum_{n-1}^2$ if $m = 2n$ and $p, q \in \sum_n^2$ if $m = 2n + 1$. Because of these formulas the conditions in (ii) imply that $\tilde{L}(f) \geq 0$. Therefore, by Proposition 1.26, $\tilde{L}$ is given by some k-atomic positive measure $\tilde{\mu}$ on $\mathcal{X}$, where $k \leq m + 1$. Continuing as in the proof of Theorem 9.15 the formulas in (9.40) are derived. □

9.5 The Hankel Rank of a Positive Semidefinite 2n-Sequence

Let $s = (s_j)_{j=0}^{2n}$ be a real sequence. We define the *kernel vector space* of s by

$$\mathcal{N}_s := \{p \in \mathbb{R}[x]_n : L_s(pq) = 0 \quad for \ q \in \mathbb{R}[x]_n \}.$$

The first assertion of the following lemma gives another description of $\mathcal{N}_s$.

Lemma 9.20

(i) *For* $p = \sum_{j=0}^{n} a_j x^j \in \mathbb{R}[x]_n$, *set* $\vec{p} := (a_0, \ldots, a_n)^T \in \mathbb{R}^{n+1}$. *Then the map* $p \mapsto \vec{p}$ *is a bijection of* $\mathcal{N}_s$ *on* $\ker H_n(s)$. *In particular,*

$$\dim \mathcal{N}_s = \dim \ker H_n(s) \quad and \quad \operatorname{rank} H_n(s) = n + 1 - \dim \mathcal{N}_s. \qquad (9.41)$$

(ii) *If s is positive semidefinite, then* $\mathcal{N}_s = \{p \in \mathbb{R}[x]_n : L_s(p^2) = 0\}$.

Proof

(i) For $p = \sum_{j=0}^{n} a_j x^j \in \mathbb{R}[x]_n$ and $q = \sum_{j=0}^{n} b_j x^j \in \mathbb{R}[x]_n$, we compute

$$L(pq) = \sum_{i,j=0}^{2n} L_s(x^{i+j}) a_i b_j = \sum_{i,j=0}^{2n} s_{i+j} a_i b_j = \vec{q}^T H_n(s) \vec{p}.$$

Therefore, $p \in \mathcal{N}_s$ if and only if $\vec{p} \in \ker H_n(s)$. Hence $p \mapsto \vec{p}$ is a bijection of $\mathcal{N}_s$ on $\ker H_n(s)$. Obviously, this implies (9.41).

(ii) The left-hand side is contained in the right-hand side by setting $p = q$.

Since s is positive semidefinite, $L_s(f^2) \geq 0$ for $f \in \mathbb{R}[x]_n$. Hence the Cauchy–Schwarz inequality (2.7) holds, that is, $L_s(pq)^2 \leq L_s(p^2) L_s(q^2)$ for $p, q \in \mathbb{R}[x]_n$. Therefore, $L_s(p^2) = 0$ implies $L_s(pq) = 0$. $\qquad\square$

Clearly, $\mathcal{N}_s$ is a linear subspace of $\mathbb{R}[x]_n$. Therefore, if $\mathcal{N}_s \neq \{0\}$, there exists a unique monic polynomial $f \in \mathbb{R}[x]_n, f_s \neq 0$, of lowest degree in $\mathcal{N}_s$.

Definition 9.21 If $\mathcal{N}_s \neq \{0\}$, we call f_s the *minimal polynomial* associated with s and the number $\operatorname{rk}(s) = \deg(f_s)$ the *Hankel rank* of s. In the case $\mathcal{N}_s = \{0\}$ we set $\operatorname{rk}(s) = n + 1$.

The Hankel rank plays a crucial role in the truncated moment problem.

Clearly, $\operatorname{rk}(s) = 0$ if and only if $f_s = 1$, or equivalently, $s_0 = \cdots = s_n = 0$. Further, by definition and Lemma 9.20(i), we have $\operatorname{rk}(s) = n + 1$ if and only if $\mathcal{N}_s = \{0\}$, or equivalently, $\operatorname{rank} H_n(s) = n + 1$.

Let us write the Hankel matrix $H_n(s)$ as

$$H_n(s) = [v_0, \ldots, v_n],$$

where $v_j = (s_j, \ldots, s_{j+n})^T, j = 0, \ldots, n$, are the column vectors of $H_n(s)$.

The second assertion of the following lemma is usually called *Frobenius' lemma* in the literature, see e.g. [Gn, Lemma X.10.1].

Lemma 9.22 *Suppose that* $1 \leq \operatorname{rk}(s) \leq n$.

(i) $\operatorname{rk}(s)$ *is the smallest integer* r, $1 \leq r \leq n$, *such that the column vector* v_r *is in the span of* $v_0, \ldots, v_{r-1}$, *or equivalently, there are reals* $\lambda_0, \ldots, \lambda_{r-1}$ *such that*

$$s_{j+r} = \lambda_{r-1} s_{j+r-1} + \cdots + \lambda_0 s_j \quad for \quad j = 0, \ldots, n. \qquad (9.42)$$

If this is true, then we have

$$f_s(x) = x^r - \lambda_{r-1}x^{r-1} - \cdots - \lambda_1 x - \lambda_0.$$

(ii) $D_{\text{rk}(s)-1}(s) \neq 0$ *and* $D_{\text{rk}(s)}(s) = 0$.

Proof

(i) Since $1 \leq \text{rk}\,(s) \leq n$, the minimal polynomial f_s is defined and we have $\deg(f_s) \geq 1$, as noted above. Let $f(x) = x^r - \lambda_{r-1}x^{r-1} - \cdots - \lambda_1 x - \lambda_0$, where $r \in \{1, \ldots, n\}$ and $\lambda_0, \ldots, \lambda_{r-1} \in \mathbb{R}$. Then

$$L_s(x^j f) = s_{j+r} - \lambda_{r-1}s_{j+r-1} - \cdots - \lambda_1 s_{j+1} - \lambda_0 s_j, \quad j = 0, \ldots, n. \qquad (9.43)$$

Hence $f \in \mathcal{N}_s$ if and only if (9.42) holds, or equivalently, $v_r = \lambda_0 v_0 + \cdots + \lambda_{r-1}v_{r-1}$. Thus the smallest r such that $v_r \in \text{span}\{v_0, \ldots, v_{r-1}\}$ is obtained if and only if the monic polynomial f has the lowest possible degree in $\mathcal{N}_s$, that is, if $f = f_s$.

(ii) Let $A_k = [v_0, \ldots, v_{k-1}]$ denote the matrix with columns $v_0, \ldots, v_{k-1}$. By (i), $\text{rk}\,(s)$ is the smallest index r such that v_r belongs to the linear span of $v_0, \ldots, v_{r-1}$.

Since v_r is in the span of $v_0, \ldots, v_{r-1}$, we have $\text{rank}\, A_{r+1} < r + 1$. Thus $D_r(s) = 0$.

Because r is the smallest such number, $v_0 \ldots, v_{r-1}$ are linearly independent and hence $\text{rank}\, A_r = r$. Further, by (i) there are real numbers $\lambda_0, \ldots, \lambda_{r-1}$ such that the equations (9.42) hold. It follows from (9.42) that each *row* of A_r is a linear combination of the r preceding rows. Therefore, each row of A_r is in the span of the first r rows of A_r. Since $\text{rank}\, A_r = r$ is also the maximal number of linearly independent *rows*, the first r rows of A_r are linearly independent. This implies $D_{r-1}(s) \neq 0$. □

Example 9.23 Let $n = 4$ and $s = (1, 1, 1, 1, 0, 0, 0, 0, 0)$. Since $D_4(s) = 0$, we easily derive from Proposition 9.25(i) that $\text{rk}\,(s) = 4$. Then $f_s = x^4$, $D_0(s) = D_3(s) = 1$, and $D_1(s) = D_2(s) = 0$. Since $L_s((1 - x^2)^2) = -1$, s is not positive semidefinite!   ○

In the rest of this section, the sequence s is positive semidefinite. Then we have $H_n(s) \succeq 0$ and $L_s(p^2) \geq 0$ for $p \in \mathbb{R}[x]_n$. From formula (9.41) and some elementary linear algebra we obtain the following lemma; we omit its simple proof.

Lemma 9.24 *If s is positive semidefinite, then the following are equivalent:*

(i) *s is positive definite.*
(ii) *$H_n(s) \succ 0$.*
(iii) *$D_n(s) \neq 0$.*
(iv) *$\text{rk}(s) = n + 1$.*

Some important properties of $\text{rk}\,(s)$ are collected in the next proposition.

Proposition 9.25 *Suppose that $s=(s_j)_{j=0}^{2n}$ is positive semidefinite and $\mathrm{rk}(s) \leq n$.*

(i) *If $n - 1 - \mathrm{rk}(s) \geq 0$, then $pf_s \in \mathcal{N}_s$ for $p \in \mathbb{R}[x]_{n-1-\mathrm{rk}(s)}$.*

(ii) *$\mathrm{rk}(s) \leq \mathrm{rank}\, H_n(s) \leq \mathrm{rk}(s) + 1$.*

(iii) *$\mathrm{rk}(s) = \mathrm{rank}\, H_n(s)$ if and only if $x^{n-\mathrm{rk}(s)}f_s \in \mathcal{N}_s$.*

(iv) *$\mathrm{rk}(s)$ is the smallest number $r \in \mathbb{N}_0$ such that $D_r(s) = 0$. More precisely,*

$$D_0(s) > 0,\ldots,D_{\mathrm{rk}(s)-1}(s) > 0,\ \ D_{\mathrm{rk}(s)}(s) = 0,\ldots,D_n(s) = 0. \qquad (9.44)$$

(In the case $\mathrm{rk}(s) = 0$, the set of inequalities should be omitted.)

Proof

(i) Since $f_s \in \mathcal{N}_s$, Corollary 9.16(i) applies, so there exists a Radon measure μ such that $\mathrm{supp}\,\mu \subseteq \mathcal{Z}(f_s)$ and $L_s(q) = \int q\, d\mu$ for $q \in \mathbb{R}[x]_{2n-1}$. Therefore, if $p \in \mathbb{R}[x]_{n-1-\mathrm{rk}(s)}$, then $(pf_s)^2 \in \mathbb{R}[x]_{2n-2}$ and $pf_s \in \mathcal{N}_s$. since

$$L_s((pf_s)^2) = \int_{\mathcal{Z}(f_s)} p(x)^2 f_s(x)^2\, d\mu(x) = 0.$$

(ii) First we note that the following set is a basis of the vector space $\mathbb{R}[x]_n$:

$$\{x^i, x^j f_s : i,j \in \mathbb{N}_0, 0 \leq i \leq \mathrm{rk}(s) - 1, 0 \leq j \leq n - \mathrm{rk}(s)\}. \qquad (9.45)$$

If $\mathrm{rk}(s) = n$, then the assertion is obvious. Hence we can assume that $\mathrm{rk}(s) < n$. By (i) we have $x^j f_s \in \mathcal{N}_s$ for $j = 0,\ldots,n - 1 - \mathrm{rk}(s)$. Thus, $\dim \mathcal{N}_s \geq n - \mathrm{rk}(s)$.

Now we prove that $\dim \mathcal{N}_s \leq n + 1 - \mathrm{rk}(s)$. Assume the contrary, that is, $\dim \mathcal{N}_s \geq n + 2 - \mathrm{rk}(s)$. Since (9.45) is a vector space basis of $\mathbb{R}[x]_n$, it follows from (i) that there exists a polynomial $f \in \mathbb{R}[x]_n, f \neq 0$, with $\deg(f) < \mathrm{rk}(s) = \deg(f_s)$ in $\mathcal{N}_s$. This contradicts the choice of f_s. Hence $\dim \mathcal{N}_s \leq n + 1 - \mathrm{rk}(s)$. Therefore, $\mathrm{rk}(s) \leq \mathrm{rank}\, H_n(s) \leq \mathrm{rk}(s) + 1$ by (9.41).

(iii) By (9.41), $\mathrm{rk}(s) = \mathrm{rank}\, H_n(s)$ if and only if $\dim \mathcal{N}_s = n + 1 - \mathrm{rk}(s)$. Since $\mathrm{rk}(s)$ is the lowest degree of nonzero polynomials in $\mathcal{N}_s$ and (9.45) is a basis of $\mathbb{R}[x]_n$, we conclude from (i) that the latter is equivalent to $x^{n-\mathrm{rk}(s)}f_s \in \mathcal{N}_s$.

(iv) Set $r := \mathrm{rk}(s)$. If $r = 0$, then $s_0 = \cdots = s_n = 0$, as noted above. Hence $s_j = 0$ for all j, since c is positive semidefinite, so that $D_k(s) = 0$ for $k = 0,\ldots,n$.

Assume now that $r \geq 1$. Then, by Lemma 9.22(ii), $D_{r-1}(s) \neq 0$ and $D_r(s) = 0$. Since $H_n(s) \succeq 0$ by assumption, $D_{r-1}(s) > 0$. Hence the matrix $H_{r-1}(s)$ is positive definite and therefore $D_k(s) > 0$ for $k = 0,\ldots,r-1$. Since $D_r(s) = 0$, the sequence $(s_j)_{j=0}^{2r}$ is not positive definite, hence neither is the sequence $(s_j)_{j=0}^{2k}$ and so $D_k(s) = 0$ for $k = r + 1,\ldots,n$. This completes the proof of (9.44). $\qquad\square$

Let us illustrate the preceding in an important special case.

Example 9.26 Suppose s has a k-atomic representing measure $\mu = \sum_{j=1}^{k} m_j \delta_{t_j}$ with $k \leq n$. Let $f \in \mathbb{R}[x]_n$. Then $f \in \mathcal{N}_s$ if and only if $L_s(f^2) = \sum_{j=1}^{k} m_j f(t_j)^2 = 0$, or equivalently, $f(t_1) = \cdots = f(t_k) = 0$. Therefore, $f_s(x) = (x - t_1) \ldots (x - t_k)$ and $\mathrm{rk}(s) = k$. By Theorem 9.27 (i)$\rightarrow$(iii) below, we also have $\mathrm{rank}\, H_n(s) = k$.   o

9.6 Truncated Hamburger and Stieltjes Moment Sequences

In this section, we settle the truncated Hamburger and Stieltjes moment problems for real sequences $s = (s_j)_{j=0}^{2n}$ (even case) and $s = (s_j)_{j=0}^{2n+1}$ (odd case).

Theorem 9.27 (The Hamburger Truncated Moment Problem; Even Case) *For any real sequence $s = (s_j)_{j=0}^{2n}$ the following are equivalent:*

  (i) *s is a truncated Hamburger moment sequence.*
 (ii) *There exist reals s_{2n+1}, s_{2n+2} such that $H_{n+1}(\tilde{s}) \succeq 0$, where $\tilde{s} := (s_j)_{j=0}^{2n+2}$.*
(iii) *$H_n(s) \succeq 0$ and $\mathrm{rank}\, H_n(s) = \mathrm{rk}(s)$.*

If $\mathrm{rk}(k) \leq n$, or equivalently, if $\mathcal{N}_s \neq \{0\}$, these conditions are equivalent to:

 (iv) *$H_n(s) \succeq 0$ and $x^{n-\mathrm{rk}(s)} f_s \in \mathcal{N}_s$.*
  (v) *$H_n(s) \succeq 0$ and there exists a polynomial of degree n in $\mathcal{N}_s$.*

Suppose that (i) is satisfied. Then s has an $\mathrm{rk}(s)$-atomic representing measure. Further, s has a unique representing measure μ if and only if $\mathrm{rk}(s) \leq n$, or equivalently, $D_n(s) = 0$. The atoms of this measure are the zeros of f_s and

$$|\mathrm{supp}\, \mu| = |\mathcal{Z}(f_s)| = \mathrm{rk}(s) = \mathrm{rank}\, H_n(s). \tag{9.46}$$

Proof
    (i)$\rightarrow$(ii) From (i) and Corollary 1.25 it follows that s has a finitely atomic representing measure μ. Hence μ has all moments. Then, setting $s_l = \int x^l\, d\mu$ for $l = 2n + 1, 2n + 2$, $\tilde{s}$ is a truncated moment sequence and hence $H_{n+1}(\tilde{s}) \succeq 0$.
    (ii)$\rightarrow$(i) Since $H_{n+1}(\tilde{s}) \succeq 0$, Corollary 9.16(i) applies with s replaced by $\tilde{s}$ and n by $n + 1$. Then Eq. (9.38) therein implies that $s_j = \int x^j\, d\mu$ for $j = 0, \ldots, 2n$.
    Next we treat the case $\mathrm{rk}(s) = n + 1$. Then $\mathcal{N}_s = \{0\}$ and $\mathrm{rank}\, H_n(s) = n + 1$ by (9.41) and Definition 9.21. The implication (i)$\rightarrow$(iii) is clear. We prove (iii)$\rightarrow$(i). Since $H_n(s) \succeq 0$ and $\mathrm{rank}\, H_n(s) = n + 1$, the sequence s is positive definite. Therefore, by Theorem 9.7, s has a one-parameter family of $(n + 1)$-atomic representing measures. Thus in the case $\mathrm{rk}(s) = n + 1$ all assertions of the theorem are proved.
    In the rest of this proof we assume that $\mathrm{rk}(s) \leq n$. Then $\mathcal{N}_s \neq \{0\}$ and the minimal polynomial f_s is defined.

(i)→(iv) By (i), s has a representing measure μ. Hence $H_n(s) \succeq 0$. Since $f_s \in \mathcal{N}_s$, we have supp $\mu \subseteq \mathcal{Z}(f_s)$ by Proposition 1.23 and therefore

$$L_s((x^{n-\mathrm{rk}(s)}f_s)^2) = \int_{\mathcal{Z}(f_s)} x^{2(n-\mathrm{rk}(s))} f_s(x)^2 \, d\mu = 0.$$

That is, $x^{n-\mathrm{rk}(s)}f_s \in \mathcal{N}_s$ and (iv) is proved.

(iv)→(v) is trivial, since $x^{n-\mathrm{rk}(s)}f_s$ belongs to $\mathcal{N}_s$ and has degree n.

(v)→(i) follows at once from Corollary 9.16(ii).

(iii)↔(iv) is clear by Proposition 9.25(iii).

This completes the proof of the equivalence of statements (i)–(v).

Suppose that (i) holds. Let μ be a representing measure for s. As noted above, supp $\mu \subseteq \mathcal{Z}(f_s)$, so μ is k-atomic with $k \leq \deg(f_s) = \mathrm{rk}(s) = \mathrm{rank}\, H_n(s)$ by (iii). By assumption, $\mathrm{rk}(s) \leq n$. Thus, $k \leq n$. Hence, as noted in Example 9.26, $k = \mathrm{rk}(s)$. The preceding proves all equalities of (9.46).

Let $\tilde{\mu}$ be another representing measure for s. Let $\mathcal{Z}(f_s) = \{t_1, \ldots, t_k\}$. Then there are nonnegative numbers m_j, n_j such that $\mu = \sum_{j=1}^k m_j \delta_{t_j}$ and $\tilde{\mu} = \sum_{j=1}^k n_j \delta_{t_j}$. Since $k \leq n$, there are interpolation polynomials $p_i \in \mathbb{R}[x]_n$ such that $p_i(t_j) = \delta_{ij}$ for $i, j = 1, \ldots, n$. Then $L_s(p_i) = \int p_i \, d\mu = m_i$ and $L_s(p_i) = \int p_i \, d\tilde{\mu} = n_i$, so that $m_i = n_i$ for all i. Hence $\mu = \tilde{\mu}$. Thus s has a unique representing measure if $\mathrm{rk}(s) \leq n$. $\qquad\square$

Remark 9.28 The proof of (i)→(iv) uses the same reasoning as the proof of Proposition 9.25(i), now applied with $p = x^{n-\mathrm{rk}(s)}$. Such arguments and the uniqueness proof based on interpolation polynomials often appear (for instance, in the proofs of Theorem 10.7 and Proposition 17.18) and in slightly different settings in this book. ○

We illustrate Theorem 9.27 with two simple examples.

Example 9.29 Let $n = 2$ and $s = (s_j)_{j=0}^4$, where $s_0 = s_1 = s_2 = s_3 = 1, s_4 = \alpha \geq 1$. Clearly, s has a representation (9.36) with $\mu = \delta_1$ and $a = \alpha - 1 \geq 0$. Hence

$$H_2(s) = \begin{pmatrix} 1 & 1 & 1 \\ 1 & 1 & 1 \\ 1 & 1 & \alpha \end{pmatrix} \succeq 0.$$

Case 1: $\alpha > 1$.

Then $f_s = x - 1$, $\mathrm{rk}(s) = 1$, and $\mathcal{N}_s = \mathbb{R} \cdot f_s$. Since $x^{2-1}f_s \notin \mathcal{N}_s$, s is not a truncated Hamburger moment sequence. Note that $(x - 1) \in \mathcal{N}_s$, but $(x - 1)^2 \notin \mathcal{N}_s$.

Case 2: $\alpha = 1$.

Then $f_s = x - 1$ and $\mathrm{rk}(s) = 1$, but $xf_s \in \mathcal{N}_s$, so s is a truncated moment sequence with unique representing measure $\mu = \delta_1$. ○

Example 9.30 Set $\mu := \delta_0 + \frac{1}{k^{2n}}\delta_k$ for fixed $k \in \mathbb{N}$ and $n \geq 2$. Clearly, the moment sequence $s = (s_j)_{j=0}^{2n}$ of μ is given by

$$s_0 = 1 + k^{-2n}, \quad s_j = k^{-(2n-j)} \text{ for } 1 \leq j < 2n, \quad s_{2n} = 1.$$

In the limit $k \rightarrow \infty$ we obtain the positive semidefinite sequence $\tilde{s} = (1, 0, \ldots, 0, 1)$. It is not a moment sequence, since $f_{\tilde{s}} = x \in \mathcal{N}(\tilde{s})$ and $x^n = x^{n-1} f_{\tilde{s}} \notin \mathcal{N}(\tilde{s})$. This shows that the set of all truncated moment $2n$-sequences is *not closed* in $\mathbb{R}^{2n+1}$. ○

The following lemma about positive block matrices is a special case of Theorem A.24. We identify $\mathbb{R}^k$ with the column matrices $M_{k,1}(\mathbb{R})$.

Lemma 9.31 *Let* $k \in \mathbb{N}$, $A \in M_k(\mathbb{R})$, $b \in \mathbb{R}^k$, $c \in \mathbb{R}$, *and suppose that* $A \succeq 0$. *Consider the block matrix* $\tilde{A} \in M_{k+1}(\mathbb{R})$ *defined by*

$$\tilde{A} = \begin{pmatrix} A & b \\ b^T & c \end{pmatrix}. \tag{9.47}$$

(i) $\tilde{A} \succeq 0$ *if and only if* $b = Au$ *for some* $u \in \mathbb{R}^k$ *and* $c \geq u^T Au$.
(ii) *Let* $b = Au$ *with* $u \in \mathbb{R}^k$ *and* $c \geq u^T Au$. *Then* $\operatorname{rank}\tilde{A} = \operatorname{rank}A$ *if and only if* $c = u^T Au$.

The three remaining existence theorems of this section use Lemma 9.31.

Theorem 9.32 (The Hamburger Truncated Moment Problem; Odd Case) *For a real sequence* $s = (s_j)_{j=0}^{2n+1}$ *the following three statements are equivalent:*

(i) *s is a truncated Hamburger moment sequence.*
(ii) *There exists a real number s_{2n+2} such that $H_{n+1}(\tilde{s}) \succeq 0$, where $\tilde{s} := (s_j)_{j=0}^{2n+2}$.*
(iii) *$H_n(s) \succeq 0$ and $(s_{n+1}, \ldots, s_{2n+1})^T \in \operatorname{range}(H_n(s))$.*

Suppose that (i) holds and set $\hat{s} := (s_j)_{j=0}^{2n}$. Then s has an $\operatorname{rk}(\hat{s})$-atomic representing measure. Further, s has a unique representing measure if and only if $\operatorname{rk}(\hat{s}) \leq n$, or equivalently, $D_n(\hat{s}) = 0$; this unique measure is $\operatorname{rk}(\hat{s})$-atomic and its set of atoms is the zero set $\mathcal{Z}(f_{\hat{s}})$ of $f_{\hat{s}}$.

Proof The proof of (i)↔(ii) is almost verbatim the same as for Theorem 9.27. In the proof of (ii)→(i) we apply Corollary 9.16 with n replaced by $n+1$. Hence Eq. (9.38) holds for $j = 0, \ldots, 2n + 1$ which gives (i).

To prove (ii)↔(iii) we consider $H_{n+1}(\tilde{s})$ as a block matrix (9.47) with

$$A = H_n(s), \quad b = (s_{n+1}, \ldots, s_{2n+1})^T, \quad c = s_{2n+2}. \tag{9.48}$$

(ii)→(iii) Since $H_{n+1}(\tilde{s}) \succeq 0$, we have $H_n(s) \succeq 0$ and $b \in \operatorname{range}(H_n(s))$ by the only if direction of Lemma 9.31(i).

(iii)→(ii) Then $b \in \operatorname{range}(H_n(s))$, say $b = H_n(s)u$ with $u \in \mathbb{R}^{n+1}$. Therefore, choosing $s_{2n+2} \geq u^T H_n(s)u$, we have $H_{n+1}(\tilde{s}) \succeq 0$ by the if part of Lemma 9.31(i).

This completes the proof of the equivalences (i)–(iii). Now suppose that (i) holds.

First we assume that $\mathrm{rk}(\hat{s}) = n + 1$. Then $\mathrm{rank}\,H_n(\hat{s}) = n + 1$ by Proposition 9.25(ii). Hence, since $H_n(s) = H_n(\hat{s})$, there exists a $v \in \mathbb{R}^{n+1}$ such that $b = H_n(s)v$. We fix v and choose $s_{2n+2} > v^T H_n(s)v$. Then, applying Lemma 9.31(ii) to the block matrix $H_{n+1}(\tilde{s})$ yields $\mathrm{rank}\,H_{n+1}(\tilde{s}) > \mathrm{rank}\,H_n(s) = \mathrm{rank}\,H_n(\hat{s}) = n + 1$. Therefore, $\mathrm{rank}\,H_{n+1}(\tilde{s}) = n + 2$. Hence, by Corollary 9.2 (or by Theorem 9.27 (iii)$\to$(i)), $\tilde{s}$ is a truncated moment sequence. Obviously, any representing measure for $\tilde{s}$ is one for s. Since these measures have the moment s_{2n+2}, they are different as s_{2n+2} varies. Thus s has a one-parameter family of representing measures.

Assume now that $\mathrm{rk}(\hat{s}) \leq n$. Any representing measure μ for s is a representing measure for $\hat{s}$. Therefore, by Theorem 9.27 applied to $\hat{s}$, μ is uniquely determined and has the properties stated above. $\square$

The next proposition shows how Lemma 9.31 can be used to obtain additional information about the structure of the extended sequence $\tilde{s}$ in Theorem 9.27(ii).

Proposition 9.33 *Let* $s = (s_j)_{j=0}^{2n}$ *be a positive semidefinite real sequence such that* $r := \mathrm{rk}(s) \leq n$. *Suppose that there exist real numbers* s_{2n+1}, s_{2n+2} *such that* $H_{n+1}(\tilde{s}) \succeq 0$, *where* $\tilde{s} := (s_0, \ldots, s_{2n}, s_{2n+1}, s_{2n+2})$. *Further, let* $\lambda_0, \ldots, \lambda_{r-1}$ *be the numbers from Eq. (9.42). Then*

$$s_{j+r} = \lambda_0 s_j + \cdots + \lambda_{r-1} s_{j+r-1} \quad \text{for} \quad j = 0, \ldots, 2n + 1 - r, \qquad (9.49)$$

$$s_{2n+2} \geq \lambda_0 s_{2n+2-r} + \cdots + \lambda_{r-1} s_{2n+1}. \qquad (9.50)$$

There is equality in (9.50) if and only if $\mathrm{rk}(s) \equiv \mathrm{rank}\,H_n(s) = \mathrm{rank}\,H_{n+1}(\tilde{s})$.

Proof We consider the Hankel matrix $H_{n+1}(\tilde{s})$ as a block matrix (9.47) with entries (9.48). Since $H_{n+1}(\tilde{s}) \succeq 0$, we have $H_n(s) \succeq 0$ and Lemma 9.31(i) applies. Therefore, we have $b = Au$ with $u \in \mathbb{R}^{n+1}$ and $c \geq u^T Au$. Equation $b = Au$ means that

$$s_{n+k+1} = \sum_{l=0}^{n} s_{k+l} u_l, \quad k = 0, \ldots, n. \qquad (9.51)$$

We prove by induction that Eq. (9.49) holds for $j = 0, \ldots, 2n + 1 - r$. If $j = 0, \ldots, n$, then (9.49) is just one of the equations (9.42). Assume that (9.49) is true for all j, where $n \leq j \leq 2n - r$. Using (9.51) and the induction hypothesis (9.49) we derive

$$s_{j+1+r} = \sum_{l=0}^{n} s_{j+r-n+l} u_l = \sum_{l=0}^{n} \sum_{i=0}^{r-1} s_{j-n+l+i} \lambda_i u_l$$

$$= \sum_{i=0}^{r-1} \left(\sum_{l=0}^{n} s_{j-n+l+i} u_l \right) \lambda_i = \sum_{i=0}^{r-1} s_{j+1+i} \lambda_i,$$

which is (9.49) for $j + 1$. This completes the induction proof of (9.49). (The preceding computation is not valid for $j = 2n + 1 - r$, since (9.51) does not hold for $k = n + 1$.)

By Lemma 9.31(i), $s_{2n+2} = c \geq u^T A u$. Using (9.51) and (9.49) we compute

$$s_{sn+2} \geq u^t A u = u^t b = \sum_{l=0}^{n} s_{n+1+l} u_l = \sum_{l=0}^{n} \sum_{i=0}^{r-1} s_{n+1-r+l+i} \lambda_i u_l = \sum_{i=0}^{r-1} s_{2n+2-r+i} \lambda_i,$$

which is the inequality (9.50).

Theorem 9.27 (ii)$\to$(iii) implies that $\mathrm{rk}(s) = \mathrm{rank}\, H_n(s)$. According to Lemma 9.31(ii), we have $\mathrm{rank}\, H_n(s) = \mathrm{rank}\, H_{n+1}(\tilde{s})$ if and only if $s_{2n+2} \equiv c = u^T A u$, or equivalently, if there is equality in (9.50). $\qquad\square$

Remark 9.34

1. Let us retain the assumptions of Proposition 9.33. Equation (9.49) is a *recursive relation* for the numbers $s_j, j \leq 2n + 1$. In particular, s_{2n+1} is determined by (9.49) for $j = 2n + 1 - r$. From Lemma 9.31(ii) it follows that *each* real number s_{sn+2} satisfying (9.50) gives an extension $\tilde{s}$ such that $H_{n+1}(\tilde{s}) \succeq 0$. In particular, we can choose $s_{2n+2} := \lambda_0 s_{2n+2-r} + \cdots + \lambda_{r-1} s_{2n+1}$; then we have $\mathrm{rank}\, H_n(s) = \mathrm{rank}\, H_{n+1}(\tilde{s})$.

2. Let $s = (s_j)_{j=0}^{2n}$ be a *positive definite* real sequence. Then $\mathrm{rank}\, H_n(s) = n + 1$. (This case is not covered by Proposition 9.33, since the assumptions of Proposition 9.33 and Theorem 9.27 imply that s is a moment sequence and $\mathrm{rank}\, H_n(s) = \mathrm{rk}(s) \leq n$.) Since $H_n(s)$ is regular, the vector b from (9.48) is in the range of $A = H_n(s)$, say $b = H_n(s)u$. Hence, by Lemma 9.31(i), for arbitrary reals s_{2n+1}, s_{2n+2} such that $s_{2n+2} \geq u^T H_n(s)u$, the extension $\tilde{s} = (s_j)_{j=0}^{2n+2}$ of s satisfies $H_{n+1}(\tilde{s}) \succeq 0$, so s is a moment sequence by Theorem 9.27. If we choose $s_{2n+2} > u^T H_n(s)u$, Lemma 9.31(ii) implies that $\mathrm{rank}\, H_n(\tilde{s}) = n + 2$, so $\tilde{s}$ is also positive definite. This recovers the extension procedure for positive definite sequences from Lemma 9.1. $\qquad\circ$

Now we turn to the truncated Stieltjes moment problem and begin with the odd case. Let $Es = (s_1, \ldots, s_m)$ denote the shifted sequence of $s = (s_0, s_1, \ldots, s_m)$.

Theorem 9.35 (The Stieltjes Truncated Moment Problem; Odd Case) *A real sequence $s = (s_j)_{j=0}^{2n+1}$ is a truncated Stieltjes moment sequence if and only if*

$$H_n(s) \succeq 0, \quad H_n(Es) \succeq 0, \quad \text{and} \quad (s_{n+1}, \ldots, s_{2n+1})^T \in \mathrm{range}\,(H_n(s)). \qquad (9.52)$$

Suppose s is a truncated Stieltjes moment sequence. Then s has an $\mathrm{rk}(\hat{s})$-atomic representing measure on $\mathbb{R}_+$. Moreover, the sequence s has a unique representing measure on $\mathbb{R}_+$ if and only if $\mathrm{rk}(s) \leq n$ or $\mathrm{rk}(Es) \leq n$, or equivalently, $D_n(s) = 0$ or $D_n(Es) = 0$.

Proof By Theorem 9.32 (i)$\leftrightarrow$(ii), s is a truncated Hamburger moment sequence if and only if $H_n(s) \succeq 0$ and $(s_{n+1}, \ldots, s_{2n+1})^T \in \mathrm{range}\,(H_n(s))$.

Suppose s has a representing measure μ supported on $\mathbb{R}_+$. Then the positive (!) measure ν given by $d\nu = x d\mu$ has the moment sequence Es, so that $H_n(Es) \succeq 0$.

Conversely, assume that $H_n(Es) \succeq 0$. By Theorem 9.32, s has an $\mathrm{rk}(\hat{s})$-atomic representing measure $\mu = \sum_{j=1}^{\mathrm{rk}(\hat{s})} m_j \delta_{t_j}$. Since $\mathrm{rk}(\hat{s}) \leq n+1$, there are Lagrange interpolation polynomials $p_i \in \mathbb{R}[x]_n$ such that $p_i(t_j) = \delta_{ij}$, $i, j = 1, \dots, \mathrm{rk}(\hat{s})$. Then

$$L(xp_i^2) = \int xp_i(x)^2 \, d\mu(x) = \sum_{j=0}^{\mathrm{rk}(\hat{s})} m_j t_j p_i^2(t_j) = m_i t_i \geq 0.$$

Since μ is $\mathrm{rk}(\hat{s})$-atomic, we have $m_i > 0$ and therefore $t_i \geq 0$. Thus μ is supported on $\mathbb{R}_+$ and s is a truncated Stieltjes moment sequence.

We turn to the proof of the uniqueness assertions. Note that $\mathrm{rk}(s) \leq n$ (resp. $\mathrm{rk}(Es) \leq n$) if and only if $D_n(s) = 0$ (resp. $D_n(Es) = 0$) by Lemma 9.24 (iii)$\leftrightarrow$(v).

First, let us assume that

$$D_n(s) \neq 0 \quad \text{and} \quad D_n(Es) \neq 0. \tag{9.53}$$

Let $\tilde{s} = (s_j)_{j=0}^{2n+3}$ with unknown s_{2n+2}, s_{2n+3}. There are polynomials f, g such that

$$D_{n+1}(\tilde{s}) = s_{2n+2} D_n(s) + f(s_0, \dots, s_{2n+1}), \tag{9.54}$$

$$D_{n+1}(E\tilde{s}) = s_{2n+3} D_n(Es) + g(s_1, \dots, s_{2n+2}). \tag{9.55}$$

Since $H_n(s) \succeq 0$ and $H_n(Es) \succeq 0$, both determinants in (9.53) are positive. Hence we can find $s_{2n+2} > 0$ such that (9.54) is positive and there exists a $c > 0$ such that (9.55) is positive for $s_{2n+3} \geq c$. Then the sequence $\tilde{s}$ fulfills (9.52) with n replaced by $n+1$. (Since $D_{n+1}(\tilde{s}) > 0$, $H_{n+1}(\tilde{s})$ is positive definite and hence its range is $\mathbb{R}^{n+2}$.) Therefore, by the existence assertion of Theorem 9.35 proved above, $\tilde{s}$ is a truncated Stieltjes moment sequence. Since $s_{2n+3} \geq c$ was arbitrary, we obtain a family of different measures on $\mathbb{R}_+$ which have the same moments $s_0, \dots, s_{2n+1}$.

Now suppose that $D_n(s) = 0$. Then, by Theorem 9.32, s has a unique representing measure on $\mathbb{R}$, so we have uniqueness on $\mathbb{R}_+$ as well.

Finally, we suppose that $D_n(Es) = 0$. Let μ_1 and μ_2 be representing measures for s. We will show that $\mu_1 = \mu_2$. Define measures ν_j by $d\nu_j = x d\mu_j$, $j = 1, 2$. Then Es is a truncated Hamburger sequence with representing measures ν_j. Since $D_n(Es) = 0$, Theorem 9.27 yields $\nu_1 = \nu_2$. Since $d\nu_j = x d\mu_j$, this implies that $\mu_1(N) = \mu_2(N)$ for all Borel sets $N \subseteq \mathbb{R} \backslash \{0\}$. We have $s_0 = \mu_j(\{0\}) + \mu_j(\mathbb{R} \backslash \{0\})$, $j = 1, 2$. Therefore, since $\mu_1(\mathbb{R} \backslash \{0\}) = \mu_2(\mathbb{R} \backslash \{0\})$, we obtain $\mu_1(\{0\}) = \mu_2(\{0\})$. Thus $\mu_1 = \mu_2$. This completes the proof of the uniqueness assertions. $\quad\square$

Theorem 9.36 (The Stieltjes Truncated Moment Problem; Even Case) *A real sequence* $s = (s_j)_{j=0}^{2n}$ *is a truncated Stieltjes moment sequence if and only if*

$$H_n(s) \succeq 0, \quad H_{n-1}(Es) \succeq 0, \quad and \quad (s_{n+1}, \dots, s_{2n})^T \in \mathrm{range}\,(H_{n-1}(Es)). \tag{9.56}$$

In this case s has an $\mathrm{rk}(s)$-atomic representing measure on $\mathbb{R}_+$.

Further, the truncated Stieltjes moment sequence s has a unique representing measure supported on $\mathbb{R}_+$ *if and only if* $\mathrm{rk}(s) \leq n$, *or equivalently, if* $D_n(s) = 0$.

Proof First assume that the three conditions (9.56) are fulfilled. Our aim is to apply Theorem 9.35 to the extended sequence $\tilde{s} = (s_j)_{j=0}^{2n+1}$ for some real number s_{2n+1}.

Since $b := (s_{n+1}, \dots, s_{2n})^T \in \mathrm{range}\,(H_{n-1}(Es))$ by (9.56), there exists a $u \in \mathbb{R}^n$ such that $b = H_{n-1}(Es)u$. Put $c = s_{2n+1} := u^T H_{n-1}(Es)u$ and $A := H_{n-1}(Es)$. Since $H_{n-1}(Es) \succeq 0$ by (9.56), the block matrix $\tilde{A}$ in (9.47) is positive semidefinite by Lemma 9.31(i). But $\tilde{A}$ is the Hankel matrix $H_n(E\tilde{s})$, where $E\tilde{s} = (s_1, \dots, s_{2n+1})$. Thus, $H_n(E\tilde{s}) \succeq 0$.

We set $w := (s_0, \dots, s_{n-1})^T$ and write $H_n(s)$ as block matrix

$$
H_n(\tilde{s}) = H_n(s) = \begin{pmatrix} s_0 & s_1 & \cdots & s_n \\ s_1 & s_2 & \cdots & s_{n+1} \\ \vdots & \vdots & \ddots & \vdots \\ s_n & s_{n+1} & \cdots & s_{2n} \end{pmatrix} = \begin{pmatrix} w & H_{n-1}(Es) \\ s_n & b^T \end{pmatrix}.
$$

Applying $H_n(\tilde{s})$ to the column vector $(0, u)^T \in \mathbb{R}^{n+1}$ we obtain

$$
H_n(\tilde{s}) \begin{pmatrix} 0 \\ u \end{pmatrix} = \begin{pmatrix} H_{n-1}(Es)u \\ b^T u \end{pmatrix} = \begin{pmatrix} b \\ u^T H_{n-1}(Es)u \end{pmatrix} = \begin{pmatrix} b \\ s_{2n+1} \end{pmatrix},
$$

that is, $(s_{n+1}, \dots, s_{2n}, s_{2n+1})^T \in \mathrm{range}\,(H_n(\tilde{s}))$. Thus, since $H_n(\tilde{s}) = H_n(s) \succeq 0$ by (9.56) and $H_n(E\tilde{s}) \succeq 0$ as shown above, $\tilde{s}$ satisfies (9.52). Therefore, by Theorem 9.35, $\tilde{s}$ is a truncated Stieltjes moment sequence which has an $\mathrm{rk}(s)$-atomic representing measure. Hence this holds for s as well.

Conversely, suppose s is a truncated Stieltjes moment sequence. It is obvious that $H_n(s) \succeq 0$ and $H_{n-1}(Es) \succeq 0$, so it remains to show the range condition in (9.56). We argue as in the proofs of Theorems 9.27 and 9.32. By Corollary 1.25, s has a finitely atomic representing measure μ with atoms in $\mathbb{R}_+$. Setting $s_l = \int x^l d\mu(x)$ for $l = 2n + 1, 2n + 2, \tilde{s} = (s_j)_{j=0}^{2n+2}$ is a truncated Stieltjes moment sequence. Since the Hankel matrix $H_n(E\tilde{s})$ is a block matrix (9.47) with

$$
A = H_{n-1}(Es), \quad b = (s_{n+1}, \dots, s_{2n})^T, \quad c = s_{2n+1}
$$

and $H_n(E\tilde{s}) \succeq 0$, Lemma 9.31(i) implies that $b \in \mathrm{range}\,(H_{n-1}(Es))$.

It remains to verify the uniqueness assertions. First, suppose that $D_n(s) = 0$. Then the representing measure of s for the even truncated Hamburger problem is unique by Theorem 9.27, so it is unique for the even Stieltjes case as well.

Now assume that $D_n(s) \neq 0$. As above we set $\tilde{s} = (s_j)_{j=1}^{2n+1}$ with unknown real s_{2n+1}. We consider $H_n(E\tilde{s})$ as a block matrix

$$
H_n(E\tilde{s}) = \begin{pmatrix} H_{n-1}(s^{(1)}) & b \\ b^T & s_{2n+1} \end{pmatrix}, \quad \text{where } b := (s_{n+1}, \dots, s_{2n})^T.
$$

Since s is a truncated moment sequence, it has a finitely atomic representing measure μ. Putting $s_{2n+1} = \int x^{2n+1} \, d\mu$, $\tilde{s}$ is an odd truncated Stieltjes moment sequence, so $H_n(E\tilde{s}) \succeq 0$ by Theorem 9.35. From Lemma 9.31(i) we obtain $H_n(E\tilde{s}) \succeq 0$ if s_{2n+1} is sufficiently large. Fix such a number s_{2n+1}. Then $\tilde{s}$ satisfies (9.52). Indeed, $H_n(\tilde{s}) = H_n(s) \succeq 0$, since s is a moment sequence, and range $(H_n(\tilde{s})) = \mathbb{R}^{n+1}$, since $D_n(\tilde{s}) = D_n(s) \neq 0$. Therefore, by Theorem 9.35, $\tilde{s}$ is a Stieltjes moment sequence. The representing measures for $\tilde{s}$ are different as s_{2n+1} varies, but all of them have the same moments $s_0, \dots, s_{2n}$. Thus, we have nonuniqueness if $D_n(s) \neq 0$. □

9.7 Exercises

1. Determine Hankel rank and minimal polynomial:

   a. $s = (3, 0, 2, 0, 2, 0, 2)$, $m = 6$.
   b. $s = (2, 2, 4, 8, 16)$, $m = 4$.
   c. Set $s_j = 1$ for $j = 0, 1, 2, 3$ and $s_j = 0$ for $j = 4, \dots, 2n$, where $n \geq 4$.

2. Decide whether or not the following sequences are truncated Hamburger moment sequences:

   a. $s = (4, 5, 9)$, $m = 3$.
   b. $s = (2, 0, 2, 0, 5)$, $m = 4$.
   c. $s = (3, 2, 6, 10, 18, 35)$, $m = 6$.
   d. $s = (3, 0, 2, 0, 2, 0, 2, 0)$, $m = 7$.

3. Decide whether or not the following sequences are truncated Stieltjes moment sequences:

   a. $s = (1, 2, 4, 8, 20)$, $m = 4$.
   b. $s = (3, 5, 9, 17, 33)$, $m = 4$.
   c. $s = (3, 2, 2, 2, 2, 3)$, $m = 5$.
   d. $s = (3, 5, 11, 29)$, $m = 3$.

9.8 Notes

Quadrature formulas of the form (9.9) were discovered by C.F. Gauss (1814) and studied by K.G.J. Jacobi [Jac] and E.B. Christoffel [Chl]. Our treatment of the positive definite case in Sect. 9.1 follows N.I. Akhiezer and M.G. Krein [AK, §1]. The representation formula (9.36) for positive semidefinite sequences was proved by E. Fischer [Fi]. The results for the truncated Hamburger and Stieltjes moment problems in Sect. 9.6 are due to R. Curto and L. Fialkow [CF1].

Chapter 10
The One-Dimensional Truncated Moment Problem on a Bounded Interval

Throughout this chapter a and b are fixed real numbers such that $a < b$ and $m \in \mathbb{N}$. We consider the truncated moment problem on the interval $[a, b]$:

Given a real m-sequence $s = (s_j)_{j=0}^m$, when is there a Radon measure μ on $[a, b]$ such that $s_j = \int_a^b x^j d\mu(x)$ for $j = 0, \ldots, m$?

In this case we say that s is a *truncated $[a, b]$-moment sequence* and μ is a representing measure for s.

In Sect. 10.1 truncated $[a, b]$-moment sequences are described in terms of positivity conditions (Theorems 10.1 and 10.2). If a solution exists, there are always atomic solutions. The main part of this chapter deals with atomic solutions of "small size". In Sect. 10.2 the cone $\mathcal{S}_{m+1}$ of all moment sequences and the index of atomic representing measures are introduced. Boundary points of $\mathcal{S}_{m+1}$ are characterized as moment sequences with unique representing measures and of index at most m (Theorem 10.7). The rest of this chapter is devoted to a detailed study of interior points of $\mathcal{S}_{m+1}$. In Sects. 10.3, 10.4, and 10.6 representing measures of index $m + 1$ and $m + 2$ are investigated. Each interior point of $\mathcal{S}_{m+1}$ has precisely two representing measures of index $m + 1$ (Theorem 10.17) and for each $\xi \in [a, b]$ a distinguished measure with root ξ and index at most $m + 2$ (Corollary 10.13). The maximal mass of a point among all representing measures is studied in Sect. 10.5 (Theorem 10.21). In Sect. 10.7 orthogonal polynomials are developed and a description of the maximal mass in terms of orthonormal polynomials is given (Theorem 10.29).

10.1 Existence of a Solution

First we collect some notation that will be used throughout this chapter.

Let $s = (s_j)_{j=0}^m$ be a real sequence. Recall that L_s is the Riesz functional on $\mathbb{R}[x]_n$ given by $L_s(x^j) = s_j$, $j = 0, \ldots, n$, and $H_k(s)$, $2k \leq m$, denotes the Hankel matrix

© Springer International Publishing AG 2017
K. Schmüdgen, *The Moment Problem*, Graduate Texts in Mathematics 277,
DOI 10.1007/978-3-319-64546-9_10

$H_k(s) := (s_{i+j})_{i,j=0}^k$. The shifted sequence Es is $Es := (s_1, \dots, s_m) = (s_{j+1})_{j=0}^{m-1}$.
Recall that $A \succeq 0$ means that the matrix $A = A^T \in M_m(\mathbb{R})$ is positive semidefinite.
The following notation differs between the two cases $m = 2n$ and $m = 2n + 1$:

$$\underline{H}_{2n}(s) := H_n(s) \equiv (s_{i+j})_{i,j=0}^n,$$

$$\overline{H}_{2n}(s) := H_{n-1}((b-E)(E-a)s) \equiv ((a+b)s_{i+j+1} - s_{i+j+2} - abs_{i+j})_{i,j=0}^{n-1},$$

$$\underline{H}_{2n+1}(s) := H_n(Es - as) \equiv (s_{i+j+1} - as_{i+j})_{i,j=0}^n,$$

$$\overline{H}_{2n+1}(s) := H_n(bs - Es) \equiv (bs_{i+j} - s_{i+j+1})_{i,j=0}^n.$$

Here the lower index always refers to the highest moment in the corresponding
matrix and the highest moment occurs only in the right lower corner. The upper and
lower bar notation and the lower indices will be seen to be useful later. An advantage
is that they allow us to treat the even and odd cases at once.

Further, we abbreviate the corresponding Hankel determinants by

$$\underline{D}_m(s) := \det \underline{H}_m(s), \qquad \overline{D}_m(s) := \det \overline{H}_m(s). \tag{10.1}$$

For $f = \sum_{j=0}^k a_j x^j \in \mathbb{R}[x]_k$ let $\vec{f} := (a_0, \dots, a_k)^T \in \mathbb{R}^{k+1}$ denote the coefficient
vector of f. Then for $p, q \in \mathbb{R}[x]_n$ and $f, g \in \mathbb{R}[x]_{n-1}$ simple computations yield

$$L_s(pq) = \vec{p}^T \underline{H}_{2n}(s)\vec{q}, \quad L_s((b-x)(a-x)fg) = \vec{f}^T \overline{H}_{2n}(s)\vec{g}, \tag{10.2}$$

$$L_s((x-a)pq) = \vec{p}^T \underline{H}_{2n+1}(s)\vec{q}, \quad L_s((b-x)pq) = \vec{p}^T \overline{H}_{2n+1}(s)\vec{q}. \tag{10.3}$$

From Proposition 3.3 we restate the formulas (3.8) and (3.9) describing the
positive polynomials $\mathrm{Pos}([a,b])_m$ on $[a,b]$ of degree at most m:

$$\mathrm{Pos}([a,b])_{2n} = \{ f + (b-x)(x-a)g : f \in \Sigma_n^2, g \in \Sigma_{n-1}^2 \}, \tag{10.4}$$

$$\mathrm{Pos}([a,b])_{2n+1} = \{ (b-x)f + (x-a)g : f, g \in \Sigma_n^2 \}. \tag{10.5}$$

The following two existence theorems rely essentially on these descriptions.

Theorem 10.1 (Truncated $[a,b]$-Moment Problem; Even Case $m = 2n$) *For a
real sequence $s = (s_j)_{j=0}^{2n}$ the following statements are equivalent:*

(i) *s is a truncated $[a,b]$-moment sequence.*
(ii) *$L_s(p^2) \geq 0$ and $L_s((b-x)(x-a)q^2) \geq 0$ for $p \in \mathbb{R}[x]_n$ and $q \in \mathbb{R}[x]_{n-1}$.*
(iii) *$\underline{H}_{2n}(s) \succeq 0$ and $\overline{H}_{2n}(s) \succeq 0$.*

Theorem 10.2 (Truncated $[a,b]$-Moment Problem; Odd Case $m = 2n+1$) *For
a real sequence $s = (s_j)_{j=0}^{2n+1}$ the following are equivalent:*

(i) *s is a truncated $[a,b]$-moment sequence.*
(ii) *$L_s((x-a)p^2) \geq 0$ and $L_s((b-x)p^2) \geq 0$ for all $p \in \mathbb{R}[x]_n$.*
(iii) *$\underline{H}_{2n+1}(s) \succeq 0$ and $\overline{H}_{2n+1}(s) \succeq 0$.*

Proofs of Theorems 10.1 and 10.2:

(i)↔(ii) We apply Proposition 1.9 to the subspace $E = \mathbb{R}[x]_m$ of $C([a, b]; \mathbb{R})$. Then L_s is a truncated $[a, b]$-moment functional if and only if $L_s(p) \geq 0$ for all $p \in E_+ = \text{Pos}([a, b])_m$. By (10.4) and (10.5) this is equivalent to condition (ii).

(ii)↔(iii) follows at once from the identities (10.2) and (10.3).   □

▸ *Remark 10.3* Since $p^2 = (b - a)^{-1}[(b - x)p^2 + (x - a)p^2]$, condition (ii) in Theorem 10.2 implies that $L_s(p^2) \geq 0$ for $p \in \mathbb{R}_n[x]$.   ○

By Theorem 1.26, we can have *finitely atomic* representing measures in Theorems 10.1 and 10.2. In the subsequent sections we study such representing measures.

10.2   The Moment Cone $\mathcal{S}_{m+1}$ and Its Boundary Points

The following notions play a crucial role in this chapter.

Definition 10.4 The *moment cone* $\mathcal{S}_{m+1}$ and the *moment curve* c_{m+1} are defined by

$$\mathcal{S}_{m+1} = \left\{ s = (s_0, s_1, \ldots, s_m) : s_j = \int_a^b t^j \, d\mu(t), j = 0, \ldots, m, \ \mu \in M_+([a, b]) \right\},$$

$$c_{m+1} = \{ \mathfrak{s}(t) := (1, t, t^2, \ldots, t^m) : t \in [a, b] \}$$

That is, $\mathcal{S}_{m+1}$ is the set of moment sequences $s = (s_0, s_1, \ldots, s_m)$ of *all* Radon measures on $[a, b]$. The curve c_{m+1} is contained in $\mathcal{S}_{m+1}$, since $\mathfrak{s}(t)$ is the moment sequence of the delta measure δ_t.

By a slight abuse of notation we consider $\mathcal{S}_{m+1}$ and c_{m+1} as subsets of $\mathbb{R}^{m+1}$ by identifying the row vectors s and $\mathfrak{s}(t)$ with the corresponding column vector s^T and $\mathfrak{s}(t)^T$ in $\mathbb{R}^{m+1}$.

We denote by $\partial \mathcal{S}_{m+1}$ the set of boundary points, by $\text{Int } \mathcal{S}_{m+1}$ the set of interior points of $\mathcal{S}_{m+1}$, and by $\mathcal{C}_{m+1}$ the conic hull of c_{m+1}. Since the polynomials $1, t, \ldots, t^m$ are linearly independent, the points $\mathfrak{s}(t) \in \mathcal{C}_{m+1}$ span $\mathbb{R}^{m+1}$. Hence the interior of $\mathcal{C}_{m+1}$ is not empty by Proposition A.33(i).

From Theorem 1.26, applied to $E = \mathbb{R}[x]_m$ and $\mathcal{X} = [a, b]$, it follows that each $s \in \mathcal{S}_{m+1}, s \neq 0$, has a k-atomic representing measure

$$\mu = \sum_{j=1}^{k} m_j \delta_{t_j}, \tag{10.6}$$

where $k \leq m+1$ and $t_j \in [a, b]$ for all j. Since μ is k-atomic, the points t_j are pairwise distinct and $m_j > 0$ for all j. The numbers t_j are called *roots* or *atoms* of μ and the numbers m_j are the *weights* of μ. We can assume without loss of generality that

$$a \leq t_1 < t_2 < \cdots < t_k \leq b. \tag{10.7}$$

That the measure μ in (10.6) is a representing measure of s means that

$$s = \sum_{j=1}^{k} m_j \mathsf{s}(t_j).$$

Thus s belongs to the conic hull of the moment curve c_{m+1}. Hence $\mathcal{S}_{m+1} = \mathcal{C}_{m+1}$. By Theorem 1.26(ii), the set of functionals L_s, where $s \in \mathcal{S}_{m+1}$, is closed in the dual space E^*. This implies that $\mathcal{C}_{m+1}$ is closed in $\mathbb{R}^{m+1}$. We state these results.

Proposition 10.5 *The moment cone $\mathcal{S}_{m+1}$ is a closed convex cone in $\mathbb{R}^{m+1}$ with nonempty interior. It is the conic hull $\mathcal{C}_{m+1}$ of the moment curve c_{m+1}.*

By Definition A.38, a convex subset B of a cone C is a *base* of C if for each $u \in C, u \neq 0$, there exists a unique $\lambda > 0$ such that $\lambda u \in B$. It is easily seen that

$$\mathcal{S}^m = \{(1, s_1, \ldots, s_m) \in \mathcal{S}_{m+1}\}$$

is a base of the cone $\mathcal{S}_{m+1}$. It is just the set of moment sequences of all *probability measures* on $[a, b]$.

Clearly, $|s_j| = |\int_a^b t^j d\mu| \leq b - a$ for $s \in \mathcal{S}^m$. Hence $\mathcal{S}^m$ is bounded in $\mathbb{R}^{m+1}$. Obviously, $\mathcal{S}^m$ is a closed subset of the closed cone $\mathcal{S}_{m+1}$. Therefore, the set $\mathcal{S}^m$ is a *convex compact base of the moment cone $\mathcal{S}_{m+1}$.*

Definition 10.6 Let $s \in \mathcal{S}_{m+1}, s \neq 0$. The *index* $\mathrm{ind}(\mu)$ of the k-atomic representing measure (10.6) for s is the sum

$$\mathrm{ind}(\mu) := \sum_{j=1}^{k} \epsilon(t_j), \quad \text{where} \quad \epsilon(t) := 2 \quad \text{for} \ t \in (a, b) \ \text{and} \ \epsilon(a) = \epsilon(b) := 1.$$

The *index* $\mathrm{ind}(s)$ of s is the minimal index of all representing measures (10.6) for s.

The reason why boundary points and interior points are counted differently lies in the following fact, which enters into many proofs given in this chapter: If $t_0 \in [a, b]$ is a zero with multiplicity k of a polynomial $p \in \mathrm{Pos}([a, b])_m$, then we have $k \geq 2$ if $t_0 \in (a, b)$, while $k = 1$ is possible if $t_0 = a$ or $t_0 = b$.

Let us recall some further notions. For $s \in \mathcal{S}_{m+1}$ let $\mathcal{M}_s$ denote the set of all representing measures for s, that is, $\mathcal{M}_s$ is the set of $\mu \in M_+([a, b])$ such that

$$s_j = \int_a^b t^j d\mu(t) \quad \text{for} \ j = 0, \ldots, m.$$

If the set $\mathcal{M}_s$ is a singleton, then s is called $[a, b]$-*determinate.*

The following theorem brings a number of important properties together.

Theorem 10.7 *For $s \in \mathcal{S}_{m+1}$, $s \neq 0$, the following statements are equivalent:*

(i) *s is a boundary point of the convex cone $\mathcal{S}_{m+1}$.*
(ii) $\mathrm{ind}(s) \leq m$.
(iii) *There exists a $p \in \mathrm{Pos}([a, b])_m$, $p \neq 0$, such that $L_s(p) = 0$.*
(iv) *$\underline{D}_m(s) = 0$ or $\overline{D}_m(s) = 0$.*
(v) *s is $[a, b]$-determinate, that is, $\mathcal{M}_s$ is a singleton.*

If p is as in (iii), then $\operatorname{supp} \mu \subseteq \mathcal{Z}(p)$ for the unique measure $\mu \in \mathcal{M}_s$.

Proof Let $\{e_0, \ldots, e_m\}$ denote the canonical vector space basis of $\mathbb{R}^{m+1}$.

(i)$\rightarrow$(iii) Let $\mu \in \mathcal{M}_s$. Since s is a boundary point of the cone $\mathcal{S}_{m+1}$, there is a supporting hyperplane of $\mathcal{S}_{m+1}$ at s (by Proposition A.34(ii)), that is, there exists a linear functional $F \neq 0$ on $\mathbb{R}^{m+1}$ such that $F(s) = 0$ and $F \geq 0$ on $\mathcal{S}_{m+1}$. Then $p(t) = F(\mathfrak{s}(t))$ is a polynomial in t of degree at most m. From $\mathfrak{s}(t) \in \mathcal{S}_{m+1}$ we get $p \in \mathrm{Pos}([a, b])$. Since $f \neq 0$, p is not the zero polynomial. Then

$$L_s(p) = L(F(\mathfrak{s}(t)) = \int_a^b F(\mathfrak{s}(t)) \, d\mu(t) = \int_a^b (F(e_0) + \cdots + F(e_m)t^m) d\mu(t)$$

$$= F(e_0)s_0 + \cdots + F(e_m)s_m = F(s) = 0.$$

This proves (iii).

(iii)$\rightarrow$(i) Let $p(t) = \sum_{j=0}^m a_j t^j$ be a polynomial as in (iii). Define a linear functional F on $\mathbb{R}^{m+1}$ by $F(e_j) = a_j$, $j = 0, \ldots, m$. Reversing the preceding reasoning, we get $F(s) = L_s(p) = 0$. Since p is not the zero polynomial, $F \neq 0$. Further, $F(\mathfrak{s}(t)) = p(t) \geq 0$ on $[a, b]$, since $p \in \mathrm{Pos}([a, b])$. Therefore, $F \geq 0$ on c_{m+1} and hence on its conic convex hull $\mathcal{C}_{m+1} = \mathcal{S}_{m+1}$. Thus F is a supporting functional to $\mathcal{S}_{m+1}$ at s. Hence s is a boundary point of $\mathcal{S}_{m+1}$ by Proposition A.34(ii).

(ii)$\rightarrow$(iii) Assume that $\mathrm{ind}(s) \leq m$. Let p denote the product of quadratic factors $(t - t_j)^2$ for all $t_j \in (a, b)$ and linear factors $(b - t)$ and $(t - a)$ provided that the end points b or a are among the points t_j, respectively. Clearly, then $p \in \mathrm{Pos}([a, b])$ and $\deg(p) = \sum_j \epsilon(t_j) = \mathrm{ind}(s) \leq m$.

(iii)$\rightarrow$(ii) Let p be a polynomial as in (iii). We choose a representing measure (10.6) for which $\mathrm{ind}\,(s) = \mathrm{ind}\,(\mu)$. Since $L_s(p) = 0$, Proposition 1.23 yields $\{t_1, \ldots, t_k\} = \operatorname{supp}\mu \subseteq \mathcal{Z}(p)$, that is, $p(t_j) = 0$ for $j = 1, \ldots, k$. Each root $t_j \in (a, b)$ has even multiplicity. (Otherwise $p(t)$ would change sign at t_j. Since $p(t) \geq 0$ on $[a, b]$, this is impossible.) The only roots of multiplicity 1 are possibly the end points a and b. Therefore, counting the roots t_j with multiplicities and adding the number $\epsilon(t_j)$ we obtain $\mathrm{ind}(s) = \sum_j \epsilon(t_j) \leq \deg(p) \leq m$.

(iii)$\leftrightarrow$(iv) We carry out the proof in the even case $m = 2n$. The proof in the odd case is similar; instead of (10.2) and (10.4) we use (10.3) and (10.5).

Let $f(x) = \sum_{i=0}^{n} a_i x^i \in \mathbb{R}[x]_n$ and $g(x) = \sum_{j=0}^{n-1} b_j x^j \in \mathbb{R}[x]_{n-1}$. As above we set $\vec{f} = (a_0, \dots, a_n)^T \in \mathbb{R}^{n+1}$ and $\vec{g} = (b_0, \dots, b_{n-1})^T \in \mathbb{R}^n$. The Hankel matrices $\underline{H}_{2n}(s)$ and $\overline{H}_{2n}(s)$ are positive semidefinite by Theorem 10.1. Therefore, by (10.2),

$$L_s(f^2) = \vec{f}^T \underline{H}_{2n}(s) \vec{f} = \| \underline{H}_{2n}(s)^{1/2} \vec{f} \|^2, \tag{10.8}$$

$$L_s((b-x)(x-a)g^2) = \vec{q}^T \overline{H}_{2n}(s) \vec{g} = \| \overline{H}_{2n}(s)^{1/2} \vec{g} \|^2, \tag{10.9}$$

where $\| \cdot \|$ denotes the Euclidean norm of $\mathbb{R}^d$. By (10.4), $\mathrm{Pos}([a,b])_{2n}$ consists of sums of polynomials of the form $p = f^2 + (b-x)(x-a)g^2$. Clearly, for such a polynomial p we have $L_s(p) = 0$ if and only if $L_s(f^2) = 0$ and $L_s((b-x)(x-a)g^2) = 0$. Therefore, by (10.8) and (10.9), $L_s(p) = 0$ for $p \in \mathrm{Pos}([a,b])_{2n}$ implies $p = 0$ if and only if both matrices $\underline{H}_{2n}(s)$ and $\overline{H}_{2n}(s)$ are regular, that is, $\underline{D}_{2n}(s) \neq 0$ and $\overline{D}_{2n}(s) \neq 0$. Hence (iii) holds if and only if $\underline{D}_{2n}(s) = 0$ or $\overline{D}_{2n}(s) = 0$.

(iii)→(v) Let $\nu \in \mathcal{M}_s$. By Proposition 1.23, $\mathrm{supp}\,\nu \subseteq \mathcal{Z}(p)$. In particular, this proves the last assertion. Let $\mathcal{Z}(p) = \{x_1, \dots, x_r\}$. Then $\nu = \sum_{i=1}^{r} n_i \delta_{x_i}$ for some numbers $n_i \geq 0$. Since $r \leq \deg(p) \leq m$, there exist Lagrange interpolation polynomials $p_j \in \mathbb{R}[x]_m$ such that $p_j(x_i) = \delta_{ij}$, $i, j = 1, \dots, r$. Then

$$L_s(p_j) = \int_a^b p_j \, d\nu = \sum_{i=1}^{r} n_i p_j(x_i) = n_j, \quad j = 1, \dots, r.$$

This shows that ν is uniquely determined by s and p.

(v)→(i) If s is not in $\partial \mathcal{S}_{n+1}$, then $s \in \mathrm{Int}\,\mathcal{S}_{m+1}$; hence s has infinitely many representing measures by Proposition 10.9 below.   □

We briefly discuss the preceding theorem. Let s be a boundary point of $\mathcal{S}_{m+1}$. Then s has a unique representing measure $\mu \in M_+([a,b])$. This measure μ has the form (10.6) and $\mathrm{ind}\,(\mu) = \mathrm{ind}(s) \leq m$. If all t_j are in the open interval (a,b), then $k \leq \frac{m}{2}$. If precisely one t_j is an end point, then $k \leq \frac{m+1}{2}$ and if both end points are among the t_j, then $k \leq \frac{m}{2} + 1$. The case $k = \frac{m}{2} + 1$ can only happen if m is even and both a and b are among the t_j. Thus, all boundary points of $\mathcal{S}_{m+1}$ can be represented by k-atomic measures with $k \leq \frac{m}{2} + 1$.

10.3   Interior Points of $\mathcal{S}_{m+1}$ and Interlacing Properties of Roots

The following result is often used in the sequel. Condition (ii) and (iii) therein should be compared with conditions (ii) and (iii) in Theorems 10.1 and 10.2.

Theorem 10.8 *For a sequence* $s \in \mathcal{S}_{m+1}$ *the following are equivalent:*

(i) *s is an interior point of* $\mathcal{S}_{m+1}$.
(ii) *The Hankel matrices* $\underline{H}_m(s)$ *and* $\overline{H}_m(s)$ *are positive definite.*
(iii) *$m = 2n$:*
   $L_s(p^2) > 0$ and $L_s((b-x)(x-a)q^2) > 0$ for $p \in \mathbb{R}[x]_n$, $q \in \mathbb{R}[x]_{n-1}, p, q \neq 0$.
   $m = 2n+1$:
   $L_s((x-a)p^2) > 0$ and $L_s((b-x)p^2) > 0$ for all $p \in \mathbb{R}[x]_n, p \neq 0$.
(iv) *$\underline{D}_m(s) > 0$ and $\overline{D}_m(s) > 0$.*
(v) *$\underline{D}_0(s) > 0, \overline{D}_0(s) > 0, \underline{D}_1(s) > 0, \overline{D}_1(s) > 0, \ldots, \underline{D}_m(s) > 0, \overline{D}_m(s) > 0$.*

Proof (i)↔(iv) is only a restatement of Theorem 10.7 (i)↔(iv). The equivalence (ii)↔(iv) is a well-known fact from linear algebra. The equivalence of (iii) and (iv) follows from formulas (10.2) and (10.3). (v)→(iv) is trivial. Obviously, $s \in$ Int $\mathcal{S}_{m+1}$ implies that $s^{(j)} := (s_0, \ldots, s_j) \in$ Int $\mathcal{S}_{j+1}$ for $j = 0, \ldots, m$. Therefore, (i)→(v) follows by applying the implication (i)→(ii) to the sequences $s^{(j)}$.   □

Proposition 10.9 *Let $s \in$ Int $\mathcal{S}_{m+1}$. For each $\xi \in [a, b]$ there exists a representing measure μ_ξ of s with ξ as an atom which has index $m + 1$ or $m + 2$ if $\xi \in (a, b)$ and index $m + 1$ if $\xi = a$ or $\xi = b$.*

Proof Clearly, $c(\xi)$ is a boundary point, because $\mathrm{ind}(c(\xi)) \leq 1$. The line through the two different points s and $c(\xi)$ of $\mathcal{S}_{m+1}$ intersects the boundary of $\mathcal{S}_{m+1}$ in a second point s'. Then $s = \lambda s' + (1 - \lambda)c(\xi)$ for some $\lambda \in (a, b)$. The unique representing measure μ' of s' satisfies $\mathrm{ind}(\mu') \leq m$ by Theorem 10.7. It is clear that $\mu = \lambda \mu' + (1 - \lambda)\delta_\xi$ is a representing measure of s. Its index is at most $m + 2$ if $\xi \in (a, b)$ and at most $m + 1$ if $\xi = a$ or $\xi = b$. We have $\mathrm{ind}(s) > m$, since otherwise s would be a boundary point by Theorem 10.7.   □

Definition 10.10 A representing measure μ of $s \in$ Int $\mathcal{S}_{m+1}$ is called *canonical* if μ is of the form (10.6) and $\mathrm{ind}(\mu) \leq m + 2$.

Each representing measure μ_ξ from Proposition 10.9 is canonical.
The following two propositions are useful for deriving interlacing properties of roots for different canonical measures of the same moment sequence $s \in$ Int $\mathcal{S}_{m+1}$.

Proposition 10.11 *Suppose that μ is a canonical representing measure of $s \in$ Int $\mathcal{S}_{m+1}$ with roots $t_1 < t_2 < \cdots < t_k$ and weights $m_1, \ldots, m_k$. Let $\tilde{\mu}$ be a representing measure of s such that $\tilde{\mu} \neq \mu$. Then we have:*

(i) supp $\tilde{\mu} \cap (t_j, t_{j+1}) \neq \emptyset$ *for* $t_j, t_{j+1} \in (a, b)$.
(ii) *If* $\mathrm{ind}(\mu) = m + 1$, *then* supp $\tilde{\mu} \cap (t_j, t_{j+1}) \neq \emptyset$ *for all* $t_j, t_{j+1} \in [a, b]$.
(iii) *If* $t_1 = a$, *then* supp $\tilde{\mu} \cap [t_1, t_2) \neq \emptyset$.
(iv) *If* $t_k = b$, *then* supp $\tilde{\mu} \cap (t_{k-1}, t_k] \neq \emptyset$.

Proof

(i) The proof is a modification of the proof of Theorem 10.7 (iii)→(iv).
   Fix $t_j, t_{j+1} \in (a, b)$. Let $p(t)$ denote the product of $(t - t_{j+1})(t - t_j)$, all factors $(t - t_i)^2$ with $t_i \in (a, b), t_i \neq t_j, t_{j+1}$, and possibly $(t - a)$ resp. $(b - t)$ if a resp. b

are among the roots. Then $p(t_j) = 0$ for all j. Since $\text{ind}(\mu) \leq m + 2$, we have $\deg(p) \leq m$. (Note that each of the two interior roots t_j, t_{j+1} is counted with 2 in the definition of $\text{ind}(s)$, but it appears only with degree 1 in p.) Hence the functional L_s applies to p and we obtain

$$L_s(p) = \int_a^b p \, d\mu = \sum_{j=1}^k m_j p(t_j) = 0.$$

Assume to the contrary that $\text{supp}\,\tilde{\mu} \cap (t_j, t_{j+1}) = \emptyset$. By the definition of p we have $p(t) \geq 0$ on $M_j := [0, t_j] \cup [t_{j+1}, 1]$. Therefore, since $0 = L_s(p) = \int_{M_j} p \, d\tilde{\mu}$, Proposition 1.23 implies that $\text{supp}\,\tilde{\mu} \subseteq \mathcal{Z}(p) = \{t_1, \ldots, t_k\}$. Thus, $\tilde{\mu} = \sum_{i=1}^k n_i \delta_{t_i}$ for some numbers $n_i \geq 0$. Since $\text{ind}(\mu) \leq m + 2$ by assumption, we have $k \leq \frac{m}{2} + 2$ and hence $k \leq m + 1$, because k is an integer and $m \geq 1$. Hence there exist interpolation polynomials p_j of degree $\deg(p_j) \leq m$ satisfying $p_j(x_i) = \delta_{ij}$, $i, j = 1, \ldots, k$. Finally, $L_s(p_j) = \int p_j \, d\tilde{\mu} = n_j$ and $L_s(p_j) = \int p_j \, d\mu = m_j$ imply that $n_j = m_j$ for all j. Thus $\mu = \tilde{\mu}$, which is the desired contradiction.

(ii) In the case $\text{ind}(\mu) = m + 1$ the preceding reasoning also works for the end points among the roots. We illustrate this for $t_1 = a$. Let $p(t)$ denote the product of $(t - a)(t - t_2)$, all factors $(t - t_i)^2$ with $t_i \in (a, b)$, $t_i \neq t_j, t_{j+1}$, and $(b - t)$ if b is a root. Then $\deg(p) = m$, because $\text{ind}(\mu) = m + 1$. Since $p(t_j) = 0$ for all j and $p(t) \geq 0$ on $[t_2, b]$, we can proceed as above and derive that $\text{supp}\,\tilde{\mu} \cap (t_1, t_2) \neq \emptyset$.

(iii) Assume to the contrary that $\text{supp}\,\tilde{\mu} \cap [a, t_2) = \emptyset$.

First we note that $t_2 \neq b$. Indeed, otherwise $\tilde{\mu}$ has support $\{b\}$ and hence $\text{ind}(\tilde{\mu}) = 1$ which contradicts the assumption $s \in \text{Int}\,\mathcal{S}_{m+1}$.

Let q be the product of factors $(t - t_2)$, $(t - t_i)^2$ if $i \geq 3$ and $t_i \neq b$, and $(t_k - t)$ if $t_k = b$. Since $t_1 = a, t_2 < b$, and $\text{ind}(\mu) \leq m + 2$, we have $\deg(q) \leq m$. The polynomial q vanishes exactly at $t_2, \ldots, t_k$. Hence $L_s(q) = \int_a^b q \, d\mu = m_1 q(a)$. By construction, $q \geq 0$ on $[t_2, b]$ and therefore $L_s(q) = \int_a^b q \, d\tilde{\mu} = \int_{t_2}^b q \, d\tilde{\mu} \geq 0$. But $q(a) < 0$, so we obtain a contradiction.

(iv) follows in a similar manner as (iii). $\qquad\square$

Proposition 10.12 *Let μ and μ' be two different canonical representing measures of $s \in \text{Int}\,\mathcal{S}_{m+1}$ with roots $t_1 < t_2 < \cdots < t_k$ and $t_1' < t_2' < \cdots < t_l'$ and weights m_i and m_j', respectively.*

(i) *The roots t_i and t_j' contained in (a, b) strictly interlace.*

(ii) *Suppose that $t_1 = t_1' = a$. Then $m_1 \neq m_1'$. Further, $m_1' > m_1$ if and only if $t_2' > t_2$.*

(iii) *If $t_k = t_l' = b$, then $m_k \neq m_l'$ and we have $m_l' > m_k$ if and only if $t_{l-1} > t_{k-1}'$.*

Proof

(i) By assumption μ and μ' have index $m + 1$ or $m + 2$. Considering the even and the odd case separately we see that the numbers of their roots in (a, b) differ by at most one. By Proposition 10.11 (i) and (iii), the smallest inner roots of μ and μ' are different. Therefore it follows from Proposition 10.11(i) that the inner roots of μ and μ' strictly interlace.

(ii) The crucial step of the proof is to show that $t_2' > t_2$ implies $m_1' > m_1$. Let q be the polynomial defined in the proof of Proposition 10.11(iii). Let us recall that $\deg(q) \le m$, $q \ge 0$ on $[t_2, b]$ and q vanishes exactly at the roots $t_2, \dots, t_k$. The root t_2' cannot be equal to all roots t_i, $i \ge 3$. (This would imply that $t_2' = t_3 = b$ and $k = 3$, so $\mathrm{ind}(\mu') = 2$ and $\mathrm{ind}(\mu) = 4$, which is impossible.) Therefore, $\int_{t_2}^b q \, d\mu' > 0$ and

$$m_1 q(a) = \int_a^b q \, d\mu = L_s(q) = \int_a^b q \, d\mu' = m_1' q(a) + \int_{t_2}^b q \, d\mu' > m_1' q(a).$$

Since $q(a) < 0$, the latter yields $m_1' > m_1$.

By interchanging the role of μ and μ' it follows that $t_2 > t_2'$ implies $m_1 > m_1'$.

Finally, we show that $m_1 \ne m_1'$. Assume to the contrary that $m_1 = m_1'$. Then we have $t_2 = t_2'$ by the preceding. Since inner roots strictly interlace by (i), this can only happen if $t_2 = t_2' = b$. But since μ and μ' have the same total mass s_0 and $m_1 = m_1'$, it follows that $\mu = \mu'$. This contradicts our assumption.

(iii) is proved in a similar manner as (ii). □

Corollary 10.13 *Let* $s \in \mathrm{Int} \, \mathcal{S}_{n+1}$. *For each* $\xi \in (a, b)$ *there exists a unique canonical representing measure* μ_ξ *of* s *with atom* ξ.

Proof The existence has been already stated in Proposition 10.9, so it remains to verify the uniqueness. If there were two different such measures, their roots contained in (a, b) would strictly interlace by Proposition 10.12(i). Since both measures have the same root $\xi \in (a, b)$, this is impossible. □

10.4 Principal Measures of Interior Points of $\mathcal{S}_{m+1}$

Now we turn to the minimal possible index $m + 1$ for interior points of $\mathcal{S}_{m+1}$. Throughout this section we suppose that $s \in \mathrm{Int} \, \mathcal{S}_{m+1}$.

Definition 10.14 A representing measure μ of the form (10.6) for s is called

- *principal* if $\mathrm{ind}(\mu) = m + 1$,
- *upper principal* if it is principal and b is an atom of μ,
- *lower principal* if it is principal and b is not an atom of μ.

That is, for principal measures the index is equal to the number of prescribed moments. The representing measures constructed in the proof of Proposition 10.9 for $\xi = 0$ and for $\xi = 1$ are principal. We will return to these measures later.

Given a fixed sequence $s = (s_0, \ldots, s_m) \in \mathcal{S}_{m+1}$ we define

$$s_{m+1}^+ = \sup_{\mu \in \mathcal{M}_s} \int_a^b x^{m+1} \, d\mu(x) = \sup_{(s_0, \ldots, s_m, s_{m+1}) \in \mathcal{S}_{m+1}} s_{m+1} \qquad (10.10)$$

$$s_{m+1}^- = \inf_{\mu \in \mathcal{M}_s} \int_a^b x^{m+1} \, d\mu(x) \quad \inf_{(s_0, \ldots, s_m, s_{m+1}) \in \mathcal{S}_{m+1}} s_{m+1}. \qquad (10.11)$$

Since $\mathcal{M}_s$ is weakly compact (by Theorem 1.19) and $\int_0^1 x^{m+1} \, d\mu$ is a continuous function on the compact space $\mathcal{M}_s$, the supremum in (10.10) and the infimum in (10.11) are attained. Thus, s_{m+1}^+ is the maximum and s_{m+1}^- is the minimum of the moment s_{m+1} over the set $\mathcal{M}_s$ of all measures which have the given moments $s_0, \ldots, s_m$. These extremal values s_{m+1}^+ and s_{m+1}^- play a crucial role in our approach to the principal measures $\mu^\pm$, but they are also of interest in themselves.

Let $s^\pm$ denote the sequence $(s_0, \ldots, s_m, s_{m+1}^\pm) \in \mathcal{S}_{m+2}$. Since s_{m+1}^+ is a maximum and s_{m+1}^- is a minimum, s^+ and s^- are not in $\mathrm{Int}\, \mathcal{S}_{m+2}$. Hence $s^\pm$ belongs to the boundary of $\mathcal{S}_{m+2}$, so by Theorem 10.7 it has a unique representing measure $\mu^\pm$ of index $\mathrm{ind}\,(\mu^\pm) \leq m + 1$. Obviously, μ^+ and μ^- are also representing measure for the sequence s. Since $s \in \mathrm{Int}\, \mathcal{S}_{m+1}$, $\mathrm{ind}\,(s) > m$ by Theorem 10.7. Hence $\mathrm{ind}\,(\mu^\pm) = m + 1$, that is, μ^+ and μ^- are principal representing measures for s.

The next proposition characterize the numbers $s_{m+1}^\pm$ in terms of determinants of Hankel matrices.

Proposition 10.15 Let $\tilde{s} = (s_0, \ldots, s_m, s_{m+1}) \in \mathcal{S}_{m+2}$. Then s_{m+1}^+ and s_{m+1}^- are the unique numbers s_{m+1} satisfying $\overline{D}_{m+1}(\tilde{s}) = 0$ and $\underline{D}_{m+1}(\tilde{s}) = 0$, respectively. Further, we have $s_{m+1}^+ > s_{m+1}^-$ and

$$\overline{D}_{m+1}(s^+) = 0, \quad \underline{D}_{m+1}(s^+) > 0, \quad \underline{D}_{m+1}(s^-) = 0, \quad \overline{D}_{m+1}(s^-) > 0. \qquad (10.12)$$

Proof We develop $\underline{D}_{m+1}(\tilde{s})$ and $\overline{D}_{m+1}(\tilde{s})$ by the last row and obtain

$$\underline{D}_{m+1}(\tilde{s}) = s_{m+1} \underline{D}_{m-1}(s) + c^+, \quad \overline{D}_{m+1}(\tilde{s}) = -s_{m+1} \overline{D}_{m-1}(s) + c^- \qquad (10.13)$$

for some numbers $c^\pm$ depending only on the given moments $s_0, \ldots, s_m$.

We prove that s_{m+1}^- is the unique number s_{m+1} for which $\underline{D}_{m+1}(\tilde{s}) = 0$. Since $s \in \mathrm{Int}\, \mathcal{S}_{m+1}$, we have $\underline{D}_{m-1}(s) > 0$ by Theorem 10.7 (i)$\leftrightarrow$(iv). Therefore, by (10.13), $\underline{D}_{m+1}(\tilde{s})$ is a strictly increasing function of s_{m+1}. Let us take some $\tilde{s} \in \mathrm{Int}\, \mathcal{S}_{m+2}$. Then $\underline{D}_{m+1}(\tilde{s}) > 0$ and $\overline{D}_{m+1}(\tilde{s}) > 0$ by Theorem 10.8. Now we decrease s_{m+1} untill s_{m+1}^- to obtain $\underline{D}_{m+1}(\tilde{s}) = 0$. Then $\overline{D}_{m+1}(\tilde{s})$ increases by (10.13) and remains positive. Hence s_{m+1}^- is the lower bound of the numbers s_{m+1} for which $\underline{D}_{m+1}(\tilde{s}) \geq 0$. Since $\overline{H}_m(s)$ is positive definite because $s \in \mathrm{Int}\, \mathcal{S}_{m+1}$, s_{m+1}^- is also the lower bound of the numbers s_{m+1} such that $\overline{H}_{m+1}(\tilde{s}) \succeq 0$, or equivalently, $\tilde{s} \in \mathcal{S}_{m+2}$. Thus, s_{m+1}^- is the number defined by (10.11).

Similarly it follows that the number s_{m+1}^+ from (10.10) is uniquely determined by the equation $\overline{D}_{m+1}(\tilde{s}) = 0$. Moreover, the proof shows that $s_{m+1}^+ > s_{m+1}^-$.   □

Before we continue we note an interesting by-product of the preceding.

Corollary 10.16 *Let* $s = (s_0, \ldots, s_m) \in \text{Int } \mathcal{S}_{m+1}$ *and let* s_{m+1} *be a real number. Then* $\tilde{s} = (s_0, \ldots, s_m, s_{m+1}) \in \mathcal{S}_{m+2}$ *if and only if* $s_{m+1}^+ \leq s_{m+1} \leq s_{m+1}^-$.

Proof The only if part is clear from the definition of $s_{m+1}^{\pm}$. Conversely, suppose that $s_{m+1}^+ \leq s_{m+1} \leq s_{m+1}^-$. Then $s_{m+1} = \lambda s_{m+1}^+ + (1 - \lambda)s_{m+1}^-$ for some $\lambda \in [0, 1]$. Therefore, $\lambda \mu^+ + (1 - \lambda)\mu^-$ is a representing measure for $\tilde{s}$, so that $\tilde{s} \in \mathcal{S}_{m+2}$.   □

Next we show that the two measures $\mu^{\pm}$ defined above are the only prinicipal representing measures of s. Let μ be an arbitrary principal measure for s. Clearly, $\tilde{s} := (s_0, \ldots, s_m, s_{m+1}) \in \mathcal{S}_{m+2}$, where $s_{m+1} := \int_a^b x^{m+1} \, d\mu$. Since $\text{ind}(\mu) = m + 1$, $\underline{D}_{m+1}(\tilde{s}) = 0$ or $\overline{D}_{m+1}(\tilde{s}) = 0$ by Theorem 10.7 (ii)→(iv). Hence, by Proposition 10.15, $s_{m+1} = s_{m+1}^-$ or $s_{m+1} = s_{m+2}^+$, which implies that $\tilde{s} = s^-$ or $\tilde{s} = s^-$. Since $\mu^{\pm}$ is the unique representing measure of $s^{\pm}$, we conclude that $\mu = \mu^-$ or $\mu = \mu^+$.

We denote by $t_j^{\pm}$ the roots and by $m_j^{\pm}$ the weights of $\mu^{\pm}$ and assume that the roots are ordered as in (10.7). Since $\text{ind}(\mu^{\pm}) = m+1$, it follows from Proposition 10.11 that the roots of μ^+ and μ^- are strictly interlacing. To describe their location further we distinguish between the even and odd cases.

Case $m = 2n$:
Since both measures $\mu^{\pm}$ have index $2n + 1$, they have exactly n roots contained in (a, b) and one end point a or b as root. Let $f(x) = \sum_{i=0}^n a_i x^i \in \mathbb{R}[x]_n$. Setting $\vec{f} = (a_0, \ldots, a_n)^T \in \mathbb{R}^{n+1}$ and using (10.3) we compute

$$\vec{f}^T \overline{H}_{m+1}(s^+)\vec{f} = L_{s^+}((b-x)f^2) = \int_a^b (b-x)f(x)^2 \, d\mu^+ = \sum_{j=1}^{n+1} m_j^+(b - t_j^+)f(t_j^+)^2.$$

Since $\overline{D}_{m+1}(s^+) = 0$, the Hankel matrix $\overline{H}_{m+1}(s^+)$ has a nontrivial kernel. Hence there exists an $f \neq 0$ such that $L_{s^+}((b-x)f^2) = 0$, that is, $(b - t_j^+)f(t_j^+) = 0$ for all $j = 1, \ldots, n + 1$. Since there are $n + 1$ roots and $\deg(f) \leq n$, this is only possible if $t_{n+1}^+ = b$. A similar reasoning using $\underline{H}_{m+1}(s^-)$ instead of $\overline{H}_{m+1}(s^+)$ shows that $t_1^+ = a$. Thus, μ^+ is upper principal and μ^- is lower principal.

Since s has only two principal measures, we conclude that μ^+ and μ^- are the principal representing measures μ_ξ from Proposition 10.9 for $\xi = b$ and $\xi = a$, that is, $\mu_+ = \mu_b$ and $\mu^- = \mu_a$.

Case $m = 2n + 1$:
Since $\text{ind}(\mu^{\pm}) = 2n + 2$, $\mu^{\pm}$ has either $n + 1$ inner roots or n inner roots and both end points as roots. Thus μ^+ has $k \geq n + 1$ roots. Let $f \in \mathbb{R}[x]_n$. Using (10.2)

we derive

$$\vec{f}^T \, \overline{H}_{m+1}(s^+)\vec{f} = L_{s^+}((b-x)(x-a)f^2) = \sum_{j=1}^{k} m_j^+ (b-t_j^+)(t_j^+ - a)f(t_j^+)^2.$$

As in the even case, $\overline{H}_{m+1}(s^+)$ has a nontrivial kernel, so there is a polynomial $f \neq 0$ such that $L_{s^+}((b-x)(x-a)f^2) = 0$. Then $(b-t_j^+)(t_j^+ - a)f(t_j^+) = 0$ for $j = 1, \ldots, k$. Since $k \geq n+1$ and $\deg(f) \leq n$, one root of μ^+ must be an end point. Therefore, since $\mathrm{ind}\,(\mu^+) = 2n + 2$, it follows that $t_1^+ = a, t_{n+2}^+ = b, k = n + 2$.

By Proposition 10.11, the roots of μ^+ and μ^- strictly interlace. Therefore, $t_j^+ < t_j^-$ and $t_i^- < t_{i+1}^+$. Hence μ^+ has both end points as roots, while μ^- has only inner roots. As in the even case, μ^+ is upper principal and μ^- is lower principal.

In contrast to the even case, μ^+ is equal to both principal measures μ_ξ from Proposition 10.9 for $\xi = b, a$, that is, $\mu_+ = \mu_b = \mu_a$. The second principal measure μ^- has only inner roots and it is not obtained from Proposition 10.9.

We illustrate the location of roots $t_j^\pm$ of the principal measures $\mu^\pm$ in the even and odd cases by the following scheme:

$$m = 2n, \qquad \mu^+ : \; a < t_1^+ < t_2^+ < \cdots < t_{n+1}^+ = b, \tag{10.14}$$

$$m = 2n, \qquad \mu^- : \; a = t_1^- < t_2^- < \cdots < t_{n+1}^- < b, \tag{10.15}$$

$$m = 2n + 1, \qquad \mu^+ : \; a = t_1^+ < t_2^+ < \cdots < t_{n+1}^+ < t_{n+2}^+ = b, \tag{10.16}$$

$$m = 2n + 1, \qquad \mu^- : \; a < t_1^- < t_2^- < \cdots < t_{n+1}^- < b. \tag{10.17}$$

Further, from Proposition 10.12 we obtain

$$m = 2n : \qquad a = t_1^- < t_1^+ < t_2^- < t_2^+ < \cdots < t_n^+ < t_{n+1}^- < t_{n+1}^+ = b, \tag{10.18}$$

$$m = 2n + 1 : \quad a = t_1^+ < t_1^- < t_2^+ < t_2^- < \cdots < t_{n+1}^+ < t_{n+1}^- < t_{n+2}^+ = b. \tag{10.19}$$

The following theorem summarizes some of the preceding results.

Theorem 10.17 *Let s be an interior point of $\mathcal{S}_{m+1}$. Then μ^+ is the unique upper principal representing measure and μ^- is the unique lower princical representing measure for s. The roots of μ^+ and μ^- are stricly interlacing.*

Now we define a distinguished canonical representing measure μ_ξ and a related polynomial $q_\xi \in \mathrm{Pos}([a, b])$ for each $\xi \in [a, b]$. Both will play a crucial role in the next sections.

Definition 10.18 Let $s \in \mathrm{Int}\,\mathcal{S}_{m+1}$. For $\xi \in (a, b)$, μ_ξ is the unique canonical measure of s which has ξ as a root (see Corollary 10.13). For $\xi = a$ or $\xi = b$, μ_ξ is the unique principal measure of s which has ξ as a root, that is, $\mu_a = \mu^-, \mu_b = \mu^+$ for $m = 2n$ and $\mu_a = \mu_b = \mu^+$ for $m = 2n + 1$.

If $\xi \in (a, b)$ is root of a principal measure $\mu^\pm$, then $\mu_\xi = \mu^\pm$ has index $m + 1$. Note that μ_a resp. μ_b is the only principal measure with root a resp. b, but there are many canonical measures with root a resp. b, see Theorems 10.25 and 10.26 below.

Definition 10.19 For $\xi \in [a, b]$ let the polynomial $q_\xi(x)$ be the product of factors $(x - t_j)^2$ for all roots $t_j \in (a, b)$, $t_j \neq \xi$, of the measure μ_ξ and $(x - a)$ resp. $(b - x)$ if a resp. b is a root of μ_ξ.

Lemma 10.20 *Let $\xi \in [a, b]$ and let $\xi, t_1, \ldots, t_k$ denote the roots of μ_ξ. Then, up to a constant positive factor, q_ξ is the unique polynomial $q \in \mathrm{Pos}([a, b])$ such that*

$$q(\xi) > 0, \ \deg(q) + \epsilon(\xi) \leq \mathrm{ind}(\mu_\xi) \ and \ q(t_j) = 0 \ for \ j = 1, \ldots, k. \quad (10.20)$$

Proof From its definition it is clear that q_ξ has these properties. Since in Definition 10.19 no factor was taken for the root ξ, we even have $\deg(q_\xi) + \epsilon(\xi) = \mathrm{ind}(\mu_\xi)$.

Let $\tilde{q}$ be another such polynomial. Since $\tilde{q} \in \mathrm{Pos}([a, b])$, its zeros in (a, b) have even multiplicities. Hence we conclude that $\deg(q_\xi) \leq \deg(\tilde{q})$. But

$$\deg(\tilde{q}) + \epsilon(\xi) \leq \mathrm{ind}(\mu_\xi) = \deg(q_\xi) + \epsilon(\xi)$$

by (10.20) implies that $\deg(\tilde{q}) \leq \deg(q_\xi)$. Thus $\deg(\tilde{q}) = \deg(q_\xi)$. Since $\tilde{q}(\xi) > 0$ and $q_\xi(\xi) > 0$, it follows that $\tilde{q}$ is a positive multiple of q_ξ. $\square$

10.5 Maximal Masses and Canonical Measures

Let $s \in \mathcal{S}_{m+1}$ and $\xi \in [a, b]$. Then we define

$$\rho_s(\xi) = \sup\{\mu(\{\xi\}) : \mu \in \mathcal{M}_s\}, \quad (10.21)$$

that is, $\rho_s(\xi)$ is the supremum of masses at ξ of all representing measures of s. We will say that a measure $\mu \in \mathcal{M}_s$ has *maximal mass* at ξ if $\mu(\{\xi\}) = \rho_s(\xi)$.

Further, we need the following number

$$\kappa_s(\xi) := \inf\{L_s(p) : p \in \mathrm{Pos}([a, b])_m, \ p(\xi) = 1\} \quad (10.22)$$

$$= \inf\left\{\frac{L_s(q)}{q(\xi)} : q \in \mathrm{Pos}([a, b])_m\right\}, \quad (10.23)$$

where $\frac{c}{0} := +\infty$ for $c \geq 0$. The equality in (10.23) is easily verified.

For arbitrary $\mu \in \mathcal{M}_s$ and $p \in \mathrm{Pos}([a,b])_m$, $p(\xi) = 1$, we obviously have

$$L_s(p) = \int_a^b p\, d\mu \geq \mu(\{\xi\}) p(\xi) = \mu(\{\xi\}).$$

Taking the infimum over p and the supremum over μ, we derive

$$\kappa_s(\xi) \geq \rho_s(\xi). \tag{10.24}$$

Theorem 10.21 *Suppose that $s \in \mathrm{Int}\, \mathcal{S}_{m+1}$. Let $\xi \in [a,b]$.*
Then the supremum in (10.22) and the infimum in (10.23) are attained and

$$\rho_s(\xi) = \kappa_s(\xi) > 0.$$

The measure μ_ξ from Definition 10.19 is the unique representing measure of s which has maximal mass $\rho_s(\xi)$ at ξ.

The infimum in (10.23) is attained at the polynomial q_ξ from Definition 10.19. If ξ is not an inner root of a principal measure of s, then q_ξ is up to a positive multiple the only polynomial $p \in \mathrm{Pos}([a,b])_m$ for which the infimum in (10.23) is attained.

Proof We denote the roots and masses of μ_ξ by $t_0 = \xi, t_1, \ldots, t_k$ and $m_0, \ldots, m_k$. For the polynomial q_ξ from Definition 10.19 we have $\deg(q_\xi) \leq \mathrm{ind}\,(\mu_\xi) - \epsilon(\xi) = m$ by (10.20), so L_s applies to q_ξ and we get

$$L_s(q_\xi) = \int_a^b q_\xi\, d\mu_\xi = \sum_{j=0}^k m_j q_\xi(t_j) = m_0 q_\xi(\xi).$$

Therefore, since $q_\xi \in \mathrm{Pos}([a,b])_m$, we obtain

$$\rho_s(\xi) \geq m_0 = \frac{L_s(q_\xi)}{q_\xi(\xi)} \geq \inf \left\{ \frac{L_s(q)}{q(\xi)} : q \in \mathrm{Pos}([a,b])_m \right\} = \kappa_s(\xi).$$

Combined with (10.24) we conclude that we have equality throughout, that is,

$$\mu_\xi(\{\xi\}) = m_0 = \rho_s(\xi) = \frac{L_s(q_\xi)}{q_\xi(\xi)} = \kappa_s(\xi). \tag{10.25}$$

The first equalities of (10.25) mean that the measure μ_ξ has maximal mass at ξ. The last equality of (10.25) says that the infimum in (10.23) is attained at q_ξ.

Let ν be an arbitrary representing measure for s which has maximal mass $\rho_s(\xi)$ at ξ. Then $\mu := \nu - \rho_s(\xi)\delta_\xi$ is a positive measure satisfying

$$\int_a^b q_\xi d\mu = \int_a^b q_\xi d\nu - \rho_s(\xi) q_\xi(\xi) = L_s(q_\xi) - \rho_s(\xi) q_\xi(\xi) = 0 \tag{10.26}$$

by (10.25). Since $q_\xi \in \mathrm{Pos}([a,b])_m$, (10.26) implies that $\mathrm{supp}\,\mu \subseteq \mathcal{Z}(q_\xi)$ by Proposition 1.23. Hence $\mathrm{supp}\,\nu \subseteq \{\xi\} \cup \mathcal{Z}(q_\xi)$. Therefore, ν is atomic and all atoms of ν are roots of μ_ξ. Thus $\mathrm{ind}(\nu) \leq \mathrm{ind}(\mu_\xi)$ and ν has ξ as an atom, since $\nu(\{\xi\}) = \rho_s(\xi) > 0$. If $\xi \in (a,b)$, then μ_ξ is canonical, hence is ν, and therefore $\nu = \mu_\xi$ by Corollary 10.13. If $\xi = a$ or $\xi = b$, then μ_ξ, hence ν, is principal. Since there is only one principal measure with root at a given endpoint, we obtain $\nu = \mu_\xi$ in this case as well.

Now let $q \in \mathrm{Pos}([a,b])_m$ be another polynomial for which the infimum $\kappa_s(\xi)$ in (10.23) is attained, so that $\kappa_s(\xi)q(\xi) = L_s(q)$ and $q(\xi) > 0$. Then, by (10.25),

$$m_0 q(\xi) = \kappa_s(\xi)q(\xi) = L_s(q) = \sum_{j=0}^{k} m_j q(t_j) = m_0 q(\xi) + \sum_{j=1}^{k} m_j q(t_j).$$

Since $q \in \mathrm{Pos}([a,b])_m$ and $m_j > 0$, we conclude that $q(t_j) = 0$ for $j = 1, \ldots, k$.

Suppose that ξ is not an inner root of a principal measure for s. Then we have $\mathrm{ind}(\mu_\xi) = m + 2$ if $\xi \in (a,b)$ and $\mathrm{ind}(\mu_\xi) = m + 1$ if $\xi = a$ or $\xi = b$. In both cases, $m = \mathrm{ind}(\mu_\xi) - \epsilon(\xi)$, so that $\deg(q) + \epsilon(\xi) \leq \mathrm{ind}(\nu_\xi)$. Thus q satisfies (10.20). Hence q is a positive constant multiple of q_ξ by Lemma 10.20. $\qquad\square$

Remark 10.22

1. The preceding proof shows that for a polynomial $q \in \mathrm{Pos}([a,b])_m$ the infimum in (10.23) is attained if and only if $q(t_j) = 0$ for all roots $t_j \neq \xi$ of μ_ξ and $q(\xi) > 0$. Let ξ be an inner root of a principal measure of s. Then $\deg(q_\xi) = m - 1$ and the infimum in (10.23) is attained at $q \in \mathrm{Pos}([a,b])_m$ if and only if $q = fq_\xi$ for some constant or linear polynomial $f \in \mathrm{Pos}([a,b])$.
2. Let s be a boundary point of $\mathcal{S}_{m+1}$. Then s has a unique representating measure μ by Theorem 10.7, so $\rho_s(\xi)$ is the corresponding weight if ξ is a root of μ and $\rho_s(\xi) = 0$ if ξ is not a root of μ. $\qquad\circ$

We derive two important consequences of the preceding theorem.

Corollary 10.23 *Let μ be a canonical representing measure of $s \in \mathrm{Int}\,\mathcal{S}_{m+1}$ with roots $t_j, j = 1, \ldots, k$. Then μ has maximal mass at each root t_j contained in (a,b). If μ is principal, μ has maximal mass at all roots.*

Proof If μ is canonical and $t_j \in (a,b)$ is a root of μ, then $\mu_{t_j} = \mu$ by Corollary 10.13 and Definition 10.18. If μ is principal and if $t_j = a$ or $t_j = b$, then also $\mu_{t_j} = \mu$ by Definition 10.18. In both cases μ has maximal mass at t_j by Theorem 10.21. $\qquad\square$

Corollary 10.24 *For each $s \in \mathcal{S}_{m+1}, s \neq 0$, there is a representing measure μ of s such that $\mathrm{ind}(\mu) \leq m + 1$ and μ has maximal mass at each root.*

Proof First let $s \in \partial\mathcal{S}_{m+1}$. Then, by Theorem 10.7, $\mathrm{ind}(\mu) \leq m$ and s is $[a,b]$-determinate, so μ obviously has maximal mass at all its roots. If $s \in \mathrm{Int}\,\mathcal{S}_{m+1}$, each principal measure has the desired properties by Corollary 10.23. $\qquad\square$

The following two theorems deal with canonical measures having prescribed masses at the end points. Because of the different behaviour of principal measures at end points we distinguish between the even and odd cases.

Theorem 10.25 *(Odd case $m = 2n + 1$) Suppose that τ and τ' are numbers such that $0 < \tau < \rho_s(a)$ and $0 < \tau' < \rho_s(b)$. Then there exist unique canonical measures $\mu^a(\tau)$ and $\mu^b(\tau')$ for s which have masses τ and τ' at a and b, respectively.*

Let $a = t_1^a(\tau) < t_2^a(\tau) < \cdots < t_k^a(\tau)$ and $t_1^b(\tau') < t_2^b(\tau') < \cdots < t_l^b(\tau') = b$ denote the roots of $\mu(\tau)$ and $\mu(\tau')$, respectively. Each root $t_j(\tau)$ is a strictly increasing continuous function of τ on the interval $(0, \rho_s(a))$, while $t_j(\tau')$ is a strictly decreasing continuous function of τ' on $(0, \rho_s(b))$. Further, $k = l = n + 2$ and

$$t_{j-1}^- = \lim_{\tau \to +0} t_j^a(\tau), \quad t_j^+ = \lim_{\tau \to \rho_s(a)-0} t_j^a(\tau), \quad j = 2, \ldots, n+2,$$

$$a = t_1^+ = t_1^a(\tau) < t_1^- < t_2^a(\tau) < t_2^+ < t_2^- < \cdots < t_{n+1}^- < t_{n+2}^a(\tau) < t_{n+2}^+ = b,$$

$$t_j^+ = \lim_{\tau' \to +0} t_j^b(\tau'), \quad t_j^- = \lim_{\tau' \to \rho_s(b)-0} t_j^b(\tau'), \quad j = 1, \ldots, n+1,$$

$$a = t_1^+ < t_1^b(\tau') < t_1^- < t_2^+ < \cdots < t_{n+1}^+ < t_{n+1}^b(\tau') < t_{n+1}^- < t_{n+2}^b(\tau') = t_{n+2}^+ = b.$$

Proof We carry out the proof for the end point a and define a sequence $r(\tau) := (s_j - \tau a^j)_{j=0}^m$. Since μ^+ has the root a by (10.16) and maximal mass at $t_1^+ = a$ by Corollary 10.23, $\mu^+ - \tau \delta_a$ is a positive measure. Obviously, its moments are $s_j - \tau a^j$, so $r(\tau)$ is a moment sequence and

$$L_{r(\tau)}(p) = L_s(p) - \tau p(a) = (\rho_s(a) - \tau)p(a) + \sum_{j=2}^{n+2} m_j p(t_j^+), \quad p \in \mathbb{R}[x]_m.$$

Since $\rho_s(a) - \tau > 0$, $L_{r(\tau)}(p) = 0$ implies that $p(t_j^+) = 0$ for all $j = 1, \ldots, n+2$, that is, $L_{r(\tau)}(p) = 0$ is equivalent to $L_s(p) = 0$. Hence, by Theorem 10.8, $s \in \text{Int } \mathcal{S}_{m+1}$ implies that $r(\tau) \in \text{Int } \mathcal{S}_{m+1}$. Thus, $r(\tau)$ has a lower principal measure $\mu(\tau)^-$ with roots $t_j(\tau)$ written as $a < t_2(\tau) < \cdots < t_{n+2}(\tau) < b$. Then $\mu(\tau) := \mu(\tau)^- + \tau \delta_a$ is a canonical representing measure of s with roots

$$a = t_1(\tau) < t_2(\tau) < \cdots < t_{n+2}(\tau) < b.$$

Since $t_1(\tau) = t_1^+ = a$ and $\tau < \rho_s(a) = \mu^+(\{a\})$, Proposition 10.12(ii) applies and yields $t_2(\tau) < t_2^+$. Applying Proposition 10.12(i) by using this fact we obtain the interlacing inequalities stated in the theorem.

Suppose that $\tilde{\mu}(\tau)$ is an arbitrary canonical measure for s with mass τ at a. Then $\tilde{\mu}(\tau) - \tau \delta_a$ is a lower principal measure for $r(\tau)$. Therefore, $\tilde{\mu}(\tau) - \tau \delta_a = \mu(\tau)^-$, which yields $\tilde{\mu}(\tau) = \mu(\tau)$.

Suppose that $0 < \tau' < \tau'' < \rho_s(a)$. Then $t_2(\tau') < t_2(\tau'')$ by Proposition 10.12(ii). By the strict interlacing of roots (proposition 10.12(i)) we get

$t_j(\tau') < t_j(\tau'')$ for $j = 2, \ldots, n + 1$. Thus, $t_j(\tau)$ is a strictly increasing function of τ on $(0, \rho_s(a))$.

Finally, we prove the continuity of $t_j(\tau)$ and the limit equalities in the theorem. Since $t_j(\tau)$ is increasing, all one-sided limits $\lim_{\tau \to \tau_0 \pm} t_j(\tau)$ for $\tau_0 \in (0, \rho_s(a))$, $\lim_{\tau \to +0} t_j(\tau)$, and $\lim_{\tau \to \rho_s(a) - 0} t_j(\tau)$ exist. It only remains to show the corresponding equalities. As a sample we assume the contrary for some $\tau_0 \in (0, \rho_s(a))$ and j and choose α, β such that

$$t_{j-1}^+ < t_{j-1}^- \leq \lim_{\tau \to \tau_0 - 0} t_j(\tau) < \alpha < \beta < \lim_{\tau \to \tau_0 + 0} t_j(\tau) \leq t_j^+. \qquad (10.27)$$

Let $\xi \in [\alpha, \beta]$. By Corollary 10.13, there is a canonical measure μ_ξ for s which has ξ as a root. Then $\tau_\xi := \mu_\xi(\{a\})$ is in $[0, \rho_s(a)]$. If $\tau_\xi < \rho_s(a)$, then $\mu_\xi = \mu^+$ by Theorem 10.21, which contradicts (10.27). We prove that $\tau_\xi > 0$. Assume to the contrary that $\tau_\xi = 0$. Since $\text{ind}(\mu_\xi) = m + 2 = 2n + 3$, then b must be a root of μ_ξ. Let $\xi_1 < \xi_2 < \ldots, < \xi_{n+1} < \xi_{n+2} = b$ be the roots of μ_ξ. Since $\xi_{n+2} = t_{n+2}^+ = b$ and $\mu_\xi(\{b\}) < \mu^+(\{b\})$ again by Theorem 10.21, it follows from Proposition 10.12(iii), applied with $\mu = \mu_\xi$ and $\mu' = \mu^+$, that $t_{n+1}^+ < \xi_{n+1}$. Proposition 10.11(iv), applied with $\mu = \mu_\xi$ and $\tilde{\mu} = \mu^-$, yields $\xi_{n+1} < t_{n+1}^-$. Thus $t_{n+1}^+ < \xi_{n+1} < t_{n+1}^-$. Therefore, by the interlacing property in Proposition 10.12(i) we get $t_{j-1}^- < \xi < t_j^+ < \xi_j < t_j^-$. Thus μ^- has no root between the inner roots ξ and ξ_j of μ_ξ. This contradicts Proposition 10.11(i) and proves that $\tau_\xi > 0$. Thus we have shown that $\tau_\xi \in (0, \rho_s(a))$.

By construction, $\xi = t_j(\tau_\xi)$. From the strict monotonicity of the function $t_j(\tau)$ and (10.27) it follows that $\tau_\xi = \tau_0$. Then $\xi = t_j(\tau_0)$. Since $\xi \in [\alpha, \beta]$ was arbitrary, this is impossible. $\qquad\qquad\qquad\qquad\qquad\qquad\qquad\qquad\qquad\qquad\qquad\qquad\quad\square$

Theorem 10.26 (Even Case $m = 2n$) *Let τ be a number such that $0 < \tau < \rho_s(a)$. Then there exists a unique canonical measure $\mu(\tau)$ for s which has mass τ at a. Let $a = t_1(\tau) < t_2(\tau) < \cdots < t_k(\tau)$ denote the roots of $\mu(\tau)$. Then $k = n + 2$. Each root $t_j(\tau)$ is a strictly increasing continuous function of τ on the interval $(0, \rho_s(a))$ and*

$$t_{j-1}^+ = \lim_{\tau \to +0} t_j(\tau), \quad t_j^- = \lim_{\tau \to \rho_s(a) - 0} t_j(\tau), \quad j = 2, \ldots, n + 1,$$

$$a = t_1^- = t_1(\tau) < t_1^+ < t_2(\tau) < t_2^- < t_2^+ < \ldots < t_{n+1}(\tau) < t_{n+1}^- < t_{n+2}(\tau) = t_{n+1}^+ = b.$$

The proof of Theorem 10.26 follows a similar pattern as the proof of Theorem 10.25; we omit the details.

However, there is an essential difference between Theorems 10.25 and 10.26: In the odd case Theorems 10.25 gives parametrizations of all canonical measures μ_ξ when ξ is not a root of a principal measure in terms of the masses $\tau \in (0, \rho_s(a))$ at a and $\tau' \in (0, \rho_s(b))$ at b. In the even case this is not true, since the measures μ_ξ with ξ contained in some interval (t_j^-, t_j^+) are not obtained in Theorem 10.26.

10.6   A Parametrization of Canonical Measures

Let us set $J := \cup_j J_j$ and $K := \cup_l K_l$, where J_j and K_l are the intervals defined by

$$m = 2n : \ K_j = (t_j^+, t_{j+1}^-) \text{ for } j=1,\dots,n, \ J_l = (t_l^-, t_l^+) \text{ for } l=1,\dots,n+1,$$

$$m = 2n+1 : \ K_j = (t_j^+, t_j^-) \text{ and } J_j = (t_j^-, t_{j+1}^+) \text{ for } j = 1,\dots,n+1.$$

That is, $J \cup K$ is the set of points $\xi \in [a, b]$ which are not roots of a principal measure, or equivalently, for which the measure μ_ξ from Definition 10.18 has index $m + 2$. Our aim in this section is to give a parametrization of measures μ_ξ, $\xi \in J \cup K$, in terms their smallest inner root.

Let us denote the roots of μ_ξ by $t^j(\xi)$ and assume that they are numbered in increasing order. Using the interlacing results of Propositions 10.11 and 10.12 and arguing similarly as in the proof of Theorem 10.25 the following properties are derived.

Even Case: $m = 2n$:
Let $\zeta \in J_1 = (a, t_1^+)$. The roots of μ_ξ are $t^1(\zeta) = \zeta$ and $t^j(\zeta) \in J_j, j = 1,\dots,n+1$. Each root $t^j(\zeta)$ is a strictly increasing continuous function on J_1 and we have

$$\zeta \in J_1 : \quad \lim_{\zeta \to t_1^- +0} t^j(\zeta) = t_j^-, \quad \lim_{\zeta \to t_1^+ -0} t^j(\zeta) = t_j^+, \ j = 1,\dots,n+1, \qquad (10.28)$$

$$a = t_1^- < t^1(\zeta) < t_1^+ < t_2^- < t^2(\zeta) < t_2^+ < \cdots < t_{n+1}^- < t^{n+1}(\zeta) < t_{n+1}^+ = b.$$

Let $\zeta \in K_1 = (t_1^+, t_2^-)$. Then μ_ξ has the roots $t^1(\zeta) = a,\ t^2(\zeta) = \zeta,\ t^{j+1}(\zeta) \in K_j$ for $j = 1,\dots,n$, and $t^{n+2}(\zeta) = b$. Each function $t^j(\zeta)$ is strictly increasing and continuous on K_1. Further,

$$\zeta \in K_1 : \quad \lim_{\zeta \to t_1^+ +0} t^j(\zeta) = t_{j-1}^+, \quad \lim_{\zeta \to t_2^- -0} t^j(\zeta) = t_j^-, \ j = 2,\dots,n+1, \qquad (10.29)$$

$$a = t_1^- = t^1(\zeta) < t_1^+ < t^2(\zeta) < t_2^- < \ldots < t_n^+ < t^{n+1}(\zeta) < t_{n+1}^- < t^{n+2}(\zeta) = t_{n+1}^+ = b.$$

Odd Case: $m = 2n + 1$:
Let $\zeta \in K_1 = (a, t_1^-) = (t_1^+, t_1^-)$. The measure μ_ξ has roots $t^1(\zeta) = \zeta,\ t^j(\zeta) \in K_j$ for $l = 1,\dots,n+1$ and $t^{n+2}(\zeta) = b$. Each root $t^l(\zeta)$ is a strictly increasing and continuous function on K_1 and

$$\zeta \in K_1 : \quad \lim_{\zeta \to t_1^+ +0} t^j(\zeta) = t_j^+, \quad \lim_{\zeta \to t_1^- -0} t^j(\zeta) = t_j^-, \ j = 1,\dots,n+1, \qquad (10.30)$$

$$a = t_1^+ < t^1(\zeta) < t_1^- < t_2^+ < \cdots < t_{n+1}^+ < t^{n+1}(\zeta) < t_{n+1}^- < t^{n+2}(\zeta) = t_{n+2}^+ = b.$$

Suppose that $\zeta \in J_1 = (t_1^-, t_2^+)$. Then μ_ζ has roots $t^1(\zeta) = a$, $t^2(\zeta) = \zeta$, and $t^{j+1}(\zeta) \in J_j$ for $j = 1, \ldots, n+1$. Each $t^j(\zeta)$ is a strictly increasing continuous function on J_1 and

$$\zeta \in J_1: \quad \lim_{\zeta \to t_1^- + 0} t^j(\zeta) = t_j^-, \quad \lim_{\zeta \to t_2^+ - 0} t^j(\zeta) = t_{j+1}^+, j = 1, \ldots, n+1, \quad (10.31)$$

$$a = t_1^+ = t^1(\zeta) < t_1^- < t^2(\zeta) < t_2^+ < t_2^- < \cdots < t_{n+1}^- < t^{n+2}(\zeta) < t_{n+2}^+ = b.$$

The preceding gives a *continuous* parametrization of the roots of the canonical measures μ_ξ for $\xi \in J \cup K$ in terms of their *first inner root* ζ contained in J_1 resp. K_1. Further, in all these cases the limits (10.28)–(10.31) show the one-sided continuity of the functions $t^j(\zeta)$ at the corresponding inner roots of the principal measure $\mu^\pm$.

Fix an interval J_i and let $\xi \in J_i$. Recall that the polynomial q_ξ from Definition 10.19 is a product of linear and quadratic factors involving the roots $t^j(\xi)$ of μ_ξ. This definition implies that the above parametrization of roots yields a continuous parametrization of the polynomials q_ξ on J_i. Here the vector space $\mathbb{R}[x]_m$ is equipped with some norm. Let $J_i = (\lambda_+, \lambda_-)$. Then λ_+ and λ_- are roots of principal measures. From the limits (10.28)–(10.31) we conclude that each one-sided limit $p_{\lambda_\pm} = \lim_{\xi \to \lambda_\pm \pm 0} q_\xi$ exists and gives a polynomial $p_{\lambda_\pm} \in \text{Pos}([a, b])_m$ which is positive at $\lambda_\pm$ and vanishes at the other roots of the corresponding principal measure. Therefore, by the remarks after the proof of Theorem 10.21, the infimum in (10.23) is attained for p_ξ at ξ and for $p_{\lambda_\pm}$ at $\lambda_\pm$, that is, we have $\kappa_s(t) = \frac{L_s(p_t)}{p_t(t)}$ for all $t \in [\lambda_+, \lambda_-]$. Thus we have a continuous parametrization of minimizing polynomials for (10.23) on the closed interval $\overline{J}_i = [\lambda_+, \lambda_-]$. (Recall that the minimizing polynomial for (10.23) is uniquely determined up to a constant factor for points in J_i, but not for inner roots of principal measures.) Therefore, since the linear functional L_s on the finite-dimensional space $\mathbb{R}[x]_m$ is continuous, the function $\kappa_s(t) = \frac{L_s(p_t)}{p_t(t)}$ is continuous on the closed interval $\overline{J}_i$. Verbatim the same proof yields the continuity of κ_s on the closure $\overline{K}_j$ of each interval K_j. Further, each inner root is a common end point of some J_i and K_j. Hence κ_s is continuous on $[a, b]$. We state this as

Theorem 10.27 *For each $s \in \text{Int } S_{m+1}$ the function $\rho_s(t) = \kappa_s(t)$ is continuous on the interval $[a, b]$.*

In the preceding proof a continuous parametrization of minimizing polynomials on intervals $\overline{J}_i$ and $\overline{K}_j$ was crucial. The following simple example shows that there is no continuous parametrization of minimizing polynomials on $[a, b]$.

Example 10.28 Let $m = 2$. Then $a = t_1^- < t_1^+ < t_2^- < t_2^+ = b$. Let us take numbers $\xi \in J_1 = (a, t_1^+)$ and $\zeta \in K_1 = (t_1^+, t_2^-)$. Clearly,

$$q_\xi(x) = (x - t^2(\xi))^2, \quad q_\zeta(x) = (b - x)(x - a).$$

Let p_ξ and p_ζ be minimizing polynomials for (10.23) at ξ and ζ, respectively. Then p_ξ and p_ζ are constant multiples of q_ξ and q_ζ, respectively. Since $t^2(\xi) \to t_2^+ = b$ as $\xi \to t_1^+ - 0$, all possible limits of these polynomials are of the form

$$\lim_{\xi \to t_1^+ - 0} p_\xi(x) = c_1(x-b)^2, \qquad \lim_{\zeta \to t_1^+ + 0} p_\zeta(x) = c_2(b-x)(x-a)$$

with $c_1 > 0$ and $c_2 > 0$. These limits are different minimizing polynomials for $\kappa_s(t_1^+)$. This shows that at the inner root t_1^+ of μ^+ one cannot have two-sided continuity of minimizing polynomials.   ○

10.7   Orthogonal Polynomials and Maximal Masses

Throughout this section, we assume that $s \in \mathrm{Int}\, \mathcal{S}_{m+1}$.

First we define and develop four sequences of orthogonal polynomials. Put

$$\underline{P}_k(x) = \begin{vmatrix} s_0 & s_1 & s_2 & \cdots & s_k \\ s_1 & s_2 & s_3 & \cdots & s_{k+1} \\ s_2 & s_3 & s_4 & \cdots & s_{k+2} \\ \cdots & & & & \\ s_{k-1} & s_k & s_{k+1} & \cdots & s_{2k-1} \\ 1 & x & x^2 & \cdots & x^k \end{vmatrix} \quad \text{for } 2k-1 \le m,$$

$$\overline{P}_k(x) = \begin{vmatrix} r_0 & r_1 & r_2 & \cdots & r_k \\ r_1 & r_2 & r_3 & \cdots & r_{k+1} \\ r_2 & r_3 & r_4 & \cdots & r_{k+2} \\ \cdots & & & & \\ r_{k-1} & r_k & r_{k+1} & \cdots & r_{2k-1} \\ 1 & x & x^2 & \cdots & x^k \end{vmatrix} \quad \text{for } 2k+1 \le m,$$

$$\underline{Q}_k(x) = \begin{vmatrix} s_1 - as_0 & s_2 - as_1 & \cdots & s_{k+1} - as_k \\ s_2 - as_1 & s_3 - as_2 & \cdots & s_{k+2} - as_{k+1} \\ s_3 - as_2 & s_4 - as_3 & \cdots & s_{k+3} - as_{k+2} \\ \cdots & & & \\ s_k - as_{k-1} & s_{k+1} - as_k & \cdots & s_{2k} - as_{2k-1} \\ 1 & x & \cdots & x^k \end{vmatrix} \quad \text{for } 2k \le m,$$

$$\overline{Q}_k(x) = \begin{vmatrix} bs_0 - s_1 & bs_1 - s_2 & \cdots & bs_k - s_{k+1} \\ bs_1 - s_2 & bs_2 - s_3 & \cdots & bs_{k+1} - s_{k+2} \\ bs_2 - s_3 & bs_3 - s_4 & \cdots & bs_{k+2} - s_{k+3} \\ \cdots & & & \\ bs_{k-1} - s_k & bs_k - s_{k+1} & \cdots & bs_{2k-1} - s_{2k} \\ 1 & x & \cdots & x^k \end{vmatrix} \quad \text{for } 2k \le m,$$

where we have set

$$r_j := (a+b)s_{j+1} - s_j - s_{j+2} \quad \text{for} \quad j = 1, \ldots, m-2.$$

The coefficients of x^k in these determinants are $D_{2k-2}, \overline{D}_{2k-2}, \underline{D}_{2k-1}$, and $\overline{D}_{2k-1}$, respectively. Since $s \in \text{Int } \mathcal{S}_{m+1}$, they are positive by Theorem 10.8. Hence each of the above polynomials has degree k with positive leading coefficient.

Further, we consider the following four sequences

$$s = (s_j)_{j=0}^{m}, \quad (b-E)(E-a)s = (r_j)_{j=0}^{m-2}, \tag{10.32}$$

$$Es - as = (s_{j+1} - as_j)_{j=0}^{m-1}, \quad bs - Es = (bs_j - s_{j+1})_{j=0}^{m-1}. \tag{10.33}$$

Their Hankel matrices are given by the formulas at the beginning of Sect. 10.1. Since $s \in \text{Int } \mathcal{S}_{m+1}$, the corresponding Hankel matrices are positive definite by Theorem 10.8, hence are the sequences in (10.32) for even m and in (10.33) for odd m.

The polynomials $\underline{P}_k, \overline{P}_k, \underline{Q}_k, \overline{Q}_k$ are orthogonal polynomials for the moment sequences (10.32)–(10.33). That is, for any polynomial $f \in \mathbb{R}[x]_{k-1}$ we have

$$L_s(\underline{P}_k f) = L_{(b-E)(E-a)s}(\overline{P}_k f) = L_{Es-as}(\underline{Q}_k f) = L_{bs-Es}(\overline{Q}_k f) = 0. \tag{10.34}$$

A simple verification can be given in a similar manner as in Sect. 5.1.

Suppose that μ is a representing measure for s. Then the moment sequences $(b-E)(E-a)s$, $Es-as$, and $bs-Es$ are represented by $(b-t)(t-a)d\mu$, $(t-a)d\mu$, and $(b-t)d\mu$, respectively. Hence the above families of polynomials are orthogonal with respect to the corresponding measures. That is, for all $k \neq j$ we have

$$\int_a^b \underline{P}_k(t) \underline{P}_j(t) \, d\mu = \int_a^b \overline{P}_k(t) \overline{P}_j(t) (b-t)(t-a) d\mu = 0 \tag{10.35}$$

$$\int_a^b \underline{Q}_k(t) \underline{Q}_j(t) (t-a) d\mu = \int_a^b \overline{Q}_k(t) \overline{Q}_j(t) (b-t) d\mu = 0.$$

It should be emphasized that the polynomial $\underline{P}_k$ is defined if $2k-1 \leq m$, while $\overline{P}_j$ is only if $2j+1 \leq m$. That is, $\overline{P}_j$ is not defined for the largest index of $\underline{P}_k$.

Now we turn to the orthonormal polynomials. Let $m = 2n$ or $m = 2n+1$. For $k = 0, \ldots, n$ and $l = 0, \ldots, n-1$ we define

$$\underline{p}_k(x) = \frac{\underline{P}_k}{\sqrt{D_{2k-2}D_{2k}}}, \quad \overline{p}_l(x) = \frac{\overline{P}_l}{\sqrt{\overline{D}_{2l}\overline{D}_{2l+2}}}, \tag{10.36}$$

$$\underline{q}_k(x) = \frac{\underline{Q}_k}{\sqrt{D_{2k-1}D_{2k+1}}}, \quad \overline{q}_k(x) = \frac{\overline{Q}_k}{\sqrt{\overline{D}_{2k-1}\overline{D}_{2k+1}}}, \tag{10.37}$$

where the determinants $\underline{D}_i$ and $\overline{D}_i$ with negative i are set to 1. All determinants $\underline{D}_j, \overline{D}_j$ occuring in (10.36)–(10.37) are positive by Theorem 10.8, since $s \in \text{Int }\mathcal{S}_{m+1}$.

Arguing as in the proof of Proposition 5.3 (see e.g. (5.3)) it follows that the polynomials in (10.36)–(10.37) have norm 1 in the corresponding norms, that is,

$$L_s((\underline{P}_k)^2) = \int_a^b \underline{p}_k(t)^2 \, d\mu = 1,$$

$$L_{(b-E)(E-a)s}((\overline{P}_l)^2) = \int_a^b \overline{p}_l(t)^2 (b-t)(t-a) \, d\mu = 1,$$

$$L_{Es-as}((\underline{q}_k)^2) = \int_a^b \underline{q}_k(t)^2 (t-a) d\mu = 1,$$

$$L_{bs-Es}((\overline{q}_k)^2) = \int_a^b \overline{q}_k(t)^2 (b-t) d\mu = 1.$$

Note that for $m = 2n + 1$ the polynomial $\underline{P}_{n+1}$ is defined (since only moments s_j with $j \leq 2n + 1$ are involved), but $\underline{p}_{n+1}$ is not (because $\underline{D}_{2n+2}$ requires s_{2n+2}).

Our next theorem gives explicit formulas for the function $\rho_s(\xi)$ in terms of the orthonormal polynomials introduced above. Recall that the intervals J_j and K_l have been defined at the beginning of Sect. 10.6, $J = \cup_j J_j$ and $K = \cup_l K_l$. As usual, $\overline{J}$ and $\overline{K}$ denote the closures of the sets J and K, respectively. Define

$$m = 2n : \qquad \underline{P}_\xi(x) = \left(\sum_{j=0}^n \underline{p}_j(\xi) \underline{p}_j(x) \right)^2, \quad \xi \in J, \tag{10.38}$$

$$\overline{P}_\xi(x) = (b-x)(x-a) \left(\sum_{j=0}^{n-1} \overline{p}_j(\xi) \overline{p}_j(x) \right)^2, \quad \xi \in K, \tag{10.39}$$

$$m = 2n + 1 : \quad \underline{Q}_\xi(x) = (x-a) \left(\sum_{j=0}^n \underline{q}_j(\xi) \underline{q}_j(x) \right)^2, \quad \xi \in J, \tag{10.40}$$

$$\overline{Q}_\xi(x) = (b-x) \left(\sum_{j=0}^n \overline{q}_j(\xi) \overline{q}_j(x) \right)^2, \quad \xi \in K. \tag{10.41}$$

Theorem 10.29 *Suppose that $s \in \text{Int }\mathcal{S}_{m+1}$.*
Even Case $m = 2n$ and $\xi \in \overline{J}$: $\underline{P}_\xi$ is a minimizing polynomial for (10.23) and

$$\rho_s(\xi) = \left(\sum_{j=0}^n \underline{p}_j(\xi)^2 \right)^{-1}.$$

Even Case m = 2n and ξ ∈ K̄: P̄_ξ is a minimizing polynomial for (10.23) and

$$\rho_s(\xi) = \frac{1}{(b-\xi)(\xi-a)}\left(\sum_{j=0}^{n-1}\overline{p}_j(\xi)^2\right)^{-1}.$$

Odd Case m = 2n + 1 and ξ ∈ J̄: Q_ξ is a minimizing polynomial for (10.23) and

$$\rho_s(\xi) = \frac{1}{\xi-a}\left(\sum_{j=0}^{n}\underline{q}_j(\xi)^2\right)^{-1}.$$

Odd Case m = 2n + 1 and ξ ∈ K̄: Q̄_ξ is a minimizing polynomial for (10.23) and

$$\rho_s(\xi) = \frac{1}{b-\xi}\left(\sum_{j=0}^{n}\overline{q}_j(\xi)^2\right)^{-1}.$$

Proof Since $s \in \text{Int } \mathcal{S}_{m+1}$, Theorem 10.8 implies $\underline{D}_j(s) \neq 0$ and $\overline{D}_j(s) \neq 0$ for $j = 0,\ldots,m$. Hence the four sequences (10.32)–(10.33) are positive definite. Recall that by Theorem 10.21 the polynomial q_ξ from Definition 10.19 is a minimizing polynomial of (10.23). Analyzing q_ξ in the various cases leads to the form stated above. We carry out the proof in the even case $m = 2n$; the odd case is treated similarly.

Let $\xi \in J_i$. From the list of atoms of μ_ξ given in Sect. 10.6 we know that neither a nor b are roots of μ_ξ. Therefore, by Definition 10.19, $q_\xi = g^2$ for some $g \in \mathbb{R}[x]_n$. Recall that $q_\xi(\xi) > 0$ by (10.20) and hence $g(\xi) \neq 0$. Setting $p_\xi := g(\xi)^{-1}g$ we have $p_\xi(x)^2 = \frac{q_\xi(x)}{q_\xi(\xi)}$, so that $\kappa_s(\xi) = \frac{L_s(q_\xi)}{q_\xi(\xi)} = L_s(p_\xi^2)$ by Theorem 10.21. Since $\kappa_s(\xi) = L_s(p_\xi^2)$ and $p_\xi(\xi) = 1$, it follows at once from the definition (10.22) of $\kappa_s(\xi)$ that p_ξ is a minimizer of $L_s(p^2)$ for $p \in \mathbb{R}[x]_n$ under the constraint $p(\xi) = 1$. This problem was settled by Proposition 9.14. By (9.34) the corresponding minimum is $(\sum_{j=0}^n \underline{p}_j(\xi)^2)^{-1}$ and by (9.35) the unique minimizer is a constant multiple of the polynomial $p(x) = \sum_{j=0}^n \underline{p}_j(\xi)\underline{p}_j(x)$. (Note that $\underline{p}_j(\xi)$ is real, since ξ is real.) Hence $p^2 \equiv \underline{P}_\xi$ is a minimizer for (10.23). This proves the assertions for $\xi \in J_i$.

By Theorem 10.27 and the discussion preceding it, ρ_s is continuous on $[a,b]$ and the continuous extension of a continuous minimizing family of polynomials for (10.23) on J_i yields minimizers for the end points of J_i. Hence the assertions remain valid for ξ in the closure $\overline{J}_i$.

Now let $\xi \in K_i$. We proceed as in the case $\xi \in J_i$ with s replaced by the positive definite sequence $(b - E)(E - a)s$. Since $\xi \in K_i$, it follows from the description in Sect. 10.6 that both end points a and b are roots of μ_ξ. Therefore, by Definition 10.19, we have $q_\xi = (b-x)(x-a)g^2$ with $g \in \mathbb{R}[x]_{n-1}$. Again $g(\xi) \neq 0$

by (10.20) and we set $p_\xi := g(\xi)^{-1} g$. For $q = (b-x)(x-a) f^2$ with $f \in \mathbb{R}[x]_{n-1}$ we compute

$$\frac{L_s(q)}{q(\xi)} = \frac{L_s((b-x)(a-x) f^2)}{q(\xi)} = [(b-\xi)(a-\xi)]^{-1} \frac{L_{(b-E)(E-a)s}(f^2)}{f(\xi)^2}. \qquad (10.42)$$

Using Theorem 10.21, Eq. (10.42), and the relation $p_\xi(\xi) = 1$ we obtain

$$\kappa_s(\xi) = \frac{L_s(q_\xi)}{q_\xi(\xi)} = [(b-\xi)(a-\xi)]^{-1} L_{(b-E)(E-a)s}(p_\xi^2).$$

Hence it follows from the definition of $\kappa_s(\xi)$ that the polynomial p_ξ is a minimizer of $L_{(b-E)(E-a)s}(f^2)$ for $f \in \mathbb{R}[x]_{n-1}$ under the constraint $f(\xi) = 1$. The orthonormal polynomials for the sequence $(b-E)(E-a)s$ are $\bar{p}_0, \bar{p}_1, \ldots, \bar{p}_{n-1}$. By Proposition 9.14, applied to the sequence $(b-E)(E-a)s$, the corresponding minimum is $(\sum_{j=0}^{n-1} \bar{p}_j(\xi)^2)^{-1}$ and the minimizer is a multiple of $p(x) = \sum_{j=0}^{n-1} \bar{p}_j(\xi) \bar{p}_j(x)$. Hence, by (10.42), the minimum $\kappa_s(\xi)$ is

$$[(b-\xi)(a-\xi)]^{-1} \left(\sum_{j=0}^{n-1} \bar{p}_j(\xi)^2 \right)^{-1}$$

and each multiple of $(b-x)(a-x) p(x)^2 \equiv \overline{P}_\xi(x)$ is a minimizer for (10.23). Thus the assertions are proved in the case $\xi \in K_i$. Arguing as in the preceding paragraph, the assertions hold for ξ in $\overline{K}_i$ as well. $\square$

We close this chapter by showing how the roots of principal measures $\mu^\pm$ and canonical measures μ_ξ can be detected from orthogonal polynomials and quasi-orthogonal polynomials, respectively.

Proposition 10.30 *The roots of the upper and lower principal representing measures μ^+ and μ^- of s are exactly the zeros of the following polynomials:*

$$m = 2n \qquad \mu^+ : (b-x)\overline{Q}_n(x), \qquad\qquad \mu^- : (x-a)\underline{Q}_n(x), \qquad (10.43)$$

$$m = 2n+1 \qquad \mu^+ : (b-x)(x-a)\overline{P}_n(x), \qquad \mu^- : \underline{P}_{n+1}(x). \qquad (10.44)$$

All these zeros are simple.

Proof We carry out the proof for $m = 2n+1$ and μ^+; the other cases can be treated similarly. Recall that $\overline{P}_0, \overline{P}_1, \ldots, \overline{P}_n$ are orthogonal polynomials for the sequence $(b-E)(E-a)s$. Hence $\overline{P}_n \in \mathbb{R}[x]_n$ is orthogonal to all polynomials $f \in \mathbb{R}[x]_{n-1}$ with respect to $(b-t)(t-a) d\mu^+$, that is, by (10.34) and (10.35) we have

$$L_{(b-E)(E-a)s}(\overline{P}_k f) = \int_a^b \overline{P}_n(t) f(t)(b-t)(t-a) d\mu^+ = 0.$$

By formula (10.16), μ^+ has n inner roots $t_2^+, \ldots, t_{n+1}^+ \in (a, b)$. Thus, we can choose f to vanish at all but one such root t_i^+ and obtain $\overline{P}_n(t_i)(b - t_i^+)(t_i^+ - a) = 0$. Hence $\overline{P}_n(t_i) = 0$. Since $t_1^+ = a$ and $t_{n+2}^+ = b$, $(b - x)(x - a)\overline{P}_n(x)$ vanishes at all $n + 2$ roots of μ^+. Because $(b - x)(x - a)\overline{P}_n(x)$ has degree $n + 2$, the roots of μ^+ exhaust the zeros of $(b - x)(x - a)\overline{P}_n(x)$ and all zeros are simple. $\qquad\square$

The roots of principal measures are described in Proposition 10.30. If $\xi \in [a, b]$ is not a root of a principal measure, then $\xi \in J \cup K$ and μ_ξ has index $m + 2$. In Sect. 10.6 these measures have been parametrized in terms of their smallest inner root $\zeta \in J_1 \cup K_1$. Then $\mu_\xi = \mu_\zeta$, so it suffices to know the roots of μ_ζ. The next proposition characterizes these roots as zeros of quasi-orthogonal polynomials.

We shall say (see also Definition 9.3 below) that a polynomial $q \in \mathbb{R}[x]_n$, $q \neq 0$, $n \geq 2$, is called *quasi-orthogonal of order n* for s if $L_s(qf) = 0$ for all $f \in \mathbb{R}[x]_{n-2}$.

Proposition 10.31 *There exist strictly increasing continuous functions φ on J_1 and ψ on K_1 with ranges $(0, +\infty)$ and $(-\infty, 0)$, respectively, such that for $\zeta \in J_1 \cup K_1$ the roots of the canonical measure μ_ζ are precisely the zeros of the polynomial g_ζ defined by*

$$m = 2n, \ \zeta \in J_1: \qquad g_\zeta(x) = (x - a)\underline{Q}_n(x) - \varphi(\zeta)(b - x)\overline{Q}_n(x), \qquad (10.45)$$

$$m = 2n, \ \zeta \in K_1: \qquad g_\zeta(x) = (b - x)(x - a)[\underline{Q}_n(x) - \psi(\zeta)\overline{Q}_n(x)], \qquad (10.46)$$

$$m = 2n + 1, \ \zeta \in J_1: \ g_\zeta(x) = (x - a)[\underline{P}_{n+1}(x) - \varphi(\zeta)(b - x)\overline{P}_n(x)], \qquad (10.47)$$

$$m = 2n + 1, \ \zeta \in K_1: g_\zeta(x) = (b - x)[\underline{P}_{n+1}(x) - \psi(\zeta)(x - a)\overline{P}_n(x)]. \qquad (10.48)$$

The polynomial g_ζ is quasi-orthogonal of order n if $m = 2n$ and $n + 1$ if $m = 2n + 1$.

Proof We carry out the proof in the case $m = 2n$ and $\zeta \in J_1 = (t_1^-, t_1^+) = (a, t_1^+)$; the other cases can be treated in similar manner with necessary modifications.

First we fix $l = 1, \ldots, n + 1$ and define a function φ_l on J_l by

$$\varphi_l(\zeta) = \frac{(\zeta - a)\underline{Q}_n(\zeta)}{(b - \zeta)\overline{Q}_n(\zeta)}, \qquad \zeta \in J_l = (t_l^-, t_l^+). \qquad (10.49)$$

First we note that $\overline{Q}_n \neq 0$ on J_l by Proposition 10.30. Hence the denominator in (10.49) is nonzero on J_l, so φ_l is continuous on J_l. Since $\underline{Q}_n$ and $\overline{Q}_n$ have degree n, positive leading terms, and no zeros in $(-\infty, a]$ by Proposition 10.30, $\frac{\underline{Q}_n(\zeta)}{\overline{Q}_n(\zeta)} > 0$ for $\zeta = a$. This holds for all $\zeta \in J_1$, because $\underline{Q}_n$ and $\overline{Q}_n$ do no vanish on J_1. The roots of μ^- and μ^+, hence the zeros $\underline{Q}_n$ and $\overline{Q}_n$ by Proposition 10.30, strictly interlace. Therefore, $\frac{\underline{Q}_n(\zeta)}{\overline{Q}_n(\zeta)} > 0$ and hence $\varphi_l(\zeta) > 0$ on J_l for all l. Since

$$\lim_{\zeta \to t_l^+ - 0} (b - \zeta)\overline{Q}_n(\zeta) = (b - t_l^+)\overline{Q}_n(t_l^+) = 0 \quad \text{and} \quad (t_l^+ - a)\underline{Q}_n(t_l^+) > 0,$$

it follows that $\lim_{\zeta \to t_l^+ - 0} \varphi_l(\zeta) = +\infty$. Similarly, $\lim_{\zeta \to t_l^- + 0} \varphi_j(\zeta) = 0$. Therefore, by the continuity of the function φ_l on J_l its the range is $(0, +\infty)$.

We show that φ_l is injective on J_l. Assume to the contrary that there exist numbers $\zeta, \zeta' \in J_l, \zeta \neq \zeta'$, such that $\varphi_l(\zeta) = \varphi_l(\zeta')$. Then the polynomial $g_\zeta(x)$ has at least $n + 2$ zeros, that is, ζ and ζ' in the interval J_l and one in each of the remaining n intervals J_i. But $\deg(g_\zeta) \leq n + 1$, so that $g_\zeta = 0$. Setting $x = b$ we get a contradiction, since b is not a root of μ^- and hence $g_\zeta(b) = (b-a)\underline{Q}_n(b) \neq 0$ by Proposition 10.30. Summarizing, we have shown that φ_l is a strictly increasing continuous function on J_l with range $(0, +\infty)$.

Now we set $\varphi := \varphi_1$. Recall that $\underline{Q}_n$ and $\overline{Q}_n$ are orthogonal polynomials with respect to $(x-a)d\mu_\zeta$ and $(b-x)d\mu_\zeta$, respectively. Hence, for $f \in \mathbb{R}[x]_{n-1}$,

$$\int_a^b g_\zeta f \, d\mu_\zeta = \int_a^b \underline{Q}_n(x)f(x)(x-a) \, d\mu_\zeta - \varphi(\zeta) \int_a^b \overline{Q}_n(x)f(x)(b-x) \, d\mu_\zeta = 0.$$

$$(10.50)$$

The measure μ_ζ has $n + 1$ roots $t^l(\zeta) \in J_l, l = 1, \dots, n+1$, where $t^1(\zeta) = \zeta$. By (10.49), we have $g_\zeta(t^1(\zeta)) = g_\zeta(\zeta) = 0$. Thus, if we take a polynomial $f \in \mathbb{R}[x]_{n-1}$ that vanishes at all $t^j(\zeta), j = 2, \dots, n+1$, except $t^i(\zeta)$, it follows from (10.50) that g_ζ vanishes at $t^i(\zeta)$. Hence all roots of μ_ζ are zeros of g_ζ. Since $\deg(g_\zeta) \leq n+1$, these $n + 1$ roots exhaust the zeros of g_ζ.

Finally, we show that g_ζ is quasi-orthogonal. As a sample we verify this in the case $m = 2n, \zeta \in K_1$. Let $f \in \mathbb{R}[x]_{n-2}$. Using again the orthogonality of $\overline{Q}_n$ and $\underline{Q}_n$ we derive

$$L_s(g_\zeta f) = L_s((b-x)(a-x)[\overline{Q}_n(x) - \psi(\zeta)\underline{Q}_n(x)]f)$$

$$= L_s((x-a)\overline{Q}_n(x)(b-x)f) - \psi(\zeta)L_s((b-x)\underline{Q}_n(x)(x-a)f)$$

$$= L_{bs-Es}(\overline{Q}_n(x)(x-a)f) - \psi(\zeta)L_{Es-as}(\underline{Q}_n(x)(b-x)f) = 0,$$

where the last equality follows from (10.34), since $(x-a)f, (b-x)f \in \mathbb{R}[x]_{n-1}$. Hence g_ζ is a quasi-orthogonal polynomial of order n. □

10.8   Exercises

1. Let $[a, b] = [0, 1]$ and $s = (1, s_1) \in \mathcal{S}_2$. Draw a picture of $\mathcal{S}_2$ and compute the numbers s_2^+ and s_2^-.
2. Let $[a, b] = [-1, 1]$, $m = 2$ and $s = (1, 0, 0)$. Compute the canonical measure μ_ξ for $\xi \in [a, b]$ and determine the principal measures μ^+ and μ^-.
3. Suppose that $a \leq 0$ and $b \geq 2$. Let $m = 4$ and $s = (8, 6, 12, 24, 48)$.

   a. Show that $s \in \mathcal{S}_5$.
   b. Show that s is determinate and compute the unique representing measure of s.

  c. Is $s' = (8, 6, 12, 24) \in S_4$ determinate?

  d. Is $s'' = (8, 6, 12, 24, 48, 96) \in S_6$ determinate?

4. Let $c > 0, d > 0$. Set $[a, b] = [-c, c], m = 3$ and $s = (2d + 2, 0, 2c^2, 0)$.

  a. Show that $s \in \text{Int } S_4$.

  b. Prove that the two principal measures μ^+ and μ_- are given by

$$\mu^+ = \delta_{-c} + 2d\delta_0 + \delta_c, \ \mu^- = (d+1)\delta_{-\gamma} + (d+1)\delta_\gamma, \text{ where } \gamma := c(d+1)^{-1/2}.$$

5. Let $\mu = \sum_{j=1}^{k} m_j \delta_{t_j}$ be a canonical representing measure for $s \in \text{Int } S_{m+1}$ and let $P \in \mathbb{R}[x]_m$ be a polynomial which has simple zeros at $t_1, \ldots, t_k$. Prove that

$$m_j = L_s\left(\frac{P(x)}{P'(t_j)(x - t_j)}\right) \quad \text{for } j = 1, \ldots, k.$$

6. Determine the extreme rays of the moment cone S_{m+1}.

10.9 Notes

According to M.G. Krein [Kr2], the ideas of this chapter go back to the Russian mathematicians P.L. Tchebycheff and A.A. Markov. Markov invented canonical measures and applied them in his study of "limiting values" of integrals [Mv2]. The theory of principal measures, canonical measures, and maximal masses presented above is taken from the fundamental paper [Kr2].

    The geometry of moment spaces S_{m+1} was elaborated by S. Karlin and L.S. Shapley [KSh]. The volume of the projection of the base S^m of the moment cone in $\mathbb{R}^m$ is computed in [KSh]. Theorem 10.29 is due to I.J. Schoenberg and G. Szegö [SSz], improving a result of Krein. The two classical monographs of S. Karlin and W.J. Studden [KSt] and of M.G. Krein and A.A. Nudelman [KN] contain further results and a detailed study of Tchebycheff systems, see also [DS]. We partly followed these books. A description of *all* solutions of various types of truncated moment problems is given by Krein [Kr3].

Chapter 11
The Moment Problem on the Unit Circle

This chapter is concerned with the *trigonometric moment problem*:

Let $s = (s_j)_{j \in \mathbb{N}_0}$ *be a complex sequence. When does there exist a Radon measure* μ *on the unit circle* $\mathbb{T}$ *such that for all* $j \in \mathbb{N}_0$,

$$s_j = \int_{\mathbb{T}} z^{-j} d\mu(z)? \tag{11.1}$$

The *truncated trigonometric moment problem* is the corresponding problem for a finite sequence $(s_j)_{j=0}^n$ of prescribed moments. The aim of this chapter is to a give a condensed treatment of some basic notions and results on these problems.

In Sect. 11.1 we prove the Fejér–Riesz theorem (Theorem 11.1) on nonnegative Laurent polynomials on $\mathbb{T}$. This is the key result for solving the trigonometric moment problem (Theorem 11.3) in Sect. 11.2. Section 11.3 deals with orthogonal polynomials on the unit circle. The Szegö recurrence relations (Theorem 11.9) and Verblunsky's theorem (Theorem 11.12) about the reflection coefficients occuring in these relations are obtained. In Sect. 11.4 the truncated trigonometric moment problem is investigated. In Sect. 11.5 we give a short digression into Carathéodory and Schur functions and the Schur algorithm and prove Geronimus' theorem (Theorem 11.31) about the equality of reflection coefficients and Schur parameters.

Throughout this chapter we adopt the following notational convention which is often used without mention: For a sequence $(s_j)_{j=0}^n$, where $n \in \mathbb{N}$ or $n = \infty$, we set $s_{-j} := \bar{s}_j$ for $j \geq 1$. The reason is that (11.1) holds for $-j$ provided it does for j.

11.1 The Fejér–Riesz Theorem

The solution of the moment problem on the unit circle is essentially based on the following *Fejér–Riesz theorem*.

© Springer International Publishing AG 2017

K. Schmüdgen, *The Moment Problem*, Graduate Texts in Mathematics 277,
DOI 10.1007/978-3-319-64546-9_11

Theorem 11.1 *Suppose that* $p(z) = \sum_{k=-n}^{n} a_k z^k \in \mathbb{C}[z, z^{-1}]$, $p \neq 0$, *is a Laurent polynomial such that* $p(z)$ *is real and nonnegative for all* $z \in \mathbb{T}$.
   Then there exists a unique polynomial $q(z) = \sum_{j=0}^{n} b_k z^k \in \mathbb{C}[z]$ *of the same degree such that* $q(z) \neq 0$ *for* $|z| < 1$, $q(0) > 0$, *and*

$$p(z) = |q(z)|^2 \quad for \quad z \in \mathbb{T}. \tag{11.2}$$

Proof Without loss of generality we can assume that $n \in \mathbb{N}$ and $a_n \neq 0$. Since $p(z)$ is real on $\mathbb{T}$, we have $a_{-k} = \bar{a}_k$ for all k. Put $f(z) := z^n p(z)$. Then $f \in \mathbb{C}[z]$ has degree $2n$, $f(0) = \bar{a}_n \neq 0$, and the nonzero zeros of f and p coincide. Clearly,

$$f(z) = a_n z^{2n} + \cdots + a_0 z^n + \cdots + \bar{a}_n = z^{2n} \overline{f(\bar{z}^{-1})}, \quad z \in \mathbb{C}, z \neq 0. \tag{11.3}$$

Equation (11.3) implies that the zeros of f are symmetric with respect to the unit circle. More precisely, if w is a zero of f, then $w \neq 0$ and $\bar{w}^{-1}$ is also a zero of f with the same multiplicity. Let $z_1, \bar{z}_1^{-1}, \ldots, z_m, \bar{z}_m^{-1}$ denote the zeros of f which are not on $\mathbb{T}$ (if there are such zeros) counted according to their multiplicities.
   Define $g(x) = p(e^{ix})$ for $x \in \mathbb{R}$. By differentiation it follows that each zero $e^{i\theta} \in \mathbb{T}$ of p has the same multiplicity as the zero $\theta \in \mathbb{R}$ of g. Since $g(x) = p(e^{ix}) \geq 0$ on $\mathbb{R}$, each zero θ of g, hence each zero $e^{i\theta} \in \mathbb{T}$ of p and so of f, is of even multiplicity. If f has zeros on $\mathbb{T}$, we denote them by $\xi_1, \xi_1, \ldots, \xi_l, \xi_l$, so that $2m + 2l = 2n$.
   Then the polynomial f and hence p factor as

$$p(z) = z^{-n} f(z) = z^{-n} a_n \prod_{k=1}^{m} (z - z_k)(z - \bar{z}_k^{-1}) \prod_{j=1}^{l} (z - \xi_j)^2$$

$$= \prod_{k=1}^{m} (z - z_k)(z^{-1} - \bar{z}_k) \prod_{j=1}^{l} (z - \xi_j)(z^{-1} - \bar{\xi}_j) \left[z^{-n} a_n \prod_{k=1}^{m} (-z)\bar{z}_k^{-1} \prod_{j=1}^{l} (-z)\xi_j \right].$$

The factor in square brackets is cz^{m+l-n} for some $c \in \mathbb{C}$. Since $m + l = n$ and $p \geq 0$ on $\mathbb{T}$, it follows that $cz^{m+l-n} = c > 0$. Setting

$$q_0(z) = \sqrt{c} \prod_{k=1}^{m} (z - z_k) \prod_{j=1}^{l} (z - \xi_j),$$

the preceding equality yields $p(z) = |q_0(z)|^2$ for $z \in \mathbb{T}$. (One of the two groups of zeros may be absent. In this case we set the corresponding product equal to one.)
   Upon multiplying q_0 by a constant of modulus one, we can have $q_0(0) > 0$. Recall that $z \mapsto \frac{1-\bar{\zeta}z}{z-\zeta}$ is a bijection of $\mathbb{T}$ for any $\zeta \in \mathbb{C}$. Therefore, if ζ is a zero of q_0, the polynomial $q_0(z) \frac{1-\bar{\zeta}z}{z-\zeta}$ has the same degree as q_0 and satisfies (11.2) as well. Continuing in this manner, we can remove all zeros of q_0 which are contained in $\mathbb{D} = \{z \in \mathbb{C} : |z| < 1\}$ and obtain a polynomial q which has the desired properties.

Let $\tilde{q}$ be another polynomial with these properties. By (11.2), $|q(z)| = |\tilde{q}(z)|$ on $\mathbb{T}$. From this it follows that the zeros of q and $\tilde{q}$ on $\mathbb{T}$ and their multiplicities coincide. Since $q(z) \neq 0$ and $\tilde{q}(z) \neq 0$ on D, the functions $f(z) := \frac{\tilde{q}(z)}{q(z)}$ and $\frac{1}{f(z)}$ are holomorphic on $\mathbb{D}$ and satisfy $|f(z)| = |\frac{1}{f(z)}| = 1$ on $\mathbb{T}$. By the maximum principle for holomorphic functions, $|f(z)| \leq 1$ and $|\frac{1}{f(z)}| \leq 1$, hence $|f(z)| = 1$, on $\mathbb{D}$. Since $|f(z)|$ attains its maximum in $\mathbb{D}$, f is constant. From $q(0) > 0$, $\tilde{q}(0) > 0$, and $|f(0)| = 1$ we get $f(0) = 1$. Hence $f(z) \equiv 1$, so that $q = \tilde{q}$. □

Remark 11.2 The polynomial q in Theorem 11.1 satisfies

$$p(z) = q(z)\,\overline{q(\overline{z}^{-1})} \quad \text{for} \quad z \in \mathbb{C}, z \neq 0. \qquad \circ$$

11.2 Trigonometric Moment Problem: Existence of a Solution

Recall that the group $*$-algebra $\mathbb{C}[\mathbb{Z}]$ of $\mathbb{Z}$ is the unital $*$-algebra $\mathbb{C}[z, z^{-1}]$ of all Laurent polynomials $p(z) = \sum_{j=-n}^{n} c_j z^j$, where $c_j \in \mathbb{C}$ and $n \in \mathbb{N}$, with involution $p \mapsto p^*(z) = \sum_{j=-n}^{n} \overline{c}_j z^{-j}$. In the $*$-algebra $\mathbb{C}[\mathbb{Z}]$ we have $z^* = z^{-1}$, that is, z is a unitary element. The character space of $\mathbb{C}[\mathbb{Z}]$ is the torus $\mathbb{T} = \{z \in \mathbb{C} : |z| = 1\}$, where $z \in \mathbb{T}$ acts on $\mathbb{C}[\mathbb{Z}]$ by the point evaluation $\chi_z(p) = p(z)$.

Note that $\mathbb{C}[\mathbb{Z}]$ can be considered as the $*$-algebra of trigonometric polynomials

$$f(\theta) = \sum_{j=-n}^{n} c_j e^{ij\theta} = c_0 + \sum_{l=1}^{n} (a_l \cos l\theta + b_l \sin l\theta), \quad \theta \in [-\pi, \pi],$$

where $c_j, a_l, b_l \in \mathbb{C}$ and $a_l = c_l + c_{-l}, b_l = i(c_l - c_{-l})$, with involution given by $f \mapsto f^*(\theta) = \sum_{j=-n}^{n} \overline{c}_j e^{-ij\theta}$.

Let $s = (s_j)_{j \in \mathbb{N}_0}$ be a sequence. We denote by L_s the linear functional on $\mathbb{C}[\mathbb{Z}]$ given by $L_s(z^{-j}) := s_j, j \in \mathbb{Z}$. Recall from Example 2.3.3 that the Hankel matrix is now the infinite *Toeplitz matrix* $H(s) = (h_{jk})_{j,k \in \mathbb{N}_0}$ with entries $h_{jk} := s_{k-j}, j, k \in \mathbb{N}_0$. Here we have set $s_{-l} = \overline{s}_l$ for $l \geq 1$ according to our notational convention.

If $s = (s_j)_{j=0}^{m}$ is a sequence and $n \leq m$, then $H_n(s) = (h_{jk})_{j,k=0}^{n}$ denotes the $(n+1) \times (n+1)$ Toeplitz matrix with entries $h_{jk} = s_{k-j}$ and we abbreviate

$$D_n := D_n(s) \equiv \det H_n(s) = \begin{vmatrix} s_0 & s_1 & \cdots & s_n \\ s_{-1} & s_0 & \cdots & s_{n-1} \\ s_{-2} & s_{-1} & \cdots & s_{n-2} \\ \cdots & \cdots & & \cdots \\ s_{-n} & s_{-n+1} & \cdots & s_0 \end{vmatrix} \qquad (11.4)$$

The next theorem, called the *Carathéodory–Toeplitz theorem*, provides a solution of the trigonometric moment problem.

Theorem 11.3 *For a complex sequence $s = (s_n)_{n \in \mathbb{N}_0}$ the following are equivalent:*

(i) *s is a moment sequence for the group $\mathbb{Z}$, that is, there exists a Radon measure μ on $\mathbb{T}$ such that*

$$s_n = \int_{\mathbb{T}} z^{-n} d\mu(z) \quad \text{for all} \quad n \in \mathbb{Z}. \tag{11.5}$$

(ii) *$L_s(q^*q) \geq 0$ for $q \in \mathbb{C}[\mathbb{Z}]$, i.e. L_s is a positive functional on the $*$-algebra $\mathbb{C}[\mathbb{Z}]$.*
(iii) *$L_s(q^*q) \geq 0$ for all $q \in \mathbb{C}[z]$.*
(iv) *The infinite Toeplitz matrix $H(s) = (s_{k-j})_{j,k=0}^{\infty}$ is positive semidefinite.*
(v) *$\sum_{j,k=0}^{\infty} s_{j-k} c_k \overline{c_j} \geq 0$ for all finite complex sequences $(c_j)_{j \in \mathbb{N}_0}$.*

The measure μ is uniquely determined by (11.5). Its support is an infinite set if and only if the Toeplitz matrix $H(s)$ is positive definite, or equivalently, the moment sequence s is positive definite, or equivalently, $D_n(s) > 0$ for all $n \in \mathbb{N}_0$.

Proof Let $q(z) = \sum_{k=0}^{n} c_k z^k \in \mathbb{C}[z]$. Then $q^*(z) = \sum_{j=0}^{n} \overline{c_j} z^{-j}$ and

$$L_s(q^*q) = \sum_{j,k=0}^{n} L_s(z^{k-j}) c_k \overline{c_j} = \sum_{j,k=0}^{\infty} s_{j-k} c_k \overline{c_j} = \sum_{j,k=0}^{\infty} h_{kj} c_k \overline{c_j}. \tag{11.6}$$

From (11.6) we conclude that (iii)$\leftrightarrow$(iv)$\leftrightarrow$(v). Further, (i)$\rightarrow$(ii) by Proposition 2.7 and (ii)$\rightarrow$(iii) is trivial. We prove the implication (iii)$\rightarrow$(i).

Let $p \in \mathbb{C}[\mathbb{Z}]$ be such that $p(z) \geq 0$ on $\mathbb{T}$. By the Fejér–Riesz theorem 11.1, there is a polynomial $q = \sum_{j=0}^{n} c_j z^j \in \mathbb{C}[z]$ such that $p = q^*q$. Then $L_s(p) = L_s(q^*q) \geq 0$ by (iii). Hence the restriction of L_s to the real subspace $E := \{ p \in \mathbb{C}[\mathbb{Z}] : p = p^* \}$ of $C(\mathbb{T}; \mathbb{R})$ is E_+-positive. Therefore, by Proposition 1.9, the restriction of L_s to E, hence also L_s on $\mathbb{C}[\mathbb{Z}]$, is given by a measure $\mu \in M_+(\mathbb{T})$. This implies (i).

Thus we have shown that the four conditions (i)–(iv) are equivalent.

Since the trigonometric polynomials are dense in $C(\mathbb{T})$ by Fejér's theorem, the measure μ is uniquely determined by (i).

We verify the last assertion. If $q(z)$ is as above, from (11.6) and (i) we obtain

$$\sum_{j,k=0}^{n} h_{kj} c_k \overline{c_j} = \sum_{j,k=0}^{n} c_k \overline{c_j} \int_{\mathbb{T}} z^{k-j} d\mu = \int_{\mathbb{T}} \left| \sum_{k=0}^{n} c_k z^k \right|^2 d\mu = \int_{\mathbb{T}} |q(z)|^2 d\mu.$$
$$\tag{11.7}$$

If μ has finite support, we can find $q \neq 0$, which vanishes on $\operatorname{supp} \mu$. Then $(c_0, \ldots, c_n) \neq 0$ and $\sum_{j,k=0}^{n} h_{kj} c_k \overline{c_j} = 0$ by (11.7). That is, the matrix $H(s)$ and the sequence s are not positive definite.

If μ has infinite support, for any vector $(c_0, \dots, c_n) \neq 0$ the polynomial q does not vanish on supp μ. Hence it follows from (11.7) that $\sum_{j,k=0}^{n} h_{kj} c_k \overline{c}_j > 0$, that is, $H(s)$ and s are positive definite.

Obviously, the infinite matrix $H(s)$ is positive definite if and only if all finite matrices $H_n(s)$ are, or equivalently, $D_n(s) > 0$ for all $n \in \mathbb{N}_0$.   □

Remark 11.4

1. The minus signs in (11.5) and in the definition $L_s(z^{-j}) = s_j$ are notational conventions following the standard literature [KN], [Sim2]. The minus sign in (11.5) also fits into the usual definition of the Fourier transform for the group $\mathbb{Z}$. It should be noted that some authors define $L_s(z^j) = s_j$ and/or $h_{jk} = s_{k-j}$.
2. In (i), we have set $s_{-n} := \overline{s}_n$ for $n \in \mathbb{N}$ by our convention. In (iii), the condition $L_s(q^*q) \geq 0$ is required only for "analytic" polynomials $q(z) = \sum_{j=0}^{n} c_j z^j$.   ○

The following theorem is the counterpart of Theorem 11.3 for the truncated trigonometric moment problem.

Theorem 11.5 *Let $n \in \mathbb{N}_0$. For a sequence $(s_j)_{j=0}^{n}$ the following are equivalent:*

(i) *There is a Radon measure μ on $\mathbb{T}$ such that*

$$s_j = \int z^{-j} d\mu(z) \quad \text{for } j = 0, \dots, n. \tag{11.8}$$

(ii) *There is a k-atomic measure μ on $\mathbb{T}$, $k \leq 2n + 1$, such that (11.8) holds.*
(iii) *The Toeplitz matrix $H_n(s)$ is positive semidefinite.*
(iv) $\sum_{j,k=0}^{n} s_{j-k} c_k \overline{c}_j \geq 0$ *for all* $(c_0, \dots, c_n)^T \in \mathbb{C}^{n+1}$.

Proof (ii)→(i) is trivial. Similarly, as in the proof of Theorem 11.3 the implications (i)→(iii)↔(iv) follow from (11.6). It suffices to prove that (iv) implies (ii). Let E be the real vector space of polynomials $p = p^* \in \mathbb{C}[\mathbb{Z}]$ such that $\deg(p) \leq n$. Clearly, $\dim E = 2n + 1$. Let $p \in E$ be such that $p(z) \geq 0$ on $\mathbb{T}$. Then the polynomial q from the Fejér–Riesz theorem also satisfies $\deg q \leq n$. Hence it follows from (11.6) and (iv) that $L_s(p) = L_s(q^*q) \geq 0$. Thus Proposition 1.26 applies and yields (ii).   □

11.3   Orthogonal Polynomials on the Unit Circle

In this section, we suppose that $s = (s_j)_{j \in \mathbb{Z}}$ is a moment sequence on $\mathbb{Z}$ such that its representing measure μ on $\mathbb{T}$ has infinite support. By Theorem 11.3 the latter holds if and only if the infinite Toeplitz matrix $H(s)$ is positive definite.

Since $H(s)$ is positive definite, there is a scalar product $\langle \cdot, \cdot \rangle_s$ on $\mathbb{C}[z]$ given by

$$\langle p, q \rangle_s := \sum_{j,k=0}^{n} a_j \overline{b}_k s_{k-j} \tag{11.9}$$

for $p = \sum_{j=0}^{n} a_j z^j \in \mathbb{C}[z]$ and $q = \sum_{k=0}^{n} b_k z^k \in \mathbb{C}[z]$. From the definition (11.9) it is immediate that $s_k = \langle 1, z^k \rangle_s$ for $k \in \mathbb{Z}$ and

$$\langle zp, zq \rangle_s = \langle p, q \rangle_s. \tag{11.10}$$

Next we define two families of polynomials P_k, P_k^* associated with s. Put $D_{-1} := 1$ and $P_0(z) := 1$. For $k \in \mathbb{N}$ we set

$$P_k(z) := \frac{1}{D_{k-1}} \begin{vmatrix} s_0 & s_1 & \dots & s_{k-1} & 1 \\ s_{-1} & s_0 & \dots & s_{k-2} & z \\ s_{-2} & s_{-1} & \dots & s_{k-3} & z^2 \\ \dots & \dots & \dots & \dots & \dots \\ s_{-k} & s_{-k+1} & \dots & s_{-1} & z^k \end{vmatrix}. \tag{11.11}$$

Since the coefficient of z^k in the determinant is D_{k-1}, P_k is a monic polynomial of degree k for $k \in \mathbb{N}_0$. The next lemma shows that $P_k, k \in \mathbb{N}_0$, are orthogonal polynomials of the unitary space $(\mathbb{C}[z], \langle \cdot, \cdot \rangle_s)$.

Lemma 11.6 $\langle P_k, P_j \rangle_s = D_{k-1}^{-1} D_k \delta_{kj}$ and $\langle P_k, z^k \rangle_s = D_{k-1}^{-1} D_k$ for $k, j \in \mathbb{N}_0$.

Proof Let $l = 0, \dots, k$. To compute $\langle P_k(z), z^l \rangle_s$ we repeat the reasoning from the proof of Proposition 5.3. Multiplying the last column by z^{-l} and applying the functional L_s the last column of the determinant will be replaced by $(s_l, s_{j-1}, \dots, s_{j-k})^T$.

If $l < k$, the last column coincides with the j-th column and hence $\langle P_k(z), z^l \rangle_s = 0$. Since P_k has degree k, this implies that $\langle P_k, P_j \rangle_s = 0$ for all $k \neq j, k, j \in \mathbb{N}_0$.

Now let $j = k$. If $j = k = 0$, then obviously $\langle P_0, P_0 \rangle_s = \langle P_0, z^0 \rangle_s = s_0 = D_0$. For $j = k \in \mathbb{N}$, the determinant becomes D_k, so that $\langle P_k, z^k \rangle_s = D_{k-1}^{-1} D_k$. Hence, since the leading coefficient of P_k is 1, we get $\langle P_k, P_k \rangle_s = \langle P_k, z^k \rangle_s = D_{k-1}^{-1} D_k$. ☐

Let $p = \sum_{j=0}^{k} c_j z^j$ be a polynomial of degree at most k. Put $\bar{p} := \sum_{j=0}^{k} \bar{c}_j z^j$. The *reciprocal polynomial* $R_k(p)$ is defined by

$$R_k(p)(z) := z^k \bar{p}(z^{-1}) = \sum_{j=0}^{k} \bar{c}_{k-j} z^j.$$

Clearly, $\deg(R_k(p)) \leq k$ and $\deg(R_k(p)) = k$ if and only if $p(0) \neq 0$. Then we have

$$(R_k(p))(z) = z^k \overline{p(\bar{z})} = z^k \overline{p(1/\bar{z})} \quad \text{for } z \in \mathbb{T}, \tag{11.12}$$

$$\langle R_k(p), R_k(q) \rangle_s = \langle q, p \rangle_s. \tag{11.13}$$

For $p = P_k$ we abbreviate $P_k^* := R_k(P_k)$. Since P_k is monic, $P_k^*(0) = 1$. Clearly,

$$P_0(z) = P_0^*(z) = 1, \quad P_1(z) = z - s_{-1} s_0^{-1}, \quad P_1^*(z) = 1 - s_1 s_0^{-1} z.$$

Combining the definition $P_k^*(z) = R_k(P_k)(z) = z^k \overline{P}_k(z^{-1})$ with (11.11) we obtain the explicit formula

$$P_k^*(z) = \frac{1}{D_{k-1}} \begin{vmatrix} s_0 & s_{-1} & \dots & s_{-k+1} & z^k \\ s_1 & s_0 & \dots & s_{-k+2} & z^{k-1} \\ s_2 & s_1 & \dots & s_{-k+3} & z^{k-2} \\ \dots & \dots & \dots & \dots & \dots \\ s_k & s_{k-1} & \dots & s_1 & 1 \end{vmatrix}, \quad k \in \mathbb{N}. \tag{11.14}$$

Remark 11.7 The notation P_k^* is standard in the literature. Note that P_k^* should not be confused with the adjoint of P_k in the $*$-algebra $\mathbb{C}[\mathbb{Z}]$! ∘

Some simple facts on these polynomials P_k^* are collected in the following lemma.

Lemma 11.8

(i) $\langle P_k^*, z^j \rangle_s = 0$ for $j = 1, \dots, k$ and $\langle P_k^*, 1 \rangle_s = D_{k-1}^{-1} D_k$.
(ii) $\|P_k\|_s^2 = \|P_k^*\|_s^2 = \langle P_k^*, 1 \rangle_s = D_{k-1}^{-1} D_k$ for $k \in \mathbb{N}_0$.

Proof All assertions follow by a repeated application of Lemma 11.6 combined with (11.13). For $j = 1, \dots, k$ we obtain

$$\langle P_k^*, z^j \rangle_s = \langle R_k(P_k), R_k(z^{k-j}) \rangle_s = \langle z^{k-j}, P_k \rangle_s = 0.$$

Further, again by (11.13), we have $\|P_k^*\|^2 = \|P_k\|^2$ and

$$\langle P_k^*, 1 \rangle = \langle R_k(P_k), R_k(z^k) \rangle = \langle z^k, P_k \rangle = \|P_k\|^2 = D_{k-1}^{-1} D_k. \qquad \square$$

The following theorem is the first main result of this section. The formulas (11.15) and (11.16) therein are called *Szegö recursion formulas*.

Theorem 11.9 *Suppose that $s = (s_j)_{j \in \mathbb{Z}}$ is a positive definite sequence on $\mathbb{Z}$. Then there exist uniquely determined complex numbers α_n for $n \in \mathbb{N}_0$ such that*

$$P_{n+1}(z) = z P_n(z) - \overline{\alpha}_n P_n^*(z), \tag{11.15}$$

$$P_{n+1}^*(z) = P_n^*(z) - \alpha_n z P_n(z). \tag{11.16}$$

Further, we have

$$\alpha_n = -\overline{P_{n+1}(0)}, \tag{11.17}$$

$$\|P_{n+1}\|_s^2 = (1 - |\alpha_n|^2) \|P_n\|_s^2 = s_0 \prod_{j=0}^{n} (1 - |\alpha_j|^2), \tag{11.18}$$

$$D_{n+1} = D_n s_0 \prod_{j=0}^{n} (1 - |\alpha_j|^2). \tag{11.19}$$

Proof Using Lemma 11.6 and formula (11.10) we obtain for $j = 1, \ldots, n$,

$$\langle P_{n+1} - zP_n, z^j \rangle = \langle P_{n+1}, z^j \rangle - \langle P_n, z^{j-1} \rangle = 0.$$

By Lemma 11.8, P_n^* is also orthogonal to $z, z^2, \ldots, z^n$. Since P_{n+1} and P_n are monic, $P_{n+1} - zP_n$ has degree at most n. Thus $P_{n+1} - zP_n$ and P_n^* are both of degree at most n and orthogonal to $z, \ldots, z^n$. Therefore, since $\langle P_n^*, 1 \rangle \neq 0$ by Lemma 11.8, setting $\alpha_n := \langle P_n^*, 1 \rangle_s^{-1} \langle P_{n+1} - zP_n, 1 \rangle_s$ we conclude that $P_{n+1} = zP_n - \overline{\alpha}_n P_n^*$ which is (11.15). Applying R_{n+1} to both sides of (11.15) yields (11.16).

Setting $z = 0$ in (11.15) and using that $P_n^*(0) = 1$ we obtain (11.17). In particular, this implies that α_n is uniquely determined by (11.15).

Next we prove (11.18). Recall that multiplication by z is unitary by (11.10), $P_{n+1} \perp P_n^*$ (since $\deg(P_n^*) \leq n$) and $\|P_n\| = \|P_n^*\|$ by Lemma 11.8(ii). Using these facts it follows from (11.15) that

$$\|P_n\|_s^2 = \|zP_n\|_s^2 = \|P_{n+1} + \overline{\alpha}_n P_n^*\|_s^2 = \|P_{n+1}\|_s^2 + |\alpha_n|^2 \|P_n\|_s^2,$$

which implies the first equality of (11.18). The second follows by repeated application of the first combined with fact that $\|P_0\|_s^2 = \langle 1, 1 \rangle_s = s_0$.

Inserting the equality $\|P_{n+1}\|_s^2 = D_n^{-1} D_{n+1}$ (from Lemma 11.8(ii)) into (11.18) we obtain (11.19). $\square$

Definition 11.10 The numbers α_n from (11.15) are called the *reflection coefficients* of s; they are also denoted by $\alpha_n(s)$, or $\alpha_n(\mu)$, where μ is the unique representing measure of s.

Remark 11.11

1. The numbers α_n also appear under the names *Verblunsky coefficients* in [Sim2], *canonical moments* in [DS], or *Schur parameters* in the literature.
2. The choice of writing $-\overline{\alpha}_n$ in (11.15) follows [Sim2]. The reason is that then $\alpha_n(\mu)$ becomes equal to the Schur parameter $\gamma_n(\mu)$ by Geronimus' theorem 11.31 below.
3. Equations (11.15) and (11.16) can be rewritten in matrix form as

$$\begin{pmatrix} P_{n+1}(z) \\ P_{n+1}^*(z) \end{pmatrix} = \begin{pmatrix} z & -\overline{\alpha}_n \\ -\alpha_n z & 1 \end{pmatrix} \begin{pmatrix} P_n(z) \\ P_n^*(z) \end{pmatrix}.$$ ∘

The second main result of this section is the following *Verblunsky theorem*. It states that sequences of numbers $\alpha_n \in \mathbb{D}$ are precisely the sequences of possible reflection coefficients of probability measures on $\mathbb{T}$ of infinite support. Since the parameters α_n appearing in the Szegö relation (11.15) are the counterpart of the Jacobi parameters a_n, b_n from Sect. 5.2, Theorem 11.12 might be called "Favard's theorem for the unit circle".

Theorem 11.12 *For a complex sequence* $(\alpha_n)_{n=0}^{\infty}$ *the following are equivalent:*

(i) *There is a positive definite sequence* $s = (s_j)_{j \in \mathbb{Z}}$, *where* $s_0 = 1$, *on* $\mathbb{Z}$ *such that* $\alpha_n = \alpha_n(s)$ *for* $n \in \mathbb{N}_0$.

(ii) *There exists a probability measure* μ *on* $\mathbb{T}$ *of infinite support such that* $\alpha_n = \alpha_n(\mu)$ *for* $n \in \mathbb{N}_0$.

(iii) $\alpha_n \in \mathbb{D}$ *for all* $n \in \mathbb{N}_0$.

By Theorem 11.3, a moment sequence on $\mathbb{Z}$ is positive definite if and only if its representing measure has infinite support. Hence Theorem 11.3 yields (i)↔(ii). (i)→(iii) follows at once from formula (11.18), since $P_{n+1} \neq 0$ and (11.18) imply that $|\alpha_n| < 1$. The main implication (iii)→(i) will be proved at the end of the next section.

We close this section by stating some facts on zeros of the polynomials P_n, P_n^*.

Proposition 11.13 *For* $n \in \mathbb{N}$ *we have:*

(i) *If* $z_0 \in \mathbb{C}$ *is a zero of* $P_n(z)$, *then* $|z_0| < 1$.

(ii) *If* $z_0 \in \mathbb{C}$ *is a zero of* $P_n^*(z)$, *then* $|z_0| > 1$.

(iii) $|P_n^*(z)| = |P_n(z)|$ *for* $z \in \mathbb{T}$.

(iv) $|P_n^*(z)| < |P_n(z)|$ *for* $z \in \mathbb{D}$.

Proof

(i) Since $P_n(z_0) = 0$, $p(z) := \frac{P_n(z)}{z - z_0}$ is a polynomial of degree $n - 1$. Hence P_n is orthogonal to $z_0 p$ in the unitary space $(\mathbb{C}[z], \langle \cdot, \cdot \rangle_s)$. Using this fact we derive

$$\|p\|_s^2 - \|zp\|_s^2 = \|(z - z_0)p + z_0 p\|_s^2 = \|P_n + z_0 p\|_s^2 = \|P_n\|_s^2 + |z_0|^2 \|p\|_s^2,$$

so that

$$(1 - |z_0|^2)\|p\|_s^2 = \|P_n\|_s^2. \tag{11.20}$$

Since $\langle \cdot, \cdot \rangle_s$ is a scalar product, we have $\|p\|_s > 0$ and $\|P_n\|_s > 0$. Therefore (11.20) implies that $|z_0| < 1$.

(ii) Recall that $P_n^*(z) = R_n(P_n)(z) = z^n P_n(z^{-1})$ and $P_n^*(0) = 1$. Hence $P_n^*(z_0) = 0$ implies $z_0 \neq 0$ and $P_n(z_0^{-1}) = 0$, so that $|z_0| > 1$ by (i).

(iii) follows from $|P_n^*(z)| = |z^n \overline{P_n(z)}| = |P_n(z)|$ for $z \in \mathbb{T}$ by (11.12).

(iv) By (ii), $P_n^*(z) \neq 0$ for $|z| \leq 1$. Hence the function $f(z) := \frac{P_n(z)}{P_n^*(z)}$ is holomorphic on $\mathbb{D}$, continuous on the closure of $\mathbb{D}$, and of modulus one on $\mathbb{T} = \partial \mathbb{D}$ by (ii). Since $n > 0$, $P_n(z)$ has a zero in $\mathbb{D}$, so in particular, $f(z)$ is not constant. Therefore, by the maximum principle for holomorphic functions, $|f(z)| < 1$ and hence $|P_n^*(z)| < |P_n(z)|$ for $z \in \mathbb{D}$. □

11.4 The Truncated Trigonometric Moment Problem

Definition 11.14 Let $n \in \mathbb{N}$. The *moment cone* $\mathcal{S}_{n+1}$ is given by

$$\mathcal{S}_{n+1} := \left\{ s = (s_0, s_1, \ldots, s_n) : s_j = \int_{\mathbb{T}} z^{-j} d\mu(z), j = 0, \ldots, n, \ \mu \in M_+(\mathbb{T}) \right\}.$$

As in Sect. 10.2, we consider $\mathcal{S}_{n+1}$ as a subset of $\mathbb{C}^{n+1}$ by identifying s with the column vector $s^T \in \mathbb{C}^{n+1}$. By a similar reasoning as in the proof to Proposition 10.5 it follows that $\mathcal{S}_{n+1}$ is a closed convex cone in $\mathbb{C}^{n+1}$ and the conic convex hull of the moment curve

$$\mathsf{c}_{n+1} = \{\mathsf{s}(z) = (1, z, z^2, \ldots, z^n) : z \in \mathbb{T}\}.$$

The latter implies that each $s \in \mathcal{S}_{n+1}$ has an atomic representing measure

$$\mu = \sum_{j=1}^{k} m_j \delta_{z_j}, \tag{11.21}$$

where $z_1, \ldots, z_k$ are pairwise different points of $\mathbb{T}$ and $m_j > 0$ for $j = 1, \ldots, k$. For such a measure μ we set $\mathrm{ind}(\mu) = 2k$. (Since the unit circle $\mathbb{T}$ has no end points, each atom of μ is counted twice.) The index $\mathrm{ind}(s)$ is defined as the minimum of indices of all such representing measures (11.21) for s.

Let $s = (s_j)_{j=0}^n \in \mathcal{S}_{n+1}$. By our notational convention, we have defined $s_{-j} := \bar{s}_j$ for $j = 1, \ldots, n$. Hence, if μ is a representing measure for s, then

$$s_j = \int_{\mathbb{T}} z^{-j} d\mu \quad \text{for } j = -n, -n+1, \ldots, n.$$

By a slight abuse of notation we denote the "double" sequence $(s_j)_{j=-n}^n$ also by s.

Recall that by Theorem 11.5 a complex sequence $s = (s_j)_{j=0}^n$ belongs to $\mathcal{S}_{n+1}$ if and only if the Toeplitz matrix $H_n(s)$ is positive semidefinite.

The following propositions characterizes boundary points and interior points of the set $\mathcal{S}_{n+1}$. The proofs are verbatim the same as for its counterparts on a bounded interval (Theorems 10.7 and 10.8) and will be omitted.

Proposition 11.15 *A sequence $s \in \mathcal{S}_{n+1}$ is a boundary point of $\mathcal{S}_{n+1}$ if and only if $\mathrm{ind}(s) \leq 2n$. In this case s is determinate, that is, it has a unique representing measure $\mu \in M_+(\mathbb{T})$.*

Proposition 11.16 *For $s \in \mathcal{S}_{n+1}$ the following statements are equivalent:*

(i) *$s \in \mathrm{Int}\, \mathcal{S}_{n+1}$, that is, s is an interior point of $\mathcal{S}_{n+1}$.*
(ii) *The Toeplitz matrix $H_n(s)$ is positive definite.*
(iii) *$D_j(s) > 0$ for $j = 0, \ldots, n$.*

Note in conditions (ii) and (iii) of Proposition 11.16 the numbers $s_{-j} = \overline{s}_j$ for $j = 1, \ldots, n$ are required.

Now we fix $s = (s_j)_{j=0}^n \in \operatorname{Int} \mathcal{S}_{n+1}$. Then, by Proposition 11.16, $D_j(s) > 0$ for $j = 0, \ldots, n$ and hence the sequence $s = (s_j)_{j=-n}^n$ is positive definite, that is,

$$\sum_{j,k=0}^n s_{j-k} c_k \overline{c_j} > 0 \quad \text{for all } (c_0, \ldots, c_n)^T \in \mathbb{C}^{n+1}, (c_0, \ldots, c_n) \neq 0.$$

This implies that Eq. (11.9) defines a scalar product on the vector space $\mathbb{C}_n[z]$. Further, proceeding as in the last section, we define the polynomials P_k and P_k^*, $k = 0, \ldots, n$, and the reflection coefficients $\alpha_j, j = 0, \ldots, n-1$, and derive the corresponding properties from Theorem 11.9.

Let us denote by C_s the set of numbers $s_{n+1} \in \mathbb{C}$ for which the extended sequence $\tilde{s} := (s_0, \ldots, s_n, s_{n+1})$ belongs to $\mathcal{S}_{n+2}$. In the proof of Proposition 11.17 below we shall use the following notation:

$$\Delta_{n+1}(z) = \begin{vmatrix} s_0 & s_1 & \cdots & s_n & z \\ s_{-1} & s_0 & \cdots & s_{n-1} & s_n \\ \cdots & \cdots & \cdots & \cdots & \cdots \\ \overline{z} & s_{-n} & \cdots & s_{-1} & s_0 \end{vmatrix}, \tag{11.22}$$

$$A_n(z) = \begin{vmatrix} s_1 & s_2 & \cdots & s_n & z \\ s_0 & s_1 & \cdots & s_{n-1} & s_n \\ \cdots & \cdots & \cdots & \cdots \\ s_{-n+1} & s_{-n+2} & \cdots & s_0 & s_1 \end{vmatrix}, \tag{11.23}$$

$$B_n(z) = \begin{vmatrix} s_{-1} & s_0 & \cdots & s_{n-2} & s_{n-1} \\ s_{-2} & s_{-1} & \cdots & s_{n-3} & s_{n-2} \\ \cdots & \cdots & \cdots & \cdots \\ \overline{z} & s_{-n+1} & \cdots & s_{-2} & s_{-1} \end{vmatrix}. \tag{11.24}$$

The next proposition is the counterpart of Corollary 10.16 for the unit circle.

Proposition 11.17 *Suppose that* $s = (s_j)_{j=0}^n \in \operatorname{Int} \mathcal{S}_{n+1}$. *Then the set* C_s *is a closed disk with radius* $r_{n+1} = \frac{D_n(s)}{D_{n-1}(s)}$ *and center* $c_{n+1} = \frac{(-1)^{n+1} A_n(0)}{D_{n-1}(s)}$.

Proof By Theorem 11.5 (i)↔(iii), C_s is the set of complex numbers s_{n+1} such that $H_{n+1}(\tilde{s})$ is positive semidefinite. Since $D_j(s) > 0$ for $= 1, \ldots, n$ by Proposition 11.16, C_s is precisely the set of numbers $z = s_{n+1} \in \mathbb{C}$ for which $\Delta_{n+1}(z) \geq 0$.

To describe this set we apply Sylvester's formula (see e.g. [Gn, p. 58]) and expand the determinant $\Delta_{n+1}(z)$ with respect to the first and last row and first and last column. Then we obtain

$$D_{n-1}(s)\Delta_{n+1}(z) = D_n(s)^2 - A_n(z)B_n(z). \tag{11.25}$$

Since $\bar{s}_j = s_{-j}$, we have $B_n(z) = \overline{A_n(z)}$. Clearly, the determinant $A_n(z)$ is a linear polynomial in z with leading coefficient $(-1)^n D_{n-1}(s)$. Therefore,

$$A_n(z) = z(-1)^n D_{n-1}(s) + A_n(0). \tag{11.26}$$

Hence, since $D_{n-1}(s) > 0$, it follows from (11.25) that $\Delta_{n+1}(z) \geq 0$ if and only if

$$D_n(s)^2 \geq |A_n(z)|^2 = |z(-1)^n D_{n-1}(s) + A_n(0)|^2,$$

or equivalently,

$$r_{n+1} \equiv \frac{D_n(s)}{D_{n-1}(s)} \geq \left| z - (-1)^{n+1}\frac{A_n(0)}{D_{n-1}(s)} \right| \equiv |z - c_{n+1}|. \tag{11.27}$$

This completes the proof of Proposition 11.17. $\qquad\square$

The closed disk C_s from Proposition 11.17 is the set of the possible $(n+1)$-th moments s_{n+1}, or more precisely, the set of numbers s_{n+1} for which the extended sequence $\tilde{s} = (s_0,\ldots,s_n,s_{n+1})$ is in $\mathcal{S}_{n+2}$. Recall that $D_{n+1}(\tilde{s}) = \Delta(s_{n+1})$ by construction and $D_j(s) = D_j(\tilde{s}) > 0$ for $j = 0,\ldots,n$ by the assumption $s \in \text{Int } \mathcal{S}_{n+1}$. Therefore, by Proposition 11.15, $\tilde{s}$ belongs to boundary of M_{n+2} if and only if $\Delta(s_{n+1}) = 0$, or equivalently, s_{n+1} lies on the circle ∂C_s.

For $\xi \in \partial C_s$ let $P_{n+1}(z;\xi)$ denote the polynomial (11.11) with $k = n+1$ and $s_{-(n+1)} = \bar{\xi}$. All other moments s_j required in (11.11) are determined by $s = (s_j)_{j=0}^n$. Let us call a representing measure μ for s canonical if $\text{ind}(\mu) = 2n+2$.

Theorem 11.18 *Suppose that $s = (s_j)_{j=0}^n \in \text{Int } \mathcal{S}_{n+1}$. For each ξ on the circle ∂C_s there exists a unique canonical representing measure μ_ξ for s such that*

$$\xi = s_{n+1}(\mu_\xi) = \int_{\mathbb{T}} z^{-(n+1)} d\mu_\xi(z). \tag{11.28}$$

The $(n+1)$ atoms of μ_ξ are precisely the roots of the polynomial $P_{n+1}(z;\xi)$. In particular, $P_{n+1}(z;\xi)$ has $n+1$ distinct simple roots, all of them lying on $\mathbb{T}$.

Proof Let $\tilde{s} = (s_0,\ldots,s_n,s_{n+1})$, where $s_{n+1} := \xi$. Because $\xi \in \partial C_s$, we have $D_{n+1}(\tilde{s}) = \Delta_{n+1}(\xi) = 0$, so $\tilde{s}$ is a boundary point of $\mathcal{S}_{n+2}$. Therefore, by Proposition 11.15, $\tilde{s}$ has a unique representing measure μ_ξ. Clearly, μ_ξ is the unique representing measure for s which satisfies (11.28).

Since $\tilde{s}$ belongs to the boundary of $\mathcal{S}_{n+2}$, $\mathrm{ind}(\mu_\xi) \leq 2n + 2$ by Proposition 11.15. Further, $\mathrm{ind}(\mu_\xi) \leq 2n$ would imply that s is in the boundary of $\mathcal{S}_{n+1}$, which contradicts the assumption $s \in \mathrm{Int}\,\mathcal{S}_{n+1}$. Thus $\mathrm{ind}(\mu_\xi) = 2n + 2$ and μ_ξ is indeed a canonical measure for s.

Since $H_{n+1}(\tilde{s})$ is positive semidefinite, Eq. (11.9), with s replaced by $\tilde{s}$, defines a nonnegative sesquilinear form $\langle \cdot, \cdot \rangle_{\tilde{s}}$ on $\mathbb{C}_{n+1}[z]$. Repeating the proof of Lemma 11.6 and using that μ_ξ is a representing measure of $\tilde{s}$ we derive

$$\int_{\mathbb{T}} |P_{n+1}(z;\xi)|^2 d\mu_\xi(z) = \langle P_{n+1}, P_{n+1} \rangle_{\tilde{s}} = D_n(s)^{-1} D_{n+1}(\tilde{s}) = 0.$$

Therefore, all $n + 1$ atoms of μ_ξ are zeros of $P_{n+1}(z;\xi)$ (by Proposition 1.23). Since $D_n(s) > 0$ and hence $\deg(P_{n+1}) = n + 1$, these atoms exhaust the zeros of P_{n+1} and all zeros are simple. The atoms of μ_ξ are in $\mathbb{T}$. Hence the zeros of P_{n+1} are in $\mathbb{T}$.   □

Next we consider extensions $\tilde{s}$ contained in $\mathrm{Int}\,\mathcal{S}_{n+2}$. Our aim is to build the bridge to the reflection coefficients. Since we assumed that $s \in \mathrm{Int}\,\mathcal{S}_{n+1}$, the sequence s is positive definite and hence (11.9) defines a scalar product on the vector space $\mathbb{C}_n[z]$. Then, proceeding as in the last section, the orthogonal polynomials $P_k, k = 0, \ldots, n$, and the reflection coefficients $\alpha_j, j = 0, \ldots, n - 1$, are defined and the corresponding properties from Theorem 11.9 remain valid. By (11.19),

$$r_{n+1} = D_{n-1}(s)^{-1} D_n(s) = s_0 \prod_{j=0}^{n-1}(1 - |\alpha_j|^2).$$

Obviously, $z = s_{n+1}$ is in the interior of C_s if and only if $|s_{n+1} - c_{n+1}| < r_{n+1}$.

Suppose that $\alpha_n \in \mathbb{D} := \{z \in \mathbb{C} : |z| < 1\}$ is given. Then, setting

$$s_{n+1} := c_{n+1} + \alpha_n r_{n+1} \equiv \frac{(-1)^{n+1} A_n(0)}{D_{n-1}(s)} + \alpha_n s_0 \prod_{j=0}^{n-1}(1 - |\alpha_j|^2), \qquad (11.29)$$

s_{n+1} belongs to the interior of C_s. Then $D_{n+1}(\tilde{s}) = \Delta(s_{n+1}) > 0$, so $\tilde{s} = (s_j)_{j=0}^{n+1}$ belongs to $\mathrm{Int}\,\mathcal{S}_{n+2}$ by Proposition 11.16. Conversely, if $\tilde{s} \in \mathrm{Int}\,\mathcal{S}_{n+2}$, then s_{n+1} is in the interior of C_s and hence s_{n+1} is of the form (11.29) for some unique $\alpha_n \in \mathbb{D}$.

Formula (11.29) describes the new moment s_{n+1} in terms of the given number α_n and of the reflection coefficients α_j and moments $s_j, j < n$, of s.

The following formula expresses α_n in terms of the moments $s_j, j = 0, \ldots, n+1$:

$$\alpha_n = \frac{(-1)^n}{D_n(s)} \begin{vmatrix} s_1 & s_2 & \cdots & s_n & s_{n+1} \\ s_0 & s_1 & \cdots & s_{n-1} & s_n \\ \cdots & \cdots & \cdots & & \cdots \\ s_{-n+1} & s_{-n+2} & \cdots & s_0 & s_1 \end{vmatrix}. \qquad (11.30)$$

We prove formula (11.30). Indeed, first using (11.29), then the formulas for r_{n+1} and c_{n+1}, and finally formula (11.26) we derive

$$\alpha_n = r_{n+1}^{-1}[s_{n+1} - c_{n+1}] = \frac{D_{n-1}(s)}{D_n(s)}\left[s_{n+1} - (-1)^{n+1}A_n(0)D_{n-1}(s)^{-1}\right]$$

$$= \frac{(-1)^n}{D_n(s)}\left[s_{n+1}(-1)^nD_{n-1}(s) + A_n(0)\right] = \frac{(-1)^n}{D_n(s)}A_n(s_{n+1}).$$

Inserting the definition (11.23) of $A_n(s_{n+1})$ into the right-hand side we obtain (11.30). This completes the proof of formula (11.30).

Finally, we prove that α_n is the n-th reflection coefficient of the sequence $\tilde{s}$. Let $\widetilde{\alpha}_n$ denote the n-th reflection coefficient of $\tilde{s}$. If P_{n+1} is the $(n + 1)$-th orthogonal polynomial for $\tilde{s}$, then $\widetilde{\alpha}_n = -\overline{P_{n+1}(0)}$ by (11.17). We develop the determinant in formula (11.11) for $P_{n+1}(0)$ by the last column and apply the complex conjugate. Then $-\overline{P_{n+1}(0)}$ becomes the right-hand side of (11.30). Hence $\widetilde{\alpha}_n = \alpha_n$, that is, α_n is the n-th reflection coefficient of $\tilde{s}$.

We summarize the preceding considerations in the following theorem.

Theorem 11.19 *Suppose that* $s = (s_j)_{j=0}^n \in \operatorname{Int} \mathcal{S}_{n+1}$. *There is a one-to-one correspondence, given by the formulas (11.29) and (11.30), between numbers* $\alpha_n \in \mathbb{D}$ *and numbers* s_{n+1} *in the interior of the disk* C_s. *This yields a one-to-one correspondence between numbers* $\alpha_n \in \mathbb{D}$ *and extensions* $\tilde{s} = (s_0, \dots, s_{n+1})$ *of* s *belonging to the interior of the moment cone* $\mathcal{S}_{n+2}$.

Remark 11.20

1. Set $\hat{s} := (s_1, \dots, s_{n+1})$. Then the determinant in (11.30) is just the determinant $D_n(\hat{s})$ for the sequence $\hat{s}$, that is, we have

$$\alpha_n = \frac{(-1)^n D_n(\hat{s})}{D_n(s)}.$$

   Therefore, if μ is a representing measure for s, then the (complex!) measure $\hat{\mu}$ defined by $d\hat{\mu}(z) = z^{-1}d\mu(z)$ has the moments $s_1, \dots, s_{n+1}$.
2. We easily compute

$$\alpha_0 = \frac{s_1}{s_0} \quad \text{and} \quad \alpha_1 = \frac{s_0 s_2 - s_1^2}{s_0^2 - s_1 s_{-1}}.$$

   In fact, the reflection coeffients α_n depend only on the quotients $\frac{s_j}{s_0}, j \in \mathbb{N}$.
3. If $s_0 = 1$, then (11.29) is a recursion formula which determines the moment sequence s *uniquely* in terms of the sequence $(\alpha_n)_{n=0}^\infty$. For this reason we restricted ourselves to probability measures in Theorem 11.12 and in Sect. 11.5.   ∘

Proof of Theorem 11.12 (iii)→(i) Let $\alpha = (\alpha_n)_{n \in \mathbb{N}_0}$ be a sequence of numbers $\alpha_n \in \mathbb{D}$. By induction we construct a positive definite sequence $s = (s_j)_{j \in \mathbb{N}_0}, s_0 = 1$, such that $\alpha_n = \alpha_n(s)$ for all $n \in \mathbb{N}_0$.

Let $n = 1$ and set $s := (1, \alpha_0)$. Then $D_1(s) = 1 - |\alpha_0|^2 > 0$, so that $s \in \text{Int} \, \mathcal{S}_2$. Further, $P_0(z) = P_0^*(z) = 1$ and $P_1(z) = z - \overline{\alpha}_0 = zP_0 - \overline{\alpha}_0 P_0^*$.

Suppose now that $s^{[n]} = (s_j)_{j=0}^{n} \in \text{Int} \, \mathcal{S}_{n+1}$ is constructed such that it has the reflection coefficients $\alpha_0, \ldots, \alpha_{n-1}$. Then, by Theorem 11.19, there exists an $s_{n+1} \in \mathbb{C}$ such that $s^{[n+1]} = (s_0, \ldots, s_{n+1}) \in \text{Int} \, \mathcal{S}_{n+2}$ has the n-th reflection coefficient α_n. By induction the preceding gives the desired positive definite sequence s. □

Summarizing the main results of this and the preceding sections we have established a one-to-one correspondence between the following three objects:

- probability measures μ on $\mathbb{T}$ of infinite support,
- positive definite sequences $s = (s_j)_{j \in \mathbb{Z}}$ on $\mathbb{Z}$, where $s_0 = 1$,
- sequences $\alpha = (\alpha_n)_{n \in \mathbb{N}_0}$ of complex numbers $\alpha_n \in \mathbb{D}$.

Indeed, for the probability measure μ on $\mathbb{T}$, s is its moment sequence (given by $s_j = \int \zeta^{-j} d\mu(\zeta)$, where $j \in \mathbb{Z}$). For the positive definite sequence s, α is its sequence of reflection coefficients from Theorem 11.9 (given by (11.30)) and μ is the unique solution of the moment problem for s from Theorem 11.3. Finally, for the sequence α, s is the positive definite sequence from Theorem 11.12 (defined inductively by (11.29) and $s_0 := 1$).

11.5 Carathéodory Functions, the Schur Algorithm, and Geromimus' Theorem

The following two notions on holomorphic functions are crucial in this section.

Definition 11.21 A holomorphic function f on $\mathbb{D} = \{z \in \mathbb{C} : |z| < 1\}$ is called a

- *Carathéodory function* if $f(0) = 1$ and $\text{Re} f(z) \geq 0$ for $z \in \mathbb{D}$,
- *Schur function* if $|f(z)| \leq 1$ for $z \in \mathbb{D}$.

For $\zeta \in \mathbb{T}$, the constant function $f(z) = \zeta$ is obviously a Schur function. For all other Schur functions f we have $|f(z)| < 1$ on $\mathbb{D}$ by the maximum principle.

The next result is *Herglotz' representation theorem* of Carathéodory functions.

Proposition 11.22 *For each probability measure μ on $\mathbb{T}$, the function F_μ defined by*

$$F_\mu(z) = \int_{\mathbb{T}} \frac{\zeta + z}{\zeta - z} \, d\mu(\zeta), \quad z \in \mathbb{D}, \tag{11.31}$$

is a Carathéodory function. Each Carathéodory function is of the form F_μ and the probability measure μ is uniquely determined by the function F_μ.

It is easily verified that F_μ is a Carathéodory function. Indeed, F_μ is holomorphic on $\mathbb{D}$, $F_\mu(0) = \mu(\mathbb{T}) = 1$ and $\operatorname{Re} F_\mu(z) > 0$ on $\mathbb{D}$, since

$$\operatorname{Re} \frac{\zeta + z}{\zeta - z} = \frac{1 - |z|^2}{|\zeta - z|^2} > 0, \quad z \in \mathbb{D}, \ \zeta \in \mathbb{T}.$$

That each Carathéodory function of the form (11.31) is proved (for instance) in [Dn, Theorem III on p. 21].

Using the representation (11.31) from Proposition 11.22 it is easy to relate the Taylor coefficients of Carathéodory functions to moment sequences.

Proposition 11.23 *Let F be a holomorphic function on $\mathbb{D}$ with Taylor expansion*

$$F(z) = 1 + 2 \sum_{n=1}^{\infty} c_n z^n. \tag{11.32}$$

Then F is a Carathéodory function if and only if there is a probability measure μ on $\mathbb{T}$ such that

$$c_n = s_n(\mu) \equiv \int_{\mathbb{T}} \zeta^{-n} d\mu(\zeta) \quad \text{for } n \in \mathbb{N}.$$

Proof Let F be a Carathéodory function. By Proposition 11.22, F is of the form (11.31). For $z \in \mathbb{D}$ and $\zeta \in \mathbb{T}$ we have the expansion

$$\frac{\zeta + z}{\zeta - z} = 1 + 2 \sum_{n=1}^{\infty} \zeta^{-n} z^n \tag{11.33}$$

which converges uniformly on $\mathbb{T}$. Hence, since $\mu(\mathbb{T}) = 1$, integrating over $\mathbb{T}$ gives

$$F(z) = 1 + 2 \sum_{n=0}^{\infty} s_n(\mu) z^n, \quad z \in \mathbb{D}.$$

Comparing the coefficients of z^n yields $c_n = s_n(\mu)$ for $n \in \mathbb{N}$.

Conversely, suppose that $c_n = s_n(\mu)$, $n \in \mathbb{N}$, for some probability measure μ. Then, using (11.32) and the uniformly converging expansion (11.33) on $\mathbb{T}$ we obtain

$$F(z) = 1 + \sum_{n=0}^{\infty} 2 \int_{\mathbb{T}} \zeta^{-n} d\mu(\zeta) z^n = \int_{\mathbb{T}} \left(1 + 2 \sum_{n=1}^{\infty} \zeta^{-n} z^n \right) d\mu(\zeta)$$

$$= \int_{\mathbb{T}} \frac{\zeta + z}{\zeta - z} d\mu(\zeta),$$

that is, $F = F_\mu$. Hence F is a Carathéodory function. $\square$

The following simple fact will be used several times.

Lemma 11.24 *If f is a Schur function such that $f(0) = 0$, then $\frac{f(z)}{z}$ is also a Schur function.*

Proof Since $f(0) = 0$, Schwarz' lemma applies and shows that $|f(z)| \leq |z|$ for $z \in \mathbb{D}$. Hence $|\frac{f(z)}{z}| \leq 1$ on $\mathbb{D}$, so that $\frac{f(z)}{z}$ is a Schur function. □

Next we note that the map

$$f \mapsto F := \frac{1 + zf(z)}{1 - zf(z)} \tag{11.34}$$

is a *bijection of the Schur functions f onto the Carathéodory functions F* with inverse

$$F \mapsto f = \frac{1}{z} \frac{F(z) - 1}{F(z) + 1}. \tag{11.35}$$

Indeed, since $w \mapsto v = \frac{1+w}{1-w}$ is a holomorphic bijection of the open unit disc $\mathbb{D}$ on the open half plane $\operatorname{Re} v > 0$, the function F in (11.34) is a Carathéodory function if f is a Schur function. The inverse $v \mapsto w = \frac{v-1}{v+1}$ maps the half plane $\operatorname{Re} v > 0$ holomorphically onto $\mathbb{D}$. Hence, if F is a Carathéodory function, the function $zf(z)$ defined by (11.35) is a Schur function and so is f by Lemma 11.24.

If F_μ is the Carathéodory function of a probability measure μ, we denote by

$$f_\mu := \frac{1}{z} \frac{F_\mu(z) - 1}{F_\mu(z) + 1} \tag{11.36}$$

the corresponding Schur function. By the preceding we have developed one-to-one correspondences between probability measures on $\mathbb{T}$, Carathéodory functions and Schur functions.

Recall that for any $\gamma \in \mathbb{D}$ the Möbius transformation

$$M_\gamma(w) := \frac{w - \gamma}{1 - \overline{\gamma}z}, \quad w \in \mathbb{D},$$

is a holomorphic bijection of $\mathbb{D}$ and a bijection of $\mathbb{T}$. Hence, finite Blaschke products

$$f(z) = e^{i\varphi} \prod_{j=1}^{n} \frac{z - \lambda_j}{1 - \overline{\lambda}_j z} \tag{11.37}$$

of order n, where $\varphi \in \mathbb{R}$ and $\lambda_1, \ldots, \lambda_n \in \mathbb{D}$, are Schur functions. Constant functions of modulus one are interpreted as Blaschke products of order 0.

Lemma 11.25 *For a probabilitiy measure μ on $\mathbb{T}$ the following are equivalent:*

(i) *μ has finite support.*
(ii) *F_μ is a rational function with all its poles in $\mathbb{T}$.*
(iii) *f_μ is a finite Blaschke product.*

Proof Clearly, the points of supp μ are the singularities of F_μ. This fact implies the equivalence of (i) and (ii).

From (11.34) and (11.35) it follows that f_μ is rational if and only if F_μ is rational; in this case f_μ has boundary values of modulus one on $\mathbb{T}$. Hence (ii) holds if and only if f_μ is an rational inner function. It is well-known (see e.g. [Y, p. 208]) that the rational inner functions are precisely the Blaschke products of finite order.   □

Let S denote the set of Schur functions and S_f the finite Blaschke products.

Now we begin to develop the *Schur algorithm*. Let f be a fixed Schur function. We define inductively $f_0(z) := f(z)$ and

$$\gamma_n := f_n(0), \quad f_{n+1}(z) := z^{-1} M_{\gamma_n}(f_n(z)) = \frac{f_n(z) - \gamma_n}{z(1 - \overline{\gamma}_n f_n(z))}, \quad n \in \mathbb{N}_0. \tag{11.38}$$

Conversely, from (11.38) we obtain

$$f_n(z) = M_{-\gamma_n}(z f_{n+1}) = \frac{\gamma_n + z f_{n+1}(z)}{1 + \overline{\gamma}_n z f_{n+1}(z)} = \gamma_n + (1 - |\gamma_n|^2) \frac{z f_{n+1}(z)}{1 + \overline{\gamma}_n z f_{n+1}(z)}.$$

Suppose that $\gamma_n \in \mathbb{D}$ and $f_n \in S$. Since M_{γ_n} is a holomorphic bijection of $\mathbb{D}$, $M_{\gamma_n}(f_n) \in S$ and so $f_{n+1} = z^{-1} M_{\gamma_n}(f_n) \in S$ by Lemma 11.24. By induction this proves that the functions $f_n(z)$ are Schur functions if $|\gamma_k| < 1$ for $k \leq n$.

Assume in the above algorithm that $\gamma_k \in \mathbb{D}$ for $k = 0, \ldots, n-1$ and $\gamma_n = f_n(0)$ is not in $\mathbb{D}$. Then $\gamma_n \in \partial\mathbb{D} = \mathbb{T}$, the algorithm terminates, and $|\gamma_k| = 1$ for $k \geq n$. It is easy to verify that this happens if and only if $f \in S_f$.

Definition 11.26 The numbers $\gamma_n = \gamma_n(f)$, $n \in \mathbb{N}_0$, are called the *Schur parameters* of the Schur function f.   .

Thus, if $f \in S$ and $f \notin S_f$, the Schur algorithm yields a sequence $(f_n)_{n \in \mathbb{N}_0}$ of Schur functions $f_n \in S \setminus S_f$ and a sequence $(\gamma_n)_{n \in \mathbb{N}_0}$ of Schur parameters $\gamma_n \in \mathbb{D}$.

We shall write $\gamma_n(\mu)$ for the Schur parameters of the Schur function f_μ given by (11.36). If the probability measure μ has infinite support, then $f_\mu \notin S_f$ by Lemma 11.25 and hence all Schur parameters $\gamma_n(\mu)$, $n \in \mathbb{N}_0$, are in $\mathbb{D}$.

For a Schur function f we denote by $a_n(f)$ its n-th Taylor coefficient, that is,

$$f(z) = \sum_{n=0}^{\infty} a_n(f) z^n.$$

It can be shown [Su] that the Schur parameter $\gamma_n(f)$ is a function of the Taylor coeffients $a_0(f), \ldots, a_n(f)$ and that the Taylor coefficient $a_k(f)$ is a function of the Schur parameters $\gamma_0(f), \ldots, \gamma_k(f)$. We shall use only the following results.

Proposition 11.27 *For $n \in \mathbb{N}$ there exists a real polynomial φ_n of $2n$ variables such that for any Schur function f we have*

$$a_n(f) = \gamma_n \prod_{j=0}^{n-1}(1 - |\gamma_j|^2) + \varphi_n(\gamma_0, \overline{\gamma}_0, \ldots, \gamma_{n-1}, \overline{\gamma}_{n-1}). \tag{11.39}$$

Proof We prove by induction on n that for all $k \in \mathbb{N}_0$ and $n \in \mathbb{N}$ there exists a polynomial $\varphi_{n,k}$ in γ_j and $\overline{\gamma}_j$ for $j = 0, \ldots, k + n - 1$ such that

$$a_n(f_k) = \gamma_{n+k} \prod_{j=k}^{n+k-1} (1 - |\gamma_j|^2) + \varphi_{n,k}(\gamma_0, \overline{\gamma}_0, \ldots, \gamma_{n+k-1}, \overline{\gamma}_{n+k-1}). \qquad (11.40)$$

Since $f = f_0$, the case $k = 0$ gives the assertion.

We compare the coefficient of z in the equation $f_k + \overline{\gamma}_k z f_k f_{k+1} = \gamma_k + z f_{k+1}$ by using that $f_j(0) = \gamma_j$. This yields $a_1(f_k) + \overline{\gamma}_k \gamma_k \gamma_{k+1} = \gamma_{k+1}$. Hence we have $a_1(f_k) = \gamma_{k+1}(1 - |\gamma_k|^2)$. This is the assertion (11.40) for $n = 1$ and $k \in \mathbb{N}_0$.

Assume that (11.40) holds for $n - 1$. Comparing the coefficient of z^n in the identity

$$f_k(z) = \gamma_k + z f_{k+1}(z) - \overline{\gamma}_k z f_{k+1}(z) f_k(z)$$

by using the definition $a_0(f_k) = f_k(0) = \gamma_k$ we derive

$$a_n(f_k) = a_{n-1}(f_{k+1}) - \overline{\gamma}_k a_{n-1}(f_{k+1} f_k)$$

$$= a_{n-1}(f_{k+1}) - \overline{\gamma}_k \sum_{j=0}^{n-1} a_{n-1-j}(f_{k+1}) a_j(f_k)$$

$$= (1 - |\gamma_k|^2) a_{n-1}(f_{k+1}) - \overline{\gamma}_k \sum_{j=1}^{n-1} a_{n-1-j}(f_{k+1}) a_j(f_k). \qquad (11.41)$$

For the terms $a_{n-1}(f_{k+1})$, $a_{n-1-j}(f_{k+1})$, $a_j(f_k)$ in (11.41) the induction hypothesis (11.40) applies with n replaced by $n-1$. Hence $a_n(f_k)$ is of the required form (11.40). This completes the induction proof. □

Corollary 11.28 *Suppose that f and g are Schur functions such that $\gamma_j(f) = \gamma_j(g)$ for $j = 0, \ldots, n$. Then*

$$|f(z) - g(z)| \le 2|z|^{n+1} \quad \text{for } z \in \mathbb{D}. \qquad (11.42)$$

If $f, g \in \mathcal{S}$ have the same Schur parameters $\gamma_k(f) = \gamma_k(g)$ for $k \in \mathbb{N}_0$, then $f = g$.

Proof First we note that $f(0) = \gamma_0(f) = \gamma_0(g) = g(0)$. By Proposition 11.27, for any $j \in \mathbb{N}$ the j-th Taylor coefficient of a Schur function is a polynomial in its Schur parameters $\gamma_0, \overline{\gamma}_0, \ldots, \gamma_j, \overline{\gamma}_j$. Hence the assumption implies that the first $n+1$ Taylor coefficients of f and g coincide, so the Schur function $h := \frac{1}{2}(f - g)$ has a zero of order $n+1$ at the origin. Therefore, by repeated application of Lemma 11.24 we obtain $|h(z)| \le |z|^{n+1}$ on $\mathbb{D}$ which gives (11.42).

If $\gamma_k(f) = \gamma_k(g)$ for $k \in \mathbb{N}_0$, then (11.42) holds for all $n \in \mathbb{N}$. Passing to the limit $n \to \infty$ yields $f(z) = g(z)$ for $z \in \mathbb{D}$. □

Now we reverse the procedure and start with Schur parameters. For $\gamma \in \mathbb{D}$, set

$$T_{\gamma,z}(w) := \frac{\gamma + zw}{a + \overline{\gamma}\, zw}.$$

Then $T_{\gamma,z}(f(z))$ is also a Schur function for any $f \in \mathcal{S}$. From (11.38) we obtain

$$f_n(z) = T_{\gamma_n,z}(f_{n+1}(z)) = \frac{\gamma_n + z f_{n+1}(z)}{1 + \overline{\gamma}_n\, z f_{n+1}(z)}.$$

Let $(\gamma_n)_{n\in\mathbb{N}_0}$ be an arbitrary sequence of numbers $\gamma_n \in \mathbb{D}$. We develop the *inverse Schur algorithm* and define the n-th *Schur approximant* $f^{[n]}(z)$ by

$$f^{[n]}(z) = T_{\gamma_0,z}(T_{\gamma_1,z}(\dots T_{\gamma_{n-1},z}(\gamma_n))), \quad n \in \mathbb{N}_0.$$

(In fact, this is a kind of continued fraction algorithm.) It is not difficult to verify that $f^{[n]}(z)$ a rational Schur function with Schur parameters given by

$$\gamma_j(f^{[n]}) = \gamma_j \text{ for } j = 0,\dots,n \text{ and } \gamma_j(f^{[n]}) = 0 \text{ for } j > n. \tag{11.43}$$

Therefore, if $n > m$, it follows from by Corollary 11.28 that

$$|f^{[n]}(z) - f^{[m]}(z)| \le 2|z|^{n+1}, \quad z \in \mathbb{D}.$$

Hence $(f^{[n]}(z))_{\in\mathbb{N}_0}$ is a Cauchy sequence for fixed $z \in \mathbb{D}$ which converges uniformly on compact subsets of $\mathbb{D}$ to some holomorphic function $f(z)$ on $\mathbb{D}$. Since $f^{[n]} \in \mathcal{S}$, it is obvious that $f \in \mathcal{S}$. Further, $\gamma_k(f) = \gamma_k(f^{[n]}) = \gamma_k$ for $n \ge k$, that is, the Schur function f has the prescribed Schur parameters $\gamma_k, k \in \mathbb{N}_0$. Since $\gamma_k \in \mathbb{D}$ for all $k \in \mathbb{N}_0$, f is not in $\mathcal{S}_f$. Therefore, by Lemma 11.25, the unique probability measure μ such that $f = f_\mu$ has infinite support.

Now let $g \in \mathcal{S}\backslash\mathcal{S}_f$ be given. Then $\gamma_n := \gamma_n(g) \in \mathbb{D}$ for $n \in \mathbb{N}_0$. If we start the inverse Schur algorithm with this sequence $(\gamma_n)_{n\in\mathbb{N}_0}$, then the corresponding Schur function $f(z)$ has the same Schur parameters as $g(z)$, so it coincides with $g(z)$ by Corollary 11.28. Hence the sequence $(f^{[n]}(z))_{\in\mathbb{N}_0}$ of Schur approximants converges to the Schur function $g(z)$ uniformly on compact subsets of $\mathbb{D}$.

For later reference we state an outcome of the preceding considerations as

Proposition 11.29 *For each sequence $(\gamma_n)_{n\in\mathbb{N}_0}$ of numbers $\gamma_n \in \mathbb{D}$ there exists a unique probability measure μ on $\mathbb{T}$ with infinite support such that*

$$\gamma_k(\mu) \equiv \gamma_k(f_\mu) = \gamma_k \quad \text{for } k \in \mathbb{N}_0.$$

The next proposition is needed in the proof of Theorem 11.31 below.

Proposition 11.30 *Let μ be a probability measure on $\mathbb{T}$ and let γ_j be its Schur parameters. For $n \in \mathbb{N}_0$ there exists a real polynomial ψ in $2n$ variables such that*

$$s_{n+1}(\mu) = \gamma_n \prod_{j=0}^{n-1}(1 - |\gamma_j|^2) + \psi_n(\gamma_0, \overline{\gamma}_0, \dots, \gamma_{n-1}, \overline{\gamma}_{n-1}). \tag{11.44}$$

Proof Recall that F_μ denotes the Carathéodory function and f_μ is the Schur function assiciated with μ. By (11.34) we then have

$$F_\mu(z) = \frac{1 + zf_\mu(z)}{1 - zf_\mu(z)} = 1 + \frac{2zf_\mu(z)}{1 - zf_\mu(z)} = 1 + \sum_{n=1}^{\infty} 2(f_\mu z)^n.$$

This formula implies that the Taylor coefficient $a_{n+1}(F_\mu)$ is a sum of the number $2a_n(f_\mu)$ and a polynomial in the lower coefficients $a_j(f_\mu)$, where $j \le n-1$. On the other hand, $a_{n+1}(F_\mu) = 2s_{n+1}(\mu)$ by Proposition 11.23. Applying formula (11.39) to the Taylor coefficients $a_n(f_\mu)$ and $a_j(f_\mu), j \le n - 1$, we obtain the assertion. $\quad\square$

The main result of this section is the following *Geronimus theorem*.

Theorem 11.31 *For each probability measure μ on $\mathbb{T}$ of infinite support we have*

$$\alpha_n(\mu) = \gamma_n(\mu) \quad for \quad n \in \mathbb{N}_0.$$

Proof The assertion will be proved by induction on n. For $n = 0$ it is easily checked that $s_1(\mu) = \alpha_0(\mu) = \gamma_0(\mu)$.

Now suppose that $\alpha_j(\mu) = \gamma_j(\mu)$ for $j = 0, \dots, n - 1$. We fix these numbers and abbreviate them by α_j. Recall that $\alpha_j \in \mathbb{D}$ for each j.

By Verblunsky's Theorem 11.12, for each number $\zeta \in \mathbb{D}$ there is a probability measure ν with infinite support such that $\alpha_n(\nu) = \zeta$ and $\alpha_j(\nu) = \alpha_j$ for $j = 0, \dots, n - 1$. By (11.29) the corresponding moment $s_{n+1}(\nu)$ is of the form

$$s_{n+1}(\nu) = \alpha_n(\nu) \prod_{j=0}^{n-1}(1 - |\alpha_j|^2) + c_{n+1}, \tag{11.45}$$

where c_{n+1} depends only on the moments $s_0, \dots, s_n$ and so on $\alpha_1, \dots, \alpha_{n-1}$ by formula (11.44). Note that c_{n+1} does not depend on $\alpha_n(\nu) = \zeta$.

On the other hand, by Proposition 11.29, for each $\zeta \in \mathbb{D}$ there exists a probability measure $\tilde{\nu}$ such that $\gamma_n(\tilde{\nu}) = \zeta$ and $\gamma_j(\tilde{\nu}) = \alpha_j$ for $j = 0, \dots, n - 1$. By Proposition 11.30 the corresponding moment $s_{n+1}(\tilde{\nu})$ is given by

$$s_{n+1}(\tilde{\nu}) = \gamma_n(\tilde{\nu}) \prod_{j=0}^{n-1}(1 - |\alpha_j|^2) + \psi_n, \tag{11.46}$$

where ψ_n depends only on $\alpha_0, \dots, \alpha_{n-1}$, but not on $\gamma_n(\tilde{\nu}) = \zeta$.

Formulas (11.45) and (11.46) describe the sets of possible $(n + 1)$-th moments when $\alpha_n(\nu) = \zeta$ and $\gamma_n(\tilde{\nu}) = \zeta$, respectively, run through $\mathbb{D}$. Both sets are open disks with centers c_{n+1} and ψ_n. Since these sets are the same, $c_{n+1} = \psi_n$.

Since $\alpha_j(\mu) = \gamma_j(\mu) = \alpha_j$ for $j = 0, \ldots, n-1$, we can set $\nu = \tilde{\nu} = \mu$ in the preceding formulas. Comparing the moment $s_{n+1}(\mu)$ in (11.45) and (11.46) by using that $c_{n+1} = \psi_n$ we obtain

$$\alpha_n(\mu) \prod_{j=0}^{n-1}(1 - |\alpha_j|^2) = \gamma_n(\mu) \prod_{j=0}^{n-1}(1 - |\alpha_j|^2).$$

Since $|\alpha_j| < 1$, this yields $\alpha_n(\mu) = \gamma_n(\mu)$. This completes the induction proof. □

11.6 Exercises

1. Let $k \in \mathbb{N}$ and $s = (s_j)_{j\in\mathbb{Z}}$, where $s_0 = 1, s_{k-n}=s_{n-k} = \frac{1}{2}, s_j = 0$ otherwise.

   a. Show that s is a moment sequence for $\mathbb{Z}$ by "guessing" the representing measure.
   b. Compute the Toeplitz determinants $H_n(s), n \in \mathbb{N}_0$.

2. Let $g(\theta) = \sum_{l=0}^{n} a_l \cos l\theta$ be a trigonometric "cosine polynomial" such that $g(\theta) \geq 0$ for $\theta \in [-\pi, \pi]$. Show that there exists a polynomial $q(z) = \sum_{j=0}^{n} c_j z^j$ with *real* coefficients c_j such that $g(\theta) = |q(e^{i\theta})|^2$ for $\theta \in [-\pi, \pi]$.
   Hint: Show that if z_j is a nonreal zero with multiplicity k_j of the polynomial $f(z)$ in the proof of Theorem 11.1, then so is its conjugate $\bar{z}_j$.

3. ([AK]) Let $s = (s_j)_{j=0}^{n}$, where $s_0 > 0$ and $n \in \mathbb{N}$, be a *real* sequence. Define $r_k = \frac{1}{2^k} \sum_{j=0}^{k} \binom{k}{j} s_{k-2j}$ for $k = 0, \ldots, n$. Show that the following are equivalent:

   (i)

$$\sum_{j,k=0}^{n} s_{j-k} \, c_k \overline{c_j} \geq 0 \quad \text{for} \quad (c_0, \ldots, c_n)^T \in \mathbb{C}^{n+1}.$$

   (ii) There are numbers $t_j \in [0, \pi]$, $m_j \geq 0$ for $j = 1, \ldots, l, l \leq 1+\frac{n}{2}$, such that

$$s_k = \sum_{j=1}^{l} m_j \cos kt_j, \quad k = 0, \ldots, n.$$

   (iii) $r = (r_k)_{k=0}^{n}$ is a truncated $[-1, 1]$-moment sequence.

   Hint: $r_k = L_s(((z + z^{-1})/2)^k)$.

4. Formulate and prove the counterpart of Exercise 3 for an infinite *real* sequence $s = (s_j)_{j\in\mathbb{N}_0}$.

More details and further results on Exercises 3 and 4 can be found in [AK, Theorems 13–17].

In the following exercises we elaborate on the moment problem for a subarc of $\mathbb{T}$, see [KN, p. 294–295]. Suppose that $-\frac{\pi}{2} \leq \alpha < \beta \leq \frac{\pi}{2}$. Let $\mathbb{T}_{\alpha,\beta}$ denote the subset of numbers e^{it}, where $t \in [2\alpha, 2\beta]$, of $\mathbb{T}$. We abbreviate $a = \tan\alpha$, $b = \tan\beta$, $c = \cos\alpha$, $d = \cos\beta$ and define a Laurent polynomial $\varphi_{\alpha,\beta}$ by

$$\varphi_{\alpha,\beta}(z) = -2\cos(\alpha - \beta) + e^{(\alpha+\beta)i}z^{-1} + e^{-(\alpha+\beta)i}z.$$

5. Show that $\varphi_{\alpha,\beta}(z) \geq 0$ for $z \in \mathbb{T}_{\alpha,\beta}$ and $\varphi_{\alpha,\beta}(z) < 0$ for $z \in \mathbb{T}\backslash\mathbb{T}_{\alpha,\beta}$.
   Hint: Verify that $\varphi_{\alpha,\beta}(e^{it}) = 4\sin(\beta - \frac{t}{2})\sin(\frac{t}{2} - \alpha)$.

6. Show that if $p \in \mathbb{C}[z, z^{-1}]$ is nonnegative on $\mathbb{T}_{\alpha,\beta}$, then there exist polynomials $q_1, q_2 \in \mathbb{C}[z]$ such that $p(z) = |q_1(z)|^2 + \varphi_{\alpha,\beta}(z)|q_2(z)|^2$ for $z \in \mathbb{T}$.
   Hints: First suppose $-\frac{\pi}{2} < \alpha < \beta < \frac{\pi}{2}$. Set $z = (1-ix)(1+ix)^{-1}$ and verify that $\varphi_{\alpha,\beta}(z) = 4(cd)^{-1}(b - x)(x - a)(1 + x^2)^{-1}$. Show that there exists a $q \in \mathbb{R}[x]$ and $n \in \mathbb{N}$ such that $p(z) = q(x)(1 + x^2)^n$ and $q(x) \geq 0$ for $x \in [a, b]$.
   Now let $-\frac{\pi}{2} < \alpha < \beta = \frac{\pi}{2}$. Set $z = (1 + i(x - a))(1 - i(x - a))^{-1}$. Show that $\varphi_{\alpha,\beta}(z) = 4c^{-1}x(1 + (x - a)^2)^{-1}$ and there are $q \in \mathbb{R}[x]$ and $n \in \mathbb{N}$ such that $p(z) = q(x)(1 + (x - a)^2)^n$ and $q(x) \geq 0$ for $x \in \mathbb{R}_+$.
   Apply Corollaries 3.24 and 3.25, respectively, to q.

7. (*Moment problem on a circular arc*)
   Let $s = (s_n)_{n\in\mathbb{N}_0}$ be a complex sequence and define a sequence $\hat{s} := (\hat{s}_n)_{n\in\mathbb{N}_0}$ by

$$\hat{s}_n := e^{(\alpha+\beta)i}s_{n+1} - 2\cos(\alpha - \beta)s_n + e^{-(\alpha+\beta)i}s_{n-1}, \quad n \in \mathbb{Z}.$$

   (Recall that $s_{-n} := \bar{s}_n$ for $n < 0$.) Verify that $L_s(\varphi_{\alpha,\beta}f) = L_{\hat{s}}(f)$ for $f \in \mathbb{C}[z, z^{-1}]$.
   Prove that s is a $\mathbb{T}_{\alpha,\beta}$- moment sequence for $\mathbb{Z}$, that is, there is a Radon measure μ on $\mathbb{T}$ supported on $\mathbb{T}_{\alpha,\beta}$ such that $s_n = \int_{\mathbb{T}} z^{-n}\,d\mu$ for $n \in \mathbb{Z}$, if and only if the two infinite Toeplitz matrices $H(s)$ and $H(\hat{s})$ are positive semidefinite.

11.7 Notes

The Fejér–Riesz Theorem 11.1 was proved in [Fj] and [Rz1]. The solution of the trigonometric moment problem in the present form is due to O. Toeplitz [To]. The Szegő recurrence relations (11.15) and (11.16) were obtained in Szegő [Sz], while Verblunsky's Theorem 11.12 was proved in [Ver]. The circle C_s and the results on the truncated trigonometric moment problem are due to N.I. Akhiezer and M.G. Krein [AK].

Carathéodory functions were first studied in [Ca1], [Ca2]. The Herglotz representation (Proposition 11.22) was obtained in [Hz]. Schur functions and Schur's algorithm were invented in I. Schur's two pioneering papers [Su], while Geronimus' Theorem 11.31 was proved in [Gs]. B. Simon's book [Sim2] has several proofs of Verblunsky's and Geronimus' theorems; the proof of Geronimus' theorem given in the text is taken from [Sim2]. Concerning Schur analysis, a collection of classical papers is [FK], a very readable discussion with many historical comments is [DK], and a deeper analysis is given in [Kv].

Part III
The Multidimensional Moment Problem

Chapter 12
The Moment Problem on Compact Semi-Algebraic Sets

In this chapter we begin the study of the multidimensional moment problem. The passage to dimensions $d \geq 2$ brings new difficulties and unexpected phenomena. In Sect. 3.2 we derived solvability criteria of the moment problem on intervals in terms of positivity conditions. It seems to be natural to look for similar characterizations in higher dimensions as well. This leads us immediately into the realm of real algebraic geometry and to descriptions of positive polynomials on semi-algebraic sets. In this chapter we treat this approach for basic closed *compact* semi-algebraic subsets of $\mathbb{R}^d$. It turns out that for such sets there is a close interaction between the moment problem and Positivstellensätze for strictly positive polynomials.

All basic notions and facts from real algebraic geometry that are needed for our treatment of the moment problem are collected in Sect. 12.1. Section 12.2 contains general facts on localizing functionals and supports of representing measures. The main existence results for the moment problem (Theorems 12.25, 12.36(ii), and 12.45) and the corresponding Positivstellensätze (Theorems 12.24, 12.36(i), and 12.44) for compact semi-algebraic sets are derived in Sects. 12.3, 12.4, and 12.6. The results in Sects. 12.3 and 12.4 are formulated in the language of preorderings and quadratic modules, that is, in terms of weighted sums of squares. In Sect. 12.6 we use another type of positivity condition which is based on the notion of a semiring.

In Sect. 12.4 we develop a fundamental technical result, the representation theorem for Archimedean quadratic modules and semirings (Theorem 12.35). In Sects. 12.6 and 12.7, the main theorems are applied to derive a number of classical results on the moment problem for concrete compact sets.

Apart from real algebraic geometry the theory of self-adjoint Hilbert space operators is our main tool for the multidimensional moment problem. In Sect. 12.5 we develop this method by studying the GNS construction and the relations to the multidimensional spectral theorem. This approach yields a very short and elegant proof of the moment problem result for Archimedean quadratic modules.

© Springer International Publishing AG 2017
K. Schmüdgen, *The Moment Problem*, Graduate Texts in Mathematics 277,
DOI 10.1007/978-3-319-64546-9_12

Throughout this chapter, A denotes a **commutative real algebra with unit element** denoted by 1. For notational simplicity we write λ for $\lambda \cdot 1$, where $\lambda \in \mathbb{R}$. Recall that $\sum A^2$ is the set of finite sums $\sum_i a_i^2$ of squares of elements $a_i \in A$.

12.1 Semi-Algebraic Sets and Positivstellensätze

The following definition collects some basic notions needed in the sequel.

Definition 12.1 A *quadratic module* of A is a subset Q of A such that

$$Q + Q \subseteq Q, \quad 1 \in Q, \quad a^2 Q \in Q \text{ for all } a \in A. \tag{12.1}$$

A quadratic module T is called a *preordering* if $T \cdot T \subseteq T$.

A *semiring* is a subset S of A satisfying

$$S + S \subseteq S, \quad S \cdot S \subseteq S, \quad \lambda \in S \text{ for all } \lambda \in \mathbb{R}, \lambda \geq 0. \tag{12.2}$$

A *cone* is a subset C of A such that $C + C \subseteq C$ and $\lambda \cdot C \subseteq C$ for $\lambda \geq 0$.

In the literature "semirings" are also called "preprimes". The name "quadratic module" stems from the last condition in (12.1) which means that Q is invariant under multiplication by squares. Setting $a = \sqrt{\lambda}$, this implies that $\lambda \cdot Q \subseteq Q$ for $\lambda \geq 0$. Hence quadratic modules are cones. While semirings and preorderings are closed under multiplication, quadratic modules are not necessarily. Semirings do not contain all squares in general. Clearly, a quadratic module is a preordering if and only if it is a semiring. In this book, we work mainly with quadratic modules and preorderings. Semirings will occur only in Theorems 12.35, 12.44, and 12.45 below.

Example 12.2 The subset $S = \{\sum_{j=0}^{n} a_j x^j : a_j \geq 0, n \in \mathbb{N}\}$ of $\mathbb{R}[x]$ is a semiring, but not a quadratic module. Clearly, $Q = \sum \mathbb{R}_d[\underline{x}]^2 + x_1 \sum \mathbb{R}_d[\underline{x}]^2 + x_2 \sum \mathbb{R}_d[\underline{x}]^2$ is a quadratic module of $\mathbb{R}_d[\underline{x}], d \geq 2$, but Q is neither a semiring nor a preordering. ○

Each cone C of A yields an ordering $\preceq$ on A by defining

$$a \preceq b \quad \text{if and only if} \quad b - a \in C.$$

Obviously, $\sum A^2$ is the smallest quadratic module of A. Since A is commutative, $\sum A^2$ is invariant under multiplication, so it is also the smallest preordering of A.

Our guiding example for A is the polynomial algebra $\mathbb{R}_d[\underline{x}] := \mathbb{R}[x_1, \ldots, x_d]$. Let $f = \{f_1, \ldots, f_k\}$ be a finite subset of $\mathbb{R}_d[\underline{x}]$. The set

$$\mathcal{K}(f) \equiv \mathcal{K}(f_1, \ldots, f_k) = \{x \in \mathbb{R}^d : f_1(x) \geq 0, \ldots, f_k(x) \geq 0\} \tag{12.3}$$

is called the *basic closed semi-algebraic set associated with* f. It is easily seen that

$$Q(\mathfrak{f}) \equiv Q(f_1, \ldots, f_k) = \left\{ \sigma_0 + f_1\sigma_1 + \cdots + f_k\sigma_k : \sigma_0, \ldots, \sigma_k \in \sum \mathbb{R}_d[\underline{x}]^2 \right\}$$
(12.4)

is the *quadratic module generated by the set* f and that

$$T(\mathfrak{f}) \equiv T(f_1, \ldots, f_k) = \left\{ \sum_{e=(e_1, \ldots, e_k) \in \{0,1\}^k} f_1^{e_1} \cdots f_k^{e_k} \sigma_e : \sigma_e \in \sum \mathbb{R}_d[\underline{x}]^2 \right\}$$
(12.5)

is the *preordering generated by the set* f. These three sets $\mathcal{K}(\mathfrak{f})$, $Q(\mathfrak{f})$, and $T(\mathfrak{f})$ play a crucial role in this chapter and the next.

By the above definitions, all polynomials from $T(\mathfrak{f})$ are nonnegative on $\mathcal{K}(\mathfrak{f})$, but in general $T(\mathfrak{f})$ does not exhaust the nonnegative polynomials on $\mathcal{K}(\mathfrak{f})$.

The following *Positivstellensatz of Krivine–Stengle* is a fundamental result of real algebraic geometry. It describes nonnegative resp. positive polynomials on $\mathcal{K}(\mathfrak{f})$ in terms of *quotients* of elements of the preordering $T(\mathfrak{f})$.

Theorem 12.3 *Let $\mathcal{K}(\mathfrak{f})$ and $T(\mathfrak{f})$ be as above and let $g \in \mathbb{R}_d[\underline{x}]$. Then we have:*

(i) *(Positivstellensatz)* $g(x) > 0$ *for all* $x \in \mathcal{K}(\mathfrak{f})$ *if and only if there exist polynomials $p, q \in T(\mathfrak{f})$ such that $pg = 1 + q$.*

(ii) *(Nichtnegativstellensatz)* $g(x) \geq 0$ *for all* $x \in \mathcal{K}(\mathfrak{f})$ *if and only if there exist $p, q \in T(\mathfrak{f})$ and $m \in \mathbb{N}$ such that $pg = g^{2m} + q$.*

(iii) *(Nullstellensatz)* $g(x) = 0$ *for* $x \in \mathcal{K}(\mathfrak{f})$ *if and only if $-g^{2n} \in T(\mathfrak{f})$ for some $n \in \mathbb{N}$.*

(iv) $\mathcal{K}(\mathfrak{f})$ *is empty if and only if -1 belongs to $T(\mathfrak{f})$.*

Proof See [PD] or [Ms1]. The orginal papers are [Kv1] and [Ste1].   □

All *"if"* assertions are easily checked and it is not difficult to show that all four statements are equivalent, see e.g. [Ms1]. Standard proofs of Theorem 12.3 as given in [PD] or [Ms1] are based on the Tarski–Seidenberg transfer principle. Assertion (i) of Theorem 12.3 will play an essential role in the proof of Proposition 12.22 below.

Now we turn to algebraic sets. For a subset S of $\mathbb{R}_d[\underline{x}]$, the real zero set of S is

$$\mathcal{Z}(S) = \{x \in \mathbb{R}^d : f(x) = 0 \quad \text{for all } f \in S\}.$$
(12.6)

A subset V of $\mathbb{R}^d$ of the form $\mathcal{Z}(S)$ is called a *real algebraic set*.

Hilbert's basis theorem [CLO, p. 75] implies that each real algebraic set is of the form $\mathcal{Z}(S)$ for some *finite* set $S = \{h_1, \ldots, h_m\}$. In particular, each real algebraic set is a basic closed semi-algebraic set, because $\mathcal{K}(h_1, \ldots, h_m, -h_1, \ldots, -h_m) = \mathcal{Z}(S)$.

Let S be a subset of $\mathbb{R}_d[\underline{x}]$ and $V := \mathcal{Z}(S)$ the corresponding real algebraic set. We denote by $\mathcal{I}$ the ideal of $\mathbb{R}_d[\underline{x}]$ generated by S and by $\hat{\mathcal{I}}$ the ideal of $f \in \mathbb{R}_d[\underline{x}]$

which vanish on V. Clearly, $\mathcal{Z}(S) = \mathcal{Z}(\mathcal{I})$ and $\mathcal{I} \subseteq \hat{\mathcal{I}}$. In general, $\mathcal{I} \neq \hat{\mathcal{I}}$. (For instance, if $d = 2$ and $S = \{x_1^2 + x_2^2\}$, then $V = \{0\}$ and $x_1^2 \in \hat{\mathcal{I}}$, but $x_1^2 \notin \mathcal{I}$.)

It can be shown [BCRo, Theorem 4.1.4] that $\mathcal{I} = \hat{\mathcal{I}}$ if and only if $\sum p_j^2 \in \mathcal{I}$ for finitely many $p_j \in \mathbb{R}_d[\underline{x}]$ implies that $p_j \in \mathcal{I}$ for all j. An ideal that obeys this property is called *real*. In particular, $\hat{\mathcal{I}}$ is real. The ideal $\mathcal{I}$ generated by a single irreducible polynomial $h \in \mathbb{R}_d[\underline{x}]$ is real if and only if h changes its sign on $\mathbb{R}^d$, that is, there are $x_0, x_1 \in \mathbb{R}^d$ such that $h(x_0)h(x_1) < 0$, see [BCRo, Theorem 4.5.1].

The quotient algebra

$$\mathbb{R}[V] := \mathbb{R}_d[\underline{x}]/\hat{\mathcal{I}} \tag{12.7}$$

is called the algebra of *regular functions* on V. Since $\hat{\mathcal{I}}$ is real, it follows that

$$\sum \mathbb{R}[V]^2 \cap \left(-\sum \mathbb{R}[V]^2\right) = \{0\}. \tag{12.8}$$

Example 12.4 Let us assume that the set f is of the form

$$f = \{g_1, \cdots, g_l, h_1, -h_1, \ldots, h_m, -h_m\}.$$

If $g := \{g_1, \ldots, g_l\}$ and $\mathcal{I}$ denotes the ideal of $\mathbb{R}_d[\underline{x}]$ generated by $h_1, \ldots, h_m$, then

$$\mathcal{K}(f) = \mathcal{K}(g) \cap \mathcal{Z}(\mathcal{I}), \quad Q(f) = Q(g) + \mathcal{I}, \quad \text{and} \quad T(f) = T(g) + \mathcal{I}. \tag{12.9}$$

We prove (12.9). The first equality of (12.9) and the inclusions $Q(f) \subseteq Q(g) + \mathcal{I}$ and $T(f) \subseteq T(g) + \mathcal{I}$ are clear from the corresponding definitions. The identity

$$ph_j = \frac{1}{4}[(p+1)^2 h_j + (p-1)^2(-h_j)] \in Q(f), \quad p \in \mathbb{R}_d[\underline{x}],$$

implies that $\mathcal{I} \subseteq Q(f) \subseteq T(f)$. Hence $Q(g) + \mathcal{I} \subseteq Q(f)$ and $T(g) + \mathcal{I} \subseteq T(f)$.   ∘

Another important concept is introduced in the following definition.

Definition 12.5 Let Q be a quadratic module or a semiring of A. Define

$$\mathsf{A}_b(Q) := \{a \in \mathsf{A} : \text{there exists a } \lambda > 0 \ \text{ such that } \lambda - a \in Q \text{and } \lambda + a \in Q\}.$$

We shall say that Q is *Archimedean* if $\mathsf{A}_b(Q) = \mathsf{A}$, or equivalently, for every $a \in \mathsf{A}$ there exists a $\lambda > 0$ such that $\lambda - a \in \mathsf{A}$.

Lemma 12.6 *Let Q be a quadratic module of A and let $a \in \mathsf{A}$. Then $a \in \mathsf{A}_b(Q)$ if and only if $\lambda^2 - a^2 \in Q$ for some $\lambda > 0$.*

Proof If $\lambda \pm a \in Q$ for $\lambda > 0$, then

$$\lambda^2 - a^2 = \frac{1}{2\lambda}[(\lambda + a)^2(\lambda - a) + (\lambda - a)^2(\lambda + a)] \in Q.$$

Conversely, if $\lambda^2 - a^2 \in Q$ and $\lambda > 0$, then

$$\lambda \pm a = \frac{1}{2\lambda} \left[(\lambda^2 - a^2) + (\lambda \pm a)^2 \right] \in Q. \qquad \square$$

Lemma 12.7 *Suppose that Q is a quadratic module or a semiring of* A.

(i) $A_b(Q)$ *is a unital subalgebra of* A.
(ii) *If the algebra* A *is generated by elements $a_1, \dots, a_n$, then Q is Archimedean if and only if each a_i there exists a $\lambda_i > 0$ such that $\lambda_i \pm a_i \in Q$.*

Proof

(i) Clearly, sums and scalar multiples of elements of $A_b(Q)$ are again in $A_b(Q)$. It suffices to verify that this holds for the product of elements $a, b \in A_b(Q)$.

First we suppose that Q is a quadratic module. By Lemma 12.6, there are $\lambda_1 > 0$ and $\lambda_2 > 0$ such that $\lambda_1^2 - a^2$ and $\lambda_2^2 - b^2$ are in Q. Then

$$(\lambda_1 \lambda_2)^2 - (ab)^2 = \lambda_2^2(\lambda_1^2 - a^2) + a^2(\lambda_2^2 - b^2) \in Q,$$

so that $ab \in A_b(Q)$ again by Lemma 12.6.

Now let Q be a semiring. If $\lambda_1 - a \in Q$ and $\lambda_2 - b \in Q$, then

$$\lambda_1 \lambda_2 \mp ab = \frac{1}{2} \left((\lambda_1 \pm a)(\lambda_2 - b) + (\lambda_2 \mp a)(\lambda_2 + b) \right) \in Q.$$

(ii) follows at once from (i). $\qquad \square$

By Lemma 12.7(ii), it suffices to check the Archimedean condition $\lambda \pm a \in Q$ for algebra generators. Often this simplifies proving that Q is Archimedean.

Corollary 12.8 *For a quadratic module Q of $\mathbb{R}_d[\underline{x}]$ the following are equivalent:*

(i) *Q is Archimedean.*
(ii) *There exists a number $\lambda > 0$ such that $\lambda - \sum_{k=1}^{d} x_k^2 \in Q$.*
(iii) *For any $k = 1, \dots, d$ there exists a $\lambda_k > 0$ such that $\lambda_k - x_k^2 \in Q$.*

Proof (i)→(ii) is clear by definition. If $\lambda - \sum_{j=1}^{d} x_j^2 \in Q$, then

$$\lambda - x_k^2 = \lambda - \sum_j x_j^2 + \sum_{j \neq k} x_j^2 \in Q.$$

This proves (ii)→(iii). Finally, if (iii) holds, then $x_k \in A_b(Q)$ by Lemma 12.6 and hence $A_b(Q) = A$ by Lemma 12.7(ii). Thus, (iii)→(i). $\qquad \square$

Corollary 12.9 *If the quadratic module $Q(\mathfrak{f})$ of $\mathbb{R}_d[\underline{x}]$ is Archimedean, then the set $\mathcal{K}(\mathfrak{f})$ is compact.*

Proof By the respective definitions, polynomials of $Q(\mathsf{f})$ are nonnegative on $\mathcal{K}(\mathsf{f})$. Since $Q(\mathsf{f})$ is Archimedean, $\lambda - \sum_{k=1}^{d} x_k^2 \in Q(\mathsf{f})$ for some $\lambda > 0$ by Corollary 12.8, so $\mathcal{K}(\mathsf{f})$ is contained in the ball centered at the origin with radius $\sqrt{\lambda}$.   □

The converse of Corollary 12.9 does not hold, as the following example shows. (However, it does hold for the preordering $T(\mathsf{f})$ as shown by Proposition 12.22 below.)

Example 12.10 Let $f_1 = 2x_1 - 1, f_2 = 2x_2 - 1, f_3 = 1 - x_1x_2$. Then the set $\mathcal{K}(\mathsf{f})$ is compact, but $Q(\mathsf{f})$ is not Archimedean (see [PD], p. 146, for a proof).   ○

The following separation result is used several times in the next sections.

Proposition 12.11 *Let Q be an Archimedean quadratic module of* A. *If $a_0 \in$ A and $a_0 \notin Q$, there exists a Q-positive linear functional φ on* A *such that $\varphi(1) = 1$ and $\varphi(a_0) \leq 0$.*

Proof Let $a \in$ A and choose $\lambda > 0$ such that $\lambda \pm a \in Q$. If $0 < \delta \leq \lambda^{-1}$, then $\delta^{-1} \pm a \in Q$ and hence $1 \pm \delta a \in Q$. This shows that 1 is an internal point of Q. Therefore, Eidelheit's separation theorem (Theorem A.27) applies and there exists a Q-positive linear functional $\psi \neq 0$ on A such that $\psi(a_0) \leq 0$. Since $\psi \neq 0$, we have $\psi(1) > 0$. (Indeed, if $\psi(1) = 0$, since ψ is Q-positive, $\lambda \pm a \in Q$ implies $\psi(a) = 0$ for all $a \in$ A and so $\psi = 0$.) Then $\varphi := \psi(1)^{-1}\psi$ has the desired properties.   □

Example 12.12 Let A $= \mathbb{R}_d[\underline{x}]$ and let K be a closed subset of $\mathbb{R}^d$. If Q is the preordering $\mathrm{Pos}(K)$ of nonnegative polynomials on K, then $\mathsf{A}_b(Q)$ is just the set of bounded polynomials on K. Hence Q is Archimedean if and only if K is compact. ○

Recall that $\hat{\mathsf{A}}$ denotes the set of characters of A. For a subset Q of A we define

$$\mathcal{K}(Q) := \{x \in \hat{\mathsf{A}} : f(x) \geq 0 \text{ for all } f \in Q\}. \tag{12.10}$$

Clearly, if Q is the quadratic module $Q(\mathsf{f})$ of A $= \mathbb{R}_d[\underline{x}]$ defined by (12.4), then $\mathcal{K}(Q)$ is just the semi-algebraic set $\mathcal{K}(\mathsf{f})$ given by (12.3).

Let Q be a quadratic module. The set $Q^{\mathrm{sat}} = \mathrm{Pos}(\mathcal{K}(Q))$ of all $f \in$ A which are nonnegative on the set $\mathcal{K}(Q)$ is obviously a preordering of A that contains Q. Then Q is called *saturated* if $Q = Q^{\mathrm{sat}}$, that is, if Q is equal to its *saturation* Q^{sat}.

Real algebraic geometry is treated in the books [BCRo, PD, Ms1]; a recent survey on positivity and sums of squares is given in [Sr3].

12.2   Localizing Functionals and Supports of Representing Measures

Haviland's Theorem 1.12 shows that there is a close link between positive polynomials and the moment problem. However, in order to apply this result reasonable desriptions of positive, or at least of strictly positive, polynomials are needed.

Recall that the moment problem for a functional L on the interval $[a, b]$ is solvable if and only if $L(p^2 + (x-a)(b-x)q^2) \geq 0$ for all $p, q \in \mathbb{R}[x]$. This condition means that two infinite Hankel matrices are positive semidefinite and this holds if and only if all principal minors of these matrices are nonnegative. In the multidimensional case we are trying to find similar solvability criteria. It is natural to consider sets that are defined by finitely many polynomial inequalities $f_1(x) \geq 0, \ldots, f_k(x) \geq 0$. These are precisely the basic closed semi-algebraic sets $\mathcal{K}(\mathfrak{f})$, so we have entered the setup of real algebraic geometry.

Let us fix a semi-algebraic set $\mathcal{K}(\mathfrak{f})$. Let L be a $\mathcal{K}(\mathfrak{f})$-moment functional, that is, L is of the form $L(p) = L^\mu(p) \equiv \int p \, d\mu$ for $p \in \mathbb{R}_d[\underline{x}]$, where μ is a Radon measure supported on $\mathcal{K}(\mathfrak{f})$. If $g \in \mathbb{R}_d[x]$ is nonnegative on $\mathcal{K}(\mathfrak{f})$, then obviously

$$L(gp^2) \geq 0 \quad \text{for all} \quad p \in \mathbb{R}_d[\underline{x}], \tag{12.11}$$

so (12.11) is a *neccesary* condition for L being a $\mathcal{K}(\mathfrak{f})$-moment functional.

The overall strategy in this chapter and the next is to solve the $\mathcal{K}(\mathfrak{f})$-moment problem by *finitely many sufficient* conditions of the form (12.11). That is, our aim is to "find" nonnegative polynomials $g_1, \ldots, g_m$ on $\mathcal{K}(\mathfrak{f})$ such that the following holds:

Each linear functional L on $\mathbb{R}_d[\underline{x}]$ which satisfies condition (12.11) for $g = g_1, \ldots, g_m$ and $g = 1$ is a $\mathcal{K}(\mathfrak{f})$-moment functional. (The polynomial $g = 1$ is needed in order to ensure that L itself is a positive functional.)

In general it is not sufficient to take only the polynomials f_j themselves as g_j. For our main results (Theorems 12.25 and 13.10), the positivity of the functional on the preordering $T(\mathfrak{f})$ is assumed. This means that condition (12.11) is required for *all* mixed products $g = f_1^{e_1} \cdots f_k^{e_k}$, where $e_j \in \{0, 1\}$ for $j = 1, \ldots, k$.

Definition 12.13 Let L be a linear functional on $\mathbb{R}_d[\underline{x}]$ and let $g \in \mathbb{R}_d[\underline{x}]$. The linear functional L_g on $\mathbb{R}_d[\underline{x}]$ defined by $L_g(p) = L(gp)$, $p \in \mathbb{R}_d[\underline{x}]$, is called the *localization* of L at g or simply the *localized functional*.

Condition (12.11) means the localized functional L_g is a positive linear functional on $\mathbb{R}_d[\underline{x}]$. Further, if L comes from a measure μ supported on $\mathcal{K}(\mathfrak{f})$ and g is nonnegative on $\mathcal{K}(\mathfrak{f})$, then

$$L_g(p) = L(gp) = \int_{\mathcal{K}(\mathfrak{f})} p(x) \, g(x) d\mu(x), \quad p \in \mathbb{R}_d[\underline{x}],$$

that is, L_g is given by the measure ν on $\mathcal{K}(\mathfrak{f})$ defined by $d\nu = g(x)d\mu$.

Localized functionals will play an important role throughout our treatment. They are used to localize the support of the measure (see Propositions 12.18 and 12.19 and Theorem 14.25) or to derive determinacy criteria (see Theorem 14.12).

Now we introduce two other objects associated with the functional L and the polynomial g. Let $s = (s_\alpha)_{\alpha \in \mathbb{N}_0^d}$ be the d-sequence given by $s_\alpha = L(x^\alpha)$ and write

$g = \sum_\gamma g_\gamma x^\gamma$. Then we define a d-sequence $g(E)s = ((g(E)s)_\alpha)_{\alpha \in \mathbb{N}_0^d}$ by

$$(g(E)s)_\alpha := \sum_\gamma g_\gamma s_{\alpha+\gamma}, \quad \alpha \in \mathbb{N}_0^d,$$

and an infinite matrix $H(gs) = (H(gs)_{\alpha,\beta})_{\alpha,\beta \in \mathbb{N}_0^d}$ over $\mathbb{N}_0^d \times \mathbb{N}_0^d$ with entries

$$H(gs)_{\alpha,\beta} := \sum_\gamma g_\gamma s_{\alpha+\beta+\gamma}, \quad \alpha, \beta \in \mathbb{N}_0^d. \tag{12.12}$$

Using these definitions for $p(x) = \sum_\alpha a_\alpha x^\alpha \in \mathbb{R}_d[\underline{x}]$ we compute

$$L_s(gp^2) = \sum_{\alpha,\beta,\gamma} a_\alpha a_\beta g_\gamma s_{\alpha+\beta+\gamma} = \sum_{\alpha,\beta} a_\alpha a_\beta (g(E)s)_{\alpha+\beta} = \sum_{\alpha,\beta} a_\alpha a_\beta H(gs)_{\alpha,\beta}.$$

$$\tag{12.13}$$

This shows that $g(E)s$ is the d-sequence for the functional L_g and $H(gs)$ is a Hankel matrix for the sequence $g(E)s$. The matrix $H(gs)$ is called the *localized Hankel matrix* of s at g.

Proposition 12.14 *Let $Q(\mathsf{g})$ be the quadratic module generated by the finite subset $\mathsf{g} = \{g_1, \ldots, g_m\}$ of $\mathbb{R}_d[\underline{x}]$. Let L be a linear functional on $\mathbb{R}_d[\underline{x}]$ and $s = (s_\alpha)_{\alpha \in \mathbb{N}_0^d}$ the d-sequence defined by $s_\alpha = L(x^\alpha)$. Then the following are equivalent:*

  (i) *L is a $Q(\mathsf{g})$-positive linear functional on $\mathbb{R}_d[\underline{x}]$.*
  (ii) *$L, L_{g_1}, \ldots, L_{g_m}$ are positive linear functionals on $\mathbb{R}_d[\underline{x}]$.*
  (iii) *$s, g_1(E)s, \ldots, g_m(E)s$ are positive semidefinite d-sequences.*
  (iv) *$H(s), H(g_1 s), \ldots, H(g_m s)$ are positive semidefinite matrices.*

Proof The equivalence of (i) and (ii) is immediate from the definition (12.4) of the quadratic module $Q(\mathsf{g})$ and Definition 12.13 of the localized functionals L_{g_j}.

By Proposition 2.7, a linear functional is positive if and only if the corresponding sequence is positive semidefinite, or equivalently, the Hankel matrix is positive semidefinite. By (12.13) this gives the equivalence of (ii), (iii), and (iv). $\square$

The solvability conditions in the existence theorems for the moment problem in this chapter and the next are given in the form (i) for some finitely generated quadratic module or preordering. This means that condition (12.11) is satisfied for finitely many polynomials g. Proposition 12.14 says there are various *equivalent* formulations of these solvability criteria: They can be expressed in the language of real algebraic geometry (in terms of quadratic modules, semirings or preorderings), of $*$-algebras (as positive functionals on $\mathbb{R}_d[\underline{x}]$), of matrices (by the positive semidefiniteness of Hankel matrices) or of sequences (by the positive semidefiniteness of sequences).

The next proposition contains a useful criterion for localizing supports of representing measures. We denote by $\mathcal{M}_+(\mathbb{R}^d)$ the set of Radon measure μ on $\mathbb{R}^d$ for which all moments are finite, or equivalently, $\int |p(x)| \, d\mu < \infty$ for all $p \in \mathbb{R}_d[\underline{x}]$.

Proposition 12.15 *Let $\mu \in \mathcal{M}_+(\mathbb{R}^d)$ and let s be the moment sequence of μ. Further, let $g_j \in \mathbb{R}_d[\underline{x}]$ and $c_j \geq 0$ be given for $j = 1, \ldots, k$. Set*

$$\mathcal{K} = \{x \in \mathbb{R}^d : |g_j(x)| \leq c_j \text{ for } j = 1, \ldots, k\}. \tag{12.14}$$

Then we have supp $\mu \subseteq \mathcal{K}$ *if and only if there exist constants $M_j > 0$ such that*

$$L_s(g_j^{2n}) \leq M_j c_j^{2n} \text{ for } n \in \mathbb{N}, j = 1, \ldots, k. \tag{12.15}$$

Proof The only if part is obvious. We prove the if direction and slightly modify the argument used in the proof of Proposition 4.1.

Let $t_0 \in \mathbb{R}^d \backslash \mathcal{K}$. Then there is an index $j = 1, \ldots, k$ such that $|g_j(t_0)| > c_j$. Hence there exist a number $\lambda > c_j$ and a ball U around t_0 such that $|g_j(t)| \geq \lambda$ for $t \in U$. For $n \in \mathbb{N}$ we then derive

$$\lambda^{2n} \mu(U) \leq \int_U g_j(t)^{2n} d\mu(t) \leq \int_{\mathbb{R}^d} g_j(t)^{2n} d\mu(t) = L_s(g_j^{2n}) \leq M_j c_j^{2n}.$$

Since $\lambda > c_j$, this is only possible for all $n \in \mathbb{N}$ if $\mu(U) = 0$. Therefore, $t_0 \notin$ supp μ. This proves that supp $\mu \subseteq \mathcal{K}$. □

We state the special case $g_j(x) = x_j$ of Proposition 12.15 separately as

Corollary 12.16 *Suppose that $c_1 > 0, \ldots, c_d > 0$. Let $\mu \in \mathcal{M}_+(\mathbb{R}^d)$ with moment sequence s. Then the measure μ is supported on the d-dimensional interval $[-c_1, c_1] \times \cdots \times [-c_d, c_d]$ if and only if there are positive constants M_j such that*

$$L_s(x_j^{2n}) \equiv s_{(0,\ldots,0,1,0,\ldots,0)}^{2n} \leq M_j c_j^{2n} \text{ for } n \in \mathbb{N}, j = 1, \ldots, d.$$

The following two propositions are basic results about the moment problem on *compact* sets. Both follow from Weierstrass' theorem on approximation of continuous functions by polynomials.

Proposition 12.17 *If $\mu \in \mathcal{M}_+(\mathbb{R}^d)$ is supported on a compact set, then μ is determinate. In particular, if K is a compact subset of $\mathbb{R}^d$, then each K-moment sequence, so each measure $\mu \in \mathcal{M}(\mathbb{R}^d)$ supported on K, is determinate.*

Proof Let $\nu \in \mathcal{M}_+(\mathbb{R}^d)$ be a measure having the same moments and so the same moment functional L as μ. Fix $h \in C_c(\mathbb{R}^d, \mathbb{R})$. We choose a compact d-dimensional interval K containing the supports of μ and h. From Corollary 12.16 it follows that supp $\nu \subseteq K$. By Weierstrass' theorem, there is a sequence $(p_n)_{n \in \mathbb{N}}$ of polynomials $p_n \in \mathbb{R}_d[\underline{x}]$ converging to h uniformly on K. Passing to the limits in the equality

$$\int_K p_n d\mu = L(p_n) = \int_K p_n d\nu$$

we get $\int h d\mu = \int h d\nu$. Since this holds for all $h \in C_c(\mathbb{R}^d, \mathbb{R})$, we have $\mu = \nu$. □

Proposition 12.18 *Suppose that $\mu \in \mathcal{M}_+(\mathbb{R}^d)$ is supported on a compact set. Let $\mathfrak{f} = \{f_1, \ldots, f_k\}$ be a finite subset of $\mathbb{R}_d[\underline{x}]$ and assume that the moment functional defined by $L^\mu(p) = \int p\, d\mu$, $p \in \mathbb{R}_d[\underline{x}]$, is $Q(\mathfrak{f})$-positive. Then $\operatorname{supp}\mu \subseteq \mathcal{K}(\mathfrak{f})$.*

Proof Suppose that $t_0 \in \mathbb{R}^d \setminus \mathcal{K}(\mathfrak{f})$. Then there exist a number $j \in \{1, \ldots, k\}$, a ball U with radius $\rho > 0$ around t_0, and a number $\delta > 0$ such that $f_j \leq -\delta$ on $2U$. We define a continuous function h on $\mathbb{R}^d$ by $h(t) = \sqrt{2\rho - \|t - t_0\|}$ for $\|t - t_0\| \leq 2\rho$ and $h(t) = 0$ otherwise and take a compact d-dimensional interval K containing $2U$ and $\operatorname{supp}\mu$. By Weierstrass' theorem, there is a sequence of polynomials $p_n \in \mathbb{R}_d[\underline{x}]$ converging to h uniformly on K. Then $f_j p_n^2 \to f_j h^2$ uniformly on K and hence

$$\lim_n L^\mu(f_j p_n^2) = \int_K (\lim_n f_j p_n^2)\, d\mu = \int_K f_j h^2\, d\mu = \int_{2U} f_j(t)(2\rho - \|t - t_0\|)\, d\mu(t)$$

$$\leq \int_{2U} -\delta(2\rho - \|t - t_0\|)\, d\mu \leq -\int_U \delta\rho\, d\mu(t) = -\delta\rho\mu(U). \qquad (12.16)$$

Since L^μ is $Q(\mathfrak{f})$-positive, we have $L^\mu(f_j p_n^2) \geq 0$. Therefore, $\mu(U) = 0$ by (12.16), so that $t_0 \notin \operatorname{supp}\mu$. This proves that $\operatorname{supp}\mu \subseteq \mathcal{K}(\mathfrak{f})$. □

The assertions of Propositions 12.17 and 12.18 are no longer valid if the compactness assumptions are omitted. But the counterpart of Proposition 12.18 for zero sets of ideals holds without any compactness assumption.

Proposition 12.19 *Let $\mu \in \mathcal{M}_+(\mathbb{R}^d)$ and let $\mathcal{I}$ be an ideal of $\mathbb{R}_d[\underline{x}]$. If the moment functional L^μ of μ is $\mathcal{I}$-positive, then L^μ annihilates $\mathcal{I}$ and $\operatorname{supp}\mu \subseteq \mathcal{Z}(\mathcal{I})$.*
(As usual, $\mathcal{Z}(\mathcal{I}) = \{x \in \mathbb{R}^d : p(x) = 0 \text{ for } p \in \mathcal{I}\}$ is the zero set of $\mathcal{I}$.)

Proof If $p \in \mathcal{I}$, then $-p \in \mathcal{I}$ and hence $L^\mu(\pm p) \geq 0$ by the $\mathcal{I}$-positivity of L^μ, so that $L^\mu(p) = 0$. That is, L^μ annihilates $\mathcal{I}$.

Let $p \in \mathcal{I}$. Since $p^2 \in \mathcal{I}$, we have $L^\mu(p^2) = \int p^2\, d\mu = 0$. Therefore, from Proposition 1.23 it follows that $\operatorname{supp}\mu \subseteq \mathcal{Z}(p^2) = \mathcal{Z}(p)$. Thus, $\operatorname{supp}\mu \subseteq \mathcal{Z}(\mathcal{I})$. □

For a linear functional L on $\mathbb{R}_d[\underline{x}]$ we define

$$\mathcal{N}_+(L) := \{f \in \operatorname{Pos}(\mathbb{R}^d) : L(p) = 0\}.$$

Proposition 12.20 *Let L be a moment functional on $\mathbb{R}_d[\underline{x}]$, that is, $L = L^\mu$ for some $\mu \in \mathcal{M}_+(\mathbb{R}^d)$. Then the ideal $\mathcal{I}_+(L)$ of $\mathbb{R}_d[\underline{x}]$ generated by $\mathcal{N}_+(L)$ is annihilated by L and the support of each representing measure of L is contained in $\mathcal{Z}(\mathcal{I}_+(L))$.*

Proof Let ν be an arbitrary representing measure of L. If $f \in \mathcal{N}_+(L)$, then we have $L(f) = \int f(x)\, d\nu = 0$. Since $f \in \operatorname{Pos}(\mathbb{R}^d)$, Proposition 1.23 applies and yields $\operatorname{supp}\nu \subseteq \mathcal{Z}(f)$. Hence $\operatorname{supp}\nu \subseteq \mathcal{Z}(\mathcal{N}_+(L)) = \mathcal{Z}(\mathcal{I}_+(L))$. In particular, the inclusion $\operatorname{supp}\nu \subseteq \mathcal{Z}(\mathcal{I}_+(L))$ implies that $L = L^\nu$ annihilates $\mathcal{I}_+(L)$. □

12.3   The Moment Problem on Compact Semi-Algebraic Sets and the Strict Positivstellensatz

The solutions of one-dimensional moment problems have been derived from desriptions of nonnegative polynomials as weighted sums of squares. The counterparts of the latter in the multidimensional case are the so-called "Positivstellensätze" of real algebraic geometry. In general these results require denominators (see Theorem 12.3), so they do not yield reasonable criteria for solving moment problems. However, for *strictly positive* polynomials on *compact* semi-algebraic sets $\mathcal{K}(\mathfrak{f})$ there are *denominator free* Positivstellensätze (Theorems 12.24 and 12.36) which provides solutions of moment problems. Even more, it turns out that there is a close interplay between this type of Positivstellensätze and moment problems on compact semi-algebraic sets, that is, existence results for the moment problem can be derived from Positivstellensätze and vice versa.

We state the main technical steps of the proofs separately as Propositions 12.21–12.23. Proposition 12.23 is also used in a crucial manner in the proof of Theorem 13.10 below.

Suppose that $\mathfrak{f} = \{f_1, \ldots, f_k\}$ is a finite subset of $\mathbb{R}_d[\underline{x}]$. Let $B(\mathcal{K}(\mathfrak{f}))$ denote the algebra of all polynomials of $\mathbb{R}_d[\underline{x}]$ which are bounded on the set $\mathcal{K}(\mathfrak{f})$.

Proposition 12.21 *Let $g \in B(\mathcal{K}(\mathfrak{f}))$ and $\lambda > 0$. If $\lambda^2 > g(x)^2$ for all $x \in \mathcal{K}(\mathfrak{f})$, then there exists a $p \in T(\mathfrak{f})$ such that*

$$g^{2n} \preceq \lambda^{2n+2}p \quad for \ \ n \in \mathbb{N}. \tag{12.17}$$

Proof By the Krivine–Stengle Positivstellensatz (Theorem 12.3(i)), applied to the positive polynomial $\lambda^2 - g^2$ on $\mathcal{K}(\mathfrak{f})$, there exist polynomials $p, q \in T(\mathfrak{f})$ such that

$$p(\lambda^2 - g^2) = 1 + q. \tag{12.18}$$

Since $q \in T(\mathfrak{f})$ and $T(\mathfrak{f})$ is a quadratic module, $g^{2n}(1 + q) \in T(\mathfrak{f})$ for $n \in \mathbb{N}_0$. Therefore, using (12.18) we conclude that

$$g^{2n+2}p = g^{2n}\lambda^2 p - g^{2n}(1 + q) \preceq g^{2n}\lambda^2 p.$$

By induction it follows that

$$g^{2n}p \preceq \lambda^{2n}p. \tag{12.19}$$

Since $g^{2n}(q + pg^2) \in T(\mathfrak{f})$, using first (12.18) and then (12.19) we derive

$$g^{2n} \preceq g^{2n} + g^{2n}(q + pg^2) = g^{2n}(1 + q + pg^2) = g^{2n}\lambda^2 p \preceq \lambda^{2n+2}p. \qquad \square$$

Proposition 12.22 *If the set $\mathcal{K}(f)$ is compact, then the associated preordering $T(f)$ is Archimedean.*

Proof Put $g(x) := (1+x_1^2)\cdots(1+x_d^2)$. Since g is bounded on the compact set $\mathcal{K}(f)$, we have $\lambda^2 > g(x)^2$ on $\mathcal{K}(f)$ for some $\lambda > 0$. Therefore, by Proposition 12.21 there exists a $p \in T(f)$ such that (12.17) holds.

Further, for any multiindex $\alpha \in \mathbb{N}_0^d$, $|\alpha| \leq k$, $k \in \mathbb{N}$, we obtain

$$\pm 2x^\alpha \preceq x^{2\alpha} + 1 \preceq \sum_{|\beta|\leq k} x^{2\beta} = g^k. \tag{12.20}$$

Hence there exist numbers $c > 0$ and $k \in \mathbb{N}$ such that $p \preceq 2cg^k$. Combining the latter with $g^{2n} \preceq \lambda^{2n+2}p$ by (12.17), we get $g^{2k} \preceq \lambda^{2k+2}2cg^k$ and so

$$(g^k - \lambda^{2k+2}c)^2 \preceq (\lambda^{2k+2}c)^2 \cdot 1.$$

Hence, by Lemma 12.6, $g^k - \lambda^{2k+2}c \in \mathsf{A}_b(T(f))$ and so $g^k \in \mathsf{A}_b(T(f))$, where $\mathsf{A} := \mathbb{R}_d[\underline{x}]$. Since $\pm x_j \preceq g^k$ by (12.20) and $g^k \in \mathsf{A}_b(T(f))$, we obtain $x_j \in \mathsf{A}_b(T(f))$ for $j = 1,\cdots,d$. Now from Lemma 12.7(ii) it follows that $\mathsf{A}_b(T(f)) = \mathsf{A}$. This means that $T(f)$ is Archimedean. $\qquad\square$

Proposition 12.23 *Suppose that L is a $T(f)$-positive linear functional on $\mathbb{R}_d[\underline{x}]$.*

(i) *If $g \in B(\mathcal{K}(f))$ and $\|g\|_\infty$ denotes the supremum of g on $\mathcal{K}(f)$, then*

$$|L(g)| \leq L(1)\,\|g\|_\infty. \tag{12.21}$$

(ii) *If $g \in B(\mathcal{K}(f))$ and $g(x) \geq 0$ for $x \in \mathcal{K}(f)$, then $L(g) \geq 0$.*

Proof

(i) Fix $\varepsilon > 0$ and put $\lambda := \| g \|_\infty + \varepsilon$. We define a real sequence $s = (s_n)_{n\in\mathbb{N}_0}$ by $s_n := L(g^n)$. Then $L_s(q(y)) = L(q(g))$ for $q \in \mathbb{R}[y]$. For any $p \in \mathbb{R}[y]$, we have $p(g)^2 \in \sum \mathbb{R}_d[\underline{x}]^2 \subseteq T(f)$ and hence $L_s(p(y)^2) = L(p(g)^2) \geq 0$, since L is $T(f)$-positive. Thus, by Hamburger's theorem 3.8, there exists a Radon measure ν on $\mathbb{R}$ such that $s_n = \int_\mathbb{R} t^n d\nu(t)$, $n \in \mathbb{N}_0$.

For $\gamma > \lambda$ let χ_γ denote the characteristic function of the set $(-\infty, -\gamma] \cup [\gamma, +\infty)$. Since $\lambda^2 - g(x)^2 > 0$ on $\mathcal{K}(f)$, we have $g^{2n} \preceq \lambda^{2n+2}p$ by Eq. (12.17) in Proposition 12.21. Using the $T(f)$-positivity of L we derive

$$\gamma^{2n}\int_\mathbb{R} \chi_\gamma(t)\,d\nu(t) \leq \int_\mathbb{R} t^{2n}d\nu(t) = s_{2n} = L(g^{2n}) \leq \lambda^{2n+2}L(p) \tag{12.22}$$

for all $n \in \mathbb{N}$. Since $\gamma > \lambda$, (12.22) implies that $\int_\mathbb{R} \chi_\gamma(t)\,d\nu(t) = 0$. Therefore, supp $\nu \subseteq [-\lambda, \lambda]$. (The preceding argument has been already used in the proof of Proposition 12.15 to obtain a similar conclusion.) Therefore, applying the

Cauchy–Schwarz inequality for L we derive

$$|L(g)|^2 \leq L(1)L(g^2) = L(1)s_2 = L(1) \int_{-\lambda}^{\lambda} t^2 \, d\nu(t)$$

$$\leq L(1)\nu(\mathbb{R})\lambda^2 = L(1)^2\lambda^2 = L(1)^2(\|g\|_\infty + \varepsilon)^2.$$

Letting $\varepsilon \to +0$, we get $|L(g)| \leq L(1) \, \| \, g \, \|_\infty$.

(ii) Since $g \geq 0$ on $\mathcal{K}(\mathfrak{f})$, we clearly have $\| \, 1 \cdot \|g\|_\infty - 2\,g\|_\infty = \|g\|_\infty$. Using this equality and (12.21) we conclude that

$$L(1)\|g\|_\infty - 2L(g) = L(1 \cdot \|g\|_\infty - 2\,g) \leq L(1)\|1 \cdot \|g\|_\infty - 2\,g\|_\infty = L(1)\|g\|_\infty,$$

which in turn implies that $L(g) \geq 0$.   □

The following theorem is the *strict Positivstellensatz* for compact basic closed semi-algebraic sets $\mathcal{K}(\mathfrak{f})$.

Theorem 12.24 *Let* $\mathfrak{f} = \{f_1, \dots, f_k\}$ *be a finite subset of* $\mathbb{R}_d[\underline{x}]$ *and let* $h \in \mathbb{R}[\underline{x}]$. *If the set* $\mathcal{K}(\mathfrak{f})$ *is compact and* $h(x) > 0$ *for all* $x \in \mathcal{K}(\mathfrak{f})$, *then* $h \in T(\mathfrak{f})$.

Proof Assume to the contrary that h is not in $T(\mathfrak{f})$. By Proposition 12.22, $T(\mathfrak{f})$ is Archimedean. Therefore, by Proposition 12.11, there exists a $T(\mathfrak{f})$-positive linear functional L on A such that $L(1) = 1$ and $L(h) \leq 0$. Since $h > 0$ on the compact set $\mathcal{K}(\mathfrak{f})$, there is a positive number δ such that $h(x) - \delta > 0$ for all $x \in \mathcal{K}(\mathfrak{f})$. We extend the continuous function $\sqrt{h(x) - \delta}$ on $\mathcal{K}(\mathfrak{f})$ to a continuous function on some compact d-dimensional interval containing $\mathcal{K}(\mathfrak{f})$. Again by the classical Weierstrass theorem, $\sqrt{h(x) - \delta}$ is the uniform limit on $\mathcal{K}(\mathfrak{f})$ of a sequence (p_n) of polynomials $p_n \in \mathbb{R}_d[\underline{x}]$. Then $p_n^2 - h + \delta \to 0$ uniformly on $\mathcal{K}(\mathfrak{f})$, that is, $\lim_n \| \, p_n^2 - h + \delta \, \|_\infty = 0$. Recall that $B(\mathcal{K}(\mathfrak{f})) = \mathbb{R}_d[\underline{x}]$, since $\mathcal{K}(\mathfrak{f})$ is compact. Hence $\lim_n L(p_n^2 - h + \delta) = 0$ by the inequality (12.21) in Proposition 12.23(i). But, since $L(p_n^2) \geq 0$, $L(h) \leq 0$, and $L(1) = 1$, we have $L(p_n^2 - h + \delta) \geq \delta > 0$ which is the desired contradiction. This completes the proof of the theorem.   □

The next result gives a solution of the $\mathcal{K}(\mathfrak{f})$-moment problem for compact basic closed semi-algebraic sets.

Theorem 12.25 *Let* $\mathfrak{f} = \{f_1, \dots, f_k\}$ *be a finite subset of* $\mathbb{R}_d[\underline{x}]$. *If the set* $\mathcal{K}(\mathfrak{f})$ *is compact, then each* $T(\mathfrak{f})$-*positive linear functional* L *on* $\mathbb{R}_d[\underline{x}]$ *is a* $\mathcal{K}(\mathfrak{f})$-*moment functional.*

Proof Since $\mathcal{K}(\mathfrak{f})$ is compact, $B(\mathcal{K}(\mathfrak{f})) = \mathbb{R}_d[\underline{x}]$. Therefore, it suffices to combine Proposition 12.23(ii) with Haviland's Theorem 1.12.   □

Remark 12.26 Theorem 12.25 was obtained from Proposition 12.23(ii) and Haviland's Theorem 1.12. Alternatively, it can derived from Proposition 12.23(i) combined with Riesz' representation theorem. Let us sketch this proof. By (12.21), the functional L on $\mathbb{R}_d[\underline{x}]$ is $\| \cdot \|_\infty$- continuous. Extending L to $C(\mathcal{K}(\mathfrak{f}))$ by the Hahn–Banach theorem and applying Riesz' representation theorem for continuous linear functionals, L is given by a signed Radon measure on $\mathcal{K}(\mathfrak{f})$. Setting $g = 1$ in

(12.21), it follows that L, hence the extended functional, has the norm $L(1)$. It is not difficult to show that this implies that the representing measure is positive.   ∘

The shortest path to Theorems 12.24 and 12.25 is probably to use Proposition 12.23 as we have done. However, in order to emphasize the interaction between both theorems and so in fact between the moment problem and real algebraic geometry we now derive each of these theorems from the other.

Proof of Theorem 12.25 (Assuming Theorem 12.24) Let $h \in \mathbb{R}_d[\underline{x}]$. If $h(x) > 0$ on $\mathcal{K}(\mathsf{f})$, then $h \in T(\mathsf{f})$ by Theorem 12.24 and so $L(h) \geq 0$ by the assumption. Therefore L is a $\mathcal{K}(\mathsf{f})$-moment functional by the implication (ii)→(iv) of Haviland's Theorem 1.12.   □

Proof of Theorem 12.24 (Assuming Theorem 12.25 and Proposition 12.22)   Suppose $h \in \mathbb{R}_d[\underline{x}]$ and $h(x) > 0$ on $\mathcal{K}(\mathsf{f})$. Assume to the contrary that $h \notin T(\mathsf{f})$. Since the preordering $T(\mathsf{f})$ is Archimedean by Proposition 12.22, Proposition 12.11 applies, so there is a $T(\mathsf{f})$-positive linear functional L on $\mathbb{R}_d[\underline{x}]$ such that $L(1) = 1$ and $L(h) \leq 0$. By Theorem 12.25, L is a $\mathcal{K}(\mathsf{f})$-moment functional, that is, there is a measure $\mu \in M_+(\mathcal{K}(\mathsf{f}))$ such that $L(p) = \int_{\mathcal{K}(\mathsf{f})} p\, d\mu$ for $p \in \mathbb{R}_d[\underline{x}]$. But $L(1) = \mu(\mathcal{K}(\mathsf{f})) = 1$ and $h > 0$ on $\mathcal{K}(\mathsf{f})$ imply that $L(h) > 0$. This is a contradiction, since $L(h) \leq 0$.   □

The preordering $T(\mathsf{f})$ was defined as the sum of sets $f_1^{e_1} \cdots f_k^{e_k} \cdot \sum \mathbb{R}_d[\underline{x}]^2$. It is natural to ask whether or not all such sets with mixed products $f_1^{e_1} \cdots f_k^{e_k}$ are really needed. To formulate the corresponding result we put $l_k := 2^{k-1}$ and let $g_1, \ldots, g_{l_k}$ denote the first l_k polynomials of the following row of mixed products:

$$f_1, \ldots, f_k, f_1 f_2, f_1 f_3, \ldots, f_1 f_k, \ldots, f_{k-1} f_k, f_1 f_2 f_3, \ldots, f_{k-2} f_{k-1} f_k, \ldots, f_1 f_2 \cdots, f_k.$$

Let $Q(\mathsf{g})$ denote the quadratic module generated by $g_1, \ldots, g_{l_k}$, that is,

$$Q(\mathsf{g}) := \sum \mathbb{R}_d[\underline{x}]^2 + g_1 \sum \mathbb{R}_d[\underline{x}]^2 + \cdots + g_{l_k} \sum \mathbb{R}_d[\underline{x}]^2.$$

The following result of T. Jacobi and A. Prestel [JP] sharpens Theorem 12.24.

Theorem 12.27 *If the set $\mathcal{K}(\mathsf{f})$ is compact and $h \in \mathbb{R}_d[\underline{x}]$ satisfies $h(x) > 0$ for all $x \in \mathcal{K}(\mathsf{f})$, then $h \in Q(\mathsf{g})$.*

We do not prove Theorem 12.27; for a proof of this result we refer to [JP]. If we take Theorem 12.27 for granted and combine it with Haviland's theorem 1.12 we obtain the following corollary.

Corollary 12.28 *If the set $\mathcal{K}(\mathsf{f})$ is compact and L is a $Q(\mathsf{g})$-positive linear functional on $\mathbb{R}_d[\underline{x}]$, then L is a $\mathcal{K}(\mathsf{f})$-moment functional.*

We briefly discuss Theorem 12.27. If $k = 1$, then $Q(\mathsf{f}) = T(\mathsf{f})$. However, for $k = 2$,

$$Q(\mathsf{f}) = \sum \mathbb{R}_d[\underline{x}]^2 + f_1 \sum \mathbb{R}_d[\underline{x}]^2 + f_2 \sum \mathbb{R}_d[\underline{x}]^2,$$

so $Q(\mathfrak{f})$ differs from the preordering $T(\mathfrak{f})$ by the summand $f_1 f_2 \sum \mathbb{R}_d[\underline{x}]^2$. If $k = 3$, then

$$Q(\mathfrak{f}) = \sum \mathbb{R}_d[\underline{x}]^2 + f_1 \sum \mathbb{R}_d[\underline{x}]^2 + f_2 \sum \mathbb{R}_d[\underline{x}]^2 + f_3 \sum \mathbb{R}_d[\underline{x}]^2 + f_1 f_2 \sum \mathbb{R}_d[\underline{x}]^2,$$

that is, the sets $g \sum \mathbb{R}_d[\underline{x}]^2$ with $g = f_1 f_3, f_2 f_3, f_1 f_2 f_3$ do not enter into the definition of $Q(\mathfrak{f})$. For $k = 4$, no products of three or four generators appear in the definition of $Q(\mathfrak{f})$. For large k, only a small portion of mixed products occur in $Q(\mathfrak{f})$ and Theorem 12.27 is an essential strengthening of Theorem 12.24.

The next corollary characterizes in terms of moment functionals when a Radon measure on a compact semi-algebraic set has a *bounded* density with respect to another Radon measure. A version for closed sets is stated in Exercise 14.11 below.

Corollary 12.29 *Suppose that the semi-algebraic set $\mathcal{K}(\mathfrak{f})$ is compact. Let μ and ν be finite Radon measures on $\mathcal{K}(\mathfrak{f})$ and let L^μ and L^ν be the corresponding moment functionals on $\mathbb{R}_d[\underline{x}]$. There exists a function $\varphi \in L^\infty(\mathcal{K}(\mathfrak{f}), \mu)$, $\varphi(x) \geq 0$ μ-a.e. on $\mathcal{K}(\mathfrak{f})$, such that $d\nu = \varphi d\mu$ if and only if there is a constant $c > 0$ such that*

$$L^\nu(g) \leq c L^\mu(g) \quad for \;\; g \in T(\mathfrak{f}). \tag{12.23}$$

Proof Choosing $c \geq \|\varphi\|_{L^\infty(\mathcal{K}(\mathfrak{f}),\mu)}$, the necessity of (12.23) is easily verified.

To prove the converse we assume that (12.23) holds. Then, by (12.23), $L := cL^\mu - L^\nu$ is a $T(\mathfrak{f})$-positive linear functional on $\mathbb{R}_d[\underline{x}]$ and hence a $\mathcal{K}(\mathfrak{f})$-moment functional by Theorem 12.25. Let τ be a representing measure of L, that is, $L = L^\tau$. Then we have $L^\tau + L^\nu = cL^\mu$. Hence both $\tau + \nu$ and $c\mu$ are representing measures of the $\mathcal{K}(\mathfrak{f})$-moment functional cL^μ. Since $\mathcal{K}(\mathfrak{f})$ is compact, $c\mu$ is determinate by Proposition 12.17, so that $\tau + \nu = c\mu$. In particular, this implies that ν is absolutely continuous with respect to μ. Therefore, by the Radon–Nikodym theorem A.3, $d\nu = \varphi d\mu$ for some function $\varphi \in L^1(\mathcal{K}(\mathfrak{f}), \mu)$, $\varphi(x) \geq 0$ μ-a.e. on $\mathcal{K}(\mathfrak{f})$. Since $\tau + \nu = c\mu$, for each Borel subset M of $\mathcal{K}(\mathfrak{f})$ we have

$$\tau(M) = c\mu(M) - \nu(M) = \int_M (c - \varphi(x)) d\mu \geq 0.$$

Therefore, $c - \varphi(x) \geq 0$ μ-a.e., so that $\varphi \in L^\infty(\mathcal{K}(\mathfrak{f}), \mu)$ and $\|\varphi\|_{L^\infty(\mathcal{K}(\mathfrak{f}),\mu)} \leq c$. $\square$

We close this section by restating Theorems 12.24 and 12.25 in the special case of compact real algebraic sets.

Corollary 12.30 *Suppose that $\mathcal{I}$ is an ideal of $\mathbb{R}_d[\underline{x}]$ such that the real algebraic set $V := \mathcal{Z}(\mathcal{I}) = \{x \in \mathbb{R}^d : f(x) = 0 \; for f \in \mathcal{I}\}$ is compact.*

(i) *If $h \in \mathbb{R}_d[\underline{x}]$ satisfies $h(x) > 0$ for all $x \in V$, then $h \in \sum \mathbb{R}_d[\underline{x}]^2 + \mathcal{I}$.*
(ii) *If $p \in \mathbb{R}_d[\underline{x}]/\mathcal{I}$ and $p(x) > 0$ for all $x \in V$, then $p \in \sum(\mathbb{R}_d[\underline{x}]/\mathcal{I})^2$.*
(iii) *If $q \in \mathbb{R}[V] \equiv \mathbb{R}_d[\underline{x}]/\hat{\mathcal{I}}$ and $q(x) > 0$ for all $x \in V$, then $q \in \sum \mathbb{R}[V]^2$.*
(iv) *Each positive linear functional on $\mathbb{R}_d[\underline{x}]$ which annihilates $\mathcal{I}$ is a V-moment functional.*

Proof Put $f_1 = 1, f_2 = h_1, f_3 = -h_1, \ldots, f_{2m} = h_m, f_{2m+1} = -h_m$, where $h_1, \ldots, h_m$ is a set of generators of $\mathcal{I}$. Then, by (12.9), the preordering $T(\mathsf{f})$ is $\sum \mathbb{R}_d[\underline{x}]^2 + \mathcal{I}$ and the semi-algebraic set $\mathcal{K}(\mathsf{f})$ is $V = \mathcal{Z}(\mathcal{I})$. Therefore, Theorem 12.24 yields (i). Since $\mathcal{I} \subseteq \hat{\mathcal{I}}$, (i) implies (ii) and (iii).

Clearly, a linear functional on $\mathbb{R}_d[\underline{x}]$ is $T(\mathsf{f})$-positive if it is positive and annihilates $\mathcal{I}$. Thus (iv) follows at once from Theorem 12.25. □

Example 12.31 (Moment problem on unit spheres) Let S^{d-1} be the unit sphere of $\mathbb{R}^d$. Then S^{d-1} is the real algebraic set $\mathcal{Z}(\mathcal{I})$ for the ideal $\mathcal{I}$ generated by $h_1(x) = x_1^2 + \cdots + x_d^2 - 1$.

Suppose that L is a linear functional on $\mathbb{R}_d[\underline{x}]$ such that

$$L(p^2) \geq 0 \quad \text{and} \quad L((x_1^2 + \cdots + x_d^2 - 1)p) = 0 \quad \text{for} \quad p \in \mathbb{R}_d[\underline{x}].$$

Then it follows from Corollary 12.30(iv) that L is an S^{d-1}-moment functional.

Further, if $q \in \mathbb{R}[S^{d-1}]$ is strictly positive on S^{d-1}, that is, $q(x) > 0$ for $x \in S^{d-1}$, then $q \in \sum \mathbb{R}[S^{d-1}]^2$ by Corollary 12.30(iii). ○

12.4 The Representation Theorem for Archimedean Modules

The main aim of this section is to derive the representation theorem for Archimedean quadratic modules (Theorem 12.35) and its application to the moment problem (Theorem 12.36). Our proof is a combination of functional-analytic and algebraic methods, but we avoid the use of Hilbert space operators! At the end of the next section we give elegant and extremely short proofs of these results based on Hilbert space methods. In view of an application given in Sect. 12.6 we also prove the representation theorem for Archimedean semirings.

Let E be a real vector space and let C be a cone in E, that is, C is a subset of E satisfying $a + b \in C$ and $\lambda a \in C$ for $a, b \in C$ and $\lambda > 0$. The cone yields an ordering " $\preceq$ " on E defined by $x \preceq y$ if and only if $y - x \in C$.

An element $e \in C$ is called an *order unit* if, given $a \in E$, there exists $\lambda > 0$ such that $\lambda e - a \in C$. Since this also holds for $-a$, there is a $\tilde{\lambda} > 0$ such that $\tilde{\lambda} e \pm a \in C$.

Suppose that $e \in C$ is an order unit. For $a \in E$ we define

$$\|a\|_e := \inf\{\lambda > 0 : -\lambda e \preceq a \preceq \lambda e\} \quad \text{and} \quad q(a) = \inf\{\lambda > 0 : a \preceq \lambda e\}.$$

Then $\| \cdot \|_e$ is a seminorm and q is a sublinear functional on E. The latter means that $q(\lambda a) = \lambda q(a)$ and $q(a + b) \leq q(a) + q(b)$ for $a, b \in E$ and $\lambda \geq 0$. Note that q is the Minkowski functional of the convex set $e - C$, that is,

$$q(a) = \inf\{\lambda > 0 : a \in \lambda(e - C)\}.$$

By definition we have $\|a\|_e = \max(q(a), q(-a))$ for $a \in E$.

Now we introduce some terminology adapted from operator algebras. Let C' denote the set of linear functionals φ on E which are C-*positive*, that is, $\varphi(c) \geq 0$ for all $c \in C$. The elements of $C'_e := \{\varphi \in C' : \varphi(e) = 1\}$ are called C-*states* and an extreme point of the convex set C'_e is called a *pure C-state* of E.

First we prove the following sharpening of Proposition 12.11.

Proposition 12.32 *Suppose that $e \in C$ is an order unit. If $a_0 \in E$ and $a_0 \notin C$, then there exists a pure C-state φ such that $\varphi(e - a_0) = q(e - a_0)$ and $\varphi(a_0) \leq 0$.*

Proof Since $a_0 \notin C$, we have $e - a_0 \notin e - C$ and hence $q(e - a_0) \geq 1$ by the definition of the Minkowski functional q of $e - C$. Hence $\varphi(e) = 1$ and $\varphi(e - a_0) = q(e - a_0)$ imply that $\varphi(a_0) \leq 0$.

Let E' denote the real vector space of all linear functionals on E and σ the weak topology $\sigma(E', E)$ on E'. In this proof we essentially use the sets

$$F := \{\varphi \in E' : \varphi(a) \leq q(a) \text{ for } a \in E\}, \quad F_0 := \{\varphi \in F : \varphi(e - a_0) = q(e - a_0)\}.$$

First we verify the inclusions

$$F_0 \subseteq C'_e \subseteq F. \tag{12.24}$$

Let $\varphi \in C'$. If $a \in E$ and $a \preceq \lambda e$, then $\varphi(a) \leq \lambda \varphi(e)$ for $\lambda > 0$ and therefore

$$\varphi(a) \leq q(a)\varphi(e) \tag{12.25}$$

by the definition of q. Hence $\psi \in F$ if $\varphi \in C'_e$. This gives the second inclusion.

Now let $\varphi \in F_0$. Then, for $c \in C$ we have $-c \preceq 0 \preceq \lambda e$ for all $\lambda > 0$, so that $q(-c) = 0$. Therefore, $-\varphi(c) = \varphi(-c) \leq q(-c) = 0$, that is, $\varphi(c) \geq 0$. Thus, $\varphi \in C'$ and the inequality (12.25) applies. By $\varphi \in F_0$ and (12.25), we have $\varphi(e) \leq q(e) \leq 1$ and $q(e - a_0) = \varphi(e - a_0) \leq q(e - a_0)\varphi(e)$. Since $q(e - a_0) \geq 1$ as noted above, we get $\varphi(e) = 1$, that is, $\varphi \in C'_e$. This proves the first inclusion of (12.24).

Next we study the set F_0. Clearly, $\varphi_0(\alpha(e - a_0)) := \alpha q(e - a_0)$, $\alpha \in \mathbb{R}$, defines a linear functional φ_0 on the vector space $E_0 := \mathbb{R} \cdot (e - a_0)$ such that $\varphi_0(b) \leq q(b)$ for $b \in E_0$. By the Hahn–Banach theorem φ_0 extends to a linear functional φ on E satisfying $\varphi(a) \leq q(a)$ for $a \in E$. Since $\varphi(e - a_0) = \varphi_0(e - a_0) = q(e - a_0)$, φ belongs to F_0. Thus F_0 is not empty. Obviously, F_0 is convex.

Now we show that F_0 is σ-compact. Let $U = \{a \in E : \|a\|_e \leq 1\}$. Recall that the polar U° of the set U is defined by $U^\circ = \{\varphi \in E' : |\varphi(a)| \leq 1 \text{ for } a \in U\}$. By the Alaoglu–Bourbaki theorem (see e.g. [Ru1, Ch. 3, Theorem 3.15]), the polar U° is σ-compact. Clearly, $U^\circ = \{\varphi \in E' : |\varphi(x)| \leq \|a\|_e, a \in E\}$. If $\varphi \in F_0$, then $\varphi(a) \leq q(a) \leq \|a\|_e$ and $-\varphi(a) \leq q(-a) \leq \|a\|_e$, so $|\varphi(a)| \leq \|a\|_e$ for $a \in E$, that is, $\varphi \in U^\circ$. Thus $F_0 \subseteq U^\circ$. Since F_0 is obviously σ-closed in U°, F_0 is σ-compact.

Thus we proved that F_0 is a *nonempty σ-compact convex* subset of E'. By the Krein–Milman theorem ([Ru1, Ch. 3, Theorem 3.22]), F_0 has an extreme point, say

φ. We show that φ is an extreme point of the larger set F. For let $\varphi = \frac{1}{2}(\varphi_1 + \varphi_2)$, where $\varphi_1, \varphi_2 \in F$. From $q(e - a_0) = \varphi(e - a_0) = \frac{1}{2}(\varphi_1(e - a_0) + \varphi_2(e - a_0)) \leq q(e - a_0)$ we conclude that $\varphi_j(e - a_0) = q(e - a_0)$, so $\varphi_j \in F_0$ for $j = 1, 2$. Hence $\varphi_1 = \varphi_2 = \varphi$, so φ is an extreme point of F. Therefore, since $\varphi \in F_0 \subseteq C'_e$ and $C'_e \subseteq F$ by (12.24), φ is also an extreme point of C'_e, that is, φ is a pure C-state.   □

Proposition 12.33 *Let Q be an Archimedean quadratic module or an Archimedean semiring of a unital commutative real algebra* A. *Then each pure Q-state φ is a character, that is,*

$$\varphi(ab) = \varphi(a)\varphi(b) \quad for \ a, b \in \mathsf{A}. \tag{12.26}$$

In the proof of this proposition we use the following technical lemma.

Lemma 12.34 *Let Q be an Archimedean quadratic module of* A *and let $\lambda > \rho > 0$. Suppose that a is an element of* A *such that $aQ \subseteq Q$ and $\rho - a \in Q$. If φ is a Q-positive linear functional on* A, *then $\varphi((\lambda - a)c) \geq 0$ for $c \in Q$.*

Proof Without loss of generality we can assume that $\lambda = 1$ and $\varphi(1) = 1$. Consider the Taylor polynomial $p_n(t)$ of the function $\sqrt{1-t}$, that is,

$$p_n(t) = \sum_{k=0}^{n} (-1)^k \binom{1/2}{k} t^k,$$

and the polynomial $q_n(t) := p_n(t)^2 - (1 - t) = \sum_k \gamma_{nk} t^k$. We compute

$$\gamma_{nk} = (-1)^k \sum_{j=k-n}^{n} \binom{1/2}{j}\binom{1/2}{k-j}$$

for $n < k < 2n$ and $\gamma_{nk} = 0$ otherwise. Since the term for the index j in the sum has sign $(-1)^{j-1}(-1)^{k-j-1} = (-1)^k$, we have $\gamma_{nk} \geq 0$ for all n and k.

Since $\rho - a \in Q$ and $aQ \subseteq Q$ by assumption, it follows from the identity

$$\rho^k - a^k = \sum_{j=0}^{k-1} \rho^{k-1-j} a^j (\rho - a)$$

that $\rho^k - a^k \in Q$ and hence $a^k \preceq \rho^k$ for all $k \in \mathbb{N}$.

Let $c \in Q$. We choose $\alpha > 0$ such that $\alpha - c \in Q$. Since $aQ \subseteq Q$ and $q_n(t)$ has only nonnegative coefficients γ_{nk}, we get $q_n(a)Q \subseteq Q$. Therefore, $q_n(a)(\alpha - c) \in Q$ and so $\varphi(q_n(a)(\alpha - c)) \geq 0$. Hence we derive

$$\varphi(q_n(a)c) \leq \alpha\varphi(q_n(a)) = \alpha \sum_{k=0}^{n} \gamma_{nk}\varphi(a^k) \leq \alpha \sum_{k=0}^{n} \gamma_{nk}\rho^k = \alpha q_n(\rho). \tag{12.27}$$

Since Q is a quadratic module and φ is Q-positive, we have $\varphi(p_n(a)^2 c) \geq 0$. Using this fact, the identity $q_n(a) = p_n(a)^2 - (1 - a)$, and (12.27) we obtain

$$\varphi((1 - a)c) = \varphi(p_n(a)^2 c) - \varphi(q_n(a)c) \geq -\varphi(q_n(a)c) \geq -\alpha q_n(\rho). \qquad (12.28)$$

Since $0 < \rho < 1$, we have $p_n(\rho) \to \sqrt{1 - \rho}$ and hence $q_n(\rho) \to 0$. Therefore, letting $n \to \infty$ in (12.28) we get $\varphi((1 - a)c) \geq 0$. $\square$

Proof of Proposition 12.33   First we suppose that Q is an *Archimedean quadratic module*. From the identity $4a = (a+1)^2 - (a-1)^2$, $a \in \mathsf{A}$, it follows that $\mathsf{A} = Q - Q$. Hence it suffices to prove (12.26) for squares $a = c^2$ and $b = d^2$, where $c, d \in \mathsf{A}$. Clearly, $\psi(c^2) \geq 0$, since $c^2 \in Q$.

Case 1: $\varphi(c^2) = 0$. There is a $\lambda > 0$ such that $\lambda - d^2 \in Q$. Since Q is a quadratic module and φ is Q-positive, $c^2(\lambda - d^2) \in Q$, so that $0 \preceq c^2 d^2 \preceq \lambda c^2$ and hence $0 \leq \varphi(c^2 d^2) \leq \lambda\varphi(c^2) = 0$. Thus, $\varphi(c^2 d^2) = 0$ and (12.26) holds for $a = c^2$ and $b = d^2$.

Case 2: $\varphi(c^2) > 0$. Clearly, $\varphi_1(\cdot) = \varphi(c^2)^{-1}\varphi(c^2 \cdot)$ is a Q-positive state. Let us choose $\lambda > 0$ such that $\frac{\lambda}{2} - c^2 \in Q$. Since φ is Q-positive, $\varphi(c^2) \leq \frac{\lambda}{2} < \lambda$ and hence $\varphi(\lambda - c^2) > 0$. Define $\varphi_2(\cdot) := \varphi(\lambda - c^2)^{-1}\varphi((\lambda - c^2) \cdot)$. Since $\frac{\lambda}{2} - c^2 \in Q$, it follows from Lemma 12.34 that the functional $\varphi((\lambda - c^2) \cdot)$ is Q-positive. Hence φ_2 is a Q-state. By construction the pure Q-state φ is the convex combination

$$\varphi = \lambda^{-1}\varphi(c^2)\,\varphi_1 + \lambda^{-1}\varphi(\lambda - c^2)\,\varphi_2$$

of the two Q-states φ_1 and φ_2. Hence $\varphi_1 = \varphi$. This yields $\varphi(c^2 b) = \varphi(c^2)\varphi(b)$ for $b \in \mathsf{A}$ which proves (12.26) when $\varphi(c^2) > 0$.

Now we assume that Q is an *Archimedean semiring*. In this case, Lemma 12.34 is not needed; the proof is very similar and even simpler. For $a \in \mathsf{A}$, there exists a $\lambda > 0$ such that $\lambda + a \in Q$, so that $a = (\lambda + a) - \lambda \in Q - Q$. Thus, $\mathsf{A} = Q - Q$. Hence it suffices to verify (12.26) for $a, b \in Q$. Then $\varphi(a) \geq 0$, since φ is Q-positive.

Case 1: $\varphi(a) = 0$.

Let $b \in Q$ and choose $\lambda > 0$ such that $\lambda - b \in Q$. Then $(\lambda - b)a \in Q$ and $ab \in Q$ (because Q is a semiring!), so that $\varphi((\lambda - b)a) = \lambda\varphi(a) - \varphi(ab) = -\varphi(ab) \geq 0$ and $\varphi(ab) \geq 0$. Hence $\varphi(ab) = 0$, so that (12.26) holds.

Case 2: $\varphi(a) > 0$.

We choose $\lambda > 0$ such that $(\lambda - a) \in Q$ and $\varphi(\lambda - a) > 0$. Since Q is a semiring, $\varphi_1(\cdot) := \varphi(a)^{-1}\varphi(a\cdot)$ and $\varphi_2(\cdot) := \varphi(\lambda - a)^{-1}\varphi((\lambda - a)\cdot)$ are Q-states satisfying $\varphi = \lambda^{-1}\varphi(a)\,\varphi_1 + \lambda^{-1}\varphi(\lambda - a)\,\varphi_2$. The latter is a convex combination of two Q-states. Since φ is a pure Q-state, $\varphi_1 = \varphi$. Hence $\varphi(ab) = \varphi(a)\varphi(b)$ for $b \in \mathsf{A}$. $\square$

The following theorem is the *representation theorem for Archimedean quadratic modules and semirings*. It has been discovered, in various versions, by a number of authors including M.H. Stone, R.V. Kadison, J.-L. Krivine, T. Jacobi, and others.

Theorem 12.35 *Suppose that Q is an Archimedean quadratic module or an Archimedean semiring of a unital commutative real algebra A. Let $a_0 \in \mathsf{A}$. If $\varphi(a_0) > 0$ for all $\varphi \in \mathcal{K}(Q)$, then $a_0 \in Q$.*

Proof Assume to the contrary that $a_0 \notin Q$. Then, by Proposition 12.32, there is a pure Q-state φ such that $\varphi(a_0) \leq 0$. Since φ is a character by Proposition 12.33, $\varphi \in \mathcal{K}(Q)$. Then $\varphi(a_0) \leq 0$ contradicts the assumption. $\square$

For our treatment of the moment problem we will need the following application of Theorem 12.35. Assertion (i) is usually called the *Archimedean Positivstellensatz*.

Theorem 12.36 *Let $\mathfrak{f} = \{f_1, \ldots, f_k\}$ be a finite subset of $\mathbb{R}_d[\underline{x}]$. Suppose that the quadratic module $Q(\mathfrak{f})$ defined by (12.4) is Archimedean.*

(i) *If $h \in \mathbb{R}_d[\underline{x}]$ satisfies $f(x) > 0$ for all $x \in \mathcal{K}(\mathfrak{f})$, then $h \in Q(\mathfrak{f})$.*
(ii) *Any $Q(\mathfrak{f})$-positive linear functional L on $\mathbb{R}_d[\underline{x}]$ is a $\mathcal{K}(\mathfrak{f})$-moment functional, that is, there exists a measure $\mu \in M_+(\mathbb{R}^d)$ supported on the compact set $\mathcal{K}(\mathfrak{f})$ such that $L(f) = \int f(x)\,d\mu(x)$ for $f \in \mathbb{R}_d[\underline{x}]$.*

Proof

(i) Set $\mathsf{A} = \mathbb{R}_d[\underline{x}]$ and $Q = Q(\mathfrak{f})$. As noted in Sect. 12.1, characters χ of A correspond to points $x_\chi \cong (\chi(x_1), \ldots, \chi(x_d))$ of $\mathbb{R}^d$ and $\mathcal{K}(Q) = \mathcal{K}(\mathfrak{f})$. Hence the assertion follows at once from Theorem 12.35.
(ii) Combine (i) and Haviland's Theorem 1.12 (ii)→(iv). $\square$

12.5 The Operator-Theoretic Approach to the Moment Problem

The spectral theory of self-adjoint operators in Hilbert space is well suited to the moment problem and provides powerful techniques for the study of this problem. The technical tool that relates the multidimensional moment problem to Hilbert space operator theory is the *Gelfand–Naimark–Segal construction*, briefly the *GNS-construction*. We develop this construction first for a general $*$-algebra (see also [Sm4, Section 8.6]) and then we specialize to the polynomial algebra.

Suppose that A is a unital (real or complex) $*$-algebra. Let $\mathbb{K} = \mathbb{R}$ or $\mathbb{K} = \mathbb{C}$.

Definition 12.37 Let $(\mathcal{D}, \langle \cdot, \cdot \rangle)$ be a unitary space. A $*$-*representation* of A on $(\mathcal{D}, \langle \cdot, \cdot \rangle)$ is an algebra homomorphism π of A into the algebra $L(\mathcal{D})$ of linear operators mapping $\mathcal{D}$ into itself such that $\pi(1)\varphi = \varphi$ for $\varphi \in \mathcal{D}$ and

$$\langle \pi(a)\varphi, \psi \rangle = \langle \varphi, \pi(a^*)\psi \rangle \quad \text{for} \quad a \in \mathsf{A}, \ \varphi, \psi \in \mathcal{D}. \tag{12.29}$$

The unitary space $\mathcal{D}$ is called the *domain* of π and denoted by $\mathcal{D}(\pi)$. A vector $\varphi \in \mathcal{D}$ is called *algebraically cyclic*, briefly *a-cyclic*, for π if $\mathcal{D} = \pi(\mathsf{A})\varphi$.

Suppose that L is a positive linear functional on A, that is, L is a linear functional such that $L(a^*a) \geq 0$ for $a \in$ A. Then, by Lemma 2.3, the Cauchy–Schwarz inequality holds:

$$|L(a^*b)|^2 \leq L(a^*a)L(b^*b) \quad \text{for} \quad a, b \in \text{A}. \tag{12.30}$$

Lemma 12.38 $\mathcal{N}_L := \{a \in \text{A} : L(a^*a) = 0\}$ *is a left ideal of the algebra* A.

Proof Let $a, b \in \mathcal{N}_L$ and $x \in$ A. Using (12.30) we obtain

$$|L((xa)^*xa)|^2 = |L((x^*xa)^*a)|^2 \leq L((x^*xa)^*x^*xa)L(a^*a) = 0,$$

so that $xa \in \mathcal{N}_L$. Applying again (12.30) we get $L(a^*b) = L(b^*a) = 0$. Hence

$$L((a+b)^*(a+b)) = L(a^*a) + L(b^*b) + L(a^*b) + L(b^*a) = 0,$$

so that $a + b \in \mathcal{N}_L$. Obviously, $\lambda a \in \mathcal{N}_L$ for $\lambda \in \mathbb{K}$.   □

Hence there exist a well-defined scalar product $\langle \cdot, \cdot \rangle_L$ on the quotient vector space $\mathcal{D}_L = \text{A}/\mathcal{N}_L$ and a well-defined algebra homomorphism $\pi_L : \text{A} \to L(\mathcal{D}_L)$ given by

$$\langle a + \mathcal{N}_L, b + \mathcal{N}_L \rangle_L = L(b^*a), \pi_L(a)(b + \mathcal{N}_L) = ab + \mathcal{N}_L, a, b \in \text{A}. \tag{12.31}$$

Let $\mathcal{H}_L$ denote the Hilbert space completion of the pre-Hilbert space $\mathcal{D}_L$. If no confusion can arise we write $\langle \cdot, \cdot \rangle$ for $\langle \cdot, \cdot \rangle_L$ and a for $a + \mathcal{N}_L$. Then we have $\pi_L(a)b = ab$, in particular $\pi_L(1)a = a$, and

$$\langle \pi_L(a)b, c \rangle = L(c^*ab) = L((a^*c)^*b) = \langle b, \pi_L(a^*)c \rangle, \quad a, b, c \in \text{A}. \tag{12.32}$$

Clearly, $\mathcal{D}_L = \pi_L(\text{A})1$. Thus, we have shown that π_L *is a* *-*representation of* A *on the domain* $\mathcal{D}(\pi_L) = \mathcal{D}_L$ *and* 1 *is an* a-*cyclic vector for* π_L. Further, we have

$$L(a) = \langle \pi_L(a)1, 1 \rangle \quad \text{for} \quad a \in \text{A}. \tag{12.33}$$

Definition 12.39 π_L is called the *GNS-representation* of A associated with L.

We show that the GNS-representation is unique up to unitary equivalence. Let π be another *-representation of A with a-cyclic vector $\varphi \in \mathcal{D}(\pi)$ on a dense domain $\mathcal{D}(\pi)$ of a Hilbert space $\mathcal{G}$ such that $L(a) = \langle \pi(a)\varphi, \varphi \rangle$ for all $a \in$ A. For $a \in$ A,

$$\|\pi(a)\varphi\|^2 = \langle \pi(a)\varphi, \pi(a)\varphi \rangle = \langle \pi(a^*a)\varphi, \varphi \rangle = L(a^*a)$$

and similarly $\|\pi_L(a)1\|^2 = L(a^*a)$. Hence there is an isometric linear map U given by $U(\pi(a)\varphi) = \pi_L(a)1, a \in$ A, of $\mathcal{D}(\pi) = \pi(\text{A})\varphi$ onto $\mathcal{D}(\pi_L) = \pi_L(\text{A})1$. Since the domains $\mathcal{D}(\pi)$ and $\mathcal{D}(\pi_L)$ are dense in $\mathcal{G}$ and $\mathcal{H}_L$, respectively, U extends by continuity to a unitary operator of $\mathcal{G}$ onto $\mathcal{H}_L$. For $a, b \in \mathsf{A}$ we derive

$$U\pi(a)U^{-1}(\pi_L(b)1) = U\pi(a)\pi(b)\varphi = U\pi(ab)\varphi = \pi_L(ab)1 = \pi_L(a)(\pi_L(b)1),$$

that is, $U\pi(a)U^{-1}\varphi = \pi_L(a)\varphi$ for $\varphi \in \mathcal{D}(\pi_L)$ and $a \in \mathsf{A}$. By definition, this means that the $*$-representations π and π_L are unitarily equivalent.

Now we specialize the preceding to the $*$-algebra $\mathbb{C}_d[\underline{x}] \equiv \mathbb{C}[x_1, \dots, x_d]$ with involution determined by $(x_j)^* := x_j$ for $j = 1, \dots, d$.

Suppose that L is a positive linear functional on $\mathbb{C}_d[\underline{x}]$. Since $(x_j)^* = x_j$, it follows from (12.32) that $X_j := \pi_L(x_j)$ is a symmetric operator on the domain $\mathcal{D}_L$. The operators X_j and X_k commute (because x_j and x_k commute in $\mathbb{C}_d[\underline{x}]$) and X_j leaves the domain $\mathcal{D}_L$ invariant (because $x_j \mathbb{C}_d[\underline{x}] \subseteq \mathbb{C}_d[\underline{x}]$). That is, $(X_1, \dots, X_d)$ is a d-tuple of *pairwise commuting symmetric operators acting on the dense invariant domain* $\mathcal{D}_L = \pi_L(\mathbb{C}_d[\underline{x}])1$ of the Hilbert space $\mathcal{H}_L$. Note that this d-tuple $(X_1, \dots, X_d)$ essentially depends on the given positive linear functional L.

The next theorem is the crucial result of the operator approach to the multidimensional moment problem and it is the counterpart of Theorem 6.1. It relates solutions of the moment problem to spectral measures of strongly commuting d-tuples $(A_1, \dots, A_d)$ of self-adjoint operators which extend our given d-tuple $(X_1, \dots, X_d)$.

Theorem 12.40 *A positive linear functional L on the $*$-algebra $\mathbb{C}_d[\underline{x}]$ is a moment functional if and only if there exists a d-tuple $(A_1, \dots, A_d)$ of strongly commuting self-adjoint operators $A_1, \dots, A_d$ acting on a Hilbert space $\mathcal{K}$ such that $\mathcal{H}_L$ is a subspace of $\mathcal{K}$ and $X_1 \subseteq A_1, \dots, X_d \subseteq A_d$. If this is fulfilled and $E_{(A_1, \dots, A_d)}$ denotes the spectral measure of the d-tuple $(A_1, \dots, A_d)$, then $\mu(\cdot) = \langle E_{(A_1, \dots, A_d)}(\cdot)1, 1\rangle_{\mathcal{K}}$ is a solution of the moment problem for L.*

Each solution of the moment problem for L is of this form.

First we explain the notions occurring in this theorem (see [Sm9, Chapter 5] for the corresponding results and more details).

A d-tuple $(A_1, \dots, A_d)$ of self-adjoint operators $A_1, \dots, A_d$ acting on a Hilbert space $\mathcal{K}$ is called *strongly commuting* if for all $k, l = 1, \dots, d, k \neq l$, the resolvents $(A_k - iI)^{-1}$ and $(A_l - iI)^{-1}$ commute, or equivalently, the spectral measures E_{A_k} and E_{A_l} commute (that is, $E_{A_k}(M)E_{A_l}(N) = E_{A_l}(N)E_{A_k}(M)$ for all Borel subsets M, N of $\mathbb{R}$). (If the self-adjoint operators are bounded, strong commutativity and "usual" commutativity are equivalent.) The spectral theorem states that, for such a d-tuple, there exists a unique spectral measure $E_{(A_1, \dots, A_d)}$ on the Borel σ-algebra of $\mathbb{R}^d$ such that

$$A_j = \int_{\mathbb{R}^d} \lambda_j \, dE_{(A_1, \dots, A_d)}(\lambda_1, \dots, \lambda_d), \ j = 1, \dots, d.$$

The spectral measure $E_{(A_1,\dots,A_d)}$ is the product of spectral measures $E_{A_1}, \cdots E_{A_d}$. Therefore, if $M_1, \dots, M_d$ are Borel subsets of $\mathbb{R}$, then

$$E_{(A_1,\dots,A_d)}(M_1 \times \cdots \times M_d) = E_{A_1}(M_1) \cdots E_{A_d}(M_d). \qquad (12.34)$$

Proof of Theorem 12.40 First assume that L is the moment functional and let μ be a representing measure of L. It is well-known and easily checked by the preceding remarks that the multiplication operators A_k, $k = 1, \dots, d$, by the coordinate functions x_k form a d-tuple of strongly commuting self-adjoint operators on the Hilbert space $\mathcal{K} := L^2(\mathbb{R}^d, \mu)$ such that $\mathcal{H}_L \subseteq \mathcal{K}$ and $X_k \subseteq A_k$ for $k = 1, \dots, d$. The spectral measure $E := E_{(A_1,\dots,A_d)}$ of this d-tuple acts by $E(M)f = \chi_M \cdot f$, $f \in L^2(\mathbb{R}^d, \mu)$, where χ_M is the characteristic function of the Borel set $M \subseteq \mathbb{R}^d$. This implies that $\langle E(M)1, 1\rangle_{\mathcal{K}} = \mu(M)$. Thus, $\mu(\cdot) = \langle E(\cdot)1, 1\rangle_{\mathcal{K}}$.

Conversely, suppose that $(A_1, \dots, A_d)$ is such a d-tuple. By the multidimensional spectral theorem [Sm9, Theorem 5.23] this d-tuple has a joint spectral measure $E_{(A_1,\dots,A_d)}$. Put $\mu(\cdot) := \langle E_{(A_1,\dots,A_d)}(\cdot)1, 1\rangle_{\mathcal{K}}$. Let $p \in \mathbb{C}_d[\underline{x}]$. Since $X_k \subseteq A_k$,

$$p(X_1, \dots, X_d) \subseteq p(A_1, \dots, A_d).$$

Therefore, since the polynomial 1 belongs to the domain of $p(X_1, \dots, X_d)$, it is also in the domain of $p(A_1, \dots, A_d)$. Then

$$\int_{\mathbb{R}^d} p(\lambda) \, d\mu(\lambda) = \int_{\mathbb{R}^d} p(\lambda) \, d\langle E_{(A_1,\dots,A_d)}(\lambda)1, 1\rangle_{\mathcal{K}} = \langle p(A_1, \dots, A_d)1, 1\rangle_{\mathcal{K}}$$

$$= \langle p(X_1, \dots, X_d)1, 1\rangle = \langle \pi_L(p(x_1, \dots, x_d))1, 1\rangle = L(p(x_1, \dots, x_d)),$$

where the second equality follows from the functional calculus and the last from (12.33). This shows that μ is a solution of the moment problem for L.   □

Proposition 12.41 *Suppose Q is an Archimedean quadratic module of a commutative real unital algebra A. Let L_0 be a Q-positive $\mathbb{R}$-linear functional on A and let π_L be the GNS representation of its extension L to a $\mathbb{C}$-linear functional on the complexification $\mathsf{A}_{\mathbb{C}} = \mathsf{A} + \mathrm{i}\mathsf{A}$. Then all operators $\pi_L(a)$, $a \in \mathsf{A}_{\mathbb{C}}$, are bounded.*

Proof Since $\sum(\mathsf{A}_{\mathbb{C}})^2 = \sum \mathsf{A}^2$ by Lemma 2.17(ii) and $\sum \mathsf{A}^2 \subseteq Q$, L is a positive linear functional on $\mathsf{A}_{\mathbb{C}}$, so the GNS representation π_L is well-defined.

It suffices to prove that $\pi_L(a)$ is bounded for $a \in \mathsf{A}$. Since Q is Archimedean, $\lambda - a^2 \in Q$ for some $\lambda > 0$. Let $x \in \mathsf{A}_{\mathbb{C}}$. By Lemma 2.17(ii), $x^*x(\lambda - a^2) \in Q$ and hence $L(x^*xa^2) = L_0(x^*xa^2) \le \lambda L_0(x^*x) = \lambda L(x^*x)$, since L_0 is Q-positive. Then

$$\|\pi_L(a)\pi_L(x)1\|^2 = \langle \pi_L(a)\pi_L(x)1, \pi_L(a)\pi_L(x)1\rangle = \langle \pi_L((ax)^*ax)1, 1\rangle$$

$$= L((ax)^*ax) = L(x^*xa^2) \le \lambda L(x^*x) = \lambda \|\pi_L(x)1\|^2,$$

where we used (12.29) and (12.33). That is, $\pi_L(a)$ is bounded on $\mathcal{D}(\pi_L)$.   □

We now illustrate the power of the operator approach to moment problems by giving short proofs of Theorems 12.35 and 12.36.

Proof of Theorem 12.35 Assume to the contrary that $a_0 \notin Q$. Since Q is Archimedean, by Proposition 12.11 there is a Q-positive $\mathbb{R}$-linear functional L_0 on A such that $L_0(1) = 1$ and $L_0(a_0) \leq 0$. Let π_L be the GNS representation of its extension to a $\mathbb{C}$-linear (positive) functional L on the unital commutative complex $*$-algebra $A_{\mathbb{C}}$.

Let $c \in Q$. If $x \in A_{\mathbb{C}}$, then $x^*xc \in Q$ by Lemma 2.17(ii), so $L_0(x^*xc) \geq 0$, and

$$\langle \pi_L(c)\pi_L(x)1, \pi_L(x)1 \rangle = L(x^*xc) = L_0(x^*xc) \geq 0 \tag{12.35}$$

by (12.33). This shows that the operator $\pi_L(c)$ is nonnegative.

For $a \in A_{\mathbb{C}}$, the operator $\pi_L(a)$ is bounded by Proposition 12.41. Let $\overline{\pi_L(a)}$ denote its continuous extension to the Hilbert space $\mathcal{H}_L$. These operators form a unital commutative $*$-algebra of bounded operators. Its completion $\mathcal{B}$ is a unital commutative C^*-algebra.

Let χ be a character of $\mathcal{B}$. Then $\tilde{\chi}(\cdot) := \chi(\overline{\pi_L(\cdot)})$ is a character of A. If $c \in Q$, then $\pi_L(c) \geq 0$ by (12.35) and so $\overline{\pi_L(c)} \geq 0$. Hence $\tilde{\chi}$ is Q-positive, that is, $\tilde{\chi} \in \mathcal{K}(Q)$. Therefore, $\tilde{\chi}(a_0) = \chi(\overline{\pi_L(a_0)}) > 0$ by the assumption of Theorem 12.35. Therefore, if we realize $\mathcal{B}$ as a C^*-algebra of continuous functions on a compact Hausdorff space, the function corresponding to $\overline{\pi_L(a_0)}$ is positive, so it has a positive minimum δ. Then $\overline{\pi_L(a_0)} \geq \delta \cdot I$ and hence

$$0 < \delta = \delta L(1) = \langle \delta 1, 1 \rangle \leq \langle \pi_L(a_0)1, 1 \rangle = L(a_0) = L_0(a_0) \leq 0,$$

which is the desired contradiction.   □

Proof of Theorem 12.36(ii) We extend L to a $\mathbb{C}$-linear functional, denoted again by L, on $\mathbb{C}_d[\underline{x}]$ and consider the GNS representation π_L. By Proposition 12.41, the symmetric operators $\pi_L(x_1), \ldots, \pi_L(x_d)$ are bounded. Hence their continuous extensions to the whole Hilbert space $\mathcal{H}_L$ are pairwise commuting bounded self-adjoint operators $A_1, \ldots, A_d$. Therefore, by Theorem 12.40, if E denotes the spectral measure of this d-tuple $(A_1, \ldots, A_d)$, then $\mu(\cdot) = \langle E(\cdot)1, 1 \rangle_{\mathcal{H}_L}$ is a solution of the moment problem for L.

Since the operators A_j are bounded, the spectral measure E, hence μ, has compact support. (In fact, supp $E \subseteq [-\|A_1\|, \|A_1\|] \times \cdots \times [-\|A_d\|, \|A_d\|]$.) Hence, since L is $Q(\mathfrak{f})$-positive by assumption, Proposition 12.18 implies that supp $\mu \subseteq \mathcal{K}(\mathfrak{f})$. This shows that L is a $\mathcal{K}(\mathfrak{f})$-moment functional.   □

The preceding proof of Theorem 12.36(ii) based on the spectral theorem is probably the most elegant approach to the moment problem for Archimedean quadratic modules. Now we derive Theorem 12.36(i) from Theorem 12.36(ii).

Proof of Theorem 12.36(i) We argue in the same manner as in the second proof of Theorem 12.24 in Sect. 12.3. Assume to the contrary that $h \notin Q(\mathfrak{f})$. Since $Q(\mathfrak{f})$ is Archimedean, Proposition 12.11 and Theorem 12.36(ii) apply to $Q(\mathfrak{f})$. By these results, there is a $Q(\mathfrak{f})$-positive linear functional L on $\mathbb{R}_d[\underline{x}]$ satisfying $L(1) = 1$ and $L(h) \leq 0$, and this functional is a $\mathcal{K}(\mathfrak{f})$-moment functional. Then there is a measure $\mu \in M_+(\mathbb{R}^d)$ supported on $\mathcal{K}(\mathfrak{f})$ such that $L(p) = \int p \, d\mu$ for $p \in \mathbb{R}_d[\underline{x}]$. (Note that $\mathcal{K}(\mathfrak{f})$ is compact by Corollary 12.9.) Again $h(x) > 0$ on $\mathcal{K}(\mathfrak{f})$, $L(1) = 1$, and $L(h) \leq 0$ lead to a contradiction. □

12.6 The Moment Problem for Semi-Algebraic Sets Contained in Compact Polyhedra

In this section, $f_1, \ldots, f_k$ are polynomials of $\mathbb{R}_d[\underline{x}]$ such that the first m polynomials $f_1, \ldots, f_m$, where $1 \leq m \leq k$, are *linear*. By a linear polynomial we mean a polynomial of degree at most one. We abbreviate

$$\hat{\mathfrak{f}} = \{f_1, \ldots, f_m\}, \quad \mathfrak{f} = \{f_1, \ldots, f_k\}.$$

Then $\mathcal{K}(\hat{\mathfrak{f}})$ is a semi-algebraic set defined by linear polynomials $f_1, \ldots, f_m$; such a set is called a *polyhedron*. The general semi-algebraic set $\mathcal{K}(\mathfrak{f})$ is contained in the polyhedron $\mathcal{K}(\hat{\mathfrak{f}})$.

Definition 12.42 Let $\mathcal{P}(\mathfrak{f})$ denote the semiring generated by $f_1, \ldots, f_k$, that is, $\mathcal{P}(\mathfrak{f})$ consists of all finite sums of terms of the form

$$\alpha f_1^{n_1} \cdots f_k^{n_k}, \quad \text{where } \alpha \geq 0, \; n_1, \ldots, n_k \in \mathbb{N}_0. \tag{12.36}$$

Note that $\mathcal{P}(\mathfrak{f})$ is not a quadratic module in general.

The following lemma goes back to H. Minkowski. In the optimization literature it is called *Farkas' lemma*. We will use it in the proof of Theorem 12.44 below.

Lemma 12.43 *Let $h, f_1, \ldots, f_m$ be linear polynomials of $\mathbb{R}_d[\underline{x}]$ such that the set $\mathcal{K}(\hat{\mathfrak{f}})$ is not empty. If $h(x) \geq 0$ on $\mathcal{K}(\hat{\mathfrak{f}})$, there exist numbers $\lambda_0 \geq 0, \ldots, \lambda_m \geq 0$ such that $h = \lambda_0 + \lambda_1 f_1 + \cdots + \lambda_m f_m$.*

Proof Let E be the vector space spanned by the polynomials $1, x_1, \ldots, x_d$ and C the cone in E generated by $1, f_1, \ldots, f_m$. It is easily shown that C is closed in E.

We have to prove that $h \in C$. Assume to the contrary that $g \notin C$. Then, by the separation of convex sets (Theorem A.26(ii)), there exists a C-positive linear functional L on E such that $L(h) < 0$. In particular, $L(1) \geq 0$, because $1 \in C$.

Without loss of generality we can assume that $L(1) > 0$. Indeed, if $L(1) = 0$, we take a point x_0 of the nonempty (!) set $\mathcal{K}(\hat{\mathfrak{f}})$ and replace L by $L' = L + \varepsilon l_{x_0}$, where l_{x_0} denotes the point evaluation at x_0 on E. Then L' is C-positive as well and $L'(h) < 0$ for small $\varepsilon > 0$.

Define a point $x := L(1)^{-1}(L(x_1), \ldots, L(x_d)) \in \mathbb{R}^d$. Then $L(1)^{-1}L$ is the evaluation l_x at the point x for the polynomials $x_1, \ldots, x_d$ and for 1, hence on the whole vector space E. Therefore, $f_j(x) = l_x(f_j) = L(1)^{-1}L(f_j) \geq 0$ for all j, so that $x \in \mathcal{K}(\hat{\mathfrak{f}})$, and $g(x)=l_x(h)=L(1)^{-1}L(h) < 0$. This contradicts the assumption.   □

Theorem 12.44 *Let $f_1, \ldots, f_k$ be polynomials of $\mathbb{R}_d[\underline{x}]$ such that the polynomials $f_1, \ldots, f_m, 1 \leq m \leq k$, are linear. Suppose that the polyhedron $\mathcal{K}(\hat{\mathfrak{f}})$ is compact and nonempty. Let $h \in \mathbb{R}_d[\underline{x}]$. If $h(x) > 0$ for all $x \in \mathcal{K}(\mathfrak{f})$, then $h \in \mathcal{P}(\mathfrak{f})$.*

Proof First we show that the semiring $\mathcal{P}(\mathfrak{f})$ is Archimedean. Let $i \in \{1, \ldots, d\}$. Since the set $\mathcal{K}(\hat{\mathfrak{f}})$ is compact, there exists a $\lambda > 0$ such that $\lambda \pm x_i \geq 0$ on $\mathcal{K}(\hat{\mathfrak{f}})$. Hence, since $\mathcal{K}(\hat{\mathfrak{f}})$ is nonempty, Lemma 12.43 implies that $(\lambda \pm x_i) \in \mathcal{P}(\mathfrak{f})$. Therefore, $\mathcal{P}(\mathfrak{f})$ is Archimedean by Lemma 12.7(ii).

Now we apply Theorem 12.35 to the Archimedean semiring $Q = \mathcal{P}(\mathfrak{f})$ of the algebra $\mathsf{A} = \mathbb{R}_d[\underline{x}]$. Clearly, each character χ of A is given by a point $x \in \mathbb{R}^d$. If χ is $\mathcal{P}(\mathfrak{f})$-positive, then $\chi(f_j) = f_j(x) \geq 0$ for all $j = 1, \ldots, k$ and therefore $x \in \mathcal{K}(\mathfrak{f})$. Thus, $\mathcal{K}(Q) = \mathcal{K}(\mathfrak{f})$ and Theorem 12.35 yields the assertion.   □

The following theorem is the main result of this section on the moment problem.

Theorem 12.45 *Retain the assumptions and the notation of Theorem 12.44. Let L be a linear functional on $\mathbb{R}_d[\underline{x}]$. Then L is a $\mathcal{K}(\mathfrak{f})$-moment functional if and only if*

$$L(f_1^{n_1} \cdots f_k^{n_k}) \geq 0 \quad \text{for all} \;\; n_1, \ldots, n_k \in \mathbb{N}_0. \tag{12.37}$$

Proof The only if part is obvious. Conversely, suppose that (12.37) is satisfied. By the definition of the semiring $\mathcal{P}(\mathfrak{f})$, (12.37) means that L is $\mathcal{P}(\mathfrak{f})$-positive. Hence, by Theorem 12.44 and Haviland's Theorem 1.12, L is a $\mathcal{K}(\mathfrak{f})$-moment functional.   □

Let us consider the important special case $m = k$. Suppose that $\mathcal{K}(\mathfrak{f})$ is a nonempty compact set. Since $k = m$, all polynomials $f_1, \ldots, f_k$ are linear. Hence $\mathcal{K}(\mathfrak{f}) = \mathcal{K}(\hat{\mathfrak{f}})$ is a polyhedron. Then, by Theorem 12.45, (12.37) is a solvability condition for the moment problem of the compact polyhedron $\mathcal{K}(\mathfrak{f})$ and Theorem 12.44 gives a description of strictly positive polynomials on $\mathcal{K}(\mathfrak{f})$.

12.7   Examples and Applications

Throughout this section, $\mathfrak{f} = \{f_1, \ldots, f_k\}$ is a finite subset of $\mathbb{R}_d[\underline{x}]$ and L denotes a linear functional on $\mathbb{R}_d[\underline{x}]$.

If L is a $\mathcal{K}(\mathfrak{f})$-moment functional, it is obviously $T(\mathfrak{f})$-positive, $Q(\mathfrak{f})$-positive, and $\mathcal{P}(\mathfrak{f})$-positive. Theorems 12.25, 12.36, and 12.45 deal with the converse implication and are the main solvability criteria for the moment problem proved in this chapter.

First we discuss Theorems 12.25 and 12.36(ii). Theorem 12.25 applies to *each* compact semi-algebraic set $\mathcal{K}(\mathfrak{f})$ and implies that L is a $\mathcal{K}(\mathfrak{f})$-moment functional if and only if it is $T(\mathfrak{f})$-positive. For Theorem 12.36(ii) the compactness of the set $\mathcal{K}(\mathfrak{f})$ is not sufficient; it requires that the quadratic module $Q(\mathfrak{f})$ is Archimedean. In this case, L is a $\mathcal{K}(\mathfrak{f})$-moment functional if and only if it is $Q(\mathfrak{f})$-positive.

Example 12.46 Let us begin with a single polynomial $f \in \mathbb{R}_d[\underline{x}]$ for which the set $\mathcal{K}(f) = \{x \in \mathbb{R}^d : f(x) \geq 0\}$ is compact. (A simple example is the d-ellipsoid given by $f(x) = 1 - a_1 x_1^2 - \cdots - a_d x_d^2$, where $a_1 > 0, \ldots, a_d > 0$.) Clearly, $T(f) = Q(f)$. Then, L is a $\mathcal{K}(f)$-*moment functional if and only if it is* $T(f)$-*positive, or equivalently, if L and L_f are positive functionals on* $\mathbb{R}_d[\underline{x}]$.

Now we add further polynomials $f_2, \ldots, f_k$ and set $\mathsf{f} = \{f, f_2, \ldots, f_k\}$. (For instance, one may take coordinate functions as $f_j = x_l$.) Since $T(f)$ is Archimedean (by Proposition 12.22, because $\mathcal{K}(f)$ is compact), so is the quadratic module $Q(\mathsf{f})$. Therefore, L is a $\mathcal{K}(\mathsf{f})$-*moment functional if and only if it is* $Q(f)$-*positive, or equivalently, if $L, L_f, L_{f_2}, \ldots, L_{f_k}$ are positive functionals on* $\mathbb{R}_d[\underline{x}]$.   ○

Example 12.47 (d-dimensional compact interval $[a_1, b_1] \times \cdots \times [a_d, b_d]$) Let $a_j, b_j \in \mathbb{R}$, $a_j < b_j$, and set $f_{2j-1} := b_j - x_j, f_{2j} := x_j - a_j$, for $j = 1, \ldots, d$. Then the semi-algebraic set $\mathcal{K}(\mathsf{f})$ for $\mathsf{f} := \{f_1, \ldots, f_{2d}\}$ is the d-dimensional interval $[a_1, b_1] \times \cdots \times [a_d, b_d]$.

Put $\lambda_j = |a_j| + |b_j|$. Then $\lambda_j - x_j = f_{2j-1} + \lambda_j - b_j$ and $\lambda_j + x_j = f_{2j} + \lambda_j + a_j$ are $Q(\mathsf{f})$, so each x_j is a bounded element with respect to the quadratic module $Q(\mathsf{f})$. Hence $Q(\mathsf{f})$ is Archimedean by Lemma 12.7(ii).

Thus, L *is a* $\mathcal{K}(\mathsf{f})$-*moment functional if and only if it is* $Q(f)$-*positive, or equivalently, if $L_{f_1}, L_{f_2}, \ldots, L_{f_k}$ are positive functionals, that is,*

$$L((b_j - x_j)p^2) \geq 0 \ \text{ and } \ L((x_j - a_j)p^2) \geq 0 \ \text{ for } \ j = 1, \ldots, d, \ p \in \mathbb{R}_d[\underline{x}].$$
(12.38)

Clearly, (12.38) implies that L itself is positive, since $L = (b_1 - a_1)^{-1}(L_{f_1} + L_{f_2})$. ○

Example 12.48 (1-dimensional interval $[a, b]$) Let $a < b$, $a, b \in \mathbb{R}$ and let $l, n \in \mathbb{N}$ be odd. We set $f(x) := (b - x)^l (x - a)^n$. Then $\mathcal{K}(f) = [a, b]$ and $T(f) = \sum \mathbb{R}[x]^2 + f \sum \mathbb{R}[x]^2$. Hence, by Theorem 12.25, *a linear functional L on $\mathbb{R}[x]$ is an $[a, b]$-moment functional if and only if L and L_f are positive functionals on* $\mathbb{R}[x]$.

This result extends Hausdorff's Theorem 3.13. It should be noted that this solvability criterion holds for arbitrary (!) odd numbers l and n, while the equality $\mathrm{Pos}([a, b]) = T(f)$ is only true if $l = n = 1$, see Exercise 3.4 b. in Chap. 3.   ○

Example 12.49 (Simplex in $\mathbb{R}^d, d \geq 2$) Let $f_1 = x_1, \ldots, f_d = x_d, f_{d+1} = 1 - \sum_{i=1}^{d} x_i, k = d + 1$. Clearly, $\mathcal{K}(\mathsf{f})$ is the simplex

$$K_d = \{x \in \mathbb{R}^d : x_1 \geq 0, \ldots, x_d \geq 0, x_1 + \cdots + x_d \leq 1\}.$$

Note that $1 - x_j = f_{d+1} + \sum_{i \neq j} f_i$ and $1 + x_j = 1 + f_j$. Therefore, $1 \pm x_j \in Q(\mathsf{f})$ and $1 \pm x_j \in \mathcal{P}(\mathsf{f})$. Hence, by Lemma 12.7(ii), the quadratic module $Q(\mathsf{f})$ and the semiring $\mathcal{P}(\mathsf{f})$ are Archimedean. Therefore, Theorem 12.36 applies to $Q(\mathsf{f})$ and Theorems 12.44 and 12.45 apply to $\mathcal{P}(\mathsf{f})$. We restate only the results on the moment problem.

By Theorems 12.36(ii) and 12.45, L is a K_d-*moment functional if and only if*

$$L(x_i p^2) \geq 0, \ i = 1, \cdots, d, \ \text{ and } \ L((1 - (x_1 + x_2 + \cdots + x_d))p^2) \geq 0 \ \text{ for } p \in \mathbb{R}_d[\underline{x}],$$

or equivalently,

$$L(x_1^{n_1} \ldots x_d^{n_d}(1 - (x_1 + \cdots + x_d))^{n_d+1}) \geq 0 \quad for \quad n_1,\ldots,n_{d+1} \in \mathbb{N}_0. \qquad \square \quad \circ$$

Example 12.50 (Standard simplex Δ_d in $\mathbb{R}^d$) Let $f_1 = x_1,\ldots, f_d = x_d, f_{d+1} = 1 - \sum_{i=1}^{d} x_i, f_{d+2} = -f_{d+1}, k = d + 2$. Then the semi-algebraic set $\mathcal{K}(\mathsf{f})$ is the standard simplex

$$\Delta_d = \{x \in \mathbb{R}^d : x_1 \geq 0,\ldots,x_d \geq 0, x_1 + \cdots + x_d = 1\}.$$

Let $\mathcal{P}_0$ denote the polynomials of $\mathbb{R}_d[\underline{x}]$ with nonnegative coefficients and $\mathcal{I}$ the ideal generated by $1 - (x_1 + \cdots + x_d)$. Then $\mathcal{P} := \mathcal{P}_0 + \mathcal{I}$ is a semiring of $\mathbb{R}_d[\underline{x}]$. Since $1 \pm x_j \in \mathcal{P}$, $\mathcal{P}$ is Archimedean. The characters of $\mathbb{R}_d[\underline{x}]$ are the evaluations at points of $\mathbb{R}^d$. Obviously, $x \in \mathbb{R}^d$ gives a $\mathcal{P}$-positive character if and only if $x \in \Delta_d$.

Let $f \in \mathbb{R}_d[\underline{x}]$ be such that $f(x) > 0$ on Δ_d. Then, $f \in \mathcal{P}$ by Theorem 12.35, so

$$f(x) = g(x) + h(x)(1 - (x_1 + \cdots + x_d)), \quad \text{where} \quad g \in \mathcal{P}_0, \ h \in \mathbb{R}_d[\underline{x}]. \qquad (12.39)$$

From Theorem 12.45 it follows that *L is a Δ_d-moment functional if and only if*

$$L(x_1^{n_1} \ldots x_d^{n_d}) \geq 0, \ L(x_1^{n_1} \ldots x_d^{n_d}(1-(x_1+ \ldots +x_d))^r) = 0, \ n_1,\ldots,n_d \in \mathbb{N}_0, r \in \mathbb{N}. \quad \circ$$

From the preceding example it is only a small step to derive an elegant proof of the following classical *theorem of G. Polya*.

Proposition 12.51 *Suppose that $f \in \mathbb{R}_d[\underline{x}]$ is a homogeneous polynomial such that $f(x) > 0$ for all $x \in \mathbb{R}^d \setminus \{0\}$, $x_1 \geq 0,\ldots,x_d \geq 0$. Then there exists an $n \in \mathbb{N}$ such that all coefficients of the polynomial $(x_1 + \cdots + x_d)^n f(x)$ are nonnegative.*

Proof We use Example 12.50. As noted therein, Theorem 12.35 implies that f is of the form (12.39). We replace in (12.39) each variable $x_j, j = 1,\ldots,d$, by $x_j(\sum_{i=1}^{d} x_i)^{-1}$. Since $(1 - \sum_j x_j(\sum_i x_i)^{-1}) = 1 - 1 = 0$, the second summand in (12.39) vanishes after this substitution. Hence, because f is homogeneous, (12.39) yields

$$\left(\sum_i x_i \right)^{-m} f(x) = g\left(x_1 \left(\sum_i x_i \right)^{-1}, \ldots, x_d \left(\sum_i x_i \right)^{-1}\right), \qquad (12.40)$$

where $m = \deg(f)$. Since $g \in \mathcal{P}_0$, $g(x)$ has only nonnegative coefficients. Therefore, after multiplying (12.40) by $(\sum_i x_i)^{n+m}$ with n sufficiently large to clear the denominators, we obtain the assertion. $\qquad \square$

Now we treat two examples which are applications of Theorem 12.45.

Example 12.52 $[-1,1]^d$

Let $k = m = 2d$ and $f_1 = 1 - x_1, f_2 = 1 + x_1,\ldots, f_{2d-1} = 1 - x_d, f_{2d} = 1 + x_d$. Then $\mathcal{K}(\hat{\mathsf{f}}) = \mathcal{K}(\mathsf{f}) = [-1,1]^d$. Therefore, by Theorem 12.45, *a linear functional*

L on $\mathbb{R}_d[x_d]$ *is a* $[-1, 1]^d$-*moment functional if and only if*

$$L((1 - x_1)^{n_1}(1 + x_1)^{n_2} \cdots (1 - x_d)^{n_{2d-1}}(1 + x_d)^{n_{2d}}) \geq 0 \quad \text{for} \quad n_1, \ldots, n_{2d} \in \mathbb{N}_0. \circ$$

Example 12.53 (Multidimensional Hausdorff moment problem on $[0, 1]^d$) Set $f_1 = x_1, f_2 = 1 - x_1, \ldots, f_{2d-1} = x_d, f_{2d} = 1 - x_d, k = 2d$. Then $\mathcal{K}(\hat{\mathfrak{f}}) = [0, 1]^d$. Let $s = (s_n)_{n \in \mathbb{N}_0^d}$ be a multisequence. We define the shift E_j of the j-th index by

$$(E_j s)_m = s_{(m_1, \ldots, m_{j-1}, m_j+1, m_{j+1}, \ldots, m_d)}, \quad m \in \mathbb{N}_0^d.$$

Proposition 12.54 *The following five statements are equivalent:*

(i) *s is a Hausdorff moment sequence on* $[0, 1]^d$.
(ii) L_s *is a* $[-1, 1]^d$-*moment functional on* $\mathbb{R}_d[\underline{x}]$.
(iii) $L_s(x_1^{m_1}(1 - x_1)^{n_1} \cdots x_d^{m_d}(1 - x_d)^{n_d}) \geq 0$ *for all* $n, m \in \mathbb{N}_0^d$.
(iv) $((I - E_1)^{n_1} \ldots (I - E_d)^{n_d} s)_m \geq 0$ *for all* $n, m \in \mathbb{N}_0^d$.
(v)

$$\sum_{j \in \mathbb{N}_0^d, j \leq n} (-1)^{|j|} \binom{n_1}{j_1} \cdots \binom{n_d}{j_d} s_{m+j} \geq 0$$

for all $n, m \in \mathbb{N}_0^d$. *Here* $|j| := j_1 + \cdots + j_d$ *and* $j \leq n$ *means that* $j_i \leq n_i$ *for* $i = 1, \ldots, d$.

Proof (i)↔(ii) holds by definition. Theorem 12.45 yields (ii)↔(iii). Let $n, m \in \mathbb{N}_0^d$. We repeat the computation from the proof of Theorem 3.15 and derive

$$L_s(x_1^{m_1}(1 - x_1)^{n_1} \cdots x_d^{m_d}(1 - x_d)^{n_d}) = ((I - E_1)^{n_1} \ldots (I - E_d)^{n_d} s)_m$$

$$= \sum_{j \in \mathbb{N}_0^d, j \leq n} (-1)^{|j|} \binom{n_1}{j_1} \cdots \binom{n_d}{j_d} s_{m+j}.$$

This identity implies the equivalence of conditions (iii)–(v).   □ ∘

12.8 Exercises

1. Suppose that Q is a quadratic module of a commutative real algebra A. Show that $Q \cap (-Q)$ is an ideal of A. This ideal is called the *support ideal* of Q.
2. Let K be a closed subset of $\mathbb{R}^d$. Show that Pos(K) is saturated.
3. Formulate solvability criteria in terms of localized functionals and in terms of d-sequences for the following sets.

    a. Unit ball of $\mathbb{R}^d$.

    b. $\{x \in \mathbb{R}^d : x_1^2 + \cdots + x_d^2 \leq r^2, \, x_1 \geq 0, \ldots, x_d \geq 0\}$.

    c. $\{(x_1, x_2, x_3, x_4) \in \mathbb{R}^4 : x_1^2 + x_2^2 \leq 1, x_3^2 + x_4^2 \leq 1\}$.

    d. $\{(x_1, x_2, x_3) \in \mathbb{R}^3 : x_1^2 + x_2^2 + x_3^2 \leq 1, x_1 + x_2 + x_3 \leq 1\}$.

    e. $\{x \in \mathbb{R}^{2d} : x_1^2 + x_2^2 = 1, \ldots, x_{2d-1}^2 + x_{2d}^2 = 1\}$.

4. Decide whether or not the following quadratic modules $Q(\mathfrak{f})$ are Archimedean.

    a. $f_1 = x_1, f_2 = x_2, f_3 = 1 - x_1 x_2, f_4 = 4 - x_1 x_2$.

    b. $f_1 = x_1, f_2 = x_2, f_3 = 1 - x_1 - x_2$.

    c. $f_1 = x_1, f_2 = x_2, f_3 = 1 - x_1 x_2$.

5. Let $f_1, \ldots, f_k, g_1, \ldots, g_l \in \mathbb{R}_d[\underline{x}]$. Set $g = (f_1, \ldots, f_k, g_1, \ldots, g_l)$, $\mathfrak{f} = (f_1, \ldots, f_k)$. Suppose that $Q(\mathfrak{f})$ is Archimedean. Show that each $Q(g)$-positive linear functional L is a determinate $\mathcal{K}(g)$-moment functional.

6. Formulate solvability criteria for the moment problem of the following semi-algebraic sets $\mathcal{K}(\mathfrak{f})$.

    a. $f_1 = x_1^2 + \cdots + x_d^2, f_2 = x_1, \ldots, f_k = x_{k-1}$, where $2 \leq k \leq d + 1$.

    b. $f_1 = x_1, f_2 = 2 - x_1, f_3 = x_2, f_4 = 2 - x_2, f_5 = x_1^2 - x_2$, where $d = 2$.

    c. $f_1 = x_1^2 + x_2^2, f_2 = ax_1 + bx_2, f_3 = x_2$, where $d = 2, a, b \in \mathbb{R}$.

7. Let $d = 2, f_1 = 1 - x_1, f_2 = 1 + x_1, f_3 = 1 - x_2, f_4 = 1 + x_2, f_5 = 1 - x_1^2 - x_2^2$ and $\mathfrak{f} = (f_1, f_2, f_3, f_4, f_5)$. Describe the set $\mathcal{K}(\mathfrak{f})$ and use Theorem 12.45 to characterize $\mathcal{K}(\mathfrak{f})$-moment functionals.

8. Find a d-dimensional version of Exercise 7, where $d \geq 3$.

9. (*Reznick's theorem* [Re2])

    Let $f \in \mathbb{R}_d[\underline{x}]$ be a homogeneous polynomial such that $f(x) > 0$ for all $x \in \mathbb{R}^d, x \neq 0$. Prove that there exists an $n \in \mathbb{N}$ such that

$$(x_1^2 + \cdots + x_d^2)^n f(x) \in \sum \mathbb{R}_d[\underline{x}]^2.$$

Hint: Mimic the proof of Proposition 12.51: Let T denote the preordering $\sum \mathbb{R}_d[\underline{x}]^2 + \mathcal{I}$, where $\mathcal{I}$ is the ideal generated by $1 - (x_1^2 + \cdots + x_d^2)$. Show that T-positive characters corresponds to points of the unit sphere, substitute $x_j(\sum_i x_i^2)^{-1}$ for x_j, apply Theorem 12.44 to T, and clear denominators.

12.9 Notes

The interplay between real algebraic geometry and the moment problem for compact semi-algebraic sets and the corresponding Theorems 12.24 and 12.25 were discovered by the author in [Sm6]. A small gap in the proof of [Sm6, Corollary 3] (observed by A. Prestel) was immediately repaired by the reasoning of the above proof of Proposition 12.22 (taken from [Sm8, Proposition 18]).

The fact that the preordering is Archimedean in the compact case was first noted by T. Wörmann [Wö]. An algorithmic proof of Theorem 12.24 was developed by M. Schweighofer [Sw1, Sw2].

The operator-theoretic proof of Theorem 12.36(ii) given above is long known among operator theorists; it was used in [Sm6]. The operator-theoretic approach to the multidimensional moment theory was investigated by F. Vasilescu [Vs1, Vs2].

The representation theorem for Archimedean modules (Theorem 12.35) has a long history. It was proved in various versions by M.H. Stone [Stn], R.V. Kadison [Kd], J.-L. Krivine [Kv1], E. Becker and N. Schwartz [BS], M. Putinar [Pu2], and T. Jacobi [Jc]. The version for quadratic modules is due to Jacobi [Jc], while the version for semirings was proved much earlier by Krivine [Kv1]. A more general version and a detailed discussion can be found in [Ms1, Section 5.4]. Lemma 12.34 appeared in [BSS]. Putinar [Pu2] has proved that a finitely generated quadratic module Q in $\mathbb{R}_d[\underline{x}]$ is Archimedean if (and only if) there exists a polynomial $f \in Q$ such that the set $\{x \in \mathbb{R}^d : f(x) \geq 0\}$ is compact.

Corollary 12.29 and its non-compact version in Exercise 14.11 below are from [Ls3]. The moment problem with bounded densities is usually called the *Markov moment problem* or *L*-moment problem. In dimension one it goes back to A.A. Markov [Mv1, Mv2], see [AK, Kr2]. An interesting more recent work is [DF]. The multidimensional case was studied in [Pu1, Pu3, Pu5, Ls3, Ls4].

For compact polyhedra with nonempty interiors Theorem 12.44 was proved by D. Handelman [Hn]. A special case was treated earlier by J.-L. Krivine [Kv2]. A related version can be found in [Cs, Theorem 4]. The general form presented above is taken from [PD, Theorem 5.4.6]. Polya's theorem was proved in [P]. Polya's original proof is elementary; the elegant proof given in the text is from [Wö]. Proposition 12.54 is a classical result obtained in [HS]. It should be noted that Reznick's theorem [Re2] is an immediate consequence of the strict Positivstellensatz, see [Sr3, 2.1.8].

Reconstructing the shape of subsets of $\mathbb{R}^d$ from its moments with respect to the Lebesgue measure is another interesting topic, see e.g. [GHPP] and [GLPR].

Chapter 13
The Moment Problem on Closed Semi-Algebraic Sets: Existence

The main subject of this chapter and the next is the moment problem on *closed* semi-algebraic sets. For a compact semi-algebraic set $\mathcal{K}(f)$ a very satisfactory solution of the existence problem in terms of the positivity on the preordering $T(f)$ was given by Theorem 12.25. This result holds for any finite set f of generators which defines the semi-algebraic set $\mathcal{K}(f)$. The representing measure is always unique and supported on $\mathcal{K}(f)$. All these features of the compact case are no longer true for noncompact sets. In this chapter we are only concerned with existence problems, while determinacy questions are studied in the next chapter.

Let us consider a semi-algebraic set $\mathcal{K}(f)$. Having Theorem 12.25 in mind it is natural to ask when the positivity of a linear functional L on the preordering $T(f)$ implies that L is a moment functional. In this case we will say that $T(f)$ has the *moment property* (MP). If at least one representing measure has support contained in $\mathcal{K}(f)$, then $T(f)$ obeys the *strong moment property* (SMP). To study when these properties hold or fail for a preordering or a quadratic module is the main theme in this chapter. The fundamental result in this respect is the *fibre theorem* (Theorem 13.10). It is stated and discussed in Sect. 13.3, but the long proof of its main implication is given only in Sect. 13.10. The fibre theorem reduces moment properties of $T(f)$ to those for fibre preorderings built by means of *bounded* polynomials on the set $\mathcal{K}(f)$. Most of the known general affirmative results on the moment problem for closed semi-algebraic sets can be derived from this theorem. In Sects. 13.4–13.7 we develop a number of applications of the fibre theorem and provide classes of preorderings satisfying (MP) or (SMP).

On the other hand, one of the new difficulties in dimensions $d \geq 2$ is that the preordering $\sum \mathbb{R}_d[\underline{x}]^2$ does not satisfy (MP), that is, there exist positive functionals which are not moment functionals. The reason for this is the existence of positive polynomials in two variables which are not sums of squares of polynomials. Section 13.1 deals with this matter. In Sect. 13.8 the concept of *stability* is used to prove the closedness of quadratic modules and to derive classes of quadratic modules for which (MP) fails. The moment problem on some cubics is studied in Sect. 13.9.

© Springer International Publishing AG 2017
K. Schmüdgen, *The Moment Problem*, Graduate Texts in Mathematics 277,
DOI 10.1007/978-3-319-64546-9_13

13.1   Positive Polynomials and Sums of Squares

In this section we begin with some simple facts on sums of squares. Then we develop Motzkin's example of a positive polynomial which is not a sum of squares and use it to construct a positive linear functional which is not a moment functional.

We begin with some simple properties of sums of squares of polynomials.

Lemma 13.1 *Let* $p, p_1, \dots, p_r \in \mathbb{R}_d[\underline{x}]$ *and* $p \neq 0$. *If* $p = \sum_{j=1}^{r} p_j^2$, *then*

$$\deg(p) = 2 \max\{\deg(p_j) : j = 1, \dots, r\}.$$

Proof Let n denote the maximum of $\deg(p_j)$, $j = 1, \dots, r$. Clearly, $\deg(p) \leq 2n$. We denote by p_{nj} the homogeneous part of degree n of p_j. (It may happen that $p_{nj} = 0$ for some j.) Then $p_{2n} := \sum_j p_{nj}^2$ is the homogeneous part of degree $2n$ of p. Since $n = \max \deg(p_j)$, there is one index, say $j = 1$, with $p_{n1} \neq 0$. Then $p_{2n}(x) \geq p_{n1}^2(x)$ on $\mathbb{R}^d$ and $p_{n1} \neq 0$. Thus $p_{2n} \neq 0$ and hence $\deg(p) \geq 2n$. $\qquad\square$

Recall that $\mathbb{R}_d[\underline{x}]_m \equiv \mathbb{R}[x_1, \dots, x_d]_m$ denotes the vector space of polynomials $p \in \mathbb{R}_d[\underline{x}]$ with $\deg(p) \leq m$ and $\sum \mathbb{R}[x]_n^2$ is the cone of sums of squares $\sum_j p_j^2$ of polynomials $p_j \in \mathbb{R}_d[\underline{x}]_n$. Then, by Lemma 13.1,

$$\mathbb{R}_d[\underline{x}]_{2n} \cap \sum \mathbb{R}_d[\underline{x}]^2 = \sum \mathbb{R}_d[\underline{x}]_n^2 \text{ for } n \in \mathbb{N}. \qquad (13.1)$$

The vector space $\mathbb{R}_d[\underline{x}]_m$ has dimension $d(m) := \binom{d+m}{m}$. A vector space basis of $\mathbb{R}_d[\underline{x}]_m$ is given by the monomials

$$x^\alpha = x_1^{\alpha_1} \cdots x_d^{\alpha_d}, \text{ where } \alpha \in N_m := \{\alpha \in \mathbb{N}_0^d : |\alpha| := \alpha_1 + \cdots + \alpha_d \leq m\}.$$

We order the basis elements of $\mathbb{R}_d[\underline{x}]_n$ in some fixed way and write them as a column vector $\underline{x}_n$. For instance, a possible "natural" ordering is

$$\underline{x}_n = (1, x_1, \dots, x_d, x_1^2, x_1 x_2, \dots, x_d^2, x_1^3, \dots, x_1^n, \dots, x_d^n)^T. \qquad (13.2)$$

Proposition 13.2 *A polynomial* $f \in \mathbb{R}_d[\underline{x}]$ *is in* $\sum \mathbb{R}_d[\underline{x}]_n^2$ *if and only if there exists a positive semidefinite matrix* $G = (a_{\alpha,\beta})_{\alpha,\beta \in N_n}$ *with real entries such that*

$$f(x) = (\underline{x}_n)^T G \underline{x}_n \equiv \sum_{\alpha, \beta \in N_n} a_{\alpha,\beta} x^{\alpha+\beta}. \qquad (13.3)$$

Proof First suppose that $f = \sum_{j=1}^{r} f_j^2$, where $f_j \in \mathbb{R}_d[\underline{x}]_n$. We write f_j as $f_j(x) = \sum_{\alpha \in N_n} f_{j,\alpha} x^\alpha$ and define

$$a_{\alpha,\beta} := \sum_{j=1}^{r} f_{j,\alpha} f_{j,\beta} \text{ and } G := (a_{\alpha,\beta})_{\alpha,\beta \in N_n}. \qquad (13.4)$$

Then

$$f(x) = \sum_{j=1}^{r} f_j(x)^2 = \sum_{j=1}^{r} \sum_{\alpha,\beta \in N_n} f_{j,\alpha} f_{j,\beta} \, x^{\alpha+\beta} = \sum_{\alpha,\beta \in N_n} a_{\alpha,\beta} x^{\alpha+\beta} = (\mathfrak{x}_n)^T G \mathfrak{x}_n$$

which proves (13.3). Obviously, the matrix G is real and symmetric. To prove that it is positive semidefinite we take a column vector $\eta \in \mathbb{R}^{d(n)}$ and compute

$$\eta^T G \eta = \sum_{\alpha,\beta \in N_n} a_{\alpha,\beta} y_\alpha y_\beta = \sum_{\alpha,\beta \in N_n} \sum_{j=1}^{r} f_{j,\alpha} f_{j,\beta} y_\alpha y_\beta = \sum_{j=1}^{r} \left(\sum_{\alpha \in N_n} f_{j,\alpha} y_\alpha \right)^2 \geq 0.$$

Conversely, assume that $f(x) = (\mathfrak{x}_n)^T G \mathfrak{x}_n$, where G is a real positive semidefinite matrix. Let $r = \operatorname{rank} G$. The assertion is trivial for $G = 0$, so we can assume that $r \in \mathbb{N}$. Let D be the diagonal matrix with nonzero, hence positive, eigenvalues $\lambda_1, \dots, \lambda_r$ of G and C the matrix with columns $u_1, \dots, u_r$ of corresponding orthonormal eigenvectors. Put $B := C\sqrt{D}$. Then $B \in M_{d(n),r}(\mathbb{R})$ and

$$G = \sum_{j=1}^{r} \lambda_j u_j (u_j)^T = CDC^T = BB^T.$$

Here the first equality holds by the spectral theorem for hermitian matrices, see e.g. (A.12). Setting $f_j(x) = \sum_{\alpha \in N_n} b_{\alpha,j} x^\alpha$, we have $f_j \in \mathbb{R}_d[\underline{x}]_n$ and

$$f(x) = (\mathfrak{x}_n)^T BB^T \mathfrak{x}_n = \sum_{\alpha,\beta \in N_n} \sum_{j=1}^{r} x^\alpha b_{\alpha,j} b_{\beta,j} x^\beta = \sum_{j=1}^{r} \left(\sum_{\alpha \in N_n} b_{\alpha,j} x^\alpha \right)^2 = \sum_{j=1}^{r} f_j(x)^2. \quad \square$$

Let $f \in \sum \mathbb{R}_d[\underline{x}]_n^2$. The matrix G defined by (13.4) is called the *Gram matrix* associated with the sum of squares (abbreviated *sos*) representation $f = \sum_{j=1}^{r} f_j^2$ and formula (13.3) is the corresponding *Gram matrix representation*. Gram matrices are a useful tool for detecting possible sos representations of polynomials, see Example 16.4. If $f(x) = \sum_{\gamma} f_\gamma x^\gamma \in \sum \mathbb{R}[x]_n^2$, comparing the coefficients of x^γ in (13.3) yields

$$\sum_{\alpha,\beta \in N_n, \alpha+\beta=\gamma} a_{\alpha,\beta} = f_\gamma, \quad \text{where} \quad \gamma \in N_{2n}. \tag{13.5}$$

The smallest number r appearing in all possible sos representations $f = \sum_{j=1}^{r} f_j^2$ of f is called the *length* of f. The preceding proof shows that the length of f is the smallest rank of all Gram matrices associated with f, so in particular, it is less than or equal to $d(n)$. That is, we have

Corollary 13.3 *Each polynomial $f \in \sum \mathbb{R}_d[\underline{x}]_n^2$ is a sum of at most $d(n) = \binom{d+n}{n}$ squares of polynomials of $\mathbb{R}_d[\underline{x}]_n$.*

The assertion of Corollary 13.3 also follows from Carathéodory's theorem A.35. Obviously, $\sum \mathbb{R}_d[\underline{x}]^2$ is a subset of the cone

$$\text{Pos}(\mathbb{R}^d) = \{p \in \mathbb{R}_d[\underline{x}] : p(x) \geq 0 \ \ \text{for} \ x \in \mathbb{R}^d\}.$$

For $d = 1$ both sets coincides as shown by Proposition 3.1, but for $d \geq 2$ we have $\sum \mathbb{R}_d[\underline{x}]^2 \neq \text{Pos}(\mathbb{R}^d)$. This was already proved in 1888 by D. Hilbert [H1], but the first explicit example was given only in 1966 by T. Motzkin [Mo]. Another famous example, the Robinson polynomial, will appear in Section 19.2.

Proposition 13.4 *Suppose that* $0 < c \leq 3$. *Then the polynomial*

$$p_c(x_1, x_2) := x_1^2 x_2^2 (x_1^2 + x_2^2 - c) + 1 \tag{13.6}$$

is in $\text{Pos}(\mathbb{R}^2) \backslash \sum \mathbb{R}[x_1, x_2]^2$, *that is,* p_c *is nonnegative on* $\mathbb{R}^2$, *but it is not a sum of squares of polynomials.*

Proof From the arithmetic-geometric mean inequality we obtain

$$x_1^4 x_2^2 + x_1^2 x_2^4 + 1 \geq 3 \sqrt[3]{x_1^4 x_2^2 \cdot x_1^2 x_2^4 \cdot 1} = 3 x_1^2 x_2^2 \geq c x_1^2 x_2^2,$$

which in turn implies that $p_c(x_1, x_2) \geq 0$ for all $(x_1, x_2) \in \mathbb{R}^2$.

Now we prove that $p_c \notin \sum \mathbb{R}[x_1, x_2]^2$. Assume to the contrary that $p_c = \sum_j q_j^2$, where $q_j \in \mathbb{R}[x_1, x_2]$. Then we have $\deg(q_j) \leq 3$ by Lemma 13.1. Since $p_c(0, x_2) = p_c(x_1, 0) = 1$, it follows that the polynomials $q_j(0, x_2)$ and $q_j(x_1, 0)$ in one variable are constant. Hence each q_j is of the form $\lambda_j + x_1 x_2 r_j$, where $\lambda_j \in \mathbb{R}$ and $r_j \in \mathbb{R}_d[\underline{x}]$ is linear. Comparing the coefficients of $x_1^2 x_2^2$ in $p_c = \sum_j q_j^2$ yields $\sum_j r_j(0)^2 = -c$. Since $c > 0$, this is a contradiction. $\square$

Motzkin's original example is p_3. Since $p_3(\pm 1, \pm 1) = 0$ and $p_3 \geq 0$ on $\mathbb{R}^2$, the mimimum of p_3 on $\mathbb{R}^2$ is zero. From the identity

$$p_c(\sqrt{c}x_1, \sqrt{c}x_2) = \frac{c^3}{27} p_3(\sqrt{3}x_1, \sqrt{3}x_2) + 1 - \frac{c^3}{27}$$

it follows that the minimum of p_c on $\mathbb{R}^2$ is $1 - \frac{c^3}{27} > 0$ if $0 < c < 3$.

Using the polynomial p_1 we now construct (by some computations) an *explicit* example of a positive linear functional on $\mathbb{R}[x_1, x_2]$ which is not a moment functional. The existence of such a functional is obtained later also by separation arguments (see Example 13.52 below).

Let L_2 denote the linear functional on $\mathbb{R}[x_1, x_2]$ defined by

$$L_2(x_1^k x_2^l) = j_{m(k/2, l/2)} \ \ \text{if} \ k \ \text{and} \ l \ \text{are even,} \ \ L_2(x_1^k x_2^l) = 0 \ \ \text{otherwise,}$$

where m is the bijection of $\mathbb{N}_0^2$ onto $\mathbb{N}$ and j_n are the numbers defined by

$$m(0,0) = 1, m(1,0) = 5, m(0,1) = 6, m(2,0) = 7,$$

$$m(0,2) = 8, m(3,0) = 9, m(0,3) = 10,$$

$$m(k,l) = l + 1 + (k+l)(k+l+1)/2 \text{ for } (k,l) \in \mathbb{N}_0^2, k+l \geq 4,$$

$$j_1 = j_2 = j_3 = 1, \ j_4 = 4, \ j_n = n!^{(n+1)!} \text{ for } n \geq 5.$$

Proposition 13.5 L_2 *is a positive functional which is not a moment functional.*

Proof First we prove that $L_2(f^2) > 0$ for all $f \in \mathbb{R}[x_1, x_2], p \neq 0$. Let us define $\alpha_{m(k,l),m(r,s)} := L_2(x_1^{k+r}x_2^{l+s})$. If $f = \sum_{k,l} c_{k,l} x_1^k x_2^l$, then

$$L_2(f^2) = \sum_{k,l,r,s} \alpha_{m(k,l),m(r,s)} c_{k,l} c_{r,s} . \tag{13.7}$$

Hence it is enough to show that the quadratic form in (13.7) is positive definite. For this it suffices to prove that $\Lambda_n := \det (\alpha_{k,l})_{k,l=1}^n > 0$ for all $n \in \mathbb{N}$.

We prove by induction that $A_n \geq 1$. We obtain $A_1 = A_2 = A_3 = 1$ and $A_4 = 4$. Assume that $n \geq 5$ and $A_{n-1} \geq 1$. A simple computation shows that

$$\max(m(k,l), m(r,s)) > m((k+r)/2, (l+s)/2) \text{ if } (k,l) \neq (r,s)$$

and the right-hand side is defined. This implies $|\alpha_{k,l}| \leq j_{n-1}$ for $k \leq n, l \leq n$, $(k,l) \neq (n,n)$ and $k,l,n \in \mathbb{N}$. Developing the determinant A_n after the n-th row by using these facts and the induction hypothesis $A_{n-1} \geq 1$ we derive

$$A_n \geq j_n A_{n-1} - (n-1)(n-1)! j_{n-1}^n \geq j_n - n! j_{n-1}^n + 1 \geq n!^{(n+1)!} - n!(n-1)!^{n!} + 1 \geq 1.$$

This completes the induction proof. Thus, in particular, $L_2(f^2) \geq 0$ for $f \in \mathbb{R}[x_1, x_2]$.

A direct verification yields $L_2(p_1) = -1$. Therefore, since $p_1 \geq 0$ on $\mathbb{R}^2$, L_2 is not a moment functional.   □

Clearly, $L_d(f) := L_2(f(x_1, x_2, 0, \ldots, 0)), f \in \mathbb{R}_d[\underline{x}]$, defines a positive linear functional on $\mathbb{R}_d[\underline{x}], d \geq 2$, which is not a moment functional. An elegant example of this kind for $d = 2$ is sketched in Exercise 13.3.

The preceding examples can be easily used to construct similar examples for the quarter plane and the Stieltjes moment problem in $\mathbb{R}^2$. Put

$$q_c(x_1, x_2) := x_1 x_2 (x_1 + x_2 - c) + 1, \ c \in (0,3]. \tag{13.8}$$

Then $p_c(x_1, x_2) = q_c(x_1^2, x_2^2)$. Since $p_c \geq 0$ on $\mathbb{R}^2$, it follows that $q_c \geq 0$ on the positive quarter plane $\mathbb{R}_+^2$. But q_c does not belong to the preordering

$$T(x_1, x_2) = \sum \mathbb{R}[x_1, x_2]^2 + x_1 \sum \mathbb{R}[x_1, x_2]^2 + x_2 \sum \mathbb{R}[x_1, x_2]^2 + x_1 x_2 \sum \mathbb{R}[x_1, x_2]^2.$$

(If q_c were in $T(x_1, x_2)$, then, replacing x_1 by x_1^2 and x_2 by x_2^2, it would follow that $p_c \in \sum \mathbb{R}[x_1, x_2]^2$, which is a contradiction.)

Define $L_2'(f) = L_2(f(x_1^2, x_2^2))$ for $f \in \mathbb{R}[x_1, x_2]$. Then L_2' is a $T(x_1, x_2)$-positive linear functional on $\mathbb{R}[x_1, x_2]$. Since $L'(q_1) = L_2(p_1) < 0$, L_2' cannot be given by a Radon measure supported on the set $\mathcal{K}(x_1, x_2) = \mathbb{R}_+^2$.

13.2   Properties (MP) and (SMP)

In this section, we suppose that A is a **finitely generated commutative real unital algebra**. Recall that in our terminology Radon measures are always nonnegative.

As discussed in Sect. 1.1.2, A is (isomorphic to) the quotient algebra $\mathbb{R}_d[\underline{x}]/\mathcal{J}$ for some ideal $\mathcal{J}$ of $\mathbb{R}_d[\underline{x}]$, the set of characters of A is the real algebraic variety $\hat{A} = \mathcal{Z}(\mathcal{J})$, and $\hat{A}$ is a locally compact Hausdorff space. For a quadratic module Q of A we recall the definition $\mathcal{K}(Q) := \{x \in \hat{A} : f(x) \geq 0 , f \in Q\}$ from (12.10). Further, let $\mathcal{M}_+(\hat{A})$ denote the Radon measures μ on $\hat{A}$ such that each $f \in A$ is μ-integrable. We will use these notions and also the basics of real algebraic geometry developed in Sect. 12.1 without mention in what follows.

Our main concepts are introduced in the following definition.

Definition 13.6 A quadratic module Q of A has the

- *moment property (MP)* if each Q-positive linear functional L on A is a moment functional, that is, there exists a Radon measure $\mu \in \mathcal{M}_+(\hat{A})$ such that

$$L(f) = \int_{\hat{A}} f(x)\, d\mu(x) \quad \text{for all } f \in A, \tag{13.9}$$

- *strong moment property (SMP)* if each Q-positive linear functional L on A is a $\mathcal{K}(Q)$–moment functional, that is, there is a Radon measure $\mu \in \mathcal{M}_+(\hat{A})$ such that $\operatorname{supp} \mu \subseteq \mathcal{K}(Q)$ and (13.9) holds.

For $d \geq 2$ the preordering $\sum \mathbb{R}_d[\underline{x}]^2$ does not satisfy (MP); an explicit example was given by Proposition 13.5. Obviously, (SMP) implies (MP). That (MP) does not imply (SMP) is shown by the following simple example in dimension $d = 1$.

Example 13.7 $T(x^3) = \sum \mathbb{R}[x]^2 + x^3 \sum \mathbb{R}[x]^2$ *satisfies (MP), but not (SMP).*

Indeed, since $\sum \mathbb{R}[x]^2 \subseteq T(x^3)$, the preordering $T(x^3)$ has (MP) by Hamburger's theorem 3.8. It remains to show that (SMP) fails.

Let s be a Stieltjes moment sequence which is Hamburger indeterminate (see e.g. the examples in Sect. 4.3). There exists an N-extremal measure μ for s such that $\operatorname{supp}\mu$ contains a negative number and $(-\varepsilon, \varepsilon) \cap \operatorname{supp}\mu \neq \emptyset$ for some $\varepsilon > 0$. (The existence of such a measure follows easily from Theorem 7.7: There is a unique N-extremal measure which has 0 in its support. Any other N-extremal measure whose support contains a negative number has the desired properties.)

We define a measure $\nu \in \mathcal{M}_+(\mathbb{R})$ by $d\nu = x^{-2}d\mu$ and a positive linear functional L on $\mathbb{R}[x]$ by $L(p) = \int p\, d\nu$. Since s is a Stieltjes moment sequence, we have $L(x^3 p^2) = \int xp^2\, d\mu = L_s(xp^2) \geq 0$ for $p \in \mathbb{R}[x]$. Thus, L is $T(x^3)$-positive.

On the other hand, since μ is N-extremal, $\mathbb{C}[x]$ is dense in $L^2(\mathbb{R}, \mu)$. Because $(1 + x^2)d\nu = (1 + x^{-2})d\mu \leq (1 + \varepsilon^{-2})d\mu$, $\mathbb{C}[x]$ is also dense in $L^2(\mathbb{R}, (1 + x^2)d\nu)$. Hence the measure ν is determinate by Corollary 6.11, that is, ν is the only representing measure for L. Since μ, hence ν, has a negative number in its support, L has no representing measure supported on $\mathbb{R}_+ = \mathcal{K}(x^3)$. This shows that $T(x^3)$ does not obey (SMP). Another proof of this fact is given in Example 13.18 below. Note that Stieltjes' theorem 3.12 implies that the functional L is not $T(x)$-positive. $\circ$

The preordering of each compact semi-algebraic set satisfies (SMP) (by Theorem 12.25) and likewise so does each Archimedean quadratic module (by Theorem 12.36). However, deciding whether or not preorderings for noncompact semi-algebraic sets obey (SMP) or (MP) is much more subtle and the fibre theorem stated below deals with this question.

Suppose that Q is a quadratic module of A. Let $\overline{Q}$ denote the closure of Q in the finest locally convex topology of the vector space A, see Appendix A.5 for the definition and some properties of this topology. Each linear functional or linear mapping is continuous in this topology. Applying the latter to the multiplication $A \times A \to A$ it follows that $\overline{Q}$ is also a quadratic module and $\overline{Q}$ is a preordering when Q is a preordering.

The next lemma gives a simple "dual" characterization of $\overline{Q}$.

Lemma 13.8 $\overline{Q}$ is the set of all $f \in A$ such that $L(f) \geq 0$ for all Q-positive linear functionals L on A.

Proof Let $\tilde{Q}$ denote the set of such elements $f \in A$. Suppose $f \in A$ and $f \notin \overline{Q}$. Then, since $\overline{Q}$ is closed in the finest locally convex topology, by the separation theorem for convex sets there is a $\overline{Q}$-positive, hence Q-positive, linear functional L on A such that $L(f) < 0$. Thus, $f \notin \tilde{Q}$. This proves that $\tilde{Q} \subseteq \overline{Q}$.

Conversely, let L be a Q-positive linear functional. Since L is continuous in the finest locally convex topology, L is also $\overline{Q}$-positive. Hence $L(f) \geq 0$ for $f \in \overline{Q}$, so that $\overline{Q} \subseteq \tilde{Q}$. $\square$

Recall that $Q^{\mathrm{sat}} = \operatorname{Pos}(\mathcal{K}(Q))$ is the saturation of Q. Clearly, $Q \subseteq \operatorname{Pos}(\mathcal{K}(Q))$ and $\operatorname{Pos}(\mathcal{K}(Q))$ is closed in the finest locally convex topology, so we have

$$Q \subseteq \overline{Q} \subseteq Q^{\mathrm{sat}} \equiv \operatorname{Pos}(\mathcal{K}(Q)).$$

Haviland's theorem 1.14 leads to the following reformulations of the properties (SMP) and (MP) in terms of the closure $\overline{Q}$ of Q in the finest locally convex topology.

Proposition 13.9

(i) *(SMP) holds if and only if* $\overline{Q} = Q^{\text{sat}} \equiv \text{Pos}(\mathcal{K}(Q))$.
(ii) *(MP) holds if and only if* $\overline{Q} = \text{Pos}(\hat{A})$.

Proof We carry out the proof of (i); (ii) is proved by the same reasoning.

Suppose that $\overline{Q} = \text{Pos}(\mathcal{K}(Q))$. Let L be a Q-positive linear functional. Then L is $\overline{Q}$-positive and hence $\text{Pos}(\mathcal{K}(Q))$-positive. Thus, by Haviland's Theorem 1.14, L comes from a measure supported on $\mathcal{K}(Q)$. This means that Q obeys (SMP).

Now assume that $\overline{Q} \neq \text{Pos}(\mathcal{K}(Q))$. Let $f_0 \in \text{Pos}(\mathcal{K}(Q)) \backslash \overline{Q}$. Then, by Lemma 13.8 there is Q-positive linear functional L such that $L(f_0) < 0$. Since $f_0 \geq 0$ on $\mathcal{K}(Q)$ and $L(f_0) < 0$, L cannot be given by a positive measure supported on $\mathcal{K}(Q)$. That is, (SMP) does not hold. $\square$

13.3   The Fibre Theorem

In this section, A is a **finitely generated commutative real unital algebra**.

Let T be a finitely generated preordering of A and let $f = \{f_1, \ldots, f_k\}$ be a set of generators of T. Further, we fix an m-tuple $h = (h_1, \ldots, h_m)$ of elements $h_k \in A$. Let $\overline{h(\mathcal{K}(T))}$ denote the closure of the subset $h(\mathcal{K}(T))$ of $\mathbb{R}^m$ defined by

$$h(\mathcal{K}(T)) = \{(h_1(x), \ldots, h_m(x)) : x \in \mathcal{K}(T)\}. \tag{13.10}$$

For $\lambda = (\lambda_1, \ldots, \lambda_r) \in \mathbb{R}^m$ we denote by $\mathcal{K}(T)_\lambda$ the subset of $\hat{A}$ given by

$$\mathcal{K}(T)_\lambda = \{x \in \mathcal{K}(T) : h_1(x) = \lambda_1, \ldots, h_m(x) = \lambda_m\}$$

and by T_λ the preordering of A generated by the sequence

$$f(\lambda) := \{f_1, \ldots, f_k, h_1 - \lambda_1, \lambda_1 - h_1, \ldots, h_m - \lambda_m, \lambda_m - h_m\}.$$

Clearly, $\mathcal{K}(T_\lambda) = \mathcal{K}(T)_\lambda$ and $\mathcal{K}(T)$ is the disjoint union of fibre set $\mathcal{K}(T)_\lambda$, where $\lambda \in h(\mathcal{K}(T))$.

Let $\mathcal{I}_\lambda$ denote the ideal of A generated by $h_1 - \lambda_1, \ldots, h_m - \lambda_m$. Then, by formula (12.9) in Example 12.4,

$$T_\lambda = T + \mathcal{I}_\lambda$$

and the preordering $T_\lambda/\mathcal{I}_\lambda$ of the quotient algebra $A/\mathcal{I}_\lambda$ is generated by

$$\pi_\lambda(f) := \{\pi_\lambda(f_1), \ldots, \pi_\lambda(f_k)\},$$

where $\pi_\lambda : A \to A/\mathcal{I}_\lambda$ denotes the canonical map.

Further, let $\hat{\mathcal{I}}_\lambda := \mathcal{I}(\mathcal{Z}(\mathcal{I}_\lambda))$ denote the ideal of all $f \in A$ which vanish on the zero set $\mathcal{Z}(\mathcal{I}_\lambda)$ of $\mathcal{I}_\lambda$. Clearly, $\mathcal{I}_\lambda \subseteq \hat{\mathcal{I}}_\lambda$ and $\mathcal{Z}(\mathcal{I}_\lambda) = \mathcal{Z}(\hat{\mathcal{I}}_\lambda)$. Set

$$\hat{T}_\lambda := T + \hat{\mathcal{I}}_\lambda.$$

Then $\hat{T}_\lambda/\hat{\mathcal{I}}_\lambda$ is a preordering of the quotient algebra $A/\hat{\mathcal{I}}_\lambda$.

In general $\mathcal{I}_\lambda \neq \hat{\mathcal{I}}_\lambda$ and equality holds if and only if the ideal $\mathcal{I}$ is *real*. The latter means that $\sum_j a_j^2 \in \mathcal{I}_\lambda$ for finitely many elements $a_j \in A$ implies that all a_j are in $\mathcal{I}_\lambda$. In real algebraic geometry the ideal $\hat{\mathcal{I}}$ is called the *real radical* of $\mathcal{I}$.

The following *fibre theorem* is the main result of this chapter. It allows us to derive (SMP) or (MP) for T from the corresponding properties of fibre preorderings.

Theorem 13.10 *Let* A *be a finitely generated commutative real unital algebra and let* T *be a finitely generated preordering of* A. *Suppose that* $h_1, \ldots, h_m$ *are elements of* A *that are bounded on the set* $\mathcal{K}(T)$. *Then the following are equivalent:*

  (i) T *satisfies property (SMP) (resp. (MP)) in* A.
 (ii) T_λ *satisfies (SMP) (resp. (MP)) in* A *for all* $\lambda \in h(\mathcal{K}(T))$.
(ii)' $\hat{T}_\lambda$ *satisfies (SMP) (resp. (MP)) in* A *for all* $\lambda \in h(\mathcal{K}(T))$.
(iii) $T_\lambda/\mathcal{I}_\lambda$ *satisfies (SMP) (resp. (MP)) in* $A/\mathcal{I}_\lambda$ *for all* $\lambda \in h(\mathcal{K}(T))$.
(iii)' $\hat{T}_\lambda/\hat{\mathcal{I}}_\lambda$ *satisfies (SMP) (resp. (MP)) in* $A/\hat{\mathcal{I}}_\lambda$ *for all* $\lambda \in h(\mathcal{K}(T))$.

Remark 13.11 The power of Theorem 13.10 can be nicely illustrated by the fact that it contains the main moment problem result (Theorem 12.25) for $A = \mathbb{R}_d[\underline{x}]$, $T = T(\mathsf{f})$, and *compact* semi-algebraic sets $\mathcal{K}(T(\mathsf{f}))$ as an immediate consequence. Indeed, since $\mathcal{K}(T(\mathsf{f}))$ is compact, the coordinate functions x_j are bounded on $\mathcal{K}(T(\mathsf{f}))$, so they can be taken as functions $h_j, j = 1, \ldots, d$. Then all fibre algebras $A/\mathcal{I}_\lambda$, $\lambda \in h(\mathcal{K}(T))$, are $\mathbb{R}$ and $T_\lambda/\mathcal{I}_\lambda = \mathbb{R}_+$ obviously has (SMP) in $A/\mathcal{I}_\lambda = \mathbb{R}$. Hence $T = T(\mathsf{f})$ obeys (SMP) in $\mathbb{R}_d[\underline{x}]$ by the implication (iii)$\to$(i) of Theorem 13.10. $\circ$

To formulate a version of the fibre theorem for quadratic modules let Q be a finitely generated quadratic module of A. We then define the quadratic module

$$Q_\lambda = Q + \mathcal{I}_\lambda \quad \text{and} \quad \hat{Q}_\lambda = Q + \hat{\mathcal{I}}_\lambda.$$

of A and the corresponding fibre set

$$\mathcal{K}(Q)_\lambda := \mathcal{K}(Q_\lambda) = \{x \in \mathcal{K}(Q) : h_1(x) = \lambda_1, \ldots, h_m(x) = \lambda_m\}.$$

Theorem 13.12 *Let* A *be a finitely generated commutative real unital algebra,* Q *a finitely generated quadratic module of* A, *and* $h_1, \ldots, h_m \in A$. *Suppose that there*

324   13 The Moment Problem on Closed Semi-Algebraic Sets: Existence

are $\alpha_j, \beta_j \in \mathbb{R}$ such that $\beta_j - h_j$ and $h_j - \alpha_j$ are in $\overline{Q}$ for $j = 1, \ldots, m$. Set $K := \prod_{j=1}^m [\alpha_j, \beta_j]$. Then the following are equivalent:

(i) Q satisfies property (SMP) (resp. (MP)) in A.
(ii) Q_λ satisfies (SMP) (resp. (MP)) in A for all $\lambda \in K$.
(ii)′ $\hat{Q}_\lambda$ satisfies (SMP) (resp. (MP)) in A for all $\lambda \in K$.
(iii) $Q_\lambda/\mathcal{I}_\lambda$ satifies (SMP) (resp. (MP)) in $A/\mathcal{I}_\lambda$ for all $\lambda \in K$.
(iii)′ $\hat{Q}_\lambda/\hat{\mathcal{I}}_\lambda$ satisfies (SMP) (resp. (MP)) in $A/\hat{\mathcal{I}}_\lambda$ for all $\lambda \in K$.

Remark 13.13

1. The fibre set $\mathcal{K}(Q)_\lambda$ may be empty for some $\lambda \in \prod_{j=1}^m [\alpha_j, \beta_j]$. For such λ we have $\overline{Q}_\lambda = $ A and (SMP) holds trivially. However, if $\lambda \in h(\mathcal{K}(T))$, say $\lambda = h(x)$ with $x \in \mathcal{K}(T)$, then $x \in \mathcal{K}(T)_\lambda$, so the fibre is not empty.
2. Suppose that $\beta_j - h_j$ and $h_j - \alpha_j$ are in $\overline{Q}$ as in Theorem 13.12. Then, since evaluations by points $x \in \mathcal{K}(Q)$ are Q-positive by definition and hence $\overline{Q}$-positive, we have $\beta_j - h_j(x) \geq 0$ and $h_j(x) - \alpha_j \geq 0$. That is, h_j is bounded on $\mathcal{K}(Q)$, so the corresponding assumption of Theorem 13.10 is satisfied.
3. In Theorem 13.10 it is only assumed that the polynomials h_j are bounded, but not that $\beta_j - h_j, h_j - \alpha_j$ are in $\overline{Q}$. Therefore, Theorem 13.10 is much stronger than Theorem 13.12.   ○

The implications (i)→(ii), the equivalence (ii)↔(ii)′ and the two equivalences (ii)↔(iii) and (ii)′ ↔(iii)′ of Theorems 13.10 and 13.12 follow immediately from Proposition 13.14 (i),(ii), and (iii), respectively, proved below.

The proofs of the remaining main implication (ii)→(i) of Theorems 13.10 and 13.12 are lengthy and technically involved. They are postponed until Sect. 13.10. Further, a crucial step in the proof of Theorem 13.10 is Proposition 12.23, which is based on the Krivine–Stengle Positivstellensatz (Theorem 12.3).

Proposition 13.14 *Let $\mathcal{I}$ be an ideal and Q a quadratic module of A. Let $\hat{\mathcal{I}}$ be the ideal of all $f \in$ A which vanish on the zero set $\mathcal{Z}(\mathcal{I})$ of $\mathcal{I}$.*

(i) *If Q satisfies (SMP) (resp. (MP)) in A, so does $Q + \mathcal{I}$ in A.*
(ii) *$Q + \mathcal{I}$ satisfies (SMP) (resp. (MP)) in A if and only if $Q + \hat{\mathcal{I}}$ does.*
(iii) *$Q+\mathcal{I}$ satisfies (SMP) (resp. (MP)) in A if and only if $(Q+\mathcal{I})/\mathcal{I}$ does in $A/\mathcal{I}$.*

Proof We only carry out the proofs for (SMP). The proofs for (MP) are even simpler, since no support conditions have to be verified. Let $\pi : \mathbb{R}_d[\underline{x}] \to \mathbb{R}_d[\underline{x}]/\mathcal{J} = $ A denote the canonical map and $\tilde{\mathcal{I}}$ the ideal $\tilde{\mathcal{I}} := \pi^{-1}(\mathcal{I})$ of $\mathbb{R}_d[\underline{x}]$.

(i) Let L be a $(Q + \mathcal{I})$-positive functional on A. Since L is Q-positive and Q obeys (SMP), L is given by a measure $\mu \in \mathcal{M}_+(\hat{A})$ supported on $\mathcal{K}(Q)$. Then $\tilde{L}(\cdot) := L(\pi(\cdot))$ is a linear functional on $\mathbb{R}_d[\underline{x}]$. Since L is $\mathcal{I}$-positive, $\tilde{L}$ is $\tilde{\mathcal{I}}$-positive and hence supp $\mu \subseteq \mathcal{Z}(\tilde{\mathcal{I}})$ by Proposition 12.19. (Note that a functional is positive on an ideal if and only if it annihilates the ideal.) But $\mathcal{Z}(\tilde{\mathcal{I}}) \subseteq \mathcal{Z}(\mathcal{I})$, so that we have supp $\mu \subseteq \mathcal{K}(Q) \cap \mathcal{Z}(\mathcal{I}) = \mathcal{K}(Q+\mathcal{I})$. That is, $Q+\mathcal{I}$ satisfies (SMP) in A.

(ii) It suffices to show that both quadratic modules $Q+\mathcal{I}$ and $Q+\hat{\mathcal{I}}$ have the same nonnegative characters and linear functionals.

For the sets of characters, using the equality $\mathcal{Z}(\mathcal{I}) = \mathcal{Z}(\hat{\mathcal{I}})$ we obtain

$$\mathcal{K}(Q+\mathcal{I}) = \mathcal{K}(Q) \cap \mathcal{Z}(\mathcal{I}) = \mathcal{K}(Q) \cap \mathcal{Z}(\hat{\mathcal{I}}) = \mathcal{K}(Q+\hat{\mathcal{I}}).$$

Since $Q+\mathcal{I} \subseteq Q+\hat{\mathcal{I}}$, a $(Q+\hat{\mathcal{I}})$-positive functional is trivially $(Q+\mathcal{I})$-positive. Conversely, let L be a $(Q+\mathcal{I})$-positive linear functional on A.

We verify that $\mathcal{Z}(\tilde{\mathcal{I}}) \subseteq \mathcal{Z}(\hat{\mathcal{I}})$. Let $x \in \mathcal{Z}(\tilde{\mathcal{I}})(\subseteq \mathbb{R}^d)$. Clearly, $\mathcal{J} \subseteq \tilde{\mathcal{I}}$ and $\mathcal{Z}(\tilde{\mathcal{I}}) \subseteq \mathcal{Z}(\mathcal{J})=\hat{A}$. Let $g \in \mathcal{I}$ and choose $\tilde{g} \in \tilde{\mathcal{I}}$ such that $g = \pi(\tilde{g})$. Since x annihilates $\mathcal{J}$ and $x \in \mathcal{Z}(\tilde{\mathcal{I}})$, we have $g(x) = \tilde{g}(x) = 0$. Thus, $x \in \mathcal{Z}(\mathcal{I}) = \mathcal{Z}(\hat{\mathcal{I}})$.

Now let $f \in \hat{\mathcal{I}}$. We choose $\tilde{f} \in \mathbb{R}_d[\underline{x}]$ such that $\pi(\tilde{f}) = f$. Then $f(x) = \tilde{f}(x)$ for $x \in \mathcal{Z}(\mathcal{J})$. Hence, since f vanishes on $\mathcal{Z}(\hat{\mathcal{I}})$ and $\mathcal{Z}(\tilde{\mathcal{I}}) \subseteq \mathcal{Z}(\hat{\mathcal{I}})$, the polynomial $\tilde{f}$ vanishes on $\mathcal{Z}(\tilde{\mathcal{I}})$. Therefore, by the real Nullstellensatz (Theorem 12.3(iii)), there are $m \in \mathbb{N}$ and $g \in \sum \mathbb{R}_d[\underline{x}]^2$ such that $p := (\tilde{f})^{2m} + g \in \tilde{\mathcal{I}}$. Upon multiplying p by some even power of $\tilde{f}$ we can assume that $2m = 2^k$ for some $k \in \mathbb{N}$. Then

$$\pi(p) = f^{2^k} + \pi(g) \in \mathcal{I}, \quad \text{where} \quad \pi(g) \in \sum A^2.$$

Being $(Q+\mathcal{I})$-positive, L annihilates $\mathcal{I}$ and is nonnegative on $\sum A^2$. Hence

$$0 = L(\pi(p)) = L(f^{2^k}) + L(\pi(g)), \quad L(\pi(g)) \geq 0, \quad L(f^{2^k}) \geq 0.$$

This implies that $L(f^{2^k}) = 0$. Since L is nonnegative on $\sum A^2$, the Cauchy–Schwarz inequality holds. By a repeated application of this inequality we derive

$$|L(f)|^{2^k} \leq L(f^2)^{2^{k-1}} L(1)^{2^{k-1}} \leq L(f^4)^{2^{k-2}} L(1)^{2^{k-2}+2^{k-1}} \leq \cdots$$
$$\leq L(f^{2^k})L(1)^{1+\cdots+2^{k-1}} = 0.$$

Thus $L(f) = 0$. That is, L annihilates $\hat{\mathcal{I}}$. Hence L is $(Q+\hat{\mathcal{I}})$-positive which completes the proof of (ii).

(iii) The assertions are only slight reformulations of Definition 13.6.

Let ρ denote the canonical map of A into $A/\mathcal{I}$. Clearly, the character set of $A/\mathcal{I}$ can be identified with $\mathcal{Z}(\mathcal{I}) = \{x \in \hat{A} : f(x) = 0 \text{ for } f \in \mathcal{I}\}$.

Suppose that $Q+\mathcal{I}$ obeys (SMP) in A. Let $\tilde{L}$ be a $(Q+\mathcal{I})/\mathcal{I}$-positive linear functional on $A/\mathcal{I}$. Then $L := \tilde{L} \circ \rho$ defines a $(Q+\mathcal{I})$-positive linear functional on A. Since $Q + \mathcal{I}$ has (SMP), the functional L, hence also $\tilde{L}$, is given by a measure of $\mathcal{M}_+(\hat{A})$ supported on $\mathcal{K}(Q+\mathcal{I}) = \mathcal{K}(Q) \cap \mathcal{Z}(\mathcal{I}) = \mathcal{K}((Q+\mathcal{I})/\mathcal{I})$. This shows that $(Q + \mathcal{I})/\mathcal{I}$ satisfies (SMP) in $A/\mathcal{I}$.

Conversely, assume that $(Q + \mathcal{I})/\mathcal{I}$ has (SMP) in $A/\mathcal{I}$ and let L be a $(Q+\mathcal{I})$-positive linear functional on A. Then L is in particular $\mathcal{I}$-positive, so it annihilates $\mathcal{I}$, and there is a well-defined linear functional $\tilde{L}$ on $A/\mathcal{I}$ such that $L = \tilde{L} \circ \rho$. Clearly, $\tilde{L}$ is $(Q + \mathcal{I})/\mathcal{I}$-positive. Since $(Q + \mathcal{I})/\mathcal{I}$ has (SMP), the functional $\tilde{L}$, and therefore also L, comes from a Radon measure with support contained in $\mathcal{K}((Q+\mathcal{I})/\mathcal{I}) = \mathcal{K}(Q+\mathcal{I})$. That is, $Q+\mathcal{I}$ obeys (SMP) in A. □

Remark 13.15 The Positivstellensatz for $\mathbb{R}_d[\underline{x}]$ (Theorem 12.3) is the only unproven result from real algebraic geometry we use in this book. In the preceding proof we derived the real Nullstellensatz for A from Theorem 12.3(iii). That the real Nullstellensatz holds for the algebra A follows from [PD, Section 4.2]. The above proof of assertion (ii) becomes much shorter if we use the latter result.   ○

We state the special case $Q = \sum A^2$ of Proposition 13.14(iii) separately as

Corollary 13.16 *If $\mathcal{I}$ is an ideal of A, then $\mathcal{I} + \sum A^2$ obeys (MP) (resp. (SMP)) on A if and only if $\sum (A/\mathcal{I})^2$ does in $A/\mathcal{I}$.*

The following simple fact is used later several times. Of course, it can also be derived directly from Hamburger's theorem 3.8.

Corollary 13.17 *If the algebra A has a single generator, then $\sum A^2$ obeys (MP).*

Proof Being single generated, A is isomorphic to a quotient algebra $\mathbb{R}[y]/\mathcal{I}$ for some ideal $\mathcal{I}$ of $\mathbb{R}[y]$. By Hamburger's theorem 3.8, $\sum \mathbb{R}[y]^2$ obeys (MP) in $\mathbb{R}[y]$ and so does $\mathcal{I} + \sum \mathbb{R}[y]^2$. Therefore, by Corollary 13.16, $\sum (\mathbb{R}[y]/\mathcal{I})^2 \cong \sum A^2$ has (MP) in $\mathbb{R}[y]/\mathcal{I} \cong A$.   □

13.4 (SMP) for Basic Closed Semi-Algebraic Subsets of the Real Line

In many applications of Theorem 13.10 the fibres are semi-algebraic subsets of the real line. In this section, we investigate (SMP) for such sets in detail.

To illustrate the corresponding phenomena we begin with two examples.

Example 13.18 $d = 1, \mathfrak{f} = \{x^3\}, \mathcal{K}(\mathfrak{f}) = \mathbb{R}_+$.

It is obvious that the polynomial x is in $\mathrm{Pos}(\mathbb{R}_+) = \mathrm{Pos}(\mathcal{K}(\mathfrak{f}))$.

We prove that $x \notin T(\mathfrak{f})$. Assume to the contrary that $x = \sum_j p_j^2 + x^3 q$, where $p_j \in \mathbb{R}[x]$ and $q \in \sum \mathbb{R}[x]^2$. Setting $x = 0$ yields $\sum_j p_j(0)^2 = 0$. Therefore, $p_j(0) = 0$ and hence $p_j(x) = x f_j(x)$ with $f_j \in \mathbb{R}[x]$ for each j. Inserting this and dividing by x we get $1 = x \sum_j f_j^2 + x^2 q$. Setting once more $x = 0$ we obtain a contradiction. Thus, $x \notin T(\mathfrak{f})$ and hence $T(\mathfrak{f}) \neq \mathrm{Pos}(\mathcal{K}(\mathfrak{f}))$.

It will be shown by Corollary 13.49 below that the preordering $T(\mathfrak{f})$ is closed in $\mathbb{R}[x]$ in the finest locally convex topology. Hence $T(\mathfrak{f}) = \overline{T(\mathfrak{f})} \neq \mathrm{Pos}(\mathcal{K}(\mathfrak{f}))$, so that $T(\mathfrak{f})$ does not obey (SMP) by Proposition 13.9(i). This was already proved in

Example 13.7. There a $T(\mathfrak{f})$-positive linear functional was constructed that cannot be given by a Radon measure supported on $\mathbb{R}_+$.

However, if we replace $\mathfrak{f} = \{x^3\}$ by $\tilde{\mathfrak{f}} = \{x\}$, then $\mathcal{K}(\mathfrak{f}) = \mathcal{K}(\tilde{\mathfrak{f}}) = \mathbb{R}_+$ and $T(\tilde{\mathfrak{f}}) = \mathrm{Pos}(\mathcal{K}(\tilde{\mathfrak{f}})) = \mathrm{Pos}(\mathbb{R}_+)$ by formula (3.4) in Proposition 3.2. Thus $T(\tilde{\mathfrak{f}})$ has (SMP) by Proposition 13.9(i) (or likewise by Stieltjes' theorem 3.12).

This shows that, in contrast to the compact case, (SMP) depends in a crucial manner on the "right" polynomials defining the noncompact semi-algebraic set!   ∘

Example 13.19 $d = 1, \mathfrak{f} = \{(1 - x^2)^3\}, \mathcal{K}(\mathfrak{f}) = [-1, 1]$.

A similar reasoning as in Example 13.18 (using the zeros ± 1 instead) shows that the polynomial $(1 - x^2) \in \mathrm{Pos}([-1, 1]) = \mathrm{Pos}(\mathcal{K}(\mathfrak{f}))$ is not in the preordering $T(\mathfrak{f})$. That is, similarly as in Example 13.18 we have $T(\mathfrak{f}) \neq \mathrm{Pos}(\mathcal{K}(\mathfrak{f})) = \mathrm{Pos}([-1, 1])$ and also $T(\tilde{\mathfrak{f}}) = \mathrm{Pos}(\mathcal{K}(\tilde{\mathfrak{f}})) = \mathrm{Pos}([-1, 1])$ when we take $\tilde{\mathfrak{f}} := \{1 - x^2\}$.

But as $\mathcal{K}(\mathfrak{f})$ is compact, $T(\mathfrak{f})$ satisfies (SMP) by Theorem 12.25. Hence we have $\overline{T(\mathfrak{f})} = \mathrm{Pos}(\mathcal{K}(\mathfrak{f}))$ by Proposition 13.9(i). In particular, $T(\mathfrak{f})$ is not closed in $\mathbb{R}[x]$.   ∘

In both examples we have $T(\mathfrak{f}) \neq \mathrm{Pos}(\mathcal{K}(\mathfrak{f}))$, but $T(\tilde{\mathfrak{f}}) = \mathrm{Pos}(\mathcal{K}(\tilde{\mathfrak{f}}))$. To overcome this difficulty we now define the "right set of generators".

Let $\mathcal{K}$ be a nonempty basic closed semi-algebraic proper subset of $\mathbb{R}$, that is, $\mathcal{K}$ is the union of finitely many closed intervals. (These intervals can be unbounded or points.)

Definition 13.20 A finite subset $\mathfrak{g}$ of $\mathbb{R}[x]$ is called a *natural choice of generators* for $\mathcal{K}$ if $\mathfrak{g}$ is the smallest set satisfying the following conditions:

- If $\mathcal{K}$ contains a least element a (that is, if $(-\infty, a) \cap \mathcal{K} = \emptyset$), then $(x - a) \in \mathfrak{g}$.
- If $\mathcal{K}$ contains a greatest element a (that is, if $(a, \infty) \cap \mathcal{K} = \emptyset$), then $(a - x) \in \mathfrak{g}$.
- If $a, b \in \mathcal{K}, a < b$, and $(a, b) \cap \mathcal{K} = \emptyset$, then $(x - a)(x - b) \in \mathfrak{g}$.

From this definition it is not difficult to see that a choice of natural generators always exists and that it is uniquely determined by the set $\mathcal{K}$. Moreover, we have $\mathcal{K} = \mathcal{K}(\mathfrak{g})$. For $\mathcal{K} = \mathbb{R}$ we set $\mathfrak{g} = \{1\}$.

Let us give some examples for the natural choice of generators:

$\mathcal{K} = \{a\} \cup [b, +\infty)$, where $a < b$: $\mathfrak{g} = \{x - a, (x - a)(x - b)\}$,
$\mathcal{K} = [a, b] \cup \{c\}$, where $a < b < c$: $\mathfrak{g} = \{x - a, (x - b)(x - c), c - x\}$,
$\mathcal{K} = \{a\} \cup \{b\}$, where $a < b$: $\mathfrak{g} = \{x - a, (x - a)(x - b), b - x\}$,
$\mathcal{K} = \{a\}$: $\mathfrak{g} = \{x - a, a - x\}$.

The next proposition shows that for the natural choice of generators the preordering contains all nonnegative polynomials on $\mathcal{K}$.

Proposition 13.21 *Suppose that $\mathcal{K}$ is a nonempty basic closed semi-algebraic subset of $\mathbb{R}$. If $\mathfrak{g}$ is the natural choice of generators for $\mathcal{K}$, then $\mathrm{Pos}(\mathcal{K}) = T(\mathfrak{g})$.*

Proof By construction, the natural generators are nonnegative on $\mathcal{K}$, so the inclusion $T(\mathfrak{g}) \subseteq \mathrm{Pos}(\mathcal{K})$ is obvious.

For the converse we prove by induction on the degree of p that $p \in \mathrm{Pos}(\mathcal{K})$ implies $p \in T(\mathfrak{g})$. If $\deg(p) = 0$, this is obvious. Assume that it is true if $\deg(p) \leq$

n. Suppose that $q \in \text{Pos}(\mathcal{K})$ and $\deg(q) = n+1$. If $q \geq 0$ on $\mathbb{R}$, then q is in $\sum \mathbb{R}[x]^2$ by Proposition 3.1 and so in $T(\mathfrak{g})$. Thus we can assume that $q(\lambda) < 0$ for some $\lambda \in \mathbb{R}$. Since $\lambda \notin \mathcal{K}$, there are three possible cases.

Case 1: $\mathcal{K}$ contains a least element a and $\lambda < a$.

Since $q \in \text{Pos}(\mathcal{K})$, q has roots in the interval $(\lambda, a]$. Let c be the least such root. Then $q = (x - c)p$ with $p \in \text{Pos}(\mathcal{K})$ and $\deg(p) = n$. Since $p \in T(\mathfrak{g})$ by the induction hypothesis and $(x-a) \in \mathfrak{g}$ by Definition 13.20, $x-c = (x-a)+(a-c) \in T(\mathfrak{g})$ and hence $q = (x - c)p \in T(\mathfrak{g})$.

Case 2: $\mathcal{K}$ contains a largest element a and $\lambda > a$.

Let d be the largest root of q in the interval $[a, \lambda)$. Then $q = (d - x)p$ with $p \in \text{Pos}(\mathcal{K})$ and $\deg(p) = n$, so $p \in T(\mathfrak{g})$ by the induction hypothesis. By Definition 13.20, $(a - x) \in \mathfrak{g}$. Therefore, $d - x = (a - x) + (d - a) \in T(\mathfrak{g})$, so that $q = (d - x)p \in T(\mathfrak{g})$.

Case 3: There exist $a, b \in \mathcal{K}$, $a < b$, such that $(a, b) \cap \mathcal{K} = \emptyset$.

In this case we take the greatest root d of p in the interval $[a, \lambda)$ and the least root c in the interval $(\lambda, b]$. Then we can write $p = (x - c)(x - d)q$ with $q \in \text{Pos}(\mathcal{K})$ and $\deg(q) = n-1$, so $q \in T(\mathfrak{g})$ by the induction hypothesis. We have $(x-a)(x-b) \in \mathfrak{g}$ by Definition 13.20. From Lemma 13.22 below it follows that $(x - c)(x - d)$ in the preordering generated by $(x-a)(x-b)$, so that $(x-c)(x-d) \in T(\mathfrak{g})$. Consequently, $p = (x - c)(x - d)q \in T(\mathfrak{g})$. $\square$

Lemma 13.22 *Suppose that $a \leq b$ and $c, d \in [a, b]$. There exists a $\gamma > 0$ such that*

$$(x - c)(x - d) - \gamma(x - a)(x - b) \in \sum \mathbb{R}[x]^2.$$

Proof By a linear transformation we can assume that $a = -1$ and $b = 1$. Set $\tau := \text{sign}(c + d)$. Since $c, d \in [-1, 1]$ and hence $(1 - \tau c)(1 - \tau d) \geq 0$, we obtain

$$2(x - c)(x - d) - (2 - |c + d|)(x^2 - 1)$$

$$= |c + d|x^2 - 2(c + d)x + 2 + 2cd - |c + d|$$

$$= |c + d|(x^2 - 2\tau x + 1) + 2 + 2cd - 2\tau(c + d)$$

$$= |c + d|(x - \tau)^2 + 2(1 - \tau c)(1 - \tau d) \geq 0.$$

Since $c, d \in [-1, 1]$, we have $\gamma := 1 - \frac{|c+d|}{2} \geq 0$ and $(x - c)(x - d) - \gamma(x^2 - 1) \geq 0$ on $\mathbb{R}$ by the preceding inequality. Therefore, $(x - c)(x - d) - \gamma(x^2 - 1) \in \sum \mathbb{R}[x]^2$ by Proposition 3.1. $\square$

The following elementary fact is used in the proof of Theorem 13.24 below. For a quadratic f we define $\mathfrak{w}(f) = |\lambda_1 - \lambda_2|$ if f has real roots λ_1, λ_2 and $\mathfrak{w}(f) = 0$ if f has no real roots.

Lemma 13.23 *If f_1 and f_2 are quadratics with positive leading coefficients, then*

$$\mathfrak{w}(f_1 + f_2) \leq \max(\mathfrak{w}(f_1), \mathfrak{w}(f_2)).$$

Proof Without loss of generality we can assume that $\mathfrak{w}(f_2) \leq \mathfrak{w}(f_1)$ and $\mathfrak{w}(f_1) > 0$. Upon translation and scaling it suffices to show the assertion for $f_1(x) = x(x - 1)$ and $f_2(x) = c(x - a)(a - (a + b))$, where $0 \leq b \leq 1, c > 0$. Then, for

$$f_1 + f_2 = (c + 1)x^2 - ((2a + b)c + 1)x + (a + b)ac$$

we have

$$\mathfrak{w}(f_1 + f_2) = \frac{\sqrt{((2a + b)c + 1)^2 - 4(c + 1)(a + b)ac}}{c + 1}.$$

Since $\mathfrak{w}(f_1) = 1$ and $\mathfrak{w}(f_2) = b \leq 1$, we have to prove that $\mathfrak{w}(f_1 + f_2) \leq 1$. After squaring and multiplying by $(c + 1)^2$, the assertion $\mathfrak{w}(f_1 + f_2) \leq 1$ is equivalent to

$$((2a + b)c + 1)^2 - 4(c + 1)(a + b)ac \leq (c + 1)^2.$$

A straightforward computation shows that this is equivalent to the inequality

$$(2a + b - 1)^2 + (1 - b^2)(c + 1) \geq 0.$$

Since $0 \leq b \leq 1$, the latter is satisfied, so the assertion $\mathfrak{w}(f_1 + f_2) \leq 1$ holds. □

The following theorem characterizes those noncompact semi-algebraic subsets $\mathcal{K}(\mathfrak{f})$ of $\mathbb{R}$ for which the preordering $T(\mathfrak{f})$ has (SMP).

Theorem 13.24 *Suppose that* $\mathfrak{f} = \{f_1, \ldots, f_k\}$ *is a finite subset of* $\mathbb{R}[x]$ *such that the semi-algebraic subset* $\mathcal{K}(\mathfrak{f})$ *of* $\mathbb{R}$ *is not compact. The following are equivalent:*

(i) *$T(\mathfrak{f})$ obeys (SMP).*
(ii) *$T(\mathfrak{f}) = \mathrm{Pos}(\mathcal{K}(\mathfrak{f}))$, that is, the preordering $T(\mathfrak{f})$ is saturated.*
(iii) *$\mathfrak{f}$ contains positive multiples of all polynomials of the natural choice of generators $\mathfrak{g}$ of $\mathcal{K}(\mathfrak{f})$.*

Proof By Proposition 13.9(i), (SMP) holds if and only if $\overline{T(\mathfrak{f})} = \mathrm{Pos}(\mathcal{K})$. Since $\mathcal{K}(\mathfrak{f})$ is not compact and semi-algebraic, it contains an unbounded interval. Therefore, $T(\mathfrak{f})$ is closed by Proposition 13.51 proved in Sect. 13.8. We take Proposition 13.51 for granted in this proof. Then (i)↔(ii).

(iii)→(ii) (iii) implies that $T(\mathfrak{g}) \subseteq T(\mathfrak{f})$. Obviously, $T(\mathfrak{f}) \subseteq \mathrm{Pos}(\mathcal{K}(\mathfrak{f}))$. Proposition 13.21 yields $\mathrm{Pos}(\mathcal{K}(\mathfrak{f})) = T(\mathfrak{g})$. Putting these facts together we obtain $T(\mathfrak{f}) = \mathrm{Pos}(\mathcal{K}(\mathfrak{f}))$.

(ii)→(iii) As already noted, $\mathcal{K}(\mathfrak{f})$ contains an unbounded interval. Upon replacing x by $-x$ we can assume that $[c, +\infty) \subseteq \mathcal{K}(\mathfrak{f})$ for some $c \in \mathbb{R}$. Also we can assume that all elements of $\mathfrak{f}$ are not constant. Each $f \in T(\mathfrak{f})$ is of the form

$$f = \sum_e f_1^{e_1} \cdots f_k^{e_k} \sigma_e \quad \text{with} \quad \sigma_e \in \sum \mathbb{R}_d[\underline{x}]^2. \tag{13.11}$$

The summation in (13.11) is over $e = (e_1 \ldots, e_k)$, where $e_1, \ldots, e_k \in \{0, 1\}$.

All f_j are nonnegative on $[c, +\infty)$, so they have positive leading coefficients. Hence the degree of f in (13.11) is equal to the maximum of the degrees of the summands.

Case 1: $\mathcal{K}(\mathfrak{f})$ contains a least element a.

Then $f := x - a \in \mathrm{Pos}(\mathcal{K}(\mathfrak{f})) = T(\mathfrak{f})$. Since f is linear, all nonzero summands in the representation (13.11) of f are multiples of linear f_j by positive constants. Since $a \in \mathcal{K}(\mathfrak{f})$, $f_j(a) \geq 0$ for each such f_j. Since $f(a) = 0$, at least one such f_j vanishes at a. Hence $f_j(x) = \lambda(x - a)$ with $\lambda > 0$, that is, $\mathfrak{f}$ contains a positive multiple of the natural generator $x - a$.

Case 2: There are $a, b \in \mathcal{K}(\mathfrak{f})$, $a < b$, such that $(a, b) \cap \mathcal{K}(\mathfrak{f}) = \emptyset$.

Then $f := (x - a)(x - b) \in \mathrm{Pos}(\mathcal{K}(\mathfrak{f})) = T(\mathfrak{f})$ and $f(x) < 0$ on (a, b). Hence the degrees of all summands in the representation (13.11) of f do not exceed 2. We omit all summands which are nonnegative on (a, b). Each linear f_j is increasing (because $f_j \geq 0$ on $[c, +\infty)$) and satisfies $f_j(a) \geq 0$ (since $a \in \mathcal{K}(\mathfrak{f})$). Hence linear f_j and their products are positive on (a, b). Since $f(x) < 0$ on (a, b), there exist quadratic polynomials $f_1, \ldots f_r$ in the set $\mathfrak{f}$ and positive numbers $\alpha_1, \ldots, \alpha_r$ such that $f(x) \geq \alpha_1 f_1(x) + \cdots \alpha_r f_r(x)$ on (a, b) and each such f_j has at least one negative value on (a, b). Since $f_j \geq 0$ on $[c, +\infty) \subseteq \mathcal{K}(\mathfrak{f})$ and at $a, b \in \mathcal{K}(\mathfrak{f})$, the polynomial f_j, hence $\alpha_j f_j$, has two real zeros in $[a, b]$, that is, $\mathfrak{w}(\alpha_j f_j) \leq b - a$. From Lemma 13.23 it follows that $\mathfrak{w}(f) = b - a$ is at most the maximum of $\mathfrak{w}(\alpha_j f_j)$, $j = 1, \ldots, r$. Thus $\mathfrak{w}(\alpha_j f_j) = b - a$ for at least one j. Hence $\alpha_j f_j = \lambda(x - a)(x - b)$ for some $\lambda > 0$, so a positive multiple of the natural generator $(x - a)(x - b)$ belongs to $\mathfrak{f}$. $\square$

13.5   Application of the Fibre Theorem: Cylinder Sets with Compact Base

Perhaps the most natural application of the fibre theorem concerns subsets of cylinders with compact base.

Proposition 13.25 *Let C be a compact set in $\mathbb{R}^{d-1}$, $d \geq 2$, and let $\mathfrak{f}$ be a finite subset of $\mathbb{R}_d[\underline{x}]$. Suppose that the semi-algebraic subset $\mathcal{K}(\mathfrak{f})$ of $\mathbb{R}^d$ is contained in the cylinder $C \times \mathbb{R}$. Then the preordering $T(\mathfrak{f})$ has (MP). If C is a semi-algebraic set in $\mathbb{R}^{d-1}$ and $\mathcal{K}(\mathfrak{f}) = C \times \mathbb{R}$, then $T(\mathfrak{f})$ satisfies (SMP).*

Proof Define $h_j(x) = x_j$ for $j = 1, \ldots, d - 1$. Since $\mathcal{K}(\mathfrak{f}) \subseteq C \times \mathbb{R}$ and C is compact, the polynomials h_j are bounded on $\mathcal{K}$, so the assumptions of Theorem 13.10 are fulfilled. Then all fibres $\mathcal{K}(\mathfrak{f})_\lambda$ are subsets of $(\lambda_1, \ldots, \lambda_{d-1}) \times \mathbb{R}$, the preordering $T(\mathfrak{f})_\lambda$ contains $\sum \mathbb{R}[x_d]^2$, and the quotient algebra $\mathbb{R}_d[\underline{x}]/\mathcal{I}_\lambda$ is an algebra of polynomials in the single variable x_d. Hence, by Corollary 13.17, $\sum(\mathbb{R}_d[\underline{x}]/\mathcal{I}_\lambda)^2$ obeys (MP) in $\mathbb{R}_d[\underline{x}]/\mathcal{I}_\lambda$ and so does $T/\mathcal{I}_\lambda$. Therefore, $T(\mathfrak{f})$ has (MP) by the implication (ii)→(i), and likewise by (iii)→(i), of Theorem 13.10.

If $\mathcal{K}(\mathfrak{f}) = C \times \mathbb{R}$, the fibres for $\lambda \in h(\mathcal{K}(\mathfrak{f}))$ are equal to $(\lambda_1, \dots, \lambda_{d-1}) \times \mathbb{R}$ and $T(\mathfrak{f})_\lambda = \sum \mathbb{R}[x_d]^2$. Hence the $T(\mathfrak{f})_\lambda$ satisfy (SMP) and so does $T(\mathfrak{f})$. $\qquad\square$

We restate this result in the special case of a strip $[a, b] \times \mathbb{R}$ in $\mathbb{R}^2$.

Example 13.26 Let $a, b \in \mathbb{R}$, $a < b$, $d = 2$, and $\mathfrak{f} = \{(x_1 - a)(b - x_1)\}$. Then

$$\mathcal{K}(\mathfrak{f}) = [a, b] \times \mathbb{R}, \quad T(\mathfrak{f}) = \sum \mathbb{R}[x_1, x_2]^2 + (x_1 - a)(b - x_1) \sum \mathbb{R}[x_1, x_2]^2.$$

By Proposition 13.25, $T(\mathfrak{f})$ obeys (SMP). That is, given a linear functional L on $\mathbb{R}[x_1, x_2]$, there exists a Radon measure μ on $\mathbb{R}^2$ supported on $[a, b] \times \mathbb{R}$ such that p is μ-integrable and

$$L(p) = \int_a^b \int_\mathbb{R} p(x_1, x_2) \, d\mu(x_1, x_2) \quad \text{for all } p \in \mathbb{R}[x_1, x_2]$$

if and only if

$$L(q_1^2 + (x_1 - a)(b - x_2)q_2^2) \geq 0 \quad \text{for all } q_1, q_2 \in \mathbb{R}[x_1, x_2]. \qquad\circ$$

Let us return to Proposition 13.25 and assume that the semi-algebraic subset $\mathcal{K}(\mathfrak{f})$ of $\mathbb{R}^d$ is only a proper subset of $C \times \mathbb{R}$. Then $T(\mathfrak{f})$ does not satisfy (SMP) in general. Recall that, by Theorem 13.10, $T(\mathfrak{f})$ obeys (SMP) if (and only if) all fibre preorderings $T(\mathfrak{f})_\lambda$, or equivalently, all preorderings $T_\lambda/\mathcal{I}_\lambda$ of the quotient algebras $\mathbb{R}_d[\underline{x}]/\mathcal{I}_\lambda$ do. If a fibre set $\mathcal{K}(\mathfrak{f})_\lambda$ is compact, we know that $T(\mathfrak{f})_\lambda = T(\mathfrak{f}(\lambda))$ has (SMP) by Theorem 12.25. Now let us look at the case when a fibre set $\mathcal{K}(\mathfrak{f})_\lambda$ is not compact. If we take the polynomials $h_j = x_j$, $j = 1, \dots, d-1$, as in the proof of Proposition 13.25, the quotient algebra $\mathbb{R}_d[\underline{x}]/\mathcal{I}_\lambda$ is (isomorphic to) the polynomial algebra $\mathbb{R}[x_d]$. Therefore, by Theorem 13.24, the preordering $T_\lambda/\mathcal{I}_\lambda$ in $\mathbb{R}[x_d]$ has (SMP) if and only if the set $\pi_\lambda(\mathfrak{f})$ contains positive constant multiples of all natural choice generators for the corresponding semi-algebraic subset $\mathcal{K}(T_\lambda/\mathcal{I}_\lambda)$ of $\mathbb{R}$. That is, in order to conclude (SMP) for $T(\mathfrak{f})$ all noncompact fibres require a careful inspection of the sequence $\pi_\lambda(\mathfrak{f})$.

We illustrate the preceding discussion with four examples. All sets are contained in the strip $[0, 1] \times \mathbb{R}$, so that (MP) is always satisfied by Proposition 13.25.

Example 13.27 $f_1(x) = x_1, f_2(x) = 1 - x_1, f_3(x) = x_2^3 - x_2^2 - x_1, f_4(x) = 4 - x_1 x_2$.

Then $h_1(x) = x_1$ is bounded and $h_1(\mathcal{K}(\mathfrak{f})) = [0, 1]$. The fibres for $\lambda \in (0, 1]$ are compact, so the preordering T_λ has (SMP). The fibre set at $\lambda = 0$ is $\{0\} \cup [1, +\infty)$ and the sequence $\pi_0(\mathfrak{f})$ is $\{0, 1, x_2^3 - x_2^2, 4\}$. Since $\pi_0(\mathfrak{f})$ does not contain multiples of all natural choice generators for $\{0\} \cup [1, +\infty)$, $T(\mathfrak{f})$ does not have (SMP). $\qquad\circ$

Example 13.28 $f_1(x) = x_1, f_2(x) = 1 - x_1, f_3(x) = 1 - x_1 x_2, f_4(x) = x_2^3$.

Taking again $h_1(x) = x_1$, we have $h_1(\mathcal{K}(\mathfrak{f})) = [0, 1]$. All fibres at $\lambda \in (0, 1]$ are compact, so they obey (SMP). The fibre set at $\lambda = 0$ is $[0, +\infty)$ and the corresponding sequence $\pi_0(\mathfrak{f}) = \{0, 1, 0, x_2^3\}$ does not contain a multiple of the natural choice generator x_2 for $[0, +\infty)$. Hence $T(\mathfrak{f})$ does not satisfy (SMP). However, if we replace f_4 by $\tilde{f}_4(x) = x_2$, then $T(\mathfrak{f})$ has (SMP).   ∘

Example 13.29 $f_1(x) = x_1, f_2(x) = 1 - x_1, f_3(x) = x_1x_2 - 1, f_4(x) = 2 - x_1x_2$.

The set $\mathcal{K}(\mathfrak{f})$ is not compact and it is the part of the strip between the two hyperbolas $x_1x_2 = 1$ and $x_1x_2 = 2$. Then $h_1(x) = x_1$ and $h_2(x) = x_1x_2$ are bounded on $\mathcal{K}(\mathfrak{f})$ and $\mathsf{h}(\mathcal{K}(\mathfrak{f})) = (0, 1] \times [0, 1]$ is not closed. All fibre sets are points. Since they are compact, all fibre preorderings obey (SMP). Therefore, $T(\mathfrak{f})$ has (SMP).   ∘

Example 13.30 $f_1(x) = x_1, f_2(x) = 1 - x_1, f_3(x) = 1 - x_1x_2$.

Then $\mathcal{K}(\mathfrak{f})$ is the part of the strip below the hyperbola $x_1x_2 = 1$. Set $h_1(x) = x_1$. Then $h_1(\mathcal{K}(\mathfrak{f})) = [0, 1]$. The fibre set at $\lambda = 0$ is the whole x_2-axis, so T_0 has (SMP). For $\lambda \in (0, 1]$, the fibre set is $(-\infty, \lambda^{-1}]$ and a multiple of its natural choice generator $\lambda^{-1} - x_2$ belongs to $\pi_\lambda(\mathfrak{f}) = \{0, 1, 1 - \lambda x_2\}$, so T_λ also has (SMP). Hence $T(\mathfrak{f})$ satisfies (SMP).   ∘

The next proposition is also about cylinder sets with compact base.

Proposition 13.31 *Suppose that* A *is a finitely generated real unital algebra with compact character space* Â. *Let* B *be the tensor product of* A *and the polynomial algebra* $\mathbb{R}[x]$ *in a single variable x. Then the preorderings* $\sum A^2$ *and* $\sum B^2$ *obey (SMP) in* A *and* B, *respectively.*

Proof Let $h_1, \ldots, h_m$ be a set of generators of A and consider a nonempty fibre for A and B, respectively. Then the corresponding fibre sets are points resp. a real line. The fibre algebra $A/\mathcal{I}_\lambda$ is $\mathbb{R}$ and the fibre algebra $B/\mathcal{I}_\lambda$ is $\mathbb{R}[x]$. In both cases $\sum (A/\mathcal{I}_\lambda)^2$ and $\sum (B/\mathcal{I}_\lambda)^2$ have (SMP) and so have $\sum A^2$ and $\sum B^2$ in A and B, respectively, by Theorem 13.10 (iii)→(i).   □

13.6 Application of the Fibre Theorem: The Rational Moment Problem on $\mathbb{R}^d$

Let us begin with some notation and preliminaries. For a subset $\mathcal{D} \subseteq \mathbb{R}_d[\underline{x}]$ we put

$$\mathcal{Z}_\mathcal{D} := \cup_{q \in \mathcal{D}} \mathcal{Z}(q), \quad \text{where} \quad \mathcal{Z}(q) = \{x \in \mathbb{R}^d : q(x) = 0\}.$$

Let $\mathbf{D}(\mathbb{R}_d[\underline{x}])$ denote the family of all multiplicative subsets $\mathcal{D}$ of $\mathbb{R}_d[\underline{x}]$ (that is, $f_1, f_2 \in \mathcal{D}$ implies $f_1f_2 \in \mathcal{D}$) such that $1 \in \mathcal{D}$ and $0 \notin \mathcal{D}$. Fix $\mathcal{D} \in \mathbf{D}(\mathbb{R}_d[\underline{x}])$. Then

$$A_\mathcal{D} := \mathcal{D}^{-1}\mathbb{R}_d[\underline{x}]$$

is a real unital algebra of rational functions which contains $\mathbb{R}_d[\underline{x}]$ as a subalgebra. Obviously, if $\mathcal{D}$ is finitely generated, so is the algebra $A_\mathcal{D}$. The elements of $A_\mathcal{D}$ are well-defined real-valued functions on $\mathbb{R}^d \backslash \mathcal{Z}_\mathcal{D}$. The following lemma shows that the characters of $A_\mathcal{D}$ are just the point evaluations on this set.

Lemma 13.32 *Define* $\chi_t(f) = f(t)$ *for* $t \in \mathbb{R}^d \setminus \mathcal{Z}_\mathcal{D}$ *and* $f \in \mathsf{A}_\mathcal{D}$. *The character set of the algebra* $\mathsf{A}_\mathcal{D}$ *is* $\widehat{\mathsf{A}_\mathcal{D}} = \{\chi_t : t \in \mathbb{R}^d \setminus \mathcal{Z}_\mathcal{D}\}$.

Proof Clearly, χ_t is a well-defined character on $\mathsf{A}_\mathcal{D}$ for $t \in \mathbb{R}^d \setminus \mathcal{Z}_\mathcal{D}$, so $\chi_t \in \widehat{\mathsf{A}_\mathcal{D}}$.

Conversely, suppose that $\chi \in \widehat{\mathsf{A}_\mathcal{D}}$. Put $t_j = \chi(x_j)$ for $j = 1, \ldots, d$. Then we have $t = (t_1, \ldots, t_d) \in \mathbb{R}^d$ and $\chi(p(x)) = p(\chi(x_1), \ldots \chi(x_d)) = p(t)$ for $p \in \mathbb{R}_d[\underline{x}]$. For $q \in \mathcal{D}$ we obtain $1 = \chi(1) = \chi(qq^{-1}) = \chi(q)\chi(q^{-1}) = q(t)\chi(q^{-1})$. Hence $q(t) \neq 0$ for all $q \in \mathcal{D}$, that is, $t \in \mathbb{R}^d \setminus \mathcal{Z}_\mathcal{D}$. Further, $\chi(q^{-1}) = q(t)^{-1}$. Therefore we derive $\chi(\frac{p}{q}) = \chi(p)\chi(q^{-1}) = p(t)q(t)^{-1} = \chi_t(\frac{p}{q})$. Thus $\chi = \chi_t$. □

Now let T be a preordering of $\mathsf{A}_\mathcal{D}$ and $\mathsf{h} = \{h_1, \ldots, h_m\}$ an m-tuple of elements $h_j \in \mathsf{A}_\mathcal{D}$. For $\lambda \in \mathbb{R}^d$ let $\mathcal{I}_\lambda$ be the ideal of $\mathsf{A}_\mathcal{D}$ generated by $h_j - \lambda_j$, $j = 1, \ldots, m$. Recall that the subset $\mathsf{h}(\mathcal{K}(T))$ of $\mathbb{R}^m$ was defined by (13.10).

We consider the following assumptions:

(i) *The functions* $h_1, \ldots, h_m \in \mathsf{A}$ *are bounded on the set* $\mathcal{K}(T)$.
(ii) *For each* $\lambda \in \mathsf{h}(\mathcal{K}(T))$ *there are a finitely generated set* $\mathcal{E}_\lambda \in \mathbf{D}(\mathbb{R}[y])$ *and a surjective algebra homomorphism*

$$\rho_\lambda : \mathcal{E}_\lambda^{-1}\mathbb{R}[y] \to \mathcal{D}^{-1}\mathbb{R}_d[\underline{x}]/\mathcal{I}_\lambda \equiv \mathsf{A}_\mathcal{D}/\mathcal{I}_\lambda.$$

Our main result of this section is the following existence theorem for the multidimensional rational moment problem.

Theorem 13.33 *Suppose that* $\mathcal{D} \in \mathbf{D}(\mathbb{R}_d[\underline{x}])$ *is finitely generated and* T *is a finitely generated preordering of the algebra* $\mathsf{A}_\mathcal{D} = \mathcal{D}^{-1}\mathbb{R}_d[\underline{x}]$. *Assume* (i) *and* (ii).

Then T *satisfies* (MP), *that is, for each* T-*positive linear functional* L *on* $\mathsf{A}_\mathcal{D}$ *there is a Radon measure* μ *on* $\widehat{\mathsf{A}_\mathcal{D}} \cong \mathbb{R}^d \setminus \mathcal{Z}_\mathcal{D}$ *such that* f *is* μ-*integrable and*

$$L(f) = \int_{\mathsf{A}_\mathcal{D}} f(x)\, d\mu(x) \quad \text{for } f \in \mathsf{A}.$$

The proof of this theorem is based on the following result which deals with the one-dimensional case. Let $\mathbb{R}[y]$ denote the real polynomials in a single variable y.

Proposition 13.34 *Suppose that* $\mathcal{E} \in \mathbf{D}(\mathbb{R}[y])$ *is finitely generated. Let* B *be the real unital algebra* $\mathsf{B} = \mathcal{E}^{-1}\mathbb{R}[y]$. *Then the preordering* $\sum \mathsf{B}^2$ *satisfies* (MP), *that is, for each positive linear functional* L *on* B *there exists a Radon measure* μ *on the locally compact Hausdorff space* $\mathcal{Y} := \mathbb{R} \setminus \mathcal{Z}_\mathcal{E}$ *such that* $\mathsf{B} \subseteq L^1(\mathcal{Y}, \mu)$ *and*

$$L(f) = \int_\mathcal{Y} f(y)\, d\mu(y) \quad \text{for } f \in \mathsf{B}. \tag{13.12}$$

Proof Since $\mathcal{E}$ is finitely generated, $\mathcal{Z}_\mathcal{E}$ is a finite set, say $\mathcal{Z}_\mathcal{E} = \{y_1, \ldots, y_k\}$. Hence $\mathcal{Y} := \mathbb{R} \setminus \mathcal{Z}_\mathcal{E}$ is a locally compact Hausdorff space in the induced topology from $\mathbb{R}$. Since $q(y) \neq 0$ for $q \in \mathcal{E}$ and $y \in \mathcal{Y}$, B is a linear subspace of $C(\mathcal{Y}; \mathbb{R})$.

We show that B is an adapted space. This means that we have to check the three conditions (i)–(iii) of Definition 1.5. Since the unital algebra B is the span of its squares, we have $\mathsf{B} = \mathsf{B}_+ - \mathsf{B}_+$, so condition (i) is satisfied. (ii) holds with $f := 1$. We verify (iii). We choose $q_j \in \mathcal{E}$ such that $q_j(y_j) = 0$. Then $h := q_1^2 \dots q_k^2 \in \mathcal{E}$ and $h(y) = 0$ for $y \in \mathcal{Z}_\mathcal{E}$. If $\mathcal{Z}_\mathcal{E}$ is empty, we set $h(y) = 1 + y^2$. Now let $f = \frac{p}{q} \in \mathsf{B}_+$, where $p \in \mathbb{R}[y]$ and $q \in \mathcal{E}$. Further, let $\varepsilon > 0$ be given. Then $g := fh \in \mathsf{B}_+$. Since h vanishes on $\mathcal{Z}_\mathcal{E}$ and $\lim_{|y| \to +\infty} h(y) = +\infty$, there exists a compact subset K_ε of $\mathcal{Y}$ such that $f(y) \le \varepsilon g(y)$ on $\mathcal{Y} \backslash K_\varepsilon$. This proves (iii). Thus, B is an adapted space.

Next we prove that $\mathsf{B}_+ \subseteq \sum \mathsf{B}^2$. Let $f \in \mathsf{B}_+$, that is, $f(y) \ge 0$ for $y \in \mathcal{Y}$. We write f as $f = \frac{p}{q} \in \mathsf{B}$, where $q \in \mathcal{E}$ and $p \in \mathbb{R}[y]$. Then $q^2 f = pq \ge 0$ on $\mathcal{Y} = \mathbb{R}\backslash\mathcal{Z}_\mathcal{E}$ and hence on $\mathbb{R}$, since $\mathcal{Z}_\mathcal{E}$ is empty or finite. The nonnegative polynomial pq in one (!) variable is a sum of squares in $\mathbb{R}[y]$ by Proposition 3.1, so that $pq = \sum_j p_j^2$ with $p_j \in \mathbb{R}[y]$. Therefore, since $q \in \mathcal{E}$, we get $f = \sum_j (\frac{p_j}{q})^2 \in \sum \mathsf{B}^2$.

Let $f \in \mathsf{B}_+$. Then $f \in \sum \mathsf{B}^2$ and hence $L(f) \ge 0$, because the functional L is positive by assumption. Thus L is a B_+-positive linear functional on the adapted subspace B of $C(\mathcal{Y}; \mathbb{R})$. Hence Theorem 1.8 applies and yields the assertion. $\square$

Proof of Theorem 13.33 Let us fix $\lambda \in h(\mathcal{K}(T))$ and abbreviate $\mathsf{B}_\lambda := \mathcal{E}_\lambda^{-1}\mathbb{R}[y]$. By Proposition 13.34, $\sum(\mathsf{B}_\lambda)^2$ has (MP) in the algebra B_λ. We denote by $\mathcal{J}^\lambda$ the kernel of the homomorphism $\rho_\lambda : \mathsf{B}_\lambda \to \mathsf{A}_\mathcal{D}/\mathcal{I}_\lambda$. Since $\sum(\mathsf{B}_\lambda)^2$ has (MP) in B_λ, it is obvious that $\mathcal{J}^\lambda + \sum(\mathsf{B}_\lambda)^2$ has (MP) in B_λ. Therefore $\sum(\mathsf{B}_\lambda/\mathcal{J}^\lambda)^2$ satisfies (MP) in $\mathsf{B}_\lambda/\mathcal{J}^\lambda$ by Corollary 13.16. Since the algebra homomorphism ρ_λ is surjective, the algebra $\mathsf{B}_\lambda/\mathcal{J}^\lambda$ is isomorphic to $\mathsf{A}_\mathcal{D}/\mathcal{I}_\lambda$. Hence $\sum(\mathsf{A}_\mathcal{D}/\mathcal{I}_\lambda)^2$ has (MP) in $\mathsf{A}_\mathcal{D}/\mathcal{I}_\lambda$ as well. Consequently, since $\sum(\mathsf{A}_\mathcal{D}/\mathcal{I}_\lambda)^2 \subseteq T/\mathcal{I}_\lambda$, the preordering $T/\mathcal{I}_\lambda$ obeys (MP) in $\mathsf{A}_\mathcal{D}/\mathcal{I}_\lambda$. Therefore, T has (MP) in $\mathsf{A}_\mathcal{D}$ by Theorem 13.10 (iii)→(i). $\square$

The general fibre theorem fits nicely to the multidimensional rational moment problem, because in general algebras of rational functions contain more *bounded* functions on $\mathcal{K}(T)$ than polynomial algebras.

The use of Theorem 13.33 is illustrated by some examples. The ideas developed therein can be combined to treat more involved examples. Throughout we suppose that the corresponding sets $\mathcal{D}$ are finitely generated.

Example 13.35 First let $d = 2$. Suppose that $\mathcal{D} \in \mathbf{D}(\mathbb{R}[x_1, x_2])$ contains $x_1 - \alpha$ and that the semi-algebraic set $\mathcal{K}(T)$ is a subset of $\{(x_1, x_2) : |x_1 - \alpha| \ge c\}$ for some $\alpha \in \mathbb{R}$ and $c > 0$. Then $h_1 := (x_1 - \alpha)^{-1} \in \mathsf{A}_\mathcal{D}$ is bounded on $\mathcal{K}(T)$, so assumption (i) holds. Let $\lambda \in h_1(\mathcal{K}(T))$. Then we have $x_1 = \lambda^{-1} + \alpha$ in the algebra $(\mathsf{A}_\mathcal{D})_\lambda$, so $(\mathsf{A}_{\mathcal{D}_1})_\lambda$ consists of rational functions in x_2 with denominators from some finitely generated set $\mathcal{E}_\lambda \in \mathbf{D}(\mathbb{R}[x_2])$. Hence assumption (ii) is also satisfied. Thus Theorem 13.33 applies to $\mathsf{A}_\mathcal{D}$.

The above setup extends at once to $d \in \mathbb{N}, d \ge 2$, if we assume that $x_j - \alpha_j \in \mathcal{D}$ and $|x_j - \alpha_j| \ge c$ on $\mathcal{K}(T)$ for some $\alpha_j \in \mathbb{R}, c > 0$, and $j = 1, \dots, d-1$. $\circ$

Example 13.36 Suppose $\mathcal{D} \in \mathbf{D}(\mathbb{R}_d[\underline{x}])$ is generated by the polynomials $q_j = 1 + x_j^2, j = 1, \dots, d$. Let $T = \sum(\mathsf{A}_\mathcal{D})^2$. Then, by Lemma 13.32, we have $\mathcal{K}(T) =$

$\widehat{A_{\mathcal{D}}} = \mathbb{R}^d$. Setting $h_j = q_j(x)^{-1}$, $h_{d+j} = x_j q_j(x)^{-1}$ for $j = 1, \ldots, d$, all $h_l \in A_{\mathcal{D}}$ are bounded on $\mathcal{K}(T)$.

Let $\lambda \in h(\mathcal{K}(T))$. Then $\lambda_j = q_j(\lambda)^{-1} \neq 0$ and $\lambda_{d+j} = \lambda_j q_j(\lambda)^{-1}$; hence we have $x_j = \lambda_{d+j} \lambda_j^{-1}$, $j = 1, \ldots, d$, in the fibre algebra $(A_{\mathcal{D}})_\lambda$. Thus, $(A_{\mathcal{D}})_\lambda = \mathbb{R}$. Taking $\mathcal{E}_\lambda = \{1\}$, we have $\mathcal{E}_\lambda^{-1} \mathbb{R}_d[\underline{x}] = \mathbb{R}_d[\underline{x}]$, so (i) and (ii) are obviously satisfied.

Therefore, $\sum(A_{\mathcal{D}})^2$ obeys (MP) by Theorem 13.33. That is, each positive linear functional on the algebra $A_{\mathcal{D}}$ is given by some positive measure on $\widehat{A_{\mathcal{D}}} = \mathbb{R}^d$.

The same conclusion and almost the same reasoning remain valid if we assume instead that $\mathcal{D}$ is generated by the single polynomial $q = 1 + x_1^2 + \cdots + x_d^2$. In this case we set $h_j = x_j q(x)^{-1}$ for $j = 1, \ldots, d$ and $h_{d+1} = q(x)^{-1}$. $\qquad$ o

Example 13.37 Suppose $\mathcal{D} \in \mathbf{D}(\mathbb{R}_d[\underline{x}])$ contains the polynomials $q_j = 1 + x_j^2$ for $j = 1, \ldots, d - 1$. Let $T = \sum(A_{\mathcal{D}})^2$. Then $h_j := q_j(x)^{-1}$ and $h_{d+j-1} := x_j q_j(x)^{-1}$ for $j = 1, \ldots, d - 1$ are in $A_{\mathcal{D}}$ and bounded on $\mathcal{K}(T) = \widehat{A_{\mathcal{D}}}$. Arguing as in Example 13.36 we conclude that we have $x_j = \lambda_{d+j-1} \lambda_j^{-1}$ for $j = 1, \ldots, d-1$ in the algebra $(A_{\mathcal{D}})_\lambda$. Therefore the fibre algebra $(A_{\mathcal{D}})_\lambda$ is an algebra $\mathcal{E}_\lambda^{-1} \mathbb{R}[x_d]$ of rational functions in the single variable x_d for some finitely generated set $\mathcal{E}_\lambda \in \mathbf{D}(\mathbb{R}[x_d])$.

Then, again by Theorem 13.33, $\sum(A_{\mathcal{D}})^2$ satisfies (MP). The same is true if we assume instead that the polynomial $q = 1 + x_1^2 + \cdots + x_{d-1}^2$ is in $\mathcal{D} \in \mathbf{D}(\mathbb{R}_d[\underline{x}])$. o

13.7 Application of the Fibre Theorem: A Characterization of Moment Functionals

In this section, we derive a theorem which characterizes moment functionals on $\mathbb{R}^d$ in terms of extensions. It will be used in the proof of Theorem 15.14 below, but it is also of interest in itself. Throughout we assume that $d \geq 2$.

Let A denote the real algebra of functions on $(\mathbb{R}^d)^\times := \mathbb{R}^d \backslash \{0\}$ generated by the polynomial algebra $\mathbb{R}_d[\underline{x}]$ and the functions

$$f_{kl}(x) := x_k x_l (x_1^2 + \cdots + x_d^2)^{-1}, \quad k, l = 1, \ldots, d. \tag{13.13}$$

Clearly, these functions satisfy the relations

$$f_{11} + f_{22} + \cdots + f_{dd} = 1, \tag{13.14}$$

$$f_{ij} f_{kl} = f_{ik} f_{jl} \quad \text{for } i, j, k, l = 1, \ldots, d. \tag{13.15}$$

Thus, the algebra A has $d + \binom{d+1}{2}$ generators $x_1, \ldots, x_d, f_{11}, f_{12}, \ldots, f_{dd}$ and the elements of A are precisely all functions of the form

$$g(p(x), f_{11}(x), \ldots, f_{dd}(x)), \tag{13.16}$$

where $p \in \mathbb{R}_d[\underline{x}]$ and g is a real polynomial in $1 + \binom{d+1}{2}$ variables. Of course, given an element of A the polynomial g is not uniquely determined.

We denote by S^{d-1} the unit sphere of $\mathbb{R}^d$ and by S^{d-1}_+ the set of all $t = (t_1, \ldots, t_d)$ of S^{d-1} for which the first nonzero coordinate t_j is positive.

The next lemma describes the character set $\hat{\mathsf{A}}$ of A.

Lemma 13.38

(i) *For $x \in (\mathbb{R}^d)^\times$ the point evaluation of functions at x is a character χ_x of A such that $x = (\chi_x(x_1), \ldots, \chi_x(x_d)) \neq 0$. Each character of A satisfying $(\chi(x_1), \ldots, \chi(x_d)) \neq 0$ is of this form.*

(ii) *For $t \in S^{d-1}$ there exists a unique character χ^t of A such that*

$$\chi^t(x_j) = 0 \quad and \quad \chi^t(f_{kl}) = f_{kl}(t) \quad for\ j, k, l = 1, \ldots, d. \tag{13.17}$$

Each character χ of A for which $\chi(x_j) = 0$ for all $j = 1, \ldots, d$ is of the form χ^t with uniquely determined $t \in S^{d-1}_+$.

(iii) *The set $\hat{\mathsf{A}}$ is the disjoint union of the set $\{\chi_x : x \in (\mathbb{R}^d)^\times\}$ and $\{\chi^t : t \in S^{d-1}_+\}$.*

Proof

(i) The first assertion is obvious.

    We prove the second assertion. For this let χ be a character of A such that $x := (\chi(x_1), \ldots, \chi(x_d)) \neq 0$. The identity $(x_1^2 + \cdots + x_d^2)f_{kl} = x_k x_l$ implies that

$$(\chi(x_1)^2 + \cdots + \chi(x_d)^2)\chi(f_{kl}) = \chi(x_k)\chi(x_l)$$

and therefore

$$\chi(f_{kl}) = (\chi(x_1)^2 + \cdots + \chi(x_d)^2)^{-1}\chi(x_k)\chi(x_l) = f_{kl}(x).$$

Thus χ acts on the generators x_j and f_{kl}, hence on the whole algebra A, by point evaluation at x, that is, we have $\chi = \chi_x$.

(ii) Fix $t \in S^{d-1}$. Since $\lim_{\varepsilon \to +0} f_{kl}(\varepsilon t) = f_{kl}(t)$ and $\lim_{\varepsilon \to +0} p(\varepsilon t) = p(0)$ for $p \in \mathbb{R}_d[\underline{x}]$, it follows that the limit

$$\chi^t(g) := \lim_{\varepsilon \to +0} g(\varepsilon t) = \lim_{\varepsilon \to +0} \chi_{\varepsilon t}(g)$$

exists for all $g \in \mathsf{A}$. Since $\chi_{\varepsilon t}$ is a character on A, so is χ^t. By construction we have $\chi^t(f_{kl}) = f_{kl}(t)$ and $\chi^t(x_j) = 0$ for all $j, k, l = 1, \ldots, d$.

    Conversely, let χ be a character of A such that $\chi(x_j) = 0$ for $j = 1, \ldots, d$. Set $\lambda_{kl} := \chi(f_{kl})$. Since χ is a character, (13.14) and (13.15) imply that

$$\lambda_{11} + \lambda_{22} + \cdots + \lambda_{dd} = 1, \tag{13.18}$$

$$\lambda_{ij}\lambda_{kl} = \lambda_{ik}\lambda_{jl} \quad for\ i, j, k, l = 1, \ldots, d. \tag{13.19}$$

First we note that $\lambda_{ii} \geq 0$ for all i. Assume to the contrary that $\lambda_{ii} < 0$ for some i. From (13.18) there exists a j such that $\lambda_{jj} > 0$. Then $\lambda_{ij}^2 = \lambda_{ii}\lambda_{jj} < 0$ by (13.19), which is a contradiction.

Set $t_i = \sqrt{\lambda_{ii}}, i = 1, \ldots, d$. Then $t = (t_1, \ldots, t_d) \in S^{d-1}$ by (13.18). From (13.19) we then obtain that $\lambda_{ij}^2 = t_i^2 t_j^2$ for all i, j. Therefore, if $t_i = 0$ for some i, then $\lambda_{ij} = 0$ and $\chi^t(f_{ij}) = \lambda_{ij} = 0 = t_i t_j$ for all j, and this remains valid if t_j is changed. By leaving out all indices with $t_i = 0$ we can assume without loss of generality that all t_i are nonzero. Further, since $\lambda_{ij}^2 = t_i^2 t_j^2$, there exists an $\varepsilon_{ij} \in \{-1, 1\}$ such that $\lambda_{ij} = \varepsilon_{ij} t_i t_j$ for $i \neq j, i, j = 1, \ldots, d$. If $\varepsilon_{1j} = -1$ for some $j = 2, \ldots, d$, we replace t_j by $-t_j$. Then $\lambda_{1j} = t_1 t_j$ for all $j = 2, \ldots, d$. Further, using (13.19) we obtain

$$\lambda_{ij}\lambda_{1j} = \varepsilon_{ij} t_i t_j t_1 t_j = \lambda_{1i}\lambda_{jj} = t_1 t_i t_j^2.$$

Since all t_k are nonzero, $\varepsilon_{ij} = 1$. Thus $\chi(f_{ij}) = \lambda_{ij} = t_i t_j$, that is, $\chi(f_{ij}) = \chi^t(f_{ij})$ for all i, j. Since χ and χ^t coincide on generators of A, we have $\chi = \chi^t$ on A. By construction, $t_1 > 0$, so that $t \in S_+^{d-1}$.

We verify the uniqueness assertion. For let $\chi^t = \chi^{\tilde{t}}$ with $t, \tilde{t} \in S_+^{d-1}$. Since $t_i^2 = \chi^t(f_{ii}) = \chi^{\tilde{t}}(f_{ii}) = \tilde{t}_i^2$, the first nonvanishing indices for t and $\tilde{t}$ are the same, say j. Then $t_j = \tilde{t}_j > 0$, since $t, \tilde{t} \in S_+^{d-1}$. Further, $t_k t_j = \chi^t(f_{kj}) = \chi^{\tilde{t}}(f_{kj}) = \tilde{t}_k \tilde{t}_j = \tilde{t}_k t_j$. Therefore, since $t_j \neq 0$, we get $\tilde{t}_k = t_k$, so that $t = \tilde{t}$.

(iii) follows at once by combining (i) and (ii).   □

Theorem 13.39 *The preordering $\sum A^2$ of the algebra A has (MP), that is, for each positive linear functional $\mathcal{L}$ on A there exist Radon measures $\nu_0 \in M_+(S^{d-1})$ and $\nu_1 \in \mathcal{M}_+(\mathbb{R}^d)$ such that $\nu_1(\{0\}) = 0$ and for all $g \in A$ of the form (13.16) we have*

$$\mathcal{L}(g(p(x), f_{11}(x), \ldots, f_{dd}(x))) = \tag{13.20}$$

$$\int_{S^{d-1}} g(p(0), f_{11}(t), \ldots, f_{dd}(t)) \, d\nu_0(t) + \int_{\mathbb{R}^d \setminus \{0\}} g(p(x), f_{11}(x), \ldots, f_{dd}(x)) \, d\nu_1(x).$$

Proof It suffices to prove that $\sum A^2$ obeys (MP). The other assertions follow from the definition of (MP) and the form of the character set given in Lemma 13.38.

From the description of $\hat{A}$ it is obvious that the functions $f_{kl}, k, l = 1, \ldots, d$, are bounded on $\hat{A}$, so we can take them as functions h_j in Theorem 13.10. Let us fix a nonempty fibre for $\lambda = (\lambda_{kl})$, where $\lambda_{kl} \in \mathbb{R}$ for all k, l. In the quotient algebra $A/\mathcal{I}_\lambda$ of A by the fibre ideal $\mathcal{I}_\lambda$ we have $\chi(f_{kl}) = \lambda_{kl}$ for all $\chi \in \hat{A}$.

First let $\chi = \chi_x$, where $x \in (\mathbb{R}^d)^\times$. Then $\chi_x(f_{kl}) = f_{kl}(x) = \lambda_{kl}$. By (13.14) we have $1 = \sum_k f_{kk}(x) = \sum_k \lambda_{kk}$, so there exists a k such that $\lambda_{kk} \neq 0$. From the equality $\lambda_{kk} = f_{kk}(x) = x_k^2(x_1^2 + \cdots + x_d^2)^{-1}$ we obtain $x_k \neq 0$. Thus $\frac{\lambda_{kl}}{\lambda_{kk}} = \frac{f_{kl}(x)}{f_{kk}(x)} = \frac{x_l}{x_k}$, so that

$$x_l = \lambda_{kl}\lambda_{kk}^{-1} x_k \quad \text{for } l = 1, \ldots, d. \tag{13.21}$$

If $\chi = \chi^t$ for $t \in S^{d-1}$, then $\chi(x_l) = \chi(x_k) = 0$, so (13.21) holds trivially. That is, in the algebra $A/\mathcal{I}_\lambda$ we have the relations (13.21) and $f_{kl} = \lambda_{kl}$. This implies that the quotient algebra $A/\mathcal{I}_\lambda$ is generated by the single polynomial x_k. Hence $\sum(A/\mathcal{I}_\lambda)^2$ satisfies (MP) in $A/\mathcal{I}_\lambda$ by Corollary 13.17. Therefore, by Theorem 13.10 (iii)→(i), the preordering $T = \sum A^2$ obeys (MP) in A. ☐

Now we are ready to prove the main result of this section.

Theorem 13.40 *A linear functional L on $\mathbb{R}_d[\underline{x}]$ is a moment functional if and only if it has an extension to a positive linear functional $\mathcal{L}$ on the larger algebra* A.

Proof Assume first that L has an extension to a positive linear functional $\mathcal{L}$ on A. By Theorem 13.39, $\mathcal{L}$ has the form described by Eq. (13.20). We define a Radon measure μ on $\mathbb{R}^d$ by

$$\mu(\{0\}) = \nu_0(S^{d-1}), \quad \mu(M\setminus\{0\}) = \nu_1(M\setminus\{0\}).$$

Then $\mu \in \mathcal{M}_+(\mathbb{R}^d)$, since $\nu_1 \in \mathcal{M}_+(\mathbb{R}^d)$. Let $p \in \mathbb{R}_d[\underline{x}]$. Setting $g(1,0,\ldots,0) = 1$, we have $g(p, f_{11}, \ldots, f_{dd}) = p$ and it follows from (13.20) that

$$L(p) = \tilde{L}(p) = \nu_0(\{0\})p(0) + \int_{\mathbb{R}^d\setminus\{0\}} p(x)\, d\nu_1(x) = \int_{\mathbb{R}^d} p(x)\, d\mu(x).$$

That is, L is a moment functional on $\mathbb{R}_d[\underline{x}]$ with representing measure μ.

Conversely, suppose that L is a moment functional on $\mathbb{R}_d[\underline{x}]$ and let μ be a representing measure. Fix a point $t \in S^{d-1}$. By Lemma 13.38(ii), χ^t is a character of A and $\chi^t(f) = f(0)$ for $f \in \mathbb{R}_d[\underline{x}]$. Therefore, for $f \in \mathbb{R}_d[\underline{x}]$, we obtain

$$L(f) = \mu(\{0\})\chi^t(f) + \int_{\mathbb{R}^d\setminus\{0\}} f(x)\, d\mu(x). \qquad (13.22)$$

For $f \in A$ we define $\mathcal{L}(f)$ by the right-hand side of (13.22). Since the functions f_{kl} are bounded on $\mathbb{R}^d\setminus\{0\}$, the integral in (13.22) exists for all $f \in A$ and it is a positive functional on A. The character χ^t is obviously a positive functional on A. Hence $\mathcal{L}$ is a positive linear functional on A which extends L by (13.22). ☐

13.8 Closedness and Stability of Quadratic Modules

While the preceding sections dealt with affirmative results for (SMP) or (MP), the aim of this section is to develop examples where (MP) *fails* and to provide some tools for proving this. By Proposition 13.9, a quadratic module Q of $\mathbb{R}_d[\underline{x}]$ has (MP) if and only if $\overline{Q} = \mathrm{Pos}(\mathbb{R}^d)$. Thus, proving that (MP) fails requires a polynomial in $\mathrm{Pos}(\mathbb{R}^d)$ that does not belong to the closure (!) of Q. This leads to the problem of when a quadratic module is closed in the finest locally convex topology. Let us look at the simplest example $\sum \mathbb{R}_d[\underline{x}]^2$. The closedness of $\sum \mathbb{R}_d[\underline{x}]^2$ can be derived from the closedness of $\sum \mathbb{R}_d[\underline{x}]_n^2$ in the finite-dimensional space $\mathbb{R}_d[\underline{x}]_{2n}$ combined with the simple, but crucial observation $\mathbb{R}_d[\underline{x}]_{2n} \cap \sum \mathbb{R}_d[\underline{x}]^2 = \sum \mathbb{R}_d[\underline{x}]_n^2$. An elaboration of the latter relation leads to the notion of stability for quadratic modules.

In this section, A is a **finitely generated commutative real unital algebra**. We equip the countable-dimensional vector space A with the finest locally convex topology (see Appendix A.5). Closedness of sets in A always refers to this topology. On each finite-dimensional subspace it inherits the unique norm topology.

Let W be a subspace of A. We denote by $\sum W^2$ the set of sums of squares of elements of W. For $g_1, \ldots, g_k \in A$, we put

$$\sum (W; g_1, \ldots, g_k) := \sum W^2 + g_1 \sum W^2 + \cdots + g_k \sum W^2. \tag{13.23}$$

If $n = \dim W < \infty$, then $\sum W^2$ and $\sum (W; g_1, \ldots, g_k)$ are contained in subspaces of dimensions at most $\frac{(n+1)n}{2}$ and $\frac{(n+1)n(k+1)}{2}$, respectively. (Indeed, if $a_1 \ldots, a_n$ is a basis of W, each square a^2 of $a \in W$ is in the span of elements $a_k a_l, k \leq l$.)

Definition 13.41 A quadratic module Q of A generated by $g_1, \ldots, g_k \in A$ is called *stable* if for each finite-dimensional subspace V of A there exists a finite-dimensional subspace W_V of A such that

$$V \cap Q \subseteq \sum (W_V; g_1, \ldots, g_k). \tag{13.24}$$

Lemma 13.42 *Definition 13.41 is independent of the choice of generators of Q.*

Proof By induction it suffices to show that the condition in Definition 13.41 is preserved if one element $g \in Q$ is added to the generators $g_1, \ldots, g_k$ of Q. If the condition holds for $g_1, \ldots, g_k$, it trivially holds for $g_1, \ldots, g_k, g$.

Conversely, suppose that it holds for $g_1, \ldots, g_k, g$. Let V be a given finite-dimensional subspace of A. Then there is a finite-dimensional subspace W_0 such that $V \cap Q \subseteq \sum (W_0; g_1, \ldots, g_k, g)$. Since $g \in Q$ and Q is generated by $g_1, \ldots, g_k$, we have $g = \sigma_0 + g_1\sigma_1 + \ldots g_k\sigma_k$ with $\sigma_j \in \sum A^2$. We choose a finite-dimensional subspace W_1 such that $1 \in W_1$ and all σ_j are in $\sum W_1^2$. If W_V is the finite-dimensional subspace spanned by $w_0 w_1$, where $w_0 \in W_0$ and $w_1 \in W_1$, then $W_0 \subseteq W_V$ and

$$gw_0^2 = \sigma_0 w_0^2 + g_1\sigma_1 w_0^2 + \ldots g_k\sigma_k w_0^2 \in \sum (W_V; g_1, \ldots, g_k)$$

for $w_0 \in W_0$. Hence $V \cap Q \subseteq \sum (W_V; g_1, \ldots, g_k)$. $\square$

For the next results we assume that $(A_n)_{n \in \mathbb{N}}$ is a sequence of linear subspaces of A such that

$$A = \bigcup_{n=1}^{\infty} A_n, \quad \text{where} \quad A_n \subseteq A_{n+1}, \quad \dim A_n < \infty \quad \text{for} \quad n \in \mathbb{N}. \tag{13.25}$$

Our standard example is $A_n = \mathbb{R}_d[\underline{x}]_n \equiv \{p \in \mathbb{R}_d[\underline{x}] : \deg(p) \leq n\}$ for $A = \mathbb{R}_d[\underline{x}]$. The main motivation for the concept of stability stems from the following fact.

Proposition 13.43 *Let Q be a finitely generated quadratic module of A with generators $g_1, \ldots, g_k$. Let $(A_n)_{n \in \mathbb{N}}$ be a sequence of subspaces of A satisfying (13.25). If Q is stable and the set $\sum(A_n; g_1, \ldots, g_k)$ is closed (in the norm topology of some finite-dimensional subspace of A which contains $\sum(A_n; g_1, \ldots, g_k)$) for each $n \in \mathbb{N}$, then Q itself is closed in the finest locally convex topology of A.*

Proof Since Q is stable, for each A_n there is a finite-dimensional subspace W_{A_n} such that $A_n \cap Q \subseteq \sum(W_{A_n}; g_1, \ldots, g_k)$. By (13.25) there is an $m_n \in \mathbb{N}$ such that A_{m_n} contains a vector space basis of W_{A_n}. Hence $W_{A_n} \subseteq A_{m_n}$. Then

$$A_n \cap Q \subseteq (W_{A_n}; g_1, \ldots, g_k) \subseteq \sum(A_{m_n}; g_1, \ldots, g_k) \subseteq Q$$

and therefore $A_n \cap Q = A_n \cap \sum(A_{m_n}; g_1, \ldots, g_k)$. Since $\sum(A_{m_n}; g_1, \ldots, g_k)$ is closed by assumption, so is $A_n \cap \sum(A_{m_n}; g_1, \ldots, g_k) = A_n \cap Q$ in A_n. Hence, by Proposition A.28, Q is closed in the finest locally convex topology of A. $\square$

Often it is convenient to rephrase the stability of quadratic modules in terms of the decomposition (13.25) of A. A quadratic module Q of A with generators $g_1, \ldots, g_k$ is stable if and only if the following condition holds:

$(*)$ *There exists a map $\mathfrak{l} : \mathbb{N}_0 \to \mathbb{N}_0$ such that, for each $n \in \mathbb{N}_0$ and $p \in Q \cap A_n$, p admits a representation $p = \sigma_0 + \sum_{j=1}^{k} g_j \sigma_j$ with $\sigma_0, \sigma_j \in \sum A^2$, $\sigma_0 \in A_{\mathfrak{l}(n)}$ and $g_j \sigma_j \in A_{\mathfrak{l}(n)}$ for $j = 1, \ldots, k$.*

Indeed, suppose that Q is stable. Applying Definition 13.41 to $V = A_n$, then W_V is contained in $A_{\mathfrak{l}(n)}$ for some $\mathfrak{l}(n) \in \mathbb{N}$. Then, by (13.23), $(*)$ holds. Conversely, assume that $(*)$ is satisfied. Then, given V we choose n such that $V \subseteq A_n$. Then $(*)$ implies that $V \cap Q \subseteq \sum(A_{\mathfrak{l}(n)}; g_1, \ldots, g_k)$, so Q is stable.

Example 13.44 The simplest example of a stable quadratic module is the preordering $\sum \mathbb{R}_d[\underline{x}]^2$ of the polynomial algebra $A = \mathbb{R}_d[\underline{x}]$.

In this case, $g_1 = 1, k = 1$ and $\mathbb{R}_d[\underline{x}]_{2n} \cap \sum \mathbb{R}_d[\underline{x}]^2 = \sum \mathbb{R}_d[\underline{x}]_n^2$ by (13.1). Thus condition $(*)$ is satisfied for $A_n = \mathbb{R}_d[\underline{x}]_n$ and $\mathfrak{l}(n) = n$, so $\sum \mathbb{R}_d[\underline{x}]^2$ is stable.   ○

Example 13.45 Let $k \in \{1, \ldots, d\}$ and $m_1, \ldots, m_k \in \mathbb{N}$. The quadratic module

$$Q = \sum \mathbb{R}_d[\underline{x}]^2 + x_1^{m_1} \sum \mathbb{R}_d[\underline{x}]^2 + \cdots + x_k^{m_k} \sum \mathbb{R}_d[\underline{x}]^2$$

of $A = \mathbb{R}_d[\underline{x}]$ is stable. Indeed, setting $A_n = \mathbb{R}_d[\underline{x}]_n$ one easily checks that condition $(*)$ is fulfilled with $\mathfrak{l}(n) = n + \max_j m_j$.   ○

The following proposition contains the crucial technical part of the proof of Theorem 13.47 below, but it is also of interest in itself.

Proposition 13.46 *Let Q be the quadratic module of A generated by $g_1, \ldots, g_k$. Suppose that $Q \cap (-Q) = \{0\}$ and (13.25) holds. Let $\mathfrak{n} = (n_0, \ldots n_k) \in \mathbb{N}_0^{k+1}$. Then*

$$Q_{\mathfrak{n}} := \sum (\mathsf{A}_{n_0})^2 + g_1 \sum (\mathsf{A}_{n_1})^2 + \cdots + g_k \sum (\mathsf{A}_{n_k})^2 \qquad (13.26)$$

is a closed subset of A_m for some $m \in \mathbb{N}$.

Proof Put $g_0 = 1$. Since A_l is finite-dimensional, each $g_i \sum (\mathsf{A}_i)^2$ is contained in some finite-dimensional subspace of A, say of dimension $r(i)$. Hence $Q_{\mathfrak{n}}$ is contained in a finite-dimensional subspace of A, so by (13.25) there is an $m \in \mathbb{N}$ such that $Q_{\mathfrak{n}} \subseteq \mathsf{A}_m$. By Carathéodory's theorem A.35, each element of the cone $g_i \sum (\mathsf{A}_i)^2$ is a sum of $r(i)$ summands $g_j b^2$, where $b \in \mathsf{A}_i$. Thus $Q_{\mathfrak{n}}$ is the image of the map

$$\Phi : \mathcal{R}_{\mathfrak{n}} := (\mathsf{A}_{n_0})^{r(n_0)} \times \cdots \times (\mathsf{A}_{n_k})^{r(n_k)} \to \mathsf{A}_m,$$

$$\Phi((f_{0j})_{j=1}^{r(n_0)} \times \cdots \times (f_{kj})_{j=1}^{r(n_k)}) := \sum_{l=0}^{k} \sum_{j=1}^{r(n_k)} g_l f_{lj}^2 .$$

We equip the finite-dimensional real vector spaces $\mathcal{R}_{\mathfrak{n}}$ and A_m with some norm. Let S be the unit sphere of $\mathcal{R}_{\mathfrak{n}}$. Since all linear and all bilinear mappings of finite-dimensional normed spaces are continuous, $\Phi : \mathcal{R}_{\mathfrak{n}} \to \mathsf{A}_m$ is continuous. Hence the image $U := \Phi(S)$ of the compact subset S of $\mathcal{R}_{\mathfrak{n}}$ is compact in A_m.

Next we show that Φ is injective. Assume that $\Phi(v) = 0$ for $v \in \mathcal{R}_{\mathfrak{n}}$. We write

$$\Phi(v) = \sum_{l=0}^{k} \sum_{j=1}^{r(n_k)} g_l f_{lj}^2$$

with $f_{lj}^2 \in \mathsf{A}_{n_j}$. Since $\Phi(v) = 0$, we have

$$g_0 f_{01}^2 = -\sum_{j=2}^{r(n_0)} g_0 f_{0j}^2 - \sum_{l=1}^{k} \sum_{j=1}^{r(n_k)} g_l f_{lj}^2 \in Q \cap (-Q).$$

Since $Q \cap (-Q) = \{0\}$ by assumption, $g_0 f_{01}^2 = 0$. Continuing this procedure by induction it follows that $g_l f_{lj}^2 = 0$ for all l, j. Thus $v = 0$ and Φ is injective.

Now we prove that $Q_{\mathfrak{n}}$ is closed in A_m. Let $(f_n)_{n \in \mathbb{N}}$ be a sequence of $Q_{\mathfrak{n}}$ which converges to some element $f \in \mathsf{A}_m$. We will show that $f \in Q_{\mathfrak{n}}$. By definition we have $f_n = \Phi(v_n)$ for some $v_n \in \mathcal{R}_{\mathfrak{n}}$. Writing $v_n = \lambda_n w_n$ with $w_n \in S$ and $\lambda_n \geq 0$, we get $f_n = \lambda_n u_n$ with $u_n \in \Phi(S) = U$. Since U is compact in A_m, the sequence $(u_n)_{n \in \mathbb{N}}$ has a subsequence which converges to some element $u \in U$. For notational simplicity we assume that the sequence (u_n) itself converges to u. Because Φ is injective, $0 \notin U$ and therefore $u \neq 0$. Then $\lim_n \lambda_n = \lim_n \frac{\|f_n\|}{\|u_n\|} = \frac{\|f\|}{\|u\|}$ and hence

$$f = \lim_n f_n = \lim_n \lambda_n u_n = (\lim_n \lambda_n)(\lim_n u_n) = \frac{\|f\|}{\|u\|}\, u \in \frac{\|f\|}{\|u\|} \cdot U \subseteq Q_n. \quad \square$$

The next theorem is our main result about the closedness of quadratic modules.

Theorem 13.47 *Let* A *be a finitely generated commutative real unital algebra and* Q *a finitely generated quadratic module of* A. *Suppose that* $Q \cap (-Q) = \{0\}$ *and* Q *is stable. Then* Q *is closed in the finest locally convex topology of* A.

Proof Since A is finitely generated, we can write A in the form (13.25). In the case $n := n_0 = \cdots = n_k$ the set Q_n in (13.26) is just the cone $\sum(A_n; g_1, \ldots, g_k)$ for $W = A_n$ defined by (13.23). Since $Q_n = \sum(A_n; g_1, \ldots, g_k)$ is closed by Proposition 13.46, the assertion follows from Proposition 13.43. $\quad \square$

The assumption $Q \cap (-Q) = \{0\}$ is fulfilled in the following important case.

Lemma 13.48 *If* Q *is a quadratic module of* $\mathbb{R}_d[\underline{x}]$ *such that* $\mathcal{K}(Q)$ *contains an interior point, then* $Q \cap (-Q) = \{0\}$.

Proof Let $p \in Q \cap (-Q)$. Then $p \geq 0$ and $-p \geq 0$ on $\mathcal{K}(Q)$. Therefore, the polynomial p vanishes on $\mathcal{K}(Q)$ and so on an open set. Hence $p = 0$. $\quad \square$

Corollary 13.49 *Let* $k \in \{1, \ldots, d\}$ *and* $m_1, \ldots, m_k \in \mathbb{N}$. *The quadratic modules*

$$\sum \mathbb{R}_d[\underline{x}]^2 + x_1^{m_1} \sum \mathbb{R}_d[\underline{x}]^2 + \cdots + x_k^{m_k} \sum \mathbb{R}_d[\underline{x}]^2$$

and $\sum \mathbb{R}_d[\underline{x}]^2$ *are stable and closed in* $\mathbb{R}_d[\underline{x}]$.

Proof Let Q be one of these quadratic modules. By Examples 13.44 and 13.45, Q is stable. Obviously, the positive d-octant $(\mathbb{R}_+)^d$ is a subset of $\mathcal{K}(Q)$, so $\mathcal{K}(Q)$ has an interior point. Therefore Q is closed by Theorem 13.47 and Lemma 13.48. $\quad \square$

The following Propositions 13.50 and 13.51 and Example 13.52 contain results about the failure of (MP) for certain quadratic modules.

Proposition 13.50 *Let* Q *be a finitely generated quadratic module of* $\mathbb{R}_d[\underline{x}]$ *such that* $\mathcal{K}(Q)$ *contains an interior point. Suppose that* $d \geq 2$ *and* Q *is stable. Then* Q *does not have (MP).*

Proof Let $g_1, \ldots, g_k$ be generators of Q. We can assume that none of the g_j is the zero polynomial. Set $g := g_1 \cdots g_k$. Since $\mathcal{K}(Q)$ has a non-empty interior U, g cannot vanish on the whole set U, so there is a point $x_0 \in U \subseteq \mathcal{K}(Q)$ such that $g(x_0) \neq 0$. Upon translation we can assume that $x_0 = 0$. Since $x_0 = 0 \in \mathcal{K}(Q)$ and $g_j(0) \neq 0$, it follows that $g_j(0) > 0$ for $j = 1, \ldots, k$. Further, we set $g_0 = 1$ and choose a polynomial $p \in \text{Pos}(\mathbb{R}^d)$ such that $p \notin \sum \mathbb{R}_d[\underline{x}]^2$; for instance, we may take the Motzkin polynomial given by (13.6).

For $n \in \mathbb{N}$ define $p_n(x) := p(2^n x)$. We prove that not all polynomials p_n are in Q. Assume to the contrary that $p_n \in Q$ for all $n \in \mathbb{N}$. Since $\deg(p_n) = \deg(p)$ and

Q is stable, there exist $m \in \mathbb{N}$ and representations

$$p_n(x) = p(2^n x) = \sum_{j=0}^{k} g_j(x)\sigma_{nj}(x),$$

where $\sigma_{nj} \in \sum \mathbb{R}_d[\underline{x}]_m^2$. Rescaling the preceding we get

$$p(x) = \sum_{j=0}^{k} g_j(2^{-n}x)\sigma_{nj}(2^{-n}x), \quad n \in \mathbb{N}. \tag{13.27}$$

Since $g_j(0) > 0$, there are $\varepsilon > 0$ and a ball $B \subseteq \mathcal{K}(Q)$ around 0 such that $g_j(x) \geq \varepsilon$, hence $g_j(2^{-n}x) \geq \varepsilon$, on B for all j and n. Then, since all polynomials in (13.27) are nonnegative on B and the left-hand side does not depend on n, the values of $\sigma_{nj}(2^{-n}x)$ are uniformly bounded on B, that is, $\sup_{x \in B} \sup_{n,j} \sigma_{nj}(2^{-n}x) < \infty$. Clearly, $\tau_{nj}(x) := \sigma_{nj}(2^{-n}x) \in \sum \mathbb{R}_d[\underline{x}]_m^2 \subseteq \mathbb{R}_d[\underline{x}]_{2m}$. The supremum over B defines a norm on the finite-dimensional space $\mathbb{R}_d[\underline{x}]_{2m}$ and the sequences $(\tau_{nj})_{n \in \mathbb{N}}$ of polynomials $\tau_{nj} \in \mathbb{R}_d[\underline{x}]_{2m}$ are bounded with respect to this norm. Hence they have converging subsequences $(\tau_{n_l j})_{l \in \mathbb{N}}$ for $j = 0, \ldots, k$. Their limits τ_j are also in $\sum \mathbb{R}_d[\underline{x}]_m^2$ by Proposition 13.46 (or Corollary 13.49). Passing to the limit $n_l \to \infty$ in (13.27) yields

$$p(x) = \sum_{j=0}^{k} g_j(0)\tau_j(x).$$

Since $\tau_j \in \sum \mathbb{R}_d[\underline{x}]^2$ and $g_j(0) > 0$, $p \in \sum \mathbb{R}_d[\underline{x}]^2$, which contradicts the choice of p.

By the preceding we proved that $p_n \notin Q$ for some n. Since Q is stable, Q is closed by Theorem 13.47 and Lemma 13.48. Therefore, $p_n \notin \overline{Q}$. But $p_n \in \text{Pos}(\mathbb{R}^d)$. Thus $\overline{Q} \neq \text{Pos}(\mathbb{R}^d)$. Hence Q does not have (MP) by Proposition 13.9(ii).   □

Let us mention a consequence of the preceding result. Let $\mathfrak{f}$ be a finite subset of $\mathbb{R}_d[\underline{x}]$ such that the semi-algebraic set $\mathcal{K}(\mathfrak{f})$ is compact. Then, by Theorem 12.25, $T(\mathfrak{f})$ has (SMP) and hence (MP). Therefore, if $d \geq 2$ and $\mathcal{K}(\mathfrak{f})$ contains an interior point, Proposition 13.50 implies that the preordering $T(\mathfrak{f})$ is *not stable*!

There is an even stronger result than Proposition 13.50, proved by C. Scheiderer [Sr1, Theorem 5.4]. We state it here without proof (for the dimension of a semi-algebraic set we refer to [BCRo, Corollary 2.8.9]):

Let Q be a finitely generated quadratic module of $\mathbb{R}_d[\underline{x}]$. If the semi-algebraic set $\mathcal{K}(Q)$ has dimension at least 2 and Q is stable, then (MP) fails.

That is, (MP) and stability exclude each other if the dimension is at least 2.

Proposition 13.51 *Let Q be a finitely generated quadratic module of $\mathbb{R}_d[\underline{x}]$ such that the semi-algebraic set $\mathcal{K}(Q)$ contains an open cone C. Then Q is stable and closed. If $d \geq 2$, then Q does not obey (MP) in $\mathbb{R}_d[\underline{x}]$.*

Proof Upon translation we can assume without loss of generality that the vertex of C is the origin. Let $p, q \in \text{Pos}(C)$. We show that

$$\max(\deg(p), \deg(q)) = \deg(p + q). \tag{13.28}$$

This is obvious if $\deg(p) \neq \deg(q)$ and if $p = 0$ or $q = 0$. Hence we can assume that $n := \deg(p) = \deg(q)$ and $pq \neq 0$. Since C has nonempty interior and $pq \neq 0$, there exists an $x_0 \in C$ such that $pq(x_0) \neq 0$. Hence $p(x_0) > 0$ and $q(x_0) > 0$. For all $\lambda > 0$ we have $\lambda x_0 \in C$ and hence $p(\lambda x_0) \geq 0$ and $q(\lambda x_0) \geq 0$. For large λ this implies that $p_n(x_0) > 0$ and $q_n(x_0) > 0$, where p_n and q_n are the corresponding homogeneous terms of degree n. Thus $(p_n + q_n)(x_0) > 0$ and hence $p_n + q_n \neq 0$. That is, $\deg(p + q) = n$, which proves (13.28).

Let $g_1, \ldots, g_k$ be a set of generators of Q. Suppose that $f = \sum_j g_j \sigma_j \in Q$ with $\sigma_j \in \sum \mathbb{R}_d[\underline{x}]^2$. Setting $p = g_j \sigma_j$ and $q = \sum_{i \neq j} g_i \sigma_i$, then p and q are nonnegative on C and hence $\deg(g_j \sigma_j) = \deg(p) \leq \deg(p + q) = \deg(f)$ by (13.28). Therefore, condition $(*)$ is satisfied with $A_n = \mathbb{R}_d[\underline{x}]_n$ and $\mathfrak{l}(n) = n$, so that Q is stable.

Since C is contained in the interior of $\mathcal{K}(Q)$, Q is closed by Theorem 13.47 and Lemma 13.48. If $d \geq 2$, then Q does not obey (MP) by Proposition 13.50. □

We illustrate the preceding with a simple, but typical example.

Example 13.52 Suppose that $d \geq 2$. The quadratic modules $Q_0 := \sum \mathbb{R}_d[\underline{x}]^2$ and

$$Q_k = \sum \mathbb{R}_d[\underline{x}]^2 + x_1 \sum \mathbb{R}_d[\underline{x}]^2 + \cdots + x_k \sum \mathbb{R}_d[\underline{x}]^2, \quad k = 1, \ldots, d,$$

do not satisfy (MP). Indeed, these quadratic modules are stable and closed by Corollary 13.49. Further, since $\mathcal{K}(Q_k)$ contains an open cone, Q_k does not obey (MP) by Proposition 13.51 (or likewise by Proposition 13.50).

Because Q_k is closed in $\mathbb{R}_d[\underline{x}]$, by the separation theorem for convex sets, each polynomial $p_0 \in \text{Pos}(\mathbb{R}^d) \backslash Q_k$ gives rise to a Q_k-positive linear functional L on $\mathbb{R}_d[\underline{x}]$ such that $L(p_0) < 0$. Each such functional L is not a moment functional. Recall that the Motzkin polynomial $p_c, 0 < c \leq 3$, defined by (13.6) is in $\text{Pos}(\mathbb{R}^2) \backslash Q_0$. ○

13.9   The Moment Problem on Some Cubics

In this section, we illustrate (SMP) and (MP) for some cubics. We use only elementary computations and do not require results from the theory of algebraic curves.

Throughout this section, we suppose that $f \in \mathbb{R}[x_1, x_2]$ is a polynomial of degree 3 and f_3 is its homogeneous part of degree 3. We denote by

$$C_f \equiv \mathcal{Z}(f) = \{(x_1, x_2) \in \mathbb{R}^2 : f(x_1, x_2) = 0\}$$

the plane real curve associated with f, by $\mathcal{I}$ the ideal generated by f and by $\hat{\mathcal{I}}$ the ideal of all $p \in \mathbb{R}[x_1, x_2]$ which vanish on C_f. Then $\mathbb{R}[C_f] = \mathbb{R}[x_1, x_2]/\hat{\mathcal{I}}$ is the algebra of regular functions on the curve C_f, see Eq. (12.7). Set

$$\mathbb{R}[C_f]_+ := \{p \in \mathbb{R}[C_f] : p(x) \geq 0 \text{ for } x \in C_f\}.$$

Proposition 13.53 *If f_3 has a nonreal zero, then $\sum \mathbb{R}[C_f]^2$ obeys (SMP).*

Proof The assumption implies that f_3 has a real zero z_0 and two nonreal complex conjugate zeros. Upon translation we can asume that $z_0 = (0, 1)$, so f_3 has the form

$$f_3(x_1, x_2) = x_1(ax_1^2 + bx_1x_2 + cx_2^2), \quad \text{where} \quad a, b, c \in \mathbb{R}, \ 4ac > b^2.$$

In particular, we have $a > 0$. Hence f has degree at most 2 in x_2, so we can write $f = \sum_{j=0}^{2} p_j(x_1)x_2^j$, where $p_j \in \mathbb{R}[x_1]$ and $\deg(p_j) \leq 3 - j$.

Let $x_1 \in \mathbb{R}$ and assume that there exists an $x_2 \in \mathbb{R}$ satisfying $(x_1, x_2) \in C_f$. This means that the quadratic equation $f(x_1, x_2) = 0$ in x_2 has a *real* solution, so that

$$g(x_1) := p_1(x_1)^2 - 4p_0(x_1)p_2(x_1) \geq 0.$$

The polynomial $g(x_1)$ has degree at most 4 and its coefficient of x_1^4 is $b^2 - 4ac < 0$. Hence there exists a $\lambda > 0$ such that $g(x_1) < 0$ for $|x_1| > \lambda$. Therefore, if $|x_1| > \lambda$, there is no $x_2 \in \mathbb{R}$ such that $(x_1, x_2) \in C_f$. That is, the curve C_f is contained in the strip $[-\lambda, \lambda] \times \mathbb{R}$ of $\mathbb{R}^2$.

Let $\mathfrak{f} = \{f, -f\}$. Then $C_f = \mathcal{Z}(f)$ is the semi-algebraic set $\mathcal{K}(\mathfrak{f})$ and the preordering $T(\mathfrak{f})$ is $\sum \mathbb{R}[x_1, x_2]^2 + \mathcal{I}$, see e.g. Example 12.4. Since $C_f \subseteq [-\lambda, \lambda] \times \mathbb{R}$, the preordering $T(\mathfrak{f})$ has (MP) by Proposition 13.25. By $\mathcal{I} \subseteq \hat{\mathcal{I}}$, each positive linear functional L on $\mathbb{R}[C_f] = \mathbb{R}[x_1, x_2]/\hat{\mathcal{I}}$ lifts to a positive linear functional $\tilde{L}$ on $\mathbb{R}[x_1, x_2]$ that vanishes on $\mathcal{I}$. Since $T(\mathfrak{f}) = \sum \mathbb{R}[x_1, x_2]^2 + \mathcal{I}$, $\tilde{L}$ is $T(\mathfrak{f})$-positive. Because $T(\mathfrak{f})$ has (MP), the functional $\tilde{L}$, hence L, is given by a Radon measure μ. Since $\tilde{L}$ vanishes on $\mathcal{I}$, μ is supported on C_f by Proposition 12.19. That is, $\sum \mathbb{R}[C_f]^2$ has (SMP). $\qquad \square$

Example 13.54 Suppose that $f_3 = x_1^3 + x_2^3$. Then f_3 has a nonreal zero, hence $\sum \mathbb{R}[C_f]^2$ satisfies (SMP) by Proposition 13.53. Examples of this kind are the Fermat curve $f = x_1^3 + x_2^3 - 1$ and the folium of Descartes $f = x_1^3 + x_2^3 - 3x_1x_2$ (Fig. 13.1). $\qquad \circ$

Proposition 13.55 *Suppose that the polynomial f is of the form*

$$f = x_1^3 + a_{20}x_1^2 + 2a_{11}x_1x_2 + a_{02}x_2^2 + a_{10}x_1 + a_{01}x_2 + a_{00}, \quad \text{where} \quad a_{02} \neq 0.$$

Then $\sum \mathbb{R}[C_f]^2$ is closed in the finest locally convex topology of the vector space $\mathbb{R}[C_f]$ and does not obey (MP).

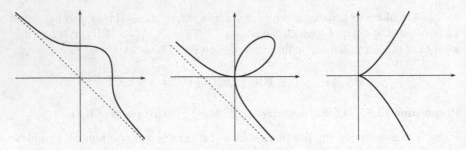

Fig. 13.1 *Left*: Fermat curve, *middle*: folium of Descartes. Both have (SMP), see Example 13.54. *Right*: Neil's parabola. It does not have (MP), see Example 13.56

Proof Replacing x_1 by $x_1 - \frac{1}{3}a_{20}$, the coefficient of x_1^2 becomes zero. Since $a_{02} \neq 0$, the coefficient of x_2 vanishes after some translation. Then f has the form

$$f(x_1, x_2) = -bx_2^2 + x_1^3 + ax_1x_2 + cx_1 + d, \quad a, b, c, d \in \mathbb{R}, \ b \neq 0. \tag{13.29}$$

From (13.29) it follows that each element of the quotient algebra $\mathbb{R}[C_f]$ is of the form

$$g = p(x_1) + x_2q(x_1), \quad \text{where} \quad p, q \in \mathbb{R}[x_1]. \tag{13.30}$$

(It suffices to note that x_2^2 can be eliminated by using (13.29), since $b \neq 0$.)

First we prove that the quadratic module $\sum \mathbb{R}[C_f]^2$ is stable. For $n \in \frac{1}{2}\mathbb{N}$, let V_n denote the vector space of elements $g \in \mathbb{R}[C_f]$, where $\deg(p) \leq n$ and $\deg(q) \leq n - 1$ in (13.30). Let $g_i = p_i(x_1) + x_2q_i(x_1)$, $i \in J$, be finitely many elements of $\mathbb{R}[C_f]$. In the algebra $\mathbb{R}[C_f]$ we compute

$$\sum_i g_i^2 = \sum_i (p_i^2 + 2x_2p_iq_i + x_2^2q_i^2)$$
$$= \sum_i \left(p_i^2 + 2x_2p_iq_i + b^{-1}(x_1^3 + ax_1x_2 + cx_1 + d)q_i^2 \right)$$
$$= \sum_i \left(p_i^2 + b^{-1}(x_1^3 + cx_1 + d)q_i^2 \right) + x_2 \sum_i \left(2p_iq_i + ab^{-1}x_1q_i^2 \right)$$
$$=: p(x_1) + x_2q(x_1). \tag{13.31}$$

Suppose that $\sum g_i^2 \in V_n$, that is, $\deg(p) \leq n$ and $\deg(q) \leq n - 1$. We shall show that $g_i \in V_{n/2}$ for all $i \in J$. Set $j := \max_i \deg(p_i)$ and $k := \max_i \deg(q_i)$.

Case 1: $j > k + 1$.

Then $2j > 2k + 3$ and hence $2j = \deg(p) \leq n$ by (13.31). Thus, $\deg(p_i) \leq j \leq n/2$ and $\deg(q_i) \leq k < j - 1 \leq n/2 - 1$, so that $g_i \in V_{n/2}$ for $i \in J$.

Case 2: $j \leq k + 1$.

Then $2j < 2k + 3$ and hence $2k + 3 = \deg(p) \leq n$ by (13.31). Therefore, we have $\deg(q_i) \leq k < n/2 - 1$ and $\deg(p_i) \leq j \leq k + 1 < n/2$, so that $g_i \in V_{n/2}$ for $i \in J$.

The preceding proves that $Q := \sum \mathbb{R}[C_f]^2$ is stable. Further, $Q \cap (-Q) = \{0\}$ by (12.8). It follows therefore from Theorem 13.47 that $Q = \sum \mathbb{R}[C_f]^2$ is closed in $\mathbb{R}[C_f]$ with respect to the finest locally convex topology.

Put $\tau := \operatorname{sign} b$. We show that there is a real number α such that

$$\tau x_1 - \alpha \in \mathbb{R}[C_f]_+ \quad \text{and} \quad \tau x_1 - \alpha \notin \sum \mathbb{R}[C_f]^2. \tag{13.32}$$

Let $(x_1, x_2) \in C_f$. Then $-bx_2^2 + ax_1x_2 + x_1^3 + cx_1 + d = 0$. This is a quadratic equation in x_2 which has a solution. Therefore,

$$a^2 x_1^2 + 4b(x_1^3 + cx_1 + d) \geq 0.$$

Suppose first that $b > 0$. Then $\lim_{x_1 \to -\infty} [a^2 x_1^2 + 4b(x_1^3 + cx_1 + d)] = -\infty$, so there is a real number α such that $a^2 x_1^2 + 4b(x_1^3 + cx_1 + d) < 0$ for $x_1 < \alpha$. Hence $x_1 \geq \alpha$ for all $(x_1, x_2) \in C_f$, that is, $\tau x_1 - \alpha \in \mathbb{R}[C_f]_+$ for $b > 0$. In the case $b < 0$ we replace x_1 by $-x_1$ and use the same reasoning.

To prove that $\tau x_1 - \alpha \notin \sum \mathbb{R}[C_f]^2$ we assume to the contrary that $\tau x_1 - \alpha = \sum g_i^2$, where $g_i \in \mathbb{R}[C_f]^2$. Since $\tau x_1 - \alpha \in V_1$, it was shown above that all g_i are in $V_{1/2}$, so they are constants. From (13.29) it is obvious that x_1 is not constant on C_f, so we have a contradiction. This completes the proof of (13.32).

By (13.32) we have $\mathbb{R}[C_f]_+ \neq \sum \mathbb{R}[C_f]^2$. Since $\sum \mathbb{R}[C_f]^2$ is closed, as shown above, it follows from Proposition 13.9(ii) (or directly from the separation theorem for convex sets) that $Q = \sum \mathbb{R}[C_f]^2$ does not obey (MP). $\square$

Example 13.56 Neil's parabola (Fig. 13.1) $f(x_1, x_2) = x_1^3 - x_2^2$.

Then $x_1 \in \mathbb{R}[C_f]_+$ and $x_1 \notin \sum \mathbb{R}[C_f]^2$. Indeed, it is obvious that $x_1 \geq 0$ on $\mathcal{Z}(f)$, that is, $x_1 \in \mathbb{R}[C_f]_+$. Arguing as in the preceding proof of Proposition 13.55 it follows that $x_1 \notin \sum \mathbb{R}[C_f]^2$.

By Proposition 13.55, $\sum \mathbb{R}[C_f]^2$ does not satisfy (MP). In fact, any positive linear functional L on $\mathbb{R}[C_f]$ such that $L(x_1 g) < 0$ for some $g \in \sum \mathbb{R}[C_f]^2$ is not a moment functional. Likewise a linear functional $\tilde{L}$ on $\mathbb{R}[x_1, x_2]$ which vanishes on $\mathcal{I}$ and satisfies $\tilde{L}(x_1 g) < 0$ for some $g \in \sum \mathbb{R}[x_1, x_2]^2$ cannot be a moment functional. In Exercise 13.3 we use this observation to "construct" a positive linear functional on $\mathbb{R}[x_1, x_2]$ which is not a moment functional.   ○

Let us return to a general cubic. Since f_3 is a homogeneous polynomial in two variables of degree 3, it is a product of 3 linear factors. If one of the zeros of f_3 is nonreal, or equivalently, f has a nonreal zero at infinity, then $\mathbb{R}[C_f]$ has (SMP) by Proposition 13.53. If all zeros of f_3 are real, it can happen that $\mathbb{R}[C_f]$ obeys (SMP) (see Example 13.57 with $p(x_1) = x_1^3$) and that $\mathbb{R}[C_f]$ does not have (MP) as shown by Proposition 13.55.

The aim of this section was to give a small glimpse into the moment problem on curves; this problem is extensively studied in [PoS, Sr2], and [Pl].

We close this section with a very simple example of curves which have (SMP).

Example 13.57 Let $p \in \mathbb{R}[x_1, x_2]$ and set $g(x_1, x_2) = x_2 - p(x_1)$. We consider the plane curve $C_g = \mathcal{Z}(g)$ and show that the preordering $\sum \mathbb{R}[C_g]^2$ has (SMP).

Let $q \in \mathbb{R}[C_g]_+$. Then $q(x_1, p(x_1)) \in \mathbb{R}[x_1]$ is nonnegative on $\mathbb{R}$. Therefore, by Proposition 3.1(i), $q(x_1, p(x_1)) = q_1(x_1)^2 + q_2(x_1)^2$ for some $q_1, q_2 \in \mathbb{R}[x_1]$, so that $q \in \sum \mathbb{R}[C_g]^2$. By Haviland's theorem 1.14, applied to the algebra $\mathsf{A} = \mathbb{R}[C_g]$, each positive linear functional on $\mathbb{R}[C_g]$ is a moment functional with representing measure on $\hat{\mathsf{A}} = C_g$. This means that $\sum \mathbb{R}[C_g]^2$ has (SMP).   ○

13.10   Proofs of the Main Implications of Theorems 13.10 and 13.12

In this section, we also use the polynomial algebra $\mathbb{R}_m[\underline{y}] := \mathbb{R}[y_1, \dots, y_m]$.

Let us begin with two technical lemmas.

Lemma 13.58 *Let μ_1 and μ_2 be Radon measures on a compact subset K of $\mathbb{R}^m$ and let A be a Borel subset of K. Then there exists a subset N of K such that $\mu_1(N) = \mu_2(N) = 0$ and a sequence $(q_n)_{n \in \mathbb{N}}$ of polynomials $q_n \in \mathbb{R}_m[\underline{y}]$ such that $\lim_{n \to \infty} q_n(\lambda) = \chi_A(\lambda)$ for $\lambda \in K \backslash N$ and $\sup_{\lambda \in K} |q_n(\lambda)| \leq 2$ for $n \in \mathbb{N}$.*

Proof Clearly, it suffices to prove the assertion for the case when K is an m-dimensional compact interval. Since the measures are regular, there exist compact subsets C_n', C_n'' and open subsets U_n', U_n'' of K such that

$$C_n' \subseteq A \subseteq U_n', \ C_n'' \subseteq A \subseteq U_n'', \ \mu_1(U_n' \backslash C_n') \leq 2^{-n}, \ \mu_2(U_n'' \backslash C_n'') \leq 2^{-n} \ \text{ for } \ n \in \mathbb{N}.$$

Then $C_n \subseteq A \subseteq U_n$, $\mu_1(U_n \backslash C_n) \leq 2^{-n}$ and $\mu_2(U_n \backslash C_n) \leq 2^{-n}$ for the compact set $C_n = C_n' \cup C_n''$ and the open set $U_n = U_n' \cap U_n''$. Further, without loss of generality we can assume $C_n \subseteq C_{n+1}$ and $U_{n+1} \subseteq U_n$, so that $U_{n+1} \backslash C_{n+1} \subseteq U_n \backslash C_n$. Applying Urysohn's lemma to the compact set C_n and the closed set $\mathbb{R}^m \backslash U_n$ of $\mathbb{R}^m$ there are exists a continuous function φ_n on K with values in $[0, 1]$ such that $\varphi_n = 1$ on C_n and $\varphi_n = 0$ on $\mathbb{R}^m \backslash U_n$. From Weierstrass' theorem, applied to the compact m-dimensional interval K, it follows that there exists a polynomial $q_n \in \mathbb{R}_m[\underline{y}]$ such that $|q_n(\lambda) - \varphi_n(\lambda)| \leq 2^{-n}$ for $\lambda \in K$. Clearly, we have $\sup_{\lambda \in K} |q_n(\lambda)| \leq 2$. Set $N = \cap_n U_n \backslash C_n$. Then $\mu_j(N) = 0$ for $j = 1, 2$, since $\mu_j(\cap_n U_n \backslash C_n) \leq \mu_j(U_k \backslash C_k) \leq 2^{-k}$ for all $k \in \mathbb{N}$.

Let $\lambda \in K \backslash N$. Then $\lambda \notin U_k \backslash C_k$ for some k. For $n \geq k$ we have $U_n \backslash C_n \subseteq U_k \backslash C_k$, so that $\lambda \notin U_n \backslash C_n$ and hence $\varphi_n(\lambda) = \chi_A(\lambda)$. Thus $\lim_n \varphi_n(\lambda) = \chi_A(\lambda)$. Since $\lim_n (q_n(\lambda) - \varphi_n(\lambda)) = 0$, we conclude that $\lim_n q_n(\lambda) = \chi_A(\lambda)$.   □

Lemma 13.59 *Suppose that μ is a Radon measure on a compact subset K of $\mathbb{R}^m$ and φ is a real-valued function of $L^1(K; \mu)$. If $\int_K q(\lambda) \varphi(\lambda) d\mu(\lambda) \geq 0$ (resp. $\int_K q(\lambda) \varphi(\lambda) d\mu(\lambda) = 0$) for all $q \in \mathbb{R}_m[\underline{y}]$, then $\varphi \geq 0$ (resp. $\varphi = 0$) μ-a.e. on K.*

Proof Given $\varepsilon > 0$, we put $A_\varepsilon := \{\lambda \in K : \varphi(\lambda) \le -\varepsilon\}$. For the Borel set A_ε we choose a sequence $(q_n)_{n \in \mathbb{N}}$ of polynomials as in Lemma 13.58. Since $|q_n^2 \varphi| \le 4|\varphi|$ on K and $\varphi \in L^1(K; \mu)$, Lebesgue's dominated convergence theorem A.2 applies and yields $\lim_n \int q_n^2 \varphi \, dv = \int \chi_{A_\varepsilon} \varphi \, d\mu$. Hence $\int \chi_{A_\varepsilon} \varphi \, d\mu \ge 0$, since $\int q_n^2 \varphi \, d\mu \ge 0$ by assumption. On the other hand, $\int \chi_{A_\varepsilon} \varphi \, d\mu \le -\varepsilon \mu(A_\varepsilon)$ by the definition of A_ε. Therefore, $\mu(A_\varepsilon) = 0$. Since $\varepsilon > 0$ was arbitrary, $\varphi(\lambda) \ge 0$ μ-a.e. on K.

The second assertion follows by applying the preceding to φ and $-\varphi$.   □

As noted in Sect. 13.3 the proofs of Theorem 13.10 and 13.12 are complete once the main implications (ii)→(i) are proved.

Proof of the Implication (ii)→(i) of Theorem 13.12 Let α_j and β_j be as in Theorem 13.12. Throughout this proof we abbreviate

$$K = \prod_{j=1}^{m} [\alpha_j, \beta_j].$$

Let us fix a Q-positive linear functional L on A. For any $p \in$ A we define a linear functional $\mathcal{L}_p$ on the polynomial algebra $\mathbb{R}_m[\underline{y}]$ by

$$\mathcal{L}_p(q) = L(q(\mathrm{h})p^2), \quad q \in \mathbb{R}_m[\underline{y}]. \tag{13.33}$$

Note that $q(\mathrm{h}) := q(h_1, \ldots, h_m)$ is a well-defined element of A, since $h_1, \ldots, h_m \in$ A by assumption.

Lemma 13.60 *For $p \in$ A there exists a Radon measure v on K such that*

$$\mathcal{L}_p(q) = L(q(\mathrm{h})p^2) = \int_K q \, dv_p, \quad q \in \mathbb{R}_m[\underline{y}]. \tag{13.34}$$

Proof The moment problem for the m-dimensional interval K was treated in Example 12.47. As shown therein (see (12.38)), for the existence of a measure v_p such that (13.34) holds it suffices to verify that for all $q \in \mathbb{R}_m[\underline{y}]$ and $j = 1, \ldots, m$,

$$\mathcal{L}_p((\beta_j - y_j)q^2) \ge 0 \text{ and } \mathcal{L}_p((y_j - \alpha_j)q^2) \ge 0. \tag{13.35}$$

By assumption, $\beta_j - h_j$ belongs to $\overline{Q}$. Because $\overline{Q}$ is also a quadratic module (as Q is), $(\beta_j - h_j)q(\mathrm{h})^2 p^2 \in \overline{Q}$. Therefore, since L is Q-positive by assumption and hence $\overline{Q}$-positive, using (13.33) we obtain

$$\mathcal{L}_p((\beta_j - \lambda_j)q^2) = L((\beta_j - h_j)q(\mathrm{h})^2 p^2) \ge 0$$

which implies the first equality of (13.35). The second one is derived in a similar manner from the assumption $(h_j - \alpha_j) \in \overline{Q}$.   □

Since K is compact, the measure ν_p is uniquely determined by p, but we will not need this here. Let ν denote the measure μ_p obtained for the element $p = 1$. A crucial step of this proof is contained in the following

Lemma 13.61 *For all $p \in A$ the measure ν_p is absolutely continuous with respect to ν, that is, each ν-null set is also a ν_p-null set.*

Proof Asssume that A is a ν-null set. For the set A, the compact set K and the measures $\mu_1 = \nu_p$, $\mu_2 = \nu$ we choose a sequence $(q_n)_{n \in \mathbb{N}}$ of polynomials $q_n \in \mathbb{R}_m[\underline{y}]$ as in Lemma 13.58. Then we have $\lim_n q_n(\lambda) = \chi_A(\lambda)$ ν_p-a.e. on K and $\lim_n q_n(\lambda)^2 = \chi_A^2(\lambda) = \chi_A(\lambda)$ ν-a.e. on K. Therefore, by Lebesgue's dominated convergence theorem A.2, we have

$$\lim_n \int_K q_n d\nu_p = \int_K \chi_A d\nu_p = \nu_p(A) \quad \text{and} \quad \lim_n \int_K q_n^2 d\nu = \int_K \chi_A d\nu = \nu(A) = 0. \tag{13.36}$$

Further, since $\sum A^2 \subseteq Q$ and $L(Q) \geq 0$, L is a positive functional on A, so the Cauchy–Schwarz inequality (2.4) holds. From this inequality and (13.34) we obtain

$$\left(\int_K q_n \, d\nu_p \right)^2 = L\big(q_n(\mathsf{h})p^2\big)^2 \leq L\big(q_n(\mathsf{h})^2\big)L(p^4) = L(p^4) \int_K q_n^2 \, d\nu. \tag{13.37}$$

Combining (13.36) and (13.37) we conclude that $\nu_p(A) = 0$. ◻

By Lemma 13.61, the assumptions of the Radon–Nikodym theorem A.3 are fulfilled, so there exists a nonnegative function $\theta_p \in L^1(K, \nu)$ such that $d\nu_p = \theta_p d\nu$. Then, by (13.34) we have

$$\mathcal{L}_p(q) = L(q(\mathsf{h})p^2) = \int_K q\theta_p \, d\nu, \quad q \in \mathbb{R}_m[\underline{y}]. \tag{13.38}$$

Now we take a basis of the form $\{p_j^2\}_{j \in \mathsf{N}}$, where $\mathsf{N} \subseteq \mathbb{N}$, of the real vector space A. (Such a basis exists: Since the algebra A is finitely generated, its vector space dimension is at most countable. If $\{f_l\}$ is a vector space basis of A, we may take elements $f_l \pm 1$ as p_j. These elements span A, since $(f_l + 1)^2 - (f_l - 1)^2 = 4f_l$.)

For $\lambda \in K$ we define a linear functional L_λ on A by

$$L_\lambda \left(\sum_j c_j p_j^2 \right) = \sum_j c_j \theta_{p_j}(\lambda). \tag{13.39}$$

Then we have the following *integral decomposition* of the functional L:

$$L(q(\mathsf{h})p) = \int_K q(\lambda)L_\lambda(p) \, d\nu(\lambda) \quad \text{for} \quad p \in A, \ q \in \mathbb{R}_m[\underline{y}]. \tag{13.40}$$

Indeed, using (13.38) and (13.39), for $p = \sum_j c_j p_j^2$ we prove (13.40) by computing

$$L(q(\mathsf{h})p) = \sum_j c_j L(q(\mathsf{h})p_j^2) = \sum_j c_j \int_K q(\lambda)\theta_{p_j}(\lambda)\, dv(\lambda)$$

$$= \int_K q(\lambda)\Big(\sum_j c_j \theta_{p_j}(\lambda)\Big)\, dv(\lambda) = \int_K q(\lambda)L_\lambda(p)\, dv(\lambda).$$

The next technical steps are collected in the following lemma.

Lemma 13.62

  (i) $L_\lambda(p) \geq 0$ v-a.e. on K for any $p \in Q$.
 (ii) $L_\lambda((h_j - \lambda_j)p) = 0$ v-a.e. on K for $p \in A$ and $j = 1, \ldots, m$.
(iii) There exists a v-null set N such that $L_\lambda \geq 0$ on $\overline{Q_{g(\lambda)}}$ for $\lambda \in K\backslash N$.

Proof

  (i) Let $p \in Q$. Then $q(\mathsf{h})^2 p$ belongs to the quadratic module Q for any $q \in \mathbb{R}_m[\underline{y}]$ and hence

$$L(q(\mathsf{h})^2 p) = \int_K q(\lambda)^2 L_\lambda(p)\, dv(\lambda) \geq 0$$

by (13.40). Therefore, since the function $\lambda \mapsto L_\lambda(p)$ is $L^1(K, v)$ (because the functions θ_{p_j} are), Lemma 13.59 applies and yields $L_\lambda(p) \geq 0$ v-a.e. on K.
 (ii) Let $q \in \mathbb{R}_m[\underline{y}]$. Applying formula (13.40) twice we derive

$$\int_K q(\lambda)L_\lambda((h_j - \lambda_j)p)\, dv(\lambda) = L(q(\mathsf{h})(h_j - \lambda_j)p) = \int_K q(\lambda)(\lambda_j - \lambda_j)L_\lambda(p\, dv(\lambda) = 0.$$

Using again Lemma 13.59 it follows that $L_\lambda((h_j - \lambda_j)p) = 0$ v-a.e. on K.
(iii) Let $g_1, \ldots, g_k$ be generators of the finitely generated quadratic module Q. From (i) and (ii) it follows that for the countable (!) subsets

$$Q' = \sum \mathbb{Q}[\underline{x}]^2 + g_1 \sum \mathbb{Q}[\underline{x}]^2 + \cdots + g_l \sum \mathbb{Q}[\underline{x}]^2,$$

$$\mathcal{I}_\lambda' = (h_1 - \lambda_1)\mathbb{Q}[\underline{x}] + \cdots + (h_m - \lambda_m)\mathbb{Q}[\underline{x}], \ \lambda \in K,$$

of Q and $\mathcal{I}_\lambda$, respectively, there exists a common v-null set N such that for all $\lambda \in K\backslash N$ we have $L_\lambda(p) \geq 0$ for $p \in Q'$ and $L_\lambda(r) = 0$ for $r \in \mathcal{I}_\lambda'$. Approximating real coefficients by rationals it follows that this holds for all elements $p \in Q$ and $r \in \mathcal{I}_\lambda$ as well. Therefore, for $\lambda \in K\backslash N$, we have $L_\lambda \geq 0$ on Q and $L_\lambda = 0$ on $\mathcal{I}_\lambda$. Thus $L_\lambda \geq 0$ on $Q_\lambda = Q + \mathcal{I}_\lambda$ and hence on the closure $\overline{Q_\lambda}$.   □

Now we can complete the proof of the implication (ii)→(i) of Theorem 13.12. Suppose that Q_λ has (SMP) for all $\lambda \in K$. Recall that L is an arbitrary Q-positive linear functional on A. Let $p \in \mathrm{Pos}(\mathcal{K}(Q))$. Then, obviously, $p \in \mathrm{Pos}(\mathcal{K}(Q_\lambda))$ for $\lambda \in K$. Since Q_λ has (SMP), $\overline{Q_\lambda} = \mathrm{Pos}(\mathcal{K}(Q_\lambda))$ by Proposition 13.9(i). Therefore, $p \in \overline{Q_\lambda}$. Therefore $L_\lambda(p) \geq 0$ ν-a.e. on K by Lemma 13.62(iii) and hence $L(p) \geq 0$ by (13.40). That is, $L \geq 0$ on $\mathrm{Pos}(\mathcal{K}(Q))$. Hence Haviland's Theorem 1.14 (i)→(iv), applies and shows that L comes from a measure $\mu \in M_+(\hat{\mathrm{A}})$ supported on $\mathcal{K}(Q)$. This proves that Q has (SMP).

The arguments in the case of (MP) are almost the same. By Proposition 13.9(iii), it suffices to replace in the preceding paragraph $\mathrm{Pos}(\mathcal{K}(Q))$ and $\mathrm{Pos}(\mathcal{K}(Q_\lambda))$ by $\mathrm{Pos}(\hat{\mathrm{A}})$. Now the proof of Theorem 13.12 is complete.   □

Remark 13.63 In fact, we have shown under the assumptions of Theorem 13.12 that

$$\overline{Q} = \cap_{\lambda \in \prod_{j=1}^{m} [\alpha_j, \beta_j]} \overline{Q_\lambda}.$$

Indeed, since $Q \subseteq Q_\lambda$ by definition, the inclusion $\overline{Q} \subseteq \cap_\lambda \overline{Q_\lambda}$ is obvious. Conversely, let $p \in \cap_\lambda \overline{Q_\lambda}$. Then, as shown at the end of the preceding proof, $L(p) \geq 0$ for each Q-positive linear functional L. Hence $p \in \overline{Q}$ by Lemma 13.8.   ○

Proof of the Implication (ii)→(i) of Theorem 13.10 In order to apply Proposition 12.23(ii) we first note that we can assume without loss of generality that $\mathrm{A} = \mathbb{R}_d[\underline{x}]$. Recall that A is (isomorphic to) the quotient algebra $\mathbb{R}_d[\underline{x}]/\mathcal{J}$ for some ideal $\mathcal{J}$ of $\mathbb{R}_d[\underline{x}]$ and $\pi : \mathbb{R}_d[\underline{x}] \to \mathrm{A}$ denotes the canonical map. We choose polynomials $\tilde{h}_j \in \mathbb{R}_d[\underline{x}]$ such that $\pi(\tilde{h}_j) = h_j$. Clearly, $\tilde{T} := \pi^{-1}(T)$ is a preordering of $\mathbb{R}_d[\underline{x}]$. Since $\mathcal{J} \subseteq \tilde{T}$, each $\tilde{T}$-positive character of $\mathbb{R}_d[\underline{x}]$ annihilates $\mathcal{J}$, so it belongs to $\hat{\mathrm{A}} = \mathcal{Z}(\mathcal{J})$. This implies that $\mathcal{K}(\tilde{T}) = \mathcal{K}(T)$ and $\tilde{h}_j(x) = h_j(x)$ for $x \in \mathcal{K}(T)$. Thus $\tilde{h}(\mathcal{K}(\tilde{T})) = h(\mathcal{K}(T))$. Since h_j is bounded on $\mathcal{K}(T)$ by (ii), so is $\tilde{h}_j$ on $\mathcal{K}(\tilde{T})$. For $\lambda \in h(\mathcal{K}(T))$, $\tilde{\mathcal{I}}_\lambda := \pi^{-1}(\mathcal{I}_\lambda)$ is an ideal of $\mathbb{R}_d[\underline{x}]$. It is easily verified that $(\tilde{T})_\lambda = \tilde{T} + \tilde{\mathcal{I}}_\lambda$. Since $\tilde{T} = \tilde{T} + \mathcal{J}$ and $(\tilde{T} + \mathcal{J})/\mathcal{J} \cong T$, by Proposition 13.14(iii), $\tilde{T}$ has (SMP) resp. (MP) in $\mathbb{R}_d[\underline{x}]$ if and only if T has in A. The same holds for T_λ. This shows that we can assume that $\mathrm{A} = \mathbb{R}_d[\underline{x}]$.

By the assumption of Theorem 13.10, each element h_j, $j = 1, \ldots, m$, of $\mathbb{R}_d[\underline{x}]$ is bounded on $\mathcal{K}(T)$, that is, there are numbers $\alpha_j, \beta_j \in \mathbb{R}$ such that $\alpha_j \leq h_j(x) \leq \beta_j$ on $\mathcal{K}(T)$. Since $\beta_j - h_j$ and $h_j - \alpha_j$ are bounded and nonnegative on $\mathcal{K}(T)$, we have $L(\beta_j - h_j) \geq 0$ and $L(h_j - \alpha_j) \geq 0$ for each T-positive linear functional L on $\mathbb{R}_d[\underline{x}]$ by Proposition 12.23(ii). Therefore, by Lemma 13.8, $\beta_j - h_j$ and $h_j - \alpha_j$ belong to the closure $\overline{T}$ of the preordering T. Thus the preordering T satisfies the assumptions of Theorem 13.12. Hence, by Theorem 13.12, T has (SMP) resp. (MP) provided that T_λ has (SMP) resp. (MP) for all $\lambda \in K = \prod_{j=1}^{m} [\alpha_j, \beta_j]$. To complete the proof of Theorem 13.10 it remains to show that it suffices to assume this for λ in the smaller set $h(\mathcal{K}(T))$.

Fix a T-positive linear functional L on $\mathbb{R}_d[\underline{x}]$. We show that the measure $\nu = \nu_1$ for the polynomial $p = 1$ in Lemma 13.60 is supported on the set $H := \overline{h(\mathcal{K}(T))}$. Suppose that $q \in \mathbb{R}_m[\underline{y}]$ is nonnegative on H. Then $q(\mathrm{h}) \in \mathbb{R}_d[\underline{x}]$ is nonnegative on $\mathcal{K}(T)$. Further, $q(\mathrm{h})$ is bounded on $\mathcal{K}(T)$, because the polynomials h_j are. Therefore, since L is T-positive, we have $\mathcal{L}_1(q) = L(q(\mathrm{h})) \geq 0$ again by Proposition 12.23(ii).

This shows that the assumptions of Haviland's Theorem 1.12 for the closed set H are satisfied. Hence, by this theorem, supp $\nu \subseteq H$. Consequently, the integration in formula (13.40) is over H rather than K. Continuing along the lines of the above proof of Theorem 13.12 we see that it suffices to have (SMP) for $\lambda \in H = \overline{\mathsf{h}(\mathcal{K}(T))}$.

Finally, let $\lambda \in H \backslash \mathsf{h}(\mathcal{K}(T))$. From its definition it follows that the fibre set $\mathcal{K}(T)_\lambda$ is empty, so that $-1 \in T_\lambda$ by Theorem 12.3(iv) and hence $\mathbb{R}_d[\underline{x}] = T_\lambda = \overline{T_\lambda}$. Therefore, the requirement $p \in \overline{T_\lambda}$ for $p \in \mathrm{Pos}(\mathcal{K}(T))$ (see the paragraph before last in the above proof of Theorem 13.12) is always fulfilled. That is, it suffices to assume (SMP) for $\lambda \in \mathsf{h}(\mathcal{K}(T))$.

The reasoning for (MP) is similar. The proof of Theorem 13.10 is complete. $\square$

Remark 13.64 Under the assumption of Theorem 13.10 we proved that

$$T = \cap_{\lambda \in \mathsf{h}(\mathcal{K}(T))} \overline{T_\lambda}.$$

$\circ$

13.11 Exercises

1. Show that the Choi–Lam polynomial

$$p(x_1, x_2, x_3) := x_1^2 x_2^2 + x_1^2 x_3^2 + x_2^2 x_3^2 + 1 - 4x_1 x_2 x_3 \tag{13.41}$$

   is nonnegative on $\mathbb{R}^3$, but it is not a sum of squares in $\mathbb{R}[x_1, x_2, x_3]$.
2. Use the Gram matrix representation (13.3) to prove that $\sum \mathbb{R}_d[\underline{x}]_n^2$ is a closed subset of $\mathbb{R}_d[\underline{x}]_{2n}$.
3. (*A positive functional on $\mathbb{R}[x_1, x_2]$ that is not a moment functional [Sm7]*)
   Let L be a $T(x^3)$-positive linear functional on $\mathbb{R}[x]$ that is not $T(x)$-positive. (Such a functional L was constructed in Example 13.7). Define linear functionals L_1 on $\mathbb{R}[x]$ and L_2 on $\mathbb{R}[x_1, x_2]$ by $L_1(x^{2n}) = L(x^n), L_1(x^{2n+1}) = 0$ for $n \in \mathbb{N}_0$ and $L_2(p) = L_1(p(x^2, x^3))$ for $p \in \mathbb{R}[x_1, x_2]$, that is, $L_2(x_1^k x_2^{2l}) = L(x^{k+3l})$ and $L_2(x_1^k x_2^{2l+1}) = 0$ for $k, l \in \mathbb{N}_0$.

   a. Show that L_2 is a positive linear functional on $\mathbb{R}[x_1, x_2]$.
      Hint: Use the $T(x^3)$-positivity of L.
   b. Show that L_2 is not a moment functional.
      Hint: Show that any representing measure would be supported on the curve $x_1^3 = x_2^2$. This is impossible, since L is not $T(x)$-positive. See Example 13.56.

4. Construct positive linear functionals on the preorderings of Examples 13.27 and 13.28 which are not $\mathcal{K}(\mathsf{f})$-moment functionals.
5. Modify the set of polynomials in Example 13.27 such that the set $\mathcal{K}(\mathsf{f})$ remains the same, but the preordering $T(\mathsf{f})$ satisfies (SMP).
6. Let $\mathcal{E} \in \mathbf{D}(\mathbb{R}[x_2])$ be finitely generated, p a nonconstant polynomial from $\mathbb{R}[x_1]$ and $\alpha > 0$. Suppose that $\mathcal{D} \in \mathbf{D}(\mathbb{R}[x_1, x_2])$ is generated by $\mathcal{E}$ and $p^2 + \alpha$. Show that the preordering $\sum(\mathsf{A}_\mathcal{D})^2$ obeys (MP) in the algebra $\mathsf{A}_\mathcal{D}$.

7. Decide whether or not the following preorderings satisfy (SMP).

   a. $f_1(x) = x_1, f_2(x) = 1 - x_1, f_3(x) = x_1x_2, f_4(x) = 1 - x_1x_2$.
   b. $f_1(x) = x_1, f_2(x) = 1 - x_1, f_3(x) = x_2, f_4(x) = 1 - x_1x_2$.
   c. $f_1(x) = x_1, f_2(x) = 1 - x_1, f_3(x) = x_2^3, f_4(x) = 1 - x_1x_2$.

8. Set $f = \{x_1, x_2, 1 - x_1x_2, x_1x_2 - 1\}$ and let $\mathcal{I}$ be the ideal generated by x_1x_2 in $\mathbb{R}[x_1, x_2]$.

   a. Does the quadratic module $\mathcal{I}$ of $\mathbb{R}[x_1, x_2]$ satisfy (SMP) or (MP)?
   b. Does the preordering $T(f)$ of $\mathbb{R}[x_1, x_2]$ satisfy (SMP) or (MP)?

9. Show that the preordering $T(x_1, x_2)$ of $\mathbb{R}[x_1, x_2]$ is closed and it does not have (MP).

10. Decide whether or not the preordering $\sum \mathbb{R}[C_f]$ of the algebra $\mathbb{R}[C_f]$ obeys (SMP) resp. (MP):

    a. $f(x_1, x_2) = x_1^3 + x_1^2x_2 + x_1x_2^2 + x_2^3 + x_1x_2 + 1$.
    b. $f(x_1, x_2) = x_1^3$.
    c. $f(x_1, x_2) = x_1^3 - 2x_2^2$.
    d. $f(x, x_2) = x_1^3 + x_1x_2^2 - x_2^2$ (cissoid of Diocles).

11. (*Moment problem on a one point set of* $\mathbb{R}$)
    Let A be the quotient algebra of $\mathbb{R}[x]$ by the ideal generated by x^2.

    a. Show that $\hat{A} = \{\chi_0\}$, where χ_0 is the point evaluation at zero.
    b. Show that $\sum A^2$ is not closed and that its closure is A_+.
    c. Show that $\sum A^2$ has (SMP).
    d. What happens with assertions a.-c. if A is replaced by the quotient algebra of $\mathbb{R}[x]$ by the ideal generated by x?

12. (*Moment problem on two intersecting lines in* $\mathbb{R}^2$)
    Consider two lines in $\mathbb{R}^2$ intersecting in one point and given by equations $a_1x_1 + b_1x_2 + c_1 = 0$ and $a_2x_1 + b_2x_2 + c_2 = 0$. Let A be the quotient algebra of $\mathbb{R}[x_1, x_2]$ by the ideal generated by $(a_1x_1 + b_1x_2 + c_1)(a_2x_1 + b_2x_2 + c_2)$.

    a. Show that for each $f \in A_+$ there exist $g_1, g_2 \in A$ such that $f = g_1^2 + g_2^2$.
    b. Conclude that $\sum A^2$ has (SMP).

13. What happens if the two lines in Exercise 12 are parallel?

14. Let $p_2 \in \mathbb{R}[x_1], p_3 \in \mathbb{R}[x_1, x_2], \ldots, p_d \in \mathbb{R}[x_1, \ldots, x_{d-1}]$, where $d \geq 2$. Consider the curve V in $\mathbb{R}^d$ given by the equations

    $$x_d = p_d(x_1, \ldots, x_{d-1}), \ldots, x_3 = p_3(x_1, x_2), x_2 = p_2(x_1).$$

    Show that $\sum \mathbb{R}[V]^2$ has (SMP).

15. a. Show that the convex hull of a compact subset of $\mathbb{R}^n$ is also compact.
    b. Suppose that M is a convex compact subset of $\mathbb{R}^n$ such that $0 \notin M$. Show that the cone generated by M is closed in $\mathbb{R}^n$.

   c. Show by an example that the assertion of b. does not hold in general if the assumption $0 \notin M$ is omitted.

16. Let A be a finitely generated (real or complex) unital $*$-algebra and let q be a seminorm on A such that $q(1) = 1$, $q(fg) \leq q(f)q(g)$ and $q(f^*) = q(f)$ for $f, g \in$ A. Let L be a positive linear functional on A and let π_L denote the GNS representation associated with L (see Definition 12.39). Suppose that there exists a constant $c > 0$ such that $|L(f)| \leq cq(f)$ for all $f \in$ A.

   a. Use the Cauchy–Schwarz inequality to show that $|L(f)| \leq L(1)q(f)$ for $f \in$ A.

   b. Show that $\pi_L(f)$ is bounded and $\|\pi_L(f)\| \leq q(f)$ for $f \in$ A.

   c. Show that there exists a compact subset K of $\hat{A}$ and a Radon measure μ on K such that $L(f) = \int_K f(x)d\mu(x)$ for $f \in$ A.

13.12 Notes

The existence of positive functionals on $\mathbb{R}_d[\underline{x}]$, $d \geq 2$, which are not moment functionals was first shown rigorously in [Sm2] and [BCJ]. The explicit functional in Proposition 13.5 is due to J. Friedrich [Fr1]. Hilbert's method of constructing positive polynomials that are not sos is elaborated in [Re4].

The fibre theorem was discovered by the author in [Sm7] for $A = \mathbb{R}_d[\underline{x}]$; the properties (SMP) and (MP) were also invented therein. The extended and more general form stated as Theorem 13.10 was obtained only recently in [Sm11]. The original proof in [Sm7] was based on the decomposition theory of states [Sm4]. An "elementary" proof was given by T. Netzer [Nt], see also [Ms1, Section 4.4]. Our approach is based on arguments from [Nt] and [Ms1]. The weaker version for quadratic modules is taken from [Ms1]. Proposition 13.14 was noted in [Sr1].

The results of Sect. 13.4 are due to S. Kuhlmann and M. Marshall [KM], see also [KMS]. Proposition 13.25 was proved in [Mt, Theorem 1.3] and later independently in [KM]; the elegant approach given in the text is from [Sm7]. The applications of the fibre theorem to the rational moment problem in Sect. 13.6 and to the extension theorem in Sect. 13.7 are taken from [Sm11]. Further results on rational moment problems can be found in [Ch, BGHN] (dimension one) and [PVs1, CMN, Sm11] (multidimensional case).

That the cone $\sum \mathbb{R}_d[\underline{x}]^2$ is closed in the finest locally convex topology was first shown in [Sm1] and independently in [BCJ] ; a general result of this kind was given in [Sm4, Theorem 11.6.3]. The stability concept of quadratic modules and its applications to the failure of (MP) in Sect. 13.8 are due to C. Scheiderer, see [PoS, Sr1]. The proof of Proposition 13.46 is taken from [Sw3]. Proposition 13.50 appeared in [Sr1], while Proposition 13.51 is from [Ms1].

The moment problem on curves was first studied in [Mt] and [St1]. In Sect. 13.9 we followed J. Matzke's thesis [Mt], where the case of cubics was completely settled. The moment problem on curves is now well-understood, see [Sr2, Pl].

Chapter 14
The Multidimensional Moment Problem: Determinacy

In this chapter, we study the determinacy problem in the multivariate case. In Sect. 14.1, we introduce several natural determinacy notions (strict determinacy, strong determinacy, ultradeterminacy) that are all equivalent to the "usual" determinacy in dimension one. In the remaining sections we develop various techniques and methods to derive *sufficient* criteria for determinacy. In Sect. 14.2, polynomial approximation is used to show that the determinacy of all marginal sequences implies the determinacy of a moment sequence (Theorem 14.6). Section 14.3 is based on operator-theoretic methods in Hilbert space. The main results (Theorems 14.12 and 14.16) show that the determinacy of appropriate 1-subsequences of a positive semidefinite d-sequence s implies that s is a (determinate) moment sequence. Section 14.4 is concerned with Carleman's condition in the multivariate case. Probably the most useful result in this chapter is Theorem 14.20; it says that if all marginal sequences of a positive semidefinite d-sequence s satisfy Carleman's condition, then s is a *determinate moment sequence*. Section 14.6 uses the disintegration of measures as a powerful method for the study of determinacy. A fibre theorem for determinacy (Theorem 14.30) states a measure is determinate if the base measure is strictly determinate and almost all fibre measures are determinate. In Sect. 14.5, we calculate the moments of the surface measure on S^{d-1} and of the Gaussian measure on $\mathbb{R}^d$.

14.1 Various Notions of Determinacy

First let us recall some results from the one-dimensional case. If $\mu \in \mathcal{M}_+(\mathbb{R})$ is a measure with moment sequence s, we know from Theorem 6.10 and Corollary 6.11 that the following statements are equivalent:

(i) *The measure μ, or equivalently, its moment sequence s, is determinate.*
(ii) *The multiplication operator X by the variable x with domain $\mathbb{C}[x]$ is essentially self-adjoint on the Hilbert space $L^2(\mathbb{R}, \mu)$.*

© Springer International Publishing AG 2017
K. Schmüdgen, *The Moment Problem*, Graduate Texts in Mathematics 277,
DOI 10.1007/978-3-319-64546-9_14

(iii) $\mathbb{C}[x]$ *is dense in* $L^2(\mathbb{R}, (1 + x^2)d\mu)$.

    Further, if μ *is determinate, then the polynomials* $\mathbb{C}[x]$ *are dense in* $L^2(\mathbb{R}, \mu)$.

In dimensions $d \geq 2$ determinacy is much more subtle and the preceding conditions lead to different determinacy notions.

Recall that $\mathcal{M}_+(\mathbb{R}^d)$ is the set of Radon measures on $\mathbb{R}^d$ for which all moments are finite. For $\mu \in \mathcal{M}_+(\mathbb{R}^d)$ we denote by $\mathcal{M}_\mu$ the set of measures $\nu \in \mathcal{M}_+(\mathbb{R}^d)$ which have the same moments as μ, or equivalently, which satisfy

$$\int_{\mathbb{R}^d} p(x)\,d\nu(x) = \int_{\mathbb{R}^d} p(x)\,d\mu(x) \quad \text{for all} \quad p \in \mathbb{R}_d[\underline{x}].$$

We write $\nu \cong \mu$ if $\nu \in \mathcal{M}_\mu$. Obviously, "$\cong$" is an equivalence relation in $\mathcal{M}_+(\mathbb{R}^d)$.

Definition 14.1 For a measure $\mu \in \mathcal{M}_+(\mathbb{R}^d)$ we shall say that

- μ is *determinate* if $\mathcal{M}_\mu$ is a singleton, that is, if $\nu \in \mathcal{M}_\mu$ implies $\mu = \nu$,
- μ is *strictly determinate* if μ is determinate and $\mathbb{C}_d[\underline{x}]$ is dense in $L^2(\mathbb{R}^d, \mu)$,
- μ is *strongly determinate* if $\mathbb{C}_d[\underline{x}]$ is dense in $L^2(\mathbb{R}^d, (1+x_j^2)d\mu)$ for $j=1,\ldots,d$,
- μ is *ultradeterminate* if $\mathbb{C}_d[\underline{x}]$ is dense in $L^2(\mathbb{R}^d, (1 + \|x\|^2)d\mu)$.

In this section we adopt the following notational convention: If μ is a measure on $\mathbb{R}^d$ and no confusion can arise, we often write $L^p(\mu)$ instead of $L^p(\mathbb{R}^d, \mu)$.

Suppose that $\mu \in \mathcal{M}_+(\mathbb{R}^d)$. Let s be the moment sequence of μ and L the moment functional of μ, that is, $L(f) = \int f(x)d\mu, f \in \mathbb{C}_d[\underline{x}]$. As in the one-dimensional case the above notions will be used synonymously for μ, s, and L. That is, we say that s, and likewise L, is *determinate, strongly determinate, strictly determinate, ultradeterminate,* if μ has this property. Let us set $\mathcal{M}_s = \mathcal{M}_L = \mathcal{M}_\mu$.

We recall the Hilbert space approach and the GNS construction from Sect. 12.5. Let L be a positive linear functional on $\mathbb{C}_d[\underline{x}]$. Then there is a scalar product $\langle \cdot, \cdot \rangle_L$ on the vector space $\mathcal{D}_L = \mathbb{C}_d[\underline{x}]/\mathcal{N}_L$ such that $\langle p + \mathcal{N}_L, q + \mathcal{N}_L \rangle_L = L(p\bar{q})$, $p, q \in \mathbb{C}_d[\underline{x}]$, where $\mathcal{N}_L = \{f \in \mathbb{C}_d[\underline{x}] : L(f\bar{f}) = 0\}$. Further, there are pairwise commuting symmetric operators $X_j, j = 1, \ldots, d$, defined by $X_j(p + \mathcal{N}_L) = x_j p + \mathcal{N}_L, p \in \mathbb{C}_d[\underline{x}]$. The Hilbert space completion of the unitary space $\mathcal{D}_L$ is denoted by $\mathcal{H}_L$. For notational simplicity we write p for $p + \mathcal{N}_L$. Then

$$\langle p, q \rangle_L = L(p\bar{q}) \quad \text{and} \quad X_j p = x_j p, \ p, q \in \mathbb{C}_d[\underline{x}],$$

that is, X_j is the multiplication operator by the variable x_j. Recall that $\overline{X}_j$ denotes the closure of operator X_j.

All four determinacy notions have been defined in terms of the measure μ. By Theorem 14.2 below, μ is strongly determinate if and only if the closures of all symmetric operators $X_1, \ldots, X_d$ are self-adjoint. Thus, strong determinacy is a natural and fundamental concept from the operator-theoretic point of view.

Our main reason for introducing strict determinacy is that it enters into the fibre Theorem 14.30 below. Another reason stems from the theory of orthogonal polynomials: Since the polynomials are dense in $L^2(\mathbb{R}^d, \mu)$, we have $\mathcal{H}_L \cong L^2(\mathbb{R}^d, \mu)$. Hence for a strictly determinate measure μ each sequence of orthonormal polynomials in $\mathcal{H}_L$ is an orthonormal basis of $L^2(\mathbb{R}^d, \mu)$.

By definition strict determinacy implies determinacy. Theorem 14.2 below shows that if μ is strongly determinate it is strictly determinate and hence determinate. Since the norm of $L^2(\mathbb{R}^d, (1 + ||x||^2)d\mu)$ is obviously stronger than that of $L^2(\mathbb{R}^d, (1 + x_j^2)d\mu)$, ultradeterminacy implies strong determinacy. Thus we have

ultradeterminate $\Rightarrow$ *strongly determinate* $\Rightarrow$ *strictly determinate* $\Rightarrow$ *determinate.*

From Weierstrass' approximation theorem it follows that each measure with *compact* support is ultradeterminate and hence all four determinacy notions are fulfilled.

As noted at the beginning of this section, in the case $d = 1$ the four concepts (determinacy, strict determinacy, strong determinacy, ultradeterminacy) are equivalent. However, in dimension $d \geq 2$ all of them are different! An example of a strongly determinate measure that is not ultradeterminate is sketched in Exercise 14.6. A strictly determinate measure which is not strongly determinate is developed in Example 14.4. In [BT] it is shown that there exist rotation invariant determinate measures $\mu \in \mathcal{M}_+(\mathbb{R}^d)$, $d \geq 2$, such that $\mathbb{C}_d[\underline{x}]$ is *not* dense in $L^2(\mathbb{R}^d, \mu)$. Such measures are not strictly determinate.

The following theorem gives an operator-theoretic characterization of strong determinacy.

Theorem 14.2 *Let L be a moment functional on $\mathbb{C}_d[\underline{x}]$ and let μ be a representing measure for L. Then μ is strongly determinate if and only if all symmetric operators $X_1, \ldots, X_d$ are essentially self-adjoint on $\mathbb{C}_d[\underline{x}]$ in $\mathcal{H}_L$, that is, we have $\overline{X_k} = (X_k)^*$ for $k = 1, \ldots, d$. If this holds, then μ is strictly determinate, that is, μ is determinate and $\mathbb{C}_d[\underline{x}]$ is dense in $L^2(\mathbb{R}^d, \mu)$.*

The following technical lemma is needed in the proof of Theorem 14.2.

Lemma 14.3 *Suppose A is a closed symmetric operator on a Hilbert space $\mathcal{K}$. Let $\mathcal{H}$ be a closed subspace of $\mathcal{K}$ and let P denote the orthogonal projection onto $\mathcal{H}$. Suppose that $\mathcal{D}$ is a dense linear subspace of $\mathcal{H}$ such that $\mathcal{D} \subseteq \mathcal{D}(A)$, $A\mathcal{D} \in \mathcal{H}$ and $X := A\lceil\mathcal{D}$ is an essentially self-adjoint operator on $\mathcal{H}$. Then $PA \subseteq AP$. Moreover, if Y denotes the restriction of A to $(I - P)\mathcal{D}(A)$, then $A = \overline{X} \oplus Y$ on $\mathcal{K} = \mathcal{H} \oplus \mathcal{H}^\perp$.*

Proof Suppose that $\varphi \in \mathcal{D}(A)$. Let $\psi \in \mathcal{D}$. Using the facts that ψ and $X\psi$ are in $\mathcal{H}$ and the relation $X \subseteq A$ it follows that

$$\langle X\psi, P\varphi \rangle = \langle pX\psi, \varphi \rangle = \langle X\psi, \varphi \rangle = \langle A\psi, \varphi \rangle = \langle \psi, A\varphi \rangle = \langle p\psi, A\varphi \rangle = \langle \psi, PA\varphi \rangle.$$

Since $\psi \in \mathcal{D}$ was arbitrary, $P\varphi \in \mathcal{D}(X^*)$ and $X^*P\varphi = AP\varphi$. Since X is essentially self-adjoint, $X^* = \overline{X}$ and hence $AP\varphi = \overline{X}P\varphi = PA\varphi$. That is, $PA \subseteq AP$.

If $\varphi \in \mathcal{D}(A)$, we have $P\varphi \in \mathcal{D}(\overline{X})$ and $(I - P)\varphi \in \mathcal{D}(A)$ as just shown. Hence $\varphi = P\varphi + (I - P)\varphi \in \mathcal{D}(\overline{X}) \oplus \mathcal{D}(Y)$ and $A\varphi = \overline{X}\varphi + Y(I - P)\varphi$ which proves that $A \subseteq \overline{X} \oplus Y$. By definition the converse inclusion is clear. $\square$

Proof of Theorem 14.2 The proof is based on the Hilbert space approach given by Theorem 12.40.

Suppose first that the symmetric operators $X_1, \ldots, X_d$ are essentially self-adjoint on $\mathcal{H}_L$. Let $(A_1, \ldots, A_d)$ be a d-tuple of strongly commuting self-adjoint operators on a larger Hilbert space $\mathcal{K}$ such that $X_j \subseteq A_j$, $j=1,\ldots,d$. By Lemma 14.3 there is a decomposition $A_j = X_j \oplus Y_j$ on $\mathcal{K} = \mathcal{H}_L \oplus \mathcal{H}_L^\perp$. Since A_j is self-adjoint, so is Y_j and for the spectral measures we obtain $E_{A_j} = E_{\overline{X_j}} \oplus E_{Y_j}$. Hence the spectral measures of $\overline{X_k}$ and $\overline{X_l}$ commute and (15.6) implies that $E_{(\overline{X_1},\ldots,\overline{X_d})} \subseteq E_{(A_1,\ldots,A_d)}$ for the corresponding joint spectral measures. Therefore, the measure $\mu(\cdot) := \langle E_{(\overline{X_1},\ldots,\overline{X_d})}(\cdot)1, 1 \rangle_{\mathcal{H}_L}$ is equal to $\langle E_{(A_1,\ldots,A_d)}(\cdot)1, 1 \rangle_{\mathcal{K}}$. Since each solution of the moment problem is of the form $\langle E_{(A_1,\ldots,A_d)}(\cdot)1, 1 \rangle_{\mathcal{K}}$ by Theorem 12.40 and so coincides with μ by the preceding considerations, μ is the unique representing measure of L. Thus μ is determinate.

Let M be a Borel set of $\mathbb{R}^d$. By definition $\mathbb{C}_d[\underline{x}]$ is dense in $\mathcal{H}_L$. Hence there is a sequence $(p_n)_{n \in \mathbb{N}}$ of polynomials $p_n \in \mathbb{C}_d[\underline{x}]$ such that $p_n(x) \to E_{(\overline{X_1},\ldots,\overline{X_d})}(M)1$ in $\mathcal{H}_L \subseteq L^2(\mu)$. Then, by the functional calculus,

$$\|(p_n(x) - E_{(\overline{X_1},\ldots,\overline{X_d})}(M))1\|_{\mathcal{H}_L}^2 = \int_{\mathbb{R}^d} |p_n(\lambda) - \chi_M(\lambda)|^2 d\langle E_{(\overline{X_1},\ldots,\overline{X_d})}(\lambda)1, 1 \rangle_{\mathcal{H}_L} \to 0,$$

so the closure of $\mathbb{C}_d[\underline{x}]$ in $L^2(\mathbb{R}^d, \mu)$ contains all characteristic functions χ_M. Hence $\mathbb{C}_d[\underline{x}]$ is dense in $L^2(\mathbb{R}^d, \mu)$. Thus $\mathcal{H}_L \cong L^2(\mathbb{R}^d, \mu)$, so μ is strictly determinate.

It is easily checked that $(X_k \pm i)\mathbb{C}_d[\underline{x}]$ is dense in $L^2(\mathbb{R}^d, \mu)$ if and only if $\mathbb{C}_d[\underline{x}]$ is dense in $L^2(\mathbb{R}^d, (1 + x_k^2)d\mu)$. Therefore, by Proposition A.42, since X_k is essentially self-adjoint, $(X_k \pm i)\mathbb{C}_d[\underline{x}]$ is dense in $\mathcal{H}_L \cong L^2(\mathbb{R}^d, \mu)$. Hence $\mathbb{C}_d[\underline{x}]$ is dense in $L^2(\mathbb{R}^d, (1 + x_k^2)d\mu)$. This proves that μ is strongly determinate.

Conversely, suppose that μ is strongly determinate. Then, by definition, for each $k = 1, \ldots, d$, $\mathbb{C}_d[\underline{x}]$ is dense in $L^2(\mathbb{R}^d, (1 + x_k^2)d\mu)$. Hence $(X_k \pm i)\mathbb{C}_d[\underline{x}]$ is dense in $L^2(\mathbb{R}^d, \mu)$ and so also in its subspace $\mathcal{H}_L$. Therefore, X_k is essentially self-adjoint on $\mathcal{H}_L$ again by Proposition A.42. $\square$

Example 14.4 (A strictly determinate measure that is not strongly determinate) Let $\nu_0 \in \mathcal{M}_+(\mathbb{R})$ be an N-extremal solution of an indeterminate moment problem. We define a measure $\nu \in \mathcal{M}_+(\mathbb{R})$ by $d\nu := (1 + x^2)^{-1}d\nu_0$. Since $\mathbb{C}[x]$ is dense in $L^2(\mathbb{R}, \nu_0) = L^2(\mathbb{R}, (1 + x^2)d\nu)$, ν is determinate by Corollary 6.11. Let μ be the image of ν under the mapping $x \mapsto (x, x^2)$ of $\mathbb{R}$ into $\mathbb{R}^2$. Clearly, $\mu \in \mathcal{M}_+(\mathbb{R}^2)$.

We prove that μ is not strongly determinate. Assume the contrary. Then $\mathbb{C}[x_1, x_2]$ is dense in $L^2(\mathbb{R}^2, (1 + x_2^2)d\mu)$. Then, by the construction of μ, $\mathbb{C}[x]$ would be dense in $L^2(\mathbb{R}, (1 + x^4)d\nu) = L^2(\mathbb{R}, (1 + x^4)(1 + x^2)^{-1}d\nu_0)$. It follows from the inequality $(1 + x^4)(1 + x^2)^{-1} \geq \frac{1}{2}(1 + x^2)$ that $\mathbb{C}[x]$ is dense in $L^2(\mathbb{R}, (1 + x^2)d\nu_0)$. Therefore ν_0 is determinate by Corollary 6.11, which contradicts the choice of ν_0.

Now we show that μ is determinate. Let $\tilde{\mu} \in \mathcal{M}_\mu$. By construction μ is supported on the graph of the parabola $x_2 = x_1^2$. Since $\int (x_2 - x_1^2)^2 d\mu = \int (x_2 - x_1^2)^2 d\tilde{\mu} = 0$, $\tilde{\mu}$ is also supported on the graph of this parabola (by Proposition 1.23). Let $\pi_1(\tilde{\mu})$ denote the projection of $\tilde{\mu}$ onto the x_1-axis. Since $\tilde{\mu} \in \mathcal{M}_\mu$, $\pi_1(\tilde{\mu})$ has the same moments as $\pi_1(\mu) = \nu$. Because ν is determinate, we obtain $\pi_1(\tilde{\mu}) = \nu = \pi_1(\mu)$. This implies that $\tilde{\mu} = \mu$, which proves that μ is determinate.

Since ν_0 is N-extremal, $\mathbb{C}[x]$ is dense in $L^2(\mathbb{R}, \nu_0) = L^2(\mathbb{R}, (1 + x^2)d\nu)$. Therefore, $\mathbb{C}[x_1, x_2]$ is dense in $L^2(\mathbb{R}^2, (1 + x_1^2)d\mu)$ and so in $L^2(\mathbb{R}^2, \mu)$. Thus μ is strictly determinate.

It is instructive to look at this example from the operator-theoretic perspective. Corollary 6.11 implies that X is essentially self-adjoint and that X^2 is not essentially self-adjoint on $L^2(\mathbb{R}, \nu)$. Using the definition of μ we verify that the multiplication operators X_1 and X_2 by the variables x_1 and x_2, respectively, on $L^2(\mathbb{R}^2, \mu)$ are unitarily equivalent to the operators X and X^2 on $L^2(\mathbb{R}, \nu)$, respectively. Hence X_2 is not essentially self-adjoint, so μ is not strongly determinate by Theorem 14.2. ∘

We close this section by proving a sufficient criterion for ultradeterminacy.

Proposition 14.5 *If $\mu \in \mathcal{M}_+(\mathbb{R}^d)$ and $\mathbb{C}_d[\underline{x}]$ is dense in $L^p(\mathbb{R}^d, \mu)$ for some $p > 2$, then μ is ultradeterminate and $\mathbb{C}_d[\underline{x}]$ is dense in $L^2(\mathbb{R}^d, (1 + \|x\|^q)d\mu)$ for any $q \geq 0$.*

Proof Let $q \geq 0$, $\varphi \in C_c(\mathbb{R}^d)$ and $f \in \mathbb{C}_d[\underline{x}]$. Note that $(\frac{p}{p-2})^{-1} + (\frac{p}{2})^{-1} = 1$. We apply the Hölder inequality and obtain

$$\int |\varphi - f|^2 (1 + \|x\|^q) \, d\mu \leq \left(\int |\varphi - f|^p d\mu \right)^{2/p} \left(\int (1 + \|x\|^q)^{\frac{p}{p-2}} d\mu \right)^{\frac{p-2}{p}}.$$

$$(14.1)$$

Since $\mathbb{C}_d[\underline{x}]$ is dense in $L^p(\mathbb{R}^d, \mu)$, there is an $f \in \mathbb{C}_d[\underline{x}]$ such that $\int |\varphi - f|^p d\mu$ is arbitrarily small. The last integral in (14.1) is finite by the assumption $\mu \in \mathcal{M}_+(\mathbb{R}^d)$. Therefore, since $C_c(\mathbb{R}^d)$ is dense in $L^2(\mathbb{R}^d, (1 + \|x\|^q)d\mu)$, so is $\mathbb{C}_d[\underline{x}]$ by (14.1). Setting $q = 2$ the latter means that μ is ultradeterminate by Definition 14.1. □

14.2 Polynomial Approximation

In this section, we develop a number of criteria for determinacy that are based on density conditions of polynomials.

For a Borel mapping $\varphi : \mathbb{R}^d \to \mathbb{R}^m$ and a Borel measure μ on $\mathbb{R}^d$, we denote by $\varphi(\mu)$ the image of μ under the mapping φ, that is, $\varphi(\mu)(M) := \mu(\varphi^{-1}(M))$ for any Borel set M of $\mathbb{R}^m$. Then the transformation formula

$$\int_{\mathbb{R}^m} f(y) \, d\varphi(\mu)(y) = \int_{\mathbb{R}^d} f(\varphi(x)) \, d\mu(x) \qquad (14.2)$$

holds for any function $f \in \mathcal{L}^1(\mathbb{R}^m, \varphi(\mu))$.

Let $\pi_j(x_1, \ldots, x_d) = x_j$ be the j-th coordinate mapping of $\mathbb{R}^d$ into $\mathbb{R}$. Then the measure $\pi_j(\mu)$ on $\mathbb{R}$ is called the j-th *marginal measure* of μ.

The following basic result is *Petersen's theorem*.

Theorem 14.6 *Let $\mu \in \mathcal{M}_+(\mathbb{R}^d)$. If all marginal measures $\pi_1(\mu), \ldots, \pi_d(\mu)$ are determinate, then μ itself is determinate.*

Proof Suppose that $\nu \in \mathcal{M}_\mu$. Let $\chi_1, \ldots, \chi_d$ be characteristic functions of Borel subsets of $\mathbb{R}$ and let $p_1, \ldots, p_d$ be polynomials in one variable. Using the Hölder inequality we derive

$$\|\chi_1(x_1) \ldots \chi_d(x_d) - p_1(x_1) \ldots p_d(x_d)\|_{L^1(\mathbb{R}^d, \nu)}$$

$$\leq \|(\chi_1(x_1) - p_1(x_1))\chi_2(x_2) \ldots \chi_d(x_d)\|_{L^1(\mathbb{R}^d, \nu)}$$

$$+ \|p_1(x_1)(\chi_2(x_2) - p_2(x_2))\chi_3(x_3) \ldots \chi_d(x_d)\|_{L^1(\mathbb{R}^d, \nu)} + \cdots$$

$$+ \|p_1(x_1) \ldots p_{d-1}(x_{d-1})(\chi_d(x_d) - p_d(x_d))\|_{L^1(\mathbb{R}^d, \nu)}$$

$$\leq \|\chi_1(x_1) - p_1(x_1)\|_{L^2(\mathbb{R}^d, \nu)} \|\chi_2(x_2) \ldots \chi_d(x_d)\|_{L^2(\mathbb{R}^d, \nu)}$$

$$+ \|\chi_2(x_2) - p_2(x_2)\|_{L^2(\mathbb{R}^d, \nu)} \|p_1(x_1)\chi_3(x_3) \ldots \chi_d(x_d)\|_{L^2(\mathbb{R}^d, \nu)} + \cdots$$

$$+ \|\chi_d(x_d) - p_d(x_d)\|_{L^2(\mathbb{R}^d, \nu)} \|p_1(x_1) \ldots p_{d-1}(x_{d-1})\|_{L^2(\mathbb{R}^d, \nu)}.$$

Let $j \in \{1, \ldots, d\}$. Clearly, $\nu \cong \mu$ implies that $\pi_j(\nu) \cong \pi_j(\mu)$. Therefore, $\pi_j(\nu) = \pi_j(\mu)$, because $\pi_j(\mu)$ is determinate by assumption. Hence

$$\|\chi_j(x_j) - p_j(x_j)\|_{L^2(\mathbb{R}^d, \nu)} = \|\chi_j(x_j) - p_j(x_j)\|_{L^2(\mathbb{R}, \pi_j(\nu))} \tag{14.3}$$

$$= \|\chi_j(x_j) - p_j(x_j)\|_{L^2(\mathbb{R}, \pi_j(\mu))}. \tag{14.4}$$

Since $\pi_j(\mu)$ is determinate, the polynomials $\mathbb{C}[x_j]$ are dense in $L^2(\mathbb{R}, \pi_j(\mu))$ by Theorem 6.10. Therefore, it follows from (14.3) and (14.4) that we can choose p_1 such that $\|\chi_1(x_1) - p_1(x_1)\|_{L^2(\mathbb{R}^d, \nu)}$ becomes arbitrarily small, then p_2 such that $\|\chi_2(x_2) - p_2(x_2)\|_{L^2(\mathbb{R}^d, \nu)}$ is small and finally p_d such that $\|\chi_d(x_d) - p_d(x_d)\|_{L^2(\mathbb{R}^d, \nu)}$ is small. Then, by the above inequality,

$$\|\chi_1(x_1) \ldots \chi_d(x_d) - p_1(x_1) \ldots p_d(x_d)\|_{L^1(\mathbb{R}^d, \nu)}$$

becomes as small as we want. Since the span of such functions $\chi_1(x_1) \ldots \chi_d(x_d)$ is dense in $L^1(\mathbb{R}^d, \nu)$, this shows that the polynomials are dense in $L^1(\mathbb{R}^d, \nu)$. Hence ν is an extreme point of $\mathcal{M}_\mu$ by Proposition 1.21. Thus each element of $\mathcal{M}_\mu$ is an extreme point. Therefore, since $\nu, \mu \in \mathcal{M}_\mu$, we have $\frac{1}{2}(\nu + \mu) \in \mathcal{M}_\mu$ and hence $\nu = \mu$. This shows that μ is determinate. $\qquad \square$

In Exercise 14.7 it is shown that the converse of Theorem 14.6 is not true, that is, there exists a (strongly) determinate measure with indeterminate marginal measures.

Now we derive some corollaries of Theorem 14.6 which provide determinacy criteria in terms of polynomial approximations.

Let K be a closed subset of $\mathbb{R}^d$. We restate Definition 2.15 in the present case.

Definition 14.7 A measure $\mu \in \mathcal{M}_+(\mathbb{R}^d)$ supported on K is called *K-determinate*, or *determinate on K*, if $\nu \in \mathcal{M}_\mu$ and $\operatorname{supp} \nu \subseteq K$ imply that $\mu = \nu$.

Let $\varphi = (\varphi_1, \dots, \varphi_m) : \mathbb{R}^d \to \mathbb{R}^m$, where $\varphi_1, \dots, \varphi_m \in \mathbb{R}_d[\underline{x}]$, be a polynomial mapping. If $\mu \in \mathcal{M}_+(\mathbb{R}^d)$, then $\varphi(\mu) \in \mathcal{M}_+(\mathbb{R}^m)$ and $\varphi_k(\mu) \in \mathcal{M}_+(\mathbb{R})$ by (14.2).

Corollary 14.8 *Let K be a closed subset of $\mathbb{R}^d$ and let $\mu \in \mathcal{M}_+(\mathbb{R}^d)$ be such that $\operatorname{supp} \mu \subseteq K$. Suppose that the map $\varphi : K \to \mathbb{R}^m$ is injective and $\varphi_k(\mu)$ is determinate for all $k = 1, \dots, m$. Then μ is K-determinate.*

Proof Let $\nu \in \mathcal{M}_+(\mathbb{R}^d)$ be such that $\nu \cong \mu$ and $\operatorname{supp} \nu \subseteq K$. Since φ is a polynomial mapping and $\nu \cong \mu$, it follows from (14.2) that $\varphi(\nu) \cong \varphi(\mu)$. By assumption, all marginal measures $\pi_k(\varphi(\mu)) = \varphi_k(\mu)$ are determinate. Therefore, $\varphi(\mu)$ is determinate by Theorem 14.6, so that $\varphi(\nu) = \varphi(\mu)$.

Let N be a Borel subset of K. Set $\tilde{N} = \varphi(N)$. Since $\varphi : K \to \mathbb{R}^m$ is injective, $\varphi^{-1}(\tilde{N}) = N$ and hence $\mu(N) = \varphi(\mu)(\tilde{N}) = \varphi(\nu)(\tilde{N}) = \nu(N)$. Thus, $\mu = \nu$. $\square$

Corollary 14.9 *Let K be a closed subset of $\mathbb{R}^d$. Assume that there are polynomials $\varphi_1 \dots, \varphi_m \in \mathbb{R}_d[\underline{x}]$ which are bounded on K and separate the points of K (that is, if $x, x' \in K$ and $\varphi_k(x) = \varphi_k(x')$ for all $k = 1, \dots, m$, then $x = x'$.) Then each measure $\mu \in \mathcal{M}_+(\mathbb{R}^d)$ satisfying $\operatorname{supp} \mu \subseteq K$ is K-determinate.*

Proof Define a mapping $\varphi = (\varphi_1, \dots, \varphi_m) : K \to \mathbb{R}^m$. Then φ is injective, because the φ_k separate the points of K. Clearly, $\operatorname{supp} \varphi_k(\mu) \subseteq \overline{\varphi_k(K)}$. Since φ_k is bounded on K, the set $\overline{\varphi_k(K)}$ is compact. Hence $\varphi_k(\mu)$ is determinate by Proposition 12.17 for $k = 1, \dots, d$. Thus Corollary 14.8 applies and gives the assertion. $\square$

For the next corollary we consider polynomial mappings $\varphi^j : \mathbb{R}^d \to \mathbb{R}^{m_j}$ and define $\varphi = (\varphi^1, \dots, \varphi^r) : \mathbb{R}^d \to \mathbb{R}^m$, where $j = 1, \dots, r$, $m_j \in \mathbb{N}$, $m = m_1 + \dots + m_r$.

Corollary 14.10 *Let μ and K be as in Corollary 14.8. If $\varphi : K \to \mathbb{R}^m$ is injective and all measures $\varphi^1(\mu), \dots, \varphi^r(\mu)$ are strictly determinate, then μ is K-determinate.*

Proof We replace in the proof of Corollary 14.8 the map π_j on a single coordinate by a map π^j on a finite set of coordinates. To be more precise, we write $y \in \mathbb{R}^m$ as $y = (y_{11}, \dots, y_{1m_1}, y_{21}, \dots, y_{rm_r})$ and define $\pi^j : \mathbb{R}^d \to \mathbb{R}^{m_j}$ by $\pi^j(y) = (y_{j1}, \dots, y_{jm_j})$. Then we have the following result: *If $\nu \in \mathcal{M}_+(\mathbb{R}^m)$ and all measures $\pi^1(\nu), \dots, \pi^r(\nu)$ are strictly determinate, then ν is determinate.*

The proof of this statement follows a similar pattern as the proof of Theorem 14.6. The polynomials $\mathbb{C}[y_{j1}, \dots, y_{jm_j}]$ are dense in $L^2(\mathbb{R}^{m_j}, \pi^j(\nu))$, because $\pi^j(\nu)$ is strictly determinate. We do not carry out the details. Using this generalization instead of Theorem 14.6 we can argue as in the proof of Corollary 14.8. $\square$

The presence of "sufficiently" many bounded polynomials on the set $\operatorname{supp} \mu$ can be used to prove stronger results than the plain determinacy. As a sample we consider subsets of cylinders with compact base sets.

Proposition 14.11 *Let K be a closed subset of $\mathbb{R}^d$, $d \geq 2$, such that K is a subset of $C \times \mathbb{R}$, where C is a compact set of $\mathbb{R}^{d-1}$. Suppose that $\mu \in \mathcal{M}_+(\mathbb{R}^d)$ is supported on K. If the marginal measure $\pi_d(\mu)$ is determinate, then μ is ultradeterminate.*

Proof We shall write $x \in \mathbb{R}^d$ as $x = (y, x_d)$ with $y \in \mathbb{R}^{d-1}$ and $x_d \in \mathbb{R}$. Suppose that $f(y)$ and $g(x_d)$ are continuous functions with compact supports. Let us abbreviate

$$M_1 = \sup\{|g(x_d)| : x_d \in \mathbb{R}\}, \ M_2 = \sup\{|p(y)| : y \in C\}, \ M_3 = \sup\{1 + \|y\| : y \in C\}.$$

We denote by $\|\cdot\|_1$ the norm of $L^2(\mathbb{R}^d, (1 + \|x\|^2)d\mu)$, by $\|\cdot\|_2$ the norm of $L^2(\mathbb{R}^d, \mu)$ and by $\|\cdot\|_3$ the norm of $L^2(\mathbb{R}, (1 + x_d^2)d\pi_d(\mu))$. Let $p \in \mathbb{C}_{d-1}[\underline{y}]$ and $q \in \mathbb{C}[x_d]$. Using the assumption $\operatorname{supp}\mu \subseteq K \subseteq C \times \mathbb{R}$ and formula (14.2) we obtain

$$\|f(y)g(x_d) - p(y)q(x_d)\|_1 \leq \|(f(y) - p(y))g(x_d)\|_1 + \|p(y)(g(x_d) - q(x_d))\|_1$$

$$\leq M_1\|f(y) - p(y)\|_1 + M_2 M_3\|(g(x_d) - q(x_d))(1 + x_d^2)\|_2$$

$$\leq M_1\mu(\mathbb{R}^d) \sup\{|f(y) - p(y)| : y \in C\} + M_2 M_3\|g(x_d) - q(x_d)\|_3.$$

By Weierstrass' approximation theorem, $p \in \mathbb{C}_{d-1}[\underline{y}]$ can be chosen such that the supremum of $|f(y) - p(y)|$ over the compact set C is arbitrary small. Since the marginal measure $\pi_d(\mu)$ is determinate, $\mathbb{C}[x_d]$ is dense in $L^2(\mathbb{R}, (1 + x_d^2)d\pi_d(\mu))$ by Theorem 6.10. Hence we find $q \in \mathbb{C}[x_d]$ such that $\|g - q\|_3$ is sufficiently small. By the preceding inequality we have shown that the function $f(y)g(x_d)$ is in the closure of $\mathbb{C}_d[\underline{x}]$ in $L^2(\mathbb{R}^d, (1 + \|x\|^2)d\mu)$. Since the span of such functions is obviously dense, $\mathbb{C}_d[\underline{x}]$ is dense in $L^2(\mathbb{R}^d, (1 + \|x\|^2)d\mu)$. This means that μ is ultradeterminate. $\square$

14.3 Partial Determinacy and Moment Functionals

The results of this section show how partial determinacy can be used to conclude that positive functionals are moment functionals. First we treat the case $d = 2$.

Theorem 14.12 *Let $Q \subseteq \mathbb{C}[x_1, x_2]$ be a set of polynomials such that $\mathbb{C}[x_1, x_2]$ is the linear span of polynomials $p(x_1)q(x_1, x_2)$, where $p \in \mathbb{C}[x_1]$ and $q \in Q$. Let L be a positive linear functional on $\mathbb{C}[x_1, x_2]$. Suppose that for each $q \in Q$ the positive linear functional $\mathcal{L}_q$ on $\mathbb{C}[x_1]$ defined by $\mathcal{L}_q(p) := L(p(x_1)(x_2^2 + 1)q\overline{q})$ is determinate. Then L is a moment functional on $\mathbb{C}[x_1, x_2]$.*

We state the important special case $Q = \{x_2^k : k \in \mathbb{N}_0\}$ of Theorem 14.12 separately as the first assertion of the following corollary.

Corollary 14.13 *Suppose that* $s = (s_{(n,k)})_{k,n\in\mathbb{N}_0}$ *is a positive semidefinite 2-sequence such that for each* $k \in \mathbb{N}_0$ *the 1-sequence*

$$(s_{(n,2k+2)}) + s_{(n,2k)})_{n\in\mathbb{N}_0} \tag{14.5}$$

is determinate. Then s is a moment sequence.

If in addition the sequence $(s_{(0,n)})_{n\in\mathbb{N}_0}$ *is determinate, then s is determinate.*

Proof Put $Q = \{x_2^k : k \in \mathbb{N}_0\}$. Since $s_{(n,m)} = L_s(x_1^n x_2^m)$, the functional $\mathcal{L}_{x_2^k}$ has the moment sequence (14.5). Therefore, by Theorem 14.12, L_s is a moment functional, so s is a moment sequence.

The determinacy of the sequence $(s_{(n,2)} + s_{(n,0)})$ implies the determinacy of $(s_{(n,0)})$ (see Exercise 6.3). Hence, if $(s_{(0,n)})$ is also determinate, both marginal sequences of s are determinate, so s is determinate by Theorem 14.6. $\qquad\square$

The proof of Theorem 14.12 is based on two operator-theoretic lemmas. In the proof of Lemma 14.14 we use the self-adjoint extension theory of symmetric operators (see [Sm9, Section 13.2] or [RS2, Section X.1]). If A is a self-adjoint operator, the unitary operator $U_A := (A - iI)(A + iI)^{-1}$ is called the *Cayley transform* of A.

Lemma 14.14 *Suppose that* $A_j, j \in J$, *is a family of pairwise strongly commuting self-adjoint operators and B is a densely defined symmetric operator on a Hilbert space* $\mathcal{H}$ *such that* $U_{A_j}B(U_{A_j})^* = B$ *for* $j \in J$. *Then there exists a family* $\tilde{A}_j, j \in J$, *of strongly commuting self-adjoint operators and a self-adjoint operator* $\tilde{B}$ *on a Hilbert space* $\mathcal{K}$ *such that* $\mathcal{H}$ *is a subspace of* $\mathcal{K}$, $B \subseteq \tilde{B}$, $A_j \subseteq \tilde{A}_j$, *and* $\tilde{A}_j$ *and* $\tilde{B}$ *strongly commute for all* $j \in J$.

Proof The proof uses the "doubling trick" from operator theory (see e.g. [Sm9, Example 13.5]). Define $\tilde{A}_j = A_j \oplus A_j$ for $j \in J$ and $B_0 = B \oplus (-B)$ on the Hilbert space $\mathcal{K} = \mathcal{H} \oplus \mathcal{H}$. Obviously, the self-adjoint operators $\tilde{A}_j, j \in J$, strongly commute as the operators $A_j, j \in J$, do by assumption. Without loss of generality we assume that the operator B is closed. Then B_0 is also closed. Let $\mathcal{N}_\pm(T) = \ker(T^* \pm iI)$ denote the deficiency spaces of a symmetric operator T. From the definition of B_0 we obtain $\mathcal{N}_\pm(B_0) = \mathcal{N}_\pm(B) \oplus \mathcal{N}_\mp(B)$. Hence $V(\varphi, \psi) := (\psi, \varphi)$, where $\varphi \in \mathcal{N}_+(B)$, $\psi \in \mathcal{N}_-(B)$, defines an isometric linear map V of $\mathcal{N}_+(B_0)$ onto $\mathcal{N}_-(B_0)$. Then, by [Sm9, Theorem 13.10]), the restriction $\tilde{B}$ of the adjoint operator $(B_0)^*$ to the domain $\mathcal{D}(\tilde{B}) := \mathcal{D}(B_0) + (I - V)\mathcal{N}_+(B_0)$ is a self-adjoint extension of B_0.

From the assumption $U_{A_j}B(U_{A_j})^* = B$ it follows that $U_{A_j}B^*(U_{A_j})^* = B^*$. This in turn implies that $U_{A_j}\mathcal{N}_\pm(B) = \mathcal{N}_\pm(B)$. From $\mathcal{N}_\pm(B_0) = \mathcal{N}_\pm(B) \oplus \mathcal{N}_\mp(B)$ and the definition of V we conclude that the Cayley transform $U_{\tilde{A}_j} = U_{A_j} \oplus U_{A_j}$ of $\tilde{A}_j$ maps $(I - V)\mathcal{N}_+(B_0)$ onto itself. Since $(U_{A_j})^*\mathcal{D}(B) = \mathcal{D}(B)$ by assumption and hence $U_{A_j}\mathcal{D}(B) = \mathcal{D}(B)$, we get $U_{\tilde{A}_j}\mathcal{D}(\tilde{B}) = \mathcal{D}(\tilde{B})$. Combined with the relation

$$U_{\tilde{A}_j}(B_0)^*(U_{\tilde{A}_j})^* = U_{A_j}B^*(U_{A_j})^* \oplus (-U_{A_j}B^*(U_{A_j})^*) = B^* \oplus (-B^*) = (B_0)^*$$

the latter implies that $U_{\tilde{A}_j}\tilde{B}(U_{\tilde{A}_j})^* = \tilde{B}$. This yields $U_{\tilde{A}_j}U_{\tilde{B}}(U_{\tilde{A}_j})^* = U_{\tilde{B}}$ and hence $U_{\tilde{A}_j}U_{\tilde{B}} = U_{\tilde{B}}U_{\tilde{A}_j}$. Since the Cayley transforms of $\tilde{A}_j$ and $\tilde{B}$ commute, so do their resolvents. Hence the self-adjoint operators $\tilde{A}_j$ and $\tilde{B}$ strongly commute. $\square$

Lemma 14.15 *Let A and B be closed symmetric linear operators on a Hilbert space $\mathcal{H}$ and let $\mathcal{D} \subseteq \mathcal{D}(A) \cap \mathcal{D}(B)$ be a linear subspace such that $A\mathcal{D} \subseteq \mathcal{D}$, $B\mathcal{D} \subseteq \mathcal{D}$, and $AB\varphi = BA\varphi$ for $\varphi \in \mathcal{D}$. Further, let Q be a subset of $\mathcal{D}$ such that the span of vectors $p(A)q$, where $p \in \mathbb{C}[x]$ and $q \in Q$, is dense in $\mathcal{H}$ and a core for B. For $q \in Q$ we denote by A_q the restriction of A to the invariant linear subspace $\mathcal{D}_q := \{p(A)(B + iI)q : p \in \mathbb{C}[x]\}$ and by $\mathcal{H}_q$ the closure of $\mathcal{D}_q$ in $\mathcal{H}$. Suppose that for each $q \in Q$ the symmetric operator A_q is essentially self-adjoint on the Hilbert space $\mathcal{H}_q$. Then the operator A on $\mathcal{H}$ is self-adjoint and we have $U_A B(U_A)^* = B$.*

Proof Let $q \in Q$ and $z \in \mathbb{C}\backslash\mathbb{R}$. Since A_q is essentially self-adjoint on $\mathcal{H}_q$, the range of $A_q + zI$ is dense in $\mathcal{H}_q$ by Proposition A.42. Hence for any $p \in \mathbb{C}[x]$ there exists a sequence $(p_n)_{n\in\mathbb{N}}$ of polynomials $p_n \in \mathbb{C}[x]$ such that

$$(A_q + zI)p_n(A)(B + iI)q = (A + zI)p_n(A)(B + iI)q \to p(A)(B + iI)q \quad \text{in} \quad \mathcal{H}_q.$$

Since B is a symmetric operator, $\|(B + iI)\varphi\|^2 = \|B\varphi\|^2 + \|\varphi\|^2$ for any $\varphi \in \mathcal{D}(B)$. We use this fact for the second equality and derive

$$\|(A + zI)Bp_n(A)q - Bp(A)q\|^2 + \|(A + zI)p_n(A)q - p(A)q\|^2$$
$$= \|B((A + zI)p_n(A)q - p(A)q)\|^2 + \|(A + zI)p_n(A)q - p(A)q\|^2$$
$$= \|(B + iI)((A + zI)p_n(A)q - p(A)q)\|^2$$
$$= \|(A + zI)p_n(A)(B + iI)q - p(A)(B + iI)q\|^2 \to 0.$$

Hence $(A + zI)Bp_n(A)q \to Bp(A)q$ and $(A + zI)p_n(A)q \to p(A)q$ as $n \to \infty$. Since the span of vectors $p(A)q$ is dense in $\mathcal{H}$, the preceding shows that the range of $A + zI$ is dense in $\mathcal{H}$, so the operator A on $\mathcal{H}$ is self-adjoint by Proposition A.42.

Because $(A + zI)^{-1}$ is bounded, it follows that $p_n(A)q \to (A + zI)^{-1}p(A)q$ and $Bp_n(A)q \to (A + zI)^{-1}Bp(A)q$. Since the operator B is closed, the latter implies that $B(A + zI)^{-1}p(A)q = (A + zI)^{-1}Bp(A)q$. By assumption, the span of vectors $p(A)q$ is a core for B. Therefore, we obtain $B(A + zI)^{-1}\psi = (A + zI)^{-1}B\psi$ for all vectors ψ of the domain $\mathcal{D}(B)$. Setting $z = i$ and $z = -i$, it follows that both operators $U_A = I - 2i(A + iI)^{-1}$ and $U_A^{-1} = I + 2i(A - iI)^{-1}$ map $\mathcal{D}(B)$ into itself. Therefore, $U_A\mathcal{D}(B) = \mathcal{D}(B)$. Let $\varphi \in \mathcal{D}(B)$. Then $\psi := U_A^{-1}\varphi \in \mathcal{D}(B)$ and

$$U_A B(U_A)^*\varphi = U_A B\psi = (I - 2i(A + iI)^{-1})B\psi = B(I - 2i(A - iI)^{-1})\psi = BU_A\psi = B\varphi.$$

This proves that $B \subseteq U_A B(U_A)^*$. If $\varphi \in \mathcal{D}(U_A B(U_A)^*)$, then $\psi := (U_A)^*\varphi \in \mathcal{D}(B)$ and therefore $\varphi = U_A\psi \in \mathcal{D}(B)$. This shows that $\mathcal{D}(U_A B(U_A)^*) \subseteq \mathcal{D}(B)$. Combined with the relation $B \subseteq U_A B(U_A)^*$ the latter yields $B = U_A B(U_A)^*$. $\square$

Proof of Theorem 14.12 Our aim is to apply Lemma 14.15. Let A and B be the closures of the operators X_1 and X_2, respectively, on $(\mathcal{H}_L, \langle \cdot, \cdot \rangle_L)$. Let $q \in Q$. Recall that $(\mathcal{H}_{\mathcal{L}_q}, \langle \cdot, \cdot \rangle_{\mathcal{L}_q})$ denotes the Hilbert space of the moment functional $\mathcal{L}_q$ on $\mathbb{C}[x_1]$. From the definitions of the functional $\mathcal{L}_q$ and the corresponding scalar products we obtain

$$\langle p_1(x_1), p_2(x_1) \rangle_{\mathcal{L}_q} = L(p_1(x_1)\bar{p}_2(x_1)(x_2^2 + 1)q\bar{q}) = \langle p_1(A)(B + i)q, p_2(A)(B + i)q \rangle_L$$

for $p_1, p_2 \in \mathbb{C}[x_1]$. From this equality it follows that the operator X_1 with domain $\mathbb{C}[x_1]$ on the Hilbert space $\mathcal{H}_{\mathcal{L}_q}$ and the operator A_q with domain $\mathcal{D}_q$ on the Hilbert space $\mathcal{H}_q$ (in the notation of Lemma 14.15) are unitarily equivalent. Since $\mathcal{L}_q$ is determinate by assumption, X_1 is essentially self-adjoint by Theorem 6.10 and so is A_q for each $q \in Q$. Therefore, by Lemma 14.15, A is self-adjoint and $U_A B(U_A)^* = B$.

From Lemma 14.14, applied in the case of a single operator A, it follows that A and B have strongly commuting self-adjoint extensions on a larger Hilbert space. Hence, by Theorem 12.40, L is a moment functional. □

Now we turn to the case when the dimension d is larger than 2. The following theorem is the counterpart of Corollary 14.13.

Theorem 14.16 *Let* $s = (s_n)_{n \in \mathbb{N}_0^d}$ *be a positive semidefinite d-sequence, where* $d \geq 3$. *Suppose that for* $j = 1, \ldots, d-1$, $k_1, \ldots, k_{j-1}, k_{j+1}, \ldots, k_d \in \mathbb{N}_0$ *the 1-sequence*

$$(s_{(k_1, \ldots, k_{j-1}, n, k_{j+1}, \ldots, k_{d-1}, 2k_d + 2)} + s_{(k_1, \ldots, k_{j-1}, n, k_{j+1}, \ldots, k_{d-1}, 2k_d)})_{n \in \mathbb{N}_0} \tag{14.6}$$

is determinate. Further, suppose that for all numbers $j, l \in \{1, \ldots, d-1\}$, $j < l$, *all sequences of one of the following two sets of 1-sequences*

$$(s_{(\ldots, k_{j-1}, n, k_{j+1}, \ldots, k_{l-1}, 2k_l + 2, k_{l+1}, \ldots)} + s_{(\ldots, k_{j-1}, n, k_{j+1}, \ldots, k_{l-1}, 2k_l, k_{l+1}, \ldots)})_{n \in \mathbb{N}_0}, \tag{14.7}$$

$$(s_{(\ldots, k_{j-1}, 2k_j + 2, k_{j+1}, \ldots, k_{l-1}, n, k_{l+1}, \ldots)} + s_{(\ldots, k_{j-1}, 2k_j, k_{j+1}, \ldots, k_{l-1}, n, k_{l+1}, \ldots)})_{n \in \mathbb{N}_0}, \tag{14.8}$$

where $k_1, \ldots, k_d \in \mathbb{N}_0$, *are determinate. Then s is a moment sequence.*

If in addition the sequence $(s_{(0, \ldots, 0, n)})_{n \in \mathbb{N}_0}$ *is determinate, then s is determinate.*

Proof The proof is given by some modifications in the proof of Theorem 14.12. Let A_j, $j = 1, \ldots, d$, denote the closure of the operator X_j on $\mathcal{H}_{L_s}$. First we fix $j \in \{1, \ldots, d-1\}$ and apply Lemma 14.15 in the case when $A = A_j$, $B = A_d$ and

$$Q = \{x_1^{k_1} \cdots x_{j-1}^{k_{j-1}} x_{j+1}^{k_{j+1}} \cdots x_d^{k_d} : k_1, \ldots, k_{j-1} k_{j+1}, \ldots, k_d \in \mathbb{N}_0\}. \tag{14.9}$$

Since all sequences (14.6) are determinate, the assumptions of Lemma 14.15 are fulfilled. Hence the operator A_j is self-adjoint and we have $U_{A_j} A_d (U_{A_j})^* = A_d$.

In order to show that the self-adjoint operators $A_1, \dots, A_{d-1}$ strongly commute we apply Lemma 14.15 once more. Let $j, l \in \{1, \dots, d-1\}, j < l$. First we assume that the sequences (14.7) are determinate. We set $A = A_j$, $B = A_l$, and define Q again by (14.9). Since all sequences (14.7) are determinate, it follows from Theorem 6.10 that the corresponding operators A_q in Lemma 14.15 are self-adjoint, so Lemma 14.15 applies and yields $U_{A_j} A_l (U_{A_j})^* = A_l$. Since A_l is self-adjoint, we get $U_{A_j} U_{A_l} (U_{A_j})^* = U_{A_l}$ and so $U_{A_j} U_{A_l} = U_{A_l} U_{A_j}$. Hence the resolvents of A_j and A_l commute, so that A_j and A_l strongly commute. In the case when the sequences (14.8) are determinate we interchange the role of j and l and proceed in a similar manner.

Thus, $A_1, \dots, A_{d-1}$ is a family of strongly commuting self-adjoint operators such that $U_{A_j} A_d (U_{A_j})^* = A_d$ for $j = 1, \dots, d-1$. That is, the assumptions of Lemma 14.14 are satisfied with $B = A_d$. From Lemma 14.14 and Theorem 12.40 we conclude that s is a moment sequence. As in the proof of Corollary 14.5, the determinacy assertion follows from Theorem 14.6. $\qquad\square$

Remark 14.17 In the preceding proof of Theorem 14.16 we have shown the following fact, which will be used in the proof of Theorem 14.20 below: If for some $j \in \{1, \dots, d\}$ all 1-sequences (14.6) are determinate, it follows from Lemma 14.15 that the operator $A_j = \overline{X}_j$ is self-adjoint on the Hilbert space $\mathcal{H}_L$. Likewise, it was shown in the proof of Theorem 14.12 that the operator $A = \overline{X}_1$ is self-adjoint on $\mathcal{H}_L$. $\qquad\circ$

14.4 The Multivariate Carleman Condition

Recall from Sect. 4.2 that Carleman's condition (4.2) for a positive semidefinite 1-sequence $(t_n)_{n \in \mathbb{N}_0}$ is that

$$\sum_{n=1}^{\infty} t_{2n}^{-\frac{1}{2n}} = +\infty. \tag{14.10}$$

For a d-sequence $s = (s_n)_{n \in \mathbb{N}_0^d}$ the 1-sequences

$$s^{[1]} := (s_{(n,0,\dots,0)})_{n \in \mathbb{N}_0}, \; s^{[2]} := (s_{(0,n,\dots,0)})_{n \in \mathbb{N}_0}, \dots, s^{[d]} := (s_{(0,\dots,0,n)})_{n \in \mathbb{N}_0}$$

are called *marginal sequences* of s. If s is the moment sequence of $\mu \in \mathcal{M}_+(\mathbb{R}^d)$, then

$$s_n^{[j]} = \int_{\mathbb{R}^d} x_j^n \, d\mu(x_1, \dots, x_d) = \int_{\mathbb{R}} y^n \, d\pi_j(\mu)(y), \quad n \in \mathbb{N}_0, j = 1, \dots, d,$$

that is, $s^{[j]}$ is the moment sequence of the j-th marginal measure $\pi_j(\mu)$ of μ. Hence Petersen's Theorem 14.6 can be rephrased by saying that if all marginal sequences $s^{[1]}, \dots, s^{[d]}$ of a d-moment sequence s are determinate, then is s itself.

Definition 14.18 Let s be a positive semidefinite d-sequence. We shall say that s, and equivalently the functional L_s, satisfy the *multivariate Carleman condition* if all marginal sequences $s^{[1]}, \ldots, s^{[d]}$ satisfy Carleman's condition (14.10), that is, if

$$\sum_{n=1}^{\infty} (s_{2n}^{[j]})^{-\frac{1}{2n}} \equiv \sum_{n=1}^{\infty} L_s(x_j^{2n})^{-\frac{1}{2n}} = \infty \quad \text{for} \quad j = 1, \ldots, d. \tag{14.11}$$

The following *theorem of Nussbaum* is the main result in this chapter. It shows the exceptional usefulness of the multivariate Carleman condition, which implies both the existence *and* uniqueness of a solution of the moment problem on $\mathbb{R}^d$!

Theorem 14.19 *Each positive semidefinite d-sequence $s = (s_n)_{n \in \mathbb{N}_0^d}$ satisfying the multivariate Carleman condition is a strongly determinate moment sequence.*

Theorem 14.19 follows at once from the following more general result.

Theorem 14.20 *Suppose that $s = (s_n)_{n \in \mathbb{N}_0^d}$ is a positive semidefinite d-sequence such that the first $d-1$ marginal sequences $s^{[1]}, \ldots, s^{[d-1]}$ fulfill Carleman's condition (14.10). Then s is a moment sequence.*

If in addition the last marginal sequence $s^{[d]}$ satisfies Carleman's condition (14.10) as well, then the moment sequence s is strongly determinate.

A crucial technical step for the proofs of Theorems 14.20 and 14.25 is the next lemma.

Lemma 14.21 *Let L be a positive linear functional on $\mathbb{C}_d[\underline{x}]$ and let $q, f \in \mathbb{R}_d[\underline{x}]$. Suppose that $\tilde{L}(\cdot) := L(f \cdot)$ is also a positive functional on $\mathbb{C}_d[\underline{x}]$. Set $t_n := L(q^n)$ and $r_n := \tilde{L}(q^n) = L(f q^n)$ for $n \in \mathbb{N}_0$. Then, if the sequence $t = (t_n)_{n \in \mathbb{N}_0}$ satisfies Carleman's condition (14.10), so does the sequence $r = (r_n)_{n \in \mathbb{N}_0}$.*

Proof Since L and $\tilde{L}$ are positive functionals, t and r are positive semidefinite sequences and the Cauchy–Schwarz inequality (2.7) holds for L. Hence, for $n \in \mathbb{N}_0$,

$$t_{2n+2}^2 = L(q^n q^{n+2})^2 \leq L(q^{2n}) L(q^{2n+4}) = t_{2n} t_{2n+4}, \tag{14.12}$$

$$r_{2n}^2 = L(q^{2n} f)^2 = L(q^{4n}) L(f^2) = t_{4n} L(f^2). \tag{14.13}$$

If $t_{2k} = 0$ for some $k \in \mathbb{N}_0$, then (14.12) implies that $t_{2n} = 0$ for $n \geq k$, so $r_{2n} = 0$ for $n \geq k$ by (14.13) and the assertion holds trivially. The assertion is also trivial if $L(f^2) = 0$. Thus we can assume that $t_{2n} > 0$ for all $n \in \mathbb{N}_0$ and $L(f^2) > 0$.

From (14.12) it follows that $\frac{t_{2j+2}}{t_{2j}} \leq \frac{t_{2j+4}}{t_{2j+2}}$ for $j \in \mathbb{N}_0$. Thus, for $k \in \mathbb{N}$, we get

$$\frac{t_{2k+2}}{t_0} = \prod_{j=1}^{k+1} \frac{t_{2j}}{t_{2j-2}} \leq \left(\frac{t_{2k+2}}{t_{2k}} \right)^{k+1},$$

that is, $t_{2k}^{k+1} \leq t_{2k+2}^{k} t_0$, so $t_{2k}^{2k+2} \leq t_{2k+2}^{2k} t_0^2$ and hence

$$\sqrt[2k]{t_{2k}} \leq \sqrt[2k+2]{t_{2k+2}} \; \sqrt[k(2k+2)]{t_0}.$$

Setting $k = 2n$, we obtain

$$t_{4n}^{-1/(4n)} \geq t_{4n+2}^{-1/(4n+2)} t_0^{-1/(2n(4n+2))} \geq t_{4n+2}^{-1/(4n+2)} (1 + t_0)^{-1}. \tag{14.14}$$

Using (14.13) and (14.14) we derive

$$r_{2n}^{-1/(2n)} \geq t_{4n}^{-1/(4n)} L(f^2)^{-1/(4n)} \geq t_{4n}^{-1/(4n)} (1 + L(f^2))^{-1}$$

$$\geq \frac{1}{2} (t_{4n}^{-1/(4n)} + t_{4n+2}^{-1/(4n+2)})(1 + t_0)^{-1} (1 + L(f^2))^{-1}.$$

Summing over $n \in \mathbb{N}$, by the Carleman condition for the sequence (t_n), the right-hand side yields $+\infty$ and so does the left hand side.   □

Applying Lemma 14.21 with $q = x_j, j = 1, \ldots, d$, yields the following

Corollary 14.22 *Let L be a positive linear functional on $\mathbb{C}_d[\underline{x}]$ and let $f \in \mathbb{R}_d[\underline{x}]$. Suppose that $\tilde{L}(\cdot) := L(f \cdot)$ is also a positive linear functional on $\mathbb{C}_d[\underline{x}]$. If L satisfies the multivariate Carleman condition, so does $\tilde{L}$.*

Proof of Theorem 14.20 Assume first that $d = 2$. Since the $(s_{(n,0)} = L(x_1^n))_{n \in \mathbb{N}_0}$ fulfills Carleman's condition (14.10) by assumption, so does the sequence

$$(s_{n,2(k+1)} + s_{(n,2k)})_{n \in \mathbb{N}_0} = (L((x_2^{k+1} + x_2^{2k})x_1^n))_{n \in \mathbb{N}_0}$$

by Lemma 14.21. Hence this 1-sequence is determinate by Carleman's Theorem 4.3. Therefore, Corollary 14.13 applies and implies that s is a moment sequence.

In the case $d \geq 3$ the proof is similar. All sequences in (14.6), (14.7), and (14.8) are of the form $(L(fx_j^n))_{n \in \mathbb{N}_0}$ or $(L(fx_l^n))_{n \in \mathbb{N}_0}$ for some polynomial $f \in \sum \mathbb{R}_d[\underline{x}]^2$. Since $s^{[1]}, \ldots, s^{[d-1]}$ satisfy Carleman's condition by assumption, these sequences do so for $j, l = 1, \ldots, d - 1$ by Lemma 14.21. Hence they are determinate by Theorem 4.3. Thus, the assumptions of Theorem 14.16 are fulfilled; hence s is a moment sequence.

Now suppose that Carleman's condition (14.10) holds for $s^{[1]}, \ldots, s^{[d]}$. Then, arguing as in the preceding paragraph, it follows that for $j = 1, \ldots, d$ all sequences (14.6) are determinate. Hence $A_j = \overline{X}_j$ is self-adjoint on $\mathcal{H}_L$ for $j =, \ldots, d$ by Remark 14.17. Therefore, by Theorem 14.2, s is strongly determinate.   □

Remark 14.23 The preceding proof of Theorems 14.20 and 14.19 was quite involved. We used Theorem 14.16 to prove s is a moment sequence. However, if we know already that s is a moment sequence, the determinacy assertion of Theorem 14.19 follows easily: Since the multivariate Carleman condition implies

Carleman's condition for the marginal sequences, the latter are determinate by Carleman's Theorem 4.3 and hence s is determinate by Petersen's Theorem 14.6. ∘

The following corollary is the multidimensional counterpart of Corollary 4.11.

Corollary 14.24 *Let $\mu \in M_+(\mathbb{R}^d)$. Suppose that there exists an $\varepsilon > 0$ such that*

$$\int_{\mathbb{R}^d} e^{\varepsilon \|x\|} \, d\mu(x) < +\infty. \tag{14.15}$$

Then $\mu \in \mathcal{M}_+(\mathbb{R}^d)$ and μ is strongly determinate.

Proof The proof is similar to the proof of Corollary 4.11. Let $j \in \{1, \ldots, d\}$ and $n \in \mathbb{N}_0$. Then $x_j^{2n} e^{-\varepsilon |x_j|} \leq \varepsilon^{-2n}(2n)!$ for $x_j \in \mathbb{R}$ (see (4.16)) and hence

$$\int_{\mathbb{R}^d} x_j^{2n} \, d\mu = \int_{\mathbb{R}^d} x_j^{2n} e^{-\varepsilon |x_j|} e^{\varepsilon |x_j|} \, d\mu \leq \varepsilon^{-2n}(2n)! \int_{\mathbb{R}^d} e^{\varepsilon \|x\|} \, d\mu < +\infty. \tag{14.16}$$

Let $p \in \mathbb{R}_d[\underline{x}]$. Then $p(x) \leq c(1 + x_1^{2n} + \cdots + x_d^{2n})$ on $\mathbb{R}^d$ for some $c > 0$ and $n \in \mathbb{N}$, so (14.16) implies that p is μ-integrable. Thus $\mu \in \mathcal{M}_+(\mathbb{R}^d)$.

Let s be the moment sequence of μ. By (14.16) there is a constant $M > 0$ such that

$$s_{2n}^{[j]} = L_s(x_j^{2n}) = \int_{\mathbb{R}^d} x_j^{2n} \, d\mu \leq M^{2n}(2n)! \quad \text{for} \quad n \in \mathbb{N}_0.$$

By Corollary 4.10, this inequality implies that $s^{[j]}$ satisfies Carleman's condition (14.10). Therefore, s and μ are strongly determinate by Theorem 14.19. □

The next theorem shows that Carleman's condition can be used to localize the support of representing measures.

Theorem 14.25 *Let s be a real d-sequence and $\mathfrak{f} = \{f_1, \ldots, f_k\}$ a finite subset of $\mathbb{R}_d[\underline{x}]$. Suppose that the corresponding Riesz functional L_s is $Q(\mathfrak{f})$-positive (that is, $L_s(p^2) \geq 0$ and $L_s(f_j p^2) \geq 0$ for all $j = 1, \ldots, k$ and $p \in \mathbb{R}_d[\underline{x}]$) and satisfies the multivariate Carleman condtion. Then s is a determinate moment sequence and its representing measure is supported on the semi-algebraic set $\mathcal{K}(\mathfrak{f})$.*

Proof By Theorem 14.20, s is a determinate moment sequence. Let μ denote its unique representing measure. It suffices to show that $L_s(f_j p^2) \geq 0$ for all $p \in \mathbb{R}_d[\underline{x}]$ implies that $f_j(x) \geq 0$ on supp μ. For simplicity we suppress the index j.

Define a linear functional $\tilde{L}$ on $\mathbb{R}_d[\underline{x}]$ by $\tilde{L}(\cdot) = L_s(f \cdot)$. Since $L_s(fp^2) \geq 0$ for $p \in \mathbb{R}_d[\underline{x}]$ by assumption, $\tilde{L}$ is a positive functional on $\mathbb{R}_d[\underline{x}]$. Therefore, since L satisfies the multivariate Carleman condition, so does $\tilde{L}$ by Corollary 14.22. Thus, by Theorem 14.20, $\tilde{L}$ is also a determinate moment functional; let τ be its representing measure. Then

$$\int_{\mathbb{R}^d} x^\alpha f(x) d\mu(x) = L_s(fx^\alpha) = \tilde{L}(x^\alpha) = \int_{\mathbb{R}^d} x^\alpha \, d\tau(x), \quad \alpha \in \mathbb{N}_0^d. \tag{14.17}$$

Put $M_+ := \{x \in \mathbb{R}^d : f(x) \geq 0\}$ and $M_- := \{x \in \mathbb{R}^d : f(x) < 0\}$. We denote by $\chi_\pm$ the characteristic function of $M_\pm$ and define measures $\mu_\pm, \nu_\pm \in M_+(\mathbb{R})$ by $d\mu_\pm = \chi_\pm d\mu$ and $d\nu_\pm = \pm f d\mu_\pm$. Then $\mu = \mu_+ + \mu_-$ and hence

$$\int x_l^{2k} \, d\mu_+(x) \leq \int x_l^{2k} \, d\mu(x), \quad k \in \mathbb{N}_0, \ l = 1, \ldots, d.$$

Therefore, since the moment sequence s of μ satisfies the multivariate Carleman condition, so does the moment sequence of μ_+ and hence the moment sequence of ν_+ by Corollary 14.22. Consequently, ν_+ is determinate by Theorem 14.20.

Since $d\nu_+ - d\nu_- = f d\mu_+ + f d\mu_- = f d\mu$, it follows from (14.17) that

$$\int x^\alpha \, d\nu_+(x) = \int x^\alpha \, d\nu_-(x) + \int x^\alpha \, d\tau(x) = \int x^\alpha d(\nu_- + \tau)(x), \quad \alpha \in \mathbb{N}_0^d.$$

Hence, since ν_+ is determinate, $\nu_+ = \nu_- + \tau$. From $0 = \nu_+(M_-) \geq \nu_-(M_-) \geq 0$ we obtain $\nu_-(M_-) = 0$ and so $\nu_- = 0$.

The latter implies that $f(x) \geq 0$ on $\text{supp } \mu$. Indeed, if we had $f(x_0) < 0$ for some $x_0 \in \text{supp } \mu$, then it would follow that $-f(x) \geq \varepsilon > 0$ on a ball B around x_0 and

$$0 = \nu_-(B) = \int_B (-f(x)) \, d\mu_-(x) = \int_B (-f(x)) \, d\mu(x) \geq \varepsilon\mu(B) > 0,$$

which is a contradiction. □

In the operator-theoretic approach Carleman's condition is closely related to the theory of *quasi-analytic vectors*. We briefly discuss this connection.

Let T be a symmetric linear operator on a Hilbert space and $x \in \cap_{n=1}^\infty \mathcal{D}(T^n)$. Since the operator T is symmetric, it is easily verified that the real sequence

$$t = (t_n := \langle T^n x, x \rangle)_{n \in \mathbb{N}_0}$$

is positive semidefinite and hence a moment sequence by Hamburger's theorem 3.8. The vector x is called *quasi-analytic* for T (see e.g. [Sm9, Definition 7.1]) if

$$\sum_{n=1}^\infty \|T^n x\|^{-1/n} = +\infty. \tag{14.18}$$

Note that $t_{2n} = \langle T^{2n}x, x \rangle = \|T^n x\|^2$ for $n \in \mathbb{N}_0$. Therefore, the vector x is quasi-analytic for T if and only if the sequence t satisfies Carleman's condition (14.10).

Now suppose that s is a positive semidefinite d-sequence and let $L = L_s$ be the corresponding positive linear functional on $\mathbb{R}_d[\underline{x}]$. Then we obtain

$$s_{2n}^{[j]} = L_s(x_j^{2n}) = \langle X_j^{2n}1, 1 \rangle_L = \|X_j^n 1\|_L^2, \quad j = 1, \ldots, d, \ n \in \mathbb{N}_0.$$

Hence the marginal sequence $s^{[j]}$ fulfills Carleman's condition (14.10) if and only if 1 is a quasi-analytic vector for the multiplication operator X_j by the variable x_j.

Quasi-analytic vectors of commuting symmetric operators are studied in detail in [Sm9, Section 7.4]. Theorem 7.18 therein gives for $d = 2$ an operator-theoretic approach to Theorem 14.19 based on quasi-analytic vectors.

14.5 Moments of Gaussian Measure and Surface Measure on the Unit Sphere

In this short section we interrupt the study of determinacy and compute the moments of two important measures. These formulas are of interest in themselves.

Let μ denote the Gaussian measure on $\mathbb{R}^d$ given by

$$d\mu = (2\pi)^{-d/2} e^{-\|x\|^2/2} \, dx,$$

where dx is the Lebesgue measure of $\mathbb{R}^d$. Obviously $\int_{\mathbb{R}^d} e^{\|x\|} d\mu(x) < \infty$. Therefore, Corollary 14.24 implies that μ is a *strongly determinate* measure of $\mathcal{M}_+(\mathbb{R}^d)$.

Further, let σ be the surface measure of the unit sphere S^{d-1} of $\mathbb{R}^d$. Recall that $s_\alpha(\mu)$ and $s_\alpha(\sigma)$, where $\alpha \in \mathbb{N}_0^d$, denote the moments of these measures.

Let $j \in \{1, \ldots, d\}$. Since both measures are invariant under the transformation $x_j \mapsto -x_j, x_i \mapsto x_i$ for $i \neq j$, we have $s_\alpha(\mu) = s_\alpha(\sigma) = 0$ if one number α_j is odd. Thus it suffices to determine the moments $s_{2\alpha}(\mu)$ and $s_{2\alpha}(\sigma)$ for $\alpha \in \mathbb{N}_0^d$.

We begin with some preliminaries. Set $(2k-1)!! := 1 \cdot 3 \cdots (2k-1)$ for $k \in \mathbb{N}$ and $(-1)!! := 1$. Using the formulas $\int_0^\infty e^{-t^2/2} \, dt = \sqrt{\frac{\pi}{2}}$ and $\int_0^\infty t e^{-t^2/2} \, dt = 1$ and integration by parts we easily compute for $k \in \mathbb{N}_0$,

$$\int_0^\infty t^{2k} e^{-t^2/2} dt = \sqrt{\frac{\pi}{2}} \, (2k-1)!! \quad \text{and} \quad \int_0^\infty t^{2k+1} e^{-t^2/2} dt = \frac{k!}{2^k}. \qquad (14.19)$$

Further, for the Gamma function $\Gamma(z) := \int_0^\infty t^{z-1} e^{-t} dt$, $\Re z > 0$, the following formulas (see e.g. [RW, p. 278]) hold for $k \in \mathbb{N}_0$:

$$\Gamma(k + 1/2) = \sqrt{\pi} \, 2^{-k} (2k-1)!! \quad \text{and} \quad \Gamma(k+1) = k!. \qquad (14.20)$$

Therefore, comparing (14.19) and (14.20) we calculate

$$\int_0^\infty t^n e^{-t^2/2} dt = 2^{(n-1)/2} \, \Gamma\big((n+1)/2\big), \quad n \in \mathbb{N}_0. \qquad (14.21)$$

This formula plays an essential role in the proof of the next proposition.

Proposition 14.26 *For* $\alpha \in \mathbb{N}_0^d$ *we have*

$$s_{2\alpha}(\mu) = (2\pi)^{-d/2} \int_{\mathbb{R}^d} x^{2\alpha} e^{-\|x\|^2/2} dx = 2^{|\alpha|} \pi^{-d/2} \, \Gamma(\alpha_1 + 1/2) \cdots \Gamma(\alpha_d + 1/2),$$

$$s_{2\alpha}(\sigma) = \int_{S^{d-1}} y^{2\alpha} d\sigma(y) = \frac{2 \, \Gamma(\alpha_1 + 1/2) \cdots \Gamma(\alpha_d + 1/2)}{\Gamma(|\alpha| + d/2)}.$$

In particular, $\mu(\mathbb{R}^d) = 1$ *and* $\sigma(S^{d-1}) = \frac{2\pi^{d/2}}{\Gamma(d/2)}$.

Proof Set $\tilde{\mu} = (2\pi)^{d/2}\mu$. Then, using formula (14.21) for $n = 2\alpha_j$, we derive

$$s_{2\alpha}(\tilde{\mu}) = \int_{\mathbb{R}^d} x^{2\alpha} e^{-\|x\|^2/2} dx = \prod_{j=1}^d \int_{\mathbb{R}} x^{2\alpha_j} e^{-x_j^2/2} dx_j = 2^d \prod_{j=1}^d \int_0^\infty x^{2\alpha_j} e^{-x_j^2/2} dx_j$$

$$= 2^d 2^{(2|\alpha|-d)/2} \prod_{j=1}^d \Gamma(\alpha_j + 1/2) = 2^{|\alpha|+d/2} \prod_{j=1}^d \Gamma(\alpha_j + 1/2). \tag{14.22}$$

This yields the formula for $s_{2\alpha}(\mu)$ stated in the proposition.

On the other hand, setting $r = \|x\|$, applying the transformation $x = ry, y \in S^{d-1}$, and using (14.21) for $n = 2|\alpha| + d - 1$, we obtain

$$s_{2\alpha}(\tilde{\mu}) = \int_{\mathbb{R}^d} x^{2\alpha} e^{-\|x\|^2/2} dx = \int_0^\infty r^{2|\alpha|+d-1} e^{-r^2/2} dr \int_{S^{d-1}} y^{2\alpha} d\sigma(y)$$

$$= 2^{(2|\alpha|+d-2)/2} \Gamma((2|\alpha| + d)/2) \, s_{2\alpha}(\sigma) = 2^{|\alpha|-1+d/2} \Gamma(|\alpha| + d/2) \, s_{2\alpha}(\sigma).$$

Inserting on the left the expression from (14.22) we get the formula for $s_{2\alpha}(\sigma)$.

The formulas for $\mu(\mathbb{R}^d)$ and $\sigma(S^{d-1})$ are obtained by letting $\alpha = 0$ and using that $\Gamma(1/2) = \sqrt{\pi}$. $\square$

14.6   Disintegration Techniques and Determinacy

The results of this section are based on the following *disintegration theorem* for measures. Recall that the image of a measure ν by a map p is denoted by $p(\nu)$.

Proposition 14.27 *Suppose that X and T are closed subsets of Euclidean spaces and ν is a finite Radon measure on X. Let $p : X \to T$ be a ν-measurable mapping and $\mu := p(\nu)$. Then there exist a mapping $t \mapsto \lambda_t$ of T into the set of Radon measures on X satisfying the following three conditions:*

(i) $\operatorname{supp} \lambda_t \subseteq p^{-1}(t)$.
(ii) $\lambda_t(p^{-1}(t)) = 1 \, \mu$–a.e.

(iii) *For each nonnegative Borel function f on X we have*

$$\int_X f(x)\, d\nu(x) = \int_T d\mu(t) \left(\int_X f(x)\, d\lambda_t(x) \right). \tag{14.23}$$

Proof This is a special case of [Bou, Proposition 2.7.13, Chapter IX]. □

We retain the assumptions and notations of Proposition 14.27 and begin with the preparations of Theorem 14.29 below.

Let A and B be countably generated complex ∗-algebras of ν-integrable functions on X and μ-integrable functions on T, respectively, which contain the constant functions. The involution is always the complex conjugation of functions. Suppose that $f(p(x)) \in A$ for all $f \in B$.

Let $g \in A$. Since (14.23) holds for $f = \bar{g}g$ and $\int f d\nu$ is finite, $|g|^2$ is λ_t-integrable μ-a.e. Because A is a countably generated, there is a common μ-null set N_0 such that $|g|^2$ is λ_t-integrable for all $g \in A$ and $t \in T \backslash N_0$. For notational simplicity we assume that $N_0 = \emptyset$ (for instance, we can set $g = 0$ on N_0 for all $g \in A$); then $|g|^2$ is λ_t-integrable on T. Since the unital ∗-algebra A is spanned by elements $|g|^2$, each $f \in A$ is λ_t-integrable on T and (14.23) holds.

We define linear functionals $\mathcal{L}_t, t \in T$, and L on A by

$$\mathcal{L}_t(f) := \int_X f(x)\, d\lambda_t(x) \quad \text{and} \quad L(f) := \int_X f(x)\, d\nu(x) \quad \text{for } f \in A.$$

Lemma 14.28 *For any $f \in$ A, the function $t \mapsto \mathcal{L}_t(f)$ is in $L^2(T, \mu)$.*

Proof We freely use the properties (i)–(iii) from Proposition 14.27. (ii) implies that $\mathcal{L}_t(1) = 1$. By the above definitions and the Cauchy–Schwarz inequality we obtain

$$\int_T d\mu(t)\, |\mathcal{L}_t(f)|^2 = \int_T d\mu(t) \left| \int_X f\, d\lambda_t(x) \right|^2$$

$$\leq \int_T d\mu(t) \left(\int_X |f|^2\, d\lambda_t(x) \right) \left(\int_X 1\, d\lambda_t(x) \right)$$

$$= \int_T d\mu(t)\, \mathcal{L}_t(f\bar{f})\mathcal{L}_t(1) = \int_T d\mu(t)\, \mathcal{L}_t(f\bar{f}) = L(f\bar{f}) < +\infty. \quad \square$$

We assume that the following three conditions are satisfied:

(1) *The measure μ is determinate on T for* B, *that is, if μ' is another Radon measure on T such that $\int f\, d\mu = \int f\, d\mu'$ for all $f \in$ B, then $\mu = \mu'$.*
(2) B *is dense in $L^2(T, \mu)$.*
(3) *For μ-almost all $t \in T$, the measure λ_t is determinate on $p^{-1}(t)$ for* A, *that is, if λ_t' is a Radon measure on the fibre $p^{-1}(t)$ such that $\int g\, d\lambda_t = \int g\, d\lambda_t'$ for all $g \in$ A, then $\lambda_t = \lambda_t'$.*

The main result of this section is the following *fibre theorem* for determinacy.

Theorem 14.29 *Let* A *and* B *be as above and retain the preceding notation. If the assumptions (1)–(3) hold, then the measure* ν *is determinate on X for* A.

Proof Suppose that ν' is a Radon measure on X such that

$$\int_X g \, d\nu = \int_X g \, d\nu' \quad \text{for} \ \ g \in \mathsf{A}. \tag{14.24}$$

Let $\mu = p(\nu)$, λ_t and $\mu' = p(\nu')$, λ'_t, respectively, be the corresponding measures from the disintegration theorem and $\mathcal{L}_t$ and $\mathcal{L}'_t$ the corresponding functionals for ν and ν', respectively. For $f \in \mathsf{B}$ we have $f(p(x)) \in \mathsf{A}$ and we compute

$$\int_T d\mu'(t) f(t) = \int_T d\mu'(t) f(t) \int_X d\lambda'_t(x) \tag{14.25}$$

$$= \int_T d\mu'(t) \int_X f(t) \, d\lambda'_t(x) = \int_X d\mu'(t) \int_X f(p(x)) d\lambda'_t(x) \tag{14.26}$$

$$= \int_X f(p(x)) d\nu'(x) = \int_X f(p(x)) d\nu(x) = \int_T d\mu(t) f(t). \tag{14.27}$$

Here the equality in (14.25) holds, since $\lambda'_t(p^{-1}(t)) = 1$ μ'-a.e. by (ii). For the second equality in (14.26) we used that $\operatorname{supp} \lambda'_t \subseteq p^{-1}(t)$ by (i) and hence $t = p(x)$ λ'_t-a.e. on $\operatorname{supp} \lambda'_t$. The first two equalities in (14.27) hold by (14.23) and (14.24). By assumption (1), μ is determinate on T for B. Hence the preceding implies that $\mu = \mu'$.

Recall that B is countably generated and dense in $L^2(T, \mu)$ by assumption (2). Hence there exist a subset N of $\mathbb{N}$ and functions $\varphi_n \in \mathsf{A}$, $n \in \mathsf{N}$, such that the subset $\{\tilde{\varphi}_n : n \in \mathsf{N}\}$ of B is an orthonormal basis of $L^2(T, \mu)$.

Fix $f \in \mathsf{A}$. We compute the Fourier coefficients of the function $\mathcal{L}_t(f)$ of $L^2(T, \mu)$ with respect to this orthonormal basis by

$$\int_T d\mu(t) \, \mathcal{L}_t(f) \, \overline{\tilde{\varphi}_n(t)} = \int_T d\mu(t) \, \overline{\tilde{\varphi}_n(t)} \int_X f(x) \, d\lambda_t(x)$$

$$= \int_T d\mu(t) \int_X \overline{\varphi_n(p(x))} f(x) \, d\lambda_t(x) = \int_X d\nu(x) \, \overline{\varphi_n(p(x))} f(x) = L(\overline{\varphi_n(p(x))} f(x)).$$

Since $\mu = \mu'$, the same reasoning with λ_t replaced by λ'_t shows that

$$\int_T d\mu(t) \, \mathcal{L}'_t(f) \, \overline{\tilde{\varphi}_n(t)} = L(\overline{\varphi_n(p(x))} f(x)).$$

Therefore, both functions $\mathcal{L}_t(f)$ and $\mathcal{L}'_t(f)$ from $L^2(T, \mu)$ (by Lemma 14.28) have the same Fourier developments

$$\mathcal{L}_t(f) = \sum_{n \in \mathsf{N}} L(\overline{\varphi_n(p)} f) \, \tilde{\varphi}_n(t), \quad \mathcal{L}'_t(f) = \sum_{n \in \mathsf{N}} L(\overline{\varphi(p)} f) \, \tilde{\varphi}_n(t)$$

in $L^2(T, \mu)$. Consequently, $\mathcal{L}_t(f) = \mathcal{L}'_t(f)$ μ-a.e. on T. That is, there is a μ-null subset M_f of T such that $\int_X f(x)\, d\lambda_t(x) = \int_X f(x)\, d\lambda'_t(x)$ for $t \in T\backslash M_f$. Since A is countably generated, there is a μ-null subset M of T such that the latter holds for *all* $f \in$ A and $t \in T\backslash M$. But then we conclude from assumption (3) that $\lambda_t = \lambda'_t$ μ-a.e. on T. Therefore, since $\mu = \mu'$ as shown above, it follows from the disintegration formula (14.23) that $\int_X \varphi\, d\nu = \int_X \varphi\, d\nu'$ for all nonnegative functions, hence for all functions, $\varphi \in C_c(X; \mathbb{R})$. This implies that $\nu = \nu'$. $\qquad\square$

We now specialize the preceding general theorem to the moment problem and develop a "reduction procedure" for proving determinacy.

Suppose that X is a closed subset of $\mathbb{R}^d$. Let $p_1, \ldots, p_m \in \mathbb{R}_d[\underline{x}]$ and define a mapping $p : X \to T$ by $p(x) = (p_1(x), \ldots, p_m(x))$, where T is a closed subset of $\mathbb{R}^m$ such that $p(X) \subseteq T$. Since the polynomials p_j are continuous and X is closed, each fibre $p^{-1}(t)$ is a closed subset of $\mathbb{R}^m$. In the case $X = \mathbb{R}^m$ the fibres $p^{-1}(t), t \in T$, are just the real algebraic varieties

$$p^{-1}(t) = \{x \in \mathbb{R}^n : p_1(x) = t_1, \ldots, p_k(x) = t_k\}.$$

Further, we set A $:= \mathbb{C}_d[\underline{x}] \equiv \mathbb{C}[x_1, \ldots, x_d]$ and B $:= \mathbb{C}_m[\underline{t}] \equiv \mathbb{C}[t_1, \ldots, t_m]$. Note that then the assumption $f(p(x)) \in$ A for $f \in$ B is fulfilled.

Suppose that $\nu \in \mathcal{M}_+(\mathbb{R}^d)$ and supp $\nu \subseteq X$. Let μ and λ_t be the corresponding measures from Proposition 14.27. For $f \in \mathbb{R}_m[\underline{t}] \subseteq$ B it follows from equations (14.25)–(14.27) that

$$\int_T f(t)\, d\mu(t) = \int_X f(p(x))\, d\nu(x) = L(f(p(x))) < +\infty.$$

Therefore, $\mu \in \mathcal{M}_+(\mathbb{R}^m)$. As noted above, there is a μ-null set N_0 such that all $g \in \mathbb{R}_d[\underline{x}] \subseteq$ A are λ_t-integrable for $t \in T\backslash N_0$. Thus, $\lambda_t \in \mathcal{M}_+(\mathbb{R}^d)$ μ-a.e. on T.

Assumptions (1) and (2) mean that μ is determinate on T (by Definition 14.7) and that $\mathbb{C}_m[\underline{t}]$ is dense in $L^2(T, \mu)$. In this case we shall call μ *strictly determinate on* T. Assumption (iii) says that λ_t is determinate on $p^{-1}(t)$ μ-a.e. on T.

Therefore, in the preceding setup Theorem 14.29 can be restated as follows.

Theorem 14.30 *If μ is strictly determinate on T and λ_t is determinate on the fibre $p^{-1}(t)$ for μ-almost all $t \in T$, then ν is determinate on X.*

Clearly, measures with compact support are strictly determinate by Weierstrass' theorem. Combined with this fact Theorem 14.30 yields the following corollaries.

Corollary 14.31 *If T is compact and the measure λ_t is determinate on the fibre $p^{-1}(t)$ for μ-almost all $t \in T$, then ν is determinate on X.*

Corollary 14.32 *If μ is strictly determinate on T and the fibre $p^{-1}(t)$ is bounded for μ-almost all $t \in T$, then ν is determinate on X.*

By specifying the sets X, T and the polynomials p_j, there are a number of applications of Theorem 14.30 and Corollaries 14.31 and 14.32. We mention three applications and retain the notation and the setup introduced before Theorem 14.30. These results are of interest in themselves and illustrate the power of the fibre theorem.

1. Let $p_1(x) = x_1, \ldots, p_m(x) = x_m$, $m < d$, and let X and T be closed subsets of $\mathbb{R}^d$ and $\mathbb{R}^m$, respectively, such that $p(X) \subseteq T$. Then Theorem 14.30 yields:

   The measure ν is determinate on X if $\mu = p(\nu)$ is strictly determinate on T for $\mathbb{C}_m[\underline{t}]$ and the measure λ_t is determinate on $p^{-1}(t)$ for μ-almost all $t \in T$.

2. Suppose that $p_1, \ldots, p_m \in \mathbb{R}_d[\underline{x}]$ are polynomials which are *bounded* on the closed subset X of $\mathbb{R}^d$. Put

$$\alpha_j = \inf \{ p_j(x) : x \in X \}, \quad \beta_j = \sup \{ p_j(x) : x \in X \}, \quad T = [\alpha_1, \beta_1] \times \cdots \times [\alpha_m, \beta_m].$$

   An immediate consequence of Corollary 14.31 is the following assertion:

   The measure ν is determinate on X if the fibre measures λ_t are determinate on $p^{-1}(t)$ for μ-almost all $t \in T$.

3. Let $p(x) = x_1^2 + \cdots + x_d^2$, $d \geq 2$, and $X = \mathbb{R}^n$, $T := p(X) = \mathbb{R}_+$. Then the fibres are circles and hence compact. That μ is determinate on $\mathbb{R}_+$ means that μ is Stieltjes determinate. In this case, μ is equal to the Friedrichs solution and hence $\mathbb{C}[t]$ is dense in $L^2(\mathbb{R}_+, \mu)$. That is, determinacy on $\mathbb{R}_+$ is the same as strict determinacy on $\mathbb{R}_+$. Thus Corollary 14.32 gives:

   The measure ν is determinate if $\mu = p(\nu)$ is determinate on $\mathbb{R}_+$, or equivalently, if μ is Stieltjes determinate.

14.7   Exercises

1. Let $\mu \in \mathcal{M}_+(\mathbb{R}^d)$ and $q \in \mathbb{C}_d[\underline{x}]$. Suppose that $q(x) \neq 0$ for $x \in \operatorname{supp} \mu$. Prove that $\mathbb{C}_d[\underline{x}]$ is dense in $L^2(\mathbb{R}^d, |q|^2 d\mu)$ if and only if $q(x)\mathbb{C}_d[\underline{x}]$ is dense in $L^2(\mathbb{R}^d, \mu)$.

2. Let $\mu, \nu \in \mathcal{M}_+(\mathbb{R}^d)$ and suppose that $\nu \cong \mu$.

   a. Let X be a real algebraic set in $\mathbb{R}^d$ such that $\operatorname{supp} \mu \subseteq X$. Show that $\operatorname{supp} \nu \subseteq X$.
   b. Suppose that $p \in \mathbb{C}_d[\underline{x}]$ is bounded on $\operatorname{supp} \mu$. Prove that p is also bounded on $\operatorname{supp} \nu$ and that $\sup \{|p(x)| : x \in \operatorname{supp} \nu\} = \sup \{|p(x)| : x \in \operatorname{supp} \mu\}$.

3. Let $\mu_1, \ldots, \mu_d \in \mathcal{M}_+(\mathbb{R})$. Show $\mu := \mu_1 \otimes \cdots \otimes \mu_d \in \mathcal{M}_+(\mathbb{R}^d)$.

4. Let $\mu = \mu_1 \otimes \cdots \otimes \mu_d$, $\mu_j \in \mathcal{M}_+(\mathbb{R})$. Prove that the following are equivalent:

   (i) Each measure $\mu_k, k = 1, \ldots, d$, is determinate.
   (ii) μ is determinate.
   (iii) μ is ultradeterminate.

5. Let $\mu \in \mathcal{M}_+(\mathbb{R}^d)$ and let L be the corresponding moment functional on $\mathbb{C}_d[\underline{x}]$. Let $p := \lambda_1 x_1 + \cdots + \lambda_d x_d$, where $\lambda_j \in \mathbb{R}$, and define a symmetric operator A_p

on $\mathcal{H}_L$ by $A_p f = p \cdot f$ for $f \in \mathbb{C}_d[\underline{x}]$. Prove that if μ is ultradeterminate, then the operator A_p is essentially self-adjoint.

6. (*A strongly determinate measure that is not ultradeterminate [Sm5]*)

   Let $\nu_0 \in \mathcal{M}_+(\mathbb{R})$ be indeterminate and N-extremal such that $\operatorname{supp} \nu \cap (-\varepsilon, \varepsilon) = \emptyset$ for some $\varepsilon > 0$ and define ν by $d\nu = (1+x^2)^{-1}d\nu_0$. Let $\mu_j = \varphi_j(\nu)$, where $\varphi_1(x) = (x, 0)$ and $\varphi_2(x) = (0, x), x \in \mathbb{R}$. Put $\mu = \mu_1 + \mu_2$.

   a. Prove that $\mu \in \mathcal{M}_+(\mathbb{R}^2)$ is strongly determinate.
   b. Prove that μ is not ultradeterminate.
      Hint: Show that the operator $A_{x_1+x_2}$ (Exercise 5) is not essentially self-adjoint.

7. (*A strongly determinate measure with indeterminate marginal measures* [Sm5]).

   Let $\nu_1 = \sum_{n=0}^{\infty} a_n \delta_{x_n}$ be an indeterminate N-extremal measure on $\mathbb{R}$ such that $x_0 = 0$ and $x_n \neq 0$ for $n \in \mathbb{N}$. Put $\mu_1 = \sum_{n=1}^{\infty} a_n \delta_{(x_n, 0)}$ and $\mu_2 = \sum_{n=1}^{\infty} a_n \delta_{(0, x_n)}$.

   a. Show that $\mu_1, \mu_2 \in \mathcal{M}_+(\mathbb{R}^2)$ are determinate.
   b. Show that $\mu = \mu_1 + \mu_2$ is strongly determinate.
   c. Show that $\pi_1(\mu)$ and $\pi_1(\mu)$ are indeterminate.
      Hint for a.: Use Exercise 7.13.

8. (*A characterization of ultradeterminacy [Fu]*)

   Let $\mu \in \mathcal{M}_+(\mathbb{R}^d)$ and let L be its moment functional. Consider the joint graph $\mathcal{G}_X := \{(f, X_1 f, \dots, X_d f) : f \in \mathbb{C}_d[\underline{x}]\}$ of operators X_j on $\mathcal{H}_L$. Prove that μ is ultradeterminate if and only if there are self-adjoint operators $A_1, \dots, A_d$ on $\mathcal{H}_L$ such that $\mathcal{G}_X$ is dense in joint graph $\mathcal{G}_A := \{(f, A_1 f, \dots, A_d f) : f \in \cap_{j=1}^d \mathcal{D}(A_j)\}$. In this case we have $A_j = \overline{X}_j$ for $j = 1, \dots, d$.

9. Let $\mu, \nu \in \mathcal{M}_+(\mathbb{R}^d)$. Suppose that $d\nu = |f(x)| d\mu$ for some bounded Borel function f on $\mathbb{R}^d$. Show that if μ is strictly determinate (strongly determinate, ultradeterminate), so is ν.

10. Let f be a polynomially bounded Borel function on $\mathbb{R}^d$, that is, $|f(x)| \leq p(x)$ on $\mathbb{R}^d$ for some $p \in \mathbb{R}_d[\underline{x}]$, and let $\varepsilon > 0$. Define $\nu_1, \nu_2 \in \mathcal{M}_+(\mathbb{R}^d)$ by

$$dv_1 = |f(x)| e^{-\varepsilon \|x\|} dx \quad \text{and} \quad dv_2 = |f(x)| e^{-\varepsilon \|x\|^2} dx,$$

where dx stands for the Lebesgue measure on $\mathbb{R}^d$. Show that ν_1 and ν_2 are strongly determinate measures of $\mathcal{M}_+(\mathbb{R}^d)$.

11. (*Radon measures with bounded density on a semi-algebraic sets $\mathcal{K}(\mathsf{f})$*)

    Prove that the assertions of Corollary 12.29 remain valid if the assumption "$\mathcal{K}(\mathsf{f})$ is compact" is replaced by the assumption "L^μ satisfies the multivariate Carleman condition (14.11)".
    Hint: Show that $L := cL^\mu - L^\nu$ also satisfies (14.11) and use Theorem 14.25.

12. Let $\varphi \in C_0^\infty(\mathbb{R}^d)$. Define a linear functional L on $\mathbb{R}_d[\underline{x}]$ by

$$L(p) = \langle p(-i\partial_1, \dots, -i\partial_d)\varphi, \varphi \rangle, \quad p \in \mathbb{R}_d[\underline{x}],$$

where $\partial_j := \frac{\partial}{\partial x_j}$ and $\langle \cdot, \cdot \rangle$ denotes the scalar product of $L^2(\mathbb{R}^d)$ with respect to the Lebesgue measure. Prove that L is an *indeterminate* moment functional.
Hint: For proving that L is a moment functional, use the Fourier transform; for the indeterminacy mimic the proof of Example 6.14.

13. (*A sufficient determinacy criterion* [PVs2, Theorem 2.1])
    Let $V_d = \{z = (z_1, \dots, z_d) \in \mathbb{C}^d : z_1^2 + \cdots + z_d^2 = -1\}$. Let L be a moment functional on $\mathbb{R}_d[\underline{x}]$. For $n \in \mathbb{N}$, we define

$$\rho_n(L) = \min \{L(p^2) : p \in \mathbb{R}_d[\underline{x}]_n, \ |p(z)| = 1 \ \text{ for } \ z \in V_d\}.$$

Since $\rho_n(L) \geq \rho_{n+1}(L)$ for $n \in \mathbb{N}$, the limit $\rho(L) := \lim_{n \to \infty} \rho_n(L)$ exists.

a. Show that L is determinate if $\rho(L) = 0$.
b. Give an example of a determinate moment functional such that $\rho(L) > 0$.

Hints: For a., reduce this to the one-dimensional case and apply Petersen's theorem 14.6. For b., use Example 14.4.

14.8   Notes

Strong determinacy and ultradeterminacy have been invented by B. Fuglede [Fu], while strict determinacy was introduced in [PSm]. The main part of Theorem 14.2 is an important classical result of A. Devinatz [Dv2]. Example 14.4 is taken from [Sm5]. Proposition 14.5 can be found in [BC1] for $d = 1$ and in [Fu] for general d. L.C. Petersen's Theorem 14.6 was proved in [Pt]. Corollary 14.13 is another classical result due to G. Eskin [Es]. Theorem 14.12 is due to J. Friedrich [Fr2].

The multivariate Carleman Theorem 14.19 was proved by A.E. Nussbaum [Nu1] using his theory of quasi-analytic vectors. The proof given in the text is taken from [PSm]. Another proof based on a localization techniques is given in [Ms3]. Extensions of Carleman's condition and various ramifications are developed in [DJ]. Theorem 14.25 is due to J.B. Lasserre [Ls3].

The fibre theorem for determinacy (Theorem 14.29) and the corresponding applications in Sect. 14.6 are due to the author; they are contained in [PSm]. The determinacy for moment problems on curves was studied in [PSr].

Chapter 15
The Complex Moment Problem

In this chapter, we give a digression into the complex moment problem on $\mathbb{C}^d$. In Sect. 15.1, we discuss the equivalence of the complex moment problem on $\mathbb{C}^d$ and the real moment problem on $\mathbb{R}^{2d}$. In Sect. 15.2, we briefly treat the moment problems for two important $*$-semigroups ($\mathbb{Z}^d$ and $\mathbb{N}_0 \times \mathbb{Z}^d$). The operator-theoretic approach to the complex moment problem (Theorem 15.6) is developed in Sect. 15.3. In Sect. 15.4, we show that each positive functional on $\mathbb{C}_d[z, \overline{z}]$ satisfying the complex multivariate Carleman condition is a determinate moment functional (Theorem 15.11). In Sect. 15.5, moment functionals on $\mathbb{C}[z, \overline{z}]$ are characterized in terms of extensions to a larger algebra (Theorem 15.14). Section 15.6 solves the two-sided complex moment problem on the complex plane (Theorem 15.15).

15.1 Relations Between Complex and Real Moment Problems

The complex moment problem on $\mathbb{C}^d$ is the moment problem for the $*$-semigroup $\mathbb{N}_0^{2d}$ with pointwise addition as semigroup operation and involution defined by $(\mathfrak{m}, \mathfrak{n})^* = (\mathfrak{n}, \mathfrak{m})$ for $\mathfrak{m}, \mathfrak{n} \in \mathbb{N}_0^d$. Recall from Example 2.3.2 that the map

$$(\mathfrak{m}, \mathfrak{n}) \equiv (m_1, \ldots, m_d, n_1, \ldots, n_d) \mapsto z^{\mathfrak{m}} \overline{z}^{\mathfrak{n}} := z_1^{m_1} \ldots z_d^{m_d} \overline{z}_1^{n_1} \ldots \overline{z}_d^{m_d}$$

extends by linearity to a $*$-isomorphism of the semigroup $*$-algebra $\mathbb{C}[\mathbb{N}_0^{2d}]$ on the polynomial algebra $\mathbb{C}_d[z, \overline{z}] := \mathbb{C}[z_1, \overline{z}_1, \ldots, z_d, \overline{z}_d]$ with involution determined by

$$(z_j)^* = \overline{z}_j, \quad (\overline{z}_j)^* = z_j, \quad j = 1, \ldots, d.$$

We will write elements of $\mathbb{C}_d[z, \overline{z}]$ as $p(z, \overline{z})$, where $z = (z_1, \ldots, z_d)$, $\overline{z} = (\overline{z}_1, \ldots, \overline{z}_d)$.

© Springer International Publishing AG 2017
K. Schmüdgen, *The Moment Problem*, Graduate Texts in Mathematics 277,
DOI 10.1007/978-3-319-64546-9_15

Let $s = (s_{m,n})_{m,n \in \mathbb{N}_0^d}$ be a complex multisequence. The corresponding Riesz functional L_s is the linear functional on $\mathbb{C}_d[\underline{z},\underline{\bar{z}}]$ defined by

$$L_s(z^m \bar{z}^n) = s_{m,n}, \quad \text{where} \quad m, n \in \mathbb{N}_0^d. \tag{15.1}$$

Then the complex moment problem asks the following:

When does there exists a measure $\mu \in M_+(\mathbb{C}^d)$ such that all $p \in \mathbb{C}_d[\underline{z},\underline{\bar{z}}]$ are μ-integrable and

$$s_{m,n} = \int_{\mathbb{C}^d} z^m \bar{z}^n \, d\mu(z) \quad \text{for} \quad m, n \in \mathbb{N}_0^d, \tag{15.2}$$

or equivalently,

$$L_s(p) = \int_{\mathbb{C}^d} p(z, \bar{z}) \, d\mu(z) \quad \text{for} \quad p \in \mathbb{C}_d[\underline{z},\underline{\bar{z}}]? \tag{15.3}$$

In this case we call s a *complex moment sequence* and L_s a *moment functional*.

It is known (by Corollary 2.16) and easily verified that a complex moment sequence $s = (s_{m,n})_{m,n \in \mathbb{N}_0^d}$ is *positive semidefinite for the $*$-semigroup* $\mathbb{N}_0^{2d}$, that is,

$$\sum_{\ell,m,\ell,n \in \mathbb{N}_0^d} s_{\ell+n, m+\ell} \, \bar{\xi}_{\ell,m} \, \bar{\xi}_{\ell,n} \geq 0$$

for each finite complex sequence $(\xi_{\ell,m})_{\ell,m \in \mathbb{N}_0^d}$.

Further, by Proposition 2.7, s is positive semidefinite for $\mathbb{N}_0^{2d}$ if and only if its Riesz functional L_s is a positive functional on the $*$-algebra $\mathbb{C}_d[\underline{z},\underline{\bar{z}}]$. Therefore, each moment functional is a positive functional on $\mathbb{C}_d[\underline{z},\underline{\bar{z}}]$.

We now discuss the relations between the complex moment problem on $\mathbb{C}^d$ and the real moment problem on $\mathbb{R}^{2d}$. For this reason we also consider the polynomial $*$-algebra $\mathbb{C}_{2d}[\underline{x}] := \mathbb{C}[x_1, y_1, \ldots, x_d, y_d]$ with involution given by

$$(x_j)^* = x_j, \ (y_j)^* = y_j, \quad j = 1, \ldots, d.$$

Let $z = (z_1, \ldots, z_d) \in \mathbb{C}^d$. We write $z_j = x_j + iy_j$, where $y_j = \operatorname{Re} z_j$ and $y_j = \operatorname{Im} z_j$, and define a bijection ϕ of $\mathbb{C}^d$ onto $\mathbb{R}^{2d}$ by

$$\phi(z) \equiv \phi(z_1, \ldots, z_d) = (x_1, y_1, \ldots, x_d, y_d).$$

There is a $*$-isomorphism Φ of the complex $*$-algebras $\mathbb{C}_d[\underline{z},\underline{\bar{z}}]$ and $\mathbb{C}_{2d}[\underline{x}]$ given by

$$\Phi(p)(x_1, y_1, \ldots, x_d, y_d) = p(x_1 + iy_1, \ldots, x_d + iy_d, x_1 - iy_1, \ldots, x_d - iy_d) \tag{15.4}$$

for $p(z, \bar{z}) \in \mathbb{C}_d[\underline{z}, \bar{\underline{z}}]$. Their Hermitian parts

$$\mathbb{C}_d[\underline{z}, \bar{\underline{z}}]_h := \{p \in \mathbb{C}_d[\underline{z}, \bar{\underline{z}}] : p^* = p\}, \quad \mathbb{C}_{2d}[\underline{x}]_h := \{p \in \mathbb{C}_{2d}[\underline{x}] : p^* = p\} = \mathbb{R}_{2d}[\underline{x}]$$

are real algebras and the complex $*$-algebras $\mathbb{C}_d[\underline{z}, \bar{\underline{z}}]$ and $\mathbb{C}_{2d}[\underline{x}]$ are the complex-ifications (as defined in Sect. 12.1) of the two real algebras $\mathbb{C}_d[\underline{z}, \bar{\underline{z}}]_h$ and $\mathbb{R}_{2d}[\underline{x}]$, respectively. Therefore, from Lemma 2.17(i) we obtain

$$\sum \mathbb{R}_{2d}[\underline{x}]^2 = \sum \mathbb{C}_{2d}[\underline{x}]^2 = \Phi\left(\sum \mathbb{C}_d[\underline{z}, \bar{\underline{z}}]^2\right) = \Phi\left(\sum (\mathbb{C}_d[\underline{z}, \bar{\underline{z}}]_h)^2\right).$$

The $*$-isomorphism Φ and its inverse Φ^{-1} map quadratic modules and preorderings of one of these real algebras $\mathbb{C}_d[\underline{z}, \bar{\underline{z}}]_h$ and $\mathbb{R}_{2d}[\underline{x}]$, respectively, onto quadratic modules and preorderings of the other. Further, the bijection ϕ of $\mathbb{C}^d$ and $\mathbb{R}^{2d}$ maps the character set $\mathbb{C}^d$ of the real algebra $\mathbb{C}_d[\underline{z}, \bar{\underline{z}}]_h$ onto the character set $\mathbb{R}^{2d}$ of the real algebra $\mathbb{R}_{2d}[\underline{x}]$. Therefore, by means of the mappings ϕ and Φ and their inverses all concepts and results from real algebraic geometry and its applications to the moment problem carry over from $\mathbb{R}^{2d}$ and $\mathbb{R}_{2d}[\underline{x}]$ to $\mathbb{C}^d$ and $\mathbb{C}_d[\underline{z}, \bar{\underline{z}}]_h$. We do not restate the complex versions of these results. Our main aim in this chapter is to develop results that are more specific to the *complex* case.

We will use the correspondence discussed in the preceding paragraph to derive two interesting facts regarding the complex moment problem.

Proposition 15.1 *There exists a positive semidefinite sequence* $s = (s_{m,n})_{m,n\in\mathbb{N}_0}$ *on the $*$-semigroup* $\mathbb{N}_0^2$, *and equivalently, a positive linear functional L_s on the $*$-algebra* $C_1[\underline{z}, \bar{\underline{z}}] \equiv \mathbb{C}[z, \bar{z}]$, *such that s is not a complex moment sequence and L_s is not a moment functional.*

Proposition 15.1 follows at once from Proposition 13.5. The following "complex" Haviland theorem is an immediate consequence of the "real" Haviland theorem 1.12 for $\mathbb{R}_{2d}[\underline{x}]$.

Proposition 15.2 *A multisequence* $s = (s_{m,n})_{m,n\in\mathbb{N}_0^d}$ *is a complex moment sequence if and only if $L_s(p) \geq 0$ for all $p \in \mathbb{C}_d[\underline{z}, \bar{\underline{z}}]$ satisfying $p(z, \bar{z}) \geq 0$ for $z \in \mathbb{C}^d$.*

As noted above, the moment problems on $\mathbb{R}^{2d}$ and $\mathbb{C}^d$ are equivalent by the isomorphism Φ of $\mathbb{C}_d[\underline{z}, \bar{\underline{z}}]$ and $\mathbb{C}_{2d}[\underline{x}]$. For $d = 1$ we state the formulas relating the corresponding moments. Let L be a linear functional on $C_1[\underline{z}, \bar{\underline{z}}] = \mathbb{C}[z, \bar{z}]$ and define a linear functional $\tilde{L}$ on $C_2[\underline{x}] = \mathbb{C}[x_1, y_1]$ and sequences $s = (s_{m,n})$, $\tilde{s} = (\tilde{s}_{m,n})$ by

$$\tilde{L}(p) = L(\Phi^{-1}(p)), \quad p \in \mathbb{C}_{2d}[\underline{x}],$$

$$s_{m,n} = L(z^m \bar{z}^n), \quad \tilde{s}_{m,n} = \tilde{L}(x_1^m y_1^n), \quad (m, n) \in \mathbb{N}_0^2.$$

Some computations yield the following formulas for $(m, n) \in \mathbb{N}_0^2$:

$$\tilde{s}_{m,n} = 2^{-m-n} \sum_{k=0}^{m} \sum_{l=0}^{n} \binom{m}{k}\binom{n}{l} i^{n-2l} s_{k+l,m+n-k-l},$$

$$s_{m,n} = \sum_{k=0}^{m} \sum_{l=0}^{n} \binom{m}{k}\binom{n}{l} i^{m-k}(-i)^{n-l} \tilde{s}_{k+l,m+n-k-l}.$$

15.2 The Moment Problems for the $*$-Semigroups $\mathbb{Z}^d$ and $\mathbb{N}_0 \times \mathbb{Z}^d$

The moment problems for the $*$-semigroups $\mathbb{Z}^d$ and $\mathbb{N}_0 \times \mathbb{Z}^d$ are moment problems on the d-torus $\mathbb{T}^d$ and on the cylinder set $\mathbb{R} \times \mathbb{T}^d$, respectively.

Let us begin with the $*$-semigroup $\mathbb{Z}^d$ from Example 2.3.3. The semigroup operation of $\mathbb{Z}^d$ is addition and the involution is given by $\mathfrak{n}^* = -\mathfrak{n}$. Further, the map $\mathfrak{n} = (n_1, \ldots, n_d) \mapsto z^{\mathfrak{n}} = z_1^{n_1} \cdots z_d^{n_d}$ gives a $*$-isomorphism of the group $*$-algebra $\mathbb{C}[\mathbb{Z}^d]$ onto the $*$-algebra of trigonometric polynomials in d variables. For notational simplicity we identify $\mathbb{C}[\mathbb{Z}^d]$ with the latter $*$-algebra, that is,

$$\mathbb{C}[\mathbb{Z}^d] = \mathbb{C}[z_1, \bar{z}_1, \ldots, z_d, \bar{z}_d : z_1\bar{z}_1 = \bar{z}_1 z_1 = 1, \ldots, z_d \bar{z}_d = \bar{z}_d z_d = 1].$$

The characters of the complex $*$-algebra $\mathbb{C}[\mathbb{Z}^d]$, or equivalently of its Hermitian part $\mathbb{C}[\mathbb{Z}^d]_h$, are the point evaluations at points of the d-torus

$$\mathbb{T}^d = \{z = (z_1, \ldots, z_d) \in \mathbb{C}^d : |z_1| = \cdots = |z_d| = 1\}.$$

Let us pass to the "real setting". Set $\mathsf{A} := \Phi(\mathbb{C}[\mathbb{Z}^d]_h)$, where Φ is defined by (15.4). Then A is a real unital algebra with $2d$ generators $x_1, y_1, \ldots, x_d, y_d$ and relations $h_j := x_j^2 + y_j^2 - 1 = 0, j = 1, \ldots, d$. That is, A of the form $\mathsf{A} = \mathbb{R}_{2d}[\underline{x}]/\mathcal{I}$, where $\mathcal{I}$ is the ideal of $\mathbb{R}_{2d}[\underline{x}]$ generated by $h_1, \ldots, h_d$. The character space $\hat{\mathsf{A}}$ is the compact real algebraic set

$$\hat{\mathsf{A}} = \mathcal{Z}(\mathcal{I}) = \phi(\mathbb{T}^d) = \{(x_1, y_1, \ldots, x_d, y_d) \in \mathbb{R}^{2d} : x_1^2 + y_1^2 = 1, \ldots, x_d^2 + y_d^2 = 1\}.$$

The following result says that *strictly positive* trigonometric polynomials on $\mathbb{T}^d$ are always sums of squares.

Proposition 15.3 *Let $p \in \mathbb{C}[\mathbb{Z}^d]$. If $p(z, \bar{z}) > 0$ for $z \in \mathbb{T}^d$, then $p \in \sum \mathbb{C}[\mathbb{Z}^d]^2$.*

Proof The assumption implies that $p \in \mathbb{C}[\mathbb{Z}^d]_h$, so $\Phi(p) \in \mathsf{A} = \mathbb{R}_{2d}[\underline{x}]/\mathcal{I}$, and $\Phi(p)(x_1, y_1, \ldots, x_d, y_d) > 0$ on $\hat{\mathsf{A}}$. Hence it follows from Corollary 12.30(ii) that $\Phi(p) \in \sum(\mathbb{R}_{2d}[\underline{x}]/\mathcal{I})^2 = \sum \mathsf{A}^2$, so that $p \in \Phi^{-1}(\sum \mathsf{A}^2) = \sum \mathbb{C}[\mathbb{Z}^d]^2$. $\qquad\square$

It is natural ask whether or not each *nonnegative* trigonometric polynomial on $\mathbb{T}^d$ is a sum of squares. This is true if and only if $d \leq 2$. For $d = 1$ this is an immediate consequence of the Fejér–Riesz theorem 11.1, while for $d = 2$ this result is much more subtle and follows (for instance) from [Ms1, Theorem 9.4.5].

The following proposition solves the moment problem for $\mathbb{Z}^d$.

Proposition 15.4 *Let* $s = (s_n)_{n \in \mathbb{Z}^d}$ *be a complex sequence and* L_s *its Riesz functional on* $\mathbb{C}[\mathbb{Z}^d]$ *defined by* $L_s(z^n) = s_n, n \in \mathbb{Z}^d$. *The following are equivalent:*

(i) *s is a positive semidefinite sequence on the ∗-semigroup $\mathbb{Z}^d$, that is,*

$$\sum_{n,m \in \mathbb{Z}^d} s_{n-m} \xi_m \overline{\xi}_n \geq 0$$

for each finite complex sequence $(\xi_n)_{n \in \mathbb{Z}^d}$.

(ii) *L_s is a positive linear functional on the ∗-algebra $\mathbb{C}[\mathbb{Z}^d]$.*

(iii) *s is a moment sequence on $\mathbb{Z}^d$, that is, there exists a Radon measure μ on $\mathbb{T}^d$ such that*

$$s_n = \int_{\mathbb{T}^d} z^{-n} \, d\mu(z) \quad \text{for } n \in \mathbb{Z}^d.$$

Proof Proposition 2.7 yields (i)↔(ii) and Lemma 2.12 gives (iii)→(ii).

The main implication (ii)→(iii) will be derived from Proposition 13.31. Indeed, we define a linear functional $\tilde{L}$ on A by $\tilde{L}(p) = L_s(\Phi(p)), p \in \mathbb{C}[\mathbb{Z}^d]_h$. By (ii), L_s is a positive functional on $\mathbb{C}[\mathbb{Z}^d]$, hence so is $\tilde{L}$ on A. Because $\hat{A}$ is compact, Proposition 13.31 applies and shows that $\tilde{L}$ is a moment functional on A. Since $A = \Phi(\mathbb{C}[\mathbb{Z}^d]_h)$ and $\hat{A} = \phi(\mathbb{T}^d)$, L_s is a moment functional on $\mathbb{C}[\mathbb{Z}^d]$. $\square$

Next we turn to the ∗-semigroup $\mathbb{N}_0 \times \mathbb{Z}^d$ with coordinatewise addition as semigroup composition and involution $(k, n)^* = (k, -n)$, where $k \in \mathbb{N}_0, n \in \mathbb{Z}^d$. It is straightforward to check that the semigroup ∗-algebra $\mathbb{C}[\mathbb{N}_0 \times \mathbb{Z}^d]$ is isomorphic to the tensor product ∗-algebra $\mathbb{C}[x] \otimes \mathbb{C}[\mathbb{Z}^d]$ and the characters of $\mathbb{C}[\mathbb{N}_0 \times \mathbb{Z}^d]$ are exactly the point evaluations at points of $\mathbb{R} \times \mathbb{T}^d$.

The next proposition provides the solution of the moment problem for $\mathbb{N}_0 \times \mathbb{Z}^d$.

Proposition 15.5 *Let* $s = (s_{k,n})_{(k,n) \in \mathbb{N}_0 \times \mathbb{Z}^d}$ *be a complex sequence and let* L_s *be its Riesz functional on* $\mathbb{C}[\mathbb{N}_0 \times \mathbb{Z}^d]$ *defined by* $L_s(x^k z^n) = s_{k,n}, (k, n) \in \mathbb{N}_0 \times \mathbb{Z}^d$. *Then the following statements are equivalent:*

(i) *s is a positive semidefinite sequence on the ∗-semigroup $\mathbb{N}_0 \times \mathbb{Z}^d$, that is,*

$$\sum_{(k,m),(l,n) \in \mathbb{N}_0 \times \mathbb{Z}^d} s_{k+l,n-m} \xi_{k,m} \overline{\xi}_{l,n} \geq 0$$

for each finite complex sequence $(\xi_{k,n})_{(k,n) \in \mathbb{N}_0 \times \mathbb{Z}^d}$.

(ii) *L_s is a positive linear functional on the ∗-algebra $\mathbb{C}[\mathbb{N}_0 \times \mathbb{Z}^d]$.*

386　　　　　　　　　　　　　　　　　　　　15 The Complex Moment Problem

(iii) *s is a moment sequence on* $\mathbb{N}_0 \times \mathbb{Z}^d$, *that is, there exists a Radon measure* μ
    on $\mathbb{R} \times \mathbb{T}^d$ *such that* $x^k z^{-\mathfrak{n}}$ *is* μ-*integrable and*

$$s_{k,\mathfrak{n}} = \int_{\mathbb{R} \times \mathbb{T}^d} x^k z^{-\mathfrak{n}} \, d\mu(x, z) \quad \text{for} \ (k, \mathfrak{n}) \in \mathbb{N}_0 \times \mathbb{Z}^d.$$

Proof The proof is almost verbatim the same as the proof of Proposition 15.4. The
implication (ii)→(iii) follows again from Proposition 13.31, but now applied to the
real algebra $\mathbb{R}[x] \otimes \mathsf{A}$, where $\mathsf{A} = \Phi(\mathbb{C}[\mathbb{Z}^d]_h)$. □

15.3　The Operator-Theoretic Approach to the Complex Moment Problem

The main aim of this section is to derive Theorem 15.6 below, which is the
counterpart of Theorem 12.40 for the complex moment problem. As Theorem 12.40
is about extensions of commuting symmetric operators to strongly commuting
self-adjoint operators, Theorem 15.6 deals with extensions of commuting *formally
normal* operators to strongly commuting *normal* operators. In order to formulate
Theorem 15.6 we need further operator-theoretic considerations. All definitions and
results used in the following discussion can be found in [Sm9, Chapters 4 and 5].

Let Z be a densely defined linear operator on a Hilbert space. The operator Z is
called *formally normal* if

$$\mathcal{D}(Z) \subseteq \mathcal{D}(Z^*) \quad \text{and} \quad \|Zf\| = \|Z^*f\| \quad \text{for} \ f \in \mathcal{D}(Z^*)$$

and Z is said to be *normal* if Z is formally normal and $\mathcal{D}(Z) = \mathcal{D}(Z^*)$. Note that Z
is normal if and only if Z is closed and $Z^*Z = ZZ^*$.

For each normal operator Z there exists a unique spectral measure E_Z on the
Borel σ-algebra of $\mathbb{C}$ such that

$$Z = \int_{\mathbb{C}} \lambda \, dE_Z(\lambda).$$

If Z_1 and Z_2 are normal operators acting on the same Hilbert space, we say that Z_1
and Z_2 *strongly commute* if the spectral measures E_{Z_1} and E_{Z_2} commute, that is,
$E_{Z_1}(M)E_{Z_2}(N) = E_{Z_2}(N)E_{Z_1}(M)$ for all Borel subsets M, N of $\mathbb{C}$. In this case,

$$Z_1 Z_2 \varphi = Z_2 Z_1 \varphi \quad \text{for} \ \varphi \in \mathcal{D}(Z_1 Z_2) \cap \mathcal{D}(Z_2 Z_1), \tag{15.5}$$

but (15.5) does not imply the strong commutativity of Z_1 and Z_2. A number of
equivalent formulations of strong commutativity are given in [Sm9, Section 5.6].

Now suppose that $N = (N_1, \ldots, N_d)$ is a fixed d-tuple of strongly commuting normal operators $N_1, \ldots, N_d$ on a Hilbert space $\mathcal{K}$. Then, by the multidimensional spectral theorem [Sm9, Theorem 5.21], there exists a unique spectral measure E_N on the Borel σ-algebra of $\mathbb{C}^d$ such that

$$N_j = \int_{\mathbb{C}^d} \lambda_j \, dE_N(\lambda_1, \ldots, \lambda_d) \quad \text{for } j = 1, \ldots, d.$$

Further, if $M_1, \ldots, M_d$ are Borel subsets of $\mathbb{C}$, then we have

$$E_N(M_1 \times \cdots \times M_d) = E_{N_1}(M_1) \cdots E_{N_d}(M_d). \tag{15.6}$$

We state basic properties of the functional calculus for the d-tuple N, see [Sm9, Theorem 4.16 and formulas (4.32), (5.32)]. For any measurable function f on $\mathbb{C}^d$ there is a normal operator $f(N) = \int_{\mathbb{C}^d} f(\lambda) \, dE_N(\lambda)$ on $\mathcal{K}$ with dense domain

$$\mathcal{D}(f(N)) = \{\varphi \in \mathcal{K} : \int_{\mathbb{C}^d} |f(\lambda)|^2 \, d\langle E_N(\lambda)\varphi, \varphi \rangle < \infty\}. \tag{15.7}$$

Let f, g be measurable functions on $\mathbb{C}^d$. If $\varphi \in \mathcal{D}(f(N))$ and $\psi \in \mathcal{D}(g(N))$, then

$$\langle f(N)\varphi, g(N)\psi \rangle = \int_{\mathbb{C}^d} f(\lambda)\overline{g(\lambda)} \, d\langle E_N(\lambda)\varphi, \psi \rangle. \tag{15.8}$$

Further, $f(N)^* = \bar{f}(N)$. For $\alpha, \beta \in \mathbb{C}$ and $\zeta \in \mathcal{D}(f(N)) \cap \mathcal{D}(g(N))$ we have

$$\zeta \in \mathcal{D}((\alpha f + \beta g)(N)) \quad \text{and} \quad (\alpha f + \beta g)(N)\zeta = \alpha f(N)\zeta + \beta g(N)\zeta.$$

If $\eta \in \mathcal{D}(g(N))$ and $\eta \in \mathcal{D}((fg)(N))$, then

$$\eta \in \mathcal{D}(f(N)g(N)) \quad \text{and} \quad f(N)g(N)\eta = (fg)(N)\eta.$$

Let $p(N, N^*)$ denote the operator $f(N)$ obtained for $f(z) = p(z, \bar{z}) \in \mathbb{C}_d[z, \bar{z}]$. In particular, $N^{\mathfrak{n}}$ is the operator for the function $z^{\mathfrak{n}}$, $\mathfrak{n} \in \mathbb{N}_0^d$. The linear subspace

$$\mathcal{D}^\infty(N) := \cap_{\mathfrak{n} \in \mathbb{N}_0^d} \mathcal{D}(N^{\mathfrak{n}}).$$

is dense in $\mathcal{K}$. From (15.7) it follows that all operators $p(N, N^*)$ are defined on $\mathcal{D}^\infty(N)$. The properties of the functional calculus stated above imply that the map

$$p \mapsto \rho_N(p) := p(N, N^*) \lceil \mathcal{D}^\infty(N)$$

is a $*$-representation ρ_N of the $*$-algebra $\mathbb{C}_d[z, \bar{z}]$ on the domain $\mathcal{D}(\rho_N) := \mathcal{D}^\infty(N)$.

Fix $\varphi \in \mathcal{D}^\infty(N)$ and define a measure $\mu_\varphi \in M_+(\mathbb{C}^d)$ by $\mu_\varphi(\cdot) := \langle E_N(\cdot)\varphi, \varphi \rangle$. Let $p \in \mathbb{C}_d[\underline{z}, \underline{\bar{z}}]$. Formula (15.8), applied with $f = p, g = 1, \varphi = \psi$, yields

$$\mathcal{L}_\varphi(p) := \langle \rho_N(p)\varphi, \varphi \rangle = \langle p(N, N^*)\varphi, \varphi \rangle = \int_{\mathbb{C}^d} p(z, \bar{z}) \, d\mu_\varphi(z). \qquad (15.9)$$

This shows that $\mathcal{L}_\varphi(\cdot) = \langle \rho_N(\cdot)\varphi, \varphi \rangle$ is a moment functional on $\mathbb{C}_d[\underline{z}, \underline{\bar{z}}]$ and μ_φ is a representing measure of $\mathcal{L}_\varphi$, see (15.3).

We now turn to a special case that is crucial for the moment problem. Suppose that μ is a Radon measure on $\mathbb{C}^d$. Let N_j denote the multiplication operator by the coordinate function z_j on the Hilbert space $\mathcal{G} := L^2(\mathbb{C}^d, \mu)$, that is,

$$(N_j\varphi)(z) := z_j\varphi(z) \quad \text{for} \quad \varphi \in \mathcal{D}(N_j) := \{\varphi \in \mathcal{G} : z_j \cdot \varphi \in \mathcal{G}\}.$$

Then the adjoint N_j^* is the multiplication operator by $\bar{z}_j$ and $\|N_j\varphi\| = \|N_j^*\varphi\|$ for $\varphi \in \mathcal{D}(N_j) = \mathcal{D}(N_j^*)$, that is, N_j is a normal operator.

For a Borel subset M of $\mathbb{C}$ set $M_j = \{z \in \mathbb{C}^d : z_j \in M\}$. Let χ_{M_j} denote the characteristic function of M_j and define $E_{N_j}(M)\varphi := \chi_{M_j} \cdot \varphi$ for $\varphi \in \mathcal{G}$. Then E_{N_j} is the spectral measure of N_j, that is, $N_j\varphi = \int_{\mathbb{C}} z_j \, dE_{N_j}(z)\varphi$ for $\varphi \in \mathcal{D}(N_j)$. Since the spectral measures E_{N_j} and E_{N_k} act as multiplications by characteristic functions, they commute, that is, N_j and N_k strongly commute for $j, k = 1, \ldots, d$. For a Borel set M of $\mathbb{C}^d$ let $E_N(M)$ be the multiplication operator by the characteristic function of M. Then E_N is the spectral measure of the d-tuple $N = (N_1, \ldots, N_d)$ of strongly commuting normal operators $N_1, \ldots, N_d$ and we have $\mu(\cdot) = \langle E_N(\cdot)1, 1 \rangle$.

The preceding facts hold for any measure $\mu \in M_+(\mathbb{C}^d)$. Now we assume in addition that all polynomials of $\mathbb{C}_d[\underline{z}, \underline{\bar{z}}]$ are in $L^2(\mathbb{C}^d, \mu)$. Then 1 is in the domain $\mathcal{D}^\infty(N)$ and the above considerations apply to the d-tuple N and $\varphi = 1$. Therefore, setting $\mu(\cdot) = \langle E_N(\cdot)1, 1 \rangle$, it follows from (15.9) that

$$L(p) := \langle \rho_N(p)1, 1 \rangle = \langle p(N, N^*)1, 1 \rangle = \int_{\mathbb{C}^d} p(z, \bar{z}) \, d\mu(z) \qquad (15.10)$$

for $p \in \mathbb{C}_d[\underline{z}, \underline{\bar{z}}]$. Therefore, L is a moment functional and $\mu(\cdot) = \langle E_N(\cdot)1, 1 \rangle$ is a representing measure of L.

Hence L is a positive linear functional on the $*$-algebra $\mathbb{C}_d[\underline{z}, \underline{\bar{z}}]$. Recall that π_L denotes the GNS representation (see Definition 12.39) associated with L. Let us describe the representation π_L. By (15.10) we have $L(pq) = \int pq \, d\mu$ for $p, q \in \mathbb{C}_d[\underline{z}, \underline{\bar{z}}]$. Comparing this with (12.31) it follows that the scalar product $\langle \cdot, \cdot \rangle_L$ on the domain $\mathcal{D}_L$ of π_L is just the scalar product of $L^2(\mathbb{C}^d, \mu)$. Thus $\mathcal{D}_L$ can be identified with the subspace $\mathbb{C}_d[\underline{z}, \underline{\bar{z}}]$ of $L^2(\mathbb{C}^d, \mu)$ and π_L acts by $\pi_L(p)q = p \cdot q$ for $p \in \mathbb{C}_d[\underline{z}, \underline{\bar{z}}], q \in \mathcal{D}_L$. In particular, we have $\pi_L(z_j) \subseteq N_j$ for $j = 1, \ldots, d$.

Now let L be an arbitrary positive linear functional on $\mathbb{C}_d[\underline{z}, \underline{\bar{z}}]$. Put $Z_j := \pi_L(z_j)$. By (12.29) we have $\langle \pi_L(z_j)\varphi, \psi \rangle = \langle \varphi, \pi_L((z_j)^*)\psi \rangle$ for $\varphi, \psi \in \mathcal{D}_L$, so that

$$\pi_L(\bar{z}_j) = \pi_L((z_j)^*) \subseteq \pi_L(z_j)^*. \qquad (15.11)$$

Using once again (12.29) and finally (15.11) we derive for arbitrary $\varphi \in \mathcal{D}_L$,

$$
\begin{aligned}
\|Z_j\varphi\|^2 &= \|\pi_L(z_j)\varphi\|^2 = \langle\pi_L(z_j)\varphi, \pi_L(z_j)\varphi\rangle = \langle\varphi, \pi_L((z_j)^*)\pi_L(z_j)\varphi\rangle \\
&= \langle\varphi, \pi_L(\bar{z}_j z_j)\varphi\rangle = \langle\varphi, \pi_L(z_j\bar{z}_j)\varphi\rangle = \langle\pi_L((z_j)^*)\varphi, \pi_L((z_j)^*)\varphi\rangle \\
&= \langle\pi_L(z_j)^*\varphi, \pi_L(z_j)^*\varphi\rangle = \|\pi_L(z_j)^*\varphi\|^2 = \|(Z_j)^*\varphi\|^2.
\end{aligned}
\tag{15.12}
$$

This proves that the operator $Z_j = \pi_L(z_j)$ is *formally normal*.

Obviously, for any $j, k = 1, \ldots, d$, the operators Z_j and Z_k commute on the domain $\mathcal{D}_L$, because z_j and z_k commute in $\mathbb{C}_d[z, \bar{z}]$ and Z_j and Z_k leave the domain $\mathcal{D}_L$ invariant. Thus, $(Z_1, \ldots, Z_d)$ is a d-tuple of *pairwise commuting formally normal operators* on the dense and invariant domain $\mathcal{D}_L$ of the Hilbert space $\mathcal{H}_L$.

The results established above contain the main technical parts for the following theorem on the operator-theoretic approach to the complex moment problem.

Theorem 15.6 *Let* $s = (s_{m,n})_{m,n\in\mathbb{N}_0^d}$ *be a complex multisequence and let* L_s *be its Riesz functional defined by (15.1). The following statements are equivalent:*

(i) s *is a complex moment sequence, or equivalently,* L_s *is a moment functional on* $\mathbb{C}_d[z, \bar{z}]$.

(ii) *There exists a* d-*tuple* $N = (N_1, \ldots, N_d)$ *of strongly commuting normal operators* $N_1, \ldots, N_d$ *on a Hilbert space* $\mathcal{G}$ *and a vector* $\varphi \in \mathcal{D}^\infty(N)$ *such that*

$$
s_{m,n} = \langle N^m\varphi, N^n\varphi\rangle \quad \text{for} \quad m, n \in \mathbb{N}_0^d. \tag{15.13}
$$

(iii) s *is a positive semidefinite sequence for the* $*$-*semigroup* $\mathbb{N}_0^{2d}$, *or equivalently,* L_s *is a positive functional on the* $*$-*algebra* $\mathbb{C}_d[z, \bar{z}]$, *and there exists a* d-*tuple* $N = (N_1, \ldots, N_d)$ *of strongly commuting normal operators* $N_1, \ldots, N_d$ *on a Hilbert space* $\mathcal{G}$ *such that* $\mathcal{H}_{L_s}$ *is a subspace of* $\mathcal{G}$ *and*

$$
\pi_{L_s}(z_1) \equiv Z_1 \subseteq N_1, \ldots, \pi_{L_s}(z_d) \equiv Z_d \subseteq N_d. \tag{15.14}
$$

If (iii) holds and E_N *is the spectral measure of the* d-*tuple* N, *then*

$$
\mu(\cdot) := \langle E_N(\cdot)1, 1\rangle \tag{15.15}
$$

is a representing measure for s *and* L_s. *Each solution of the complex moment problem for* s *resp.* L_s *is obtained in this manner.*

Proof

(i)→(iii) Since L_s is a moment functional, it is a positive functional. Let μ be a representing measure for L_s. Let N_j denote the multiplication operator by z_j on $\mathcal{G} := L^2(\mathbb{C}^d, \mu)$. Then, as shown above, $N = (N_1, \ldots, N_d)$ is a d-tuple of strongly commuting normal operators on $\mathcal{G}$ such that $\pi_{L_s}(z_j) \subseteq N_j, j = 1, \ldots, d$, and $\mu(\cdot) =$

$\langle E_N(\cdot)1, 1\rangle$. This proves (iii) and the latter shows that each representing measure μ of L_s is of the form (15.15).

(iii)→(ii) By (iii), $\pi_{L_s}(z_j) \subseteq N_j$ for $j = 1, \dots, d$. Hence

$$\pi_{L_s}(z^n) = \pi_{L_s}(z_1)^{n_1} \dots \pi_{L_s}(z_d)^{n_d} \subseteq N_1^{n_1} \dots N_d^{n_d} \subseteq N^n, \quad n \in \mathbb{N}_0^d. \tag{15.16}$$

Using (15.1), the definition of the scalar product in $\mathcal{D}_{L_s}$, and (15.16) we obtain

$$s_{m,n} = L_s(\bar{z}^n z^m) = \langle \pi_{L_s}(z^m)1, \pi_{L_s}(z^n)1 \rangle = \langle N^m 1, N^n 1 \rangle, \quad m, n \in \mathbb{N}_0^d, \tag{15.17}$$

which is statement (ii) with $\varphi = 1$.

(ii)→(i) Using first (15.13) and then (15.8) we obtain

$$s_{m,n} = \langle N^m \varphi, N^n \varphi \rangle = \int z^m \bar{z}^n \, d\langle E_N(z)\varphi, \varphi \rangle, \quad m, n \in \mathbb{N}_0^d, \tag{15.18}$$

which proves that s is a moment sequence.

Finally, suppose that (iii) holds. From (15.17) and (15.18) with $\varphi = 1$ it follows that $\mu(\cdot) = \langle E_N(\cdot)1, 1\rangle$ is a representing measure for s. This completes the proof of Theorem 15.6. □

Remark 15.7 In Theorem 15.6(ii), $N_1, \dots, N_d$ are strongly commuting normal operators and $\varphi \in \mathcal{D}^\infty(N)$. Hence Theorem 15.6 remains valid if (15.13) is replaced by $s_{m,n} = \langle N^m (N^*)^n \varphi, \varphi \rangle$ for $m, n \in \mathbb{N}_0^d$. ○

We restate the main part of Theorem 15.6 in the special case $d = 1$ separately as

Corollary 15.8 *A positive linear functional L on $\mathbb{C}_1[z, \bar{z}] = \mathbb{C}[z, \bar{z}]$ is a moment functional if and only if the formally normal operator $Z = \pi_L(z)$ has a normal extension acting on a possibly larger Hilbert space $\mathcal{G}$, that is, there exists a normal operator N on a Hilbert space $\mathcal{G}$ such that $\mathcal{H}_L$ is a subspace of $\mathcal{G}$ and $Z \subseteq N$.*

If this holds and E_N denotes the spectral measure of the normal operator N, then $\mu(\cdot) = \langle E(\cdot)1, 1\rangle$ is a representing measure for L. All solutions of the moment problem for L are of this form.

A by-product of the study of the complex moment problem is the following operator-theoretic result, which is of interest in itself.

Corollary 15.9 *There exists a formally normal operator on a Hilbert space which has no normal extension on a possibly larger Hilbert space.*

Proof From Proposition 15.1, there exists a positive functional L on $C[z, \bar{z}]$ that is not a moment functional. By Corollary 15.8, $Z = \pi_L(z)$ has the desired properties. □

Remark 15.10 A densely defined linear operator T on a Hilbert space $\mathcal{H}$ is called *subnormal* if there exists a normal operator N on a Hilbert space $\mathcal{G}$ such that $\mathcal{H}$ is a subspace of $\mathcal{G}$ and $T \subseteq N$. Using this notion Corollary 15.8 says that a positive linear functional L on $\mathbb{C}[z, \bar{z}]$ is a moment functional if and only if the (formally normal) operator $Z = \pi_L(z)$ is subnormal.   ○

15.4   The Complex Carleman Condition

A complex moment sequence s, likewise its Riesz functional L_s, is called *determinate* if s, or equivalently L_s, has a *unique* representing measure.

The next theorem is a fundamental result on the complex moment problem.

Theorem 15.11 *Let* $s = (s_{m,n})_{m,n \in \mathbb{N}_0^d}$ *be a positive semidefinite complex sequence for the $*$-semigroup* $\mathbb{N}_0^{2d}$. *If the complex multivariate Carleman condition*

$$\sum_{n=1}^{\infty} L_s(z_j^n \bar{z}_j^n)^{-\frac{1}{2n}} = \infty \quad for \quad j = 1, \ldots, d \tag{15.19}$$

holds, then s is a determinate complex moment sequence and its Riesz functional L_s *is a determinate moment functional on* $\mathbb{C}_d[z, \bar{z}]$.

Proof For notational simplicity we identify the $*$-algebras $\mathbb{C}_d[z, \bar{z}]$ and $\mathbb{C}_{2d}[x]$ by the $*$-isomorphism Φ from Sect. 15.1. Then the complex moment problem for L_s becomes a real moment problem on $\mathbb{C}_{2d}[x]$ and it suffices to prove the corresponding assertions for this real moment problem. Our aim is to apply Theorem 14.19.

Recall that $z_j = x_j + iy_j$, where $(x_j)^* = x_j$ and $(y_j)^* = y_j$, and $Z_j = \pi_{L_s}(z_j)$. Using the basic property (12.33) of the GNS construction we obtain

$$L_s(x_j^{2n}) = \langle \pi_{L_s}(x_j^{2n})1, 1 \rangle = \langle \pi_{L_s}(x_j)^n 1, \pi_{L_s}(x_j)^n 1 \rangle = \|\pi_{L_s}(x_j)^n 1\|^2, \tag{15.20}$$

$$L_s(z_j^n \bar{z}_j^n) = \langle \pi_{L_s}(\bar{z}_j^n z_j^n)1, 1 \rangle = \langle \pi_{L_s}(z_j)^n 1, \pi_{L_s}(z_j)^n 1 \rangle = \|(Z_j)^n 1\|^2. \tag{15.21}$$

Further, since $2x_j = z_j + \bar{z}_j$, we derive

$$2^n \|\pi_{L_s}(x_j)^n 1\| = \|(\pi_{L_s}(z_j) + \pi_{L_s}(\bar{z}_j))^n 1\| = \|(Z_j + (Z_j)^*)^n 1\|$$

$$= \left\| \sum_{k=0}^{n} \binom{n}{k} (Z_j)^{*k}(Z_j)^{n-k} 1 \right\| \leq \sum_{k=0}^{n} \binom{n}{k} \|(Z_j)^{*k}(Z_j)^{n-k} 1\|$$

$$= \sum_{k=0}^{n} \binom{n}{k} \|(Z_j)^k(Z_j)^{n-k} 1\| = \sum_{k=0}^{n} \binom{n}{k} \|(Z_j)^n 1\| = 2^n \|(Z_j)^n 1\|.$$

Here for the third equality we used that the operators Z_j and Z_j^* commute, so the binomial theorem applies, while the fourth equality follows from the fact that the operator Z_j is formally normal, that is, from Eq. (15.12).

Combining the preceding estimate with (15.20) and (15.21) we conclude that

$$L_s(x_j^{2n}) \leq L_s(z_j^n \bar{z}_j^n), \quad n \in \mathbb{N}_0.$$

Almost the same reasoning shows that

$$L_s(y_j^{2n}) \leq L_s(z_j^n \bar{z}_j^n), \quad n \in \mathbb{N}_0.$$

Therefore, assumption (15.19) implies that the functional L_s on $\mathbb{C}_{2d}[\underline{x}]$ satisfies the multivariate Carleman condition (14.11) of the real case, so the assertions follow from Theorem 14.19.   □

We illustrate the power of Theorem 15.11 by deriving two results for the moment problem on *bounded* subsets of $\mathbb{C}^d$. For a sequence $s = (s_{m,n})_{m,n \in \mathbb{N}_0^d}$ we set

$$s_n^{(j)} := s_{(0,\ldots,0,n,\ldots,0),(0,\ldots,0,n,\ldots,0)},$$

where the number n stands at the places j and $d + j$. Further, for $r = (r_1, \ldots, r_d)$, where $r_1 > 0, \ldots, r_d > 0$, let $\overline{\mathbb{D}}_r$ denote the closed polydisc

$$\overline{\mathbb{D}}_r = \{z = (z_1, \ldots, z_d) \in \mathbb{C}^d : |z_j| \leq r_j, \; j = 1, \ldots, d\}.$$

Corollary 15.12 *A sequence* $s = (s_{m,n})_{m,n \in \mathbb{N}_0^d}$ *is a complex moment sequence which has a representing measure supported on the polydisc* $\overline{\mathbb{D}}_r$ *if and only if* s *is positive semidefinite for the* *-semigroup* $\mathbb{N}_0^{2d}$ *and there are numbers* $M_j > 0$ *such that*

$$s_n^{(j)} \leq M_j r_j^{2n} \quad \text{for} \quad n \in \mathbb{N}, j = 1, \ldots, r. \tag{15.22}$$

Proof It is clear that a complex moment sequence is positive semidefinite for $\mathbb{N}_0^{2d}$. If s has a representing measure μ with support contained in $\overline{\mathbb{D}}_r$, then

$$s_n^{(j)} = \int_{\mathbb{C}^d} z_j^n \bar{z}_j^n \, d\mu(z) \leq r_j^{2n} \mu(\mathbb{C}^d),$$

which gives (15.22).

Now we prove the if direction. From the definition of $s_n^{(j)}$ we get $s_n^{(j)} = L_s(z_j^n \bar{z}_j^n)$. For sufficiently large n we have $M_j \leq 2^{2n}$ and therefore by (15.22),

$$L_s(z_j^n \bar{z}_j^n)^{\frac{1}{2n}} = (s_n^{(j)})^{\frac{1}{2n}} \leq 2r_j.$$

Hence the Carleman condition (15.19) is fulfilled, so s is a complex moment sequence by Theorem 15.11. Let μ be a representing measure of s.

To show that μ is supported on $\overline{\mathbb{D}}_r$ we use the correspondence between moment problems of $\mathbb{C}^d$ and $\mathbb{R}^{2d}$ and apply Proposition 12.15 with $g_j(x) = x_j^2 + y_j^2 \cong z_j\overline{z}_j$ for $j = 1, \ldots, d$. Then (15.22) implies (12.15) and the set $\mathcal{G}$ defined by (12.14) is just the polydisc $\overline{\mathbb{D}}_r$. Hence $\operatorname{supp}\mu \subseteq \overline{\mathbb{D}}_r$ by Proposition 12.15. $\square$

Corollary 15.13 *A sequence $s = (s_{m,n})_{m,n \in \mathbb{N}_0^d}$ is a complex moment sequence on $\mathbb{C}^d$ which has a representing measure with compact support if and only if s is positive semidefinite for the $*$-semigroup $\mathbb{N}_0^{2d}$ and there is a constant $c > 0$ such that the sequences $(c^n s_n^{(j)})_{n \in \mathbb{N}_0}, j = 1, \ldots, d,$ are bounded.*

Proof The assertion is easily derived from Corollary 15.12. It suffices to note that condition (15.22) holds if and only if all sequences $(c^n s_n^{(j)})_{n \in \mathbb{N}_0}$ are bounded, where $c := (\max_j r_j^2)^{-1}$. $\square$

15.5 An Extension Theorem for the Complex Moment Problem

This section deals with the complex moment problem for $d = 1$, that is, with the moment problem for the $*$-semigroup $\mathbb{N}_0^2$ with involution $(m, n)^* = (n, m)$. Clearly, $\mathbb{N}_0^2$ is a $*$-subsemigroup of the larger $*$-semigroup

$$\mathsf{N}_+ = \{(m, n) \in \mathbb{Z}^2 : m + n \geq 0\} \quad \text{with involution} \quad (m, n)^* = (n, m).$$

Hence $\mathbb{C}[\mathbb{N}_0^d]$ is a $*$-subalgebra of the larger $*$-algebra $\mathbb{C}[\mathsf{N}_+]$.

The following result is the *Stochel–Szafraniec extension theorem*.

Theorem 15.14 *A linear functional L on $\mathbb{C}[\mathbb{N}_0^d] = \mathbb{C}[z, \overline{z}]$ is a moment functional if and only if L has an extension to a positive linear functional on $\mathbb{C}[\mathsf{N}_+]$.*

Proof First we describe the semigroup $*$-algebra $\mathbb{C}[\mathsf{N}_+]$. From its definition it is clear that $\mathbb{C}[\mathsf{N}_+]$ is the complex $*$-algebra generated by the functions $z^m\overline{z}^n$ on $\mathbb{C}\backslash\{0\}$, where $m, n \in \mathbb{Z}$ and $m + n \geq 0$. If $r(z)$ denotes the modulus and $u(z)$ the phase of z, then $z^m\overline{z}^n = r(z)^{m+n}u(z)^{m-n}$. Then, setting $k = m + n$, it follows that

$$\mathbb{C}[\mathsf{N}_+] = \operatorname{Lin}\{r(z)^k u(z)^{2m-k} : k \in \mathbb{N}_0, m \in \mathbb{Z}\}.$$

The two functions $r(z)$ and $u(z)$ itself are not in $\mathbb{C}[\mathsf{N}_+]$, but the functions $r(z)u(z) = z$ and $v(z) := u(z)^2 = z\overline{z}^{-1}$ are in $\mathbb{C}[\mathsf{N}_+]$ and they generate the $*$-algebra $\mathbb{C}[\mathsf{N}_+]$. Writing $z = x_1 + ix_2$ with $x_1, x_2 \in \mathbb{R}$, we get

$$1 + v(z) = 1 + \frac{x_1 + ix_2}{x_1 - ix_2} = 2\,\frac{x_1^2 + ix_1x_2}{x_1^2 + x_2^2}, \quad 1 - v(z) = 2\,\frac{x_2^2 - ix_1x_2}{x_1^2 + x_2^2}.$$

This implies that the complex algebra $\mathbb{C}[\mathsf{N}_+]$ is generated by the five functions

$$x_1, \ x_2, \ \frac{x_1^2}{x_1^2 + x_2^2}, \ \frac{x_2^2}{x_1^2 + x_2^2}, \ \frac{x_1 x_2}{x_1^2 + x_2^2}. \tag{15.23}$$

The Hermitian part $\mathbb{C}[\mathsf{N}_+]_h$ of the complex $*$-algebra $\mathbb{C}[\mathsf{N}_+]$ is the *real* algebra generated by the functions (15.23). This real algebra is the special case $d = 2$ of the $*$-algebra A treated in Sect. 13.7. Thus, if we identify $\mathbb{C}$ with $\mathbb{R}^2$ in the obvious way, the assertion of Theorem 15.14 follows at once from Theorem 13.40.   □

In terms of sequences Theorem 15.14 can be rephrased in the following manner:

A positive semidefinite sequence $c = (c_{m,n})_{(m,n)\in\mathbb{N}_0^2}$ *for the $*$-semigroup* $\mathbb{N}_0^2$ *is a moment sequence for* $\mathbb{N}_0^2$ *if and only if there is a positive semidefinite sequence* $\tilde{c} = (\tilde{c}_{m,n})_{(m,n)\in\mathsf{N}_+}$ *for the $*$-semigroup* N_+ *such that* $\tilde{c}_{m,n} = c_{m,n}$ *for* $(m, n) \in \mathbb{N}_0^2$.

Using this reformulation we now give a proof of Theorem 15.14 in the context of $*$-semigroups.

Second Proof of Theorem 15.14 This proof makes essential use of the $*$-semigroup $\mathbb{N}_0 \times \mathbb{Z}$ with involution $(k, l)^* = (k, -l)$ and its $*$-subsemigroup $\mathcal{S}$ given by $\mathcal{S} = \{(k, l) \in \mathbb{N}_0 \times \mathbb{Z} : k + l \text{ even}\}$. It is straightforward to check that the map

$$\omega : \mathsf{N}_+ \to \mathcal{S}, \quad \omega(m, n) := (m + n, m - n)$$

is a $*$-isomorphism of N_+ and $\mathcal{S}$.

First we suppose that c has an extension to a positive semidefinite sequence $\tilde{c}$ for N_+. We define a function $\varphi : \mathbb{N}_0 \times \mathbb{Z} :\to \mathbb{R}$ by

$$\varphi((k, l)) := \tilde{c}_{\omega^{-1}(k,l)} \quad \text{if} \quad (k, l) \in \mathcal{S}, \tag{15.24}$$

$$\varphi((k, l)) := 0 \qquad \text{if} \quad (k, l) \in \mathbb{N}_0 \times \mathbb{Z}, (k, l) \notin \mathcal{S}. \tag{15.25}$$

We show that φ is a positive semidefinite function on the $*$-semigroup $\mathbb{N}_0 \times \mathbb{Z}$. Suppose that finitely many elements $s_i \in \mathbb{N}_0 \times \mathbb{Z}$ and complex numbers ξ_i are given. Set $u := (0, 1)$. For $s, t \in \mathbb{N}_0 \times \mathbb{Z}$ we easily verify that $s + t \in \mathcal{S}$ if and only if $s, t \in \mathcal{S}$ or $s, t \notin \mathcal{S}$. Further, $s + u \in \mathcal{S}$ if $s \notin \mathcal{S}$. Since the sequence $\tilde{c}$ is positive semidefinite on N_+ and ω^{-1} is a $*$-isomorphism of $\mathcal{S}$ onto N_+, the restriction of φ to $\mathcal{S}$ is positive semidefinite on $\mathcal{S}$. Using the preceding facts we derive

$$\sum_{i,j} \varphi(s_i^* + s_j) \, \overline{\xi}_i \xi_j = \sum_{s_i^* + s_j \in \mathcal{S}} \varphi(s_i^* + s_j) \, \overline{\xi}_i \xi_j$$

$$= \sum_{s_i, s_j \in \mathcal{S}} \varphi(s_i^* + s_j) \, \overline{\xi}_i \xi_j + \sum_{s_i, s_j \notin \mathcal{S}} \varphi(s_i^* + s_j) \, \overline{\xi}_i \xi_j$$

$$= \sum_{s_i, s_j \in \mathcal{S}} \varphi(s_i^* + s_j) \, \overline{\xi}_i \xi_j + \sum_{s_i, s_j \notin \mathcal{S}} \varphi((s_i + u)^* + (s_j + u)) \, \overline{\xi}_i \xi_j \geq 0,$$

that is, φ is positive semidefinite on $\mathbb{N}_0 \times \mathbb{Z}$.

Hence, by Proposition 15.5, $(\varphi(k, l))_{(k,l) \in N_0 \times \mathbb{Z}}$ is a moment sequence for $\mathbb{N}_0 \times \mathbb{Z}$. Let ν be a representing measure for this sequence and let μ denote its image under the mapping $\mathbb{R} \times \mathbb{T} \ni (r, z) \mapsto rz \in \mathbb{C}$. Using (15.24) and (15.25) we obtain

$$c_{m,n} = \varphi(\omega(m, n)) = \int_{\mathbb{R} \times \mathbb{T}} r^{m+n} z^{n-m} d\nu(r, z)$$

$$= \int_{\mathbb{R} \times \mathbb{T}} (rz)^m (\overline{rz})^n d\nu(r, z) = \int_{\mathbb{C}} z^m \overline{z}^n d\mu(z), \quad m, n \in \mathbb{N}_0,$$

that is, c is a moment sequence for $\mathbb{N}_0^2$.

Conversely, assume that c is a moment sequence for $\mathbb{N}_0^2$ and let μ be a representing measure. Then, for $m, n \in \mathbb{N}_0$, we have

$$c_{m,n} = \mu(\{0\})\delta_{m+n,0} + \int_{\mathbb{C} \setminus \{0\}} z^m \overline{z}^n d\mu(z). \tag{15.26}$$

Since μ is a representing measure for c, the function $z^m \overline{z}^n$ is μ-integrable on $\mathbb{C} \setminus \{0\}$ for $m + n \geq 0$, that is, for $(m, n) \in N_+$. Therefore, the right-hand side of (15.26) is well-defined for $(m, n) \in N_+$; let $\tilde{c}_{m,n}$ denote the corresponding number. It is easily verified that the second summand in (15.26) defines a positive semidefinite sequence for N_+. Obviously, the map $(m, n) \mapsto \delta_{m+n,0}$ defines a character of N_+. Hence the first summand in (15.26) is a positive semidefinite sequence for N_+ as well. Thus, $\tilde{c} = (\tilde{c}_{(m,n)})$ is a positive semidefinite sequence for the $*$-semigroup N_+. $\square$

15.6   The Two-Sided Complex Moment Problem

The two-sided complex moment problem is the moment problem for the $*$-semigroup $\mathbb{Z}^2$ with involution $(m, n)^* = (n, m)$.

It is easily verified that the map $(m, n) \mapsto z^m \overline{z}^n$ gives a $*$-isomorphism of the semigroup $*$-algebra $\mathbb{C}[\mathbb{Z}^2]$ on the $*$-algebra $\mathbb{C}[z, \overline{z}, z^{-1}, \overline{z}^{-1}]$ of complex Laurent polynomials in z and $\overline{z}$. For notational simplicity we identify these $*$-algebras. The character space of $\mathbb{C}[\mathbb{Z}^2]$ consists of evaluation functionals at points of $\mathbb{C}^\times := \mathbb{C} \setminus \{0\}$.

Let $s = (s_{m,n})_{(m,n) \in \mathbb{Z}^2}$ be a complex sequence. As usual, L_s is the Riesz functional on $\mathbb{C}[\mathbb{Z}^2]$ defined by $L_s(z^m \overline{z}^n) = s_{m,n}, (m, n) \in \mathbb{Z}^2$.

Then the two-sided complex moment problem is the following question:

When does there exist a Radon measure μ on $\mathbb{C}^\times$ such that the function $z^m \overline{z}^n$ on $\mathbb{C}^\times$ is μ-integrable and

$$s_{mn} = \int_{\mathbb{C}^\times} z^m \overline{z}^n d\mu(z) \quad for \quad (m, n) \in \mathbb{Z}^2,$$

or equivalently,

$$L_s(p) = \int_{\mathbb{C}^\times} p(z,\bar{z})\,d\mu(z) \quad for \quad p \in \mathbb{C}[\mathbb{Z}^2] = \mathbb{C}[z,\bar{z},z^{-1},\bar{z}^{-1}]\,?$$

Note that this requires conditions for the measure μ at infinity *and* at zero. In the affirmative case we call s a moment sequence for $\mathbb{Z}^2$ and L_s a moment functional.

The following fundamental result is *Bisgaard's theorem.*

Theorem 15.15 *A linear functional L on $\mathbb{C}[\mathbb{Z}^2]$ is a moment functional if and only if L is a positive functional, that is, $L(f^*f) \geq 0$ for all $f \in \mathbb{C}[\mathbb{Z}^2]$.*

In terms of $*$-semigroups the main assertion of this theorem says that *each positive semidefinite sequence on $\mathbb{Z}^2$ is a moment sequence on $\mathbb{Z}^2$.* This result is really surprising, since $\mathbb{C}^\times$ has dimension 2 and no additional condition (in terms of positivity or of some appropriate extension) is required.

Proof The only if part is obvious. We prove the if part.

First we describe the semigroup $*$-algebra $\mathbb{C}[\mathbb{Z}^2]$ in terms of generators. Clearly, a vector space basis of $\mathbb{C}[\mathbb{Z}^2] = \mathbb{C}[z,\bar{z},z^{-1},\bar{z}^{-1}]$ is the set $\{z^k\bar{z}^l : k,l \in \mathbb{Z}\}$. Writing $z = x_1 + ix_2$ with $x_1,x_2 \in \mathbb{R}$ we get

$$z^{-1} = \frac{x_1 - ix_2}{x_1^2 + x_2^2} \quad and \quad \bar{z}^{-1} = \frac{x_1 + ix_2}{x_1^2 + x_2^2}.$$

Hence $\mathbb{C}[\mathbb{Z}^2]$ is the complex unital algebra generated by the four functions

$$x_1, \quad x_2, \quad y_1 := \frac{x_1}{x_1^2 + x_2^2}, \quad y_2 := \frac{x_2}{x_1^2 + x_2^2} \tag{15.27}$$

on $\mathbb{R}^2\backslash\{0\}$. Note that all four functions are unbounded on $\mathbb{R}^2\backslash\{0\}$.

Let A be the Hermitian part of the complex $*$-algebra $\mathbb{C}[\mathbb{Z}^2]$. Then A is a real algebra and its character set $\hat{A}$ is given by the point evaluations χ_x at $x \in \mathbb{R}^2\backslash\{0\}$. (Obviously, χ_x is a character for $x \in \mathbb{R}^2\backslash\{0\}$. Since $(y_1 + iy_2)(x_1 - ix_2) = 1$, there is no character χ on A for which $\chi(x_1) = 0$ and $\chi(x_2) = 0$.)

The three functions

$$h_1(x) = x_1y_1 = \frac{x_1^2}{x_1^2 + x_2^2}, \quad h_2(x) = x_2y_2 = \frac{x_2^2}{x_1^2 + x_2^2}, \quad h_3(x) = x_1y_2 = x_2y_1 = \frac{x_1x_2}{x_1^2 + x_2^2}$$

are elements of A and they are bounded on $\hat{A} = \{\chi_x : x \in \mathbb{R}^2, x \neq 0\}$.

To apply Theorem 13.10 we consider a nonempty fibre set given by $h_j(x) = \lambda_j$, where $\lambda_j \in \mathbb{R}$ for $j = 1,2,3$. Then $\lambda_1 + \lambda_2 = 1$, so we can assume without loss of generality that $\lambda_1 \neq 0$. In the quotient algebra $A/\mathcal{I}_\lambda$ we then have $x_1y_1 = \lambda_1 \neq 0$, so that $y_1 = \lambda_1x_1^{-1}$, and $x_2y_1 = x_1y_2 = \lambda_3$, so that $y_2 = \lambda_3x_1^{-1}$ and $x_2 = \lambda_3\lambda_1^{-1}x_1$. Thus the algebra $A/\mathcal{I}_\lambda$ is generated by x_1 and x_1^{-1} and hence a quotient of the algebra $\mathbb{R}[x_1, x_1^{-1}]$ of Laurent polynomials. Since $\sum \mathbb{R}[x, x^{-1}]^2$ obeys (MP) by Theorem 3.16, so does the preordering $\sum (A/\mathcal{I}_\lambda)^2$ of its quotient algebra $A/\mathcal{I}_\lambda$ by Corollary 13.16. Therefore, $\sum A^2$ satisfies (MP) by Theorem 13.10. By definition, this means that each positive functional on A, hence on $\mathbb{C}[\mathbb{Z}^2]$, is a moment functional.   □

Remark 15.16 The generators x_1, x_2, y_1, y_2 of the algebra A satisfy the relations

$$x_1 y_1 + x_2 y_2 = 1 \quad \text{and} \quad (x_1^2 + x_2^2)(y_1^2 + y_2^2) = 1. \qquad \circ$$

15.7 Exercises

1. Let $s = (s_{m,n})_{m,n \in \mathbb{N}_0^d}$ be a positive semidefinite sequence for $\mathbb{N}_0^{2d}$. Show that
   $|s_{\ell+m,\mathfrak{l}+n}|^2 \le s_{\ell+\mathfrak{l},\ell+\mathfrak{l}} \, s_{m+n,m+n}$ for $\ell, \mathfrak{l}, m, n \in \mathbb{N}_0^d$.
2. Let $s = (s_{m,n})_{m,n \in \mathbb{N}_0}$ be a positive semidefinite sequence for $\mathbb{N}_0^2$. Show that there is a measure $\mu \in \mathcal{M}_+(\mathbb{R})$ supported on $\mathbb{R}_+$ such that $s_{n,n} = \int x^n \, d\mu$ for $n \in \mathbb{N}_0$.
3. Let $\nu \in M_+(\mathbb{C}^d)$. Suppose that all polynomials $p \in \mathbb{C}_d[\underline{z}, \overline{z}]$ are ν-integrable. Define $\mu \in M_+(\mathbb{C}^d)$ by $d\mu = (1 + \|z\|^2)^{-1} d\nu$, where $\|z\|^2 := z_1 \overline{z}_1 + \cdots + z_d \overline{z}_d$. Show that the complex moment sequence of μ is determinate.
4. Suppose that $\mu \in M_+(\mathbb{C}^d)$ satisfies $\int e^{\varepsilon \|z\|^2} \, d\mu < \infty$. Show that all moments of μ are finite and the moment sequence of μ is determinate.
5. Let $s = (s_{m,n})_{m,n \in \mathbb{N}_0}$ be a complex sequence and let $R > r > 0$. Give necessary and sufficient conditions for s to be a complex moment sequence with a representing measure supported on the following set K:

   a. $K := \{z \in \mathbb{C} : r \le |z| \le R\}$.
   b. $K := \{z \in \mathbb{C} : \mathrm{Im}\, z \ge 0, \, r \le |z| \le R\}$.

6. Let s be a positive semidefinite sequence for the $*$-semigroup $\mathbb{N}_0^4$. Show that s is a complex moment sequence on $\mathbb{C}^2$ with representing measure supported on the closed ball $\{(z_1, z_2) \in \mathbb{C} : |z_1|^2 + |z_2|^2 \le r^2\}$ if and only if there is a constant $M > 0$ such that

$$\sum_{k=0}^{n} \binom{2n}{k} s_{2k,4n-2k,2k,4n-2k} \le M r^{4n} \quad \text{for } n \in \mathbb{N}_0.$$

7. Let $s = (s_{m,n})_{m,n \in \mathbb{N}_0}$ be a complex moment sequence with representing measure μ supported on $\overline{\mathbb{D}} = \{z \in \mathbb{C} : |z| \le 1\}$. What can be said about the support of μ if

   a. $s_{1,1} = s_{0,0}$,
   b. $s_{m,n} = s_{m-n,0}$ for $m \ge n$,
   c. $s_{1,1} = s_{2,0}$,
   d. $s_{m,n} = s_{m+n,0}$ for $m, n \in \mathbb{N}_0$?

8. Let s be a complex moment sequence and let μ be a representing measure of s. For $p \in \mathbb{C}[z, \bar{z}]$ we define a Radon measure μ_p on $\mathbb{C}$ by $d\mu_p(z) = |p(z, \bar{z})|^2 d\mu(z)$. Express the moment sequence of μ_p in terms of p and s.

15.8   Notes

The interplay between the complex moment problem and subnormality has been known and investigated for a long time by operator-theorists [Br, Fo]. Theorem 15.6 for $d = 1$ goes back to Y. Kilpi [Ki], see also [StS2, Proposition 3]. Normal extensions of formally normal unbounded operators have been extensively studied by J. Stochel and F.H. Szafraniec [StS1, StS2, StS3].

The two formulas at the end of Sect. 15.1 are taken from [StS4]. Proposition 15.5 is due to A. Devinatz [Dv1]. The existence of formally normal operators without normal extensions (Corollary 15.9) was discovered in [Cd], see [Sm3] and [St1] for simple explicit examples. The complex Carleman condition was investigated in [StS1]; Theorem 15.11 can be found therein. A solution of the moment problem for discs without using Carleman's condition is given in [Sf].

Theorem 15.14 was proved in [StS4]; our second proof is the original proof in [StS4], while our first proof is taken from [Sm11]. Theorem 15.15 is due to T.M. Bisgaard [Bi]; the very short proof in the text is also from [Sm11].

Chapter 16
Semidefinite Programming and Polynomial Optimization

"Finding" the minimum or infimum $p^{\min}$ of a real polynomial p over a semi-algebraic set $\mathcal{K}(\mathsf{f})$ is a basic optimization problem. Sum of squares decompositions of polynomials by means of Positivstellensätze and moment problem methods provide powerful tools for polynomial optimization. The aim of this chapter is to give a short digression into these applications by outlining the main ideas.

In Sect. 16.2, we introduce two relaxations p_n^{mom} and p_n^{sos} for $p^{\min}$ in terms of Hankel matrices and the quadratic module $Q(\mathsf{f})$ and formulate them as a semidefinite program and its dual. In Sects. 16.3 and 16.4, these relaxations are investigated on $\mathcal{K}(\mathsf{f})$ and $\mathbb{R}^d$, respectively. If the quadratic module $Q(\mathsf{f})$ is Archimedean and hence $\mathcal{K}(\mathsf{f})$ is compact, a simple application of the Archimedean Positivstellensatz shows that both relaxations converge to the minimum $p^{\min}$ (Theorem 16.6). Section 16.1 contains a brief introduction to semidefinite programming.

In this chapter we use some results of positive semidefinite matrices from Appendix A.3.

16.1 Semidefinite Programming

Let Sym_n denote the vector space of real symmetric $n \times n$-matrices and $\langle \cdot, \cdot \rangle$ the scalar product on Sym_n defined by $\langle A, B \rangle = \mathrm{Tr}\, AB$. Recall that $A \succeq 0$ means that the matrix A is positive semidefinite and $A \succ 0$ that A is positive definite.

Now we define a semidefinite program and its dual program.

Suppose that a vector $b \in \mathbb{R}^m$ and $m + 1$ matrices $A_0, \ldots, A_m \in \mathrm{Sym}_n$ are given. Then the primal *semidefinite program* (SDP) is the following:

$$p_* = \inf_{y \in \mathbb{R}^m} \{ b^T y : A(y) := A_0 + y_1 A_1 + \cdots + y_m A_m \succeq 0 \}. \tag{16.1}$$

© Springer International Publishing AG 2017
K. Schmüdgen, *The Moment Problem*, Graduate Texts in Mathematics 277,
DOI 10.1007/978-3-319-64546-9_16

That is, one minimizes the linear function $b^T y = \sum_{j=1}^{m} b_j y_j$ in a vector variable $y = (y_1, \ldots, y_m)^T \in \mathbb{R}^m$ subject to the *linear matrix inequality* (LMI) constraint

$$A(y) := A_0 + y_1 A_1 + \cdots + y_m A_m \succeq 0. \tag{16.2}$$

The set of points $y \in \mathbb{R}^m$ satisfying (16.2) is called a *spectrahedron*. By Proposition A.18, the matrix $A(y)$ is positive semidefinite if and only if all its principal minors are nonnegative. Since each such minor is a polynomial, the set of points for which $A(y) \succeq 0$ is described by finitely many polynomial inequalities. Hence each spectrahedron is a basic closed semi-algebraic set. By (16.2), a spectrahedron is convex. But a convex basic closed semi-algebraic set is not necessarily a spectrahedron.

Let $p \in \mathbb{R}_m[y]$. Further, let $e \in \mathbb{R}^m$ be such that $p(e) > 0$. Then the polynomial p is called *hyperbolic* with respect to e if for each $y \in \mathbb{R}^m$ the univariate polynomial $p(e + ty) \in \mathbb{R}[t]$ has only real zeros. In this case, the closure of the connected component of the set $\{y \in \mathbb{R}^m : p(y) > 0\}$ containing e is called *rigidly convex*. If $A(e) \succ 0$, it can be shown that $p(y) := \det A(y)$ is a hyperbolic polynomial with respect to e and the corresponding spectrahedron is rigidly convex. Polyhedra and ellipsoids are spectrahedra. The set $\{(y_1, y_2) \in \mathbb{R}^2 : y_1^4 + y_2^4 \leq 1\}$ is not rigidly convex, hence it is not a spectrahedron. Spectrahedra form an interesting class of sets, but their study is outside the scope of this book (see e.g. [BTB]).

If all matrices A_j are diagonal, the constraint $A(y) \succeq 0$ consists of inequalities of linear functions, so (16.1) is a linear program. Conversely, each linear problem becomes a semidefinite program by writing the linear constraints as an LMI with a diagonal matrix. Thus, linear programs are special cases of semidefinite programs.

The *dual program* associated with (16.1) is defined by

$$p^* = \sup_{Z \in \mathrm{Sym}_n} \{-\langle A_0, Z \rangle : Z \succeq 0 \text{ and } \langle A_j, Z \rangle = b_j, \, j = 1, \ldots, m\}. \tag{16.3}$$

Thus, one maximizes the linear function $-\langle A_0, Z \rangle = -\mathrm{Tr}\, A_0 Z$ in a matrix variable $Z \in \mathrm{Sym}_n$ subject to the constraints $Z \succeq 0$ and $\langle A_j, Z \rangle = \mathrm{Tr}\, A_j Z = b_j$, $j = 1, \ldots, m$. It can be shown that the dual program (16.3) is also a semidefinite program.

A vector $y \in \mathbb{R}^m$ resp. a matrix $Z \in \mathrm{Sym}_n$ is called *feasible* for (16.1) resp. (16.3) if it satisfies the corresponding constraints. If there are no feasible points, we set $p_* = +\infty$ resp. $p^* = -\infty$. A program is called *feasible* if it has a feasible point.

Proposition 16.1 *If y is feasible for (16.1) and Z is feasible for (16.3), then*

$$b^T y \geq p_* \geq p^* \geq -\langle A_0, Z \rangle. \tag{16.4}$$

Proof Since y is feasible for (16.1) and Z is feasible for (16.3), $A(y) \succeq 0$ and $Z \succeq 0$. Therefore, $\langle A(y), Z \rangle \geq 0$ by Proposition A.21(i), so using (16.3) we derive

$$b^T y = \sum_{j=1}^{m} y_j \langle A_j, Z \rangle = \langle A(y), Z \rangle - \langle A_0, Z \rangle \geq -\langle A_0, Z \rangle.$$

Taking the infimum over all feasible vectors y and the supremum over all feasible matrices Z we obtain (16.4). □

In contrast to linear programming, p_* is not equal to p^* in general (see Exercise 15.4). The number $p_* - p^*$ is called the *duality gap*. The next proposition shows that, under the stronger assumption of *strict feasibility*, the duality gap is zero.

Proposition 16.2

(i) *If $p_* > -\infty$ and (16.1) is strictly feasible (that is, there exists a vector $y \in \mathbb{R}^m$ such that $A(y) \succ 0$), then $p_* = p^*$ and the supremum in (16.3) is a maximum.*
(ii) *If $p^* < +\infty$ and (16.3) is strictly feasible (that is, there exists a matrix $Z \in \mathrm{Sym}_n$ such that $Z \succ 0$ and $\langle A_j, Z \rangle = b_j$ for $j = 1, \ldots, m$), then $p_* = p^*$ and the infimum in (16.1) is a minimum.*

Proof We carry out the proof of (i); the proof of (ii) is similar.

The subset $\mathcal{U}$ of positive definite matrices is an open convex cone in the vector space Sym_{n+1} (in any norm topology). We define matrices $\tilde{A}_i \in \mathrm{Sym}_{n+1}$

$$\tilde{A}_0 = \begin{pmatrix} p_* & 0 \\ 0 & A_0 \end{pmatrix} \quad \text{and} \quad \tilde{A}_j = \begin{pmatrix} -b_j & 0 \\ 0 & A_j \end{pmatrix}, \quad j = 1, \ldots, m,$$

and consider the convex subset

$$\mathcal{C} := \left\{ \tilde{A}_0 + y_1 \tilde{A}_1 + \cdots + y_m \tilde{A}_m : y_1, \ldots, y_m \in \mathbb{R} \right\}$$

of Sym_{n+1}. Then $\mathcal{U} \cap \mathcal{C} = \emptyset$. (Indeed, otherwise $A_0 + y_1 A_1 + \cdots + y_m A_m \succ 0$ and $p_* - \sum_j b_j y_j > 0$ which contradicts the definition of p_*.) Therefore, by separation of convex sets (Theorem A.26(i)), there are a linear functional L on Sym_{n+1} and a real number a such that $L(A) \leq a < L(B)$ for $A \in \mathcal{C}$ and $B \in \mathcal{U}$. Clearly, $L \neq 0$. Since $\mathcal{U}$ is a cone, it follows that $a \leq 0$. Hence $L \leq 0$ on $\mathcal{C}$ and $L > 0$ on $\mathcal{U}$. Since $L \leq 0$ on $\mathcal{C}$, we have $L(\tilde{A}_0) \leq 0$ and $L(\tilde{A}_j) = 0$ for $j = 1, \ldots, n$.

By Riesz' theorem for the Hilbert space $(\mathrm{Sym}_{n+1}, \langle \cdot, \cdot \rangle)$, there exists a matrix $\tilde{Z} \in \mathrm{Sym}_{n+1}$ such that $L(\cdot) = \langle \cdot, \tilde{Z} \rangle$. Since the closure of $\mathcal{U}$ is the cone of positive semidefinite matrices and $L > 0$ on $\mathcal{U}$, it follows that $L(B) = \langle B, \tilde{Z} \rangle = \langle \tilde{Z}, B \rangle \geq 0$ for all $B \succeq 0$. Therefore, $\tilde{Z} \succeq 0$ by Proposition A.21(ii). We write $\tilde{Z}$ as

$$\tilde{Z} = \begin{pmatrix} z_0 & z \\ z^T & Z \end{pmatrix},$$

where $z_0 \in \mathbb{R}, z \in \mathbb{R}^n, Z \in \mathrm{Sym}_n$. Then $L(\tilde{A}_0) \leq 0$ and $L(\tilde{A}_j) = 0, j = 1, \ldots, m$, yield

$$\langle A_0, Z \rangle + z_0 p_* \leq 0 \quad \text{and} \quad \langle A_j, Z \rangle = z_0 b_j, \; j = 1, \ldots, m. \tag{16.5}$$

A crucial step is to prove that $z_0 \neq 0$. Assume to the contrary that $z_0 = 0$. Then $z = 0$, because $\tilde{Z} \succeq 0$. Since $L \neq 0$, we have $\tilde{Z} \neq 0$, so $Z \neq 0$. By $z_0 = 0$, (16.5) implies $\langle A_0, Z \rangle \leq 0$ and $\langle A_j, Z \rangle = 0$, so that $\langle A(y), Z \rangle \leq 0$ for all $y \in \mathbb{R}^m$. But by the assumption of strict feasibility there exists a $y \in \mathbb{R}^m$ such that $A(y) \succ 0$. Fix such a y. Since $\tilde{Z} \succeq 0$, we have $Z \succeq 0$ and hence $\langle A(y), Z \rangle \geq 0$ by Proposition A.21(i). Thus, $\langle A(y), Z \rangle = 0$, so that $A(y)Z = 0$ by Proposition A.21(i). Now $A(y) \succ 0$ implies that $A(y)$ is invertible. Hence $Z = 0$, which is a contradiction. This proves that $z_0 \neq 0$.

From $\tilde{Z} \succeq 0$ we get $z_0 > 0$. Upon scaling we can assume that $z_0 = 1$. From (16.5) we then obtain $\langle A_j, Z \rangle = b_j$ for $j = 1, \ldots, m$, so Z is feasible for the dual program (16.3), and $-\langle A_0, Z \rangle \geq p_*$. Therefore, $p^* \geq -\langle A_0, Z \rangle \geq p_*$. Since $p_* \geq p^*$ by Proposition 16.1, we get $p_* = p^* = -\langle A_0, Z \rangle$. This shows that the supremum in (16.3) is a maximum. $\qquad \square$

A large number of problems in various mathematical fields can be formulated in terms of semidefinite programming, see e.g. [VB]. We give only two examples.

Example 16.3 (Largest eigenvalue of a symmetric matrix) Let $\lambda_{\max}(B)$ denote the largest eigenvalue of $B \in \mathrm{Sym}_n$. Then

$$\lambda_{\max}(B) = \min_{y \in \mathbb{R}} \; \{y : (yI - B) \succeq 0\}$$

gives $\lambda_{\max}(B)$ by a semidefinite program. Since this program and its dual are strictly feasible (take $y \in \mathbb{R}$ such that $(yI - B) \succ 0$ and $Z = I$), Proposition 16.2 yields

$$\lambda_{\max}(B) = \max_{Z \in \mathrm{Sym}_n} \; \{\langle B, Z \rangle : Z \succeq 0, \langle I, Z \rangle = 1\}.$$

A semidefinite program for the sum of the j largest eigenvalues of B is developed in [OW]. $\qquad \circ$

Example 16.4 (Sos representation of a polynomial) By Proposition 13.2, a polynomial $f(x) = \sum_\alpha f_\alpha x^\alpha \in \mathbb{R}_d[\underline{x}]_{2n}$ is in $\sum \mathbb{R}_d[\underline{x}]_n^2$ if and only if there exists a positive semidefinite matrix G such that $f(x) = (\mathfrak{x}_n)^T G \mathfrak{x}_n$, where $\mathfrak{x}_n$ is given by (13.2). If we write $\mathfrak{x}_n(\mathfrak{x}_n)^T = \sum_{\alpha \in N_{2n}} A_\alpha x^\alpha$ with $A_\alpha \in \mathrm{Sym}_{d(n)}$, Eq. (13.5) means that $\langle G, A_\alpha \rangle = \mathrm{Tr}\, G A_\alpha = f_\alpha$ for $\alpha \in \mathbb{N}_0^d, |\alpha| \leq 2n$. Therefore, $f \in \sum \mathbb{R}_d[\underline{x}]_n^2$ if and only if there exists a matrix $G \in \mathrm{Sym}_{d(n)}$ such that

$$G \succeq 0 \quad \text{and} \quad \langle Q, A_\alpha \rangle = f_\alpha \quad \text{for } \alpha \in \mathbb{N}_0^d, |\alpha| \leq 2n. \tag{16.6}$$

Hence f is a sum of squares if and only if the *feasibility condition* (16.6) of the corresponding semidefinite program is satisfied. That is, testing whether or not f is a sum of squares means checking the feasibility of a semidefinite program.   ∘

16.2 Lasserre Relaxations of Polynomial Optimization with Constraints

In this section and the next, $\mathsf{f} = \{f_0, \ldots, f_k\}$ is a fixed finite subset of $\mathbb{R}_d[\underline{x}]$ and $p \in \mathbb{R}_d[\underline{x}], p \neq 0$. For simplicity we assume that $f_0 = 1$ and $f_j \neq 0$ for all j.

Our aim is to minimize the polynomial p over the semi-algebraic set $\mathcal{K}(\mathsf{f})$. That is, we want to "compute"

$$p^{\min} := \inf\{p(x) : x \in \mathcal{K}(\mathsf{f})\}. \tag{16.7}$$

Let $n \in \mathbb{N}_0$. We denote by n_j the largest integer such that $n_j \leq \frac{1}{2}(n - \deg(f_j))$, where $j = 0, \ldots, k$, and set

$$Q(\mathsf{f})_n := \left\{ \sum_{j=0}^{k} f_j \sigma_j : \sigma_j \in \sum \mathbb{R}_d[\underline{x}]^2, \deg(f_j \sigma_j) \leq n \right\}$$

$$= \left\{ \sum_{j=0}^{k} f_j \sigma_j : \sigma_j \in \sum \mathbb{R}_d[\underline{x}]_{n_j}^2 \right\}.$$

Let $Q(\mathsf{f})_n^*$ denote the set of linear functionals L on $\mathbb{R}_d[\underline{x}]_n$ satisfying $L(1) = 1$ and $L(g) \geq 0$ for all $g \in Q(\mathsf{f})_n$.

Now we define two *relaxations* of (16.7), called *Lasserre relaxations*, by

$$p_n^{\mathrm{mom}} := \inf\{L(p) : L \in Q(\mathsf{f})_n^*\}, \tag{16.8}$$

$$p_n^{\mathrm{sos}} := \sup\{\lambda \in \mathbb{R} : p - \lambda \in Q(\mathsf{f})_n\}. \tag{16.9}$$

Here we set $p_n^{\mathrm{sos}} = -\infty$ if there is no $\lambda \in \mathbb{R}$ such that $p - \lambda \in Q(\mathsf{f})_n$.

Let us motivate these relaxations. Obviously, $p^{\min}$ is the supremum of numbers $\lambda \in \mathbb{R}$ such that $p - \lambda \geq 0$ on $\mathcal{K}(\mathsf{f})$. Clearly, each $f \in Q(\mathsf{f})$ is nonnegative on $\mathcal{K}(\mathsf{f})$. Replacing $p - \lambda \geq 0$ on $\mathcal{K}(\mathsf{f})$ by the stronger requirement $p - \lambda \in Q(\mathsf{f})_n$ gives the number p_n^{sos}, which is less then or equal to $p^{\min}$. Further, by definition, $p^{\min}$ is the infimum of all evaluations of p at points of $\mathcal{K}(\mathsf{f})$. Taking the infimum over the larger set of functionals $Q(\mathsf{f})_n^*$ yields the number p_n^{mom}, which is also less than or equal to $p^{\min}$.

Next we want to reformulate the two relaxations (16.8) and (16.9) in terms of a semidefinite program and its dual. Let us begin with (16.8).

In order to describe the functionals L of $Q(\mathsf{f})_n^*$ we introduce the $\binom{d+n}{n}$ variables $y_\alpha := L(x^\alpha)$, where $\alpha \in \mathbb{N}_0^d$, $|\alpha| \leq n$. If the functional L is given by a Radon measure μ, then y_α is just the α-th moment of μ. But in general $L \in Q(\mathfrak{f})_n^*$ does not imply that L is given by a Radon measure, so y_α is only a variable here.

Let L be a linear functional on $\mathbb{R}_d[\underline{x}]_n$. Then, by definition, L belongs to $Q(\mathfrak{f})_n^*$ if and only if $L(1) = 1$ and for $j = 0, \ldots, k$ we have

$$L(f_j q^2) \geq 0 \quad \text{for} \quad q \in \mathbb{R}_d[\underline{x}]_{n_j}. \tag{16.10}$$

Now we reformulate the conditions (16.10) in terms of the variables y_α. Clearly, $L(1) = 1$ means that $y_0 = 1$. Let us fix $j = 0, \ldots, k$ and write $f_j = \sum_\alpha f_{j,\alpha} x^\alpha$. We denote by $H_{n_j}(f_j y)$ the type $\binom{d+n_j}{n_j} \times \binom{d+n_j}{n_j}$-matrix with entries

$$H_{n_j}(f_j y)_{\alpha,\beta} := \sum_\gamma f_{j,\gamma} y_{\alpha+\beta+\gamma}, \quad \text{where} \quad |\alpha| \leq n_j, |\beta| \leq n_j. \tag{16.11}$$

That is, $H_{n_j}(f_y)$ is a truncation of the localized infinite Hankel matrix $H(f_j y)$ whose entries are defined by (12.12). For $q = \sum_\alpha a_\alpha x^\alpha \in \mathbb{R}_d[\underline{x}]_{n_j}$ repeating the computation of (12.13) we derive

$$L(f_j q^2) = \sum_{\alpha,\beta,\gamma} f_{j,\gamma} a_\alpha a_\beta L(x^{\alpha+\beta+\gamma}) = \sum_{\alpha,\beta,\gamma} f_{j,\alpha} a_\beta a_\gamma y_{\alpha+\beta+\gamma} = \sum_{\alpha,\beta} H_{n_j}(f_j y)_{\alpha,\beta} a_\alpha a_\beta. \tag{16.12}$$

That is, (16.10) holds if and only if the matrix $H_{n_j}(f_j y)$ is positive semidefinite. Let $H(\mathfrak{f})(y)$ denote the block diagonal matrix with diagonal blocks $H_{n_0}(f_0 y), H_{n_1}(f_1 y)$, $\ldots, H_{n_k}(f_k y)$. Then, by the preceding, condition (16.10) is satisfied for all $j = 0, \ldots, k$ if and only if the matrix $H(\mathfrak{f})(y)$ is positive semidefinite.

Note that $H(\mathfrak{f})(y)$ is a symmetric $N \times N$-matrix, where $N := \sum_{j=0}^k \binom{d+n_j}{n_j}$. By (16.11), all matrix entries of $H(\mathfrak{f})(y)$ are linear functions with real coefficients of the variables y_α, where $\alpha \in \mathbb{N}_0^d$, $|\alpha| \leq n$. Hence, inserting the condition $y_0 = 1$, there are (constant!) real symmetric $N \times N$-matrices A_α such that

$$H(\mathfrak{f})(y) = A_0 + \sum_{\alpha \in \mathbb{N}_0^d, 0 < |\alpha| \leq n} y_\alpha A_\alpha. \tag{16.13}$$

We write the polynomial p as $p(x) = \sum_\alpha p_\alpha x^\alpha$. Then $L(p) = p_0 + \sum_{\alpha \neq 0} p_\alpha y_\alpha$. By (16.8), p_n^{mom} is the infimum of the function $L(p)$ in the $M := \binom{d+n}{n} - 1$ variables y_α, where $0 < |\alpha| \leq n$, subject to the condition that the matrix $H(\mathfrak{f})(y)$ given by (16.13) is positive semidefinite.

Summarizing, the relaxation (16.8) leads to the *primal semidefinite program*

$$p_n^{\text{mom}} - p_0 = \inf_{(y_\alpha) \in \mathbb{R}^M} \left\{ \sum_{0 < |\alpha| \leq n} p_\alpha y_\alpha : A_0 + \sum_{0 < |\alpha| \leq n} y_\alpha A_\alpha \succeq 0 \right\}. \tag{16.14}$$

Our next aim is to show that (16.9) yields the corresponding dual program. Suppose that $\lambda \in \mathbb{R}$ and $p - \lambda \in Q(f)_n$, that is,

$$p - \lambda = \sum_{j=0}^{k} f_j \sigma_j, \quad \text{where} \quad \sigma_j \in \sum \mathbb{R}_d[x]_{n_j}^2. \tag{16.15}$$

By Proposition 13.2, $\sigma_j \in \sum \mathbb{R}_d[x]_{n_j}^2$ if and only if there is a positive semidefinite matrix $Z(j) = (Z(j)_{\alpha,\beta})_{|\alpha|,|\beta| \le n_j}$ of type $\binom{d+n_j}{n_j} \times \binom{d+n_j}{n_j}$ such that

$$\sigma_j(x) = \sum_{|\alpha|,|\beta| \le n_j} Z(j)_{\alpha,\beta} x^{\alpha+\beta}.$$

Let Z be the block diagonal matrix of type $N \times N$ with blocks $Z(0), \ldots, Z(k)$. Clearly, $Z \succeq 0$ if and only if $Z(j) \succeq 0$ for $j = 0, \ldots, k$.

Recall that $f_j = \sum_\alpha f_{j,\alpha} x^\alpha$. From (16.13) it follows that the j-th diagonal block of the matrix A_α has the (β, γ)-matrix entry $\sum_\delta f_{j,\delta}$, where the summation is over all δ satisfying $\alpha = \beta + \gamma + \delta$. Hence, equating coefficients in (16.15) yields

$$p_0 - \lambda = \sum_{j=0}^{k} f_{j,0} Z(j)_{00} = \operatorname{Tr} A_0 Z = \langle A_0, Z \rangle,$$

$$p_\alpha = \sum_{j=0}^{k} \sum_{\beta+\gamma+\delta=\alpha} f_{j,\delta} Z(j)_{\beta,\gamma} = \operatorname{Tr} A_\alpha Z = \langle A_\alpha, Z \rangle, \quad \alpha \ne 0.$$

Thus, taking the supremum of λ in (16.9) is equivalent to taking the supremum of $\lambda - p_0 = -\langle A_0, Z \rangle$ subject to the conditions $p_\alpha = \langle A_\alpha, Z \rangle$, $\alpha \in \mathbb{N}_0^d$, $0 < |\alpha| \le n$.

By the preceding, the relaxation (16.9) leads to the corresponding *dual program*

$$p_n^{sos} - p_0 = \sup_{Z \in \operatorname{Sym}_N} \left\{ -\langle A_0, Z \rangle : Z \succeq 0, \ p_\alpha = \langle A_\alpha, Z \rangle \ \text{for} \ 0 < |\alpha| \le n \right\}.$$

$$\tag{16.16}$$

16.3 Polynomial Optimization with Constraints

Let us retain the notation from the preceding section. Some simple properties of the numbers (16.8) and (16.9) are given in the next lemma.

Lemma 16.5

(i) $p_n^{sos} \le p_{n+1}^{sos}$ and $p_n^{mom} \le p_{n+1}^{mom}$ for $n \in \mathbb{N}_0$.
(ii) $p_n^{sos} \le p_n^{mom} \le p^{min}$ for $n \in \mathbb{N}_0$.

Proof

(i) Obviously, $Q(\mathfrak{f})_n$ is a subspace of $Q(\mathfrak{f})_{n+1}$. Therefore, $p - \lambda \in Q(\mathfrak{f})_n$ implies that $p - \lambda \in Q(\mathfrak{f})_{n+1}$, so that $p_n^{sos} \leq p_{n+1}^{sos}$. Since the restrictions of functionals from $Q(\mathfrak{f})_{n+1}^*$ belong to $Q(\mathfrak{f})_n^*$, it follows that $p_n^{mom} \leq p_{n+1}^{mom}$.

(ii) Since polynomials of $Q(\mathfrak{f})$ are nonnegative on $\mathcal{K}(\mathfrak{f})$, each point evaluation at $x \in \mathcal{K}(\mathfrak{f})$ is in $Q(\mathfrak{f})_n^*$. Hence $p_n^{mom} \leq p(x)$ for all $x \in \mathcal{K}(\mathfrak{f})$. This implies that $p_n^{mom} \leq p^{min}$.

Let $L \in Q(\mathfrak{f})_n^*$. If $p - \lambda \in Q(\mathfrak{f})_n$, then $L(p - \lambda) = L(p) - \lambda L(1) = L(p) - \lambda \geq 0$, that is, $\lambda \leq L(p)$. Taking the supremum over λ and the infimum over L we get $p_n^{sos} \leq p_n^{mom}$. □

As an application of the Archimedean Positivstellensatz we show that for an Archimedean module $Q(\mathfrak{f})$ both relaxations converge to the minimum of p.

Theorem 16.6 *Suppose that the quadratic module $Q(\mathfrak{f})$ is Archimedean. Then the set $\mathcal{K}(\mathfrak{f})$ is compact, so p attains its minimum over $\mathcal{K}(\mathfrak{f})$, and we have*

$$\lim_{n \to \infty} p_n^{sos} = \lim_{n \to \infty} p_n^{mom} = p^{min}. \tag{16.17}$$

Proof By Corollary 12.9, the semi-algebraic set $\mathcal{K}(\mathfrak{f})$ is compact. Fix $\lambda \in \mathbb{R}$ such that $\lambda < p^{min}$. Then $p(x) - \lambda > 0$ on $\mathcal{K}(\mathfrak{f})$ and hence $p - \lambda \in Q(\mathfrak{f})$ by the Archimedean Positivstellensatz (Theorem 12.36(i)), that is, $p - \lambda = \sum_j f_j \sigma_j$ for some elements $\sigma_j \in \sum \mathbb{R}_d[x]^2$. We choose $n \in \mathbb{N}$ such that $n \geq \deg(f_j \sigma_j)$ for all $j = 0, \ldots, k$. Then we have $p - \lambda = \sum_j f_j \sigma_j \in Q(\mathfrak{f})_n$ and hence $p_n^{sos} \geq \lambda$ by the definition (16.9) of p_n^{sos}. Since $\lambda < p^{min}$ was arbitrary and we have $p_n^{sos} \leq p_n^{mom} \leq p^{min}$ and $p_n^{sos} \leq p_{n+1}^{sos}$ by Lemma 16.5, this implies (16.17). □

The following propositions describe two interesting situations. The first shows that the supremum in (16.16) is attained if $\mathcal{K}(\mathfrak{f})$ has interior points and the second says that we have finite convergence if the quadratic module $Q(\mathfrak{f})$ is stable.

Proposition 16.7 *Suppose that $\mathcal{K}(\mathfrak{f})$ has a nonempty interior. Then $p_n^{sos} = p_n^{mom}$ for $n \in \mathbb{N}$. If $p_n^{mom} > -\infty$, then the supremum in (16.16) is a maximum.*

Proof Clearly, $Q(\mathfrak{f})_n$ is of the form (13.26). Therefore, since $\mathcal{K}(\mathfrak{f})$ has an interior point, it follows from Lemma 13.48 and Proposition 13.46 that $Q(\mathfrak{f})_n$ is closed in $\mathbb{R}_d[x]_n$.

Suppose that $\lambda \in \mathbb{R}$ and $\lambda > p_n^{sos}$. Then $p - \lambda \notin Q(\mathfrak{f})_n$ by the definition p_n^{sos}. Therefore, because $Q(\mathfrak{f})_n$ is a *closed* convex set in $\mathbb{R}_d[x]_n$, Theorem A.26(ii) applies, so there exists a linear functional L on $\mathbb{R}_d[x]_n$ such that $L(p - \lambda) < 0$ and $L(g) \geq 0$ for all $g \in Q(\mathfrak{f})_n$. Pick $x \in \mathcal{K}(\mathfrak{f})$ and define $L_\varepsilon(g) = L(g) + \varepsilon g(x)$ for $g \in \mathbb{R}_d[x]_n$ and $\varepsilon > 0$. Then, $L_\varepsilon \geq 0$ on $Q(\mathfrak{f})_n$ and $L_\varepsilon(p - \lambda) < 0$ if $\varepsilon > 0$ is sufficiently small. Further, $L_\varepsilon(1) \geq \varepsilon > 0$. Hence, upon replacing L_ε by $L_\varepsilon(1)^{-1} L_\varepsilon$, we can assume that $L_\varepsilon(1) = 1$. Then $L_\varepsilon \in Q(\mathfrak{f})_n^*$ and hence

$$p_n^{mom} \leq L_\varepsilon(p) = L_\varepsilon(p - \lambda) + L_\varepsilon(\lambda) < L_\varepsilon(\lambda) = \lambda.$$

Thus we have shown that $\lambda > p_n^{sos}$ implies $\lambda > p_n^{mom}$. Therefore, since $p_n^{sos} \leq p_n^{mom}$ by Lemma 16.5(ii), the equality $p_n^{sos} = p_n^{mom}$ holds.

Assume now in addition that $p_n^{mom} > -\infty$. The set $\mathcal{K}(\mathfrak{f})$ has an interior point, so it contains an open ball U. Define $L(g) = \int_U g(x)dx$ for $g \in \mathbb{R}_d[\underline{x}]_n$, where dx means the Lebesgue integration on $\mathbb{R}^d$. Since $U \subseteq \mathcal{K}(\mathfrak{f})$, we have $L \in Q(\mathfrak{f})_n^*$. Let $q = \sum_\alpha a_\alpha x^\alpha \in \mathbb{R}_d[\underline{x}]_{n_j}$, $q \neq 0$. Since $f_j \neq 0$, hence $f_j q^2 \neq 0$, and U is open, using (16.12) we obtain

$$\sum_{\alpha,\beta} H_{n_j}(f_j y)_{\alpha,\beta} a_\alpha a_\beta = L(f_j q^2) = \int_U (f_j q^2)(x)\, dx > 0. \qquad (16.18)$$

Thus $H_{n_j}(y) \succ 0$ for each j, so that $H(\mathfrak{f})(y) \succ 0$. This means that the semidefinite program (16.14) is strictly feasible. Therefore, by Proposition 16.2(i), its dual program (16.16) attains its maximum.   □

Proposition 16.8 *Suppose that the quadratic module $Q(\mathfrak{f})$ is stable. Then there exists an $n_0 \in \mathbb{N}_0$, depending only on $\deg(p)$, such that $p_n^{sos} = p_{n_0}^{sos}$ for all $n \geq n_0$.*

Proof We use the characterization of stability given by condition $(*)$ in Sect. 13.8. It says that for any $n \in \mathbb{N}_0$ there exists a number $\mathfrak{l}(n) \in \mathbb{N}_0$ such that each $q \in Q(\mathfrak{f})_n$ can be represented as $q = \sum_j f_j \sigma_j$ with $\sigma_j \in \sum \mathbb{R}_d[\underline{x}]^2$ such that $\deg(f_j\sigma_j) \leq \mathfrak{l}(n)$. Recall that $f_1 = 1$. Set $n_0 := \max(\deg(p), \mathfrak{l}(\deg(p)))$.

Suppose that $n \geq n_0$. First let $p_n^{sos} = -\infty$. Then, by definition, there is no $\lambda \in \mathbb{R}$ such that $p - \lambda \in Q(\mathfrak{f})_n$. Hence there is no $\lambda \in \mathbb{R}$ such that $p - \lambda \in Q(\mathfrak{f})_{n_0}$, so that $p_{n_0}^{sos} = -\infty$. Now assume that $p_{n_0}^{sos} > -\infty$. Let λ be a real number such that $\lambda < p_n^{sos}$. Then $p - \lambda \in Q(\mathfrak{f})_n$ by definition. Since $\deg(p - \lambda) \leq \deg(p)$, it follows from the definition of n_0 that $p - \lambda \in Q(\mathfrak{f})_{n_0}$, so that $p_{n_0}^{sos} \geq \lambda$. The preceding proves that $p_{n_0}^{sos} \geq p_n^{sos}$. It is obvious that $p_n^{sos} \geq p_{n_0}^{sos}$ for $n \geq n_0$. Thus, $p_n^{sos} = p_{n_0}^{sos}$.   □

Remark 16.9 Suppose that $\mathcal{K}(\mathfrak{f})$ is compact and $Q(\mathfrak{f})$ is the preorder $T(\mathfrak{f})$. Then $Q(\mathfrak{f})$ is Archimedean by Proposition 12.22 and has property (MP) by Theorem 12.25. Therefore, if $d \geq 2$ and $\mathcal{K}(\mathfrak{f})$ has interior points, Proposition 13.50 implies that $Q(\mathfrak{f})$ is not stable. Thus, in this case, Proposition 16.7 applies, but Proposition 16.8 does not. Likewise, if $\dim \mathcal{K}(\mathfrak{f}) \geq 2$, it can be shown that the assumptions of Theorem 16.6 and Proposition 16.8 exclude each other.   ∘

Remark 16.10 It is natural to look for conditions which imply finite convergence for the limit $\lim_{n\to\infty} p_n^{mom} = p^{min}$. The flat extension theory developed in Sect. 17.6 below yields such a result:

Let $m := \max\{1, \deg(f_j) : j = 0, \dots, k\}$ and $n > m$. If the infimum in (16.8) is attained at L and $\operatorname{rank} H_{n-m}(L) = \operatorname{rank} H_n(L)$, then $p_k^{mom} = p^{min}$ for all $k \geq n$.

To prove this we take Theorem 17.38 for granted and apply this result with n replaced by $n - m$ and $\tilde{L}$ by L. Since $L \in Q(\mathfrak{f})_n^*$, the positivity condition in Theorem 17.38(ii) is fulfilled. Therefore, by Theorem 17.38(i), L has an r-atomic representing measure μ, where $r = \operatorname{rank} H_{n-m}(L)$. Using that $L(p) = p_n^{mom}$

(because L attains the infimum (16.8)) and $L(1) = \int 1 d\mu = 1$ (by $L \in Q(\mathfrak{f})_n^*$) we derive

$$p_n^{\mathrm{mom}} \le p^{\min} = p^{\min} \int 1 d\mu \le \int p(x) d\mu = L(p) = p_n^{\mathrm{mom}}, \qquad (16.19)$$

so that $p_n^{\mathrm{mom}} = p^{\min}$ and hence $p_k^{\mathrm{mom}} = p^{\min}$ for $k \ge n$. Further, from (16.19) it follows easily that each atom of μ minimizes the polynomial p over the set $\mathcal{K}(\mathfrak{f})$. ∘

16.4 Global Optimization

In this section, we specialize the setup of Sect. 16.2 to the case $k = 1, f_1 = 1$. Then $\mathcal{K}(\mathfrak{f}) = \mathbb{R}^d$ and (16.7) becomes a global optimization problem on $\mathbb{R}^d$:

$$p^{\min} := \inf\{p(x): x \in \mathbb{R}^d\}. \qquad (16.20)$$

First we rewrite the two relaxations (16.8) and (16.9) in this case. For this we suppose that $n \ge \deg(p)$. Obviously, $Q(\mathfrak{f}) = \sum \mathbb{R}_d[\underline{x}]^2$, so $Q(\mathfrak{f})_n$ is the set of $\sigma \in \sum \mathbb{R}_d[\underline{x}]^2$ such that $\deg(\sigma) \le n$. Since we assumed that $n \ge \deg(p)$, it follows from Lemma 13.1 that $p - \lambda \in Q(\mathfrak{f})_n$ if and only if $p - \lambda \in \sum \mathbb{R}_d[\underline{x}]^2$. Thus

$$p_n^{\mathrm{sos}} = \sup\left\{\lambda \in \mathbb{R}: p - \lambda \in \sum \mathbb{R}_d[\underline{x}]^2\right\} \qquad (16.21)$$

does not depend on $n \ge \deg(p)$. Let us denote the number from (16.21) by p^{sos}.

Recall that $\lfloor \frac{n}{2} \rfloor$ is the largest integer not greater than $\frac{n}{2}$. Then $Q(\mathfrak{f})_n^*$ is the set of linear functionals L on $\mathbb{R}_d[\underline{x}]_n$ such that $L(1) = 1$ and $L(g^2) \ge 0$ for $g \in \mathbb{R}_d[\underline{x}]_{\lfloor \frac{n}{2} \rfloor}$. Set $y_\alpha = L(x^\alpha)$ for $|\alpha| \le n$. The matrix $H(\mathfrak{f})(y)$ from Sect. 16.2 is the truncated Hankel matrix $H_{\lfloor \frac{n}{2} \rfloor}(L)$ with entries $H_{\lfloor \frac{n}{2} \rfloor}(L)_{\alpha,\beta} = y_{\alpha+\beta}, |\alpha|, |\beta| \le \lfloor \frac{n}{2} \rfloor$. Clearly, a linear functional L on $\mathbb{R}_d[\underline{x}]_n$ is in $Q(\mathfrak{f})_n^*$ if and only if $H_{\lfloor \frac{n}{2} \rfloor}(L) \succeq 0$, so that

$$p_n^{\mathrm{mom}} = \inf\{L(p): H_{\lfloor \frac{n}{2} \rfloor}(L) \succeq 0, y_0 = 1\}.$$

From Proposition 16.7 and Lemma 16.5(ii) we conclude that

$$p^{\mathrm{sos}} \equiv p_n^{\mathrm{sos}} = p_n^{\mathrm{mom}} \le p^{\min} \quad \text{for} \quad n \ge \deg(p),$$

that is, both semidefinite programs (16.14) and (16.16) yield the same *lower bound* p^{sos} for the polynomial $p(x)$ on $\mathbb{R}^d$.

It is obvious that $p^{\min} = -\infty$ if $\deg(p)$ is odd. Now assume that $2k := \deg(p)$ is even. It may happen that this lower bound is trivial, that is, $p^{\mathrm{sos}} = -\infty$. The nontrivial case $p^{\mathrm{sos}} > -\infty$ holds if the semidefinite program (16.16) for (16.21) is feasible. This means that p is, upon adding a constant, a sum of squares in $\mathbb{R}_d[\underline{x}]$.

The following proposition gives a necessary condition and a sufficient condition for $p^{sos} > -\infty$. Let p_{2k} denote the homogeneous part of p of degree $2k = \deg(p)$, $k \in \mathbb{N}$.

Proposition 16.11 *If $p^{sos} > -\infty$, then we have $p_{2k} \in \sum \mathbb{R}_d[\underline{x}]^2$. Conversely, if*

$$p_{2k} - c(x_1^2 + \cdots + x_d^2)^k \in \sum \mathbb{R}_d[\underline{x}]^2 \tag{16.22}$$

for some constant $c > 0$, then $p^{sos} > -\infty$.

Proof First suppose that $p^{sos} > -\infty$. Then there exists a real number λ such that $(p - \lambda) \in \sum \mathbb{R}_d[\underline{x}]^2$, say $p - \lambda = \sum_i q_i^2$. Comparing the parts of degree $2k > 0$ on both sides it follows that $p_{2k} \in \sum \mathbb{R}_d[\underline{x}]^2$.

To prove the second assertion we assume that (16.22) holds. In this proof we write $f \geq g$ if $f - g \in \sum \mathbb{R}_d[\underline{x}]^2$ and abbreviate $\Delta := x_1^2 + \cdots + x_d^2$. Let

$$p(x) - p_{2k}(x) = \sum_{|\alpha| < 2k} a_\alpha x^\alpha.$$

For $\alpha \in \mathbb{N}_0^d$, $|\alpha| < 2k$, we write $x^\alpha = x^\beta x^\gamma$ with $|\beta| < k$ and $|\gamma| \leq k$. Since

$$2a_\alpha x^\alpha + |a_\alpha|(\varepsilon^{-2}x^{2\beta} + \varepsilon^2 x^{2\gamma}) = |a_\alpha|(\varepsilon^{-1}x^\beta + (\operatorname{sign} a_\alpha)\varepsilon x^\gamma)^2 \geq 0$$

for $\varepsilon > 0$, we conclude that there are real numbers b_α such that

$$p(x) \geq p_{2k}(x) - \varepsilon^2 \sum_{|\alpha|=k} b_\alpha x^{2\alpha} + \sum_{|\alpha|<k} b_\alpha x^{2\alpha}. \tag{16.23}$$

The multinomial theorem implies that $\Delta^k \geq x^{2\alpha}$ for $\alpha \in \mathbb{N}_0^d$, $|\alpha| = k$. Therefore, $\frac{c}{2}\Delta^k \geq \varepsilon^2 \sum_{|\alpha|=k} b_\alpha x^{2\alpha}$ for sufficiently small $\varepsilon > 0$. Since $p_{2k} \geq c\Delta^k$ by (16.22), it follows from (16.23) that

$$p(x) \geq \frac{c}{2}\Delta^k + \sum_{|\alpha|<k} b_\alpha x^{2\alpha}. \tag{16.24}$$

If $k = 1$, the sum in (16.24) is a constant λ, so that $(p(x) - \lambda) \in \sum \mathbb{R}_d[\underline{x}]^2$.

Now assume that $k \geq 2$. For $a > 0$ set $q(y) = a^k - 2^{-(k-1)}(a + y)^k + y^k$. It can be verified that the polynomial $q(x)$ is nonnegative on $\mathbb{R}_+$ (in fact, it attains its minimum 0 at $y = a$). Hence $q \in \sum \mathbb{R}[y]^2 + y\sum \mathbb{R}[y]^2$ by Proposition 3.2. Therefore, setting $y = \Delta$, we get $q(\Delta) \in \sum \mathbb{R}_d[\underline{x}]^2$. Expanding $(a + \Delta)^k$ yields

$$q(\Delta) = (1 - 2^{-(k-1)})(a^k + \Delta^k) - 2^{-(k-1)}\sum_{j=1}^{k-1}\binom{k}{j}a^j\Delta^{k-j} \geq 0. \tag{16.25}$$

Dividing by $1 - 2^{-(k-1)} > 0$, estimating Δ^k by (16.25), and using (16.24) we obtain

$$p(x) \geq \frac{c}{2(2^{(k-1)} - 1)} \sum_{j=1}^{k-1} \binom{k}{j} a^j \Delta^{k-j} + \sum_{1 \leq |\alpha| < k} b_\alpha x^{2\alpha} + b_0 - \frac{c}{2} a^k. \qquad (16.26)$$

If $|\alpha| < k$, then $\Delta^{|\alpha|} \geq x^{2\alpha}$ by the multinomial theorem. Hence, if a is sufficiently large, the first summand in (16.26) is greater than the second sum. Thus, we have $p(x) \geq b_0 - \frac{c}{2} a^k$. Setting $\lambda = b_0 - \frac{c}{2} a^k$, this yields $(p(x) - \lambda) \in \sum \mathbb{R}_d[x]^2$. $\qquad \square$

We illustrate the preceding with two examples.

Example 16.12 Consider the Motzkin polynomial $p = x_1^2 x_2^2 (x_1^2 + x_2^2 - 3) + 1$, see (13.6). Then $p^{\min} = 0$. Further, there is no $\lambda \in \mathbb{R}$ such that $(p - \lambda) \in \sum \mathbb{R}[x_1, x_2]^2$. Hence $p^{sos} = -\infty$. But $p_6 = x_1^4 x_2^2 + x_1^2 x_2^4 \in \sum \mathbb{R}[x_1, x_2]^2$. This shows the necessary condition given in Proposition 16.11 is not sufficient. $\qquad\circ$

Example 16.13 Let $p_\delta := x_1^2 x_2^2 (x_1^2 + x_2^2 - 3) + 1 + \delta(x_1^6 + x_2^6)$ for $\delta \leq 0$. It can be shown that $(p_\delta)^{\min} = \frac{\delta}{1+\delta}$. If $\delta > 0$, then $(p_\delta)_6 = x_1^4 x_2^2 + x_1^2 x_2^4 + \delta(x_1^6 + x_2^6)$ satisfies condition (16.22). Hence, by Proposition 16.11, we have $(p_\delta)^{sos} > -\infty$ for all $\delta > 0$.

Note that $\lim_{\delta \to +0}(p_\delta)^{\min} = 0$, but $\lim_{\delta \to +0}(p_\delta)^{sos} = -\infty$. (Indeed, otherwise there exist a $\lambda \in \mathbb{R}$ and a positive null sequence (δ_n) such that $p_{\delta_n} - \lambda$ is in $\sum \mathbb{R}[x_1, x_2]^2$ for all n. Since $\sum \mathbb{R}[x_1, x_2]^2$ is closed, then $p_0 - \lambda = \lim_n(p_{\delta_n} - \lambda)$ would be in $\sum \mathbb{R}[x_1, x_2]^2$, which contradicts Example 16.12.) $\qquad\circ$

16.5 Exercises

1. Show that the closed unit disc in $\mathbb{R}^2$ and the set $\{(x_1, x_2) \in \mathbb{R}^2 : x_1 x_2 \geq 1\}$ are spectrahedra.
2. Find a semidefinite program that is strictly feasible such that its dual program is not feasible. What can be said about p_* for such a program?
3. Consider the problem of minimizing x_1 subject to the constraints $x_1 \geq 0$ and $x_1 x_2 \geq 1$, where $(x_1, x_2) \in \mathbb{R}^2$.

   a. Write this as a semidefinite program.
   b. Show that $p_* = 0$ and the infimum in (16.1) is not attained.

4. (*A semidefinite program with positive finite duality gap [VB]*)
   Consider the problem of minimizing x_1 subject to the constraint

   $$\begin{pmatrix} 0 & x_1 & 0 \\ x_1 & x_2 & 0 \\ 0 & 0 & x_1 + 1 \end{pmatrix} \succeq 0.$$

   Show that this is a semidefinite program such that $p_* = 0$ and $p^* = -1$.

5. Let $\lambda_{\min}(B)$ the smallest eigenvalue of $B \in \mathrm{Sym}_n$. Describe $-\lambda_{\min}(B)$ by a semidefinite program and determine the dual program. Show that both programs are strictly feasible.

6. Prove by induction on d that $x_1^{2k} \cdots + x_d^{2k} - \frac{1}{2^{(d-1)(k-1)}}(x_1^2 + \cdots + x_d^2)^k \in \mathbb{R}_d[x]^2$ for $k \in \mathbb{N}$.

7. Find an example $p \in \sum \mathbb{R}[x_1, x_2]^2$ such that $p^{\min} = p^{\text{sos}} = 0$, but p_{2k} does not satisfy condition (16.22) in Proposition 16.11.

16.6 Notes

The semidefinite relaxations and corresponding results are due to J.B. Lasserre [Ls1] who first applied Positivstellensätze and moment methods in polynomial optimization. Proposition 16.11 is taken from Marshall [Ms2] and parts of our approach follow [Ms1]. Semidefinite programming is treated in [V] and [BN].

As noted in this chapter's introduction, we wanted to give only a small glimpse into the main ideas. There is now an extensive literature on this area. We refer to the books [Ms1, Ls2, BTB, AL] and the articles [Pa, Sw3, La2, Nie].

An important result on finite convergence (that is, $p_n^{\text{mom}} = p^{\min}$ for large n) of the relaxation (16.8) was obtained by J. Nie [Nie]. He proved finite convergence under the assumptions that the quadratic module is Archimedean and standard sufficient optimality conditions (linear independence of gradients, strict complementarity, second order sufficient condition) hold at each global minimizer. Since these conditions hold generically, Nie's theorem implies that (16.8) has finite convergence generically if the quadratic module is Archimedean.

Part IV
The Multidimensional Truncated Moment Problem

Chapter 17
Multidimensional Truncated Moment Problems: Existence

This chapter and the next two are devoted to the *truncated $\mathcal{K}$-moment problem*:

> *Given a subset* N *of* $\mathbb{N}_0^d$, *a sequence* $s = (s_\alpha)_{\alpha \in N}$, *and a closed subset* $\mathcal{K}$ *of* $\mathbb{R}^d$, *when does a Radon measure* μ *on* $\mathcal{K}$ *exist such that* $s_\alpha = \int x^\alpha \, d\mu$ *for all* $\alpha \in N$?

In the affirmative case we call s a truncated $\mathcal{K}$-moment sequence and the corresponding Riesz functional L_s a truncated $\mathcal{K}$-moment functional.

The existence results developed in this chapter provide important theoretical insights into the truncated moment problem, but for proving that a given sequence is a truncated $\mathcal{K}$-moment sequence their usefulness is limited. While the main results in previous chapters (for instance, Theorems 10.1, 10.2 and 12.25, 12.36) contain tractable criteria in terms of Hankel matrices, the solvability conditions of this chapter are difficult to verify. One reason is that useful descriptions of strictly positive polynomials up to a fixed degree $2n$ are missing. If the strict Positivstellensatz (Theorem 12.24) is applied, the degrees of the involved polynomials of the preordering exceed $2n$ in general (see e.g. [Ste2] for an elaborated example in dimension one).

In Sects. 17.1 and 17.2, we formulate the truncated $\mathcal{K}$-moment problem and its projective version and derive basic existence criteria in terms of positivity conditions (Theorems 17.3, 17.9, 17.10, and 17.15).

Hankel matrices are useful tools for the truncated moment problem. We introduce and study them in Sects. 17.3 and 17.4. In Sect. 17.5, we solve the full moment problem on $\mathbb{R}_d[\underline{x}]$ with finite rank Hankel matrix. The positive semidefiniteness of the Hankel matrix is not sufficient for being a truncated moment functional, but combined with a flatness condition it is. This is the flat extension theorem of Curto and Fialkov (Theorems 17.35 and 17.36). It is stated and discussed in Sect. 17.6 and proved in Sect. 17.7.

In this chapter, if not stated otherwise, N denotes a nonempty (finite or infinite) subset of $\mathbb{N}_0^d$, $\mathcal{A} = \{x^\alpha : \alpha \in N\}$ is the set of monomials, $\mathsf{A} = \mathrm{Lin}\,\{x^\alpha : \alpha \in N\}$ is their linear span, and $\mathcal{K}$ is a **closed** subset of $\mathbb{R}^d$. In Section 17.2, $\mathcal{K}$ also denotes a

© Springer International Publishing AG 2017
K. Schmüdgen, *The Moment Problem*, Graduate Texts in Mathematics 277,
DOI 10.1007/978-3-319-64546-9_17

closed subset of the projective space $\mathbb{P}^d(\mathbb{R})$. Further, we assume the following:

$$\textbf{There exists an element } e \in \mathsf{A} \textbf{ such that } e(x) \geq 1 \textbf{ for } x \in \mathcal{K}. \tag{17.1}$$

There are at least two important special cases where condition (17.1) is satisfied. First, if $0 \in \mathsf{N}$, then (17.1) holds with $e(x) = 1 \in \mathsf{A}$. The second case is when A is the vector space of homogeneous polynomials of degree $2n$ and $\mathcal{K}$ is a subset of the unit sphere S^{d-1}; then $e(x) := (x_1^2 + \cdots + x_d^2)^n \in \mathsf{A}$ and $e(x) = 1$ on $\mathcal{K}$.

17.1 The Truncated $\mathcal{K}$-Moment Problem and Existence Criteria

First we recall two standard notations. For a real sequence $s = (s_\alpha)_{\alpha \in \mathsf{N}}$ the associated Riesz functional is the real-valued linear functional L_s on A given by $L(x^\alpha) = s_\alpha,\ \alpha \in \mathsf{N}$. For a measure $\mu \in M_+(\mathbb{R}^d)$ such that $\mathsf{A} \subseteq \mathcal{L}^1(\mathbb{R}^d, \mu)$, L^μ denotes the linear funtional on A defined by

$$L^\mu(p) = \int_{\mathbb{R}^d} p(x)\, d\mu(x), \quad p \in \mathsf{A}. \tag{17.2}$$

Definition 17.1 A real sequence $s = (s_\alpha)_{\alpha \in \mathsf{N}}$ is called a *truncated $\mathcal{K}$-moment sequence* if there exists a measure $\mu \in M_+(\mathbb{R}^d)$ supported on $\mathcal{K}$ such that

$$x^\alpha \in \mathcal{L}^1(\mathbb{R}^d, \mu) \quad \text{and} \quad s_\alpha = \int_{\mathbb{R}^d} x^\alpha\, d\mu(x) \quad \text{for } \alpha \in \mathsf{N}. \tag{17.3}$$

A real-valued linear functional L on A is a *truncated $\mathcal{K}$-moment functional* if there is a measure $\mu \in M_+(\mathbb{R}^d)$ supported on $\mathcal{K}$ such that $L = L^\mu$, that is,

$$p \in \mathcal{L}^1(\mathbb{R}^d, \mu) \quad \text{and} \quad L(p) = \int_{\mathbb{R}^d} p(x)\, d\mu(x) \quad \text{for } p \in \mathsf{A}. \tag{17.4}$$

Each such measure μ is called a *representing measure* of L; the set of representing measures is denoted by $\mathcal{M}_{L,\mathcal{K}}$.

For $\mathcal{K} = \mathbb{R}^d$ we call a truncated $\mathcal{K}$-moment sequence simply a *truncated moment sequence* and a truncated $\mathcal{K}$-moment functional a *truncated moment functional*, and we denote the set of representing measures by $\mathcal{M}_L$.

Obviously, s satisfies (17.3) if and only if L_s does (17.4). That is, s is a truncated $\mathcal{K}$-moment sequence if and only if L_s is a truncated $\mathcal{K}$-moment functional. We often prefer to work with functionals rather than sequences.

Note that each measure μ satisfing (17.2), (17.3), or (17.4) is *finite*. Indeed, using the element $e \in$ A from condition (17.1) we obtain

$$\mu(\mathbb{R}^d) = \int 1 \, d\mu \le \int e \, d\mu = L(e) < \infty.$$

Thus, the *truncated $\mathcal{K}$-moment problem* asks: when is a sequence $s = (s_\alpha)_{\alpha \in \mathsf{N}}$ a truncated $\mathcal{K}$-moment sequence, or equivalently, when is a linear functional L on A a truncated $\mathcal{K}$-moment functional? Equations (17.3) and (17.4) are equivalent versions of the truncated $\mathcal{K}$-moment problem.

In the special case $\mathsf{N} = \mathbb{N}_0^d, \mathsf{A} = \mathbb{R}_d[x]$ we obtain the (full) multidimensional $\mathcal{K}$-moment problem. But here our main emphasis will be on *finite* sets N.

Let us relate the truncated $\mathcal{K}$-moment problem to the moment problem on a finite-dimensional space E treated in Sect. 1.2. We set $\mathcal{X} = \mathcal{K}$ and consider the linear subspace $E := \mathsf{A} \lceil \mathcal{K}$ of $C(\mathcal{K}; \mathbb{R})$ spanned by the restrictions of functions $f \lceil \mathcal{K}, f \in \mathsf{A}$. That is, E is the quotient vector space of A by the equivalence relation "$f \sim g$ if and only if $f(x) = g(x)$ for $x \in \mathcal{K}$".

Obviously, if $\tilde{L}$ is a linear functional on E, then $L(f) := \tilde{L}(f \lceil \mathcal{K}), f \in \mathsf{A}$, defines a linear functional L on A such that

$$L(f) = 0 \quad \text{for } f \in \mathsf{A}, \ f \lceil \mathcal{K} = 0. \tag{17.5}$$

Conversely, if L is a linear functional on A satisfying (17.5), then there is a well-defined (!) linear functional $\tilde{L}$ on E given by

$$\tilde{L}(f \lceil \mathcal{K}) := L(f), \quad f \in \mathsf{A}, \tag{17.6}$$

and it is clear that L is a truncated $\mathcal{K}$-moment functional according to Definition 17.1 if and only if $\tilde{L}$ is a moment functional on E by Definition 1.1. Further, it is obvious that L and $\tilde{L}$ have the same representing measures. This correspondence can be used to restate results from Sect. 1.2 in the present setting. However, one has to be careful: While $f \in$ A is only zero if f is the null polynomial, $g \in E$ is zero if and only if $g(x) = 0$ for all $x \in \mathcal{X}$. This occasionally leads to slight modifications in the formulations of results.

There are two important cases where (17.5) is satisfied. First, if $f \lceil \mathcal{K} = 0$ implies $f = 0$; this happens if $\mathcal{K}$ has a nonempty interior in $\mathbb{R}^d$. Secondly, by Lemma 17.4, if condition (17.8) holds, in particular, if L is a truncated $\mathcal{K}$-moment functional.

Theorem 17.2 *If the set N is finite, then each truncated $\mathcal{K}$-moment functional L on A has a k-atomic representing measure $\mu \in \mathcal{M}_{L,\mathcal{K}}$, where $k \le |\mathsf{N}| = \dim \mathsf{A}$ and all atoms of μ are in $\mathcal{K}$.*

Proof Since L is a truncated $\mathcal{K}$-moment functional, (17.5) is satisfied. Hence the assertion follows from Corollary 1.25, applied to $\tilde{L}$ and E.   □

The following standard notation is often used in the sequel:

$$\text{Pos}(\mathsf{A}, \mathcal{K}) := \{p \in \mathsf{A} : p(x) \geq 0 \quad \text{for} \quad x \in \mathcal{K}\}. \tag{17.7}$$

Theorem 17.3 *Suppose that $\mathcal{K}$ is a compact subset of $\mathbb{R}^d$. A linear functional L on A is a truncated $\mathcal{K}$-moment functional if and only if*

$$L(p) \geq 0 \quad \text{for all} \quad p \in \text{Pos}(\mathsf{A}, \mathcal{K}). \tag{17.8}$$

In this case the set $\mathcal{M}_{L,\mathcal{K}}$ of representing measures is vaguely compact.

The following simple fact is used several times.

Lemma 17.4 *If a functional L on A satisfies (17.8) (for instance, if L is a truncated $\mathcal{K}$-moment functional), then (17.5) holds, so the functional $\tilde{L}$ on E given by (17.6) is well-defined.*

Proof Let $f \in \mathsf{A}$ and $f \lceil \mathcal{K} = 0$. Then $\pm f \in \text{Pos}(\mathsf{A}, \mathcal{K})$ and hence $L(\pm f) \geq 0$ by (17.8), so that $L(f) = 0$. This means that (17.5) is satisfied.   □

Proof of Theorem 17.3. The necessity of condition (17.8) is obvious. Conversely, suppose that (17.8) holds. Then the functional $\tilde{L}$ is well-defined by Lemma 17.4. Since $\mathcal{X} := \mathcal{K}$ is compact and condition (17.1) is assumed, it follows from Proposition 1.9, that $\tilde{L}$ on E, hence L on A, has a representing measure with support in $\mathcal{X} = \mathcal{K}$.

We prove that the set $\mathcal{M}_{L,\mathcal{K}} = \mathcal{M}_{\tilde{L}}$ is vaguely compact. Let $f \in C_c(\mathcal{K}; \mathbb{R})$. Then there is a constant M_f such that $|f(x)| \leq M_f e(x)$ for $x \in \mathcal{K}$ and therefore

$$\sup_{\mu \in \mathcal{M}_{L,\mathcal{K}}} \left| \int_{\mathcal{K}} f \, d\mu \right| \leq M_f \sup_{\mu \in \mathcal{M}_{L,\mathcal{K}}} \int_{\mathcal{K}} e \, d\mu = M_f L(e) = M_f \tilde{L}(e \lceil \mathcal{X}) < \infty.$$

Hence, $\mathcal{M}_{\tilde{L}} = \mathcal{M}_{L,\mathcal{K}}$ is relatively vaguely compact by Theorem A.6.

Let $(\mu_i)_{i \in I}$ be a net of measures $\mu_i \in \mathcal{M}_{L,\mathcal{K}}$ converging vaguely to $\mu \in M_+(\mathcal{K})$. Since $\mathcal{K}$ is compact, $\mathsf{A} \lceil \mathcal{K} \subseteq C_c(\mathcal{K}; \mathbb{R})$ and hence

$$\int_{\mathcal{K}} f \, d\mu = \lim_i \int_{\mathcal{K}} f \, d\mu_i = \lim_i L(f) = L(f) \quad \text{for } f \in \mathsf{A},$$

so that $\mu \in \mathcal{M}_{L,\mathcal{K}}$. This shows that $\mathcal{M}_{L,\mathcal{K}}$ is vaguely closed. Being relatively vaguely compact and vaguely closed, $\mathcal{M}_{L,\mathcal{K}}$ is vaguely compact.   □

If the set $\mathcal{K}$ is only closed and not compact, A is not necessarily an adapted subspace of $C(\mathcal{K}, \mathbb{R})$ and in contrast to Haviland's theorem 1.12 condition (17.8) is not sufficient for being a truncated $\mathcal{K}$-moment functional. A simple counterexample in dimension one was given in Example 9.30. Counterparts of Theorem 17.3 for closed sets $\mathcal{K}$ will be obtained in the next section (Theorems 17.13 and 17.15).

For $m \in \mathbb{N}$ we abbreviate

$$\mathbb{N}_{0,m}^d := \{\alpha = (\alpha_1, \ldots, \alpha_d) \in \mathbb{N}_0^d : |\alpha| := \alpha_1 + \cdots + \alpha_d \leq m\}. \qquad (17.9)$$

Recall that $\mathbb{R}_d[x]_m$ are the polynomials $p \in \mathbb{R}[x_1, \ldots, x_d]$ such that $\deg(p) \leq m$.

Our main guiding example for the preceding setup is the following. The reader might always think of this important special case. Let $n \in \mathbb{N}$ and set $\mathsf{N} := \mathbb{N}_{0,2n}^d$. Then $\mathsf{A} = \mathbb{R}_d[x]_{2n}$ and Definition 17.1 has the following form.

Definition 17.5 A sequence $s = (s_\alpha)_{\alpha \in \mathbb{N}_{0,2n}^d}$ is a *truncated $\mathcal{K}$-moment sequence* if there is a measure $\mu \in M_+(\mathbb{R}^d)$ supported on $\mathcal{K}$ such that x^α is μ-integrable and

$$s_\alpha = \int_{\mathbb{R}^d} x^\alpha \, d\mu(x) \quad \text{for } \alpha \in \mathbb{N}_{0,2n}^d.$$

A linear functional L on $\mathbb{R}_d[x]_{2n}$ is a *truncated $\mathcal{K}$-moment functional* if there exists a measure $\mu \in M_+(\mathbb{R}^d)$ with support in $\mathcal{K}$ such that p is μ-integrable and

$$L(p) = \int_{\mathbb{R}^d} p(x) \, d\mu(x) \quad \text{for } p \in \mathbb{R}_d[x]_{2n}.$$

The next theorem restates Theorems 17.2 and 17.3 in this special case. Put

$$\mathrm{Pos}(\mathcal{K})_{2n} := \{p \in \mathbb{R}_d[x]_{2n} : p(x) \geq 0 \quad \text{for} \quad x \in \mathcal{K}\}. \qquad (17.10)$$

Theorem 17.6 *Suppose that $\mathcal{K}$ is a compact subset of $\mathbb{R}^d$. A linear functional L on $\mathbb{R}_d[x]_{2n}$ is a truncated $\mathcal{K}$-moment functional if and only if*

$$L(p) \geq 0 \quad \text{for all} \quad p \in \mathrm{Pos}(\mathcal{K})_{2n}. \qquad (17.11)$$

In this case, $\mathcal{M}_{L,\mathcal{K}}$ is a vaguely compact subset of $M_+(\mathcal{K})$ and L has a k-atomic representing measure, where $k \leq \binom{2n+d}{d}$ and all atoms of μ are in $\mathcal{K}$.

The next result is *Stochel's theorem*. It says that solving the *truncated $\mathcal{K}$-moment problem* on $\mathbb{R}_d[x]_{2n}$ for all $n \in \mathbb{N}$ leads to a solution of the *full $\mathcal{K}$-moment problem*.

Theorem 17.7 *Let $\mathcal{K}$ be a closed subset of $\mathbb{R}^d$ and L a linear functional on $\mathbb{R}_d[x]$. Suppose that for each $n \in \mathbb{N}$ the restriction $L_n := L \lceil \mathbb{R}_d[x]_{2n}$ is a truncated $\mathcal{K}$-moment functional on $\mathbb{R}_d[x]_{2n}$, that is, there exists a measure $\mu_n \in M_+(\mathbb{R}^d)$ supported on $\mathcal{K}$ such that*

$$L_n(p) \equiv L(p) = \int p(x) \, d\mu_n(x) \quad \text{for } p \in \mathbb{R}_d[x]_{2n}. \qquad (17.12)$$

Then L is a $\mathcal{K}$-moment functional. Further, L has a representing measure μ which is the limit of a subsequence $(\mu_{n_k})_{k \in \mathbb{N}}$ in the vague convergence of $M_+(\mathcal{K})$.

Proof Apply Theorem 1.20 to $\mathcal{X} = \mathcal{K}, E = \mathsf{A}\lceil\mathcal{K}, E_n = \mathbb{R}_d[x]_{2n}$. $\qquad\qquad\square$

The following corollary rephrases Theorem 17.7 in terms of sequences.

Corollary 17.8 *Let* $s = (s_\alpha)_{\alpha\in\mathbb{N}_0^d}$ *be a real multisequence. If for each* $n \in \mathbb{N}$ *the truncation* $s^{(n)} := (s_\alpha)_{\alpha\in\mathbb{N}_{0,2n}^d}$ *has a representing measure supported on* $\mathcal{K}$*, the sequence s does as well.*

17.2 The Truncated Moment Problem on Projective Space

In this section, we study truncated moment problems on the real projective space $\mathbb{P}^d(\mathbb{R})$ and apply this to the truncated $\mathcal{K}$-moment problem for *closed* sets in $\mathbb{R}^d$.

Let $\mathbb{P}^d(\mathbb{R})$ be the d-dimensional real projective space. The points of $\mathbb{P}^d(\mathbb{R})$ are equivalence classes of $(d+1)$-tuples $(t_0,\dots,t_d) \neq 0$ of reals under the equivalence relation

$$(t_0,\dots,t_d) \sim (t_0',\dots,t_d') \quad\text{if}\quad (t_0,\dots,t_d) = \lambda(t_0',\dots,_d') \tag{17.13}$$

for some $\lambda \neq 0$. The equivalence class is denoted by $[t_0 : \cdots : t_d]$ and $t_0,\dots,t_d$ are called homogeneous coordinates of the point $t = [t_0 : \cdots : t_d] \in \mathbb{P}^d(\mathbb{R})$. That is, $\mathbb{P}^d(\mathbb{R}) = (\mathbb{R}^{d+1}\setminus\{0\})/\sim$. The map

$$\varphi : (t_1,\dots,t_d) \mapsto [1 : t_1 : \cdots : t_d] \tag{17.14}$$

is an injection of $\mathbb{R}^d$ into $\mathbb{P}^d(\mathbb{R})$. We identify $t \in \mathbb{R}^d$ with its image $\varphi(t)$ in $\mathbb{P}^d(\mathbb{R})$. In this manner $\mathbb{R}^d$ becomes a subset of $\mathbb{P}^d(\mathbb{R})$. The complement of $\mathbb{R}^d$ in $\mathbb{P}^d(\mathbb{R})$ is the hyperplane $\mathbb{H}_\infty^d = \{[0 : t_1 : \cdots : t_d] \in \mathbb{P}^d(\mathbb{R})\}$ at infinity. Note that $\mathbb{H}_\infty^d$ can be identified with $\mathbb{P}^{d-1}(\mathbb{R})$.

We denote by $\mathcal{H}_{d+1,2n}$ the *homogeneous* polynomials from $\mathbb{R}[x_0,x_1,\dots,x_d]$ of degree $2n$. Recall that $\mathbb{R}_d[x]_{2n}$ are the polynomials from $\mathbb{R}[x_1,\dots,x_d]$ of degree at most $2n$. It is not difficult to verify that the map

$$\phi : p(x_0,\dots,x_d) \mapsto \hat{p}(x_1,\dots,x_d) := p(1,x_1,\dots,x_d)$$

is a bijection of the vector spaces $\mathcal{H}_{d+1,2n}$ and $\mathbb{R}_d[x]_{2n}$ with inverse given by

$$\phi^{-1} : q(x_1,\dots,x_d) \mapsto \check{q}(x_0,\dots,x_d) := x_0^{2n}q\left(\frac{x_1}{x_0},\dots,\frac{x_n}{x_0}\right). \tag{17.15}$$

If $\deg(q) = 2n$, then $\check{q}$ is the homogenization of q. Clearly, the mappings Φ and Φ^{-1} preserve the positivity of polynomials on corresponding sets.

There is a unique topology on the space $\mathbb{P}^d(\mathbb{R})$ for which the maps

$$(t_0,\ldots,t_{j-1},t_{j+1},\ldots,t_d) \mapsto [t_0 : \cdots : t_{j-1} : 1 : t_{j+1} : \cdots : t_d], \quad j = 1,\ldots,d,$$

of $\mathbb{R}^d$ into $\mathbb{P}^d(\mathbb{R})$ are homeomorphisms. Then $\mathbb{P}^d(\mathbb{R})$ is a compact topological Hausdorff space and $\mathbb{R}^d$ is dense in $\mathbb{P}^d(\mathbb{R})$. In fact, $\mathbb{P}^d(\mathbb{R})$ is even a C^∞-manifold.

Each homogeneous polynomial $q \in \mathcal{H}_{d+1,2n}$ can be considered as a continuous function, denoted by $\tilde{q}$, on the projective space by

$$\tilde{q}(t) := \frac{q(t_0,\ldots,t_d)}{(t_0^2 + \cdots + t_d^2)^n}, \quad t = [t_0 : \cdots : t_d] \in \mathbb{P}^d(\mathbb{R}). \tag{17.16}$$

(Indeed, the fraction in (17.16) is invariant under the equivalence relation (17.13), so $\tilde{q}(t)$ is well-defined. It is easily verified that $\tilde{q}(t)$ is continuous on $\mathbb{P}^d(\mathbb{R})$.)

If we replace $\mathbb{R}^d$ by the projective space $\mathbb{P}^d(\mathbb{R})$ and consider *homogeneous* polynomials, then for each closed subset $\mathcal{K}$ of $\mathbb{P}^d(\mathbb{R})$ truncated $\mathcal{K}$-moment functionals can be defined almost verbatim in the same manner as for closed subsets of $\mathbb{R}^d$. We will restate the corresponding definition in Theorem 17.9.

To formulate Theorem 17.9 we assume that N is a nonempty subset of the set

$$\{\alpha = (\alpha_0,\ldots,\alpha_d) \in \mathbb{N}_0^{d+1} : \alpha_0 + \alpha_1 + \cdots + \alpha_d = 2n\}.$$

Then $A := \mathrm{Lin}\,\{x_0^\alpha : \alpha \in N\}$ is a subspace of the vector space $\mathcal{H}_{d+1,2n}$. Further, suppose that $\mathcal{K}$ is a closed subset of the projective space $\mathbb{P}^d(\mathbb{R})$ and define

$$\mathrm{Pos}(A,\mathcal{K}) = \{p \in A : \tilde{p}(x) \geq 0 \quad \text{for} \quad x \in \mathcal{K}\}.$$

Theorem 17.9 *Let L be a linear functional on* A. *Assume that there exists an* $e \in A$ *such that* $\tilde{e}(x) \geq 1$ *for* $x \in \mathcal{K}$. *Then L is a truncated $\mathcal{K}$-moment functional, that is, there exists a Radon measure μ on $\mathbb{P}^d(\mathbb{R})$ with support in $\mathcal{K}$ such that*

$$L(p) = \int_{\mathbb{P}^d(\mathbb{R})} \tilde{p}(t)\,d\mu(t) \quad \text{for} \quad p \in A, \tag{17.17}$$

if and only if

$$L(p) \geq 0 \quad \text{for all} \quad p \in \mathrm{Pos}(A,\mathcal{K}). \tag{17.18}$$

In this case, the set $\mathcal{M}_{L,\mathcal{K}}$ of such measures μ is compact in the vague topology of $M_+(\mathcal{K})$ and L has a k-atomic representing measure, where $k \leq |N| = \dim A$ and all atoms of μ are in $\mathcal{K}$.

Proof The proof is almost verbatim the same as the proof of Theorem 17.3. Note that the closed subset $\mathcal{K}$ of the compact space $\mathbb{P}^d(\mathbb{R})$ is compact. The assumption on e is needed in order to apply Proposition 1.26 with $\mathcal{X} = \mathcal{K}$. $\qquad\square$

It is well-known that the real projective space $\mathbb{P}^d(\mathbb{R})$ can be identified with the quotient space $S^d/\sim$ of the unit sphere

$$S^d = \{(x_0, \ldots, x_d) \in \mathbb{R}^{d+1} : x_0^2 + x_1^2 \cdots + x_d^2 = 1\}$$

under the equivalence relation "$\sim$" on S^d : "$x \sim y$ if and only if $x = y$ or $x = -y$". The space $S^d/\sim$ is obtained from S^d by identifying antipodal points. (The map $x \mapsto x\|x\|^{-1}$ gives a topological homeomorphism of $\mathbb{R}^{d+1}/\sim$ on $S^d/\sim$.) One could even take $S^d/\sim$ with the quotient topology as the definition of the space $\mathbb{P}^d(\mathbb{R})$.

Let S_+^d denote the set of points of S^d for which the first nonzero coordinate is *positive*. Clearly, there is a canonical bijection of $S^d/\sim$ on S_+^d. We can equip S_+^d with the compact topology induced from $S^d/\sim$ under this bijection and treat the truncated projective moment problem on S_+^d. This topology on S_+^d is different from the topology induced from S^d. But, since moment functionals on finite-dimensional spaces have finitely atomic representing measures, this does not cause any difficulty.

Now we avoid the use of the projective space $\mathbb{P}^d(\mathbb{R})$ and look directly for integral representations by measures on S^d. The next theorem restates the assertions of Theorem 17.9 in a slightly different form.

Theorem 17.10 *Let* A *be as above and let* $\mathcal{K}$ *be a closed subset of* S^d. *For a linear functional* $L \neq 0$ *on* A *the following statements are equivalent:*

(i) $L(p) \geq 0$ *for all* p *in* $\mathrm{Pos}(\mathsf{A}, \mathcal{K}) \equiv \{f \in \mathsf{A} : f(x) \geq 0 \text{ for } x \in \mathcal{K}\}$.

(ii) L *is a truncated* $\mathcal{K}$-*moment functional, that is, there exists a measure* $\mu \in M_+(S^d)$ *supported on* $\mathcal{K}$ *such that*

$$L(p) = \int_{S^d} p(x) \, d\mu(x) \quad for \quad p \in \mathsf{A}. \tag{17.19}$$

(iii) *There is a* k-*atomic measure* $\mu \in M_+(S^d)$, $k \leq \dim \mathsf{A}$, *with all atoms in* $\mathcal{K}$ *for which (17.19) holds, that is, there are points* $x_1, \ldots, x_k \in \mathcal{K}$ *and positive numbers* $c_1, \ldots, c_k$ *such that*

$$L(p) = \sum_{j=1}^{k} c_j p(x_j) \quad for \quad p \in \mathsf{A}. \tag{17.20}$$

Proof Obviously, (iii)$\rightarrow$(ii)$\rightarrow$(i). The main implication (i)$\rightarrow$(iii) follows from Proposition 1.26, applied to $\mathcal{X} = \mathcal{K}$ and the subspace $E := \mathsf{A}{\upharpoonright}\mathcal{X}$ of $C(\mathcal{X}; \mathbb{R})$. $\quad\square$

Remark 17.11 Obviously, the polynomial $e(x) := (x_0^2 + \cdots + x_d^2)^n$ satisfies $\tilde{e}(t) = 1$ for $t \in \mathbb{P}^d(\mathbb{R})$ and $e(x) = 1$ for $x \in S^d$. Therefore, if e is in A, it can be taken in Theorems 17.9 and 17.10 to satisfy assumption (17.1). $\quad\circ$

In the remaining part of this section we return to the truncated $\mathcal{K}$-moment problem for $\mathsf{A} = \mathbb{R}_d[\underline{x}]_{2n}$ and study the case of a *closed* subset $\mathcal{K}$ of $\mathbb{R}^d$.

We consider $\mathcal{K}$ as a subset of the projective space $\mathbb{P}^d(\mathbb{R})$ by the injection φ defined by (17.14). Let $\mathcal{K}^-$ denote the closure of $\mathcal{K}$ in $\mathbb{P}^d(\mathbb{R})$. Then $\mathcal{K}^-$ is the disjoint union of $\mathcal{K}$ and the intersection $\mathcal{K}_\infty$ of $\mathcal{K}^-$ with $\mathbb{H}^d_\infty$.

For $p \in \mathbb{R}_d[\underline{x}]_{2n}$ let p_{2n} denote its homogeneous part of degree $2n$ and put

$$\widetilde{p_{2n}}(t) := \frac{p_{2n}(t_1, \cdots, t_d)}{(t_1^2 + \cdots + t_d^2)^n} \quad \text{for} \quad t = (0 : t_1 : \cdots : t_d) \in \mathbb{H}^d_\infty \cong \mathbb{P}^{d-1}(\mathbb{R}).$$

Note that the fraction is a well-defined continuous function on $\mathbb{H}^d_\infty$.

Lemma 17.12 *Let μ_∞ be a Radon measure on $\mathbb{H}^d_\infty$ supported on $\mathcal{K}_\infty$. Then*

$$L_\infty(p) := \int_{\mathcal{K}_\infty} \widetilde{p_{2n}}(t)\, d\mu_\infty(t), \quad p \in \mathbb{R}_d[\underline{x}]_{2n}, \tag{17.21}$$

defines a linear functional L_∞ on $\mathbb{R}_d[\underline{x}]_{2n}$ such that $L_\infty(p) \geq 0$ for $p \in \mathrm{Pos}(\mathcal{K})_{2n}$.

Proof Let $p \in \mathrm{Pos}(\mathcal{K})_{2n}$. Since $\check{p}(1,x) = p(x) \geq 0$ for $x \in \mathcal{K}$, $\check{p} \geq 0$ on $\varphi(\mathcal{K}) \cong \mathcal{K}$. Therefore, $\check{p} \geq 0$ on the closure $\mathcal{K}^- = \mathcal{K} \cup \mathcal{K}_\infty$ of $\mathcal{K}$, so that $\check{p}(0,t) = p_{2n}(t) \geq 0$ and hence $\widetilde{p_{2n}}(t) \geq 0$ for $(0,t) \in \mathcal{K}_\infty$. Then $L_\infty(p) \geq 0$ by (17.21). □

This functional L_∞ satisfying condition (17.11) is the main new ingredient of the next theorem. The following result characterizes those linear functionals on $\mathbb{R}_d[\underline{x}]_{2n}$ for which the positivity condition (17.11) is fulfilled.

Theorem 17.13 *Let $\mathcal{K}$ be a closed subset of $\mathbb{R}^d$ and let L be a linear functional on $\mathbb{R}_d[\underline{x}]_{2n}$. Then L satisfies (17.11) if and only if there are Radon measures μ on $\mathbb{R}^d$ with support in $\mathcal{K}$ and μ_∞ on $\mathbb{H}^d_\infty$ with support in $\mathcal{K}_\infty$ such that*

$$L(p) = \int_{\mathcal{K}} p(t)\, d\mu(t) + \int_{\mathcal{K}_\infty} \widetilde{p_{2n}}(t)\, d\mu_\infty(t) \quad \text{for} \quad p \in \mathbb{R}_d[\underline{x}]_{2n}. \tag{17.22}$$

Proof Using Lemma 17.12 we conclude that both summands of (17.22) are nonnegative on $\mathrm{Pos}(\mathcal{K})_{2n}$, so L satisfies condition (17.11).

Conversely, assume that (17.11) holds. Define a linear functional $\check{L}$ on $\mathcal{H}_{d+1,2n}$ by

$$\check{L}(\check{p}) = L(p), \quad p \in \mathbb{R}_d[\underline{x}]_{2n}. \tag{17.23}$$

Suppose that $\widetilde{p} \geq 0$ on $\mathcal{K}^-$. Then we have $p \geq 0$ on $\mathcal{K}$ by (17.15) and (17.16) and hence $\check{L}(\check{p}) = L(p) \geq 0$ by (17.11). That is, Theorem 17.9 applies to the compact subset $\mathcal{X} := \mathcal{K}^-$ of $\mathbb{P}^d(\mathbb{R})$ and the functional $\check{L}$ on $E := \mathcal{H}_{d+1,2n}\lceil\mathcal{K}^-$. Therefore, there exists a Radon measure $\tilde{\mu}$ on $\mathbb{P}^d(\mathbb{R})$ supported on $\mathcal{K}^-$ such that

$$\check{L}(\check{p}) = \int_{\mathcal{K}^-} \widetilde{p}(t)\, d\tilde{\mu}, \quad p \in \mathbb{R}_d[\underline{x}]_{2n}. \tag{17.24}$$

Define Radon measures μ_∞ on $\mathbb{H}_\infty^d$ by $\mu_\infty(M) = \tilde{\mu}(M), M \subseteq \mathbb{H}_\infty^d$, and μ on $\mathbb{R}^d$ by

$$d\mu(t_1, \ldots, t_d) = (1 + t_1^2 + \cdots + t_d^2)^{-n} d\tilde{\mu}([1 : t_1 : \cdots : t_d]). \tag{17.25}$$

Since $\operatorname{supp} \tilde{\mu} \subseteq \mathcal{K}^-$, μ_∞ and μ are supported on $\mathcal{K}_\infty$ and $\mathcal{K}$, respectively.

First let $q \in \mathbb{R}_d[\underline{x}]_{2n-1}$. Then, by (17.15) and (17.16), we have

$$\widetilde{q}(t) = \frac{t_0^{2n} q\left(\frac{t_1}{t_0}, \ldots, \frac{t_d}{t_0}\right)}{(t_0^2 + \cdots + t_d^2)^n} = \frac{t_0 \, q_0(t_1, \ldots, t_d)}{(t_0^2 + \cdots + t_d^2)^n}, \quad t = [t_0 : \cdots : t_d] \in \mathbb{P}^d(\mathbb{R}),$$

for some polynomial q_0. Therefore, if $t \in \mathcal{K}_\infty$, then $t_0 = 0$, so that

$$\widetilde{q}(t) = 0 \quad \text{for} \quad t \in \mathcal{K}_\infty. \tag{17.26}$$

Now let $p \in \mathbb{R}_d[\underline{x}]_{2n}$. Then $q := p - p_{2n} \in \mathbb{R}_d[\underline{x}]_{2n-1}$. Again by (17.15) and (17.16),

$$\widetilde{p}(t) = \frac{p(t_1, \ldots, t_d)}{(1 + t_1^2 + \cdots + t_d^2)^n} \quad \text{for} \quad t = [1 : t_1 : \cdots : t_d] \in \mathcal{K}. \tag{17.27}$$

Hence, using the formulas (17.23)–(17.27) we derive

$$L(p) = \check{L}(\check{p}) = \int_{\mathcal{K}^-} \widetilde{p}(t) \, d\tilde{\mu}(t) = \int_{\mathcal{K}} \widetilde{p}(t) \, d\tilde{\mu}(t) + \int_{\mathcal{K}_\infty} \widetilde{p_{2n}}(t) \, d\tilde{\mu}(t) + \int_{\mathcal{K}_\infty} \widetilde{q}(t) \, d\tilde{\mu}(t)$$

$$= \int_{\mathcal{K}} \frac{p(t_1, \ldots, t_d)}{(1 + t_1^2 + \cdots + t_d^2)^n} \, d\tilde{\mu}([1 : t_1 : \cdots : t_d]) + \int_{\mathcal{K}_\infty} \widetilde{p_{2n}}(t) \, d\mu_\infty(t)$$

$$= \int_{\mathcal{K}} p(t) \, d\mu(t) + \int_{\mathcal{K}_\infty} \widetilde{p_{2n}}(t) \, d\mu_\infty(t),$$

which proves (17.22). $\qquad\qquad\qquad\qquad\qquad\qquad\qquad\qquad\qquad\qquad\qquad\qquad\qquad \square$

Remark 17.14

1. If the set $\mathcal{K}$ in Theorem 17.13 is compact, then $\mathcal{K}_\infty$ is empty, so the second summand in (17.22) does not occur and we obtain Theorem 17.6.
2. Theorem 17.13 explains why for noncompact sets $\mathcal{K}$ the positivity condition (17.11) is not sufficient for L to be a truncated $\mathcal{K}$-moment functional. The reason is that there may be a functional L_∞ given by some measure at "infinity".
3. Most of the results and notions developed in this chapter and the next have their counterparts for the truncated moment problem on $\mathbb{P}^d(\mathbb{R})$. We leave it to the reader to state the corresponding projective versions.
4. The truncated moment problem on a closed subset $\mathcal{K}$ of $\mathbb{P}^d(\mathbb{R})$ has several advantages. First, since $\mathbb{P}^d(\mathbb{R})$ and hence $\mathcal{K}$ is compact, truncated $\mathcal{K}$-moment

functionals are characterized by the positivity condition (17.18). As noted in Remark 2, for the truncated moment problem on $\mathbb{R}^d$ condition (17.18) is not sufficient. Secondly, homogeneous polynomials are more convenient to deal with and additional technical tools such as the apolar scalar product (see Sect. 19.1) are available.   ∘

Finally, we use the truncated moment problem on $\mathbb{P}^d(\mathbb{R})$ to derive the following result on the truncated moment problem for *closed* subsets of $\mathbb{R}^d$.

Theorem 17.15 *Let $\mathcal{K}$ be a closed subset of $\mathbb{R}^d$ and L_0 a linear functional on $\mathbb{R}_d[x]_{2n-2}, n \in \mathbb{N}$. Then L_0 is a truncated $\mathcal{K}$-moment functional on $\mathbb{R}_d[x]_{2n-2}$ if and only if L_0 admits an extension to a linear functional L on $\mathbb{R}_d[x]_{2n}$ such that*

$$L(p) \geq 0 \quad \text{for} \quad p \in \text{Pos}(\mathcal{K})_{2n}. \tag{17.28}$$

Proof Let L_0 be a truncated $\mathcal{K}$-moment functional. Then, by Theorem 17.2, L_0 has a finitely atomic representing measure μ. Clearly, $\mathbb{R}_d[x]_{2n} \subseteq \mathcal{L}^1(\mathcal{K}, \mu)$ and the functional L defined by $L(p) = \int f\, d\mu, p \in \mathbb{R}_d[x]_{2n}$, satisfies (17.28).

Conversely, assume that (17.28) holds. Then, by Theorem 17.13, L is of the form (17.22). If $p \in \mathbb{R}_d[x]_{2n-2}$, then $p_{2n} = 0$. Hence the second summand in (17.22) vanishes and we obtain $L_0(p) = \int_{\mathcal{K}} p(t)\, d\mu(t)$ for $p \in \mathbb{R}_d[x]_{2n-2}$.   □

17.3 Hankel Matrices

Recall that N is a nonempty subset of $\mathbb{N}_0^d$ and $A = \text{Lin}\{x^\alpha : \alpha \in N\}$. Throughout this section, we suppose that *L is a linear functional on the linear subspace*

$$A^2 := \text{Lin}\{pq : p, q \in A\} = \text{Lin}\{x^\beta : \beta \in N + N\}$$

of $\mathbb{R}_d[x]$. Note that in contrast to Sects. 17.1–17.2 we now consider functionals on A^2 rather than A.

The following definition contains the basic notions studied in this section.

Definition 17.16 The *Hankel matrix* of L is the symmetric matrix

$$H(L) = (h_{\alpha,\beta})_{\alpha,\beta \in N}, \quad \text{where } h_{\alpha,\beta} := L(x^{\alpha+\beta}), \ \alpha, \beta \in N.$$

The (cardinal) number $\text{rank}\, L := \text{rank}\, H(L)$ is called the *rank* of L. The *kernel* $\mathcal{N}_L$ and the real algebraic set $\mathcal{V}_L$ are defined by

$$\mathcal{N}_L = \{f \in A : L(fg) = 0 \text{ for all } g \in A\}, \tag{17.29}$$

$$\mathcal{V}_L = \{t \in \mathbb{R}^d : f(t) = 0 \text{ for all } f \in \mathcal{N}_L\}. \tag{17.30}$$

Hankel matrices are fundamental tools for the study of truncated moment problems. Many problems and properties of the functional L are easily translated into those for the matrix $H(L)$, see e.g. Proposition 17.17 below.

Further, we introduce a symmetric bilinear form $\langle \cdot, \cdot \rangle'$ on $\mathsf{A} \times \mathsf{A}$ by

$$\langle f, g \rangle' := L(fg), \quad f, g \in \mathsf{A}.$$

Equation (17.29) means that $\mathcal{N}_L$ is just *the kernel of the bilinear form* $\langle \cdot, \cdot \rangle'$. This is one reason for the importance of the vector space $\mathcal{N}_L$. We define a symmetric bilinear form $\langle \cdot, \cdot \rangle$ on the quotient space $\mathcal{D}_L := \mathsf{A}/\mathcal{N}_L$ by

$$\langle f + \mathcal{N}_L, g + \mathcal{N}_L \rangle := \langle f, g \rangle' = L(fg), \quad f, g \in \mathsf{A}. \tag{17.31}$$

Clearly, this bilinear form $\langle \cdot, \cdot \rangle$ is non-degenerate, that is, its kernel is trivial. Let us write $\tilde{f}$ for the equivalence class $f + \mathcal{N}_L$. Then, by (17.31) we have

$$L(fg) = \langle \tilde{f}, \tilde{g} \rangle, \quad f, g \in \mathsf{A}. \tag{17.32}$$

Since each element of A^2 is a sum of products fg with $f, g \in \mathsf{A}$, all numbers s_α, $\alpha \in \mathsf{N} + \mathsf{N}$, and hence the functional L can be recovered from the space $(\mathcal{D}_L, \langle \cdot, \cdot \rangle)$ by using Eq. (17.32).

Let us fix an ordering of the index set N; for instance, we can take the lexicographic ordering. For $f = \sum_{\alpha \in \mathsf{N}} f_\alpha x^\alpha \in \mathsf{A}$ we let $\vec{f} = (f_\alpha)^T$ denote the coefficient vector of f written as a column according to the ordering of N. Since $\vec{f}$ has only finitely many nonzero terms f_α, products such as $\vec{f}^T H(L)$ and $H(L)\vec{f}$ are well-defined. If the set N is finite, then $\vec{f} \in \mathbb{R}^{|\mathsf{N}|}$.

A number of basic facts on Hankel matrices are collected in the next proposition. In dimension one some assertions have been already noted in Lemma 9.20.

Proposition 17.17

(i) *For $f \in \mathsf{A}$ and $g \in \mathsf{A}$ we have*

$$L(fg) = \vec{f}^T H(L) \vec{g}. \tag{17.33}$$

(ii) *A polynomial $f \in \mathsf{A}$ belongs to $f \in \mathcal{N}_L$ if and only if $\vec{f} \in \ker H(L)$. In particular,* $\dim \mathcal{N}_L = \dim \ker H(L)$.

(iii) $\operatorname{rank} L \equiv \operatorname{rank} H(L) = \dim(\mathsf{A}/\mathcal{N}_L) \equiv \dim \mathcal{D}_L$.

(iv) *L is a positive functional (that is, $L(f^2) \geq 0$ for $f \in \mathsf{A}$ by Definition 2.2) if and only if the Hankel matrix $H(L)$ is positive semidefinite.*

(v) *If the functional L is positive, then $\mathcal{N}_L = \{f \in \mathsf{A} : L(f^2) = 0\}$.*

(vi) *If L is a truncated moment functional, then $\operatorname{supp} \mu \subseteq \mathcal{V}_L$ for each representing measure μ of L.*

Proof

(i) Let $f = \sum_{\alpha \in N} f_\alpha x^\alpha \in A$ and $g = \sum_{\alpha \in N} g_\alpha x^\alpha \in A$. Clearly, then we have $fg = \sum_{\alpha, \beta \in N} f_\alpha g_\beta x^{\alpha + \beta}$ and therefore

$$L(fg) = \sum_{\alpha, \beta \in N} f_\alpha g_\beta L(x^{\alpha + \beta}) = \sum_{\alpha, \beta \in N} h_{\alpha, \beta} f_\alpha g_\beta = (f_\alpha)^T H(L)(g_\beta) = \vec{f}^T H(L) \vec{g}.$$

(ii) follows at once by combining the definition (17.29) of $\mathcal{N}_L$ and (17.33).

(iii) Let us take $m \in \mathbb{N}$ column vectors

$$h^{(1)} = (h_{\alpha^{(1)}, \beta})_{\beta \in N}, \ldots, h^{(m)} = (h_{\alpha^{(m)}, \beta})_{\beta \in N}, \quad \text{where } \alpha^{(1)}, \ldots, \alpha^{(m)} \in N,$$

of the Hankel matrix $H(L)$ and consider a linear combination of these vectors with real coefficients $c_1, \ldots, c_m$. Put $p(x) = \sum_i c_i x^{\alpha^{(i)}}$. Then $p \in A$ and

$$\sum_{i=1}^m c_i h_{\alpha^{(i)}, \beta} = L\left(\sum_{i=1}^m c_i x^{\alpha^{(i)}} x^\beta \right) = L(p(x) x^\beta), \quad \beta \in N.$$

Hence this linear combination of $h^{(1)}, \ldots, h^{(m)}$ is zero if and only if $p \in \mathcal{N}_L$. Thus, $\{h^{(1)}, \ldots, h^{(m)}\}$ is a maximal set of linearly independent column vectors of $H(L)$ if and only if $\{x^{\alpha^{(1)}}, \ldots, x^{\alpha^{(m)}}\}$ is a maximal set of linearly independent monomials in the quotient vector space $\mathcal{D}_L = A/\mathcal{N}_L$. This implies $\operatorname{rank} H(L) = \dim \mathcal{D}_L$.

If the set N is finite, this equality can be obtained by a shorter reasoning: Using that $\dim \ker H(L) = \dim \mathcal{N}_L$ by (ii) and the rank-nullity characterization we get

$$\operatorname{rank} H(L) = \dim \mathbb{R}^{|N|} - \dim \ker H(L)$$

$$= \dim A - \dim \mathcal{N}_L = \dim(A/\mathcal{N}_L) = \dim(\mathcal{D}_L).$$

(iv) is obtained from (17.33) by setting $f = g$.

(v) If $f \in \mathcal{N}_L$, then $L(f^2) = 0$ by setting $f = g$ in (17.33). We prove the converse. Since L is a positive functional, the Cauchy–Schwarz inequality (2.7) holds:

$$L(fg)^2 \leq L(f^2) L(g^2) \quad \text{for } f, g \in A. \tag{17.34}$$

Therefore, if $L(f^2) = 0$, then $L(fg) = 0$ for all $g \in A$, so that $f \in \mathcal{N}_L$.

(vi) follows at once from Proposition 1.23. $\qquad \square$

Suppose that the functional L is positive on A^2, that is, $L(f^2) \geq 0$ for $f \in A$. Then, by (17.32), the non-degenerate symmetric bilinear form $\langle \cdot, \cdot \rangle$ on $\mathcal{D}_L$ is positive definite and hence a scalar product. Thus, $(\mathcal{D}_L, \langle \cdot, \cdot \rangle)$ is a real unitary space. Further, by Proposition 17.17(iii), $\mathcal{D}_L$ is finite-dimensional if and only if the Hankel matrix $H(L)$ has finite rank. In this case the unitary space $(\mathcal{D}_L, \langle \cdot, \cdot \rangle)$ is obviously complete, that is, $(\mathcal{D}_L, \langle \cdot, \cdot \rangle)$ is a real finite-dimensional Hilbert space. In particular, $\mathcal{D}_L$ has finite dimension if the set N is finite.

The next proposition shows that $\mathcal{N}_L$ obeys some ideal-like properties.

Proposition 17.18 *Let $p \in \mathcal{N}_L$ and $q \in$ A. Suppose that $pq \in$ A.*

(i) *If L is a positive functional and $pq^2 \in$ A, then $pq \in \mathcal{N}_L$.*
(ii) *If L is a truncated $\mathcal{K}$-moment functional, then $pq \in \mathcal{N}_L$.*

Proof

(i) Since the functional L is positive, the Cauchy–Schwarz inequality (17.34) holds. Using that $pq^2 \in$ A and $L(p^2) = 0$ we obtain

$$L((pq)^2)^2 = L(p\,pq^2)^2 \le L(p^2)L((pq^2)^2) = 0,$$

so that $pq \in \mathcal{N}_L$ by Proposition 17.17(iv).
(ii) Let $\mu \in \mathcal{M}_{L,\mathcal{K}}$. Recall that $\operatorname{supp}\mu \subseteq \mathcal{V}_L$ by Proposition 17.17(vi). Therefore, since $p \in \mathcal{N}_L$ and hence $p(x) = 0$ on $\mathcal{V}_L$, we get

$$L((pq)^2) = \int_{\mathcal{K}} (pq)^2(x)\,d\mu(x) = \int_{\mathcal{V}_L \cap \mathcal{K}} p(x)^2 q(x)^2\,d\mu(x) = 0.$$

Thus, $pq \in \mathcal{N}_L$ again by Proposition 17.17(iv).   □

We restate the preceding results in the case of our standard example.

Corollary 17.19 *Let N be the set $\mathbb{N}_{0,2n}^d$ (see (17.9)) and let L be a linear functional on $A = \mathbb{R}_d[x]_{2n}$. Suppose that $p \in \mathcal{N}_L$ and $q \in \mathbb{R}_d[x]_n$.*

(i) *If L is a positive functional and $pq \in \mathbb{R}_d[x]_{n-1}$, then $pq \in \mathcal{N}_L$.*
(ii) *If L is a truncated $\mathcal{K}$-moment functional and $pq \in \mathbb{R}_d[x]_n$, then $pq \in \mathcal{N}_L$.*

Proof

(i) Proposition 17.18(i) yields the assertion in the case $q = x_j, j = 1,\dots,d$. Since $\mathcal{N}_L$ is a vector space, repeated applications give the general case.
(ii) follows from Proposition 17.18(ii).   □

Remark 17.20 Example 9.29, Case 1, shows that the assertion $pq \in \mathcal{N}_L$ in Corollary 17.19(i) is no longer valid if $pq \in \mathbb{R}_d[x]_n$. This provides an important difference between positive functionals and moment functionals!   ○

17.4 Hankel Matrices of Functionals with Finitely Atomic Measures

Throughout this section, μ is a finitely atomic signed measure

$$\mu = \sum_{j=1}^{k} m_j \delta_{x_j}, \quad \text{where} \quad m_j \in \mathbb{R},\ x_j \in \mathbb{R}^d \ \text{for}\ j = 1, \dots, k, \tag{17.35}$$

and L denotes the corresponding functional on A^2 defined by

$$L(f) = \int f(x)\, d\mu = \sum_{j=1}^{k} m_j f(x_j), \quad f \in \mathsf{A}^2. \tag{17.36}$$

If $m_j \geq 0$ for all j, then μ is a Radon measure and L is a truncated moment functional. Clearly, μ is k-atomic if all $m_j > 0$ and the points x_j are pairwise distinct.

For $x \in \mathbb{R}^d$, we denote the column vector $(x^\alpha)_{\alpha \in \mathbb{N}}$ by $\mathfrak{s}_{\mathsf{N}}(x)$ or simply by $\mathfrak{s}(x)$ if no confusion can arise. Note that $\mathfrak{s}_{\mathsf{N}}(x)$ is the moment vector of the delta measure δ_x for A, not for A^2!

Proposition 17.21 *For the Hankel matrix $H(L)$ of the functional L we have*

$$H(L) = \sum_{j=1}^{k} m_j \mathfrak{s}_{\mathsf{N}}(x_j)\, \mathfrak{s}_{\mathsf{N}}(x_j)^T, \tag{17.37}$$

$$\operatorname{rank} H(L) \leq k = |\operatorname{supp} \mu| \leq |\mathcal{V}_L|. \tag{17.38}$$

Suppose that $m_j \neq 0$ for $j = 1, \dots, k$. Then the following are equivalent:

(i) $\operatorname{rank} H(L) = k$.
(ii) *The point evaluations $l_{x_1}, \dots, l_{x_k}$ on A (!) are linearly independent.*
(iii) *The vectors $\mathfrak{s}_{\mathsf{N}}(x_1), \dots, \mathfrak{s}_{\mathsf{N}}(x_k)$ are linearly independent.*

Proof Clearly, for $x \in \mathbb{R}^d$ the (α, β)-entry of the matrix $\mathfrak{s}(x)\mathfrak{s}(x)^T$ is just $x^{\alpha+\beta}$. This means that $\mathfrak{s}(x)\mathfrak{s}(x)^T$ is the Hankel matrix of the point evaluation l_x on A^2. Since $L = \sum_j m_j l_{x_j}$, this yields the formula (17.37) for $H(L)$.

By (17.37), the matrix $H(L)$ is a sum of k matrices $m_j \mathfrak{s}(x_j)\mathfrak{s}(x_j)^T$. These matrices are of rank one (if $m_j \neq 0$) or rank zero (if $m_j = 0$). Hence $\operatorname{rank} H(L) \leq k$. Since $\operatorname{supp} \mu \subseteq \mathcal{V}_L$, it is obvious that $|\operatorname{supp} \mu| \leq |\mathcal{V}_L|$.

Now we assume that $m_j \neq 0$ for all j and prove the equivalence of (i)–(iii).

(i)$\leftrightarrow$(ii) Let $f = \sum_{\alpha \in \mathbb{N}} f_\alpha x^\alpha \in \mathsf{A}$. Recall that $\vec{f}$ is the column vector $(f_\alpha)^T$. Let $h^{(\alpha)}$ be the α-th column of $H(L)$ and $e_\alpha = (\delta_{\alpha,\beta})_{\beta \in \mathbb{N}}$ the α-th basis vector. Then

$$H(L)\vec{f} = \sum_{\alpha \in \mathbb{N}} f_\alpha H(L) e_\alpha = \sum_{\alpha \in \mathbb{N}} f_\alpha h^{(\alpha)} = \sum_{\alpha \in \mathbb{N}} f_\alpha \sum_{j=1}^{k} m_j x_j^\alpha \mathfrak{s}(x_j) = \sum_{j=1}^{k} m_j l_{x_j}(f) \mathfrak{s}(x_j).$$

$$\tag{17.39}$$

Since $m_j \neq 0$ and $\operatorname{im} H(L)$ is contained in the span of vectors $\mathfrak{s}(x_1), \ldots, \mathfrak{s}(x_k)$, it follows from (17.39) that $\operatorname{rank} H(L) = \dim \operatorname{im} H(L)$ is k if and only if the restrictions of point evaluations $l_{x_1}, \ldots, l_{x_k}$ to A are linearly independent.

(ii)$\leftrightarrow$(iii) Clearly, $\mathfrak{s}(x_j)$ is the column $(l_{x_j}(x^\alpha))_{\alpha \in N}$. Thus, if $c_1, \ldots, c_k \in \mathbb{R}$, then $\sum_j c_j \mathfrak{s}(x_j)$ is the column vector $\left((\sum_j c_j l_{x_j})(x^\alpha)\right)_{\alpha \in N}$. Therefore, $\sum_j c_j \mathfrak{s}(x_j) = 0$ if and only if $\sum_j c_j l_{x_j} = 0$ on A. Hence (ii) and (iii) are equivalent. $\qquad\square$

Corollary 17.22 *For each truncated $\mathcal{K}$-moment functional L on A^2 we have*

$$\operatorname{rank} H(L) \leq |\mathcal{V}_L \cap \mathcal{K}|. \tag{17.40}$$

Proof By Theorem 17.2 and Proposition 17.17(vi), there is a k-atomic measure $\mu \in \mathcal{M}_{L,\mathcal{K}}$ and $\operatorname{supp} \mu \subseteq \mathcal{V}_L \cap \mathcal{K}$. Since $\operatorname{rank} H(L) \leq |\operatorname{supp} \mu|$ by (17.38), this yields (17.40). $\qquad\square$

Example 17.23 Let $d = 1, \mathsf{N} = \{0, 1\}$, so that $\mathsf{A} = \{a + bx : a, b \in \mathbb{R}\}$. The functionals l_{-1}, l_0, l_1 are linearly independent on A^2, but they are linearly dependent on A, since $2l_0 = l_{-1} + l_1$ on A. For $L = l_{-1} + l_0 + l_1$ we have $\operatorname{rank} H(L) = 2$. $\quad\circ$

Note that (17.40) is a necessary condition for truncated $\mathcal{K}$-moment functionals.

There are several *necessary conditions* for a linear functional L on A^2 to be a truncated moment functional. These are

• the *positivity condition:*

$$L(f^2) \geq 0 \quad \text{for } f \in \mathsf{A}, \tag{17.41}$$

• the *rank-variety condition:*

$$\operatorname{rank} H(L) \leq |\mathcal{V}_L|, \tag{17.42}$$

• the *consistency condition:*

$$p \in \mathcal{N}_L, q \in \mathsf{A} \text{ and } pq \in \mathsf{A} \quad \text{imply} \quad pq \in \mathcal{N}_L. \tag{17.43}$$

The positivity condition is obvious. Conditions (17.42) and (17.43) follow from (17.40) and Proposition 17.18(ii), applied with $\mathcal{K} = \mathbb{R}^d$.

Another necessary condition is given in Proposition 18.15(ii) below.

Remark 17.24 Let $h^{(\alpha)} = (h_{\alpha,\beta})_{\beta \in N}$ denote the α-th column vector and $CH(L)$ the linear span of column vectors $h^{(\alpha)}$, $\alpha \in N$, of the Hankel matrix $H(L)$. We define a surjective linear mapping $\varphi : \mathsf{A} \to CH(L)$ by

$$f = \sum_{\alpha \in N} f_\alpha x^\alpha \in \mathsf{A} \mapsto \varphi(f) = \sum_{\alpha \in N} f_\alpha h^{(\alpha)}.$$

These "functions" $\varphi(f)$ of column vectors are another tool. By (17.39) we have

$$H(L)\vec{f} = \sum_{\alpha \in \mathbb{N}} f_\alpha h^{(\alpha)} = \varphi(f) \quad \text{for } f = \sum_{\alpha \in \mathbb{N}} f_\alpha x^\alpha \in \mathsf{A}.$$

Hence $\operatorname{rank} H(L) = \dim \operatorname{im} H(L) = \dim CH(L)$. By Proposition 17.17(ii), $\mathcal{N}_L$ is the kernel of φ. Hence the consistency condition (17.43) can be reformulated as

$$p, q, pq \in \mathsf{A} \quad \text{and} \quad \varphi(p) = 0 \implies \varphi(pq) = 0. \qquad \circ$$

Example 17.28 below shows that it may happen that $|\mathcal{V}_L| > \operatorname{rank} H(L)$. Now we turn to functionals for which equality holds in the rank-variety condition (17.42).

Definition 17.25 A linear functional L on A^2 is called *minimal* if $\operatorname{rank} H(L)$ is finite and $\operatorname{rank} H(L) = |\mathcal{V}_L|$.

Our next aim is to derive two simple but very useful formulas   (17.45) and (17.47).

Let μ and L be as above and let $\mathcal{F} = \{f_1, \dots, f_n\}$ be a finite subset of A. Define

$$M_\mathcal{F} = \begin{pmatrix} f_1(x_1) \, f_1(x_2) \cdots \, f_1(x_k) \\ f_2(x_1) \, f_2(x_2) \cdots \, f_2(x_k) \\ \cdots \quad \cdots \quad \cdots \quad \cdots \\ f_n(x_1) \, f_n(x_2) \cdots \, f_n(x_k) \end{pmatrix} \equiv \begin{pmatrix} l_{x_1}(f_1) \, l_{x_2}(f_1) \cdots \, l_{x_k}(f_1) \\ l_{x_1}(f_2) \, l_{x_2}(f_2) \cdots \, l_{x_k}(f_2) \\ \cdots \quad \cdots \quad \cdots \quad \cdots \\ l_{x_1}(f_n) \, l_{x_2}(f_n) \cdots \, l_{x_k}(f_n) \end{pmatrix}.$$

Then, using (17.36) we compute

$$M_\mathcal{F} (m_1, \dots, m_k)^T = (L(f_1), \dots, L(f_n))^T. \qquad (17.44)$$

For minimal functionals L we will use (17.44) to derive the masses m_j from the points x_i and the functional L. If L is minimal, then $\operatorname{rank} H(L) = |\mathcal{V}_L|$ and hence (17.38) and Proposition 17.33(iii) imply that $k = \operatorname{rank} H(L) = \dim(\mathsf{A}/\mathcal{N}_L)$.

Lemma 17.26 *Let μ and L be defined by (17.35) and (17.36), respectively, and suppose that L is minimal. Let $\{f_1, \dots, f_k\}$ be a subset of A such that the representatives $\tilde{f}_j = f_j + \mathcal{N}_L$ form a basis of the quotient space $\mathsf{A}/\mathcal{N}_L$. Then the $k \times k$-matrix $M_\mathcal{F}$ is invertible and*

$$(m_1, \dots, m_k)^T = (M_\mathcal{F})^{-1}(L(f_1), \dots, L(f_k))^T. \qquad (17.45)$$

Proof Assume to the contrary that the matrix $M_\mathcal{F}$ is not invertible. Then the column rank of $M_\mathcal{F}$ is less than k, so there exist reals $\alpha_1, \dots, \alpha_k$, not all zero, such that $\alpha_1 f_1(x_j) + \cdots + \alpha_k f_k(x_j) = 0$ for all j. From the definition (17.36) of L we compute that $f := \alpha_1 f_1 + \cdots + \alpha_k f_k \in \mathsf{A}$ satisfies $L(f^2) = 0$, so that $f \in \mathcal{N}_L$. This contradicts the linear independence of the set $\{\tilde{f}_1, \dots, \tilde{f}_k\}$ in $\mathsf{A}/\mathcal{N}_L$. Thus, $M_\mathcal{F}$ is invertible and (17.45) follows from (17.44). $\square$

Now we introduce the matrix

$$
H_{\mathcal{F}}(L) = \begin{pmatrix} L(f_1^2) & L(f_1 f_2) & \cdots & L(f_1 f_n) \\ L(f_1 f_2) & L(f_2^2) & \cdots & L(f_2 f_n) \\ \cdots & \cdots & \cdots & \cdots \\ L(f_1 f_n) & L(f_2 f_n) & \cdots & L(f_n^2) \end{pmatrix}. \tag{17.46}
$$

If N is finite and $\mathcal{F}$ is our standard basis $\{x^\alpha : \alpha \in \mathsf{N}\}$, then $H_{\mathcal{F}}(L)$ is just the "usual" Hankel matrix $H(L)$ from Definition 17.16. Let $D(m)$ denote the diagonal matrix with diagonal entries $m_1, \ldots, m_k$. Then a simple computation yields

$$
\begin{pmatrix} L(f_1^2) & L(f_1 f_2) & \cdots & L(f_1 f_n) \\ L(f_1 f_2) & L(f_2^2) & \cdots & L(f_2 f_n) \\ \cdots & \cdots & \cdots & \cdots \\ L(f_1 f_n) & L(f_2 f_n) & \cdots & L(f_n^2) \end{pmatrix} =
$$

$$
\begin{pmatrix} f_1(x_1) & f_1(x_2) & \cdots & f_1(x_k) \\ f_2(x_1) & f_2(x_2) & \cdots & f_2(x_k) \\ \cdots & \cdots & \cdots & \cdots \\ f_n(x_1) & f_n(x_2) & \cdots & f_n(x_k) \end{pmatrix} \begin{pmatrix} m_1 & 0 & \cdots & 0 \\ 0 & m_2 & \cdots & 0 \\ \cdots & \cdots & \cdots & \cdots \\ 0 & 0 & \cdots & m_k \end{pmatrix} \begin{pmatrix} f_1(x_1) & f_2(x_1) & \cdots & f_n(x_1) \\ f_1(x_2) & f_2(x_2) & \cdots & f_n(x_2) \\ \cdots & \cdots & \cdots & \cdots \\ f_1(x_k) & f_2(x_k) & \cdots & f_n(x_k) \end{pmatrix}
$$

which means that

$$
H_{\mathcal{F}}(L) = M_{\mathcal{F}} D(m)(M_{\mathcal{F}})^T. \tag{17.47}
$$

Proposition 17.27 *If N is finite and L is a minimal truncated moment functional on A^2, then L is determinate and its unique representing measure is* rank $H(L)$-*atomic.*

Proof Set $k := \operatorname{rank} H(L)$. Since L is minimal, $k = |\mathcal{V}_L|$. We write $\mathcal{V}_L = \{x_1, \ldots, x_k\}$. Let ν be an arbitrary representing measure of L. Since $\operatorname{supp} \nu \subseteq \mathcal{V}_L$ by Proposition 17.17(vi), ν is of the form $\nu = \sum_{j=1}^{k} m_j \delta_{x_j}$ with $m_j \geq 0$ for all j.

Now we apply (17.47) to the standard Hankel matrix $H(L)$. If $m_j = 0$ for one j, then $\operatorname{rank} D(m) < k$ and hence $\operatorname{rank} H(L) < k$ by (17.47), which is a contradiction. Thus, $m_j > 0$ for all j, so ν is k-atomic.

We choose a set $\{f_1, \ldots, f_k\}$ as in Lemma 17.26. Then the numbers m_j are given by (17.45), so the measure ν is uniquely determined and L is determinate. $\square$

Example 17.28 (An example for which $|\mathcal{V}_L| > \operatorname{rank} H(L)$) Suppose that $d = 2$, $n \geq 3$, and $\mathsf{A} = \mathbb{R}[x_1, x_2]_n$. Set

$$
p(x_1, x_2) = (x_1 - \alpha_1) \cdots (x_1 - \alpha_n), \quad q(x_1, x_2) = (x_2 - \beta_1) \cdots (x_2 - \beta_n),
$$

where α_i, β_j are real numbers such that $\alpha_1 < \cdots < \alpha_n$ and $\beta_1 < \cdots < \beta_n$. Then $\mathcal{Z}(p) \cap \mathcal{Z}(q) = \{(\alpha_i, \beta_j) : i, j = 1, \ldots, n\}$.

We define an n^2-atomic measure μ such that all atoms are in $\mathcal{Z}(p) \cap \mathcal{Z}(q)$ and a truncated moment functional L on A^2 by $L(f) = \int f \, d\mu, f \in \mathsf{A}^2$. Then we have $L(p^2) = L(q^2) = 0$, so that $p, q \in \mathcal{N}_L$. Clearly, p and q are linearly independent in $\mathcal{N}_L$. Hence rank $H(L) = \dim (\mathsf{A}/\mathcal{N}_L) \leq \binom{n+2}{2} - 2$. From supp $\mu \subseteq \mathcal{V}_L$ and $p, q \in \mathcal{N}_L$ we obtain $n^2 = |\text{supp}\,\mu| \leq |\mathcal{V}_L| \leq |\mathcal{Z}(p) \cap \mathcal{Z}(q)| = n^2$. Thus, $\mathcal{V}_L = \mathcal{Z}(p) \cap \mathcal{Z}(q)$ and

$$|\mathcal{V}_L| - \text{rank}\,H(L) \geq n^2 - \binom{n+2}{2} + 2 = \binom{n-1}{2}. \tag{17.48}$$

We show that L is *determinate*. Let ν be a representing measure of L. Since supp $\nu \subseteq \mathcal{V}_L = \mathcal{Z}(p) \cap \mathcal{Z}(q)$, ν can be written as $\nu = \sum_{i,j=1}^{n} m_{ij} \delta_{(\alpha_i, \beta_j)}$. Let us fix $i, j \in \{1, \ldots, n\}$. Put $f_{ij} := (x_1 - \alpha_i)^{-1}(x_2 - \beta_j)^{-1} p \, q$. Then $L(f_{ij}) = \int f_{ij} \, d\nu = m_{ij}$. Hence the measure ν is uniquely determined by L, that is, L is determinate.

From the assumption $n \geq 3$ and (17.48) it follows that $|\mathcal{V}_L| > \text{rank}\,H(L)$ and the difference $|\mathcal{V}_L| - \text{rank}\,H(L)$ is large as n becomes large. Since L is determinate, L has no (rank L)-atomic representing measure, so L is *not minimal*. ○

17.5 The Full Moment Problem with Finite Rank Hankel Matrix

Throughout this section, we assume that $\mathsf{N} = \mathbb{N}_0^d$. Then $\mathsf{A} = \mathsf{A}^2 = \mathbb{R}_d[\underline{x}]$ and we are concerned with the *full* moment problem.

Let L be a positive functional on the real $*$-algebra $\mathbb{R}_d[\underline{x}]$ with identity involution (that is, $f^* = f$ for $f \in \mathbb{R}_d[\underline{x}]$). We consider the GNS construction associated with L, see Sect. 12.5. By Proposition 17.17(iv), the vector space $\mathcal{N}_L$ defined in (17.29) coincides with the left ideal $\mathcal{N}_L$ from Lemma 12.38. Therefore, by (17.32), $(\mathcal{D}_L, \langle \cdot, \cdot \rangle)$ is the domain of the GNS representation π_L of $\mathbb{R}_d[\underline{x}]$. Recall that, by the definition of the GNS representation, $\pi_L(p)$ acts on $\mathcal{D}_L$ by multiplication with p.

Theorem 17.29 *Suppose that L is a positive linear functional on $\mathbb{R}_d[\underline{x}]$ such that* rank $H(L) = r$, *where* $r \in \mathbb{N}$. *Then L is a moment functional with unique representing measure μ. This measure μ has r atoms and* supp $\mu = \mathcal{V}_L$.

Proof By Proposition 17.17, $\dim \mathcal{D}_L = \text{rank}\,H(L) = r$. As noted before the theorem, $\mathcal{D}_L$ is the representation space of the GNS representation π_L. Therefore, $X_k := \pi_L(x_k), k = 1, \ldots, d$, are commuting self-adjoint operators acting on the r-dimensional unitary space $\mathcal{D}_L$. From linear algebra it is known that these operators have a common set of eigenvectors, say $e_1, \ldots, e_r$, which form an orthonormal basis of $\mathcal{D}_L$. Let $X_k e_j = t_{kj} e_j$ and put $t_j = (t_{1j}, \ldots, t_{dj})^T$. Since all eigenvalues of X_k are real, $t_j \in \mathbb{R}^d$. We develop the vector $1 \in \mathcal{D}_L$ with respect to this orthonormal basis and obtain $1 = \sum_{j=1}^r m_j e_j$. Set $\mu := \sum_{j=1}^r m_j^2 \delta_{t_j}$. For $p \in \mathbb{R}_d[\underline{x}]$ we derive

$$L(p) = \langle \pi_L(p)1, 1 \rangle = \langle p(\pi_L(x_1), \ldots, \pi_L(x_d))1, 1 \rangle = \langle p(X_1, \ldots, X_d)1, 1 \rangle$$

$$= \sum_{i,j=1}^r \langle p(t_{1j}, \ldots, t_{dj}) m_j e_j, m_i e_i \rangle = \sum_{j=1}^r p(t_j) m_j^2 = \int p(t)\, d\mu(t). \qquad (17.49)$$

Here the first equality holds by formula (12.33), while the second equality follows from the fact that π_L is an algebra homomorphism. (This is just the finite-dimensional version of the proof of Theorem 12.40.) By (17.49), μ is a representing measure of L.

By the properties of the GNS construction, $\pi_L(\mathbb{R}_d[\underline{x}])1 = \mathcal{D}_L$. Hence $m_j \neq 0$ for all j. Since rank $H(L) = r$, the point evaluations $l_{t_1} \ldots, l_{t_r}$ on $\mathsf{A} = \mathbb{R}_d[\underline{x}]$ are linearly independent by Proposition 17.21. Therefore, $t_i \neq t_j$ for $i \neq j$. By the preceding we have shown that $\mu = \sum_{j=1}^r m_j^2 \delta_{t_j}$ is r-atomic. Because μ has finite, hence compact, support, Proposition 12.17 implies that L has only one representing measure. (This fact also follows from the uniqueness of the spectral decomposition of $X_1, \ldots, X_d$.)

By Proposition 17.17(vi), supp $\mu \subseteq \mathcal{V}_L$. We prove that $\mathcal{V}_L \subseteq$ supp μ. Let $t_0 \in \mathcal{V}_L$. Assume to the contrary that $t_0 \notin$ supp $\mu = \{t_1, \ldots, t_r\}$. By Lagrange interpolation there exists a $p \in \mathbb{R}_d[\underline{x}]$ such that $p(t_0) = 1$ and $p(t_j) = 0$ for $j = 1, \ldots, r$. Then

$$L(p^2) = \int p^2\, d\mu = \sum_{j=1}^r m_j p(t_j)^2 = 0,$$

so that $p \in \mathcal{N}_L$. Since $p(t_0) = 1$, $t_0 \notin \mathcal{V}_L$, which is a contradiction. Thus we have shown that $\mathcal{V}_L =$ supp μ. $\qquad \square$

17.6 Flat Extensions and the Flat Extension Theorem

In this section we derive another result, called the flat extension theorem, which provides a sufficient condition for the solvability of the truncated moment problem. To formulate this theorem we begin with some preliminaries.

We suppose in this section that N is a finite subset of $\mathbb{N}_0^d$, $\mathcal{A} = \{x^\alpha : \alpha \in \mathsf{N}\}$ and $\mathsf{A} = \mathrm{Lin}\{x^\alpha : \alpha \in \mathsf{N}\}$. Now let N_0 be another subset of $\mathbb{N}_0^d$. We denote by $\mathcal{B} = \{x^\alpha : \alpha \in \mathsf{N}_0\}$ the corresponding monomials and by B the linear span of elements of $\mathcal{B}$.

Definition 17.30 We say the set $\mathcal{B}$ is connected to 1 if $1 \in \mathcal{B}$ and each $p \in \mathcal{B}$, $p \neq 1$, can be written as $p = x_{i_1} \ldots x_{i_k}$ with $x_{i_1}, x_{i_1} x_{i_2}, \ldots, x_{i_1} \cdots x_{i_k} \in \mathcal{B}$.

Example 17.31 Clearly, $\{1, x_1, x_2, x_1 x_2\}$ and $\{1, x_2, x_2 x_3, x_1 x_2 x_3\}$ are connected to 1, but $\{1, x_1, x_2, x_1 x_2 x_3\}$ is not. $\qquad \circ$

From now on we assume that $\mathcal{B}$ is a subset of $\mathcal{A}$. Then B^2 is a subspace of A^2.

Definition 17.32 A linear functional L on A^2 is called *flat* with respect to B if

$$\operatorname{rank} H(L) = \operatorname{rank} H(L_0),$$

where L_0 denotes the restriction to B^2 of L.

To motivate this definition we write the Hankel matrix $H(L)$ as a block matrix

$$H(L) = \begin{pmatrix} H(L_0) & X_{12} \\ X_{21} & X_{22} \end{pmatrix}.$$

The matrix $H(L)$ is called a flat extension (see Definition A.22) of the matrix $H(L_0)$ if $\operatorname{rank} H(L) = \operatorname{rank} H(L_0)$. Hence the functional L is flat with respect to B if and only if the Hankel matrix $H(L)$ is a flat extension of the Hankel submatrix $H(L_0)$.

The next proposition contains a reformulation of the flatness condition that can be used in noncommutative settings as well. Another proof based on Hankel matrices is sketched in Exercise 17.6. Recall that $\mathcal{N}_L$ is defined by (17.29).

Proposition 17.33 *A linear functional L on A^2 is flat with respect to B if and only if $\mathsf{A} = \mathsf{B} + \mathcal{N}_L$. In this case, $\mathcal{N}_{L_0} = \mathcal{N}_L \cap \mathsf{B}$.*

Proof Clearly, $\mathcal{N}_L \cap \mathsf{B} \subseteq \mathcal{N}_{L_0}$ and there are canonical maps σ_1 and σ_2:

$$\mathsf{B}/\mathcal{N}_{L_0} \xleftarrow{\sigma_1} \mathsf{B}/(\mathcal{N}_L \cap \mathsf{B}) \xrightarrow{\sigma_2} \mathsf{A}/\mathcal{N}_L.$$

Here the dimensions are increasing from left to right, so by Proposition 17.17(iii),

$$\operatorname{rank} H(L_0) = \dim\,(\mathsf{B}/\mathcal{N}_{L_0}) \leq \dim\,(\mathsf{B}/(\mathcal{N}_L \cap \mathsf{B})$$

$$\leq \dim\,(\mathsf{A}/\mathcal{N}_L) = \operatorname{rank} H(L).$$

Therefore, L is flat with respect to B if and only if we have equality throughout, that is, σ_1 and σ_2 are bijections, or equivalently, $\mathsf{A} = \mathsf{B} + \mathcal{N}_L$ and $\mathcal{N}_{L_0} = \mathcal{N}_L \cap \mathsf{B}$.

To complete the proof we show that $\mathsf{A} = \mathsf{B} + \mathcal{N}_L$ implies that $\mathcal{N}_{L_0} = \mathcal{N}_L \cap \mathsf{B}$. Obviously, $\mathcal{N}_L \cap \mathsf{B} \subseteq \mathcal{N}_{L_0}$. To prove the converse inclusion, let $f \in \mathcal{N}_{L_0}$. We write $g \in \mathsf{A}$ as $g = g_1 + g_2$ with $g_1 \in \mathsf{B}$ and $g_2 \in \mathcal{N}_L$. Then $fg_1 \in \mathsf{B}^2$. Since $f \in \mathcal{N}_{L_0}$ and $g_2 \in \mathcal{N}_L$, we get $L(fg) = L(fg_1) + L(fg_2) = L_0(fg_1) = 0$, so that $f \in \mathcal{N}_L \cap \mathsf{B}$. □

Proposition 17.34 *Let L be a linear functional on A^2 which is flat with respect to B. If $L(p^2) \geq 0$ for all $p \in \mathsf{B}^2$, then $L(q^2) \geq 0$ for all $q \in \mathsf{A}^2$.*

Proof Let L_0 denote the restriction to B^2 of L. Since L_0 is positive on B^2 by assumption, $H(L_0) \succeq 0$. Because L is flat with respect to B, $H(L)$ is a flat extension of $H(L_0)$. Then $H(L) \succeq 0$ by Theorem A.24(iii). Hence L is positive on A^2. □

The next result is the *flat extension theorem*.

Theorem 17.35 *Suppose that $\mathcal{B}$ is a finite set of monomials of $\mathbb{R}_d[\underline{x}]$ such that $\mathcal{B}$ is connected to 1 and*

$$\mathcal{A} = \mathcal{B} \cup x_1 \cdot \mathcal{B} \cup \cdots \cup x_d \cdot \mathcal{B}. \tag{17.50}$$

Let A and B be the linear spans of $\mathcal{A}$ and $\mathcal{B}$, respectively. Suppose that L is a linear functional on A^2 which is flat with respect to B.

Then L has a unique extension to a linear functional $\tilde{L}$ of $\mathbb{R}_d[\underline{x}]$ such that $\tilde{L}$ is flat with respect to A. If $L(p^2) \geq 0$ for all $p \in B$, then $\tilde{L}(f^2) \geq 0$ for all $f \in \mathbb{R}_d[\underline{x}]$. Further, if $\mathcal{I}(\mathcal{N}_L)$ denotes the ideal of $\mathbb{R}_d[\underline{x}]$ generated by $\mathcal{N}_L$, then

$$\mathcal{N}_{\tilde{L}} = \mathcal{I}(\mathcal{N}_L) \quad \text{and} \quad \mathcal{V}_{\tilde{L}} = \mathcal{V}_L. \tag{17.51}$$

The functional $\tilde{L}$ on $\mathbb{R}_d[\underline{x}]$ is also flat with respect to B, because the functional L on A^2 is flat with respect to B.

The proof of Theorem 17.35 is lengthy and will be given in the next section. The following theorem of Curto and Fialkow contains the main applications concerning the truncated moment problem.

Theorem 17.36 *Let $\mathcal{A}, \mathcal{B}, A, B$ satisfy the assumptions of Theorem 17.35 and let L be a linear functional on A^2 which is flat with respect to B. Suppose that $L(p^2) \geq 0$ for $p \in B$. Then L is a minimial truncated moment functional.*

In particular, L is determinate and its unique representing measure μ is r-atomic, where $r := \operatorname{rank} H(L) \leq |\mathcal{B}|$, and satisfies $\operatorname{supp} \mu = \mathcal{V}_L$.

Proof Let $\tilde{L}$ be the positive linear functional on $\mathbb{R}_d[\underline{x}]$ from Theorem 17.35. Since both L and $\tilde{L}$ are flat with respect to B, we have

$$r = \operatorname{rank} H(L) = \operatorname{rank} H(\tilde{L}) = \operatorname{rank} H(L_0) \leq |\mathcal{B}|.$$

Therefore, it follows from Theorem 17.29 that $\tilde{L}$, hence L, has an r-atomic representing measure μ and $\operatorname{supp} \mu = \mathcal{V}_{\tilde{L}}$. Since $\mathcal{V}_L = \mathcal{V}_{\tilde{L}}$ by (17.51), we have $r = \operatorname{rank} H(L) = |\mathcal{V}_L|$, so L is minimal and hence determinate by Proposition 17.27. $\square$

The most important application of Theorem 17.36 concerns the following case:

$$\mathsf{N}_0 = \mathbb{N}_{0,n-1}^d, \quad \mathsf{N} = \mathbb{N}_{0,n}^d, \quad \mathcal{B} = \{x^\alpha : |\alpha| \leq n-1\}, \quad \mathcal{A} = \{x^\alpha : |\alpha| \leq n\}.$$

Then it is obvious that $\mathcal{B}$ is connected to 1, assumption (17.50) holds,

$$\mathsf{B}^2 = \mathbb{R}_d[\underline{x}]_{2n-2}, \quad \text{and} \quad \mathsf{A}^2 = \mathbb{R}_d[\underline{x}]_{2n}.$$

Let L be a linear functional on $\mathbb{R}_d[\underline{x}]_{2n}$. Recall from Definition 17.16 that the Hankel matrix of L is the matrix $H_n(L) := H(L)$ with entries

$$h_{\alpha,\beta} := L(x^{\alpha+\beta}), \quad \text{where} \quad \alpha, \beta \in \mathbb{N}_0^d, |\alpha|, |\beta| \leq n.$$

Clearly, the Hankel matrix of the restriction L_0 of L to $\mathbb{R}_d[x]_{2n-2}$ is the matrix $H_{n-1}(L) := H(L_0)$ with entries $h_{\alpha,\beta}$, where $\alpha, \beta \in \mathbb{N}_0^d$, $|\alpha|, |\beta| \leq n - 1$. Further, $|\mathcal{B}| = \dim \mathbb{R}_d[x]_{n-1} = \binom{n-1+d}{d}$. Then Theorem 17.36 has the following special case.

Theorem 17.37 *Suppose that L is a linear functional on $\mathbb{R}_d[x]_{2n}, n \in \mathbb{N}$, such that*

$$L(p^2) \geq 0 \text{ for } p \in \mathbb{R}_d[x]_{n-1} \text{ and } r := \operatorname{rank} H_n(L) = \operatorname{rank} H_{n-1}(L). \quad (17.52)$$

Then L is a determinate truncated moment functional and its unique representing measure is r-atomic with $r \leq \binom{n-1+d}{d}$.

The next theorem deals with the truncated $\mathcal{K}$-moment problem, where

$$\mathcal{K} := \mathcal{K}(\mathfrak{f}) = \{x \in \mathbb{R}^d : f_0(x) \geq 0, \dots, f_k(x) \geq 0\} \quad (17.53)$$

is the semi-algebraic set associated to a finite subset $\mathfrak{f} = \{f_0, \dots, f_k\}$ of $\mathbb{R}_d[x]$ with $f_0 = 1$. Let us abbreviate $m := \max\{1, \deg(f_j) : j = 1, \dots, k\}$.

Theorem 17.38 *Let L be a linear functional on $\mathbb{R}_d[x]_{2n}$, $n \in \mathbb{N}$, and $r := \operatorname{rank} H_n(L)$. Retaining the preceding notation, the following are equivalent:*

(i) *L is a truncated $\mathcal{K}$-moment functional which has an r-atomic representing measure μ with all atoms in $\mathcal{K}$.*

(ii) *L extends to a linear functional $\tilde{L}$ on $\mathbb{R}_d[x]_{2(n+m)}$ such that $\operatorname{rank} H_{n+m}(\tilde{L}) = \operatorname{rank} H_n(L)$ and $\tilde{L}(f_j p^2) \geq 0$ for all $p \in \mathbb{R}_d[x]_n$ and $j = 0, \dots, k$.*

Proof
(i)$\rightarrow$(ii) Let $\tilde{L}$ be the truncated $\mathcal{K}$-moment functional on $\mathbb{R}_d[x]_{2(n+m)}$ given by a measure μ as in (i). Then, by (17.38),

$$r = |\operatorname{supp} \mu| \geq \operatorname{rank} H_{n+m}(\tilde{L}) \geq \operatorname{rank} H_n(L) = r,$$

so that $\operatorname{rank} H_{n+m}(\tilde{L}) = \operatorname{rank} H_n(L)$. The positivity condition is obvious.

(ii)$\rightarrow$(i) Since $\operatorname{rank} H_n(L) = \operatorname{rank} H_{n+m}(\tilde{L})$, $\tilde{L}$ is flat with respect to $\mathbb{R}_d[x]_n$. Therefore, since $L(p^2) \geq 0$ for $p \in \mathbb{R}_d[x]_n$ by assumption (with $f_0 = 1$), Proposition 17.34 implies that $\tilde{L}(q^2) \geq 0$ for $q \in \mathbb{R}_d[x]_{n+m}$. Further, $\tilde{L}$ is flat with respect to $\mathbb{R}_d[x]_{n+m-1}$. (This is where $m \geq 1$ is used!) Hence, by Theorem 17.37, $\tilde{L}$, hence L, is a truncated moment functional with r-atomic representing measure $\mu = \sum_{i=1}^{r} m_i \delta_{x_i}$, where all $m_i > 0$. It remains to prove that all atoms $x_j, j = 1, \dots, r$, are in $\mathcal{K}$.

Since $|\operatorname{supp} \mu| = r = \operatorname{rank} H_n(L)$, it follows from Proposition 17.21 (i)$\leftrightarrow$(ii) that the functionals $l_{x_1}, \dots, l_{x_r}$ are linearly independent on $\mathbb{R}_d[x]_n$. Hence there are polynomials $p_1, \dots, p_r \in \mathbb{R}_d[x]_n$ such that $l_{x_i}(p_j) \equiv p_j(x_i) = \delta_{ij}$ for all i, j. Let

$j, k \in \{0, \ldots, r\}$. Then we have $f_k p_j^2 \in \mathbb{R}_d[\underline{x}]_{2(n+m)}$ and

$$\tilde{L}(f_k p_j^2) = \int f_k p_j^2 \, d\mu = \sum_{i=1}^{r} m_i f_k(x_i) p_j(x_i)^2 = m_j f_k(x_j) \geq 0.$$

Since $m_j > 0$, we conclude that $f_k(x_j) \geq 0$ for all k. Hence $x_j \in \mathcal{K}$ by (17.53). $\quad\square$

Remark 17.39

1. The proof of the implication (ii)→(i) shows that μ is also a representing measure for the functional $\tilde{L}$ on $\mathbb{R}_d[\underline{x}]_{2(n+m)}$.
2. As we have discussed in Remark 16.10, Theorem 17.38 has an interesting application to polynomial optimization over the set $\mathcal{K}$. $\quad\circ$

17.7 Proof of Theorem 17.36

The following "truncated ideal-like" property of $\mathcal{N}_L$ will be used twice below.

Lemma 17.40 *If $f \in \mathcal{N}_L$ and $x_i f \in \mathsf{A}$, then $x_i f \in \mathcal{N}_L$.*

Proof Since $x_i f \in \mathsf{A}$, we have $x_i f = p + q$ with $p \in \mathsf{B}$ and $q \in \mathcal{N}_L$ by the equality $\mathsf{A} = \mathsf{B} + \mathcal{N}_L$ from Proposition 17.33. Therefore, by $f \in \mathcal{N}_L$ and by the assumption (17.50), for all $g \in \mathsf{B} \subseteq \mathsf{A}$ we have $x_i g \in x_i \mathsf{B} \subseteq \mathsf{A}$ and

$$0 = L(f(x_i g)) = L(pg) + L(qg) = L(pg) = L_0(pg),$$

so that $p \in \mathcal{N}_{L_0}$. Hence $p \in \mathcal{N}_L$ and $x_i f = p + q \in \mathcal{N}_L$. $\quad\square$

By Proposition 17.33 the flatness implies that $\mathsf{A} = \mathsf{B} + \mathcal{N}_L$. We choose a subspace $\mathcal{D}$ of B such that A is the direct sum of vector spaces $\mathcal{D}$ and $\mathcal{N}_L$.

Without loss of generality we assume that $1 \in \mathcal{D}$. (If such a choice is impossible, then $1 \in \mathcal{N}_L$. But then a repeated application of Lemma 17.40 yields $\mathsf{A} \in \mathcal{N}_L$, so that $L \equiv 0$ and the assertion holds trivially.)

Let π denote the projection of A onto $\mathcal{D} \subseteq \mathsf{B}$ with respect to the direct sum

$$\mathsf{A} = \mathcal{D} \oplus \mathcal{N}_L. \tag{17.54}$$

Further, let $X_i : \mathcal{D} \to \mathcal{D}, i = 1, \ldots, d$, denote the operator $X_i : \mathcal{D} \to \mathcal{D}$ defined by $X_i(f) := \pi(x_i f), f \in \mathcal{D}$. The crucial step of the proof is the following

Lemma 17.41 *If $f \in \mathcal{D}$, then $X_i X_j f = X_j X_i f$ for $i, j = 1, \ldots, d$ and $f \in \mathcal{D}$.*

Proof Let $f \in \mathcal{D}$. By the definition of the operators X_j and X_i we obtain

$$X_i X_j(f) = X_i(\pi(x_j f)) = \pi(x_i \pi(x_j f)) = x_i x_j f - x_i(I - \pi)(x_j f) - (I - \pi)(x_i(\pi(x_j f))).$$

Therefore, $(X_iX_j - X_jX_i)f = g_0 + g_1 + g_2$, where

$$g_0 := (I - \pi)(x_j(\pi(x_if))) - (I - \pi)(x_i(\pi(x_jf))),$$
$$g_1 := x_j(I - \pi)(x_if), \ g_2 := -x_i(I - \pi)(x_jf).$$

Clearly, $g_0 \in \mathcal{N}_L$ by the definition of the projection π. Since $(I - \pi)(x_if) \in \mathcal{N}_L$ and $g_1 = x_j(I - \pi)(x_jf) \in x_j\mathsf{B} \subseteq \mathsf{A}$, Lemma 17.40 implies that $g_1 \in \mathcal{N}_L$. The same reasoning shows that $g_2 \in \mathcal{N}_L$. Thus $g_0 + g_1 + g_2 \in \mathcal{N}_L$.

On the other hand, the sum $g_0 + g_1 + g_2 = (X_iX_j - X_jX_i)f$ belongs to $\mathcal{D}$. Since $\mathcal{D} \cap \mathcal{N}_L = \{0\}$ by (17.54), we conclude that $X_iX_jf = X_jX_if$. □

Let $p \in \mathbb{R}_d[x]$. Since the operators X_i and X_j on $\mathcal{D}$ commute by Lemma 17.41, it follows that $p(X) := p(X_1, \ldots, X_d)$ is a well-defined linear operator mapping $\mathcal{D}$ into itself. Recall that $1 \in \mathcal{D}$. Set $\varphi(p) := P(X)1 \in \mathcal{D}$. Clearly, the map $p \mapsto p(X)$ is an algebra homorphism of $\mathbb{R}_d[x]$ into the linear operators of $\mathcal{D}$. Hence we have

$$\varphi(pq) = p(X)\varphi(q), \quad p, q \in \mathbb{R}_d[x]. \tag{17.55}$$

In particular, this implies that $\ker \varphi$ is an ideal of $\mathbb{R}_d[x]$.

Lemma 17.42

(i) $\varphi(p) = \pi(p)$ for $p \in \mathsf{A}$.
(ii) $L(pq) = L(\varphi(pq))$ for $p, q \in \mathsf{A}$.

Proof

(i) The proof is given by induction on the degree of p. Obviously, $\varphi(1) = \pi(1) = 1$. Suppose that the assertion holds for all p of degree k. Let $q \in \mathsf{A}$ and $\deg(q) = k + 1$. It suffices to prove the assertion for monomials $q \in \mathcal{A}$. Since $\mathcal{A} = \mathcal{B}^+$, we have $q = x_ip$ for some i and $p \in \mathsf{B}$ or $q \in \mathcal{B}$. In the latter case, q is also of the form $q = x_ip$ for some i and $p \in \mathcal{B}$, because $\mathcal{B}$ is connected to 1. Thus, in any case, $q = x_ip$. Then $\deg(p) = k$ and hence $\varphi(p) = \pi(p)$ by the induction assumption. From (17.55) and (17.54) we obtain

$$\varphi(q) = X_i(\varphi(p)) = X_i(\pi(p)) = \pi(x_i\pi(p)) = x_i\pi(p) + g$$

for some $g \in \mathcal{N}_L$. Similarly, by (17.54), $q = x_ip = x_i(\pi(p) + h)$ for some $h \in \mathcal{N}_L$. Thus we have $x_i\pi(p) = q - x_ih = \varphi(q) - g$, which implies that $x_ih \in \mathsf{A}$. Therefore, since $h \in \mathcal{N}_L$, it follows from Lemma 17.40 that $x_ih \in \mathcal{N}_L$. Applying the projection π to the identity $q - x_ih = \varphi(q) - g$ we get $\pi(q) = \pi(\varphi(q)) = \varphi(q)$. This completes the induction.
(ii) First we show that

$$L(pq) = L(\varphi(pq)) \quad \text{for} \quad q \in \mathsf{B}, \ p \in \mathsf{A}. \tag{17.56}$$

We proceed by induction on $\deg(p)$. For constant p this follows from (i), since $L(q) = L(\pi(q)) = L(\varphi(q))$ for $q \in$ A. Assume that (17.56) holds for all $p \in$ A with $\deg(p) = k$. Suppose that $f \in$ A and $\deg(f) = k + 1$. Arguing as in the proof of (i), we can assume without loss of generality that $f = x_i p$ with $p \in \mathcal{B}$. Thus we obtain

$$L(fq) = L(px_iq) = L(p\varphi(x_iq)) = L(\varphi(px_iq)) \tag{17.57}$$

by applying first (i) and then the induction hypothesis. Using the definitions of $\varphi(g)$ and $g(X)$ for $g \in \mathbb{R}_d[\underline{x}]$ and applying (17.55) twice we derive

$$\varphi(p\varphi(x_iq)) = p(X)(\varphi(x_iq)) = P(X)(X_i(q))$$
$$= (x_ip)(X)(q) = f(X)(q) = f(X)\varphi(q) = \varphi(fq).$$

Inserting this into (17.57) we get $L(pq) = L(\varphi(pq))$. This proves (17.56).

Let $p, q \in$ A. From (i) we obtain $L(pq) = L(p\varphi(q))$. Using first this equality, then (17.56), and finally (17.55) we derive

$$L(pq) = L(p\varphi(q)) = L(\varphi(p\varphi(q))) = L(P(X)\varphi(q)) = L(\varphi(pq))$$

which completes the proof.   □

Now we define a linear functional $\tilde{L}$ on $\mathbb{R}_d[\underline{x}]$ by

$$\tilde{L}(f) = L(\varphi(f)), \quad f \in \mathbb{R}_d[\underline{x}]. \tag{17.58}$$

Lemma 17.42(ii) implies that $\tilde{L}(f) = L(f)$ for $f = pq$, where $p, q \in$ A, and hence for all $f \in$ A^2. Therefore, $\tilde{L}$ is an extension of L.

The first statement of the next lemma shows that $\tilde{L}$ is flat with respect to A.

Lemma 17.43

(i) $\operatorname{rank} H(\tilde{L}) = \operatorname{rank} H(L) = \dim \mathcal{D}$.
(ii) $\mathcal{N}_{\tilde{L}} = \mathcal{I}(\mathcal{N}_L) = \ker \varphi$ and $\mathcal{V}_{\tilde{L}} = \mathcal{V}_L$.

Proof First we show that

$$\mathcal{I}(\mathcal{N}_L) \subseteq \ker \varphi \subseteq \mathcal{N}_{\tilde{L}}. \tag{17.59}$$

If $p \in \mathcal{N}_L$, then $\pi(p) = 0$ and hence $\varphi(p) = p(X)1 = 0$, so that $p \in \ker \varphi$. From (17.55) we conclude that $\ker \varphi$ is an ideal of $\mathbb{R}_d[\underline{x}]$. Hence $\mathcal{I}(\mathcal{N}_L) \subseteq \ker \varphi$.

Now we verify the second inclusion of (17.59). Let $f \in \ker \varphi$. Then, by (17.55), for $p \in \mathbb{R}_d[\underline{x}]$ we obtain

$$\tilde{L}(fp) = L(\varphi(fp)) = L(p(X)\varphi(f)) = 0,$$

so that $f \in \mathcal{N}_{\tilde{L}}$. This proves (17.59).

From (17.54) and Lemma 17.42(i) it follows that $\mathbb{R}_d[\underline{x}] = \mathcal{D} + \mathcal{I}(\mathcal{N}_L)$. Using this equality, the corresponding definitions, Proposition 17.17(ii), and (17.59) we derive

$$\dim \mathcal{D} = \dim(\mathsf{A}/\mathcal{N}_L) = \operatorname{rank} H(L) \leq \operatorname{rank} H(\tilde{L}),$$

$$\operatorname{rank} H(\tilde{L}) = \dim(\mathbb{R}_d[\underline{x}]/\mathcal{N}_{\tilde{L}}) \leq \dim(\mathbb{R}_d[\underline{x}]/\ker \varphi)$$

$$\leq \dim(\mathbb{R}_d[\underline{x}]/\mathcal{I}(\mathcal{N}_L)) \leq \dim \mathcal{D}.$$

Hence we have equalities throughout and therefore also in (17.59). This proves (i) and the first assertion of (ii). Since $\mathcal{N}_L$ and $\mathcal{I}(\mathcal{N}_L)$ define the same real algebraic set $\mathcal{V}_L$, the equality $\mathcal{N}_{\tilde{L}} = \mathcal{I}(\mathcal{N}_L)$ implies that $\mathcal{V}_{\tilde{L}} = \mathcal{V}_L$.   □

Lemma 17.44 *If L' is another linear functional on $\mathbb{R}_d[\underline{x}]$ such that L' is an extension of L and L' is flat with respect to* A, *then $L' = \tilde{L}$.*

Proof Since L' is flat with respect to A, $\mathcal{N}_L = \mathcal{N}_{L'} \cap \mathsf{A}$ by Proposition 17.33. Obviously, $\mathcal{N}_{L'}$ is an ideal. Hence $\ker \varphi = \mathcal{J}(\mathcal{N}_L) \subseteq \mathcal{N}_{L'}$ by Lemma 17.43(ii).

Let $f \in \mathbb{R}_d[\underline{x}]$. Then $\varphi(f) \in \mathcal{D} \subseteq \mathsf{A}$ and hence $\varphi(\varphi(f)) = \pi(\varphi(f)) = \varphi(f)$ by Lemma 17.42(i), so that $(f - \varphi(f)) \in \ker \varphi \subseteq \mathcal{N}_{L'}$. Therefore, $L'(f - \varphi(f)) = 0$. Since $L'\lceil \mathcal{D} = L$, we get $L'(f) = L'(\varphi(f)) = L(\varphi(f)) = \tilde{L}(f)$. Thus, $L' = \tilde{L}$.   □

Lemma 17.45 *Suppose that $L(p^2) \geq 0$ for $p \in$ B. Then $\tilde{L}$ is positive on $\mathbb{R}_d[\underline{x}]$.*

Proof By Proposition 17.34, L is positive on A^2. Since $\tilde{L}$ is flat with respect to A by Lemma 17.43(i), the assertion follows by repeated application of Proposition 17.34. Alternatively, one can also argue as follows.

Let $f \in \mathbb{R}_d[\underline{x}]$. Then $g := (f - \varphi(f)) \in \ker \varphi = \mathcal{N}_{\tilde{L}}$ as noted in the proof of Lemma 17.44. Hence $\tilde{L}(2\varphi(f)g + g^2) = 0$. Since L is positive on A^2 and $\varphi(f) \in \mathsf{A}$,

$$\tilde{L}(f^2) = \tilde{L}((\varphi(f) + g)^2) = \tilde{L}(\varphi(f)^2) + \tilde{L}(2\varphi(f)g + g^2) = L(\varphi(f)^2) \geq 0.$$ □

Putting the preceding lemmas together we have proved Theorem 17.35.   □

17.8 Exercises

1. Give an example of a truncated $\mathcal{K}$-moment functional and a closed set $\mathcal{K}$ such that $\mathcal{M}_{L,\mathcal{K}}$ is not vaguely compact.
2. Let $p \in \mathbb{R}_d[\underline{x}]_{2n}$ and let $\check{p} \in \mathcal{H}_{d+1,2n}$ be defined by (17.15).

   a. Show that $p \in \sum \mathbb{R}[\underline{x}]_n^2$ if and only if $\check{p} \in \sum (\mathcal{H}_{d+1,n})^2$.
   b. Show that the following are equivalent:

      (i) $p(x) \geq 0$ for all $x \in \mathbb{R}^d$.
      (ii) $\check{p}(y) \geq 0$ for all $y \in \mathbb{R}^{d+1}$.
      (iii) $\check{p}(y) \geq 0$ for all $y \in S^d$.

3. Show that for a linear functional L on $\mathcal{H}_{d+1,2n}$ the following are equivalent:

   (i) There exists a Radon measure on $\mathbb{R}^{d+1}$ such that $p \in \mathcal{L}^1(\mathbb{R}^{d+1}, \mu)$ and $L(p) = \int p \, d\mu$ for $p \in \mathcal{H}_{d+1,2n}$.

   (ii) There is a Radon measure on S^d such that $L(p) = \int p \, d\mu$ for $p \in \mathcal{H}_{d+1,2n}$.

   (iii) $L(p) \geq 0$ for all $p \in \mathcal{H}_{d+1,2n}$ such that $p \geq 0$ on S^d.

4. Let L be a linear functional on $\mathcal{H}_{d+1,2n}$ given by a k-atomic measure on $\mathbb{R}^{d+1}$. Show that L has a k'-atomic representing measure on S^d, where $k' \leq k$. Find examples where $k' = k$ and examples where $k' = k$ is impossible.

5. Prove that for a linear functional L on $\mathbb{R}_d[\underline{x}]_{2n}$ the following are equivalent:

   (i) L is a truncated moment functional.

   (ii) There exist a number $k \in \mathbb{N}$ and an extension $\tilde{L}$ of L to a positive linear functional on $\mathbb{R}_d[\underline{x}]_{2n+2k}$ such that rank $H_{n+k}(L) = $ rank $H_{n+k-1}(L)$.

   Hint: For instance, combine Theorems 17.15 and 17.29 and Proposition 17.21.

6. Let L be a positive functional on A^2. Give a second proof for both assertions of Proposition 17.33 by applying Proposition A.25 to Hankel matrices.

7. Show that Proposition 17.27 remains valid for arbitrary sets N. Use this result to give another proof of the determinacy assertion of Theorem 17.29.

8. ([CFM, Lemma 2.5]) Let $m_1, \ldots, m_k \in \mathbb{R}$ and let $x_1, \ldots, x_k \in \mathbb{R}^d$ be pairwise distinct points. Define a functional L on A^2 by $L(f) = \sum_{j=1}^k m_j f(x_j), f \in \mathsf{A}^2$. Suppose that $k = $ rank $H(L)$. Use formula (17.47) to prove that the following are equivalent:

   (i) $L(f^2) \geq 0$ for all $f \in \mathsf{A}$.

   (ii) $m_1 \geq 0, \ldots, m_k \geq 0$.

   (iii) $m_1 > 0, \ldots, m_k > 0$.

17.9　Notes

The multidimensional truncated moment problem was first investigated in the (unpublished) Thesis of J. Matzke [Mt] and independently by R. Curto and L. Fialkow [CF2, CF3]. It is now an active research area, see e.g. [Pu4, La1, CF5, FN1, FN2, BlLs, Bl2, Sm10, F1, DSm1, DSm2]. Multidimensional trigonometric truncated moment problems and related extension problems are treated in the monograph [BW].

Theorem 17.7 is from J. Stochel [St2], while Theorem 17.13 was proved in [Sm10]. Theorem 17.15 was obtained in [Mt] and independently in [CF5]. Theorem 17.29 is contained in [CF2]. Minimal moment functionals have been first studied in [CFM] where they are called "extremal".

The flat extension Theorem 17.36 was obtained by R. Curto and L. Fialkow [CF2]. Theorem 17.35 and the proof presented in the text are taken from [LM]; see [MS] for a general result that applies to noncommutative ∗-algebras as well. Another approach to the flat extension theory is given in [Vs3].

The following result was proved in [CF4] (see [F1] for a correction): Suppose that $N = \mathbb{N}_{0,n}^2$, $n \in \mathbb{N}, n \geq 2$, and $p \in \mathbb{R}[x_1, x_2]$, $\deg(p) \leq 2$. If L is a linear functional on $\mathbb{R}[x_1, x_2]_{2n}$ satisfying the positivity condition (17.41), the consistency condition (17.43), and $p \in \mathcal{N}_L$, then L is a truncated moment functional.

Chapter 18
Multidimensional Truncated Moment Problems: Basic Concepts and Special Topics

While the preceding chapter dealt with existence questions, this chapter is devoted to fundamental notions and various special topics concerning the structure of solutions.

In Sect. 18.1, we begin the study of the cone of truncated $\mathcal{K}$-moment functionals. We investigate its extreme points, introduce its Carathéodory number and consider extreme points of the solution set.

In Sects. 18.2 and 18.3, we define two real algebraic sets $\mathcal{V}_+(L, \mathcal{K})$ and $\mathcal{V}(L)$ associated with a truncated $\mathcal{K}$-moment functional resp. a linear functional L on A. These sets are fundamental concepts for truncated moment problems. They contain the supports of representing measures. The set $\mathcal{V}(L)$ is called the core variety of L. If L is a moment functional, then $\mathcal{V}(L)$ coincides with the set of atoms of representing measures (Theorem 18.21) and L is determinate if and only if $|\mathcal{V}(L)| \leq \dim(A\lceil\mathcal{V}(L))$ (Theorem 18.23). We show that a linear functional $L \neq 0$ on A is a truncated moment functional if and only if $L(e) > 0$ and $\mathcal{V}(L)$ is not empty (Theorem 18.22).

Section 18.4 deals with the maximal mass $\rho_L(t)$ of the one point set $\{t\}$ among all representing measures of L. In Sect. 18.5, we construct ordered maximal mass representations (Theorem 18.38). In Sect. 18.6, we use evaluation polynomials for the study of truncated moment problems (Theorems 18.42, 18.43, and 18.46).

In this chapter, N denotes a fixed **finite** subset of $\mathbb{N}_0^d$, A is the linear span of monomials x^α, $\alpha \in$ N, and, if not specified otherwise, $\mathcal{K}$ is a **closed** subset of $\mathbb{R}^d$. Further, we retain condition (17.1) from Chap. 17 and assume the following:

There exists an element $e \in$ A such that $e(x) \geq 1$ for $x \in \mathcal{K}$. $\qquad$ (18.1)

18.1 The Cone of Truncated Moment Functionals

In this section, we use some concepts and results on convex sets from Appendix A.6. The main objects of this chapter are introduced in the following definition.

© Springer International Publishing AG 2017
K. Schmüdgen, *The Moment Problem*, Graduate Texts in Mathematics 277,
DOI 10.1007/978-3-319-64546-9_18

Definition 18.1 The *moment cone* $\mathcal{S}(\mathsf{A}, \mathcal{K})$ is the set of truncated $\mathcal{K}$-moment sequences $(s_\alpha)_{\alpha \in \mathsf{N}}$ and $\mathcal{L}(\mathsf{A}, \mathcal{K})$ is the set of truncated $\mathcal{K}$-moment functionals on A.

Clearly, $\mathcal{L}(\mathsf{A}, \mathcal{K})$ is a cone in the dual space A^*, $\mathcal{S}(\mathsf{A}, \mathcal{K})$ is a cone in $\mathbb{R}^{|\mathsf{N}|}$ and the map $s \mapsto L_s$ is a bijection of $\mathbb{R}^{|\mathsf{N}|}$ and A^* which maps $\mathcal{S}(\mathsf{A}, \mathcal{K})$ onto $\mathcal{L}(\mathsf{A}, \mathcal{K})$.

Set $E = \mathsf{A}{\restriction}\mathcal{K}$ and $\mathcal{X} = \mathcal{K}$. Then we are in the setup of Section 1.2. Clearly, $f \mapsto f{\restriction}\mathcal{K}$ maps $\mathrm{Pos}(\mathsf{A}, \mathcal{K})$ onto E_+. As discussed in Sect. 17.1, $L \mapsto \tilde{L}$ is a bijection of $\mathcal{L}(\mathsf{A}, \mathcal{K})$ onto the cone $\mathcal{L}$ of moment functionals of $E \subseteq C(\mathcal{K}; \mathbb{R})$. Here $\tilde{L}$ is the linear functional on E defined by $\tilde{L}(f{\restriction}\mathcal{K}) = L(f), f \in \mathsf{A}$, see (17.6) and Lemma 17.4. Repeating verbatim the reasoning from the proof of Proposition 1.27 we obtain

$$\mathcal{L}(\mathsf{A}, \mathcal{K}) \subseteq \mathrm{Pos}(\mathsf{A}, \mathcal{K})^{\wedge} = \overline{\mathcal{L}(\mathsf{A}, \mathcal{K})}. \tag{18.2}$$

If $L \in \mathcal{L}(\mathsf{A}, \mathcal{K}), L \neq 0$, condition (18.1) implies that $L(e) > 0$. Thus, $L \mapsto L(e)$ is a strictly $\mathcal{L}(\mathsf{A}, \mathcal{K})$-positive linear functional on A^*. Hence, by (A.24), the convex set

$$\mathcal{L}_1(\mathsf{A}, \mathcal{K}) := \{L \in \mathcal{L}(\mathsf{A}, \mathcal{K}) : L(e) = 1\} \tag{18.3}$$

is a base of the cone $\mathcal{L}(\mathsf{A}, \mathcal{K})$, see Definition A.38. If the set $\mathcal{K}$ is compact, there are further and stronger results.

Proposition 18.2 *Suppose that $\mathcal{K}$ is compact.*

(i) $\mathcal{L}(\mathsf{A}, \mathcal{K}) = \mathrm{Pos}(\mathsf{A}, \mathcal{K})^{\wedge}$.
(ii) $\mathcal{S}(\mathsf{A}, \mathcal{K})$ *is closed in $\mathbb{R}^{|\mathsf{N}|}$ and $\mathcal{L}(\mathsf{A}, \mathcal{K})$ is closed in the unique norm topology of the finitedimensional vector space A^*.*
(iii) $\mathcal{L}_1(\mathsf{A}, \mathcal{K})$ *is compact base of the cone $\mathcal{L}(\mathsf{A}, \mathcal{K})$ in A^*.*

Proof

(i)   is a restatement of Theorem 17.3.
(ii) Since $\mathcal{L}$ is closed in E^* by Proposition 1.26(ii) and $L \mapsto \tilde{L}$ is a linear bijection of $\mathcal{L}(\mathsf{A}, \mathcal{K})$ onto $\mathcal{L}$, so is $\mathcal{L}(\mathsf{A}, \mathcal{K})$ in A^*. Therefore, since the homeomorphism $s \mapsto L_s$ of $\mathbb{R}^{|\mathsf{N}|}$ on A^* maps $\mathcal{S}(\mathsf{A}, \mathcal{K})$ onto $\mathcal{L}(\mathsf{A}, \mathcal{K})$, the cone $\mathcal{S}(\mathsf{A}, \mathcal{K})$ is closed in $\mathbb{R}^{|\mathsf{N}|}$.
(iii) Let K be a ball in $\mathbb{R}^d$ containing $\mathcal{K}$. We equip A with the supremum norm $\|\cdot\|_K$ on K. The norm topology of A^* is given by the dual norm of $\|\cdot\|_K$ on A^*. Let $L \in \mathcal{L}_1(\mathsf{A}, \mathcal{K})$. Then, by Theorem 17.2, L has a k-atomic representing measure $\mu = \sum_{j=1}^{k} c_j \delta_{x_j}$, where $k \leq |\mathsf{N}| = \dim \mathsf{A}$ and $c_j > 0$, $x_j \in \mathcal{K}$. Using condition (18.1) we obtain $L(e) = 1 = \sum_j c_j e(x_j) \geq \sum_j c_j$, so that $c_j \leq 1$ for all j. Therefore,

$$|L(f)| \leq \sum_{j=1}^{k} c_j |f(x_j)| \leq \sum_{j=1}^{k} \|f\|_K \leq |\mathsf{N}| \, \|f\|_K, \quad f \in \mathsf{A}.$$

Hence $\mathcal{L}_1(A, \mathcal{K})$ is bounded in A^*. Since the linear functional $e \mapsto L(e)$ is continuous on A^*, $\mathcal{L}_1(A, \mathcal{K})$ is obviously closed in A^*. Thus, $\mathcal{L}_1(A, \mathcal{K})$ is compact. □

Example 9.30 shows that for noncompact closed subsets $\mathcal{K}$ the cone $\mathcal{S}(A, \mathcal{K})$ is not closed in general.

Proposition 18.3 *Let* $(\sum A^2)^\wedge$ *denote the cone of positive functionals on* A^2.

(i) *The cone* $(\sum A^2)^\wedge$ *is pointed, that is,* $(\sum A^2)^\wedge \cap (-(\sum A^2)^\wedge) = \{0\}$.
(ii) *For each* $x \in \mathbb{R}^d$, *the point evaluation* l_x *at* x *spans an extreme ray of* $(\sum A^2)^\wedge$.

Proof

(i) Let $L \in (\sum A^2)^\wedge \cap (-(\sum A^2)^\wedge)$. Then $L(f^2) \geq 0$ and $-L(f^2) \geq 0$, hence $L(f^2) = 0$, for all $f \in A$. Therefore, it follows at once from the Cauchy–Schwarz inequality (2.7) that $L(fg) = 0$ for $f, g \in A$. Hence $L = 0$ on A^2.
(ii) Let $L_1, L_2 \in (\sum A^2)^\wedge$. Assume that there are numbers $c_1 > 0, c_2 > 0$ such that $l_x = c_1 L_1 + c_2 L_2$. Set $q := e(x)^{-1} e$. Then $l_x(q) = 1$.

Fix $j \in \{1, 2\}$. Let $f, g, h \in A$. Since L_j is a positive functional on A^2, the Cauchy–Schwarz inequality (2.7) holds. Using this inequality we derive

$$0 \leq c_j L_j((f - f(x)q)h)^2 \leq L_j(h^2) c_j L_j(f - f(x)q)^2$$

$$\leq L_j(h^2)[c_1 L_1(f - f(x)q)^2 + c_2 L_2(f - f(x)q)^2] = L_j(h^2) l_x(f - f(x)q)^2 = 0.$$

Here the last two equalities hold by the relations $l_x = c_1 L_1 + c_2 L_2$ and $l_x(q) = 1$ and by the definition of the evaluation functional l_x. Hence, since $c_j > 0$, we have $L_j((f - f(x)q)h) = 0$, so that

$$L_j(fh) = f(x) L_j(qh) \quad \text{for } f, h \in A. \tag{18.4}$$

Put $a_j := L_j(q^2)$. Setting $h = g - g(x)q$ in (18.4) we obtain

$$L_j(fg) - g(x) L_j(fq) = f(x) L_j(gq) - f(x)g(x)a_j. \tag{18.5}$$

For $h = q$, (18.4) yields $L_j(fq) = f(x)a_j$. Replacing f by g, we get $L_j(gq) = g(x)a_j$. Inserting these two facts into (18.5) it follows that $L_j(fg) = f(x)g(x)a_j = a_j l_x(fg)$ for arbitrary $f, g \in A$. Therefore, $L_j = a_j l_x$ on A^2, that is, L_j belongs to the ray spanned by l_x. This proves that l_x spans an extreme ray. □

Example 18.4 (An extreme ray of $(\sum A^2)^\wedge$ *which is not spanned by a point evaluation)* Let $A^2 = \mathbb{R}_2[\underline{x}]_6$ be the polynomials in two variables of degree at most 6.

We denote by Exr the set of elements of all extreme rays of the cone $(\sum A^2)^\wedge$. By Lemma 18.3(i), $(\sum A^2)^\wedge$ is pointed. The cone $(\sum A^2)^\wedge$ is obviously closed in the norm topology of the dual of $\mathbb{R}_2[\underline{x}]_6$. Thus, Proposition A.37 applies and shows that each element of $(\sum A^2)^\wedge$ is a finite sum of elements of Exr.

On the other hand, by Proposition 13.5, there exists a positive linear functional L_2 on $\mathbb{R}_2[\underline{x}]$ such that $L_2(p_c) < 0$ for the Motzkin polynomial p_c. The restriction $L = L_2\lceil \mathbb{R}_2[\underline{x}]_6$ is in $(\sum \mathsf{A}^2)^\wedge$, but L is not a truncated moment functional, since $p_c \geq 0$ on $\mathbb{R}^2$. Hence L is not a positive combination of point evaluations. Since $L \in (\sum \mathsf{A}^2)^\wedge$ is a sum of elements of Exr, we conclude that there exists an extreme ray of $(\sum \mathsf{A}^2)^\wedge$ that is not spanned by a point evaluation. As shown in [Bl1, Theorem 1.6], the Hankel matrix of each nonzero functional contained in such an extreme ray has rank 7.   ∘

Example 18.4 shows that extreme rays of the cone $(\sum \mathsf{A}^2)^\wedge$ are not necessarily spanned by point evaluations. But all extreme rays of the smaller cone of truncated $\mathcal{K}$-moment functionals do, as shown in the next proposition.

Proposition 18.5 *For a truncated $\mathcal{K}$-moment functional $L \neq 0$ on A^2 the following statements are equivalent:*

(i) *L spans an extreme ray of the cone $\mathcal{L}(\mathsf{A}^2, \mathcal{K})$.*
(ii) *L has a 1-atomic representing measure.*
(iii) *There are a point $x \in \mathcal{K}$ and a number $c > 0$ such that $L = cl_x$.*
(iv) *$\operatorname{rank} H(L) = 1$.*

Proof
(i)→(ii) Let $\mu = \sum_{j=1}^{k} m_j \delta_{x_j} \in \mathcal{M}_L$ with the smallest number k of atoms. Thus, $L = \sum_{j=1}^{k} m_j l_{x_j}$. Assume to the contrary that (ii) is not true. Then we have $k \geq 2$ and $L = m_1 L_1 + L_2$, where $L_1 := l_{x_1}$ and $L_2 := \sum_{j=2}^{k} m_j l_{x_j}$. Clearly, $m_1 > 0$ and $L_2 \neq 0$. Since L spans an extreme ray, $L_j = a_j L$ for some $a_j > 0, j = 1, 2$. Thus, $l_{x_1} = L_1 = a_1 a_2^{-1} L_2$. Hence μ has less than k atoms, which is a contradiction.

(ii)↔(iii) is obvious.

(iii)→(i) By Proposition 18.3, l_x generates an extreme ray of the cone of positive functionals and hence also of the subcone $\mathcal{L}(\mathsf{A}, \mathcal{K})$.

(ii)→(iv) Let $\mu = c\delta_x \in \mathcal{M}_L$. Recall that $\mathfrak{s}(x) \equiv \mathfrak{s}_\mathsf{N}(x)$ denotes the vector $(x^\alpha)_{\alpha \in \mathsf{N}} \in \mathbb{R}^{|\mathsf{N}|}$. By formula (19.29), we have $H(L) = c\,\mathfrak{s}(x)\mathfrak{s}(x)^T$, which implies that $\operatorname{rank} H(L) \leq 1$. Since $L \neq 0$, $H(L) \neq 0$. Hence $\operatorname{rank} H(L) = 1$.

(iv)→(iii) Since $H(L) \succeq 0$ has rank one, there is a nonzero vector $\zeta \in \mathbb{R}^{|\mathsf{N}|}$ such that $H(L) = \zeta\zeta^T$. Let $\mu = \sum_{j=1}^{k} m_j \delta_{x_j}$ be a representing measure of L, where $m_j > 0$ for all j. Then, $H(L) = \sum_{j=1}^{k} m_j \mathfrak{s}(x_j)\mathfrak{s}(x_j)^T$ by (19.29). For $\eta \in \mathbb{R}^{|\mathsf{N}|}$ we obtain

$$\eta^T H(L)\eta = \sum_{j=1}^{k} m_j |\langle \eta, \mathfrak{s}(x_j)\rangle|^2 = |\langle \eta, \zeta\rangle|^2, \tag{18.6}$$

where $\langle \cdot, \cdot \rangle$ is the canonical scalar product of $\mathbb{R}^{|\mathsf{N}|}$. From (18.6) and $m_j > 0$ it follows that $\eta \perp \zeta$ implies $\eta \perp \mathfrak{s}(x_j)$. Hence all vectors $\mathfrak{s}(x_j)$ are multiples of ζ. Therefore, since $\zeta \neq 0$, there exists an i such that ζ, hence all $\mathfrak{s}(x_j)$ are multiplies of $\mathfrak{s}(x_i)$. Then $H(L) = c\,\mathfrak{s}(x_i)\mathfrak{s}(x_i)^T$ with $c > 0$.

We prove that $L = cl_{x_i}$. Let $f = \sum_{\alpha \in \mathsf{N}} f_\alpha x^\alpha \in \mathsf{A}$ and set $\vec{f} = (f_\alpha)^T \in \mathbb{R}^{|\mathsf{N}|}$. Then $\langle \vec{f}, \mathfrak{s}_{x_i} \rangle = \sum_\alpha f_\alpha (x_i)^\alpha = f(x_i)$. Therefore, using (17.33) for $f, g \in \mathsf{A}$ we obtain

$$L(fg) = \vec{f}^T H(L)\vec{g} = c \langle \vec{f}, \mathfrak{s}(x_i) \rangle \langle \vec{g}, \mathfrak{s}(x_i) \rangle = cf(x_i)g(x_i) = cl_{x_i}(fg).$$

Since A^2 is spanned by products fg with $f, g \in \mathsf{A}$, we get $L = cl_{x_i}$.   □

For general vector spaces not of the form A^2, point evaluations do not necessarily span extreme rays, as shown by the following simple example.

Example 18.6 (A point evaluation on A which does not span an extreme ray) Let $d = 1, \mathsf{N} = \{0, 1\}$, so that $\mathsf{A} = \mathrm{Lin}\{1, x\}$. Then $l_0 = \frac{1}{2}(l_1 + l_{-1})$ on A, but $l_1 \neq al_0$ on A for all $a \geq 0$. Hence l_0 does not span an extreme ray of the cone $\mathcal{L}(\mathsf{A}, \mathbb{R})$.   ○

The next definition contains two other important notions.

Definition 18.7 Let $s \in \mathcal{S}(\mathsf{A}, \mathcal{K})$ and $L \in \mathcal{L}(\mathsf{A}, \mathcal{K})$. The *Carathéodory number* $\mathsf{C}(s)$ resp. $\mathsf{C}(L)$ is the smallest number $k \in \mathbb{N}_0$ such that s resp. L has a k-atomic representing measure $\mu \in M_+(\mathcal{K})$.
   The *Carathéodory number* $\mathsf{C}(\mathsf{A}, \mathcal{K})$ is the largest $\mathsf{C}(s)$, where $s \in \mathcal{S}(\mathsf{A}, \mathcal{K})$.

Thus, $\mathsf{C}(\mathsf{A}, \mathcal{K})$ the smallest number $n \in \mathbb{N}$ such that each sequence $s \in \mathcal{S}(\mathsf{A}, \mathcal{K})$ has a k-atomic representing measure, where $k \leq n$. Obviously, $\mathsf{C}(s) = \mathsf{C}(L_s)$. By definition we have $\mathsf{C}(L) = 0$ for $L = 0$.
   Recall from Corollary 1.25 that each moment sequence $s \in \mathcal{S}(\mathsf{A}, \mathcal{K})$ has a k-atomic representing measure, where $k \leq |\mathsf{N}|$. Therefore, the Carathéodory numbers $\mathsf{C}(s)$, $\mathsf{C}(L)$, and $\mathsf{C}(\mathsf{A}, \mathcal{K})$ are well-defined and satisfy

$$\mathsf{C}(\mathsf{A}, \mathcal{K}) = \sup\{\mathsf{C}(s) : s \in \mathcal{S}(\mathsf{A}, \mathcal{K})\} = \sup\{\mathsf{C}(L) : L \in \mathcal{L}(\mathsf{A}, \mathcal{K})\} \leq |\mathsf{N}| = \dim \mathsf{A}.$$

Remark 18.8 Let P be a cone in a finite-dimensional real vector space $E \neq \{0\}$. Let $\mathrm{Exr}(P)$ denote the set of elements of extreme rays of P. We assume that each element of P is a finite sum of elements of $\mathrm{Exr}(P)$. (By Proposition A.37, this holds if P is pointed and closed. It is also true for the cones $\mathcal{S}(\mathsf{A}^2, \mathcal{K})$ and $\mathcal{L}(\mathsf{A}^2, \mathcal{K})$ by Theorem 17.2 and Proposition 18.5, even though these cones are not necessarily closed.)
   For $c \in P$, let $\mathsf{C}(c)$ denote the smallest number k such that c is a sum of k elements from $\mathrm{Exr}(P)$. By Carathéodory's theorem, $\mathsf{C}(c) \leq \dim E$. The largest number $\mathsf{C}(c)$ for $c \in P$ is called the Carathéodory number $\mathsf{C}(P)$ of the cone P.
   Now let $P = \mathcal{S}(\mathsf{A}^2, \mathcal{K})$ and let E be the vector space spanned by $\mathcal{S}(\mathsf{A}^2, \mathcal{K})$. Then $\mathrm{Exr}(P) = \{\lambda l_x : \lambda \geq 0, x \in \mathcal{K}\}$ by Proposition 18.5. Thus, in this case the Carathéodory number $\mathsf{C}(P)$ of the cone P is just the Carathéodory number $\mathsf{C}(\mathsf{A}, \mathcal{K})$ according to Definition 18.7.   ○

Example 18.9 (Moment cone and Carathéodory number in dimension one on $\mathbb{R}[x]_m$) Let $d = 1, m \in \mathbb{N}, \mathsf{N} = \{0, 1, \ldots, m\}$, and $\mathcal{K} := [a, b]$, where $a, b \in \mathbb{R}, a < b$.

Then $A = \mathbb{R}[x]_m$ and $\mathcal{S}(A, \mathcal{K})$ is just the moment cone $\mathcal{S}_{m+1}$ from Definition 10.4. For the Carathéodory number we have

$$C(A, \mathcal{K}) = n + 1 \quad \text{if } m = 2n \text{ or } m = 2n + 1, \ n \in \mathbb{N}.$$

That is, if $\lfloor \frac{m}{2} \rfloor$ denotes the largest integer k satisfying $k \leq \frac{m}{2}$, we can write

$$C(A, \mathcal{K}) = \left\lfloor \frac{m}{2} \right\rfloor + 1. \tag{18.7}$$

We prove formula (18.7) in the even case $m = 2n$; the odd case is similar. For this we use some notions and facts from Sects. 10.2 and 10.3. Let $s \neq 0$ be a sequence of the moment cone $\mathcal{S}_{m+1}$ and choose a representing measure $\mu = \sum_{j=1}^{k} m_j \delta_{x_j}$ of s such that $\mathrm{ind}\,(\mu) = \mathrm{ind}\,(s)$. Recall that for the definition of $\mathrm{ind}\,(\mu)$ atoms in (a, b) are counted twice, while possible atoms a or b are counted only once. If s is a boundary point of $\mathcal{S}_{m+1}$, then $\mathrm{ind}\,(\mu) \leq m$ by Theorem 10.7. Hence μ has at most $n + 1$ atoms. If s is an interior point of $\mathcal{S}_{m+1}$, it has a principal representing measure μ^+ with $n + 1$ atoms by (10.14). Since $\mathrm{ind}\,(s) \geq m + 1 = 2n + 1$ (again by Theorem 10.7), s has no representing measure with fewer atoms than $n + 1$. This proves that $C(A, \mathcal{K}) = n + 1 = \lfloor \frac{m}{2} \rfloor + 1$. $\quad$°

Determining the Carathéodory number in the multivariate case is a difficult problem and only a few results are known, see Theorem 18.43(ii) and Sect. 19.1 below.

Our next result characterizes the extreme points of the set $\mathcal{M}_{L,\mathcal{K}}$.

Proposition 18.10 *Suppose that $L \neq 0$ is a truncated $\mathcal{K}$-moment functional on A and $\mu \in \mathcal{M}_{L,\mathcal{K}}$. Let J_μ denote the canonical embedding of A into $L^1_{\mathbb{R}}(\mathbb{R}^d, \mu)$. The following statements are equivalent:*

(i) *μ is an extreme point of the set $\mathcal{M}_{L,\mathcal{K}}$.*
(ii) *$J_\mu(A) = L^1_{\mathbb{R}}(\mathbb{R}^d, \mu)$.*
(iii) *μ is k-atomic with atoms $t_1, \ldots, t_k \in \mathcal{K}$ and there exist elements $p_1, \ldots, p_k \in A$ such that $p_j(t_i) = \delta_{ij}$ for $i, j = 1, \ldots, k$.*

Proof
(i)↔(ii) By Proposition 1.21, μ is an extreme point of the set $\mathcal{M}_{L,\mathcal{K}}$ if and only if the image of A is dense in $L^1_{\mathbb{R}}(\mathbb{R}^d, \mu)$. Since A is finite-dimensional, the latter holds if and only if $J_\mu(A) = L^1_{\mathbb{R}}(\mathbb{R}^d, \mu)$. Thus (i) and (ii) are equivalent.

(iii)→(ii) is obvious.

(ii)→(iii) Because N is finite, $J_\mu(A) = L^1_{\mathbb{R}}(\mathbb{R}^d, \mu)$ (by (ii)) is finite-dimensional. Hence the measure μ is k-atomic, where $k = \dim L^1_{\mathbb{R}}(\mathbb{R}^d, \mu)$. Let $\mu = \sum_{l=1}^{k} m_l \delta_{t_l}$ with $t_1, \ldots, t_k \in \mathcal{K}$. Since $J_\mu(A) = L^1_{\mathbb{R}}(\mathbb{R}^d, \mu)$, the characteristic function χ_{t_j} of each singleton $\{t_j\}, j = 1, \ldots, k$, is in the image $J_\mu(A)$. Thus $\chi_{t_j} = J_\mu(p_j)$ for some polynomial $p_j \in A$. Then $p_j(t_i) = \delta_{ij}$ for $i, j = 1, \ldots, k$. $\quad\square$

Condition (iii) in Proposition 18.10 suggests the following definition.

Definition 18.11 We say that points $t_1, \ldots, t_k$ of $\mathcal{K}$ satisfy the *separation property* $(SP)_A$ if there exist elements $p_1, \ldots, p_k \in A$ such that $p_j(t_i) = \delta_{ij}, i, j = 1, \ldots, k$.

Proposition 18.10 (i)$\leftrightarrow$(iii) says for each truncated $\mathcal{K}$-moment functional $L \neq 0$ the extreme points of $\mathcal{M}_{L,\mathcal{K}}$ are precisely those k-atomic measures in $\mathcal{M}_{L,\mathcal{K}}$ for which all atoms are in $\mathcal{K}$ and satisfy property $(SP)_A$.

We close this section with simple but useful determinacy criterion.

Proposition 18.12 Let $p \in \text{Pos}(A, \mathcal{K})$. Suppose that $\mathcal{Z}(p) \cap \mathcal{K} = \{t_1, \ldots, t_k\}$ and $t_1, \ldots, t_k \in \mathcal{K}$ satisfy $(SP)_A$. Let $\mu = \sum_{j=1}^{k} m_j \delta_{t_j}$, where $m_j \geq 0$ for all j. If L is a truncated $\mathcal{K}$-moment functional and $\mu \in \mathcal{M}_{L,\mathcal{K}}$, then the set $\mathcal{M}_{L,\mathcal{K}} = \{\mu\}$, that is, L is a determinate truncated $\mathcal{K}$-moment functional.

Proof Let ν be another measure from $\mathcal{M}_{L,\mathcal{K}}$. Since $p \in \text{Pos}(A, \mathcal{K})$ and

$$\int_{\mathcal{K}} p \, d\nu = L(p) = \int p \, d\mu = \sum_j m_j p(t_j) = 0,$$

Proposition 1.23 implies that $\text{supp } \nu \subseteq \mathcal{Z}(p) \cap \mathcal{K} = \{t_1, \ldots, t_k\}$. Hence ν is of the form $\nu = \sum_{j=1}^{k} n_j \delta_{t_j}$ for some numbers $n_j \geq 0$.

Now fix $i \in \{1, \ldots, k\}$. Since $t_1, \ldots, t_k$ satisfy $(SP)_A$, there exists a $q \in A$ such that $q(t_j) = \delta_{ij}$ for $j = 1, \ldots, k$. Then $L(q) = \int q \, d\mu = m_i$ and $L(q) = \int q \, d\nu = n_i$, which implies that $m_i = n_i$. This proves that $\mu = \nu$, that is, L is determinate. $\qquad\square$

18.2 Inner Moment Functionals and Boundary Moment Functionals

Throughout this section, L is a **truncated $\mathcal{K}$-moment functional on** A.

Definition 18.13

$$\mathcal{N}_+(L, \mathcal{K}) := \{p \in \text{Pos}(A, \mathcal{K}) : L(p) = 0\},$$

$$\mathcal{V}_+(L, \mathcal{K}) := \{x \in \mathbb{R}^d : p(x) = 0 \text{ for } p \in \mathcal{N}_+(L, \mathcal{K})\}.$$

Clearly, $\mathcal{N}_+(L, \mathcal{K})$ is a subcone of the cone $\text{Pos}(A, \mathcal{K})$ and $\mathcal{V}_+(L, \mathcal{K})$ is a real algebraic set in $\mathbb{R}^d$. In the case $\mathcal{K} = \mathbb{R}^d$ we write

$$\mathcal{N}_+(L) := \mathcal{N}_+(L, \mathbb{R}^d), \quad \mathcal{V}_+(L) := \mathcal{V}_+(L, \mathbb{R}^d). \tag{18.8}$$

Remark 18.14 We compare $\mathcal{N}_+(L, \mathcal{K})$ and $\mathcal{V}_+(L, \mathcal{K})$ with their counterparts $\mathcal{N}_L$ and $\mathcal{V}_L$ from Definition 17.16. For this we suppose that the linear functional L is defined on A^2. Then the definition of $\mathcal{N}_+(L, \mathcal{K})$ refers to elements from A^2, while

$\mathcal{N}_L$ contains only polynomials from A by (17.29). Obviously, if $p \in \mathcal{N}_L$, then $p^2 \in \mathcal{N}_+(L, \mathcal{K})$. Therefore, for any closed set $\mathcal{K}$ we have

$$V_+(L, \mathcal{K}) \subseteq V_L, \tag{18.9}$$

but in general $V_+(L, \mathcal{K}) \neq V_L$. In Sect. 19.6 we develop examples for which $V_L \neq V_+(L)$. ○

The next proposition is the counterpart of Proposition 17.18 in the present setting.

Proposition 18.15

(i) *For each representing measure $\mu \in \mathcal{M}_{L,\mathcal{K}}$ we have* $\operatorname{supp} \mu \subseteq V_+(L, \mathcal{K}) \cap \mathcal{K}$.
(ii) *Let $p \in \mathcal{N}_+(L, \mathcal{K})$ and $q \in$ A be such that $pq \in$ A. Then $L(pq) = 0$. If in addition $q \in \operatorname{Pos}(A, \mathcal{K})$, then $pq \in \mathcal{N}_+(L, \mathcal{K})$.*

Proof

(i) follows at once from Proposition 1.23.
(ii) The proof is almost verbatim the same as the proof of Proposition 17.18. Take a measure $\mu \in \mathcal{M}_{L,\mathcal{K}}$. Using that $\operatorname{supp} \mu \subseteq V_+(L, \mathcal{K}) \cap \mathcal{K}$ by (i) and $p(x) = 0$ on $V_+(L, \mathcal{K})$ by definition, we obtain

$$L(pq) = \int_{\mathcal{K}} (pq)(x)\, d\mu(x) = \int_{V_+(L,\mathcal{K}) \cap \mathcal{K}} p(x)q(x)\, d\mu(x) = 0. \tag{18.10}$$

If q is in $\operatorname{Pos}(A, \mathcal{K})$, so is pq and therefore $pq \in \mathcal{N}_+(L, \mathcal{K})$. □

The set $\mathcal{N}_+(L, \mathcal{K})$ is closely related to the set $N_+(L)$ from Definition 1.39. Since L is a truncated $\mathcal{K}$-moment functional, there is a moment functional $\tilde{L}$ on $E := A \lceil \mathcal{K}$ defined by $\tilde{L}(f \lceil \mathcal{K}) := L(f), f \in$ A. Clearly, we have $f \in \mathcal{N}_+(L, \mathcal{K})$ if and only if $f \lceil \mathcal{K} \in N_+(\tilde{L})$, that is, $N_+(\tilde{L}) = \mathcal{N}_+(L, \mathcal{K}) \lceil \mathcal{K}$. Therefore, Proposition 1.42 may be restated as the following.

Proposition 18.16

(i) *Let $p \in \mathcal{N}_+(L, \mathcal{K}), p \lceil \mathcal{K} \neq 0$. Then $\varphi_p(L') = L'(p), L' \in$ A*, defines a supporting functional φ_p of the cone $\mathcal{L}(A, \mathcal{K})$ at L. Each supporting functional of $\mathcal{L}(A, \mathcal{K})$ at L is of this form.*
(ii) *L is a boundary point of the cone $\mathcal{L}(A, \mathcal{K})$ if and only if $\mathcal{N}_+(L, \mathcal{K}) \lceil \mathcal{K} \neq \{0\}$.*
(iii) *L is an inner point of the cone $\mathcal{L}(A, \mathcal{K})$ if and only if $\mathcal{N}_+(L, \mathcal{K}) \lceil \mathcal{K} = \{0\}$.*

Note that if $p \in$ A vanishes on $\mathcal{K}$, then obviously $p \in \mathcal{N}_+(L, \mathcal{K})$, but $\varphi_p \equiv 0$, so φ_p is not a supporting hyperplane of $\mathcal{L}(A, \mathcal{K})$ at L. That is, the requirement $p \lceil \mathcal{K} \neq 0$ cannot be omitted in Proposition 18.16(i).

Since $L \in \operatorname{Pos}(A, \mathcal{K})^\wedge$, $\mathcal{N}_+(L, \mathcal{K})$ is an exposed face of the cone $\operatorname{Pos}(A, \mathcal{K})$. The nonempty exposed faces of the cone $\mathcal{L}(A, \mathcal{K})$ are exactly the sets

$$F_p := \{L' \in \mathcal{L}(A, \mathcal{K}) : \varphi_p(L') \equiv L'(p) = 0\}, \quad \text{where } p \in \operatorname{Pos}(A, \mathcal{K}).$$

Example 18.17 (Finitely atomic representing measure) Let $\mu = \sum_{j=1}^{k} c_j \delta_{x_j}$ be a representing measure of L, where $x_j \in \mathcal{K}$ and $c_j > 0$ for all j. Clearly, then we have

$$\mathcal{N}_+(L, \mathcal{K}) = \{p \in \mathrm{Pos}(\mathsf{A}, \mathcal{K}) : p(x_1) = \cdots = p(x_k) = 0\}. \qquad (18.11)$$

Whether or not $\mathcal{V}_+(L, \mathcal{K})$ is "small" or "large" depends very sensitively on the size of A and the number and the location of atoms x_j. Roughly speaking, if sufficiently many atoms are well distributed in a real algebraic set, this set is contained in $\mathcal{V}_+(L, \mathcal{K})$. We illustrate this by the following very simple example. Similar results for $\mathbb{R}_2[\underline{x}]_4$ and $\mathbb{R}_2[\underline{x}]_6$ will be given by Propositions 19.35 and 19.37 below.   ∘

Example 18.18 Let $d = 2, \mathsf{N} = \{(n_1, n_2) \in \mathbb{N}_0^2 : n_1 + n_2 \leq 2\}, \mathcal{K} = \mathbb{R}^2$. Then A is the space of polynomials in two variables of degree at most 2. Let L be a truncated moment functional and $\mu = \sum_{j=1}^{k} c_j \delta_{x_j}$ a k-atomic representing measure of L.

Clearly, each element of $\mathrm{Pos}(\mathsf{A}, \mathcal{K})$ is a sum of squares of linear polynomials. If $k = 1$, then $\mathcal{V}_+(L) = \{x_1\}$. Suppose that $k \geq 2$. Then, if all points x_j are on a line, then $\mathcal{V}_+(L)$ is this line, and if the points x_j are not on a line, then $\mathcal{V}_+(L) = \mathbb{R}^2$.   ∘

18.3   The Core Variety of a Linear Functional

In this section, L is a **linear functional** on A and $\mathcal{K} = \mathbb{R}^d$, that is, we consider the truncated moment problem on $\mathbb{R}^d$. Our aim is to study the core variety $\mathcal{V}(L)$ of L.

We define inductively linear subspaces $\mathcal{N}_k(L), k \in \mathbb{N}$, of A and real algebraic sets $\mathcal{V}_j(L), j \in \mathbb{N}_0$, by $\mathcal{V}_0(L) = \mathbb{R}^d$,

$$\mathcal{N}_k(L) := \{p \in \mathrm{Pos}(\mathsf{A}, \mathcal{V}_{k-1}(L)) : L(p) = 0\}, \qquad (18.12)$$

$$\mathcal{V}_j(L) := \{t \in \mathbb{R}^d : p(t) = 0 \text{ for } p \in \mathcal{N}_j(L)\}. \qquad (18.13)$$

If $\mathcal{V}_k(L)$ is empty for some k, we set $\mathcal{V}_j(L) = \mathcal{V}_k(L) = \emptyset$ for all $j \geq k, j \in \mathbb{N}$.

For $k = 1$ these notions coincide with those defined in Eq. (18.8), that is, $\mathcal{N}_1(L) = \mathcal{N}_+(L) \equiv \mathcal{N}_+(L, \mathbb{R}^d)$ and $\mathcal{V}_1(L) = \mathcal{V}_+(L) \equiv \mathcal{V}_+(L, \mathbb{R}^d)$.

Definition 18.19 The *core variety* $\mathcal{V}(L)$ of the linear functional L on A is

$$\mathcal{V}(L) := \bigcap_{j=0}^{\infty} \mathcal{V}_j(L). \qquad (18.14)$$

Since $\mathcal{V}(L)$ is the zero set of real polynomials, $\mathcal{V}(L)$ is a real algebraic set in $\mathbb{R}^d$.

The preceding definitions resemble the corresponding definitions in Sect. 1.2.5. More precisely, if we set $\mathcal{X} = \mathbb{R}^d, E = \mathsf{A} \subseteq C(\mathbb{R}^d; \mathbb{R})$ and L is a moment functional on E, then formulas (1.22), (1.23), (1.24) coincide with (18.12), (18.13), (18.14), respectively, that is, we have $\mathcal{N}_k(L) = N_k(L), \mathcal{V}_j(L) = V_j(L), \mathcal{V}(L) = V(L)$.

Note that L is now an arbitrary linear functional on A, while it was a moment functional in Subsection 1.2.5.

Proposition 18.20

(i) $\mathcal{N}_k(L) \subseteq \mathcal{N}_{k+1}(L)$ and $\mathcal{V}_k(L) \subseteq \mathcal{V}_{k-1}(L)$ for $k \in \mathbb{N}$.
(ii) If L is a truncated moment functional on A and $\mu \in \mathcal{M}_L$, then $\operatorname{supp} \mu \subseteq \mathcal{V}(L)$.
(iii) There exists a $k \in \mathbb{N}_0$ such that $\mathcal{V}_k(L) = \mathcal{V}_{k+1}(L)$; for any such k we have

$$\mathcal{V}(L) = \mathcal{V}_k(L) = \mathcal{V}_n(L), \quad n \geq k. \tag{18.15}$$

Proof The proofs of (i) and (ii) are verbatim the same as the proofs of assertions (i) and (ii) of Proposition 1.48.

(iii) Define an ideal $\mathcal{I}_j$ of $\mathbb{R}_d[\underline{x}]$ by $\mathcal{I}_j = \{p \in \mathbb{R}_d[\underline{x}] : p(x) = 0$ for $x \in \mathcal{V}_j(L)\}$ for $j \in \mathbb{N}_0$. Since $\mathcal{V}_{j+1}(L) \subseteq \mathcal{V}_j(L)$ by (i), we have $\mathcal{I}_j \subseteq \mathcal{I}_{j+1}$. Thus,

$$\mathcal{I}_0 \subseteq \mathcal{I}_1 \subseteq \cdots \subseteq \mathcal{I}_j \subseteq \mathcal{I}_{j+1} \subseteq \cdots$$

is an ascending chain of ideals in $\mathbb{R}_d[\underline{x}]$. Since $\mathbb{R}_d[\underline{x}]$ is a Noetherian ring (see e.g. [CLO, p. 77]), there exists a $k \in \mathbb{N}_0$ such that $\mathcal{I}_k = \mathcal{I}_n$ for $n \geq k$. Let $\mathcal{U}_j$ denote the real algebraic set defined by $\mathcal{I}_j$. Since $\mathcal{I}_j$ is the vanishing ideal of $\mathcal{V}_j(L)$, it follows that $\mathcal{U}_j = \mathcal{V}_j(L)$. Clearly, $\mathcal{I}_k = \mathcal{I}_n$ implies $\mathcal{U}_k = \mathcal{U}_n$, so that $\mathcal{V}_k(L) = \mathcal{V}_n(L)$ for $n \geq k$.

Assume that $\mathcal{V}_k(L) = \mathcal{V}_{k+1}(L)$ for some $k \in \mathbb{N}_0$. Then $\mathcal{N}_{k+1}(L) = \mathcal{N}_{k+2}(L)$ by (18.12) and hence $\mathcal{V}_{k+1}(L) = \mathcal{V}_{k+2}(L)$ by (18.13). Proceeding by induction we get $\mathcal{V}_n(L) = \mathcal{V}_{n+1}(L)$ for $n \geq k$, which implies (18.15). $\square$

For a truncated moment functional L on A we define the set of possible atoms:

$$\mathcal{W}(L) := \{x \in \mathbb{R}^d : \mu(\{x\}) > 0 \quad \text{for some} \quad \mu \in \mathcal{M}_L\}. \tag{18.16}$$

The importance of the core variety stems from the following three theorems.

The main assertion of the first theorem is only a restatement of Theorem 1.49. It says that the set $\mathcal{W}(L)$ of atoms is equal to the real algebraic set $\mathcal{V}(L)$; in particular, $\mathcal{W}(L)$ is a closed subset of $\mathbb{R}^d$.

Theorem 18.21 *Let L be a truncated moment functional on A. Then*

$$\mathcal{W}(L) = \mathcal{V}(L). \tag{18.17}$$

Each representing measure μ of L is supported on $\mathcal{V}(L)$. For each point $x \in \mathcal{V}(L)$ there is a finitely atomic representing measure μ of L which has x as an atom.

Proof Theorem 1.49 applies with $E = $ A, $\mathcal{X} = \mathbb{R}^d$ and yields the equality (18.17), since $\mathcal{V}(L) = V(L)$ and $\mathcal{W}(L) = W(L)$. The other assertions follow from Proposition 18.20(ii) and Corollary 1.25. $\square$

The next results provide existence and determinacy criteria in terms of the core variety $\mathcal{V}(L)$; they are based on Theorems 1.30 and 1.36.

Recall that by condition (18.1) there exists an $e \in$ A such that $e(x) \geq 1$ on $\mathcal{K} = \mathbb{R}^d$.

Theorem 18.22 *A linear functional $L \neq 0$ on A is a truncated moment functional if and only if $L(e) \geq 0$ and the core variety $\mathcal{V}(L)$ is not empty.*

Proof In this proof we abbreviate $\mathcal{N}_j = \mathcal{N}_j(L)$, $\mathcal{V}_j = \mathcal{V}_j(L)$, and $\mathcal{V} = \mathcal{V}(L)$.

First suppose L is a truncated moment functional. Since each representing measure is supported on $\mathcal{V}$ by Theorem 18.21 and $L \neq 0$, we have $\mathcal{V} \neq \emptyset$ and $L(e) > 0$.

Now we prove the converse implication. If $L(e) = 0$, then $e \in \mathcal{N}_1$, so $\mathcal{V}$ is empty, which contradicts the assumption. Therefore, $L(e) > 0$. By Proposition 18.20(iii), there is a number $k \in \mathbb{N}_0$ such that $\mathcal{V} = \mathcal{V}_k = V_{k+1}$.

Let $f \in \text{Pos}(A, \mathcal{V})$. We prove that $L(f) \geq 0$. Assume to the contrary that $L(f) < 0$. Set $q := f - L(e)^{-1}L(f)e \in$ A. Since $L(e) > 0$, $L(f) < 0$, and $e(x) \geq 1$ on $\mathbb{R}^d$, we have $q(x) > 0$ on $\mathcal{V} = \mathcal{V}_k$ and $L(q) = 0$. Hence $q \in \mathcal{N}_{k+1}$ and $\mathcal{V} = \mathcal{V}_{k+1} \subseteq \mathcal{Z}(q) = \emptyset$. This is a contradiction, since $\mathcal{V}$ is not empty. Thus we have shown that $L(f) \geq 0$.

Therefore, by Lemma 17.4, condition (17.5) holds. Hence there exists a well-defined linear functional $\tilde{L}$ on $E = A\lceil \mathcal{V}$ given by $\tilde{L}(f\lceil) := L(f), f \in$ A.

We prove that $\tilde{L}$ is strictly E_+–positive. Let $f \in$ A be such that $f\lceil \mathcal{V} \in E_+$ and $f\lceil \mathcal{V} \neq 0$. Then $f \in \text{Pos}(A, \mathcal{V})$ and hence $L(f) \geq 0$ as shown in the paragraph before last. Since $f\lceil \mathcal{V} \neq 0$ and $\mathcal{V}$ is not empty, there exists an $x_0 \in \mathcal{V}$ such that $f(x_0) > 0$. If $L(f)$ were zero, then $f \geq 0$ on $\mathcal{V}_k$ implies $f \in \mathcal{N}_{k+1}$, so $\mathcal{V} = \mathcal{V}_{k+1} \subseteq \mathcal{Z}(f)$. This is impossible, because $x_0 \in \mathcal{V}$ and $f(x_0) > 0$. Thus $\tilde{L}$ is strictly E_+–positive.

Therefore, by Theorem 1.30, $\tilde{L}$ is a moment functional on E and so is L on A. $\square$

Theorem 18.23 *For any truncated moment functional L on A the following are equivalent:*

  (i) *L is not determinate.*
 (ii) *$|\mathcal{V}(L)| > \dim(A\lceil \mathcal{V}(L))$.*
(iii) *There is a representing measure μ of L such that $|\text{supp}\,\mu| > \dim(A\lceil \mathcal{V}(L))$.*

In particular, L is not determinate if $|\mathcal{V}(L)| > \dim A = |\mathsf{N}|$ or if L has a representing measure μ such that $|\text{supp}\,\mu| > \dim A = |\mathsf{N}|$.

Proof Since $\mathcal{W}(L) = \mathcal{V}(L)$ by Theorem 18.21, the assertions are only restatements of Theorem 1.36 and Corollary 1.37, applied with $E = A$ and $\mathcal{X} = \mathbb{R}^d$. $\square$

Suppose that L is a truncated moment functional on A. Let k be the smallest integer for which $\mathcal{V}_k(L) = \mathcal{V}_{k+1}(L)$. Then, by Proposition 18.20 and (18.17),

$$\mathcal{W}(L) = \mathcal{V}(L) = \mathcal{V}_n(L) = \mathcal{V}_k(L) \subsetneq \mathcal{V}_j(L) \subsetneq \mathcal{V}_1(L) \tag{18.18}$$

for $n \geq k$, $1 < j < k$. In general, it is difficult to determine the set $\mathcal{V}(L) = \mathcal{W}(L)$. It would be important to have useful criteria ensuring that $\mathcal{V}(L) = \mathcal{V}_1(L) \equiv \mathcal{V}_+(L)$; see also Theorem 1.45.

We close this section with two illustrating examples.

Example 18.24 ($\mathcal{V}(L) \neq \emptyset$ and L is not a truncated moment functional) Let $d = 1$ and $\mathsf{N} = \{0, 2\}$. Then $\mathsf{A} = \{a + bx^2 : a, b \in \mathbb{R}\}$, so condition (18.1) is satisfied with $e(x) = 1$. Define a linear functional L on A by $L(a + bx^2) = -a$. Clearly, L is not a truncated moment functional, because $L(1) = -1$.

Then $\text{Pos}(\mathsf{A}, \mathbb{R}) = \{a + bx^2 : a \geq 0, b \geq 0\}$, so $\mathcal{N}_1(L) = \mathbb{R}_+ \cdot x^2$ and $\mathcal{V}_1(L) = \{0\}$. Hence $\text{Pos}(\mathsf{A}, \mathcal{V}_1(L)) = \{a + bx^2 : a \geq 0, b \in \mathbb{R}\}$, $\mathcal{N}_2(L) = \mathbb{R} \cdot x^2$, and $\mathcal{V}_2(L) = \{0\}$. Therefore, $\mathcal{V}(L) = \{0\} \neq \emptyset$.   ∘

Example 18.25 (A truncated moment functional with $\mathcal{V}(L) = \mathcal{V}_2(L) \neq \mathcal{V}_1(L)$) Let $d = 1$ and $\mathsf{N} = \{0, 2, 4, 5, 6, 7, 8\}$. Then $\mathsf{A} = \text{Lin}\{1, x^2, x^4, x^5, x^6, x^7, x^8\}$. We fix a real number $\alpha > 1$ and define $\mu = \delta_{-1} + \delta_1 + \delta_\alpha$. For the corresponding moment functional $L \equiv L^\mu = l_{-1} + l_1 + l_\alpha$ on A we shall show that

$$\mathcal{W}(L) = \mathcal{V}(L) = \mathcal{V}_2(L) = \{1, -1, \alpha\} \subset \{1, -1, \alpha, -\alpha\} = \mathcal{V}_1(L). \tag{18.19}$$

Let us prove (18.19). Put $p(x) := (x^2 - 1)^2(x^2 - \alpha^2)^2$. Clearly, $p \in \mathsf{A}$, $L(p) = 0$ and $p \in \text{Pos}(\mathbb{R})$, so that $p \in \mathcal{N}_1(L)$. Conversely, let $f \in \mathcal{N}_1(L)$. Then $L(f) = 0$ implies that $f(\pm 1) = f(\alpha) = 0$. Since $f \in \text{Pos}(\mathbb{R})$, the zeros $1, -1, \alpha$ have even multiplicities. Hence $f(x) = (x-1)^2(x+1)^2(x-\alpha)^2(ax^2 + bx + c)$ with $a, b, c \in \mathbb{R}$. The coefficients of x and x^3 of f are $\alpha^2 b - 2\alpha c$ and $4\alpha c + (1 - 2\alpha^2)b - 2\alpha a$. Since x and x^3 are not in A, these coefficients have to be zero. This yields $b = 2\alpha a, c = \alpha^2 a$. Therefore, $ax^2 + bx + c = a(x + \alpha)^2$. Clearly, $a \geq 0$, so that $f = ap \in \mathbb{R}_+ \cdot p$. Thus we have shown that $\mathcal{N}_1(L) = \mathbb{R}_+ \cdot p$. Hence $\mathcal{V}_1(L) = \mathcal{Z}(p) = \{1, -1, \alpha, -\alpha\}$.

Now we set $q(x) = x^4(x^2 - 1)(\alpha - x)$. Then $q \in \mathsf{A}$. Since $q(\pm 1) = q(\alpha) = 0$ and $q(-\alpha) = \alpha^4(\alpha^2 - 1)2\alpha > 0$, we have $q \in \text{Pos}(\mathsf{A}, \mathcal{V}_1(L))$ and $L(q) = 0$. Thus, $q \in \mathcal{N}_2(L)$ and hence $\mathcal{V}_2(L) \subseteq \mathcal{V}_1(L) \cap \mathcal{Z}(q) = \{1, -1, \alpha\}$.

Since $1, -1, \alpha$ are atoms of μ, $\{1, -1, \alpha\} \subseteq \mathcal{W}(L)$. Now (18.19) follows from (18.18) and the preceding. Note that the moment functional L is determinate.   ∘

18.4   Maximal Masses

In this section, we suppose that $L \neq 0$ is a **truncated $\mathcal{K}$-moment functional** on A.

The next definition contains the main notions that will be studied in this section.

Definition 18.26

$$\rho_L(x) := \sup\{\mu(\{x\}) : \mu \in \mathcal{M}_{L,\mathcal{K}}\}, \quad x \in \mathcal{K}, \tag{18.20}$$

$$\mathcal{W}(L, \mathcal{K}) := \{x \in \mathcal{K} : \mu(\{x\}) > 0 \text{ for some } \mu \in \mathcal{M}_{L,\mathcal{K}}\}.$$

That is, $\mathcal{W}(L,\mathcal{K})$ is the set of atoms and $\rho_L(x)$ is the supremum of point masses at x of all solutions of the truncated $\mathcal{K}$-moment problem for L. Note that $\mathcal{W}(L,\mathbb{R}^d)$ coincides with the set $\mathcal{W}(L)$ defined by (18.16).

Since L always has an atomic representing measure by Theorem 17.2 and $L \neq 0$ by assumption, $\mathcal{W}(L,\mathcal{K})$ is not empty. It is obvious that

$$\mathcal{W}(L,\mathcal{K}) = \{x \in \mathcal{K} : \rho_L(x) > 0\}.$$

By the inequality (18.25) below, $\rho_L(x) < \infty$ for all $x \in \mathcal{K}$. From Proposition 18.15(i),

$$\mathcal{W}(L,\mathcal{K}) \subseteq \mathcal{V}_+(L,\mathcal{K}) \cap \mathcal{K}. \tag{18.21}$$

In general, $\mathcal{W}(L,\mathcal{K}) \neq \mathcal{V}_+(L,\mathcal{K}) \cap \mathcal{K}$, as shown by Example 1.41 with $\mathcal{K} = \mathbb{R}^2$. It is an interesting problem to determine when we have equality in (18.21).

Further, we define another important nonnegative number $\kappa_L(x)$ by

$$\kappa_L(x) = \inf\{ L(p) : p \in \mathrm{Pos}(\mathsf{A},\mathcal{K}), \ p(x) = 1\} \tag{18.22}$$

$$= \inf\left\{ \frac{L(p)}{p(x)} : p \in \mathrm{Pos}(\mathsf{A},\mathcal{K})\right\}, \tag{18.23}$$

where we set $\frac{c}{0} := +\infty$ for $c \geq 0$. The equality in (18.23) is obvious. The polynomial $p_x := e(x)^{-1}e$ is in $\mathrm{Pos}(\mathsf{A},\mathcal{K})$ and satisfies $p_r(x) = 1$, so the corresponding set in (18.22) is not empty and $\kappa_L(x)$ is well defined.

Note that both quantities $\rho_L(x)$ and $\kappa_L(x)$ depend on the set $\mathcal{K}$ as well; for simplicity we have suppressed this dependence.

The number $\kappa_L(x)$ is defined by a conic optimization problem (A.25) for the cone $C = \mathrm{Pos}(\mathsf{A},\mathcal{K})$ in A and functionals $L_1 = L$ and $L_0 = l_x$, see Appendix A.6. The corresponding dual problem (A.26) is

$$\kappa_L(x)^* := \sup\{c \in \mathbb{R} : (L - cl_x) \in \mathrm{Pos}(\mathsf{A},\mathcal{K})^{\wedge}\}. \tag{18.24}$$

A standard fact of the duality theory of conic optimization (see e.g. [BN]) states that if L is an interior point of the dual cone $C^{\wedge}$, then the infimum in (A.25) is attained. Proposition 18.28(iii) below is the corresponding result in the present context.

Proposition 18.27 *Let $x \in \mathcal{K}$. Then*

$$\rho_L(x) = c_L(x) := \sup\{c \in \mathbb{R}_+ : (L - cl_x) \in \mathcal{L}(\mathsf{A},\mathcal{K})\} \leq \kappa_L(x) < +\infty \tag{18.25}$$

and the functional $L - \rho_L(x)l_x$ belongs to the boundary of the cone $\mathcal{L}(\mathsf{A},\mathcal{K})$.

If $\mathcal{K}$ is compact, then we have $\kappa_L(x) = \rho_L(x)$, the supremum in Eq. (18.20) is a maximum, and $L - \rho_L(x)l_x$ is a truncated $\mathcal{K}$-moment functional.

Proof In this proof, we abbreviate $\mathcal{L} := \mathcal{L}(A, \mathcal{K})$.

Let $c \in \mathbb{R}_+$ be such that $\tilde{L} := (L - cl_x) \in \mathcal{L}$. Such c exists; for instance, $c = 0$. If $\tilde{\mu}$ is a representing measure of $\tilde{L}$, then $\mu = \tilde{\mu} + c\delta_x$ is a positive measure representing L. Hence $c \leq \mu(\{x\}) \leq \rho_L(x)$. Taking the supremum over c yields $c_L(x) \leq \rho_L(x)$.

Let us assume that $c_L(x) < \rho_L(x)$. By the definition of $\rho_L(x)$, there exist a number $c \in (c_L(x), \rho_L(x))$ and a representing measure μ of s such that $\mu(\{x\}) = c$. Then $\tilde{\mu} := \mu - c \cdot \delta_x$ is a positive (!) Radon measure representing $\tilde{L} = L - cl_x$. Hence $\tilde{L} \in \mathcal{L}$, so that $c \leq c_L(x)$, which is a contradiction. This proves that $c_L(x) \geq \rho_L(x)$. Combining the preceding two paragraphs, we have shown that $c_L(x) = \rho_L(x)$.

Let $\nu \in \mathcal{M}_{L,\mathcal{K}}$. For any $p \in \text{Pos}(A, \mathcal{K}), p(x) = 1$, we have

$$L(p) = \int p(y)\, d\nu(y) \geq \nu(\{x\})p(x) = \nu(\{x\}).$$

Taking the infimum over such p and the supremum over $\nu \in \mathcal{M}_{L,\mathcal{K}}$ we obtain $\kappa_L(x) \geq \rho_L(x)$. This completes the proof of (18.25).

Set $L_0 := L - \rho_L(x)l_x = L - c_L(x)l_x$. From the definition of $c_L(x)$ in (18.25) it follows at once that the functional L_0 belongs to the boundary of $\mathcal{L}(A, \mathcal{K})$.

For the rest of this proof we assume that $\mathcal{K}$ is compact.

Since $\mathcal{K}$ is compact, we have $\mathcal{L}(A, \mathcal{K}) = \text{Pos}(A, \mathcal{K})^\wedge$ by Proposition 18.2(i). Hence $c_L(x) = \kappa_L(x)^*$ by (18.24). Lemma A.40 yields $\kappa_L(x) = \kappa_L(x)^*$. Therefore, $\kappa_L(x) = c_L(x) = \rho_L(x)$. Further, $\mathcal{L}(A, \mathcal{K})$ is closed in A^* by Proposition 18.2(ii). Therefore, the boundary functional L_0 belongs to the cone $\mathcal{L}(A, \mathcal{K})$.

By the definition of $\rho_L(x)$, there is a sequence $(\mu_n)_{n\in\mathbb{N}}$ of $\mathcal{M}_{L,\mathcal{K}}$ such that $\lim_n \mu_n(\{x\}) = \rho_L(x)$. By Proposition 17.3(ii), $\mathcal{M}_{L,\mathcal{K}}$ is compact in the vague topology. Hence there exists a subnet $(\mu_i)_{i\in J}$ of the sequence $(\mu_n)_{n\in\mathbb{N}}$ which converges vaguely to some measure $\mu \in \mathcal{M}_{L,\mathcal{K}}$. Then

$$\rho_L(x) = \lim_n \mu_n(\{x\}) = \lim_i \mu_i(\{x\}) \leq \mu(\{x\}),$$

where the last inequality holds by Proposition A.5, applied to $K = \{x\}$. By definition, $\mu(\{x\}) \leq \rho_L(x)$. Thus, $\rho_L(x) = \mu(\{x\})$. This shows that the supremum in (18.20) is attained at μ.   □

Proposition 18.28 *Suppose that $x \in \mathcal{K}$.*

(i) *If $\rho_L(x) > 0$, then $x \in \mathcal{V}_+(L, \mathcal{K})$.*

(ii) *Suppose that $\mathcal{K}$ is compact and assume that the infimum in (18.23), or equivalently in (18.22), is a minimum. Then $x \in \mathcal{V}_+(L, \mathcal{K})$ implies that $\rho_L(x) > 0$.*

(iii) *If L is a relatively interior point of the cone $\mathcal{L}(A, \mathcal{K})$, or equivalently, $L(f) > 0$ for all $f \in \text{Pos}(A, \mathcal{K}), f \lceil \mathcal{K} \neq 0$, then the infimum in (18.23) is a minimum.*

Proof

(i) Suppose that $\rho_L(x) > 0$ and assume to the contrary that $x \notin \mathcal{V}_+(L, \mathcal{K})$. Then there exists a $p_0 \in \mathcal{N}_+(L, \mathcal{K})$ such that $p_0(x) \neq 0$. Upon scaling we can assume that $p_0(x) = 1$. Since $L(p_0) = 0$ by $p_0 \in \mathcal{N}_+(L, \mathcal{K})$, then $\kappa_L(x) = 0$ by

(18.22). Hence $\rho_L(x) = 0$, because $\kappa_L(x) \geq \rho_L(x) \geq 0$ by (18.25). This is a contradiction.

(ii) Let $x \in \mathcal{V}_+(L, \mathcal{K})$. Let us assume that the infimum in (18.22) is attained at $p_0 \in \text{Pos}(A, \mathcal{K})$, that is, $L(p_0) = \kappa_L(x)$. If $L(p_0)$ were zero, then we would have $p_0 \in \mathcal{N}_+(L, \mathcal{K})$ and hence $p_0(x) = 0$, since $x \in \mathcal{V}_+(L, \mathcal{K})$. This contradicts $p_0(x) = 1$ by (18.22). Thus $L(p_0) > 0$. Since $\mathcal{K}$ is compact, $\kappa_L(x) = \rho_L(x)$ by Proposition 18.27. Therefore, $\rho_L(x) = \kappa_L(x) = L(p_0) > 0$.

(iii) We will derive the assertion from Theorem 1.30. Let $E = A\lceil\mathcal{K}, \mathcal{X} = \mathcal{K}, \tilde{f} = f\lceil\mathcal{K}$. Recall that $L \mapsto \tilde{L}$ is a bijection of $\mathcal{L}(A, \mathcal{K})$ onto the cone $\mathcal{L}$ of the moment functional on E, where $\tilde{L}$ is defined by (17.6), that is, $\tilde{L}(\tilde{f}) = L(f), f \in A$. The second assumption on L means that $\tilde{L}$ is strictly E_+-positive. By Lemma 1.29, this holds if and only $\tilde{L}$ is an inner point of $\mathcal{L}$, or equivalently, L is a relatively interior point of $\mathcal{L}(A, \mathcal{K})$. Then, by Theorem 1.30(iii), there exists $\tilde{f}_0 \in E_+$, $\tilde{f}_0(x) = 1$, such that $\tilde{L}(\tilde{f}_0) = \inf\{\tilde{L}(\tilde{f}) : \tilde{f} \in E_+, \tilde{f}(x) = 1\}$. Since $\tilde{L}(\tilde{f}) = L(f)$ and $\tilde{f} \in E_+$ if and only if $f \in \text{Pos}(A, \mathcal{K})$, we conclude that $f_0 \in A$ attains the infimum in (18.23). □

The following simple example shows that the supremum in (18.20) is not necessarily attained if the set $\mathcal{K}$ is not compact.

Example 18.29 Let $\mathcal{K} = \mathbb{R}$ and $\mathsf{N} = \{0, 1, 2\}$. Define $L(p) = \frac{1}{2}(p(1) + p(-1))$ for $p \in A = \mathbb{R}[x]_2$. For $c \in [0, 1)$, we set $\mu_c := c\delta_0 + \frac{1}{2}(1 - c)(\delta_{y(c)} + \delta_{-y(c)})$, where $y(c) = (1 - c)^{-1/2}$. Then μ_c is a representing measure of L and $\mu_c(\{0\}) = c$.

One verifies that $\kappa_L(0) = c_L(0) = \rho_L(0) = 1$. While the infimum for $\kappa_L(0)$ in (18.22) is attained at $p = 1$, the suprema for $\rho_L(0)$ in (18.20) and $c_L(0)$ in (18.25) are not attained. That is, there is no measure $\mu \in \mathcal{M}_L$ such that $\mu(\{0\}) = 1$ and $L' := L - \rho_L(0)l_0 \neq 0$ is not a truncated moment functional on A, since $L'(1) = 0$. ∘

Definition 18.30 Let $\mu = \sum_{j=1}^{k} m_j\delta_{x_j}$ be a measure in $\mathcal{M}_{L,\mathcal{K}}$, where $k \in \mathbb{N}$ and $x_j \in \mathcal{K}$ for $j = 1, \ldots, k$, and let $i \in \{1, \ldots, k\}$. We shall say that

- μ has *maximal mass* at x_i if $m_i = \rho_L(x_i)$,
- μ is a *maximal mass measure* for the functional L, or briefly, μ is *maximal mass*, if $m_j = \rho_L(x_j)$ for all $j = 1, \ldots, k$.

Recall that a measure $\mu = \sum_{j=1}^{k} m_j\delta_{x_j}$ is called *k-atomic* if $m_j > 0$ for all j and $x_i \neq x_j$ for all $i \neq j$.

The following proposition and Theorem 18.33 below deal with the question of when masses of atoms of atomic representing measures are maximal masses.

Proposition 18.31 *Suppose that $\mu = \sum_{j=1}^{k} m_j\delta_{x_j}$ is a representing measure of L, where $k \in \mathbb{N}$ and $m_j > 0$ and $x_j \in \mathcal{K}$ for all j. Let $i \in \{1, \ldots, k\}$. Then:*

(i) *If $m_i = \kappa_L(x_i)$, then $m_i = \rho_L(x_i)$, that is, μ has maximal mass at x_i.*

(ii) *$m_i = \kappa_L(x_i)$ if and only there exists a sequence $(p_n)_{n \in \mathbb{N}}$ with $p_n \in \text{Pos}(A, \mathcal{K})$ such that $p_n(x_i) = 1$ and $\lim_n p_n(x_j) = 0$ for all $j \neq i$.*

(iii) *If there exists a $p \in \text{Pos}(A, \mathcal{K})$ such that $p(x_i) = 1$ and $p(x_j) = 0$ for $j \neq i$, then $m_i = \kappa_L(x_i) = \rho_L(x_i)$.*

(iv) *Assume that the infimum (18.22) for $x = x_i$ is a minimum. If $m_i = \kappa_L(x_i)$, then there exists a $p \in \mathrm{Pos}(A, \mathcal{K})$ such that $p(x_i) = 1$ and $p(x_j) = 0$ for $j \neq i$.*

(v) *If $m_j = \kappa_L(x_j)$ for all $j = 1, \ldots, k$, then $k \leq \dim A$ and μ is k-atomic.*

Proof By the definition of $\rho_L(x_i)$ and (18.25) we always have

$$m_i \leq \rho_L(x_i) \leq \kappa_L(x_i). \tag{18.26}$$

(i) follows at once from (18.26).

(ii) First suppose that there exists a sequence $(p_n)_{n \in \mathbb{N}}$ as stated above. Since $p_n(x_i) = 1$, we have $m_i \leq \kappa_L(x_i) \leq L(p_n)$ by (18.23). Combined with the equality

$$\lim_n L(p_n) = \sum_{j=1}^{k} m_j \left(\lim_n p_n(x_j) \right) = \sum_{j=1}^{k} m_j \delta_{ji} = m_i$$

we conclude that $m_i = \kappa_L(x_i)$.

Conversely, suppose that $m_i = \kappa_L(x_i)$. By the definition of $\kappa_L(x_i)$, there exists a sequence $(p_n)_{n \in \mathbb{N}}$ of elements $p_n \in \mathrm{Pos}(A, \mathcal{K})$ such that $p_n(x_i) = 1$ and $\kappa_L(x_i) = \lim_n L(p_n)$. Clearly, $0 \leq p_n(x_j) \leq m_j^{-1} L(p_n)$ for all $j = 1, \ldots, k$. Since the sequence $(L(p_n))_{n \in \mathbb{N}}$ converges, each sequence $(p_n(x_j))_{n \in \mathbb{N}}$ is bounded. By passing to a subsequence if necessary, we can assume that $a_j := \lim_n p_n(x_j)$ exists for all j. From $p_n \in \mathrm{Pos}(A, \mathcal{K})$ we get $a_j \geq 0$. Then

$$m_i = \kappa_L(x_i) = \lim_n L(p_n) = m_i + \sum_{j=1, j \neq i}^{k} m_j a_j. \tag{18.27}$$

Since $m_j > 0$ for all j by assumption, this implies that $a_j = \lim_n p_n(x_j) = 0$ for all $j = 1, \ldots, k, j \neq i$. Thus the sequence $(p_n)_{n \in \mathbb{N}}$ has the desired properties.

(iii) From (ii), with $p_n = p$, we get $m_i = \kappa_L(x_i)$. Hence $m_i = \rho_L(x_i)$ by (i).

(iv) Suppose that for $x = x_i$ the infimum in (18.23) is attained at $p \in A$, that is, $L(p) = \kappa_L(x_i)$ and $p(x_i) = 1$. Thus we can set $p_n = p$ for $n \in \mathbb{N}$ in the second half of the proof of (ii). From (18.27) we conclude that $a_j = p(x_j) = 0$ for all $j \neq i$.

(v) Fix $j \in \{1, \ldots, k\}$. Since $m_j = \kappa_L(x_j)$, there is a sequence $(p_{jn})_{n \in \mathbb{N}}$ of elements as stated in (i). To prove that $k \leq \dim A$ it suffices to show that the linear functionals $l_{t_1}, \ldots, l_{t_k}$ on A are linearly independent. Assume that there are numbers $\lambda_1, \ldots, \lambda_k \in \mathbb{R}$ such that $0 = \sum_{i=1}^{k} \lambda_i l_{x_i}(f)$ for all $f \in A$. Setting $f = p_{jn}$ and passing to the limit $n \to \infty$, we obtain

$$0 = \lim_n \sum_{i=1}^{k} \lambda_i l_{x_i}(p_{jn}) = \sum_{i=1}^{k} \lambda_i \left(\lim_n p_{jn}(x_i) \right) = \sum_{i=1}^{k} \lambda_i \delta_{ij} = \lambda_j.$$

Thus, $\lambda_j = 0$ for all j, so the functionals $l_{t_1}, \ldots, l_{t_k}$ are linearly independent. In particular, the points x_j are pairwise distinct. Hence μ is k-atomic. $\square$

The following definition is suggested by Proposition 18.31 (iii) and (iv).

Definition 18.32 We shall say that points $x_1, \ldots, x_k \in \mathcal{K}$ have the *positive separation property* $(PSP)_{A,\mathcal{K}}$ if the following holds:
There are elements $q_1, \ldots, q_k \in \text{Pos}(A, \mathcal{K})$ such that $q_j(x_i) = \delta_{ij}, i, j = 1, \ldots, k$.

We summarize some of the preceding results in the following theorem.

Theorem 18.33 *Let L be a truncated $\mathcal{K}$-moment functional on A and let $\mu = \sum_{j=1}^{k} m_j \delta_{x_j}$ be a k-atomic representing measure of L, where $k \in \mathbb{N}$ and $x_1, \ldots, x_k \in \mathcal{K}$.*

(i) *If the points $x_1, \ldots, x_k$ satisfy the positive separation property $(PSP)_{A,\mathcal{K}}$, then $m_j = \rho_L(x_j) = \kappa_L(x_j)$ for all $j = 1, \ldots, k$ and the measure μ is maximal mass.*
(ii) *Suppose that $\mathcal{K}$ is compact and the infimum (18.23) is attained at each point $x = x_j, j = 1, \ldots, k$. If μ is maximal mass, then the points $x_1, \ldots, x_k$ have property $(PSP)_{A,\mathcal{K}}$.*
(iii) *Suppose that $\mathcal{K}$ is compact and L is a relatively inner point of the cone $\mathcal{L}(A, \mathcal{K})$. Then μ is maximal mass if and only if the points $x_1, \ldots, x_k$ obey the positive separation property $(PSP)_{A,\mathcal{K}}$.*

Proof

(i) Fix $j = 1, \ldots, k$. Let q_j be as in Definition 18.32. Setting $p = q_j$ in Proposition 18.31(iii) we obtain $m_j = \kappa_L(x_j) = \rho_L(x_j)$.
(ii) Since $\mathcal{K}$ is compact, $\rho_L(x_j) = \kappa_L(x_j)$ for all j by Proposition 18.27. Hence $m_j = \kappa_L(x_j)$, because μ is maximal mass. Therefore, Proposition 18.31(iv) implies that the points $x_1, \ldots, x_k$ have property $(PSP)_{A,\mathcal{K}}$.
(iii) Since L is a relatively inner point of $\mathcal{L}(A, \mathcal{K})$, it follows from Proposition 18.28(iii) that the infimum in (18.23) is attained at each point $x = x_j$. Now the assertion follows from (i) and (ii). $\qquad\square$

The next example gives a simple recipe for constructing maximal mass measures.

Example 18.34 (Squares of polynomials of degree one) Let $x_1, \ldots, x_{d+1} \in \mathbb{R}^d$. Assume that these points are not contained in a $(d-1)$-dimensional affine subspace, or equivalently, for any $i \in \{1, \ldots, d+1\}$ the d vectors $x_k - x_i, k \neq i$, do not lie in a $(d-1)$-dimensional linear subspace of $\mathbb{R}^d$.

Fix $j \in \{1, \ldots, d+1\}$ and take an index $i \in \{1, \ldots, d+1\}$ such that $i \neq j$. We choose a vector $a_j \in \mathbb{R}^d, a_j \neq 0$, which is orthogonal to the $d - 1$ vectors $x_k - x_i$, where $k = 1, \ldots, d+1, k \neq j, i$. Then $a_j \cdot x_k = a_j \cdot x_i$ for $k = 1, \ldots, d+1, k \neq j, i$, where $\cdot$ is the scalar product of $\mathbb{R}^d$. Further, a_j is not orthogonal to $x_j - x_i$, since otherwise the vectors $x_k - x_i, k \neq i$, would be contained in a $(d - 1)$-dimensional linear subspace. Upon scaling a_j we can assume that $a_j \cdot (x_j - x_i) = 1$. Putting

$$q_j(x) := (a_j \cdot x - a_j \cdot x_i)^2, \tag{18.28}$$

we have $q_j(x_i) = \delta_{ij}$ for $i, j = 1, \ldots, d+1$ by construction.

Suppose that $x_1, \ldots, x_{d+1} \in \mathcal{K}$ and $\mathbb{R}_d[\underline{x}]_2 \subseteq \mathsf{A}$. We define $\mu = \sum_{j=1}^{d+1} m_j \delta_{x_j}$ and $L = \sum_{j=1}^{d+1} m_j l_{x_j}$, where $m_j > 0$. Since $x_1, \ldots, x_{d+1}$ obey the positive separation property $(PSP)_{\mathsf{A},\mathcal{K}}$, the measure $\mu \in \mathcal{M}_{L,\mathcal{K}}$ is maximal mass by Theorem 18.33(i). ∘

Example 18.35 Let $t_j = (t_{j1}, \ldots, t_{jd})$, $j = 1, \ldots, k$, be $k \geq 2$ pairwise different points of $\mathbb{R}^d$. Since $t_i \neq t_j$ for $i \neq j$, there is a number $n_{ij} \in \{1, \ldots, d\}$ such that $t_{i,n_{ij}} \neq t_{j,n_{ij}}$. Then the Langrange interpolation polynomials

$$p_j(x) = \prod_{i=1, i \neq j}^{k} \frac{x_{n_{ij}} - t_{i,n_{ij}}}{t_{j,n_{ij}} - t_{i,n_{ij}}} \, , j = 0, \ldots, k, \tag{18.29}$$

satisfy $p_j \in \mathbb{R}_d[\underline{x}]_{k-1}$ and $p_j(t_i) = \delta_{ij}$, $i, j = 1, \ldots, k$.

Suppose that $\mathcal{K}$ contains $t_1, \ldots, t_k$. Define $\mu = \sum_{j=1}^{k} m_j \delta_{t_j}$ and $L = \sum_{j=1}^{k} m_j l_{t_j}$, where $m_j > 0$ for $j = 1, \ldots, k$. Obviously, $\mu \in \mathcal{M}_{L,\mathcal{K}}$.

If $\mathbb{R}_d[\underline{x}]_{k-1} \subseteq \mathsf{A}$, then μ is an extreme point of $\mathcal{M}_{L,\mathcal{K}}$ by Proposition 18.10.

Suppose that $\mathbb{R}_d[\underline{x}]_{2k-2} \subseteq \mathsf{A}$. Then, since $p_j^2 \in \mathrm{Pos}(\mathsf{A}, \mathcal{K})$ and $p_j^2(x_i) = \delta_{ij}$ for $i, j = 1, \ldots, k$, the atoms $t_1, \ldots, t_k$ has the positive separation property $(PSP)_{\mathsf{A},\mathcal{K}}$. Therefore, μ is maximal mass by Theorem 18.33(i).   ∘

18.5   Constructing Ordered Maximal Mass Measures

In this section, L is a **truncated $\mathcal{K}$-moment functional on A such that $L \neq 0$**.

Finding maximal mass representing measures is a difficult task even in the case when $\mathcal{K}$ is compact. But there is a weaker notion of ordered maximal mass measures for which a simple general construction method exists.

Definition 18.36 A k-atomic measure $\mu = \sum_{i=1}^{k} m_i \delta_{t_i} \in \mathcal{M}_{L,\mathcal{K}}$ with all atoms t_i in $\mathcal{K}$ is called *ordered maximal mass* for L if

$$m_j = \rho_{L_{j-1}}(t_j) \quad \text{for } j = 1, \ldots, k,$$

where L_{j-1} is the functional on A defined by $L_{j-1} = \sum_{i=j}^{k} m_i l_{t_i}$ and $L_0 = L$.

Obviously, each maximal mass measure is ordered maximal mass. The converse is not true; the measure μ in Proposition 19.25 below is ordered maximal mass, but not maximal mass.

From now on suppose that $\mathcal{K}$ is **compact**. Then, from Proposition 18.27 we recall that for each functional $L' \in \mathcal{L}(\mathsf{A}, \mathcal{K})$ and $t \in \mathcal{K}$ we have $\rho_{L'}(t) = \kappa_{L'}(t) = c_{L'}(t)$ and this number is the largest $c \in \mathbb{R}_+$ such that $(L' - c l_x) \in \mathcal{L}(\mathsf{A}, \mathcal{K})$.

Since $L \neq 0$ and there is an atomic measure in $\mathcal{M}_{L,\mathcal{K}}$ by Theorem 17.2, we have $\rho_L(t) > 0$ for some $t \in \mathcal{K}$. Set $L_0 := L$. We choose a point $t_1 \in \mathcal{K}$ such that $\rho_{L_0}(t_1) > 0$ and define $L_1 = L - \rho_{L_0}(t_1) l_{x_1}$. Then $L_1 \in \mathcal{L}(\mathsf{A}, \mathcal{K})$ and $\rho_{L_1}(x_1) = 0$.

(Indeed, if $\rho_{L_1}(x_1) > 0$, there would exist a $c > \rho_{L_0}(t_1) = c_{L_0}(t_1)$ such that $(L_0 - cl_x) \in \mathcal{L}(A, \mathcal{K})$, a contradiction.) If $L_1 = 0$, we set $L_j = 0$ for all $j \in \mathbb{N}, j \geq 2$, and we are done. If $L_1 \neq 0$, we apply the preceding construction with L replaced by L_1. Continuing this procedure, we obtain functionals $L_n \in \mathcal{L}(A, \mathcal{K})$ such that $L_n = 0$ or there are points $t_j \in \mathcal{K}$ satisfying $\rho_{L_{j-1}}(t_j) > 0$, $\rho_{L_j}(t_j) = 0$ for $j = 1, \ldots, n$, and

$$L_n = L - \sum_{j=1}^{n} \rho_{L_{j-1}}(t_j)l_{t_j}. \tag{18.30}$$

Let $j \leq i \leq n$. By construction, we have $L_i(f) \leq L_j(f)$ for $f \in \text{Pos}(A, \mathcal{K})$. Hence $0 \leq \kappa_{L_i}(t_j) \leq \kappa_{L_j}(t_j) = \rho_{L_j}(t_j) = 0$ by (18.22), so that $\kappa_{L_i}(t_j) = \rho_{L_i}(t_j) = 0$. Therefore, if $j < i$, then $j \leq i - 1$ and hence $\rho_{L_{i-1}}(t_j) = 0$, but $\rho_{L_{i-1}}(t_i) > 0$, so that $t_i \neq t_j$. In particular, this shows that the points $t_1, \ldots, t_n$ are pairwise distinct.

Lemma 18.37 $L_k = 0$ for all $k \in \mathbb{N}, k \geq |\mathsf{N}|$.

Proof Assume to the contrary that $L_k \neq 0$ for some $k \geq |\mathsf{N}|$. Then $L_{|\mathsf{N}|} \neq 0$. Set $m := |\mathsf{N}| + 1$. Since $m > |\mathsf{N}| = \dim A$, the point evaluations $l_{t_j}, j = 1, \ldots, m$, on A are linearly dependent. Hence there are reals $\lambda_1, \ldots, \lambda_m$, not all zero, such that

$$\sum_{j=1}^{m} \lambda_j l_{t_j}(f) = 0, \quad f \in A. \tag{18.31}$$

Let r be the smallest index for which $\lambda_r \neq 0$. Since $L_{|\mathsf{N}|} \neq 0$ and $r - 1 \leq |\mathsf{N}|$, we have $L_{r-1} \neq 0$ and $\rho_{L_{i-1}}(t_i) > 0$ for $i = 1, \ldots, m$ by construction. From (18.30) it follows that $L_{r-1} = L_m + \sum_{i=r}^{m} \rho_{L_{i-1}}(t_i)l_{t_i}$. Then, if μ_m is a finitely atomic representing measure for L_m, so is $\mu_{r-1} := \mu_m + \sum_{i=r}^{m} \rho_{L_{i-1}}(t_i)\delta_{t_i}$ for L_{r-1} and $\mu_{r-1}(\{t_r\}) \geq \rho_{L_{r-1}}(t_r)$. Hence $\mu_{r-1}(\{t_r\}) = \rho_{L_{r-1}}(t_r) = \kappa_{L_{r-1}}(t_r) > 0$. Thus, Proposition 18.31(ii) applies to $\mu_{r-1}, L_{r-1}, x_i := t_r$, so there exists a sequence $(p_n)_{n \in \mathbb{N}}$ from $\text{Pos}(A, \mathcal{K})$ such that $p_n(t_r) = 1$ and $\lim_n p_n(t) = 0$ for all atoms $t \neq t_r$ of μ_{r-1}. Since $\rho_{L_{i-1}}(t_i) > 0$ for $i = 1, \ldots, m$, the points $t_j, r < j \leq m$, are atoms of μ_{r-1} and they are different from t_r, as noted above. Setting $f = p_n$ in (18.31) and passing to the limit we obtain

$$0 = \lim_{n \to \infty} \sum_{j=1}^{m} \lambda_j l_{t_j}(p_n) = \sum_{j=r}^{m} \lambda_j \left(\lim_{n \to \infty} p_n(t_j) \right) = \sum_{j=r}^{m} \lambda_j \delta_{jr} = \lambda_r.$$

This contradicts the choice of r and proves the assertion. $\square$

By Lemma 18.37, the above construction terminates. Let $k \in \mathbb{N}$ be the smallest number such that $L_{k-1} \neq 0$ and $L_k = 0$. By (18.30) we have for $j = 1, \ldots, k$,

$$L = \sum_{i=1}^{k} \rho_{L_{i-1}}(t_i)l_{t_i}, \tag{18.32}$$

$$L_{j-1} = \sum_{i=j}^{k} \rho_{L_{i-1}}(t_i)l_{t_i}. \tag{18.33}$$

Note that $k \leq |\mathsf{N}|$ by Lemma 18.37. From (18.32) it follows that the measure

$$\mu := \sum_{j=1}^{k} m_j \delta_{t_j}, \quad \text{where} \quad m_j := \rho_{L_{j-1}}(t_j), \ j = 1, \ldots, k,$$

belongs to $\mathcal{M}_{L,\mathcal{K}}$. Since the points $t_1, \ldots, t_k$ are pairwise distinct, as shown above, and $\rho_{L_{j-1}}(t_j) > 0$ for all j by construction, μ is k-atomic. The functionals L_{j-1} in (18.33) are precisely the linear functionals L_{j-1} associated with μ in Definition 18.36. Therefore, μ is an *ordered maximal mass* representing measure for L according to Definition 18.36. Moreover, μ can be chosen so that an arbitrary point $t_1 \in \mathcal{K}$ satisfying $\rho_L(t_1) > 0$ is an atom of μ. Then $t_1 \in \mathcal{V}_+(L, \mathcal{K})$ by Proposition 18.28(i). We summarize the preceding in the following theorem.

Theorem 18.38 *Suppose that $\mathcal{K}$ is compact and $L \neq 0$ is a truncated $\mathcal{K}$-moment functional on A. Let $t_1 \in \mathcal{K}$ be such that $\rho_L(t_1) > 0$. Then the above construction gives a k-atomic measure $\mu = \sum_{j=1}^{k} m_j \delta_{x_j} \in \mathcal{M}_{L,\mathcal{K}}, \ k \leq |\mathsf{N}|$, such that all atoms of μ are in $\mathcal{K}$ and μ is ordered maximal mass.*

Illustrative examples for this theorem are given in Sect. 19.4 and in Exercise 19.11.

18.6 Evaluation Polynomials

Throughout this section, we assume that L is a fixed **positive** functional on A^2, that is, L is a real linear functional on A^2 such that $L(f^2) \geq 0$ for $f \in \mathsf{A}$.

From Proposition 17.17(v) and formula (17.30) we recall

$$\mathcal{N}_L = \{f \in \mathsf{A} : L(f^2) = 0\}, \quad \mathcal{V}_L = \{x \in \mathbb{R}^d : f(x) = 0 \ \text{for all} f \in \mathcal{N}_L\}.$$

For $f \in \mathsf{A}$ we denote by $\tilde{f} = f + \mathcal{N}_L$ the equivalence class of $f \in \mathsf{A}$ in $\mathcal{D}_L = \mathsf{A}/\mathcal{N}_L$. Recall from Section 17.3 that there is a scalar product $\langle \cdot, \cdot \rangle$ on the vector space $\mathcal{D}_L$ defined by $L(fg) = \langle \tilde{f}, \tilde{g} \rangle, f, g \in \mathsf{A}$.

Let $t \in \mathcal{V}_L$. Since $f(t) = 0$ for $f \in \mathcal{N}_L$, there is a well-defined linear functional F_t on $\mathcal{D}_L = \mathsf{A}/\mathcal{N}_L$ given by $F_t(\tilde{f}) := f(t), f \in \mathsf{A}$. Each linear functional on the finite-dimensional real Hilbert space $(\mathcal{D}_L, \langle \cdot, \cdot \rangle)$ is continuous, so by Riesz' theorem there is a unique element $\widetilde{E}_t \in \mathcal{D}_L, \ E_t \in \mathsf{A}$, such that

$$f(t) \equiv F_t(\tilde{f}) = \langle \tilde{f}, \widetilde{E}_t \rangle = L(f E_t), \quad f \in \mathsf{A}. \tag{18.34}$$

Definition 18.39 $E_t \in \mathsf{A}$ is called an *evaluation polynomial* at the point $t \in \mathcal{V}_L$.

Note that the polynomial E_t is not uniquely determined by t. In fact, it is only determined up to a summand from the vector space $\mathcal{N}_L$, but the values of $E_t(\cdot)$ on the set $\mathcal{V}_L$ do not depend on the particular choice of the element of the equivalence class $\widetilde{E}_t$. In what follows we will fix a choice of the polynomial E_t.

Let $\{\tilde{f}_i : i = 1,\ldots,k\}$ be an orthonormal basis of the Hilbert space $\mathcal{D}_L$. Using (18.34) we develop $\widetilde{E}_t$ with respect to this basis and obtain

$$\widetilde{E}_t = \sum_{i=1}^{k} \langle \widetilde{E}_t, \tilde{f}_i \rangle\, \tilde{f}_i = \sum_{i=1}^{k} f_i(t)\tilde{f}_i. \tag{18.35}$$

As noted above, $\tilde{f}(x) = f(x)$ for $x \in V_L$ and $f \in \mathsf{A}$. Therefore,

$$E_t(x) = \sum_{i=1}^{k} f_i(t)f_i(x) \quad \text{for} \quad t, x \in V_L. \tag{18.36}$$

Since the f_i are polynomials, it follows from (18.36) that $E_t(x)$ is jointly continuous in (t, x) for $t, x \in V_L$. Using (18.34) we derive

$$E_t(t) = L(E_t E_t) = \langle \tilde{E}_t, \tilde{E}_t \rangle = \|\tilde{E}_t\|^2 \quad \text{for} \quad t \in V_L, \tag{18.37}$$

where $\|\cdot\|$ denotes the norm of the Hilbert space $\mathcal{D}_L$. In particular, $\widetilde{E}_t \neq 0$ in $\mathcal{D}_L$ and hence $E_t(t) \neq 0$, since $\langle e, \widetilde{E}_t \rangle = e(t) > 0$ by (18.34) and (18.1).

For $t \in \mathbb{R}^d$ we now define another quantity $\tau_L(t)$ by a variational problem:

$$\tau_L(t) = \inf\{L(p^2) : p \in \mathsf{A},\ p(t) = 1\} = \inf\left\{ \frac{L(p^2)}{p(x)^2} : p \in \mathsf{A}\right\}, \tag{18.38}$$

where we set $\frac{c}{0} := +\infty$ for $c \geq 0$. The following result is similar to Proposition 9.12.

Proposition 18.40 *For $t \in V_L$ the infimum in (18.38) is a minimum and equal to $\|\widetilde{E}_t\|^{-2}$. Further, $\|\widetilde{E}_t\|^{-2}\widetilde{E}_t$ is the unique element $\tilde{p} \in \mathcal{D}_L$ such that the infimum in (18.38) is attained at $p \in \mathsf{A}$.*

Proof The proof is almost verbatim the same as the proof of Proposition 9.12.

Let $p \in \mathsf{A}$. We develop $\tilde{p}$ with respect to the orthonormal basis $\{\tilde{f}_1,\ldots,\tilde{f}_k\}$:

$$\tilde{p} = \sum_i c_i\tilde{f}_i, \quad \text{where} \quad c_i = \langle \tilde{p}, \tilde{f}_i \rangle, \quad \text{and} \quad L(p^2) = \|\tilde{p}\|^2 = \sum_i c_i^2.$$

Hence $p(t) = \sum_i c_i f_i(t)$. Therefore, the problem in (18.38) is to minimize $\sum_i c_i^2$ for $(c_1,\ldots,c_k)^T \in \mathbb{R}^k$ under the constraint $\sum_i c_i f_i(t) = 1$. This problem is solved by Lemma 9.13, now applied to $\mathbb{R}^k$ with $g = (c_1,\ldots,c_k)^T$ and $f = (f_1(t),\ldots,f_k(t))^T$. Since $\{\tilde{f}_1,\ldots,\tilde{f}_k\}$ is an orthonormal basis, (18.35) yields $\|\widetilde{E}_t\|^2 = \sum_i f_i(t)^2$. Then, by Lemma 9.13, the infimum is attained at p if and only if $c_i = f_i(t)\|\widetilde{E}_t\|^{-2}$, that is,

$$\tilde{p} = \sum_i f_i(t)\|\widetilde{E}_t\|^{-2}\tilde{f}_i = \|\widetilde{E}_t\|^{-2}\widetilde{E}_t,$$

and the minimum is $\left(\sum_i f_i(t)^2\right)^{-1} = \|\widetilde{E}_t\|^{-2}$. $\qquad\square$

Note that if L is a truncated $\mathcal{K}$-moment functional, then we have

$$\tau_L(t) \geq \kappa_L(t) \geq \rho_L(t) \quad \text{for} \quad t \in \mathcal{V}_L \cap \mathcal{K}. \tag{18.39}$$

Indeed, the first inequality follows at once by comparing the corresponding definitions (18.22) and (18.38), while the second inquality holds by (18.25).

Definition 18.41 A family $\{E_{x_j} : j = 1, \ldots, k\}$ of evaluation polynomials E_{x_j}, $x_j \in \mathcal{V}_L$, is called *orthogonal* if $\langle \widetilde{E_{x_i}}, \widetilde{E_{x_j}} \rangle = 0$ for $i, j = 1, \ldots k, i \neq j$.

As noted above, $\widetilde{E_{x_i}} \neq 0$. Hence a set $\{E_{x_j} : j = 1, \ldots, k\}$ of evaluation polynomials is orthogonal if and only if there are numbers $c_i > 0$, $i = 1, \ldots, k$, such that

$$\langle \widetilde{E_{x_i}}, \widetilde{E_{x_j}} \rangle = c_i^{-1} \delta_{ij}, \quad i, j = 1, \ldots, k. \tag{18.40}$$

Here the inverse c_i^{-1} is taken in order to obtain formula (18.41) below. Note that combining (18.34) and (18.40) yields $E_{x_i}(x_j) = c_i^{-1} \delta_{ij}$.

The remaining results of this section deal with truncated moment functionals L that have $(\operatorname{rank} L)$-atomic representing measures. The next theorem characterizes such functionals in terms of evaluation polynomials.

Theorem 18.42 *Let L be a positive functional on A^2. Set $k := \operatorname{rank} L$. There is a one-to-one correspondence between orthogonal families $\{E_{x_j} : j = 1, \ldots, k\}$ of evaluation polynomials E_{x_j}, $x_j \in \mathcal{V}_L$, and k-atomic representing measures μ of L. It is given by $\mu = \sum_{j=1}^{k} c_j \delta_{x_j}$, where c_j are the numbers from (18.40). In this case,*

$$\tau_L(x_j) = c_j, \quad j = 1, \ldots, k. \tag{18.41}$$

Proof Let $\{E_{x_j} : j = 1, \ldots, k\}$ be an orthogonal family of evaluation polynomials. Put $\mu = \sum_{j=1}^{k} c_j \delta_{x_j}$. Since $k = \operatorname{rank} L = \dim \mathcal{D}_L$ by Proposition 17.17(iii), it follows from (18.40) that $\{c_j^{1/2} \widetilde{E_{x_j}} : j = 1, \ldots, k\}$ is an orthonormal basis of the Hilbert space $\mathcal{D}_L$. Therefore, using Parseval's identity and (18.35) we derive

$$L(pq) = \langle \tilde{p}, \tilde{q} \rangle = \sum_j \langle \tilde{p}, c_j^{1/2} \widetilde{E_{x_j}} \rangle \langle \tilde{q}, c_j^{1/2} \widetilde{E_{x_j}} \rangle$$

$$= \sum_j c_j \langle \tilde{p}, \widetilde{E_{x_j}} \rangle \langle \tilde{q}, \widetilde{E_{x_j}} \rangle = \sum_j c_j p(x_j) q(x_j) = \int (pq)(x) \, d\mu(x)$$

for $p, q \in \mathsf{A}$. Since A^2 is the span of products pq with $p, q \in \mathsf{A}$, the preceding implies that $L(f) = \int f \, d\mu$ for all $f \in \mathsf{A}^2$, that is, $\mu \in \mathcal{M}_L$.

Conversely, suppose that $\mu = \sum_{i=1}^{k} m_i \delta_{x_i}$ is a k-atomic measure of $\mathcal{M}_L$. By Proposition 17.17(vi), each atom x_j is in $\mathcal{V}_L$. Let J denote the embedding of A into $L^2(\mathbb{R}^d, \mu)$ and $\langle \cdot, \cdot \rangle_\mu$ the scalar product of $L^2(\mathbb{R}^d, \mu)$. Then

$$\langle \tilde{p}, \tilde{q} \rangle = L(pq) = \int pq \, d\mu = \langle J(p), J(q) \rangle_\mu, \quad p, q \in \mathsf{A}. \tag{18.42}$$

Hence $\tilde{p} \mapsto J(p)$ is a well-defined isometric linear map of $\mathcal{D}_L$ into $L^2(\mathbb{R}^d, \mu)$. Since μ is k-atomic and hence $\dim L^2(\mathbb{R}^d, \mu) = k = \dim \mathcal{D}_L$, J is bijective. Hence for any $j = 1, \ldots, k$ there is a $p_j \in \mathsf{A}$ such that $J(p_j)$ is the characteristic function of the point x_j, that is, $J(p_j)(x_i) = p_j(x_i) = \delta_{ji}$ for $i, j = 1, \ldots, k$. By (18.42),

$$\langle \tilde{f}, m_j^{-1} \widetilde{p_j} \rangle = m_j^{-1} L(fp_j) = m_j^{-1} \int fp_j \, d\mu = m_j^{-1} \sum_i m_i f(x_i) p_j(x_i) = f(x_j)$$

for $f \in \mathsf{A}$, that is, $E_{x_j} := m_j^{-1} p_j$ is an evaluation polynomial at x_j. Further,

$$\langle m_i^{-1} \widetilde{p_i}, m_j^{-1} \widetilde{p_j} \rangle = m_i^{-1} m_j^{-1} \int p_i p_j \, d\mu = m_i^{-1} m_j^{-1} \sum_n m_n p_i(x_n) p_j(x_n) = m_i^{-1} \delta_{ij}.$$

This shows that $\{E_{x_j} : j = 1, \ldots, k\}$ is an orthogonal family of evaluation polynomials for L with constants $c_j = m_j$ in (18.40). The corresponding measure is $\sum_j m_j \delta_{x_j} = \mu$. This proves the asserted one-to-one correspondence.

We verify (18.41). Take $p \in \mathsf{A}$ such that $p(x_j) = 1$. Since $k = \dim \mathcal{D}_L$, the set $\{E_{x_j} : j = 1, \ldots, k\}$ is an orthogonal basis of $\mathcal{D}_L$, so $\tilde{p} = \sum_{i=1}^k \alpha_i \widetilde{E}_{x_i}$ with $\alpha_i \in \mathbb{R}$. By (18.40), we have $\| \widetilde{E}_{x_i} \|^2 = E_{x_i}(x_i) = c_i^{-1}$ for each i. Using this formula we derive $1 = p(x_j) = \sum_{i=1}^k \alpha_i E_{x_i}(x_j) = \alpha_j c_j^{-1}$, so that $\alpha_j = c_j$ for all j. Therefore,

$$L(p^2) = \| \tilde{p} \|^2 = \sum_i \alpha_i^2 \| \widetilde{E}_{x_i} \|^2 = \sum_i \alpha_i^2 c_i^{-1}.$$

Hence the infimum in Eq. (18.38) is obviously obtained when we set $\alpha_i = 0$ for all $i \neq j$. Thus, $\tau_L(x_j) = \alpha_j^2 c_j^{-1} = c_j$. □

The next theorem deals with the "nice" but rare case when $\mathrm{Pos}(\mathsf{A}^2, \mathcal{K}) = \sum \mathsf{A}^2$, see e.g. Exercise 18.9 for a very simple example.

Theorem 18.43 *Suppose that the set $\mathcal{K}$ is compact and*

$$\mathrm{Pos}(\mathsf{A}^2, \mathcal{K}) = \sum \mathsf{A}^2 := \Big\{ \sum_i f_i^2 : f_i \in \mathsf{A} \Big\}.$$

(i) *If L is a positive linear functional on A^2, then L has a (rank L)-atomic maximal mass representing measure μ with all atoms contained in $\mathcal{V}_L \cap \mathcal{K}$. Moreover,*

$$\rho_L(x) = \tau_L(x) \quad for \quad x \in \mathcal{V}_L \cap \mathcal{K}. \tag{18.43}$$

(ii) *Assume that $f \lceil \mathcal{K} = 0$ for $f \in \mathsf{A}$ implies $f = 0$. Then $\mathcal{C}(\mathsf{A}^2, \mathcal{K}) = |\mathsf{N}|$.*

Proof

(i) First we note that L is a truncated $\mathcal{K}$-moment functional by Theorem 17.10, since $\mathrm{Pos}(\mathsf{A}^2, \mathcal{K}) = \sum \mathsf{A}^2$ and $\mathcal{K}$ is compact.

We verify (18.43). Define $L_1 := L - \tau_L(x)l_x$. By the second equality in (18.38), we have $L_1(f^2) = L(f^2) - \tau_L(x)f(x)^2 \geq 0$ for $f \in \mathsf{A}$, that is, L_1 is a positive functional on A^2. Thus, again by Theorem 17.10, L_1 is a truncated $\mathcal{K}$-moment functional. If $\mu_1 \in \mathcal{M}_{L_1,\mathcal{K}}$, then $\mu := \mu_1 + \tau_L(x)\delta_x$ is a representing measure of $L = L_1 + \tau_L(x)l_x$. Hence $\rho_L(x) \geq \mu(\{x\}) \geq \tau_L(x)$. Since $\rho_L(x) \leq \tau_L(x)$ by (18.39), (18.43) is proved.

Next we prove the following assertion: If $\tau_L(x) > 0$, then

$$\operatorname{rank} L = 1 + \operatorname{rank} L_1. \tag{18.44}$$

Indeed, since $L(f^2) \geq L_1(f^2) \geq 0$ for $f \in \mathsf{A}$, we have $H_n(L) \succeq H_n(L_1) \succeq 0$ and hence $\operatorname{rank} L \geq \operatorname{rank} L_1$. By Proposition 18.40 there is a unique $\widetilde{p} \in \mathcal{D}_L$ such that the infimum in (18.38) is attained at p, that is, $p(x) = 1$ and $\tau_L(x) = L(p^2)$. Then $L_1(p^2) = 0$ and $L(p^2) \neq 0$. From the uniqueness of $\widetilde{p}$ and (17.29) we conclude that $\mathcal{N}_L$ is a subspace of $\mathcal{N}_{L_1}$ of codimension 1. Hence (18.44) follows from Proposition 17.17.

To prove the remaining assertions on L we shall proceed by induction.

First assume that $\tau_L(x) = 0$ for all $x \in \mathcal{V}_L \cap \mathcal{K}$. Then $\rho_L(x) = 0$ on $\mathcal{V}_L \cap \mathcal{K}$ by (18.43). Since each $\mu \in \mathcal{M}_{L,\mathcal{K}}$ is supported on $\mathcal{V}_L \cap \mathcal{K}$, Theorem 17.2 implies that $L = 0$, so the assertions on L hold trivially.

Now we assume that $\tau_L(x_1) > 0$ for some $x_1 \in \mathcal{V}_L \cap \mathcal{K}$. Then, as shown above, $L_1 := L - \tau_L(x_1)l_{x_1}$ is a truncated $\mathcal{K}$-moment functional satisfying (18.44). We replace L by L_1 and proceed by induction. After $k := \operatorname{rank} L$ steps we obtain a positive functional L_k such that $\operatorname{rank} L_k = 0$. Then $L_k = 0$ and

$$L = L_k + \sum_{j=1}^{k} \tau_{L_{j-1}}(x_j)l_{x_j} = \sum_{j=1}^{k} \tau_{L_{j-1}}(x_j)l_{x_j}, \quad \text{where} \quad L_0 := L.$$

Hence $\mu := \sum_{j=1}^{k} \tau_{L_{j-1}}(x_j)\delta_{x_j}$ is a representing measure of L such that

$$|\operatorname{supp} \mu| \leq k = \operatorname{rank} L = \operatorname{rank} H(L).$$

By (17.38), $\operatorname{rank} H(L) \leq |\operatorname{supp} \mu|$. Thus $\operatorname{rank} L = |\operatorname{supp} \mu|$, so μ is $(\operatorname{rank} L)$-atomic. Therefore Theorem 18.42 applies and formula (18.41) yields $\tau_{L_{j-1}}(x_j) = \tau_L(x_j)$ for $j = 1, \ldots, k$. Since $\tau_L(x_j) = \rho_L(x_j)$ by (18.43), we have $\tau_{L_{j-1}}(x_j) = \rho_L(x_j)$, that is, μ has maximal mass at each atom x_j. Thus, μ is maximal mass.

(ii) By (i), each $L \in \mathcal{L}(\mathsf{A}^2, \mathcal{K})$ has a $(\operatorname{rank} L)$-atomic representing measure. Hence $\mathsf{C}(\mathsf{A}^2, \mathcal{K}) \leq |\mathsf{N}|$, since $\operatorname{rank} L \leq |\mathsf{N}|$. From the assumption of (ii) it follows that there are points $x_1, \ldots, x_{|\mathsf{N}|} \in \mathcal{K}$ such that the functionals $l_{x_1}, \ldots, l_{x_{|\mathsf{N}|}}$ are linearly independent on A. Define $L = \sum_{j=1}^{|\mathsf{N}|} l_{x_j}$ and $\mu = \sum_{j=1}^{|\mathsf{N}|} \delta_{x_j}$. Then $L \in \mathcal{L}(\mathsf{A}^2, \mathcal{K})$ and $\mu \in \mathcal{M}_{L,\mathcal{K}}$. From Proposition 17.21, $\operatorname{rank} L = \operatorname{rank} H(L) = |\mathsf{N}|$. By (17.38), L has no representing measure with fewer atoms. Hence $|\mathsf{N}| \leq \mathsf{C}(\mathsf{A}^2, \mathcal{K})$. Thus, $\mathsf{C}(\mathsf{A}^2, \mathcal{K}) = |\mathsf{N}|$. $\qquad\square$

Remark 18.44 In condition (18.1) we assumed that A contains an element e such that $e(x) \geq 1$ on $\mathcal{K}$. If L is a truncated $\mathcal{K}$-moment functional on A^2, all results of this section hold if $e \in A^2$ rather than $e \in A$. Indeed, the proof of Theorem 18.43 remains valid, since Theorem 17.10 was applied to A^2. Since $e \in A^2$, for each $x \in \mathcal{K}$ there is an $f_x \in A$ such that $f_x(x) \neq 0$, so $\langle f_x, \widetilde{E_x} \rangle = f_x(x) \neq 0$, and hence $\widetilde{E_x} \neq 0$. ∘

There is a similar result about positive functionals of "small rank". It is based on the following theorem due to G. Blekherman [B12, Theorem 2.3].

Theorem 18.45 *If L is a positive linear functional on $\mathbb{R}_d[\underline{x}]_{2n}$ such that*

$$\operatorname{rank} L \leq 3n - 3 \quad \text{if} \quad n \geq 3, \quad \operatorname{rank} L \leq 6 \quad \text{if} \quad n = 2, \tag{18.45}$$

then L is a truncated moment functional.

The proof in [B12] is based on tools from algebraic geometry (Cayley–Bacharach duality, see e.g. [EGH]) which are beyond the scope of this book. Taking Theorem 18.45 for granted we obtain the following result.

Theorem 18.46 *Let $d, n \in \mathbb{N}$ and let L be a positive linear functional on $\mathbb{R}_d[\underline{x}]_{2n}$, that is, $L(f^2) \geq 0$ for $f \in \mathbb{R}_d[\underline{x}]_n$. Suppose that condition (18.45) is satisfied. Then L is a truncated moment functional with $(\operatorname{rank} L)$-atomic representing measure and the equality (18.43) holds with $\mathcal{K} = \mathbb{R}^d$.*

The proof of Theorem 18.46 follows the same lines as the proof of Theorem 18.43 with Theorem 17.10 replaced by Theorem 18.45. We omit the details.

18.7 Exercises

1. Suppose that $\mathcal{K}$ is compact and let $f \in A$. Prove the following:

   a. f is an interior point of $\operatorname{Pos}(A, \mathcal{K})$ if and only if $f(x) > 0$ for all $x \in \mathcal{K}$.
   b. f is a boundary point of $\operatorname{Pos}(A, \mathcal{K})$ if and only if $f(x_0) = 0$ for some $x_0 \in \mathcal{K}$.
   c. The assertion of a. is no longer valid in general if $\mathcal{K}$ is not compact.
      Hint: $f(x) = 1 + x^2, \mathcal{K} = \mathbb{R}, A = \mathbb{R}[x]_n$ with $n \geq 3$.

2. Let $A = \mathbb{R}_d[\underline{x}]_{2n}$ and let $\mathcal{K}$ be a closed subset of $\mathbb{R}^d$. Discuss whether or not (depending on $\mathcal{K}$) the following statements are true:

   a. Each boundary point of $\mathcal{S}(A, \mathcal{K})$ is determinate.
   b. Each interior point of $\mathcal{S}(A, \mathcal{K})$ is indeterminate.

3. ($\mathcal{N}_+(L_s, \mathcal{K})$ and $\mathcal{V}_+(L_s, \mathcal{K})$ *for one-dimensional bounded intervals*)
   Let $d = 1, \mathcal{K} = [0, 1], m \in \mathbb{N}$, and $s \in \mathcal{S}_{m+1}, s \neq 0$. Suppose that $\operatorname{ind}(s) \leq m$, see Definitions 10.4 and 10.6. Use Theorem 10.7 to determine $\mathcal{N}_+(L_s, \mathcal{K})$ and $\mathcal{V}_+(L_s, \mathcal{K})$.

4. Let $d = 1, \mathcal{K} = [0, 1], \mathsf{A} = \mathbb{R}[x]_2$, and $x_1 = 0, x_2 = \frac{1}{2}, x_3 = 1$. Define $L = \sum_{j=1}^{3} l_{x_j}$. Show that $\rho_L(0) = \frac{6}{5}$.

5. Let $d = 1, \mathcal{K} = [0, 1], \mathsf{A} = \mathbb{R}[x]_n$. Consider a k-atomic measure $\mu = \sum_{j=1}^{k} m_j \delta_{x_j}$ and $L = \sum_{j=1}^{k} m_j l_{x_j}$.

    a. Decide when $m_j = \rho_L(x_j)$ for one j.
    b. Decide when μ is maximal mass.
    c. Give an example such that $m_1 = \rho_L(x_1)$, but $m_2 \neq \rho_L(x_2)$.
    d. Discuss these problems in the case $\mathcal{K} = \mathbb{R}$.

6. Let $d = 2, x_1 = (0,0), x_2 = (1,0), x_3 = (0,1), x_4 = (1,1)$. Set $\mu := \sum_{j=1}^{4} m_j \delta_{x_j}$ and $L := \sum_{j=1}^{4} m_j \delta_{x_j}$, where $m_j > 0, j = 1, 2, 3, 4$. Suppose that $x_1, x_2, x_3, x_4 \in \mathcal{K}$.

    a. Suppose that $\mathbb{R}_2[\underline{x}]_2 \subseteq \mathsf{A}$. Show that μ is an extreme point of $\mathcal{M}_{L,\mathcal{K}}$.
    b. Let $\mathcal{K}$ be a rectangle and $\mathsf{A} = \mathbb{R}_2[\underline{x}]_2$. Find a 3-atomic measure $\nu \in \mathcal{M}_{L,\mathcal{K}}$.
    c. Suppose that $\mathbb{R}_2[\underline{x}]_4 \subseteq \mathsf{A}$. Show that μ is determinate.

7. Suppose that $\mathcal{K}$ is compact. Show that $C(\mathsf{A}, \mathcal{K}) \leq 1 + \sup_{s \in \partial \mathcal{S}(\mathsf{A}, \mathcal{K})} C(s)$.
    Hint: Use Proposition 1.43. Write $s \in \mathcal{S}(\mathsf{A}, \mathcal{K})$ as $s = s_0 + c_{L_s}(x)\mathfrak{s}(x), s_0 \in \partial \mathcal{S}(\mathsf{A}, \mathcal{K})$.

8. Extend the construction of maximal mass measures in Example 18.34 by taking products of squares of linear polynomials in (18.28).

9. Let $d \geq 2, \mathsf{N} = \{(1, 0, \ldots, 0), \ldots, (0, \ldots, 0, 1)\}$. Let $\mathcal{K}$ be a compact subset of $\mathbb{R}^d$ such that $S^{d-1} \subseteq \mathcal{K}$. Then A^2 are the homogeneous polynomials in $\mathbb{R}_d[\underline{x}]$ of degree 2. Show that $\mathrm{Pos}(\mathsf{A}^2, \mathcal{K}) = \sum \mathsf{A}^2$ and $C(\mathsf{A}^2, \mathcal{K}) = d$.

18.8 Notes

The real algebraic set $\mathcal{V}_+(L)$ first appeared in the Thesis of J. Matzke [Mt], while the core variety $\mathcal{V}(L)$ was invented by L. Fialkow [F2]. Theorem 18.22 is due to G. Blekherman and L. Fialkow (personal communication from L. Fialkow, June 2016). Theorems 18.21 and 18.23 were proved in [DSm1]. Most of the results on maximal masses in Sects. 18.4 and 18.5 are taken from [Sm10]. A number of results of this chapter (for instance, Theorems 18.42 and 18.43) are new.

Chapter 19
The Truncated Moment Problem
for Homogeneous Polynomials

Let $\mathcal{H}_{d,m}$ denote the real homogeneous polynomials in d variables of degree m. In the final chapter of the book we deal with the truncated moment problem for $\mathcal{H}_{d,2n}$ on $\mathbb{R}^d$, on the unit sphere S^{d-1}, and on S^{d-1}_+. Since S^{d-1}_+ is a realization of the projective space $\mathbb{P}^{d-1}(\mathbb{R})$, the latter is in fact the truncated projective moment problem. The existence problem in this case was considered in Sect. 17.2.

In Sects. 19.1 and 19.2, we treat the apolar scalar product on the vector space $\mathcal{H}_{d,m}$ and its relations to the actions of differential operators. These are powerful technical tools for the study of homogeneous polynomials. They are applied in Sect. 19.3 to the truncated projective moment problem and to Carathéodory numbers. In Sect. 19.4, we develop some properties of the Robinson polynomial and use them to construct interesting examples on the truncated moment problem. In Sect. 19.5, some results on zeros of positive polynomials (Theorem 19.32) are derived. Section 19.6 contains some applications to the truncated moment problem on $\mathbb{R}^2$.

In this chapter, N is the set

$$\mathsf{N}_{d,m} := \{\alpha \in \mathbb{N}_0^d : \alpha_1 + \cdots + \alpha_d = m\}, \quad \text{where} \quad m \in \mathbb{N},$$

and the corresponding span $\mathsf{A} = \mathrm{Lin}\,\{x^\alpha : \alpha \in \mathsf{N}_{d,m}\}$ is the vector space $\mathcal{H}_{d,m}$. The elements of $\mathcal{H}_{d,m}$ are also called d-forms of degree m, or briefly, *forms*.

19.1 The Apolar Scalar Product

First we note that the vector space $\mathcal{H}_{d,m}$ has the dimension

$$\dim \mathcal{H}_{d,m} = |\mathsf{N}_{d,m}| = \binom{m+d-1}{d-1}. \tag{19.1}$$

© Springer International Publishing AG 2017
K. Schmüdgen, *The Moment Problem*, Graduate Texts in Mathematics 277,
DOI 10.1007/978-3-319-64546-9_19

Our first aim is to define the apolar scalar product $[\cdot,\cdot]$ on $\mathcal{H}_{d,m}$. We abbreviate

$$\binom{m}{\alpha} = \frac{m!}{\alpha_1!\ldots\alpha_d!} \quad \text{for } \alpha = (\alpha_1,\ldots,\alpha_d) \in \mathsf{N}_{d,m}.$$

Let us consider homogeneous polynomials

$$p(x) = \sum_{\alpha \in \mathsf{N}_{d,m}} \binom{m}{\alpha} a_\alpha x^\alpha \quad \text{and} \quad q(x) = \sum_{\alpha \in \mathsf{N}_{d,m}} \binom{m}{\alpha} b_\alpha x^\alpha \tag{19.2}$$

of $\mathcal{H}_{d,m}$ and define

$$[p,q] := \sum_{\alpha \in \mathsf{N}_{d,m}} \binom{m}{\alpha} a_\alpha b_\alpha. \tag{19.3}$$

(The reason for including the multinomial coefficients $\binom{m}{\alpha}$ in (19.2) and (19.3) will be seen below when we derive a number of nice formulas.)

Definition 19.1 $[\cdot,\cdot]$ is called the *apolar scalar product* on $\mathcal{H}_{d,m}$.

It is obvious that $[\cdot,\cdot]$ is a scalar product on $\mathcal{H}_{d,m}$. Thus, $(\mathcal{H}_{d,m}, [\cdot,\cdot])$ is a finite-dimensional real Hilbert space. Note that the coefficient a_α of $p \in \mathcal{H}_{d,m}$ in (19.2) can be recovered from the apolar scalar product by

$$[p,x^\alpha] = a_\alpha, \quad \alpha \in \mathsf{N}_{d,m}.$$

As usual, $a \cdot b = a_1 b_1 + \cdots + a_d b_d$ denotes the standard scalar product of $\mathbb{R}^d$.

Let $y = (y_1,\ldots,y_d) \in \mathbb{R}^d$. We denote by $(y\cdot)^m$ the element of $\mathcal{H}_{d,m}$ defined by

$$(y\cdot)^m(x) := (y \cdot x)^m \equiv \left(\sum_{j=1}^d y_j x_j\right)^m = \sum_{\alpha \in \mathsf{N}_{d,m}} \binom{m}{\alpha} y^\alpha x^\alpha, \tag{19.4}$$

where the last equality holds by the multimonomial theorem. The map $y \mapsto (y\cdot x)^{2n}$, called the $2n$-th *Veronese embedding*, plays an important role in algebraic geometry.

Now suppose that $f(x) = \sum_{j=1}^k c_j (y_j \cdot x)^m \in \mathcal{H}_{d,m}$, where $y_j \in \mathbb{R}^d$ and $c_j \in \mathbb{R}$, and let $p(x) \in \mathcal{H}_{d,m}$ be as in (19.2). From (19.4) it follows that

$$[p,f] = \sum_{j=1}^k \binom{m}{\alpha} a_\alpha c_j y_j^\alpha = \sum_{j=1}^k c_j p(y_j) = \sum_{j=1}^k c_j l_{y_j}(p), \quad p \in \mathcal{H}_{d,m}. \tag{19.5}$$

Thus, the scalar product with f is just the corresponding linear combination of point evaluations l_{y_j} at y_j. This gives the link to the truncated moment problem. Further,

Eq. (19.5) shows the reproducing kernel property of the apolar scalar product:

$$[p, (y \cdot x)^m] = l_y(p) = p(y), \quad y \in \mathbb{R}^d, \, p \in \mathcal{H}_{d,m}. \tag{19.6}$$

That is, the point evaluation at y is the linear functional given by $(y \cdot)^m$.

In particular, setting $p = (a \cdot)^m$ and $y = b$ in (19.6), we obtain

$$[(a \cdot)^m, (b \cdot)^m] = (a \cdot b)^m, \quad a, b \in \mathbb{R}^d, \tag{19.7}$$

where $a \cdot b$ means the usual scalar product in $\mathbb{R}^d$. Therefore, if $a \perp b$ in $\mathbb{R}^d$, then $(a \cdot)^m \perp (b \cdot)^m$ in the Hilbert space $(\mathcal{H}_{d,m}, [\cdot, \cdot])$. Further, if the set $\{y_1, \ldots, y_r\}$ is orthonormal in $\mathbb{R}^d$, so is $\{(y_1 \cdot)^m, \ldots, (y_r \cdot)^m\}$ in $(\mathcal{H}_{d,m}, [\cdot, \cdot])$.

Next we derive some useful facts on bases and spanning sets of $\mathcal{H}_{d,m}$.

Lemma 19.2 *For each open subset U of $\mathbb{R}^d$, the polynomials $(y \cdot)^m(x), y \in U$, span the vector space $\mathcal{H}_{d,m}$.*

Proof Let $p \in \mathcal{H}_{d,m}$. Since $(\mathcal{H}_{d,m}, [\cdot, \cdot])$ is a Hilbert space, it suffices to show that $[p, (y \cdot x)^m] = 0$ for all $y \in U$ implies $p = 0$. By (19.6), $[p, (y \cdot x)^m] = p(y) = 0$ on the open set U. Therefore, $p = 0$.   □

The following is a classical result of O. Biermann (1903).

Proposition 19.3 *The set $\{(\alpha \cdot)^m(x) : \alpha \in \mathsf{N}_{d,m}\}$ is a basis of the vector space $\mathcal{H}_{d,m}$.*

Proof For $\beta \in \mathsf{N}_{d,m}$ we define

$$f_\beta(x) = \prod_{j=1}^{d} \prod_{i=0}^{\beta_j - 1} (m x_j - i(x_1 + \cdots + x_d)). \tag{19.8}$$

Since f_β is a product of $|\beta| = m$ homogeneous linear polynomials, $f_\beta \in \mathcal{H}_{d,m}$.

Let $\alpha \in \mathsf{N}_{d,m}$. First suppose that $\alpha \neq \beta$. There is an index j such that $\alpha_j < \beta_j$. Then there is a factor with indices $j, i = \alpha_j$ in (19.8). Evaluated at the point $x = \alpha$ this factor is $m\alpha_j - \alpha_j(\alpha_1 + \cdots + \alpha_d) = 0$. Hence $f_\beta(\alpha) = 0$. If $\beta = \alpha$, then

$$f_\beta(\beta) = \prod_{j=1}^{d} \prod_{i=0}^{\beta_j - 1} (m\beta_j - im) = m^m \prod_{j=1}^{d} \beta_j! \neq 0.$$

Thus, $[f_\beta, (\alpha \cdot x)^m] = f_\beta(\alpha) = 0$ for $\alpha \neq \beta$ and $[f_\beta, (\beta \cdot x)^m] = f_\beta(\beta) \neq 0$. This in turn implies that the polynomials $(\alpha \cdot)^m(x)$, where $\alpha \in \mathsf{N}_{d,m}$, are linearly independent. Since the vector space $\mathcal{H}_{d,m} = \mathrm{Lin}\{x^\alpha : \alpha \in \mathsf{N}_{d,m}\}$ has dimension $|\mathsf{N}_{d,m}|$, they form a vector space basis of $\mathcal{H}_{d,m}$.   □

Definition 19.4 A subset $\{y_1, \ldots, y_r\}$ of $\mathbb{R}^d$ is called a *basic set of nodes* if the polynomials $\{(y_1 \cdot)^m, \ldots, (y_r \cdot)^m\}$ form a basis of the vector space $\mathcal{H}_{d,m}$.

Obviously, if $\{y_1, \ldots, y_r\}$ is a basic set of nodes, then $r = |N_{d,m}| = \dim \mathcal{H}_{s,m}$. Proposition 19.3 says that the set $N_{d,m}$ is a basic set of nodes.

Remark 19.5 By Lemma 19.2, each polynomial $f \in \mathcal{H}_{d,m}$ can be written in the form $f(x) = \sum_{j=1}^{k} c_j (y_j \cdot x)^m$, where $c_j \in \mathbb{R}$ and $y_j \in \mathbb{R}^d$. The smallest number k among all such representations of f is called the *real Waring rank* of f. Note that here the numbers c_j are real, while for the *width* $w(f)$ of $f \in Q_{d,2n}$ (see Definition 19.17 below) the numbers c_j are positive.   ○

Let $\{f_i : i \in I\}$ and $\{g_i : i \in I\}$ be subsets of $\mathcal{H}_{d,m}$, where I is an index set such that $|I| = \dim \mathcal{H}_{d,m} = \binom{m+d-1}{d-1}$. We say that the sets are *dual bases* of $\mathcal{H}_{d,m}$ if

$$[f_i, g_j] = \delta_{i,j} \quad \text{for } i, j \in I. \tag{19.9}$$

It is well-known from linear algebra and it is easy to verify that in this case the sets $\{f_i : i \in I\}$ and $\{g_i : i \in I\}$ are indeed vector space bases of $\mathcal{H}_{d,m}$ and

$$f = \sum_{i \in I} [f_i, f] \, g_i = \sum_{i \in I} [g_i, f] f_i \quad \text{for } f \in \mathcal{H}_{d,m}. \tag{19.10}$$

Proposition 19.6 *Two subsets* $\{f_i : i \in I\}$ *and* $\{g_i : i \in I\}$, $|I| = \binom{m+d-1}{d-1}$, *of* $\mathcal{H}_{d,m}$ *form dual bases of* $\mathcal{H}_{d,m}$ *if and only if the following Marsden identity holds:*

$$(y \cdot x)^m = \sum_{i \in I} f_i(y) g_i(x), \quad x, y \in \mathbb{R}^d. \tag{19.11}$$

Proof Suppose that these sets are dual bases. Setting $f = (y \cdot)^m$ in (19.10) and using that $[f_i, (y \cdot)^m] = f_i(y)$ by (19.6) we obtain

$$(y \cdot)^m = \sum_{i} f_i(y) g_i. \tag{19.12}$$

Evaluating both sides of this equation at x yields (19.11).

Conversely, assume that (19.11) holds. Then (19.12) holds for $y \in \mathbb{R}^d$. Since $\mathcal{H}_{d,m}$ is spanned by the functions $(y \cdot)^m$, $y \in \mathbb{R}^d$, (19.12) implies that the functions $g_i, i \in I$, span the whole vector space $\mathcal{H}_{d,m}$. Therefore, since $|I| = \binom{m+d-1}{d-1} = \dim \mathcal{H}_{d,m}$, it follows that $\{g_i : i \in I\}$ is a basis of $\mathcal{H}_{d,m}$. Now we take y as a variable and apply the apolar scalar product with $g_j(y)$ in (19.11). Then we obtain

$$g_j(x) = [g_j(y), (y \cdot x)^m] = \sum_{i \in I} [f_i, g_j] g_i(x), \quad x \in \mathbb{R}^d.$$

Since $\{g_i : i \in I\}$ is a basis, we conclude that $[f_i, g_j] = \delta_{ij}$ for $i, j \in I$. This means that the corresponding sets are dual bases.   □

19.2 The Apolar Scalar Product and Differential Operators

The apolar scalar product is a valuable tool for the study of forms and truncated moment problems. One reason for this is its nice interplay with differential operators.

Let $i \in \{1, \ldots, d\}$ and $\alpha = (\alpha_1, \ldots, \alpha_d) \in \mathbb{N}_0^d$. Then we abbreviate $\partial_i = \frac{\partial}{\partial x_i}$ and $\partial^\alpha = (\partial_1)^{\alpha_1} \cdots (\partial_d)^{\alpha_d}$. Further, we define

$$p(\partial) := \sum_\alpha a_\alpha \partial^\alpha \quad \text{for} \quad p(x) = \sum_\alpha a_\alpha x^\alpha \in \mathbb{R}_d[\underline{x}]$$

and consider $p(\partial)$ as a differential operator acting on polynomials. Since the operators $\frac{\partial}{\partial x_i}$ and $\frac{\partial}{\partial x_j}$ commute on polynomials, we have

$$(pg)(\partial)f = p(\partial)q(\partial)f = q(\partial)p(\partial)f \quad \text{for} \quad p, q, f \in \mathbb{R}_d[\underline{x}]. \tag{19.13}$$

Lemma 19.7 *Let $y_1, \ldots, y_k \in \mathbb{R}^d$, $c_1, \ldots, c_k \in \mathbb{R}$, and define*

$$f(x) = \sum_{j=1}^k c_j (y_j \cdot x)^m \in \mathcal{H}_{d,m}.$$

If $p \in \mathcal{H}_{n,d}$ and $n \leq m$, then

$$(p(\partial)f)(x) = m(m-1)\ldots(m+1-n) \sum_{j=1}^k c_j p(y_j)(y_j \cdot x)^{m-n}. \tag{19.14}$$

Let $y \in \mathbb{R}^d$ and $p \in \mathcal{H}_{d,m}$. Then

$$p(\partial)(y \cdot x)^m = m! \, p(y). \tag{19.15}$$

In particular, if $p(y) = 0$, then $p(\partial)(y \cdot x)^m = 0$.

Proof For $\alpha \in \mathrm{N}_{d,n}$ we derive

$$\partial^\alpha (y \cdot x)^m = \left(\frac{\partial}{\partial x_1}\right)^{\alpha_1} \cdots \left(\frac{\partial}{\partial x_d}\right)^{\alpha_d} (y \cdot x)^m$$

$$= m(m-1)\ldots(m+1-n) \, y_1^{\alpha_1} \cdots y_d^{\alpha_d} (y \cdot x)^{m-n}$$

$$= m(m-1)\ldots(m+1-n) \, y^\alpha (y \cdot x)^{m-n}.$$

By linearity this implies (19.14). Formula (19.15) follows from (19.14) by setting $f(x) = (y \cdot x)^m$. $\qquad\square$

Lemma 19.8 *If $p, q \in \mathcal{H}_{d,m}$, then $[p, q] = \frac{1}{m!} p(\partial)q = \frac{1}{m!} q(\partial)p$.*

Proof By the symmetry of the scalar product it suffices to prove the first formula. By linearity we can restrict ourselves to $p = \binom{m}{\alpha} x^{\alpha}$ and $q = x^{\beta}$, where $\alpha, \beta \in \mathsf{N}_{d,m}$.

First suppose that $\alpha \neq \beta$. Then $[x^{\alpha}, x^{\beta}] = 0$ by (19.3). Since $\alpha, \beta \in \mathsf{N}_{d,m}$ and $\alpha \neq \beta$, there is an index $j \in \{1, \ldots, d\}$ such that $\alpha_j > \beta_j$. Then $(\frac{\partial}{\partial x_j})^{\alpha_j} x_j^{\beta_j} = 0$ and hence $\partial^{\alpha} x^{\beta} = 0$. That is, $[\binom{m}{\alpha} x^{\alpha}, x^{\beta}] = \frac{1}{m!} \binom{m}{\alpha} \partial^{\alpha} x^{\beta} = 0$ for $\alpha \neq \beta$.

Now assume that $\alpha = \beta$. Then we have $\alpha_j = \beta_j$ and hence $(\frac{\partial}{\partial x_j})^{\alpha_j} x_j^{\beta_j} = \alpha_j!$ for all $j = 1, \ldots, d$. This yields $\partial^{\alpha} x^{\beta} = \alpha_1! \ldots \alpha_d!$ and

$$\frac{1}{m!} \binom{m}{\alpha} \partial^{\alpha} x^{\beta} = \frac{1}{m!} \frac{m!}{\alpha_1! \ldots \alpha_d!} \alpha_1! \ldots \alpha_d! = 1 = \left[\binom{m}{\alpha} x^{\alpha}, x^{\beta} \right].$$

This completes the proof.   □

Combining formula (19.15) with Lemma 19.8 we obtain the following

Corollary 19.9 *If* $f(x) = \sum_{j=1}^{k} c_j (y_j \cdot x)^m \in \mathcal{H}_{d,m}$ *with* $y_j \in \mathbb{R}^d$, $c_j \in \mathbb{R}$, *then*

$$\frac{1}{m!} f(\partial) p = [p, f] = \sum_{j=1}^{k} c_j p(y_j) = \sum_{j=1}^{k} c_j l_{y_j}(p) \quad \text{for } p \in \mathcal{H}_{d,m}. \tag{19.16}$$

Formula (19.16) expresses the apolar scalar product in terms of differential operators. In Theorems 19.14 and 19.16 below we apply this to the moment problem.

Lemma 19.10 *Let* $n \in \mathbb{N}, n < m$. *If* $p \in \mathcal{H}_{d,m}$, $q \in \mathcal{H}_{d,n}$ *and* $f \in \mathcal{H}_{d,m-n}$, *then*

$$m! [p, fq]_m = (m - n)! [f, q(\partial) p]_{m-n},$$

where $[\cdot, \cdot]_m$ *and* $[\cdot, \cdot]_{m-n}$ *are the apolar scalar products of* $\mathcal{H}_{d,m}$ *and* $\mathcal{H}_{d,m-n}$, *respectively.*

Proof Using Lemma 19.8 and Eq. (19.13) we derive

$$m! [p, fq]_m = (fq)(\partial) p = f(\partial)(q(\partial) p) = (m - n)! [f, q(\partial) p]_{m-n}. \quad □$$

Remark 19.11 Let $n, m \in \mathbb{N}, n \leq m$. From formula (19.14) it follows that

$$[p, f]_{n,m} := (m(m-1) \ldots (m+1-n))^{-1} p(\partial) f, \quad p \in \mathcal{H}_{d,n}, f \in \mathcal{H}_{d,m}, \tag{19.17}$$

defines a bilinear mapping, called the *apolar pairing*, of $\mathcal{H}_{d,n} \times \mathcal{H}_{d,m}$ to $\mathcal{H}_{d,m-n}$. In the special case $m = n$ the right-hand side of (19.17) becomes the scalar $m! p(\partial) f$ and we obtain the apolar scalar product $[\cdot, \cdot]$ on $\mathcal{H}_{d,m}$.   ○

We close this section by deriving the following version of Sylvester's *apolarity lemma*. It illustrates nicely the usefulness of the apolar scalar product, but will not be needed later for the study of moment problems.

Proposition 19.12 *Let $k, m \in \mathbb{N}, k \leq m$, and $(a_j, b_j) \in \mathbb{R}^2, j = 1, \ldots, k$. Suppose that $a_j b_i \neq a_i b_j$ for $i \neq j$. Set $q = (b_1 x_1 - a_1 x_2) \ldots (b_k x_1 - a_k x_2)$. Let $f \in \mathcal{H}_{2,m}$. Then we have $q(\partial)f = 0$ if and only if f is of the form*

$$f = \sum_{j=1}^{k} c_j (a_j x_1 + b_j x_2)^m \quad with \quad c_1, \ldots, c_m \in \mathbb{R}. \tag{19.18}$$

Proof Let E be the subspace of $\mathcal{H}_{2,m}$ spanned by $h_j := (a_j x_1 + b_j x_2)^m, j = 1, \ldots, k$.

We describe the orthogonal compliment $E^{\perp}$ of E with respect to the apolar scalar product. Let $g \in \mathcal{H}_{2,m}$. Then $g \in E^{\perp}$ if and only if $[h_j, g] = g(a_j, b_j) = 0$ for $j = 1, \ldots, k$. Thus, $E^{\perp}$ is the set of $g \in \mathcal{H}_{2,m}$ vanishing at all (a_j, b_j). Therefore, since g is a homogeneous polynomial in 2 variables, each polynomial $b_j x_1 - a_j x_2$ divides g. The assumption $a_j b_i \neq a_i b_j$ for $i \neq j$ implies that $b_i x_1 - a_i x_2$ and $b_j x_1 - a_j x_2$ are relatively prime for $i \neq j$. Hence the product q of all factors $b_j x_1 - a_j x_2$ divides g. That is, $g = hq$ for some $h \in \mathcal{H}_{2,m-k}$. Conversely, each g of the form $g = hq$ vanishes at all points (a_j, b_j), so it is in $E^{\perp}$. Summarizing, we have shown that

$$E^{\perp} = \{hq : h \in \mathcal{H}_{2,m-k}\}. \tag{19.19}$$

First let $q(\partial)f = 0$. Then $m![g,f] = g(\partial)f = h(\partial)q(\partial)f = 0$ for $g = hq \in E^{\perp}$ by (19.19), so that $f \in E^{\perp\perp} = E$. This means that f is of the form (19.18).

Conversely, assume that f has a representation (19.18). Then $f \in E$. By (19.19), $m![hq, f] = (hq)(\partial)f = h(\partial)q(\partial)f = 0$ for all $h \in \mathcal{H}_{2,m-k}$. If $k = m$, then $q(\partial)f = 0$ and we are finished. Now suppose $k < m$. Note that $q(\partial)f \in \mathcal{H}_{2,m-k}$. If $[\cdot, \cdot]'$ denotes the apolar scalar product of $\mathcal{H}_{2,m-k}$, we have $(m - k)![h, q(\partial)f]' = h(\partial)q(\partial)f = 0$ for all $h \in \mathcal{H}_{2,m-k}$, that is, $q(\partial)f \in (\mathcal{H}_{2,m-k})^{\perp} = \{0\}$. Thus, $q(\partial)f = 0$. $\qquad\square$

19.3 The Apolar Scalar Product and the Truncated Moment Problem

Let us begin by defining three cones in the vector space $\mathcal{H}_{d,m}$ resp. $\mathcal{H}_{d,2n}$:

$$Q_{d,m} := \left\{f \in \mathcal{H}_{d,m} : f = \sum_{j=1}^{k} (y_j \cdot)^m, \text{ where } y_1, \ldots, y_k \in \mathbb{R}^d, k \in \mathbb{N}\right\}, \tag{19.20}$$

$$\sum_{d,2n}^{2} := \left\{f \in \mathcal{H}_{d,2n} : f = \sum_{j=1}^{k} f_j^2, \text{ where } f_1, \ldots, f_k \in \mathcal{H}_{d,n}, k \in \mathbb{N}\right\}, \tag{19.21}$$

$$P_{d,2n} := \text{Pos}(\mathcal{H}_{d,2n}, \mathbb{R}^d) \equiv \{f \in \mathcal{H}_{d,2n} : f(x) \geq 0 \quad \text{for } x \in \mathbb{R}^d\}. \tag{19.22}$$

Since $\lambda(y \cdot x)^m = (\sqrt[m]{\lambda} y \cdot x)^m$ for $\lambda \geq 0$, $Q_{d,m}$ is a cone in $\mathcal{H}_{d,m}$. Obviously, $P_{d,2n}$ and $\sum_{d,2n}^2$ are cones in $\mathcal{H}_{d,2n}$. Clearly,

$$Q_{d,2n} \subseteq \sum\nolimits_{d,2n}^2 \subseteq P_{d,2n}. \tag{19.23}$$

In general, we have $Q_{d,2n} \neq P_{d,2n}$. Exercise 19.6 contains an example of a polynomial $p \in \sum_{2,4}^2$ such that $p(x) > 0$ for all $x \in \mathbb{R}^2 \setminus \{0\}$, but $p \notin Q_{2,4}$. At the end of this section we discuss briefly when the equality $\sum_{d,2n}^2 = P_{d,2n}$ holds.

Recall from Lemma 19.2 that the polynomials $(y \cdot)^m \in Q_{d,m}$, $y \in \mathbb{R}^d$, span $\mathcal{H}_{d,m}$. Thus $Q_{d,m}$, hence $\sum_{d,2n}^2$ and $P_{d,2n}$ by (19.23), span $\mathcal{H}_{d,m}$ resp. $\mathcal{H}_{d,2n}$. Therefore, all three cones have nonempty interiors in $\mathcal{H}_{d,m}$ resp. $\mathcal{H}_{d,2n}$ by Proposition A.33(i).

Proposition 19.13 $Q_{d,2n}$ and $P_{d,2n}$ are closed in the norm topology of $\mathcal{H}_{d,2n}$ and they are dual cones to each other with respect to the apolar scalar product, that is,

$$Q_{d,2n} = \{f \in \mathcal{H}_{d,2n} : [p,f] \geq 0 \text{ for } p \in P_{d,2n}\}, \tag{19.24}$$

$$P_{d,2n} = \{p \in \mathcal{H}_{d,2n} : [p,f] \geq 0 \text{ for } f \in Q_{d,2n}\}. \tag{19.25}$$

Proof Obviously, $P_{d,2n}$ is closed. We prove that $Q_{d,2n}$ is closed. Set $r := \dim \mathcal{H}_{d,2n}$. Let $(f_k)_{k \in \mathbb{N}}$ be a sequence from $Q_{d,2n}$ converging to $f \in \mathcal{H}_{d,2n}$. From Carathéodory's theorem (Proposition A.35) and the definition of $Q_{d,2n}$ it follows that each f_k is of the form

$$f_k(x) = \sum_{j=1}^r (y_j^{(k)} \cdot x)^{2n}, \quad \text{where } y_j^{(k)} = (y_{kj1}, \ldots, , y_{kjd})^T \in \mathbb{R}^d. \tag{19.26}$$

Fix $i \in \{1, \ldots, d\}$. The coefficient of x_i^{2n} in f_k is $\sum_{j=1}^r (y_{kji})^{2n}$. Clearly, as $k \to \infty$, these numbers converge to the coefficient of x_i^{2n} in f. In particular, these sequences are bounded, so there exists a $C > 0$ such that $|y_{kji}| \leq 1 + \sum_{l=1}^r (y_{kli})^{2n} \leq C$ for all k, j, i. Hence we can choose a subsequence $(k_m)_{m \in \mathbb{N}}$ such that $y_{ji} := \lim_{m \to \infty} y_{k_m ji}$ exists. Setting $y_j = (y_{j1}, \ldots, y_{jd})^T$ and passing to the limit $k \to \infty$ in (19.26) we get $f(x) = \sum_{j=1}^r (y_j \cdot)^{2n}$, so that $f \in Q_{d,2n}$. This proves that $Q_{d,2n}$ is closed.

We verify (19.24) and (19.25). We identify the dual of $\mathcal{H}_{d,2n}$ with the functionals $[\cdot, f]$ and $[p, \cdot]$, respectively; then the sets on the right-hand sides of (19.24) and (19.25) are the dual cones $(P_{d,2n})^\wedge$ and $(Q_{d,2n})^\wedge$, respectively. Let $f = \sum_{j=1}^r (y_j \cdot)^{2n}$ and $p \in P_{d,2n}$. Then, by (19.5), $[p,f] = \sum_{j=1}^r p(y_j) \geq 0$. Therefore, $Q_{d,2n} \subseteq (P_{d,2n})^\wedge$ and $P_{d,2n} \subseteq (Q_{d,2n})^\wedge$, so that $(Q_{d,2n})^\wedge \supseteq (P_{d,2n})^{\wedge\wedge}$ and $(P_{d,2n})^\wedge \supseteq (Q_{d,2n})^{\wedge\wedge}$.

On the other hand, since $P_{d,2n}$ and $Q_{d,2n}$ are closed, $(P_{d,2n})^{\wedge\wedge} = P_{d,2n}$ and $(Q_{d,2n})^{\wedge\wedge} = Q_{d,2n}$ by the bipolar theorem (Proposition A.32). By the preceding we have proved that $Q_{d,2n} = (P_{d,2n})^\wedge$ and $P_{d,2n} = (Q_{d,2n})^\wedge$, or equivalently, that (19.24) and (19.25) are satisfied. $\qquad \square$

Let us call a truncated $\mathbb{R}^d$-moment functional on $\mathcal{H}_{d,m}$ (as defined in Definition 17.1) simply a *moment functional*. Our first main result in this section is the following.

Theorem 19.14 *Let $f \in Q_{d,m}$ and*

$$f(x) = \sum_{j=1}^{k} c_j(y_j \cdot x)^m \quad \text{with } y_1,\ldots,y_k \in \mathbb{R}^d,\ c_1 > 0,\ldots,c_k > 0,\ k \in \mathbb{N}.$$

$$(19.27)$$

Then there is a moment functional L_f on $\mathcal{H}_{d,m}$ defined by

$$L_f(p) := \frac{1}{m!}f(\partial)p = [p,f] = \sum_{j=1}^{k} c_j l_{y_j}(p), \quad p \in \mathcal{H}_{d,m},$$

$$(19.28)$$

with representing measure $\mu = \sum_{j=1}^{k} c_j\delta_{y_j}$. Each moment functional on $\mathcal{H}_{d,m}$ is of the form L_f with $f \in Q_{d,m}$ uniquely determined.

Proof Equation (19.16) gives (19.28). The latter means that L_f is a moment functional and μ is a representing measure of L_f.

Conversely, let L be a moment functional on $\mathcal{H}_{d,m}$. If $L = 0$, then $L = L_f$ for $f(x) := (0 \cdot x)^m = 0$. Now let $L \neq 0$. By Theorem 17.2, L has a k-atomic representing measure $\mu = \sum_{j=1}^{k} c_j\delta_{y_j}$, where $c_j > 0$ and $k \in \mathbb{N}$. Then $f(x) := \sum_{j=1}^{k} c_j(y_j \cdot x)^m$ belongs to $Q_{d,m}$ and formula (19.16) implies that $L = L_f$.

Suppose that $L_f = L_g$ for $f, g \in Q_{d,m}$. Then $[p,f] = [p,g]$ by (19.28) and hence $[p,f-g] = 0$ for all $p \in \mathcal{H}_{d,m}$. Because $[\cdot,\cdot]$ is a scalar product on $\mathcal{H}_{d,m}$, we conclude that $f = g$. □

Obviously, $c(y \cdot x)^m = ((\sqrt[m]{c}\,y)\cdot x)^m$ for $c \geq 0$. Therefore, since the polynomials in $\mathcal{H}_{d,m}$ are homogeneous, the representation (19.27) of $f \in Q_{d,m}$ in Theorem 19.14 and the representing measure μ of the truncated moment functional L_f are not unique.

There are several natural ways of normalizing these representations. Without loss of generality let us assume that $f \neq 0$ and $y_j \neq 0$ for all j.

First, upon replacing y_j by $\sqrt[m]{c_j}\,y_j$ and allowing points of $\mathbb{R}^d$, we can always assume that $c_j = 1$ for all j. Secondly, replacing y_j by $\|y_j\|^{-1}y_j$ and c_j by $\|y_j\|^m c_j$ and allowing coefficients $c_j > 0$, we can assume that all points y_j belong to the unit sphere S^{d-1} of $\mathbb{R}^d$. Further, thirdly, upon replacing y_j by $-y_j$ if necessary, we can assume that all y_j are in S_+^{d-1}. Recall that S_+^{d-1} denotes the set of all $t = (t_1,\ldots,t_d)$ of the unit sphere S^{d-1} for which the first nonzero coordinate t_j is positive. These three normalizations mean to study the truncated moment problem for the vector space $\mathcal{H}_{d,m}$ on $\mathbb{R}^d$, S^{d-1}, and S_+^{d-1}, respectively.

Since S_+^{d-1} is a realization of the projective space $\mathbb{P}^{d-1}(\mathbb{R})$ (see Sect. 17.2), the normalization by S_+^{d-1} refers to the truncated moment problem for $\mathcal{H}_{d,m}$ on $\mathbb{P}^{d-1}(\mathbb{R})$. Hence the third normalization is most important among the three.

Let us retain the notation of Theorem 19.14. Then the Hankel matrix $H(L_f)$ of the moment functional L_f has the entries $h_{\alpha,\beta}, \alpha, \beta \in N(d,n)$, given by

$$h_{\alpha,\beta} = L_f(x^{\alpha+\beta}) = [x^{\alpha+\beta}, f] = \sum_{j=1}^{k} c_j x^{\alpha+\beta}(y_j) = \sum_{j=1}^{k} c_j y_{j1}^{\alpha_1+\beta_1} \cdots y_{jd}^{\alpha_d+\beta_d}.$$

If $s(y)$ is the column vector $(y^\alpha)_{\alpha \in N(d,n)}$ for $y \in \mathbb{R}^d$, then by Proposition 17.21,

$$H(L_f) = \sum_{j=1}^{k} c_j s(y_j) s(y_j)^T. \tag{19.29}$$

Our next theorem is a classical result of D. Hilbert (1909).

Theorem 19.15 *Let $d \in \mathbb{N}$ and $n \in \mathbb{N}$. Then the polynomial*

$$h_{d,2n}(x) := \|x\|^{2n} \equiv (x_1^2 + \cdots + x_d^2)^n$$

is in $Q_{d,2n}$, that is, there exist $k \in \mathbb{N}$ and vectors $y_1, \ldots, y_k \in \mathbb{R}^d \setminus \{0\}$ such that

$$h_{d,2n}(x) = \sum_{j=1}^{k} (y_j \cdot x)^{2n}. \tag{19.30}$$

The points y_j can be chosen such that $y_1 = \|y_1\|u$, where $u \in S^{d-1}$ is arbitrary.

Proof Let σ denote the surface measure of the unit sphere S^{d-1} and define

$$h(x) := \int_{S^{d-1}} (y \cdot x)^{2n} d\sigma(y), \quad x \in \mathbb{R}^d.$$

We expand $(y \cdot x)^{2n}$ by the multinomial theorem and conclude that $h(x)$ is a homogeneous polynomial of degree $2n$, that is, $h \in \mathcal{H}_{d,2n}$. Fix $x \in \mathbb{R}^d$. For $y \in S^{d-1}$ we abbreviate $z_1(y) := \|x\|^{-1}(y \cdot x)$. Since $|z_1(y)| \le 1$, there is an orthogonal transformation of variables $y \mapsto z(y) = (z_1(y), \ldots, z_d(y))$ on S^{d-1}. Using the invariance of σ under an orthogonal change of variables and the relation $y \cdot x = z_1 \|x\|$ we derive

$$h(x) = \int_{S^{d-1}} (z_1 \|x\|)^{2n} d\sigma(z) = \|x\|^{2n} \int_{S^{d-1}} z_1^{2n} d\sigma(z) = \|x\|^{2n} c \tag{19.31}$$

for some $c > 0$ depending on d, n. (For the explicit value of c, see Sect. 14.5.)

On the other hand, we define a truncated S^{d-1}-moment functional on $\mathcal{H}_{d,2n}$ by

$$L(f) = \int_{S^{d-1}} f(y) d\sigma(y), \quad f \in \mathcal{H}_{d,2n}.$$

Obviously, if $f = 0$ on S^{d-1} for some $f \in \mathcal{H}_{d,2n}$, then $f = 0$ on $\mathbb{R}^d$ and so $f = 0$. That is, the map $f \mapsto f \lceil S^{d-1}$ of $\mathcal{H}_{d,2n}$ is injective. Setting $\mathcal{X} = S^{d-1}$ and $E \cong \mathcal{H}_{d,2n}$ we are in the setup of Sect. 1.2 and L is a moment functional on E. Clearly, if $L(f) = 0$ for some $f \in E_+ \cong \text{Pos}(\mathcal{H}_{d,2n}, S^{d-1})$, then $f = 0$. That is, L is strictly E_+-positive. Therefore, by Theorem 1.30(ii), L has a k-atomic representing measure $\mu = \sum_{j=1}^{k} m_j \delta_{t_j}$ which has u as an atom, say $t_1 = u$. Then

$$L(f) = \sum_{j=1}^{k} m_j f(t_j) = \sum_{j=1}^{k} f(m_j^{1/2n} t_j). \tag{19.32}$$

Specializing to $f(y) = (y \cdot x)^{2n}$ we get $L(f) = h(x) = c\|x\|^{2n} = \sum_{j=1}^{k} (m_j^{1/2n} t_j \cdot x)^{2n}$ by (19.31) and (19.32). Setting $y_j = c^{-1/2n} m_j^{1/2n} t_j$ the last equation yields (19.30). By construction, $y_1 \|y_1\|^{-1} = t_1 = u$. $\qquad\square$

For many pairs (d, n) explicit representations (19.30) of $h_{d,2n}$ are known by classical formulas [Re1, Nat]. For instance, the following polynomial identities hold:

$$12 \, h_{3,4} = \quad 8x_1^4 + 8x_2^4 + 8x_3^4 +$$
$$(x_1 + x_2 + x_3)^4 + (x_1 - x_2 + x_3)^4 + (x_1 + x_2 - x_3)^4 + (x_1 - x_2 - x_3)^4,$$

$$6 \, h_{4,4} = \quad (x_1 + x_2)^4 + (x_1 - x_2)^4 + (x_1 + x_3)^4 + (x_1 - x_3)^4 + (x_1 + x_4)^4 + (x_1 - x_4)^4 +$$
$$(x_2 + x_3)^4 + (x_2 - x_3)^4 + (x_2 + x_4)^4 + (x_2 - x_4)^4 + (x_3 + x_4)^4 + (x_3 - x_4)^4,$$

$$60 \, h_{3,6} = \quad 40x_1^6 + 40x_2^6 + 40x_3^6 +$$
$$4(x_1 + x_2)^6 + 4(x_1 - x_2)^6 + 4(x_1 + x_3)^6 + 4(x_1 - x_3)^6 +$$
$$(x_1 + x_2 + x_3)^6 + (x_1 - x_2 + x_3)^6 + (x_1 + x_2 - x_3)^6 + (x_1 - x_2 - x_3)^6.$$

As usual, let $\Delta = (\frac{\partial}{\partial x_1})^2 + \cdots + (\frac{\partial}{\partial x_d})^2$ denote the Laplacian.

Theorem 19.16 *Let $d, n \in \mathbb{N}$. There is a truncated S^{d-1}-moment functional on $\mathcal{H}_{d,2n}$ defined by*

$$L(p) = \frac{1}{(2n)!} \Delta^n p = [p, (x_1^2 + \cdots + x_d^2)^n], \quad p \in \mathcal{H}_{d,2n}. \tag{19.33}$$

For each representation (19.30) of $h_{d,2n}(x) \in Q_{d,2n}$ we have

$$L(p) = \sum_{j=1}^{k} p(y_j), \quad p \in \mathcal{H}_{d,2n}. \tag{19.34}$$

The functional L is strictly $\mathrm{Pos}(\mathcal{H}_{d,2n}, S^{d-1})$-positive, that is, $L(f) > 0$ for each polynomial $f \in \mathrm{Pos}(\mathcal{H}_{d,2n}, S^{d-1}), f \neq 0$.

Proof Since $h_{d,2n} \in Q_{d,2n}$ by Theorem 19.15, it follows from Theorem 19.14 that $L \equiv L_{h_{d,2n}}$ is a truncated S^{d-1}-moment functional and (19.30) implies (19.34).

We show that L is strictly positive. Suppose that $f \in \mathrm{Pos}(\mathcal{H}_{d,2n}, S^{d-1}), f \neq 0$. Then there is a $u \in S^{d-1}$ such that $f(u) > 0$. By Theorem 19.15, there exists a representation (19.30) such that $y_1 = \|y_1\|u \neq 0$. Then, by (19.30),

$$L(f) = \sum_{j=1}^{k} f(y_j) \geq f(y_1) = \|y_1\|^{2n} f(u) > 0. \qquad \square$$

At the end of this section, we turn briefly to Carathéodory numbers.

Definition 19.17 For $f \in Q_{d,m}$ the *width* $w(f)$ of f is the smallest number k among all possible representations (19.27) of f, where we set $w(f) = 0$ for $f = 0$.

For instance, $w(h_{3,6}) = 11$ and $w(h_{3,8}) = 16$, see [Re1, Theorems 9.28 and 9.37].

Let L be a moment functional on $\mathcal{H}_{m,d}$. Then, by Theorem 19.14, $L = L_f$ for some unique function $f \in Q_{d,m}$. We define $w(L) := w(f)$. Further, let $\mathsf{C}(L)$ denote the smallest number k for all k-atomic representing measures $\mu \in M_+(\mathbb{R}^d)$ of L.

From Definition 18.7 we recall that the *Carathéodory number*

$$\mathsf{C}_{d,m} := \mathsf{C}(\mathcal{H}_{d,m}, \mathbb{R}^d)$$

is the maximum of the numbers $\mathsf{C}(L)$ for all moment functionals L on $\mathcal{H}_{d,m}$.

Proposition 19.18 *For each moment functional L on $\mathcal{H}_{d,m}$ we have $w(L) = \mathsf{C}(L)$. The number $\mathsf{C}_{d,m}$ is the largest width $w(L)$ for all moment functionals L on $\mathcal{H}_{d,m}$.*

Proof By definition, $\mathsf{C}(L) = w(L) = 0$ if $L = 0$. Hence we can assume that $L \neq 0$. Throughout this proof, we abbreviate $k := \mathsf{C}(L)$ and $r := w(L)$.

Let μ be a k-atomic representing measure of L, say $\mu = \sum_{j=1}^{k} c_j \delta_{y_j}$, where $c_j > 0$, $y_j \in \mathbb{R}^d$ for $j = 1, \ldots, k$. Then $f(x) := \sum_{j=1}^{k} c_j (y_j \cdot x)^m \in Q_{d,m}$ and

$$L(p) = \int p(x) d\mu = \sum_{j=1}^{k} c_j p(y_j) = L_f(p) \quad \text{for } p \in \mathcal{H}_{d,m},$$

so that $L = L_f$. Hence $r = w(f) \leq k$.

Conversely, let L be a moment functional. By Theorem 19.14, $L = L_f$ for some $f \in Q_{d,m}$. We can write $f(x) = \sum_{j=1}^{r} c_j (y_j \cdot x)^m$ with $y_j \in \mathbb{R}^d, j = 1, \ldots, w(f) = r$. Then, by Theorem 19.14, $\mu := \sum_{j=1}^{r} c_j \delta_{y_j}$ is a representing measure for $L_f = L$ and μ is k-atomic with $k \leq r$. Hence $k \leq r$. Thus we have proved that $r = k$.

The second assertion follows at once from the equality $w(L) = \mathsf{C}(L)$. $\qquad \square$

As noted above, each moment functional of $\mathcal{H}_{d,m}$ is a truncated S^{d-1}-moment functional and $\mathrm{Pos}(\mathcal{H}_{d,m}, \mathbb{R}^d) = \mathrm{Pos}(\mathcal{H}_{d,m}, S^{d-1})$. Hence, Theorem 18.43(ii), applied with $\mathcal{K} = S^{d-1}$, and (19.1) yield the following result: *If the equality*

$$\mathrm{Pos}(\mathcal{H}_{d,2n}, \mathbb{R}^d) = \sum \mathcal{H}_{d,n}^2, \tag{19.35}$$

that is, $\mathsf{P}_{d,2n} = \sum_{d,2n}^2$, *is satisfied, then* $\mathsf{C}_{d,2n} = |\mathsf{N}_{d,n}| = \binom{n+d-1}{d-1}$.

Unfortunately, (19.35) holds only in very few cases. D. Hilbert [H1] has shown that the equality (19.35) is fulfilled if and only if $n = 1$ or $d = 2$ or $(d,n) = (3,2)$.

It is not difficult to verify (see Exercises 19.7–19.9) that (19.35) holds if $n = 1$ or $d = 2$ and that (19.35) does not hold if $d \geq 3, n \geq 3$ or $d \geq 4, n \geq 2$.

The remaining case $(d,n) = (3,2)$ is more difficult. A simple proof of the fact that polynomials of $\mathrm{Pos}(\mathcal{H}_{3,4}, \mathbb{R}^3)$ are sums of squares can be found in [CL].

Taking these results for granted we get

$$\mathsf{C}_{d,2} = d, \quad \mathsf{C}_{2,2n} = n + 1, \quad \mathsf{C}_{3,4} = 6.$$

The general formula for $\mathsf{C}_{d,2n}$ is not yet known.

19.4 Robinson's Polynomial and Some Examples

For typographical simplicity we write x, y, z instead of x_0, x_1, x_2 in this section. Our aim is to investigate the (homogeneous) *Robinson polynomial*

$$R(x, y, z) := x^6 + y^6 + z^6 - x^4 y^2 - x^4 z^2 - x^2 y^4 - y^4 z^2 - x^2 z^4 - y^2 z^4 + 3x^2 y^2 z^2.$$

Clearly, $R \in \mathcal{H}_{3,6}$ and R is symmetric in all three variables x, y, z. It is convenient to write R in two other forms:

$$R = x^2(x^2 - z^2)^2 + y^2(y^2 - z^2)^2 - (x^2 - z^2)(y^2 - z^2)(x^2 + y^2 - z^2) \tag{19.36}$$

$$= (x^2 + y^2 - z^2)(x^2 - y^2)^2 + (x^2 - z^2)(y^2 - z^2)z^2. \tag{19.37}$$

Proposition 19.19

(i) $R \in \mathrm{Pos}(\mathbb{R}^3)$.
(ii) *Let* $(x, y, z)^T \in \mathbb{R}^3$. *Then we have* $R(x, y, z) = 0$ *if and only if* $x = 0, y^2 = z^2$ *or* $y = 0, x^2 = z^2$ *or* $z = 0, x^2 = z^2$ *or* $x^2 = y^2 = z^2$.
(iii) *R is not a sum of squares in* $\mathbb{R}[x, y, z]$.

Proof

(i) If $(x^2 - z^2)(y^2 - z^2)(x^2 + y^2 - z^2) \leq 0$, then it is obvious from (19.36) that $R(x, y, z) \geq 0$. Thus it remains to consider the case when

$$(x^2 - z^2)(y^2 - z^2)(x^2 + y^2 - z^2) \geq 0.$$

Then we have $x^2 + y^2 - z^2 \geq 0$ and $(x^2 - z^2)(y^2 - z^2) \geq 0$. But in this case (19.37) implies that $R(x, y, z) \geq 0$. (Another way to prove (i) is to verify the identity

$$(x^2 + y^2)R = x^2 z^2 (x^2 - z^2)^2 + y^2 z^2 (y^2 - z^2)^2 + (x^2 - y^2)^2 (x^2 + y^2 - z^2)^2.$$

Obviously, this implies that $R \geq 0$ on $\mathbb{R}^3$.)

(ii) The if part is clear from (19.36). Conversely, suppose that $R(x, y, z) = 0$. If $x = 0$, then (19.36) yields $y^2(y^2 - z^2)^2 + z^2(y^2 - z^2)^2 = 0$ and hence $y^2 = z^2$. Similarly, since R is symmetric, $y = 0$ implies $x^2 = z^2$ and $z = 0$ implies $x^2 = y^2$.

Suppose now that $x \neq 0, y \neq 0, z \neq 0$. Assume to the contrary that $x^2 = y^2 = z^2$ does not hold. By symmetry we can assume without loss of generality that $x^2 > z^2$. Then $(x^2 - z^2)(x^2 + y^2 - z^2) > 0$. Therefore, since $x^2(x^2 - z^2)^2 > 0$, the last term in (19.36) must be negative, which implies that $z^2 < y^2$. Then it follows from (19.37) that $R(x, y, z) > 0$, which is a contradiction.

(iii) Assume to the contrary that $R = \sum_j f_j^2$. Clearly, $\deg(f_j) \leq 3$. We write f_j as $f_j = xp_j + q_j$, where p_j and q_j contain only *even* powers of x. Then

$$R = \sum_j \; (x^2 p_j^2 + 2xp_j q_j + q_j^2). \tag{19.38}$$

Comparing the coefficients of x^6 we conclude that

$$1 = \sum_j \; p_j(1, 0, 0)^2. \tag{19.39}$$

Since R has only even powers of x, the odd powers of x on the right-hand side of (19.38) vanish. Thus, $\sum_j 2xp_j q_j = 0$. Then (19.38) implies that $R \geq x^2 p_j^2$ on $\mathbb{R}^3$ for all j. Setting $y = x$ and writing $p_j(x, x, z) = a_i x^2 + b_i xz + c_i z^2$ this yields

$$R(x, x, z) = z^2(z^2 - x^2)^2 \geq x^2 p_j(x, x, z)^2 = x^2(a_i x^2 + b_i xz + c_i z^2)^2.$$

Choosing x large, it follows that $a_i = 0$. Then $z^2(z^2 - x^2)^2 \geq x^2(b_i x + c_i z)^2$. Hence $b_i x + c_i z = 0$ if $x \pm z = 0, x \neq 0$. Therefore, $b_i = c_i = 0$, so that $p_j(x, x, z) = 0$. Similarly, $p(x, -x, z) = 0$. These relations imply that $p_j(x, y, z)$ is divisible by $x - y$ and $x + y$. Replacing y by z in the preceding, p_j is divisible by $x - z$ and $x + z$. Thus,

$$p_j(x, y, z) = (x^2 - y^2)(x - z^2)g_j.$$

Since $\deg(p_j) \leq 2$, $g_j \equiv 0$ and $p_j(1, 0, 0) = 0$ for all j which contradicts (19.39).

An elegant and short proof of assertion (iii) can be given by using a basic result from the theory of cubics, see Example 19.36 below. $\square$

As discussed in Sect. 17.2, the subset S_+^2 of the unit sphere S^2 can be considered as a realization of the projective space $\mathbb{P}^2(\mathbb{R})$. From the description of the zero set of R given in Proposition 19.19(ii) it follows that R has exactly 10 zeros in the projective space $\mathbb{P}^2(\mathbb{R})$. The corresponding points of S_+^2 are

$$t_1 = \frac{1}{\sqrt{3}}(1,1,1), t_2 = \frac{1}{\sqrt{3}}(1,1,-1), t_3 = \frac{1}{\sqrt{3}}(1,-1,1), t_4 = \frac{1}{\sqrt{3}}(1,-1,-1),$$

$$t_5 = \frac{1}{\sqrt{2}}(1,0,1), t_6 = \frac{1}{\sqrt{2}}(1,0,-1), t_7 = \frac{1}{\sqrt{2}}(1,1,0), t_8 = \frac{1}{\sqrt{2}}(1,-1,0),$$

$$t_9 = \frac{1}{\sqrt{2}}(0,1,1), t_{10} = \frac{1}{\sqrt{2}}(0,1,-1).$$

Proposition 19.20 *Let $p(x,y,z) \in \mathcal{H}_{3,6}$. Then we have $p \in \mathrm{Pos}(\mathbb{R}^3)$ and $p(t_j) = 0$ for $j = 1,2,\ldots,9$ if and only if $p \in \mathbb{R}_+ \cdot R + \mathbb{R}_+ \cdot h_{10}^2$, where*

$$h_{10}(x,y,z) = y^3 - z^3 + (z-y)x^2 = (y-z)(y^2 + yz + z^2 - x^2).$$

Proof The if part is easily verified. It suffices to prove the only if part.
Suppose that $p \in \mathrm{Pos}(\mathbb{R}^3)$ and $p(t_j) = 0$ for $j = 1,2,\ldots,9$. Write

$$p(x,y,z) = \sum_{j,k,l=0}^{6} a_{jkl}x^j y^k z^l.$$

Since $p \geq 0$ on $\mathbb{R}^3$, each zero t_j is a local minimum of p, so all first-order partial derivatives of p at t_j vanish as well (see also Lemma 19.26 below). Thus

$$p(t_j) = \frac{\partial p}{\partial x}(t_j) = \frac{\partial p}{\partial y}(t_j) = \frac{\partial p}{\partial z}(t_j) = 0 \quad \text{for } j = 1,\ldots,9. \tag{19.40}$$

These are 36 homogeneous linear equations for the 28 variables a_{jkl}. It can be shown (for instance, by using a computer program such as *Mathematica*) that the corresponding matrix has rank 26. Hence the dimension of the solution space is 2. Since R and h_{10}^2 are in $\mathrm{Pos}(\mathbb{R}^3)$ and vanish at $t_1,\ldots,t_9$, the coefficient vectors of R and h_{10}^2 are solutions of (19.40). Hence $p = \lambda_1 R + \lambda_2 h_{10}^2$ with $\lambda_1, \lambda_2 \in \mathbb{R}$. Since $R(t_{10}) = 0$ and $h_{10}(t_{10})^2 > 0$, we get $p(t_{10}) = \lambda_2 h_{10}(t_{10})^2 \geq 0$, so that $\lambda_2 \geq 0$. For $t = (2,1,1)$ we have $h_{10}(t) = 0$ and $R(t) \neq 0$. Then, $p(t) = \lambda_1 R(t) \geq 0$ and hence $\lambda_1 \geq 0$. $\square$

Remark 19.21 Let $i \in \{1,\ldots,10\}$. The assertion of Proposition 19.20 remains valid if t_{10} is replaced by t_i: There is a polynomial $h_i \in \mathcal{H}_{3,3}$ such that $p \in \mathrm{Pos}(\mathcal{H}_{3,6}, \mathbb{R}^3)$ vanish on t_j for $j = 1,\ldots,10, j \neq i$, if and only if $p \in \mathbb{R}_+ \cdot R + \mathbb{R}_+ \cdot h_i^2$. We have

$$h_1 = x(y^2 + z^2 - x^2) + y(z^2 + x^2 - y^2) + z(x^2 + y^2 - z^2) - xyz.$$

All other polynomials h_i can be obtained by symmetry from h_{10} and h_1; for instance, $h_5(x,y,z) = h_{10}(y,x,-z)$, $h_2(x,y,z) = h_1(x,y,-z)$ etc.   ∘

Corollary 19.22 *Let $p \in \mathrm{Pos}(\mathcal{H}_{3,6}, \mathbb{R}^3)$. The following are equivalent:*

(i) $p(t_j) = 0$ *for* $j = 1, \ldots, 10$.
(ii) $p = \lambda R$ *for some number* $\lambda \geq 0$.
(iii) $(\gamma R - p) \in \mathrm{Pos}(\mathbb{R}^3)$ *for some number* $\gamma \geq 0$.

Proof
(i)→(ii) By Proposition 19.20, (i) implies $p = \lambda_1 R + \lambda_2 h_{10}^2$ with $\lambda_1 \geq 0, \lambda_2 \geq 0$. Since $h_{10}(t_{10}) \neq 0$ and $R(t_{10}) = 0$, we obtain $\lambda_2 = 0$, so that $p = \lambda_1 R$.
(ii)→(iii) is trivial by setting $\gamma = \lambda$.
(iii)→(i) Since $0 \leq p \leq \gamma R$ on $\mathbb{R}^3$ and $R(t_j) = 0$ for $j = 1, \ldots, 10$, p vanishes at $t_1, \ldots, t_{10}$ as well.   □

From Corollary 19.22 (iii)→(ii) it follows that the Robinson polynomial R spans an extreme ray of the cone $\mathrm{Pos}(\mathcal{H}_{3,6}, \mathbb{R}^3)$.
Now we turn to the truncated moment problem on $\mathbb{P}^2(\mathbb{R}) \cong S_+^2$ and set

$$N := \{\alpha = (\alpha_0, \alpha_1, \alpha_2) \in \mathbb{N}_0^3 : \alpha_0 + \alpha_1 + \alpha_2 = 6\}.$$

Then $\mathrm{Lin}\{x^\alpha : \alpha \in N\}$ is the vector space $\mathcal{H}_{3,6}$ of 3-forms of degree 6.
We fix a point $t_0 \in S_+^2$ such that $t_0 \neq t_j$ for $j = 1, \ldots, 10$. Then $R(t_0) \neq 0$. Put

$$\nu = \sum_{j=1}^{10} m_j \delta_{t_j}, \quad \mu = \nu + m_0 \delta_{t_0} = \sum_{j=0}^{10} m_i \delta_{t_i}, \quad \text{where } m_j \geq 0, \quad j = 0, \ldots, 10,$$

and let L^ν and L^μ denote the corresponding truncated moment functionals on the projective space $\mathbb{P}^2(\mathbb{R}) \cong S_+^2$. The following simple fact will be used several times.

Lemma 19.23 *Let $p \in \mathcal{H}_{3,6}$. If $p(x) \geq 0$ on S_+^2, then $p(x) \geq 0$ on $\mathbb{R}^3$.*

Proof The equalities $\mathbb{R}^3 = \cup_{c \in \mathbb{R}} c\, S_+^2$ and $p(cx) = c^6 p(x)$ give the assertion.   □

In the following proofs we use the projective versions of the notions and results.

Proposition 19.24 *L^ν is a determinate truncated moment functional on $\mathbb{P}^2(\mathbb{R}) \cong S_+^2$ and ν is a maximal mass measure. Further, if $m_j > 0$ for $j = 1, \ldots, 10$, then*

$$\mathcal{N}_+(L^\nu, S_+^2) = \mathbb{R}_+ \cdot R \quad and \quad \mathcal{V}_+(L^\nu, S_+^2) = \mathrm{supp}\,\nu = \{t_1, \ldots, t_{10}\}. \tag{19.41}$$

Proof By straightforward computations we verify that the polynomials

$$q_1 = x^2 y(x+y)z(x+z), \qquad q_2 = x^2(x-y)yz(x-z),$$
$$q_3 = x^2 y(x+y)z(x-z), \qquad q_4 = x^2 y(x-y)z(x+z),$$

$$q_5 = x^2(x^2 - y^2)z(x + z), \qquad q_6 = x^2(x^2 - y^2)z(x - z),$$

$$q_7 = x^2(x - y)(x^2 - z^2), \qquad q_8 = x^2(x + y)(x^2 - z^2),$$

$$q_9 = (x^2 - y^2)(x^2 - z^2)(y - z)^2, \quad q_{10} = (x^2 - y^2)(x^2 - z^2)(y + z)^2$$

of $\mathcal{H}_{3,6}$ satisfy $q_j(t_i) = 0$ for $j \neq i$ and $q_j(t_j) \neq 0$, where $i, j = 1 \ldots, 10$. From this it follows that the points $t_1, \ldots, t_{10}$ obey property $(SP)_{\mathcal{H}_{3,6}}$, see Definition 18.11. Hence L^ν is determinate by Proposition 18.12. In particular, ν is maximal mass.

We abbreviate $\mathcal{N}_+ := \mathcal{N}_+(L^\nu, S_+^2)$ and prove that $\mathcal{N}_+ = \mathbb{R}_+ \cdot R$. Since $m_j > 0$ for $j = 1, \ldots, 10$, $\mathcal{N}_+$ consists of $p \in \mathrm{Pos}(\mathcal{H}_{3,6}, S_+^2)$ that vanish at $t_1, \ldots, t_{10}$, see (18.11). Hence $R \in \mathcal{N}_+$. Conversely, let $p \in \mathcal{N}_+$. Then $p \geq 0$ on S_+^2. Hence $p \geq 0$ on $\mathbb{R}^3$ by Lemma 19.23, so $p \in \mathbb{R}_+ \cdot R$ by Corollary 19.22 (i)$\rightarrow$(ii). Thus, $\mathcal{N}_+ = \mathbb{R}_+ \cdot R$.

Since $\{t_1, \ldots, t_{10}\}$ is the zero set of R on the projective space, the second equality of (19.41) follows at once from the first. $\qquad\square$

Proposition 19.25 *Suppose that $m_j > 0$ for $j = 0, 1, \ldots, 10$. Then the measure μ is ordered maximal mass. If $h(t_0) \neq 0$ and $R(t_0) > 0$ (for instance, if $t_0 = (1, 0, 0)$), then μ is not maximal mass.*

Proof In this proof we abbreviate $L = L^\mu$.

Put $p := R(t_0)^{-1}R$. Then $p \geq 0$ on S_+^2, $p(t_0) = 1$, and $p(t_j) = 0$ for $j = 1, \ldots, 10$. Hence $\rho_L(t_0) = m_0$ by Proposition 18.31(iii). Since $\mu - m_0\delta_{t_0} = \nu$ by definition and ν is maximal mass by Proposition 19.24, μ is ordered maximal mass.

Let $f \in \mathrm{Pos}(\mathcal{H}_{3,6}, S_+^2)$ and suppose that $L(f) = 0$. Then $f \geq 0$ on S_+^2 and hence $f \geq 0$ on $\mathbb{R}^3$ by Lemma 19.23. From $L(f) = 0$ and $m_j > 0$ we obtain $f(t_j) = 0$ for $j = 0, 1, \ldots, 10$. Therefore, by Corollary 19.22, $f = \lambda R$ with $\lambda \geq 0$. Since $f(t_0) = 0$ and $R(t_0) \neq 0$, $f = 0$. This shows that L is strictly $\mathrm{Pos}(\mathcal{H}_{3,6}, S_+^2)$-positive. Therefore, by Proposition 18.28(iii), the infimum in (18.23) is attained for each point x.

Assume to the contrary that μ is maximal mass. Then Theorem 18.33(ii) applies to the compact set $\mathcal{K} = \mathbb{P}^2(\mathbb{R}) \cong S_+^2$, so there exists a polynomial $p \in \mathcal{H}_{3,6}$ such that $p \geq 0$ on S_+^2, hence $p \geq 0$ on $\mathbb{R}^3$ by Lemma 19.23, and $p(t_{10}) = 1$, $p(t_j) = 0$ for $j = 0, \ldots, 9$. From Proposition 19.20 we obtain $p = \lambda_1 R + \lambda_2 h_{10}^2$ with $\lambda_1, \lambda_2 \in \mathbb{R}_+$. Since $R(t_0) > 0$ and $h_{10}(t_0) \neq 0$ by assumption, $p(t_0) = \lambda_1 R(t_0) + \lambda_2 h(t_0)^2 = 0$ implies that $\lambda_1 = \lambda_2 = 0$. Thus, $p = 0$ and $p(t_{10}) = 0$. This is a contradiction. $\qquad\square$

19.5 Zeros of Positive Homogeneous Polynomials

Zeros of positive polynomials play a crucial role for the truncated moment problem. They appeared in the definition of the subcone $\mathcal{N}_+(L, \mathcal{K})$ (see (18.11)) and in the study of maximal masses (see Definition 18.32). In this section, we develop some basic results on zeros of positive polynomials of $\mathcal{H}_{3,2n}$.

We begin with two simple preliminary lemmas.

Lemma 19.26 *Let U be an open subset of $\mathbb{R}^d$, $u \in U$, and $f \in \mathbb{R}_d[\underline{x}]$. Suppose that $f(u) = 0$ and $f(x) \geq 0$ for all $x \in U$. Then we have $\frac{\partial f}{\partial x_i}(u) = 0$ for $i = 1, \ldots, d$ and the Hessian matrix $\left(\frac{\partial^2 f}{\partial x_i \partial x_j}(u) \right)_{i,j=1}^d$ is positive semidefinite.*

Proof Upon translation we can assume that $u = 0$. Let $t = (t_1, \ldots, t_d) \in U$. Since $f(0) = 0$, the Taylor expansion of f at 0 is

$$f(t) = \sum_{i=1}^d t_i \frac{\partial f}{\partial x_i}(0) + \sum_{i,j=1}^d t_i t_j \frac{\partial^2 f}{\partial x_i \partial x_j}(0) + \sum_{i,j,k=1}^d t_i t_j t_l \frac{\partial^3 f}{\partial x_i \partial x_j \partial x_k}(0) + \cdots .$$

We set $t_j = 0$ for $j \neq i$ and consider small $|t_i|$. Since $f(t) \geq 0$ on U, this expansion implies that $\frac{\partial f}{\partial x_i}(u) = 0$. Thus the linear part of the Taylor expansion disappears. Using again that $f(t) \geq 0$ on U and taking all $|t_i|$ small we conclude that the quadratic part is nonnegative. Hence the Hessian is a positive semidefinite matrix. $\qquad \square$

Lemma 19.27 *Let $f \in \mathbb{R}_d[\underline{x}]$. Suppose that f is irreducible in $\mathbb{R}_d[\underline{x}]$, but reducible in $\mathbb{C}_d[\underline{x}]$. Then there exist polynomials $g_1, g_2 \in \mathbb{R}_d[\underline{x}]$ such that $f = \epsilon(g_1^2 + g_2^2)$, where $\epsilon = 1$ or $\epsilon = -1$.*

Proof Since f is reducible in $\mathbb{C}_d[\underline{x}]$, there is a nontrivial factorization $f = qp$ with $p, q \in \mathbb{C}_d[\underline{x}]$. Taking complex conjugates and using that $f \in \mathbb{R}_d[\underline{x}]$ we get $f = \overline{q}\,\overline{p}$. Hence $f \cdot f = (q\overline{q}) \cdot (p\overline{p})$. Since f is irreducible in $\mathbb{R}_d[\underline{x}]$ and $q\overline{q}, p\overline{p} \in \mathbb{R}_d[\underline{x}]$, the uniqueness of irreducible factorizations implies that $f = a(q\,\overline{q})$ for some nonzero real number a. We write $\sqrt{|a|}\, q = g_1 + ig_2$ with $g_1, g_2 \in \mathbb{R}_d[\underline{x}]$ and set $\epsilon = a|a|^{-1}$. Then $f = a(q\,\overline{q}) = \epsilon(g_1^2 + g_2^2)$. $\qquad \square$

Next we collect some facts on plane curves. We will need only a few elementary notions and some basic results stated as Lemmas 19.28, 19.29, and 19.38.

Let $f \in \mathcal{H}_{3,m}$. Then the set

$$\mathcal{Z}_{\mathbb{P}}(f) = \{ t \in \mathbb{P}^2(\mathbb{R}) : f(t) = 0 \}$$

is called the *real projective curve* associated with f, or briefly, the curve $f = 0$. Note that the equation $f(t) = 0$ is well-defined for $t \in \mathbb{P}^2(\mathbb{R})$, since f is homogeneous.

A point t of the curve $\mathcal{Z}_{\mathbb{P}}(f)$ is called *singular* if $\frac{\partial f}{\partial x_1}(t) = \frac{\partial f}{\partial x_2}(t) = \frac{\partial f}{\partial x_3}(t) = 0$.

Let $t = [t_1 : t_2 : t_3] \in \mathcal{Z}_{\mathbb{P}}(f)$. If $f \geq 0$ in some neighbourhood $U \subseteq \mathbb{R}^3$ of $(t_1, t_2, t_3) \in \mathbb{R}^3$, Lemma 19.26 implies that t is a singular point of the curve $\mathcal{Z}_{\mathbb{P}}(f)$.

Lemma 19.28 *If $f \in \mathcal{H}_{3,m}$ is irreducible in $\mathbb{C}_3[\underline{x}]$, then the curve $\mathcal{Z}_{\mathbb{P}}(f)$ has at most $\frac{1}{2}m(m-1)$ singular points.*

Proof See e.g. [Wr, II, Theorem 4.4]. $\qquad \square$

Let $f \in \mathcal{H}_{3,m}$ and $g \in \mathcal{H}_{3,n}$. For each point $t \in \mathcal{Z}_{\mathbb{P}}(f) \cap \mathcal{Z}_{\mathbb{P}}(g)$ the *intersection multiplicity* $I_t(f, g) \in \mathbb{N}$ of the curves $f = 0$ and $g = 0$ at t is defined in

[Wr, III, Section 2.2] or in [Fn, Chapter 3, Section 3.3]. We do not restate the precise definition here. In what follows we use only the fact that $I_t(f, g) \geq 2$ if $t \in \mathcal{Z}_\mathbb{P}(f) \cap \mathcal{Z}_\mathbb{P}(g)$ is a singular point of one of the curves $f = 0$ or $g = 0$.

Also, we will need *Bezout's theorem*. For *complex* projective curves we even have equality in (19.42), but for our applications the following simple version is sufficient. The symbol $|Z|$ denotes the number of points of a set Z.

Lemma 19.29 *Suppose that the polynomials* $f \in \mathcal{H}_{3,n}$ *and* $g \in \mathcal{H}_{3,m}$ *are relatively prime in* $\mathbb{R}_3[\underline{x}]$, *that is,* f *and* g *have no common factor of positive degree. Then* $|\mathcal{Z}_\mathbb{P}(f) \cap \mathcal{Z}_\mathbb{P}(g)| \leq nm$. *More precisely,*

$$\sum\nolimits_{t \in \mathcal{Z}_\mathbb{P}(f) \cap \mathcal{Z}_\mathbb{P}(g)} I_t(f, g) \leq nm. \tag{19.42}$$

Proof See e.g. [Wr, p. 59] or [Fn, p. 57]. □

Having finished the preparations we turn to zeros of positive polynomials.

Proposition 19.30 *Let* $f \in \mathcal{H}_{3,2n}$ *be irreducible in* $\mathbb{R}_3[\underline{x}]$. *If* $f \in \mathrm{Pos}(\mathbb{R}^3)$, *then*

$$|\mathcal{Z}_\mathbb{P}(f)| \leq \alpha(n) := \max(n^2, (2n-1)(n-1)). \tag{19.43}$$

Proof First assume that f is also irreducible in $\mathbb{C}_3[\underline{x}]$. Since $f \in \mathrm{Pos}(\mathbb{R}^3)$, by Lemma 19.26 all zeros of f are singular points of the curve $f = 0$. Since f is irreducible in $\mathbb{C}_3[\underline{x}]$, this curve has at most $\frac{1}{2}(2n-1)(2n-2) = (2n-1)(n-1)$ singular points by Lemma 19.28. Therefore, $|\mathcal{Z}_\mathbb{P}(f)| \leq (2n-1)(n-1)$.

Now let f be reducible in $\mathbb{C}_3[\underline{x}]$. Then Lemma 19.27 applies. Since $f \geq 0$ on $\mathbb{R}^3$, we have $f = g_1^2 + g_2^2$ with $g_1, g_2 \in \mathbb{R}_3[\underline{x}]$. Clearly, g_1, g_2 are also homogeneous, $\mathcal{Z}_\mathbb{P}(f) = \mathcal{Z}_\mathbb{P}(g_1) \cap \mathcal{Z}_\mathbb{P}(g_2)$, and $\deg(g_j) \leq n$ for $j = 1, 2$. We verify that g_1 and g_2 are relatively prime in $\mathbb{C}_3[\underline{x}]$. Indeed, if $h \in \mathbb{C}_3[\underline{x}]$ is a common factor of g_1 and g_2, then $h \cdot \overline{h}$ is a common factor of g_1^2 and g_2^2 and so of f. Therefore, $h \cdot \overline{h}$, and hence h, is constant, because f is irreducible in $\mathbb{R}_3[\underline{x}]$. That is, the assumptions of Bezout's theorem (Lemma 19.29) are satisfied, so the real projective curves defined by $g_1 = 0$ and $g_2 = 0$ intersect in at most $n \cdot n$ points. Thus, $|\mathcal{Z}_\mathbb{P}(f)| \leq n^2$. □

Lemma 19.31 *The function* $\alpha(n)$ *on* $\mathbb{N}$ *defined by (19.43) is subadditive, that is,*

$$\alpha(n_1) + \alpha(n_2) + \cdots + \alpha(n_r) \leq \alpha(n_1 + \cdots + n_r) \quad \text{for } n_1, \ldots, n_r \in \mathbb{N}. \tag{19.44}$$

Proof Note that $\frac{\alpha(n)}{n}$ is monotone increasing on $\mathbb{N}$. Therefore, $\frac{\alpha(n_j)}{n_j} \leq \frac{\alpha(n_1 + n_2)}{n_1 + n_2}$ for $j = 1, 2$ and hence

$$\alpha(n_1) + \alpha(n_2) \leq \frac{n_1 \alpha(n_1 + n_2)}{n_1 + n_2} + \frac{n_2 \alpha(n_1 + n_2)}{n_1 + n_2} = \alpha(n_1 + n_2).$$

Formula (19.44) follows by induction on r. □

Clearly, $\alpha(1) = 1, \alpha(2) = 4, \alpha(3) = 10$, and $\alpha(n) = (2n-1)(n-1)$ for $n \geq 3$. Note that the Robinson polynomial $R \in \mathcal{H}_{3,6}$ has exactly $\alpha(3) = 10$ zeros in $\mathbb{P}^2(\mathbb{R})$.

Let us call a polynomial $f \in \mathbb{R}_d[\underline{x}]$ *indefinite* if neither f nor $-f$ is in $\mathrm{Pos}(\mathbb{R}^d)$, that is, there exist points $t_1, t_2 \in \mathbb{R}^d$ such that $f(t_1) < 0$ and $f(t_2) > 0$.

Our main result in this section is the following theorem.

Theorem 19.32 *Let* $n \in \mathbb{N}$ *and* $f \in \mathcal{H}_{3,2n}$. *Suppose that* $f(x) \geq 0$ *on* $\mathbb{R}^3$ *and* $|\mathcal{Z}_\mathbb{P}(f)| > \alpha(n)$. *Then* $|\mathcal{Z}_\mathbb{P}(f)|$ *is infinite and there are polynomials* $p \in \mathcal{H}_{3,2n_1}$, $q \in \mathcal{H}_{3,n_2}$ *such that* $f = pq^2$, *where* $n_1 + n_2 = n$, $p \in \mathrm{Pos}(\mathbb{R}^3)$, $|\mathcal{Z}_\mathbb{P}(p)| < \infty$, q *is indefinite, and* $|\mathcal{Z}_\mathbb{P}(q)|$ *is infinite. Moreover,* q *is a product of indefinite irreducible polynomials in* $\mathbb{R}_3[\underline{x}]$ *and* p *and* q *are relatively prime in* $\mathbb{R}_3[\underline{x}]$. *(It is possible that* p *is a positive real constant; in this case* $n_1 = 0$ *and we set* $\mathcal{H}_{3,0} := \mathbb{R}$.)

Proof Since the assertion trivially holds for $f = 0$, we can assume that $f \neq 0$. Let $f = f_1 \cdots f_r$ be a factorization of f as a product of irreducible factors in $\mathbb{R}_3[\underline{x}]$. Since f is homogeneous, so are all factors f_j. Set $m_j = \deg(f_j)$.

Assume first that no factor f_j is indefinite. Then $m_j = 2n_j$ is even. Multiplying by -1 if necessary, we can suppose that $f_j \in \mathrm{Pos}(\mathbb{R}^3)_{2n_j}$. Then, by (19.43) and (19.44),

$$|\mathcal{Z}_\mathbb{P}(f)| \leq \sum_j |\mathcal{Z}_\mathbb{P}(f_j)| \leq \sum_j \alpha(n_j) \leq \alpha(n_1 + \cdots + n_r) = \alpha(n),$$

which contradicts the assumption $|\mathcal{Z}_\mathbb{P}(f)| > \alpha(n)$.

As shown in the preceding paragraph, there is at least one indefinite factor, say f_1. It suffices to show that $|\mathcal{Z}_\mathbb{P}(f_1)|$ is infinite and that f_1^2 divides f. Indeed, then $f = f_1^2 h$ with $h \in \mathcal{H}_{3,2k}, k < n$. Clearly, $h \geq 0$ on $\mathbb{R}^3$. If $\mathcal{Z}_\mathbb{P}(h)$ is finite, we are finished; if not, we proceed by induction.

Set $g = f_2 \cdots f_r$. Then $f = f_1 g$. Since f_1 is indefinite, there are $t_1, t_2 \in \mathbb{R}^3$ such that $f_1(t_1) < 0$ and $f_1(t_2) > 0$. After affine changes of coordinates we can assume that $t_1 = (1, 0, a), t_2 = (1, 0, b), a < b$. By continuity, there is a $\delta > 0$ such that $f_1(1; u, a) < 0$ and $f_1(1, u, b) > 0$ for $|u| < \delta$. Fix $u \in (-\delta, \delta)$. Since $f_1(1, u, y)$ is a nonconstant polynomial in y, by the intermediate value theorem there exists a $c_u \in (a, b)$ such that $f_1(1, u, c_u) = 0$ and $f_1(1, u, y) < 0$ for $y < c_u, f_1(1, u, y) > 0$ for $y > c_u$ in a neighbourhood of c_u. From $f(1, u, y) = f_1(1, u, y)g(1, u, y) \geq 0$ we get $g(1, u, y) \leq 0$ if $y < u_c, g(1, u, y) \geq 0$ if $y > u_c$ in a neighbourhood of c_u. Hence $g(1, u, c_u) = 0$, so that $(1 : u : c_u) \in \mathcal{Z}_\mathbb{P}(f_1) \cap \mathcal{Z}_\mathbb{P}(g)$. Thus, $\mathcal{Z}_\mathbb{P}(f_1) \cap \mathcal{Z}_\mathbb{P}(g)$, hence $\mathcal{Z}_\mathbb{P}(f_1)$, is infinite. Clearly, f_1 and g are not relatively prime, because this would imply that their curves have only finitely many common points by Bezout's theorem. Since f_1 is irreducible, f_1 divides g. Because $f = f_1 g$, then f_1^2 divides f.

By construction, q is a product of irreducible indefinite factors f_i. Since $\mathcal{Z}_\mathbb{P}(p)$ is finite and each $\mathcal{Z}_\mathbb{P}(f_i)$ is infinite, p and q are relatively prime in $\mathbb{R}_3[\underline{x}]$. $\square$

The reasoning concerning f_1 in the preceding proof yields the following.

Corollary 19.33 *For each indefinite polynomial* $f \in \mathcal{H}_{3,m}$ *the set* $\mathcal{Z}_\mathbb{P}(f)$ *is infinite.*

Remark 19.34 As noted in [CLR, p. 13], Theorem 19.32 remains valid if $\alpha(n)$ is replaced by $\hat{\alpha}(n) := \frac{3}{2}n(n-1) + 1$. This stronger statement is based on deep results of I.G. Petrovskii and O.A. Oleinik on ovals of plane curves [PO]. Let $\beta(n)$ denote the maximum of $|\mathcal{Z}_{\mathbb{P}}(f)|$, where $f \in \mathcal{H}_{3,2n}$ and $\mathcal{Z}_{\mathbb{P}}(f)$ is finite. Then

$$n^2 \leq \beta(n) \leq \hat{\alpha}(n) = \frac{3}{2}n(n-1) + 1.$$

○

19.6   Applications to the Truncated Moment Problem on $\mathbb{R}^2$

In this section, we suppose that $t_1, \ldots, t_k$ are pairwise distinct points of $\mathbb{R}^2$, $c_1, \ldots, c_k$ are positive numbers and L denotes the truncated moment functional on $\mathbb{R}_2[\underline{x}]_{2n}$, where $n = 2, 3$, given by the k-atomic measure $\mu = \sum_{j=1}^{k} c_j \delta_{t_j}$:

$$L(f) = \int f \, d\mu \equiv \sum_{j=1}^{k} c_j f(t_j), \quad f \in \mathbb{R}_2[\underline{x}]_{2n}. \tag{19.45}$$

Then the vector space $\mathcal{N}_L$ and the cone $\mathcal{N}_+(L)$ from Definitions 17.16 and 18.13 are

$$\mathcal{N}_L = \{p \in \mathbb{R}_2[\underline{x}]_n : p(t_1) = \cdots = p(t_k) = 0\}, \tag{19.46}$$

$$\mathcal{N}_+(L) = \{p \in \mathrm{Pos}(\mathbb{R}^2)_{2n} : p(t_1) = \cdots = p(t_k) = 0\}, \tag{19.47}$$

and $\mathcal{V}_L$ and $\mathcal{V}_+(L)$ are the real algebraic sets in $\mathbb{R}^2$ defined by $\mathcal{N}_L$ and $\mathcal{N}_+(L)$, respectively.

   Our aim is to illustrate the usefulness of Theorem 19.32 by developing some results on $\mathcal{V}_L$ and $\mathcal{V}_+(L)$ for $n = 2, 3$. Since Theorem 19.32 deals with forms, we have to switch between polynomials of $\mathbb{R}_2[\underline{x}]_{2n}$ and forms of $\mathcal{H}_{3,2n}$.

   Let $q \in \mathbb{R}_2[\underline{x}]_m$ be of degree m and let $p \in \mathcal{H}_{3,m}$. We define the *homogenization* $q_h \in \mathcal{H}_{3,m}$ and the *dehomogenization* $\hat{p} \in \mathbb{R}_2[\underline{x}]_m$ by

$$q_h(x_1, x_2, x_3) := x_1^m q\left(\frac{x_2}{x_1}, \frac{x_3}{x_1}\right) \quad \text{and} \quad \hat{p}(x_1, x_2) := p(1, x_1, x_2). \tag{19.48}$$

Then $\widehat{q_h} = q$ and $\deg(q_h) = \deg(q) = m$. Clearly, $\deg(\hat{p}) \leq m$, but it may happen that $\deg(\hat{p}) < m$; for instance, if $p(x_1, x_2, x_3) = x_1^2$ and $m = 2$, then $\hat{p} = 1$. If $\deg(\hat{p}) = m$, then $(\hat{p})_h = p$. If $p_j \in \mathcal{H}_{3,m_j}$ for $j = 1, 2$, then $p_1 p_2 \in \mathcal{H}_{3,m_1+m_2}$ and $\widehat{p_1 p_2} = \hat{p_1} \hat{p_2}$. For even m, it is obvious that $q \in \mathrm{Pos}(\mathbb{R}^2)$ if and only if $q_h \in \mathrm{Pos}(\mathbb{R}^3)$.

   The zero set $\mathcal{Z}(q)$ in $\mathbb{R}^2$ is a subset of the zero set $\mathcal{Z}_{\mathbb{P}}(q_h)$ in $\mathbb{P}^2(\mathbb{R})$. In general, $\mathcal{Z}_{\mathbb{P}}(q_h)$ is larger than $\mathcal{Z}(q)$, because it contains possible zeros of q at "infinity".

   We will use the following version of Bezout's theorem for real curves in $\mathbb{R}^2$ which follows from Lemma 19.29: *Let $f, g \in \mathbb{R}_2[\underline{x}]$, $\deg(f) = m_1$, and $\deg(g) = m_2$. If f and g are relatively prime in $\mathbb{R}_2[\underline{x}]$, then $|\mathcal{Z}(f) \cap \mathcal{Z}(g)| \leq m_1 m_2$ and*

$$\sum_{t \in \mathcal{Z}(f) \cap \mathcal{Z}(g)} I_t(f, g) \leq m_1 m_2. \tag{19.49}$$

By a *quadratic* or a *cubic* we mean a real curve $\mathcal{Z}(p)$ in $\mathbb{R}^2$ for some polynomial $p \in \mathbb{R}_2[\underline{x}]$ of degree 2 or 3, respectively.

Proposition 19.35 *Let $n = 2$ and $k = 5$. Suppose that $t_1, \ldots, t_5$ are distinct points of $\mathbb{R}^2$, no three are collinear. There exists a unique, up to a constant multiple, polynomial $f \in \mathbb{R}_2[\underline{x}]$ of degree 2 such that $t_1, \ldots, t_5$ lie on the quadratic $\mathcal{Z}(f)$. Then*

$$\mathcal{N}_+(L) = \mathbb{R}_+ \cdot f^2 \quad and \quad \mathcal{V}_+(L) = \mathcal{Z}(f). \tag{19.50}$$

Proof The existence and uniqueness of $f \in \mathbb{R}_2[\underline{x}]_2$ satisfying $t_1, \ldots, t_5 \in \mathcal{Z}(f)$ is a classical result in curve theory, see e.g. [Bx, Theorem 5.10].

Let $g \in \mathcal{N}_+(L)$, $g \neq 0$. Then, by (19.47), $g \in \mathrm{Pos}(\mathbb{R}^2)$ and g vanishes at $t_1, \ldots, t_5$. If $\deg(g) = 2$, then g is a sum of squares of linear polynomials and all t_j are collinear, a contradiction. Hence $\deg(g) = 4$, so that $g_\mathrm{h} \in \mathcal{H}_{3,4}$. We have $g_\mathrm{h} \in \mathrm{Pos}(\mathbb{R}^3)$ and $|\mathcal{Z}_\mathbb{P}(g_\mathrm{h})| \geq |\mathcal{Z}(g)| \geq 5 > 4 = \alpha(2)$. Therefore, by Theorem 19.32, we can write $g_\mathrm{h} = pq^2$, where q is indefinite, $p \in \mathrm{Pos}(\mathbb{R}^3)$ and $\mathcal{Z}_\mathbb{P}(p)$ is finite. Then $g = \hat{p}\,\hat{q}^2$.

Since q is indefinite, q is not constant. Assume that $\deg(q) = 1$. Then $p \in \mathcal{H}_{3,2}$ and $\deg(\hat{q}) \leq 1$. Thus, $\hat{q} \neq 0$ (by $g \neq 0$) is either constant or $\hat{q} = 0$ is a line. Since $g(t_j) = \hat{p}(t_j)\hat{q}(t_j)^2 = 0$ for $j = 1, \ldots, 5$ and no three points of the t_j are on a line, $\hat{p}$ vanishes on at least three points, so that $|\mathcal{Z}_\mathbb{P}(\hat{p})| \geq 3 > \alpha(1) = 1$. From Theorem 19.32, applied to p, it follows that $\mathcal{Z}_\mathbb{P}(p)$ is infinite, which is a contradiction.

Thus, $q \in \mathcal{H}_{3,2}$ and p is a nonzero constant. Then g_h is a multiple of q^2, so g is a multiple of $\hat{q}^2$. Since $g \in \mathcal{N}_+(L)$ vanishes on $t_1, \ldots, t_5$, so does $\hat{q}^2$ and hence $\hat{q}$. The uniqueness assertion concerning f implies that $\hat{q} = cf$ for some $c \in \mathbb{R}$. Therefore, $g \in \mathbb{R}_+ \cdot f^2$. This proves that $\mathcal{N}_+(L) \subseteq \mathbb{R}_+ \cdot f^2$. The converse inclusion and the equality $\mathcal{V}_+(L) = \mathcal{Z}(f)$ are obvious.   □

The case of cubics is more subtle. In the following discussion we use some classical results on cubics (see e.g. [Bx]) and we state some facts without proof.

Case 1: $k = 8$.

Suppose $Z_8 := \{t_1, \ldots, t_8\}$ is set of pairwise distinct points of $\mathbb{R}^2$ such that no four of them are collinear and no seven lie on a quadratic. Then $\mathcal{N}_L$ is the vector space of polynomials $p \in \mathbb{R}_2[\underline{x}]_3$ of degree 3 vanishing on Z_8 and this space has dimension 2. There is a point t_9 in the projective zero set of $\mathcal{N}_L$ which has the following property: For each $t_{10} \in \mathbb{R}^2$, different from $t_1, \ldots, t_9$, there is a unique, up to a constant, cubic which contains the points $t_1, \ldots, t_8, t_{10}$. Note t_9 may or may not be in the set Z_8. (All preceding results follow from Theorems 13.4, 13.6, and 13.7 in [Bx].)

Assume that $t_9 \in \mathbb{R}^2$ and $t_9 \notin Z_8$. Let the cubics f, g form a basis of $\mathcal{N}_L$. Then

$$\mathcal{Z}(f) \cap \mathcal{Z}(g) = Z_9 := \{t_1, \ldots, t_9\}, \tag{19.51}$$

that is, both cubics intersect in exactly 9 points of $\mathbb{R}^2$, and $\mathcal{V}_L = \mathcal{Z}(f) \cap \mathcal{Z}(g) = Z_9$. It is not difficult to verify that (19.51) implies that f and g are relatively prime.

As proved by B. Reznick [Re4, Theorem 3.4 and 4.1], there exists a polynomial $p \in \operatorname{Pos}(\mathbb{R}^2)$ such that $p(t_j) = 0$ for $j = 1, \ldots, 8$ and $p(t_9) > 0$. Then $p \in \mathcal{N}_+(L)$ and $t_9 \notin \mathcal{V}_+(L)$. From $Z_8 = \operatorname{supp} \mu \subseteq \mathcal{V}_+(L) \subseteq \mathcal{V}_L = Z_9$ we get $\mathcal{V}_+(L) = Z_8$. Thus,

$$\mathcal{V}_+(L) = Z_8 \subsetneqq Z_9 = \mathcal{V}_L. \tag{19.52}$$

We illustrate this with an example.

Example 19.36 The cubics $f = x_1(1 - x_1^2)$, $g = x_2(1 - x_2^2)$ satisfy (19.51), where the points $t_1, \ldots, t_8$ are $(\pm 1, \pm 1), (\pm 1, 0), (0, \pm 1)$ and $t_9 = (0, 0)$. The dehomogenized Robinson polynomial $p := \hat{R}$ is in $\operatorname{Pos}(\mathbb{R}^2)$, vanishes at $t_1, \ldots, t_8$, and satisfies $p(t_9) > 0$. Hence (19.52) holds.

By Proposition 19.19(iii), Robinson's polynomial $R \in \mathcal{H}_{3,6}$ is not a sum of squares. The preceding yields a very short proof of this fact. Indeed, assume that $R = \sum_j f_j^2$ with $f_j \in \mathbb{R}_3[\underline{x}]$. Then $\hat{R} = \sum_j (\hat{f}_j)^2$ and $\hat{f}_j \in \mathbb{R}_2[\underline{x}]_3$. Clearly, each $\hat{f}_j$ vanish at all zeros of $\hat{R}$. Therefore, since $\hat{R}$ vanishes at Z_8, $\hat{f}_j \in \mathcal{N}_L$, so that $\hat{f}_j \in \operatorname{Lin}\{f, g\}$. Then, (19.51) implies that $\hat{f}_j(t_9) = 0$ for all j. Hence $\hat{R}(t_9) = 0$. But $\hat{R}(t_9) = R(1, 0, 0) = 1$, a contradiction. ○

Case 2: $k = 9$.

Retain the assumptions and the notation of Case 1. Since $\mathcal{N}_L$ is spanned by f, g, (19.51) yields $\mathcal{V}_L = Z_9$. From $Z_9 = \operatorname{supp} \mu \subseteq \mathcal{V}_+(L) \subseteq \mathcal{V}_L$ we get $Z_9 = \mathcal{V}_+(L) = \mathcal{V}_L$.

Case 3: $k \geq 10$.

There exists a truncated moment functional L such that $\mathcal{V}_+(L)$ consists of 10 points, see (19.41). (This example is on $\mathbb{P}^2(\mathbb{R})$, but by a linear transformation we obtain such a measure on $\mathbb{R}^2$, see Exercise 19.12.)

In the next proposition the real algebraic set $\mathcal{V}_+(L)$ is a cubic.

Proposition 19.37 *Suppose that $k \geq 10$. Let $t_j, j = 1, \ldots, k$, be pairwise distinct points in $\mathbb{R}^2$ of a cubic $\mathcal{Z}(f)$, $f \in \mathbb{R}_2[\underline{x}]_3$. Suppose that no four of the points $t_1, \ldots, t_8$ are collinear and no seven of them are on a quadratic. Let $\mu = \sum_{j=1}^{k} c_j \delta_{t_j}$ be a k-atomic measure and let L be its moment functional defined by (19.45). Then,*

$$\mathcal{N}_+(L) = \mathbb{R}_+ \cdot f^2 \quad and \quad \mathcal{V}_+(L) = \mathcal{Z}(f).$$

Proof As in the proof of Proposition 19.35 it suffices to show that $\mathcal{N}_+(L) \subseteq \mathbb{R}_+ \cdot f^2$.

Let $g \in \mathcal{N}_+(L), g \neq 0$. Then $g \in \operatorname{Pos}(\mathbb{R}^2)$, so that $g_h \in \operatorname{Pos}(\mathbb{R}^3)$, and $g(t_j) = 0$ for $j = 1, \ldots, k$. By Lemma 19.26, each t_j is a singular point of $\mathcal{Z}(g)$, so that $I_{t_j}(f, g) \geq 2$. Therefore, $\sum_j I_{t_j}(f, g) \geq k \cdot 2 \geq 20 > 18 \geq \deg(g) \deg(f)$. Hence Bezout's theorem (19.49) implies that f and g are not relatively prime.

Let r be a nonconstant common factor of f and g, say $f = ru$ and $g = rv$. We show that $\mathcal{Z}(g_h)$ is infinite. Since $g_h = r_h v_h$, by Corollary 19.33 it suffices to show that r_h is indefinite. This is obvious if $\deg(r)$ is odd. Let $\deg(r) = 2$. Then $\deg(u) = 1$. We have $f(t_j) = r(t_j)u(t_j) = 0$ for $j = 1, \ldots, 9$. Since no four of $t_1, \ldots, t_9$ lie on the line $u = 0$, r vanishes at 6 points t_i. The polynomial r is not in $\mathrm{Pos}(\mathbb{R}^2)$, since then $r_h \in \mathrm{Pos}(\mathbb{R}^3)$ and hence $6 \leq |\mathcal{Z}(r)| \leq |\mathcal{Z}_{\mathbb{P}}(r_h)| \leq \alpha(2) = 4$ by Theorem 19.32, a contradiction. Similarly, $-r$ is not in $\mathrm{Pos}(\mathbb{R}^2)$. Thus, r, and hence r_h, is indefinite.

Since $\mathcal{Z}_{\mathbb{P}}(g_h)$ is infinite, by Theorem 19.32 there is a factorization $g_h = q^2 p$, where q is indefinite, $p \in \mathrm{Pos}(\mathbb{R}^3)$ and $\mathcal{Z}_{\mathbb{P}}(p)$ is finite. Note that q is not constant, since $g \neq 0$ and hence $g_h \neq 0$. Further, $g = \hat{p}\hat{q}^2$. Thus, $\hat{q} \neq 0$, because $g \neq 0$.

Case I: $\deg(q) = 1$.

Then $\deg(p) \leq 4$, so that $p \in \mathcal{H}_{3,2k}$ for $k \in \{0, 1, 2\}$. Since $p \in \mathrm{Pos}(\mathbb{R}^3)$ and $\mathcal{Z}_{\mathbb{P}}(p)$ is finite, we conclude from Theorem 19.32 that $|\mathcal{Z}(\hat{p})| \leq |\mathcal{Z}_{\mathbb{P}}(p)| \leq \alpha(2) = 4$. We have $g(t_j) = \hat{p}(t_j)\hat{q}(t_j)^2 = 0$ for $j = 1, \ldots, 8$. Therefore, at least four of the points $t_1, \ldots, t_8$ lie on the curve $\hat{q} = 0$. Since $\deg(\hat{q}) \leq \deg(q) = 1$, $\hat{q}$ is either a nonzero constant or $\hat{q} = 0$ is a line and we obtain a contradiction to our assumption.

Case II: $\deg(q) = 2$.

Then $\deg(p) \leq 2$, so that $|\mathcal{Z}(\hat{p})| \leq \alpha(1) = 1$. Hence, since $g(t_j) = \hat{p}(t_j)\hat{q}(t_j)^2 = 0$ for $j = 1, \ldots, 8$, seven of the points $t_1, \ldots, t_8$ lie on the curve $\hat{q} = 0$. Since $\deg(\hat{q}) \leq 2$, $\hat{q}$ is a nonzero constant or $\hat{q} = 0$ is a line or a quadratic, again a contradiction.

From the preceding two cases it follows that $\deg(q) = 3$ and p is constant. Therefore $\deg(\hat{q}) = 3$. Since $g = \hat{p}\hat{q}^2$, $\hat{q}$ vanishes at all 10 points $t_1, \ldots, t_{10}$. These points determine the cubic uniquely [Bx, Theorem 13.7], so $\hat{q}$ is a constant multiple of f. Hence $g = \hat{p}\hat{q}^2$ is a nonnegative multiple of f^2, that is, $g \in \mathbb{R}_+ \cdot f^2$. $\square$

The next lemma is *Chasles' theorem*. In the literature this result or its generalization to curves of higher degrees is also called the *Cayley–Bacharach theorem*.

Lemma 19.38 *Let $t_1, \ldots, t_9$ be pairwise distinct points of $\mathbb{R}^2$ and suppose that $f, g \in \mathbb{R}_2[\underline{x}]_3$ define two cubics satisfying $\mathcal{Z}(f) \cap \mathcal{Z}(g) = \{t_1, \ldots, t_9\}$. Let $h \in \mathbb{R}_2[\underline{x}]_3$. If the points $t_1, \ldots, t_8$ lie on the cubic $\mathcal{Z}(h)$, then also $t_9 \in \mathcal{Z}(h)$.*

Proof [EGH, Theorem CB3], see also [SR, Chapter V, Section 1.1] and [RRS]. From the assumptions and Bezout's theorem one easily derives that no four of the points t_j are collinear and no seven are on a quadratic. Using this fact the assertion of Lemma 19.38 follows from [Bx, Theorem 13.7] as well. $\square$

Recall that a positive functional on $\mathbb{R}_2[\underline{x}]$ that is not a moment functional was given in Proposition 13.5. Now we use the preceding setup to construct another positive functional on $\mathbb{R}_2[\underline{x}]_6$ that is not a truncated moment functional. Let

$$t_1 = (1, 1), \ t_2 = (1, -1), \ t_3 = (-1, 1), \ t_4 = (-1, -1), \tag{19.53}$$

$$t_5 = (0, 1), \ t_6 = (0, -1), \ t_7 = (1, 0), \ t_8 = (-1, 0), \ t_9 = (0, 0). \tag{19.54}$$

The first 8 points $t_1, \ldots, t_8$ are zeros of the dehomogenized Robinson polynomial

$$\hat{R}(x_1, x_2) := R(1, x_1, x_2) = x_1^6 + x_2^6 + 1 - x_1^4(x_2^2 + 1) - x_2^4(x_1^2 + 1) - x_1^4 - x_2^4 + 3x_1^2 x_2^2$$

and $\hat{R}(t_9) = 1$. For $c \in \mathbb{R}$, we define $L_c = \sum_{j=1}^{8} l_{t_j} - c l_{t_9}$.

Proposition 19.39 *Let* $0 < c \leq \frac{4}{5}$. *Then* L_c *is a positive functional on* $\mathbb{R}_2[\underline{x}]_6$ *(that is,* $L_c(p^2) \geq 0$ *for* $p \in \mathbb{R}_2[\underline{x}]_3$*) which is not a truncated moment functional.*

Proof Recall that $t_1, \ldots, t_9$ are the points satisfying (19.36) for the two cubics $f = x_1(1 - x_1^2)$ and $g = x_2(1 - x_2^2)$ from Example 19.36. Hence Lemma 19.38 applies.

First we show that the point evaluations $l_{t_1}, \ldots, l_{t_9}$ are linearly dependent on $\mathbb{R}_2[\underline{x}]_3$. Assume the contrary and let $\mathcal{L}$ be the linear span of these functionals. Then there is a linear functional F on $\mathcal{L}$ such that $F(l_{t_j}) = 0, j = 1, \ldots, 8$, and $F(l_{t_9}) = 1$. Since $\mathcal{L} \subseteq (\mathbb{R}_2[\underline{x}]_3)^*$, the functional F is given by some polynomial $p \in \mathbb{R}_2[\underline{x}]_3$, that is, $F(l) = l(p)$ for $l \in \mathcal{L}$. Then $F(l_{t_j}) = p(t_j)$. Hence $t_1, \ldots, t_8 \in \mathcal{Z}(p)$. Since $t_1, \ldots, t_8$ do not lie on a quadratic, $\deg(p) = 3$ and $t_1, \ldots, t_8$ are on the cubic $\mathcal{Z}(p)$. Hence, by Chasles' theorem (Lemma 19.38), $p(t_9) = 0$ which contradicts $p(t_9) = F(l_{t_9}) = 1$.

Since the functionals $l_{t_1}, \ldots, l_{t_9}$ are linearly dependent, there are real numbers $\gamma_1, \ldots, \gamma_9$, not all zero, such that

$$\sum_{j=1}^{9} \gamma_j l_{t_j}(p) = \sum_{j=1}^{9} \gamma_j p(t_j) = 0 \quad \text{for} \quad p \in \mathbb{R}_2[\underline{x}]_3. \tag{19.55}$$

Setting $p = (1 - x_1^2)(1 \pm x_2)$ and $p = (1 - x_1)x_2(1 \pm x_2)$ in (19.55) yields $2\gamma_6 + \gamma_9 = 2\gamma_5 + \gamma_9 = 0$ and $2\gamma_3 + \gamma_5 = 2\gamma_4 + \gamma_6 = 0$. If we interchange the role of x_1 and x_2, we have $2\gamma_7 + \gamma_9 = 2\gamma_8 + \gamma_9 = 0$ and $2\gamma_2 + \gamma_7 = 0$. For $p = (1 + x_1)x_2(1 + x_2)$ we get $2\gamma_1 + \gamma_5 = 0$. Therefore, putting $\gamma_9 = 1$, we obtain on $\mathbb{R}_2[\underline{x}]_3$,

$$l_{t_9} + \frac{1}{4}(l_{t_1} + l_{t_2} + l_{t_3} + l_{t_4}) - \frac{1}{2}(l_{t_5} + l_{t_6} + l_{t_7} + l_{t_8}) = 0. \tag{19.56}$$

(We could have avoided Chasles' theorem by verifying this identity on $\mathbb{R}_2[\underline{x}]_3$ by a direct computation.) Let $p \in \mathbb{R}_2[\underline{x}]_3$. By (19.56),

$$p(t_9)^2 = \left(-\frac{1}{4} \sum_{j=1}^{4} p(t_j) + \frac{1}{2} \sum_{j=5}^{8} p(t_j) \right)^2 \leq \frac{5}{4} \sum_{j=1}^{8} p(t_j)^2. \tag{19.57}$$

Then L_c is a positive functional, since $0 < c \leq \frac{4}{5}$ and hence

$$L_c(p^2) = \sum_{j=1}^{8} p(t_j)^2 - c p(t_9)^2 \geq (1 - 5c/4) \sum_{j=1}^{8} p(t_j)^2 \geq 0.$$

Since $t_1, \cdots, t_8$ are zeros of $\hat{R}$ and $\hat{R}(t_9) = 1$, we have $L_c(\hat{R}) = -c < 0$. Thus, L_c is not a truncated moment functional, since $R \in \text{Pos}(\mathbb{R}^3)$ and hence $\hat{R} \in \text{Pos}(\mathbb{R}^2)$. $\qquad\qquad\qquad\qquad\qquad\qquad\qquad\qquad\qquad\qquad\qquad\qquad\qquad\qquad\square$

19.7 Exercises

1. (*Covariance of the apolar scalar product*)
   Let $p, q \in \mathcal{H}_{d,m}$ and $A \in M_d(\mathbb{R})$. Define $(p \circ A)(x) = p(Ax), x \in \mathbb{R}^d$, where $x \in \mathbb{R}^d$ is a column vector. Prove that $p \circ A \in \mathcal{H}_{d,m}$ and $[p \circ A, q] = [p, q \circ A^T]$.
   Hint: It suffices to verify the latter for $p = (a \cdot)^m, q = (b \cdot)^m$, where $a, b \in \mathbb{R}^d$.
2. Let $p \in \mathcal{H}_{d,2n}$. Show that $p \circ U = p$ for all orthogonal transformations U if and only if there is a number $c \in \mathbb{R}$ such that $p(x) = c\|x\|^{2n}$.
3. Define $D_a := a_1 \partial_1 + \cdots + a_d \partial_d$ for $a = (a_1, \ldots, a_d) \in \mathbb{R}^d$. Let $y_1, \ldots, y_m \in \mathbb{R}^d$. Show that $\frac{1}{m!} D_{y_1} \cdots D_{y_m} f = [f, (y_1 \cdot) \cdots (y_m \cdot)]$ for $f \in \mathcal{H}_{d,m}$.
4. Let $p \in \mathcal{H}_{d,m}, m \in \mathbb{N}$, and $x \in \mathbb{R}^d$. Suppose that $\frac{\partial p}{\partial x_j}(x) = 0$ for $j = 1, \ldots, d$. Use Euler's identity to show that $p(x) = 0$.
5. Show that the polynomial $f(x_1, x_2) = x_1^2 x_2^2 \in \mathcal{H}_{2,4}$ is not in $Q_{2,4}$.
6. Let $f_\varepsilon(x_1, x_2) = x_1^2 x_2^2 + \varepsilon(x_1^2 + x_2^2)^2 \in \mathcal{H}_{2,4}$ for $\varepsilon > 0$. Note that f_ε is positive on $\mathbb{R}^2 \setminus \{0\}$. Show that f_ε is not in $Q_{2,4}$ if $0 < \varepsilon < \frac{1}{4}$.
   Hint: Assume the contrary. Write $f_\varepsilon = \sum_{j=1}^5 (a_j x_1 + b_j x_2)^4$ by Carathéodory's theorem. Compare first the coefficients of x_1^4, x_2^4, and finally of $x_1^2 x_2^2$.
7. Let $d = 2$ or $n = 1$. Let p be a homogeneous polynomial of $\mathcal{H}_{d,2n}$ such that $p \geq 0$ on $\mathbb{R}^d$. Show that p is a sum of squares of elements of $\mathcal{H}_{d,n}$.
8. Find polynomials $p \in \mathcal{H}_{3,6}$ and $q \in \mathcal{H}_{4,4}$ which are not sums of squares of elements of $\mathcal{H}_{3,3}$ and $\mathcal{H}_{4,2}$, respectively.
   Hint: Use homogenized Motzkin and Choi–Lam polynomials, (13.6) and (13.41).
9. Suppose that $d \geq 3, n \geq 3$ or $d \geq 4, n \geq 2$. Show that $\sum \mathcal{H}_{d,n}^2 \neq \text{Pos}(\mathcal{H}_{d,2n}, \mathbb{R}^d)$.
10. Find a polynomial $p \in \mathcal{H}_{2,2}$ such that p^2 does not span an extreme ray of $\sum_{2,4}^2$.
11. Let $t_1, \ldots, t_8$ be the points of $\mathbb{R}^2$ given by (19.53) and (19.54) and let $L = \sum_{j=1}^8 m_j l_{t_j}$, where $m_j \geq 0$ for $j = 1, \ldots, 8$.

    a. Show the truncated moment functional L on $\mathbb{R}_2[\underline{x}]_6$ is determinate.
    b. Let $t_9 \in \mathbb{R}^2$ be such that $t_9 \neq t_j$ for $j = 1, \ldots, 8$. Show that $\mu = \delta_{t_9} + \sum_{j=1}^8 m_j \delta_{t_j}$ is an ordered maximal mass measure.

12. Let R be the Robinson polynomial. Find a linear transformation A of $\mathbb{R}^3$ such that $p \in \mathcal{H}_{3,6}$ defined by $p(x) := R(Ax)$ has exactly 10 zeros in $\mathbb{R}^2$.

13. (*Hessians at zeros for sums of squares*)
Let $f_1, \ldots, f_k \in \mathcal{H}_{2,n}$ and $f = f_1^2 + \cdots + f_k^2$. Suppose that $t \in \mathcal{Z}(f)$. Show that

$$\sum_{i,j=1}^d \zeta_i \zeta_j \frac{\partial^2 f}{\partial x_i \partial x_j}(t) = 2 \sum_{l=1}^k \left(\sum_i^d \zeta_i \frac{\partial f_l}{\partial x_i}(t) \right)^2 \equiv 2 \sum_{l=1}^k (\zeta \cdot \nabla f_l(t))^2$$

for $\zeta = (\zeta_1, \ldots, \zeta_d)^T \in \mathbb{R}^d$, where $\nabla = (\frac{\partial}{\partial x_1}, \ldots, \frac{\partial}{\partial x_d})^T$.
Hint: First verify that $\frac{\partial^2 f_l^2}{\partial x_i \partial x_j}(t) = 2 \frac{\partial f_l}{\partial x_i}(t) \frac{\partial f_l}{\partial x_j}(t)$.

14. Let $t_1, \ldots, t_9$ be pairwise distinct points of $\mathbb{R}^2$ and let $f, g \in \mathbb{R}_2[\underline{x}]_3$ be two cubics satisfying $\mathcal{Z}(f) \cap \mathcal{Z}(g) = \{t_1, \ldots, t_9\}$. Show that no four of the points $t_1, \ldots, t_9$ are collinear and no seven are on a quadratic.

15. ([Re1]) Suppose that $a > 0, a \neq \sqrt{2}$. Let L be the moment functional of the measure $\mu = \sum_{j=1}^8 \delta_{t_j}$ on $\mathbb{R}_2[\underline{x}]_6$, where the eight points t_j are $(\pm 1, \pm 1), (\pm a, 0), (0, \pm a)$.

   a. Show that $f = x_1(x_1^2 - a^2 + (a^2 - 1)x_2^2)$, $g = x_2(x_2^2 - a^2 + (a^2 - 1)x_1^2)$ form a basis of the vector space $\mathcal{N}_L$.
   b. What happens to $\mathcal{N}_L$ in the case $a = \sqrt{2}$?
   c. Show that $\mathcal{Z}(f) \cap \mathcal{Z}(g) = \{t_1, \ldots, t_8, t_9\}$, where $t_9 := (0, 0)$.
   d. Let $L' := L + l_{t_9}$ and $L'' = L' + \delta_{t_{10}}$, where $t_{10} = (2, 1 - 3(1 - a^2)^{-1})$ if $a \neq 1$ or $t_{10} = (3, 3)$. Determine $\mathcal{N}_{L'}, \mathcal{N}_{L''}, \mathcal{N}_+(L'), \mathcal{N}_+(L''), \mathcal{V}_+(L')$, and $\mathcal{V}_+(L'')$.

19.8 Notes

The apolar scalar product goes back to the 19th century, see [Re1] for some historical discussion and [Veg], [Re3]. Representing homogeneous polynomials of degree m as finite sums of m-th powers of linear forms is called Waring's problem, see [El1, El2]. Hilbert's Theorem 19.15 was proved in [H2]. It is developed in [Nat] where explicit formulas for $h_{4,2n}, n = 2, 3, 4, 5$, are given.

Robinson's polynomial was discovered by R.M. Robinson in [Rb]. Its extremality (Corollary 19.22) was shown in [CL]. Proposition 19.25 is taken from [Sm10].

Theorem 19.32 is due to M.D. Choi, T.Y. Lam and B.Reznick [CLR]. Recent results on Carathéodory numbers can be found in [RiS] and [DSm2].

The construction method of a positive functional that is not a moment functional in Proposition 19.39 (with a more complicated polynomial) was invented by the author [Sm2]. It was elaborated in [Bl1]. Formulas (19.56) and (19.57) appeared in [BW, Exercise 1.6.28].

Appendix

A.1 Measure Theory

For all notions and results on measure theory we refer to [Ba, BCR1], and [Ru2].

Throughout this appendix, $\mathcal{X}$ is a **locally compact Hausdorff space**.

We denote by $C_c(\mathcal{X})$ the continuous functions on $\mathcal{X}$ with compact support, by $C_c(\mathcal{X}; \mathbb{R})$ the real-valued functions in $C_c(\mathcal{X})$, by $C_c(\mathcal{X})_+$ the nonnegative functions in $C_c(\mathcal{X})$, and by $C_0(\mathcal{X})$ the functions $f \in C(\mathcal{X})$ which vanish at infinity, or equivalently, for which the set $\{x \in \mathcal{X} : |f(x)| \geq \varepsilon\}$ is compact for all $\varepsilon > 0$.

The *Borel algebra* $\mathfrak{B}(\mathcal{X})$ is the σ-algebra generated by the open subsets of $\mathcal{X}$. A $\mathfrak{B}(\mathcal{X})$-measurable function on $\mathcal{X}$ is called a *Borel function*.

Definition A.1 A *Radon measure* on $\mathcal{X}$ is a measure $\mu : \mathfrak{B}(\mathcal{X}) \to [0, +\infty]$ such that $\mu(K) < \infty$ for each compact subset K of $\mathcal{X}$ and

$$\mu(M) = \sup \{\mu(K) : K \subseteq M, \ K \text{ compact}\} \quad \text{for all } M \in \mathfrak{B}(\mathcal{X}). \tag{A.1}$$

Thus, in our terminology Radon measures are always nonnegative!

Condition (A.1) is the *regularity* of μ. If the locally compact space $\mathcal{X}$ is σ-*compact* (that is, $\mathcal{X}$ is a countable union of compact sets), then each Radon measure is also *outer regular* (see e.g. [Ba, Corollary 29.7])

$$\mu(M) = \inf \{\mu(U) : M \subseteq U, \ U \text{ open}\} \quad \text{for } M \in \mathfrak{B}(\mathcal{X}).$$

Closed subsets of $\mathbb{R}^d$ are obviously σ-compact. All measures for moment problems occuring in this book are Radon measures on σ-compact locally compact Hausdorff spaces; in most cases they are supported on closed subsets of $\mathbb{R}^d$.

Let $M_+(\mathcal{X})$ denote the set of Radon measures on $\mathcal{X}$, $M_+^1(\mathcal{X})$ the subset of probability measures in $M_+(\mathcal{X})$, and $M_+^b(\mathcal{X})$ the subset of finite measures in $M_+(\mathcal{X})$.

© Springer International Publishing AG 2017
K. Schmüdgen, *The Moment Problem*, Graduate Texts in Mathematics 277,
DOI 10.1007/978-3-319-64546-9

A *complex Radon measure* on $\mathcal{X}$ is a map $\mu : \mathfrak{B}(\mathcal{X}) \to \mathbb{C}$ which is of the form $\mu = \mu_1 - \mu_2 + i(\mu_3 - \mu_4)$, where $\mu_1, \mu_2, \mu_3, \mu_4 \in M_+^b(\mathcal{X})$. The set of complex Radon measures on $\mathcal{X}$ is denoted by $M(\mathcal{X})$.

For $\mu \in M_+(\mathcal{X})$ we denote the μ-integrable Borel functions on $\mathcal{X}$ by $\mathcal{L}^1(\mathcal{X}, \mu)$, that is, a Borel function f on $\mathcal{X}$ is in $\mathcal{L}^1(\mathcal{X}, \mu)$ if and only if $\int |f(x)| \, d\mu < \infty$.

We say that a measure $\mu \in M_+(\mathcal{X})$ is *supported* on a set $M \in \mathfrak{B}(\mathcal{X})$ if $\mu(\mathcal{X} \backslash M) = 0$. The *support* of $\mu \in M_+(\mathcal{X})$, denoted by supp μ, is the smallest closed subset M of $\mathcal{X}$ such that $\mathcal{X}$ is supported on M. By this definition, $x_0 \in \mathcal{X}$ is in supp μ if and only if $\mu(U) > 0$ for each open subset U of $\mathcal{X}$ containing x_0.

Theorem A.2 (Lebesgue's Dominated Convergence Theorem) *Suppose that $\mu \in M_+(\mathcal{X})$. Let $f_n, n \in \mathbb{N}$, and f be complex Borel functions on $\mathcal{X}$ and let $g : \mathcal{X} \to [0, +\infty]$ be a μ-integrable function. Suppose that*

$$\lim_{n \to \infty} f_n(x) = f(x) \ \ and \ \ |f_n(x)| \le g(x), \ n \in \mathbb{N}, \ \mu\text{-}a.e. \ on \ \mathcal{X}. \tag{A.2}$$

Then the functions f and f_n are μ-integrable,

$$\lim_{n \to \infty} \int_\Omega |f_n - f| \, d\mu = 0, \ \ and \ \ \lim_{n \to \infty} \int_\Omega f_n \, d\mu = \int_\Omega f \, d\mu.$$

Proof [Ru2, Theorem 1.34]. $\qquad\qquad\qquad\qquad\qquad\qquad\qquad\qquad\qquad\qquad\qquad\qquad\qquad$ □

Theorem A.3 (Radon–Nikodym Theorem) *Let $\mu, \nu \in M_+(\mathcal{X})$. Suppose that μ is σ-finite (that is, $\mathcal{X}$ is a countable union of sets which have finite μ-measure). If ν is absolutely continuous with respect to μ (that is, each μ-null set is also a ν-null set), then there exists a nonnegative function $h \in \mathcal{L}^1(\mathcal{X}, \mu)$ such that $d\nu = h(x)d\mu$.*

Proof [Ba, Theorem 17.10] or [Ru2, Theorem 6.9]. $\qquad\qquad\qquad\qquad\qquad\qquad\qquad$ □

Theorem A.4 (Riesz' Representation Theorem) *For each linear functional L on $C_c(\mathcal{X}; \mathbb{R})$ such that $L(f) \ge 0$ for all $f \in C_c(\mathcal{X})_+$ there exists a unique (positive) Radon measure μ on $\mathcal{X}$ such that*

$$F(f) = \int_\mathcal{X} f \, d\mu, \ f \in C_c(\mathcal{X}; \mathbb{R}).$$

Proof [Ru2, Theorem 2.14], [Ba, Theorems 29.1 and 29.3], or [BCRl, Chapter 2, Corollary 2.3]. $\qquad\qquad\qquad\qquad\qquad\qquad\qquad\qquad\qquad\qquad\qquad\qquad\qquad\qquad\qquad\qquad\qquad$ □

Now we develop some basics on the vague convergence of measures, see e.g. [Ba, § 30 and 31] or [BCRl, Chapter 2, § 3 and 4] for more details.

The *vague topology* on $M_+(\mathcal{X})$ is the coarsest topology for which all mappings $\mu \mapsto \int f d\mu$, where $f \in C_c(\mathcal{X})$, are continuous. A net $(\mu_i)_{i \in I}$ from $M_+(\mathcal{X})$ *converges vaguely* to $\mu \in M_+(\mathcal{X})$ if and only if for all $f \in C_c(\mathcal{X}; \mathbb{R})$ we have

$$\lim_i \int f d\mu_i = \int f d\mu. \tag{A.3}$$

Proposition A.5 *A net* $(\mu_i)_{i \in I}$ *of measures* $\mu_i \in M_+(\mathcal{X})$ *converges vaguely to* $\mu \in M_+(\mathcal{X})$ *if and only if*

$$\limsup_i \mu_i(K) \le \mu(K) \quad \text{and} \quad \liminf_i \mu_i(A) \ge \mu(A)$$

for every compact subset K and every relatively compact open subset A of $\mathcal{X}$.

Proof [Ba, Theorem 30.2], see also [BCR1, Chapter 2, Exercise 4.11].   □

Theorem A.6 *A subset $\mathcal{M}$ of $M_+(\mathcal{X})$ is relatively compact in the vague topology (that is, its closure is vaguely compact) if and only if it is vaguely bounded, that is,*

$$\sup_{\mu \in \mathcal{M}} \left| \int f \, d\mu \right| < \infty \quad \text{for all} \quad f \in C_c(\mathcal{X}; \mathbb{R}).$$

For any $a > 0$ the set $\{\mu \in M_+(\mathcal{X}) : \mu(\mathcal{X}) \le a\}$ is vaguely compact.

Proof [Ba, Theorem 31.2 and Corollary 31.3] or [BCR1, Chapter 2, Theorem 4.5 and Proposition 4.6].   □

Proposition A.7 *If $C_c(\mathcal{X}; \mathbb{R})$ is separable with respect to the supremum norm, then the vague topology on $M_+(\mathcal{X})$ is metrizable.*

Proof [Ba, Lemma 31.4 and Theorem 31.5].   □

A function $h : \mathcal{X} \to [-\infty, +\infty]$ is called *upper semicontinuous* (resp. *lower semicontinuous*) if for each number $a \in \mathbb{R}$ the set $\{x \in \mathcal{X} : f(x) < a\}$ (resp. $\{x \in \mathcal{X} : f(x) > a\}$) is open in $\mathcal{X}$, or equivalently, if for each convergent net $(x_i)_{i \in I}$ in $\mathcal{X}$ we have

$$\limsup_i f(x_i) \le f(\lim_i x_i) \quad (\text{resp.} \quad f(\lim_i x_i) \le \liminf_i f(x_i)).$$

The following result is called the *portmanteau theorem*.

Proposition A.8 *Let $(\mu_i)_{i \in I}$ be a net of measures $\mu_i \in M_+^b(\mathcal{X})$ and $\mu \in M_+^b(\mathcal{X})$. Then the following five statements are equivalent:*

(i) $\limsup_i \mu_i = \mu$ *in the vague convergence and* $\lim_i \mu_i(\mathcal{X}) = \mu(\mathcal{X})$.
(ii) $\limsup_i \mu_i(A) \le \mu(A)$ *for all closed subsets A of $\mathcal{X}$ and* $\lim_i \mu_i(\mathcal{X}) = \mu(\mathcal{X})$.
(iii) $\limsup_i \int h \, d\mu_i \le \int h \, d\mu$ *for all upper semicontinuous bounded real functions h on $\mathcal{X}$.*
(iv) $\liminf_i \mu_i(A) \ge \mu(A)$ *for all open subsets A of $\mathcal{X}$ and* $\lim_i \mu_i(\mathcal{X}) = \mu(\mathcal{X})$.
(v) $\liminf_i \int h \, d\mu_i \ge \int h \, d\mu$ *for all lower semicontinuous bounded real functions h on $\mathcal{X}$.*

Proof By [BCR1, Chapter 2, Proposition 4.2] or [Ba, Theorem 30.8], (i) holds if and only if the net (μ_i) converges weakly to μ. Using this result the assertion follows from [BCR1, Chapter 2, Theorem 3.1]; for (i) and (iii), see [Ba, Theorem 30.10].   □

Proposition A.9 *Let $(\mu_i)_{i \in I}$ be a net from $M_+^b(\mathcal{X})$ satisfying $\sup_{i \in I} \mu_i(\mathcal{X}) < \infty$. If this net converges vaguely to $\mu \in M_+^b(\mathcal{X})$, then (A.3) holds for all $f \in C_0(\mathcal{X})$.*

Proof [Ba, Theorem 30.6] or [BCR1, Chapter 2, Proposition 4.4].   □

A.2   Pick Functions and Stieltjes Transforms

Let $\mathbb{C}_+ = \{z \in \mathbb{C} : \operatorname{Im} z > 0\}$ denote the upper half-plane.

Definition A.10 A holomorphic function $f : \mathbb{C}_+ \to \mathbb{C}$ is a *Pick function* if

$$\operatorname{Im} f(z) \geq 0 \quad \text{for all} \quad z \in \mathbb{C}_+.$$

We denote the set of Pick functions by $\mathfrak{P}$. In the literature Pick functions are also called *Nevanlinna functions*.

Examples of Pick functions are $\tan z$ or the principal logarithm $\operatorname{Log} z$.

If $f \in \mathfrak{P}$ and $f(z_0)$ is real for some point $z_0 \in \mathbb{C}_+$, then f is not open and hence a constant. That is, all nonconstant Pick functions map $\mathbb{C}_+$ into $\mathbb{C}_+$.

Each Pick function f can be extended to a holomorphic function on $\mathbb{C}\backslash\mathbb{R}$ by setting $f(\bar{z}) := \overline{f(z)}$ for $z \in \mathbb{C}_+$.

Theorem A.11 (Canonical Integral Representation of Pick Functions) *For each Pick function f there exist numbers $a, b \in \mathbb{R}$, $b \geq 0$, and a finite Radon measure ν on the real line such that*

$$f(z) = a + bz + \int_{\mathbb{R}} \frac{1 + zt}{t-z} \, d\nu(t), \quad z \in \mathbb{C}/\mathbb{R}, \tag{A.4}$$

where the numbers a, b and the measure ν are uniquely determined by f. Conversely, any function f of this form is a Pick function.

Proof [AG, Nr. 69, Theorem 2] or [Dn, p. 20, Theorem 1].   □

Often the canonical representation (A.4) is written in the form

$$f(z) = a + bz + \int_{\mathbb{R}} \left(\frac{1}{t-z} - \frac{t}{1+t^2} \right) d\tau(t), \quad z \in \mathbb{C}/\mathbb{R}, \tag{A.5}$$

where τ is a Radon measure on $\mathbb{R}$ such that $\int (1 + t^2)^{-1} d\tau(t) < \infty$. The two canonical representations (A.4) and (A.5) are related by the formula

$$d\nu(t) = (1 + t^2)^{-1} d\tau(t).$$

Moreover, $a = \operatorname{Re}(f(i))$ and $b = \lim_{y \to \infty} \frac{f(iy)}{iy}$. It should be emphasized that the function $t(1 + t^2)^{-1}$ in (A.5) is not necessarily τ-integrable.

Definition A.12 The *Stieltjes transform* of a complex Radon measure $\mu \in M(\mathbb{R})$ is

$$I_\mu(z) = \int_\mathbb{R} \frac{1}{t-z}\, d\mu(t), \ \ z \in \mathbb{C}\backslash\mathbb{R}. \tag{A.6}$$

Stieltjes transforms are also called *Cauchy transforms* or *Borel transforms*.

Theorem A.13 (Stieltjes–Perron Inversion Formula) *Each complex Radon measure $\mu \in M(\mathbb{R})$ is uniquely determined by the values of its Stieltjes transform $I_\mu(z)$ on $\mathbb{C}\backslash\mathbb{R}$. In fact, for $a, b \in \mathbb{R}, a < b$, we have*

$$\mu((a,b)) + \tfrac{1}{2}\mu(\{a\}) + \tfrac{1}{2}\mu(\{b\}) = \lim_{\varepsilon\to+0} \tfrac{1}{2\pi i} \int_a^b [I_\mu(t+i\varepsilon) - I_\mu(t-i\varepsilon)]\, dt,$$
$$\tag{A.7}$$

$$\mu(\{a\}) = \lim_{\varepsilon\to+0} -i\varepsilon\, I_\mu(a+i\varepsilon). \tag{A.8}$$

In particular, if $I_\mu(z) = 0$ for all $z \in \mathbb{C}\backslash\mathbb{R}$, then $\mu = 0$.

Proof By definition $\mu \in M(\mathbb{R})$ is a linear combination of measures from $M_+^b(\mathbb{R})$. Hence one can assume without loss of generality that $\mu \in M_+^b(\mathbb{R})$. In this case proofs are given (for instance) in [AG, Nr. 69] and [Wei, Appendix B]. □

Note that there exist complex Radon measures $\mu \neq 0$ for which $I_\mu = 0$ on $\mathbb{C}_+$.

Now suppose that $\mu \in M_+(\mathbb{R})$. Then $\overline{I_\mu(z)} = I_\mu(\bar z)$ for all $z \in \mathbb{C} \backslash \mathbb{R}$. Hence the measure μ is uniquely determined by the values of its Stieltjes transform $I_\mu(z)$ on $\mathbb{C}_+$ and the formulas (A.7) and (A.8) can be written as

$$\mu((a,b)) + \tfrac{1}{2}\mu(\{a\}) + \tfrac{1}{2}\mu(\{b\}) = \lim_{\varepsilon\to+0} \tfrac{1}{\pi} \int_a^b \operatorname{Im} I_\mu(t+i\varepsilon)\, dt, \tag{A.9}$$

$$\mu(\{a\}) = \lim_{\varepsilon\to+0} \varepsilon \operatorname{Im} I_\mu(a+i\varepsilon). \tag{A.10}$$

The Stieltjes transform I_μ of $\mu \in M_+(\mathbb{R})$ is a Pick function, since

$$\operatorname{Im} I_\mu(z) = \operatorname{Im} z \int_\mathbb{R} \frac{1}{|t-z|^2}\, d\mu(t), \ \ z \in \mathbb{C}\backslash\mathbb{R}.$$

The next result characterizes these Stieltjes transforms among Pick functions.

Theorem A.14 *A Pick function f is the Stieltjes transform I_μ of a measure $\mu \in M_+(\mathbb{R})$ if and only if*

$$\sup \{ |yf(iy)| : y \in \mathbb{R}, y \geq 1 \} < \infty. \tag{A.11}$$

Proof [AG, Nr. 69, Theorem 3]. □

Proposition A.15 *Let K be a closed subset of $\mathbb{R}$ and $\mu \in M_+(\mathbb{R})$. The Stieltjes transform $I_\mu(z)$ has a holomorphic extension to $\mathbb{C}\backslash K$ if and only if supp $\mu \subseteq K$.*

Proof [Dn, Lemma 2, p. 26].   □

Let us sketch the proof of Proposition A.15:

If supp $\mu \subseteq K$, then (A.6) with $z \in \mathbb{C}\backslash K$ defines a holomorphic extension of I_μ to $\mathbb{C}\backslash K$. Conversely, suppose that I_μ has a holomorphic extension, say f, to $\mathbb{C}\backslash K$. Then $\lim\limits_{\varepsilon \to +0} I_\mu(t \pm i\varepsilon) = f(t)$ and hence $\lim\limits_{\varepsilon \to +0} \operatorname{Im} I_\mu(t+i\varepsilon) = 0$ for $t \in \mathbb{R}\backslash K$. Therefore, by (A.9) and (A.10), we have supp $\mu \subseteq K$.

A.3 Positive Semidefinite and Positive Definite Matrices

Let $\mathbb{K}$ be either the real field $\mathbb{R}$ or the complex field $\mathbb{C}$. If not stated otherwise, $\langle \cdot, \cdot \rangle$ and $\| \cdot \|$ denote the standard scalar product and the Euclidean norm of $\mathbb{K}^n$.

Let $M_{n,m}(\mathbb{K})$ be the (n, m)-matrices over $\mathbb{K}$ and $M_n(\mathbb{K}) := M_{n,n}(\mathbb{K})$. For a matrix $A = (a_{jk}) \in M_{n,m}(\mathbb{K})$ we denote by $\overline{A}$ the matrix $\overline{A} = (\overline{a}_{jk})$ and by A^T the transposed matrix. A matrix $A = (a_{jk})_{j,k=1}^n$ is called *Hermitian* if $\overline{a}_{jk} = a_{kj}$ for $j, k = 1, \dots, n$.

Proofs of the following Propositions A.16 and A.18–A.20 and the corresponding notions can be found in standard texts on matrices such as [Gn, Zh].

Proposition A.16 (Spectral theorem for Hermitian matrices) *Let $A \in M_n(\mathbb{K})$ be Hermitian. There exist an orthonormal basis $u_1, \dots, u_n$ of $\mathbb{K}^n$ and reals $\lambda_1, \dots, \lambda_n$ such that $Au_j = \lambda_j u_j$, $j = 1, \dots, n$. If D denotes the diagonal matrix with diagonal entries $\lambda_1, \dots, \lambda_n$ and U the matrix with columns $u_1, \dots, u_n$, then*

$$A = \sum_{j=1}^n \lambda_j u_j (\overline{u}_j)^T = UDU^{-1}. \tag{A.12}$$

Given a Hermitian matrix $A = (a_{jk})_{j,k=1}^n$, the expression

$$Q_A(\xi) = \langle A\xi, \xi \rangle = \sum_{j,k=1}^n a_{jk} \xi_j \overline{\xi}_k \quad \text{for} \quad \xi = (\xi_1, \dots, \xi_n)^T \in \mathbb{K}^n$$

is called the *Hermitian form associated with A* or briefly a *Hermitian form*.

Definition A.17 A Hermitian matrix A and the Hermitian form Q_A are called

- *positive semidefinite* if $Q_A(\xi) \geq 0$ for all $\xi \in \mathbb{K}^n$,
- *positive definite* if $Q_A(\xi) > 0$ for all $\xi \in \mathbb{K}^n$, $\xi \neq 0$.

We write $A \succeq 0$ if A is positive semidefinite and $A \succ 0$ if A is positive definite.

Given integers i_j, $j = 1, \dots, k \leq n$, such that $1 \leq i_1 < i_2 < \dots < i_k \leq n$ let $D(i_1, \dots, i_k)$ denote the determinant of the $k \times k$-submatrix of A formed by the

i_j-th rows and columns of A. Then $D(i_1, \ldots, i_k)$ is called a *principal minor* of A and $D(1, \ldots, k)$ is a *main principal minor* of A.

Proposition A.18 *For a Hermitian matrix $A \in M_n(\mathbb{K})$ the following are equivalent:*

(i) *A is positive semidefinite.*
(ii) *All principal minors $D(i_1, \ldots, i_k)$ are nonnegative.*
(iii) *All eigenvalues of A are nonnegative.*
(iv) *There exist vectors $v_1, \ldots, v_n \in \mathbb{K}^n$ such that $A = \sum_{j=1}^n v_j(\bar{v}_j)^T$.*

Proposition A.19 *For a Hermitian matrix $A \in M_n(\mathbb{K})$ the following are equivalent:*

(i) *A is positive definite.*
(ii) *All main principal minors $D(1, \ldots, k)$, $k = 1, \ldots, n$, are positive.*
(iii) *All eigenvalues of A are positive.*
(iv) *There are linearly independent vectors $v_1, \ldots, v_n \in \mathbb{K}^n$ such that $A = \sum_{j=1}^k v_j(\bar{v}_j)^T$.*
(v) *A is positive semidefinite and $\det A \neq 0$.*
(vi) *A is positive semidefinite and $\operatorname{rank} A = n$.*

For each positive semidefinite matrix A there is a unique positive semidefinite matrix B, denoted by $A^{1/2}$, such that $B^2 = A$. If A is of the form (A.12) and $D^{1/2}$ denotes the diagonal matrix with entries $\lambda_1^{1/2}, \ldots, \lambda_n^{1/2}$, then $A^{1/2} = UD^{1/2}U^{-1}$.

The *trace* $\operatorname{Tr} A$ of a matrix $A \in M_n(\mathbb{K})$ is the sum of all diagonal entries of A.

Proposition A.20 *For $A, B \in M_n(\mathbb{K})$ we have:*

(i) $\operatorname{Tr} AB = \operatorname{Tr} BA$.
(ii) $\operatorname{Tr} A \geq 0$ *if $A \succeq 0$. In particular, $\operatorname{Tr} \bar{A}^T A \geq 0$.*
(iii) *$A \succeq 0$ and $\operatorname{Tr} A = 0$ imply that $A = 0$. In particular, $\operatorname{Tr} \bar{A}^T A = 0$ implies $A = 0$.*

Let Sym_n denote the real vector space of symmetric matrices in $M_n(\mathbb{R})$. There is a scalar product $\langle \cdot, \cdot \rangle$ on Sym_n defined by

$$\langle A, B \rangle := \operatorname{Tr} AB = \sum_{i,j=1}^n a_{ij}b_{ij} \quad \text{for} \quad A = (a_{ij}), \ B = (b_{ij}) \in \operatorname{Sym}_n.$$

For matrices from Sym_n we have the following basic properties.

Proposition A.21

(i) *If $A \succeq 0$ and $B \succeq 0$, then $\langle A, B \rangle = \operatorname{Tr} AB \geq 0$.*
(ii) *If $A \succeq 0$, $B \succeq 0$, and $\langle A, B \rangle = 0$, then $AB = 0$.*
(iii) *If $\langle A, B \rangle \geq 0$ for all $B \succeq 0$, then $A \succeq 0$.*

Proof

(i) As noted above, each positive semidefinite matrix $C \in \mathrm{Sym}_n$ has a unique positive semidefinite square root $C^{1/2}$. Then we derive

$$\mathrm{Tr}\, AB = \mathrm{Tr}\, A^{1/2}A^{1/2}B^{1/2}B^{1/2} = \mathrm{Tr}\, B^{1/2}A^{1/2}A^{1/2}B^{1/2} = \mathrm{Tr}\, (A^{1/2}B^{1/2})^T(A^{1/2}B^{1/2}).$$

Since $(A^{1/2}B^{1/2})^T(A^{1/2}B^{1/2}) \succeq 0$, this yields $\mathrm{Tr}\, AB \geq 0$.

(ii) By the preceding equality, we have $0 = \langle A, B \rangle = \mathrm{Tr}\, (A^{1/2}B^{1/2})^T(A^{1/2}B^{1/2})$. Therefore, $(A^{1/2}B^{1/2})^T(A^{1/2}B^{1/2}) = 0$ and hence $A^{1/2}B^{1/2} = 0$. Thus we obtain $AB = A^{1/2}(A^{1/2}B^{1/2})B^{1/2} = 0$.

(iii) Assume to the contrary that A is not positive semidefinite. Then, by Proposition A.18, A admits an eigenvalue $\lambda < 0$, so $Au = \lambda u$ with $u \in \mathbb{R}^n$, $u \neq 0$. Then $B := uu^T \succeq 0$ and $\mathrm{Tr}\, AB = \mathrm{Tr}\, (\lambda uu^T) = \lambda \mathrm{Tr}\, uu^T$. Since $uu^T \succeq 0$ and $uu^T \neq 0$, $\mathrm{Tr}\, uu^T > 0$ and hence $\mathrm{Tr}\, AB < 0$, which contradicts the assumption. $\qquad\square$

A.4 Positive Semidefinite Block Matrices and Flat Extensions

Suppose that $A = A^T \in M_n(\mathbb{R})$, $B \in M_{n,m}(\mathbb{R})$, $C = C^T \in M_m(\mathbb{R})$, where $n, m \in \mathbb{N}$. We consider the symmetric block matrix $X \in M_{n+m}(\mathbb{R})$ defined by

$$X = \begin{pmatrix} A & B \\ B^T & C \end{pmatrix}. \tag{A.13}$$

Definition A.22 The block matrix X is called a *flat extension* of A if

$$\mathrm{rank}\, X = \mathrm{rank}\, A.$$

Proposition A.23 *The block matrix X is a flat extension of A if and only if there exists a matrix $U \in M_{n,m}(\mathbb{R})$ such that $B = AU$ and $C = U^T AU$.*

Proof The proof is based essentially on the well-known fact that the rank of a matrix is the maximal number of linearly independent columns. First suppose that $\mathrm{rank}\, X = \mathrm{rank}\, A$. Then each column of $\left(\begin{smallmatrix} B \\ C \end{smallmatrix}\right)$ is in the span of columns of $\left(\begin{smallmatrix} A \\ B^T \end{smallmatrix}\right)$. Hence $\left(\begin{smallmatrix} B \\ C \end{smallmatrix}\right)$ is of the form $\left(\begin{smallmatrix} AU \\ B^T U \end{smallmatrix}\right)$, so that $B = AU$ and $C = B^T U = U^T AU$.

Conversely, if X is of the prescribed form, one easily verifies that the dimensions of the ranges of X and A are equal; hence $\mathrm{rank}\, X = \mathrm{rank}\, A$. $\qquad\square$

The following description of positive semidefinite block matrices is due to J.L. Shmuljan [Sh].

Theorem A.24 *Suppose that $A \succeq 0$.*

(i) $X \succeq 0$ *if and only if $B = AU$ for some matrix $U \in M_{n,m}(\mathbb{R})$ and $C \succeq U^T AU$.*

(ii) *Suppose that $B = AU$ with $U \in M_{n,m}(\mathbb{R})$. Then X is a flat extension of A if and only if $C = U^T AU$.*

(iii) *If X is an arbitrary flat extension of A, then $X \succeq 0$.*

Proof

(i) First suppose that $B = AU$ with $U \in M_{n,m}(\mathbb{R})$. Then we have the identity

$$X = \begin{pmatrix} A & AU \\ U^TA & C \end{pmatrix} = \tag{A.14}$$

$$\begin{pmatrix} A^{1/2} & A^{1/2}U \\ 0 & 0 \end{pmatrix}^T \begin{pmatrix} A^{1/2} & A^{1/2}U \\ 0 & 0 \end{pmatrix} + \begin{pmatrix} 0 & 0 \\ 0 & C - U^TAU \end{pmatrix}. \tag{A.15}$$

If $C \succeq U^TAU$, both summands in (A.15) are positive semidefinite, so that $X \succeq 0$.

Conversely, assume that $X \succeq 0$. We show that B can be written as $B = AU$ with $U \in M_{n,m}(\mathbb{R})$. Since $X \succeq 0$, it has a positive square root. If X_1 and X_2 denote the rows of $X^{1/2}$ we write

$$X^{1/2} = \begin{pmatrix} X_1 \\ X_2 \end{pmatrix}.$$

Comparing the square of $X^{1/2}$ with X yields $A = X_1 X_1^T, B = X_1 X_2^T, C = X_2 X_2^T$.

For $v \in \mathbb{R}^n$ we have $\langle Av, v \rangle = \langle X_1 X_1^T v, v \rangle = \|X_1^T v\|^2$. Therefore, $Av = 0$ implies that $X_1^T v = 0$. Hence there exists a well-defined (!) linear operator $V_1 : \mathbb{R}^n \to \mathbb{R}^{n+m}$ such that $V_1 Av = X_1^T v$ for $v \in \mathbb{R}^n$. Then $V_1 A = X_1^T$. Similarly, there is a linear operator $V_2 : \mathbb{R}^m \to \mathbb{R}^{n+m}$ such that $V_2 C = X_2^T$. Then $V_1^T V_2 C : \mathbb{R}^m \to \mathbb{R}^n$ is given by a matrix $U \in M_{n,m}(\mathbb{R})$ and we compute $B = X_1 X_2^T = (V_1 A)^T V_2 C = A V_1^T V_2 C = AU$.

For $v \in \mathbb{R}^m$, we have $\langle (C - U^TAU)v, v \rangle = \langle X \begin{pmatrix} 0 \\ v \end{pmatrix}, \begin{pmatrix} 0 \\ v \end{pmatrix} \rangle \geq 0$. Thus, $C \succeq U^TAU$.

(ii) By the assumption of (ii), $B = AU$. Therefore, if $C = U^TAU$, then X is a flat extension of A by Proposition A.23.

Conversely, suppose $C \neq U^TAU$. Let $P : \mathbb{R}^{n+m} \to \mathbb{R}^n$ be the canonical projection given by $P(x, y) = x, x \in \mathbb{R}^n, y \in \mathbb{R}^m$. From (A.14) it follows that P is a surjection of the range of X on the range of A. We choose $v \in \mathbb{R}^m$ such that $Cv \neq U^TAUv$. Set $u := -Uv$ and $w := Cv - U^TAUv$. From (A.14) we derive $X \begin{pmatrix} u \\ v \end{pmatrix} = \begin{pmatrix} 0 \\ w \end{pmatrix}$. Since $w \neq 0$, P is not injective. Therefore, $\operatorname{rank} X = \dim \operatorname{im} X > \dim \operatorname{im} A = \operatorname{rank} A$, so X is not a flat extension of A.

(iii) By Proposition A.23, each flat extension X of A is of the form (A.13) with $B = AU$ and $C = U^TAU$. Hence $X \succeq 0$ by (A.14) and (A.15). $\qquad\square$

Proposition A.25 *Suppose that $A \succeq 0$. Then the matrix X given by (A.13) is a flat extension of A if and only if*

$$\mathbb{R}^{n+m} = (\mathbb{R}^n, 0) + \ker X. \tag{A.16}$$

If this holds and $x \in \mathbb{R}^n$, then $x \in \ker A$ if and only if $(x, 0) \in \ker X$.

Proof First suppose that X is a flat extension of A. Then, by Proposition A.23, there is a matrix $U \in M_{n,m}(\mathbb{R})$ such that $B = AU$ and $C = U^T A U$. Let $x \in \mathbb{R}^n$ and $y \in \mathbb{R}^m$. Then $X(-Uy, y) = 0$, so that $(x, y) = (x + Uy, 0) + (-Uy, y) \in \mathbb{R}^{n+m} + \ker X$. This proves (A.16).

Now assume that (A.16) holds. By reordering the canonical basis of $\mathbb{R}^n$, we may assume that the first n' canonical basis elements of $\mathbb{R}^n$ are linearly independent of $\ker X$ and that, together with a basis of $\ker X$, they form a basis of $\mathbb{R}^{n+m}$. Then

$$\mathbb{R}^{n+m} = (\mathbb{R}^{n'}, 0) \oplus \ker X. \tag{A.17}$$

Let $m' := \dim \ker X$. Clearly, $n' + m' = n + m$ by (A.17). From (A.16) it follows that $n' \leq n$. We write X as block matrix of the form (A.13) with respect to the decomposion (A.17). Let $A', B', (B')^T, C'$ denote the corresponding matrix entries. The matrix of change of bases between the canonical basis of $\mathbb{R}^{n+m}$ and this new basis is of the form

$$\begin{pmatrix} I & -U' \\ 0 & I \end{pmatrix}$$

for some matrix $U' \in M_{n',m'}(\mathbb{R})$. Then the matrix of the symmetric operator corresponding to X on $\mathbb{R}^{n+m}$ in the new basis is

$$\begin{pmatrix} A' & 0 \\ 0 & 0 \end{pmatrix} = \begin{pmatrix} I & 0 \\ -(U')^T & I \end{pmatrix} \begin{pmatrix} A' & B' \\ (B')^T & C' \end{pmatrix} \begin{pmatrix} I & -U' \\ 0 & I \end{pmatrix}$$

$$= \begin{pmatrix} A' & B' - A'U' \\ (B')^T - (U')^T A' & C' - (U')^T B' - (B')^T U' + (U')^T A' U' \end{pmatrix}.$$

Thus, $B' = A'U'$, $(B')^T = (U')^T A'$, and $C' - (U')^T B' - (B')^T U' + (U')^T A' U' = 0$. We insert the first two relations into the last and obtain $C' = (U')^T A' U'$. Hence, by Proposition A.23, X is a flat extension of A', so that $\operatorname{rank} X = \operatorname{rank} A'$. Since $n' \leq n$ as noted above, A' corresponds to a submatrix of A in the new basis. Therefore, $\operatorname{rank} A' \leq \operatorname{rank} A \leq \operatorname{rank} X$. Hence $\operatorname{rank} X = \operatorname{rank} A$, so X is a flat extension of A.

We verify the last assertion. We write X as in Proposition A.23 and compute $X(x, 0) = (Ax, B^T x) = (Ax, U^T Ax)$. Hence $(x, 0) \in \ker X$ if and only if $x \in \ker A$. $\qquad\qquad\square$

A.5   Locally Convex Topologies

Introductions to locally convex spaces are given in [Ru1] and [Cw].

In this section, $\mathbb{K} = \mathbb{R}$ or $\mathbb{K} = \mathbb{C}$ and V is a $\mathbb{K}$-vector space.

A map $p : V \to \mathbb{R}_+$ is called a *seminorm* on the vector space V if $p(\lambda x) = |\lambda| p(x)$ and $p(x + y) \leq p(x) + p(y)$ for all $\lambda \in \mathbb{K}$ and $x, y \in V$.

Let P be a family of seminorms on V such that $p(x) = 0$ for $p \in P$ implies $x = 0$. The *locally convex topology* defined by P is the topology on V for which the sets

$$\{y \in V : p_1(x - y) \leq \varepsilon, \ldots, p_k(x - y) \leq \varepsilon\}$$

where $p_1, \ldots, p_k \in P, k \in \mathbb{N}, \varepsilon > 0$, form a base of neighbourhoods of $x \in V$.

Theorem A.26 (Separation Theorem for Convex Sets) *Suppose that V is real vector space equipped with a locally convex topology. Let X and Y be nonempty disjoint convex subsets of V.*

(i) *If Y is open, then there exist a linear functional $L : V \to \mathbb{R}$ and an $a \in \mathbb{R}$ such that $L(x) \leq a < L(y)$ for $x \in X$ and $y \in Y$.*

(ii) *If X is compact and Y is closed, then there are a linear functional $L : V \to \mathbb{R}$ and numbers $a, b \in \mathbb{R}$ such that $L(x) \leq a < b \leq L(y)$ for all $x \in X$ and $y \in Y$.*

If Y is a cone, then $a < 0$ in (i) and (ii).

Proof [Ru1, Part 1, Theorem 3.4].   □

The following algebraic version of the separation theorem is *Eidelheit's theorem*. A point x_0 of a set X in a real vector space V is called an *internal point* if, given $v \in V$, there exists a $\delta_v > 0$ such that $x_0 + \delta v \in X$ for all $\delta \in \mathbb{R}, |\delta| \leq \delta_v$.

Theorem A.27 *Let X and Y be nonempty disjoint convex sets in a real vector space V. Suppose that Y has at least one internal point. Then there exists a linear functional L on V such that $L(x) \leq L(y)$ for $x \in X$, $y \in Y$. If Y is a cone, then L is Y-positive.*

Proof [KNa, Theorem 3.8] or [Kö], § 17, (3), p. 187.   □

If the vector space V is finite-dimensional, it has a unique locally convex topology; we call it the *natural topology* of V. It can be given by any norm on V.

Let τ_f denote the finest locally convex topology on V. It is obtained by taking the family of *all* seminorms on V as P. A neighbourhood base of zero is given by the absolutely convex absorbing subsets of V. All linear functionals on V and all seminorms on V are continuous in the topology τ_f. The topology τ_f induces the natural topology on each finite-dimensional linear subspace. If V has countable vector space dimension, we have the following characterization of closed subsets.

Proposition A.28 *Suppose that V is a $\mathbb{K}$-vector space of at most countable dimension. A subset C of V is closed in the topology τ_f if and only $C \cap E$ is closed in V for each finite-dimensional subspace E of V.*

Proof The assertion follows from the Krein–Smuljan theorem [KNa, Theorem 22.6] applied to the dual locally convex space of V. We include a direct "elementary" proof following [Ms1, Section 3.6].

If C is τ_f-closed, then $C \cap E$ is closed in the induced topology. Since the finite-dimensional space E has a unique locally convex topology, the induced topology of τ_f is the natural topology of E.

We prove the converse implication. Since the assertion is trivial if V has finite dimension, we can assume that V has a countable basis, say $\{v_n : n \in \mathbb{N}\}$. Let V_n be the linear span of $v_1, \dots, v_n$. For numbers $\varepsilon_1 > 0, \dots, \varepsilon_n > 0$ we abbreviate

$$U(\varepsilon_1, \dots, \varepsilon_n) := \left\{ v = \sum_{j=1}^n \lambda_j e_j : |\lambda_j| \le \varepsilon_j \text{ for } j = 1, \dots, n \right\}.$$

It suffices to show that the complement $M := V \backslash C$ of C is open, that is, for each point $v \in M$ there exists a τ_f-neighbourhood U such that $U \subseteq M$. Upon translation we can assume without loss of generality that $v = 0$.

We prove by induction that there is a positive sequence $\varepsilon = (\varepsilon_n)_{n \in \mathbb{N}}$ such that

$$U(\varepsilon_1, \dots, \varepsilon_n) \subseteq M \cap V_n. \tag{A.18}$$

Clearly, since $M \cap V_1$ is open, there is an $\varepsilon_1 > 0$ such that $U(\varepsilon_1) \subseteq M \cap V_1$. Suppose that $\varepsilon_1, \dots, \varepsilon_n$ are constructed such that (A.18) is satisfied.

Assume to the contrary that there is no number $\varepsilon_{n+1} > 0$ such that (A.18) holds for $n + 1$. Then for each $k \in \mathbb{N}$ there exists an $x_k = \sum_{j=1}^{n+1} \lambda_{k,j} v_j \in U(\varepsilon_1, \dots, \varepsilon_n, \frac{1}{k})$ such that $x_k \notin M$. Since all "coordinates" λ_{kj} of x_k are bounded, the sequence $(x_k)_{k \in \mathbb{N}}$ has a convergent subsequence in V_{n+1}. Let x be the limit of such a subsequence. By $x_k \in U(\varepsilon_1, \dots, \varepsilon_n, \frac{1}{k})$, we have $|\lambda_{k,j}| \le \varepsilon_j$ for $j = 1, \dots, n$ and $|\lambda_{k,n+1}| \le \frac{1}{k}$. Therefore, $x \in V_n$ and $x \in U(\varepsilon_1, \dots, \varepsilon_n)$, so that $x \in M \cap V_n$ by (A.18) and hence $x \notin C \cap V_n$. But $x_k \in C \cap V_n$, since $x_k \notin M$. Since x is the limit of a subsequence of (x_k), this contradicts the assumption that $C \cap V_n$ is closed.

Thus, by induction there exists a positive sequence ε satisfying (A.18). Since $U = \bigcup_{n=1}^\infty U(\varepsilon_1, \dots, \varepsilon_n)$ is an absolutely convex and absorbing set, U is a neighbourhood of zero in the topology τ_f. By (A.18) we have $U \subseteq M$. $\square$

Remark A.29 The preceding proof shows the following result: Let $(V_n)_{n \in \mathbb{N}}$ be a sequence of finite-dimensional subspaces of V such that $V_n \subseteq V_{n+1}$ for $n \in \mathbb{N}$ and $V = \bigcup_{n=1}^\infty V_n$. Then M is closed in the finest locally convex topology of V if and only if $M \cap V_n$ is closed in V_n for all $n \in \mathbb{N}$. $\circ$

A.6 Convex Sets and Cones

The basics of convex sets are developed in the monographs [Sn, Rf], and [Bv].

Let E be a real vector space. We denoted its dual vector space by E^*.

Definition A.30 A subset C of E is called

- *convex* if $\lambda x + (1 - \lambda)y \in C$ for $x, y \in C$ and $\lambda \in [0, 1]$,
- a *cone* if $x + y \in C$ and $\lambda x \in C$ for $x, y \in C$ and $\lambda \geq 0$.

Definition A.31 For a cone C in E the *dual cone* of C is the set

$$C^\wedge = \{F \in E^* : L(x) \geq 0 \quad \text{for} \quad x \in C\}. \tag{A.19}$$

Clearly, $C^\wedge$ is a cone in the real vector space E^*.

By the *convex hull* resp. *conic hull* of a subset $\mathcal{X}$ of E we mean the smallest convex set resp. cone which contains $\mathcal{X}$.

From now on we assume that E is a **finite-dimensional** real vector space. All topological notions refer to the unique norm topology of E.

Proposition A.32 *If C is a closed cone of E, then $C = C^{\wedge\wedge}$.*

Proof This is a special case of the bipolar theorem [Cw, Theorem V.1.8]. It follows at once from separation of convex sets. Obviously, $C \subseteq C^{\wedge\wedge}$ by definition. Assume to the contrary that there exists an $x \in C^{\wedge\wedge}\backslash C$. Since C is closed, by Theorem A.26(ii) there is a linear functional L on E such that $L(x) < 0$ and $L(y) \geq 0$ for $y \in C$. Then $L \in C^\wedge$ and hence $L(x) \geq 0$, because $x \in C^{\wedge\wedge}$. This is a contradiction. □

A point $x \in E$ is called a *relatively interior point* of a set C if x is in the interior of C in the vector space spanned by C.

Proposition A.33 *Suppose that C is a non-empty convex subset of E.*

(i) *S has relatively interior points. If E is the span of C, then C has interior points.*
(ii) *The set C and its closure $\overline{C}$ have the same interior points.*

Proof

(i) [Rf, Theorem 6.2] or [Sn, Theorem 1.1.13].
(ii) [Rf, Theorem 6.3] or [Sn, Theorem 1.1.15(a)]. □

Proposition A.34 *Suppose that C is a cone in E.*

(i) *If C is closed and $x_0 \in E$ is not in C, then there exists a linear functional F on E such that $F(x_0) < 0$ and $F(x) \geq 0$ for $x \in C$.*
(ii) *A point $x_0 \in E$ is a boundary point of C if and only if there exists a linear functional $F \neq 0$ on E such that $F(x_0) = 0$ and $F(x) \geq 0$ for $x \in C$. Such a functional is called a supporting functional of C at x_0.*

Proof

(i) [Sn, Theorem 1.3.9] or [Rf, Corollary 11.4.2].
(ii) The if part is almost obvious; for the only if part we refer to [Sn, Theorems 1.3.2 and 1.3.9] or [Rf, Theorem 11.6].

Clearly, (i) is a special case of Theorem A.26(ii). In [Rf] both results are stated for convex sets; since C is a cone, the corresponding hyperplanes pass the origin which yields the results as stated above.   □

The following result is *Carathéodory's theorem* [Ca1]. Since it is used in a crucial manner in this book, we include a proof of it.

Proposition A.35 *Let $d = \dim E$ and let X be a non-empty subset of E.*

(i) *Each point of the conic hull of X is a nonnegative combination of at most d points of X.*
(ii) *Each point of the convex hull of X is a convex combination of at most $d + 1$ points of X.*

Proof

(i) Each point x in the conic hull of X can be represented as

$$x = \sum_{j=1}^{k} \lambda_j x_j \quad \text{with } \lambda_j \geq 0, \ x_j \in X, \ j = 1, \ldots, k. \tag{A.20}$$

Suppose that (A.20) is a corresponding representation of x with minimal k. Assume to the contrary that $k > d = \dim E$. Then $x_1, \ldots, x_k$ are linearly dependent. Hence there are reals $\gamma_1, \ldots, \gamma_k$ such that at least one is positive and

$$\sum_{j=1}^{k} \gamma_j x_j = 0. \tag{A.21}$$

Upon renumbering the points we can assume $\lambda_1 \gamma_1^{-1} > 0$ is the minimum of all $\lambda_j \gamma_j^{-1}$, where $\gamma_j > 0$. By (A.20) and (A.21) we have

$$x = \sum_{j=1}^{k} \lambda_j x_j - \lambda_1 \gamma_1^{-1} \sum_{j=1}^{k} \gamma_j x_j = \sum_{j=2}^{k} (\lambda_j - \lambda_1 \gamma_1^{-1} \gamma_j) x_j. \tag{A.22}$$

Then $\lambda_j - \lambda_1 \gamma_1^{-1} \gamma_j \geq 0$ for all j. Indeed, if $\gamma_j > 0$, then $\lambda_1 \gamma_1^{-1} \leq \lambda_j \gamma_j^{-1}$ by construction. If $\gamma_j < 0$, this is trivial. Hence (A.22) is a representation (A.20) with $k - 1$ summands which contradicts the minimality of k.
(ii) The assertion follows by a similar reasoning as (i) (employing the linear dependence of $x_j - x_1, j = 2, \ldots, k$) or directly from (i) by using that x is in the convex hull of X if and only if $(1, x)^T$ is in the conic hull of $(1, X)^T$ in $\mathbb{R} \oplus E$.   □

Next we turn to extreme rays, faces, and extreme points.

Definition A.36 A subset F of the cone C is called

- a *face* of C if F is a subcone such that $x = y + z \in F$ with $y, z \in C$ implies that $y, z \in F$.
- an *extreme ray* of C if F is a face of the form $F = \mathbb{R}_+ \cdot x$ for some $x \in C, x \neq 0$.
- an *exposed face* of C if there exists a linear functional $L \in C^\wedge$ such that

$$F = F_L := \{c \in C : L(c) = 0\}. \tag{A.23}$$

It is easily seen that an exposed face is indeed a face. The empty set $\emptyset$ and the cone C are faces. A face $F \neq \emptyset, C$ is called a proper face. In general, not all proper faces are exposed faces, but they are if the cone C is finitely generated.

Let $L \in C^\wedge$ and let c_0 be an element of the exposed face F_L defined by (A.23). Then $L(c) \geq 0$ for all $c \in C$ (by $L \in C^\wedge$) and $L(c_0) = 0$ (by $c_0 \in F_L$). Therefore, if $L \neq 0$, then L is a supporting hyperplane of C at c_0 and c_0 is a boundary point of C.

The extreme rays are the one-dimensional faces. Restating the definition, an *extreme ray* of C is a subset $F = \mathbb{R}_+ \cdot x$, where $x \in C, x \neq 0$, such that $x = x_1 + x_2$ with $x_1, x_2 \in C$ implies that $x_1, x_2 \in \mathbb{R}_+ \cdot x$. In this case, we say that the vector x spans an extreme ray of C. Let $\mathrm{Exr}(C)$ denote the set of elements of all extreme rays of C.

A cone C is called *pointed* if $C \cap (-C) = \{0\}$, or equivalently, if C does not contain a line.

Proposition A.37 *Suppose* $C \neq \{0\}$ *is a pointed closed cone of a finite-dimensional real vector space. Then each point of* C *is a finite sum of elements of* $\mathrm{Exr}(C)$.

Proof [Sn, Theorem 1.4.3] or [Bv, II, Corollary 8.5 and problem 3]; see also [Rf, Corollary 18.7.1]. □

Often it is convenient to study cones by means of bases.

Definition A.38 A *base* of a cone C is a convex subset B of C such that for each $x \in C, x \neq 0$, there exists a unique number $\lambda_x > 0$ satisfying $\lambda_x x \in B$.

A linear functional L on E is *strictly C-positive* if $L(c) > 0$ for all $c \in C, c \neq 0$.

Let L be a strictly C-positive linear functional. Then it is easily verified that

$$B(L) := \{c \in C : L(c) = 1\} \tag{A.24}$$

is a base of C. It can be shown that each base of C is of this form. For $x \in C, x \neq 0$, the corresponding number $L(x)$ is given by $L(x) = \lambda_x^{-1}$, where $\lambda_x > 0$ and $\lambda_x x \in B$. If $E = C - C$, then the map $L \mapsto B(L)$ yields a one-to-one correspondence between bases of C and strictly C-positive linear functionals on E.

Definition A.39 A point $x \in B$ is called an *extreme point* of a convex set B if whenever $x = \lambda y + (1 - \lambda)z$ with $y, z \in B$ and $0 < \lambda < 1$, then $y = z = x$.

Suppose that B is a base of C. We record some basic facts. A point $x \in B$ is an extreme point of the convex set B if and only if x spans an extreme ray of C, see [Bv, II, Lemma 8.4]. Further, if B is compact, C is closed [Bv, II, Lemma 8.6].

By a classical result of H. Minkowski, each *compact* convex set B in the finite-dimensional real vector space E is the convex hull of its extreme points.

Finally, we turn to conic optimization. Let L_1 and L_0 be linear functionals on E. A *linear conic optimization problem* is given by

$$c := \inf \{L_1(f) : f \in C, L_0(f) = 1\} \tag{A.25}$$

and the corresponding dual problem to (A.25) is

$$c^* := \sup \{\lambda \in \mathbb{R} : (L_1 - \lambda L_0) \in C^\wedge\}. \tag{A.26}$$

Lemma A.40 *Suppose that $L_1, L_0 \in C^\wedge$ and there exists an element $f_0 \in C$ such that $L_0(f_0) > 0$. Then $c = c^*$.*

Proof The set $V := \{f \in C : L_0(f) = 1\}$ is not empty, since $L_0(f_0)^{-1}f_0 \in V$. Hence c is defined. Let $f \in C$. From $L_1(g) \geq c$ for $g \in V$ and $L_0(f) \geq 0$ (by $L_0 \in C^\wedge$) we conclude that $L_1(f) \geq cL_0(f)$. Therefore, $(L_1 - cL_0) \in C^\wedge$ and hence $c \leq c^*$.

Since $L_1 \in C^\wedge$, $(L_1 - 0 \cdot L_0) \in C^\wedge$, so the corresponding set in (A.26) is not empty. Fix $\lambda \leq c^*$ such that $(L_1 - \lambda L_0) \in C^\wedge$. Then, for $f \in C$, we have $L_1(f) - \lambda L_0(f) \geq 0$, so that $L_1(g) \geq \lambda$ for $g \in V$. Taking the infimum over $g \in V$ and then the supremum over such λ we get $c \geq c^*$. $\square$

A.7   Symmetric and Self-Adjoint Operators on Hilbert Space

All operator-theoretic facts in this appendix can be found in most books on Hilbert space operators such as [RS2], [AG], and [Sm9].

Suppose that $\mathcal{H}$ is a complex Hilbert space with scalar product $\langle \cdot, \cdot \rangle$.

By an *operator* on $\mathcal{H}$ we mean a linear mapping T of a linear subspace $\mathcal{D}(T)$, called the *domain* of T, into $\mathcal{H}$. The *kernel* of T is $\mathcal{N}(T) = \{\varphi \in \mathcal{D}(T) : T\varphi = 0\}$. The bounded operators defined on $\mathcal{H}$ are denoted by $\mathbf{B}(\mathcal{H})$.

An operator T is called *symmetric* if $\langle T\varphi, \psi \rangle = \langle \varphi, T\psi \rangle$ for $\varphi, \psi \in \mathcal{D}(T)$.

If T_1 and T_2 are linear operators on $\mathcal{H}$, we say that T_2 is an *extension* of T_1 and write $T_1 \subseteq T_2$ if $\mathcal{D}(T_1) \subseteq \mathcal{D}(T_2)$ and $T_1\varphi = T_2\varphi$ for $\varphi \in \mathcal{D}(T_1)$.

An operator T is called *closed* if for each sequence (φ_n) from $\mathcal{D}(T)$ such that $\varphi_n \to \varphi$ and $T\varphi_n \to \psi$ in $\mathcal{H}$ then $\varphi \in \mathcal{D}(T), \psi = T\varphi$. If T has a closed extension, it has a smallest closed extension, called the *closure* of T and denoted by $\overline{T}$.

For a closed operator T the *resolvent set* $\rho(T)$ is the set of numbers $z \in \mathbb{C}$ for which the operator $T - zI$ has a bounded inverse $(T - zI)^{-1}$ that is defined on the whole Hilbert space $\mathcal{H}$. The set $\sigma(T) := \mathbb{C} \backslash \rho(T)$ is the *spectrum* of T.

Suppose that $\mathcal{D}(T)$ is dense in $\mathcal{H}$. Then the *adjoint operator* T^* of T is defined: Its domain $\mathcal{D}(T^*)$ is the set of vectors $\psi \in \mathcal{H}$ for which there exists an $\eta \in \mathcal{H}$ such that $\langle T\varphi, \psi \rangle = \langle \varphi, \eta \rangle$ for all $\varphi \in \mathcal{D}(T)$; in this case $T^*\psi = \eta$.

The operator T is symmetric if and only $T \subseteq T^*$.

A densely defined operator T is called *self-adjoint* if $T = T^*$ and *essentially self-adjoint* if its closure $\overline{T}$ is self-adjoint, or equivalently, if $\overline{T} = T^*$.

Let us turn to self-adjoint extensions of a densely defined symmetric operator T. The *deficiency indices* of T is the pair $(d_+(T), d_-(T))$ of (cardinal) numbers

$$d_+(T) = \dim \mathcal{N}(T^* - zI) \quad \text{for } \operatorname{Im} z > 0, \quad d_-(T) = \dim \mathcal{N}(T^* - zI) \quad \text{for } \operatorname{Im} z < 0.$$

These numbers are independent of the choice of z satisfying $\operatorname{Im} z > 0$ resp. $\operatorname{Im} z < 0$.

Proposition A.41 *Let T be a densely defined symmetric operator on $\mathcal{H}$. Then*

$$\dim (\mathcal{D}(T^*)/\mathcal{D}(\overline{T})) = d_+(T) + d_-(T). \tag{A.27}$$

The operator T has a self-adjoint extension on $\mathcal{H}$ if and only if it has equal deficiency indices, that is, $d_+(T) = d_-(T)$.

Proof [Sm9, Proposition 3.7 and Theorem 13.10]. □

If $d_+(T) = d_-(T) \neq 0$, then T has "many" different self-adjoint extension on $\mathcal{H}$, see [Sm9, Theorem 13.10]. Further, each densely defined symmetric operator has a self-adjoint extension on a possibly *larger* Hilbert space [Sm9, Proposition 3.17].

Proposition A.42 *For any densely defined symmetric operator T on $\mathcal{H}$ the following are equivalent:*

(i) *T is essentially self-adjoint.*
(ii) *T has a unique self-adjoint extension on $\mathcal{H}$.*
(iii) *$(T - z_+I)\mathcal{D}(T)$ and $(T - z_-I)\mathcal{D}(T)$ are dense in $\mathcal{H}$ for some (then for all) $z_+, z_- \in \mathbb{C}$ such that $\operatorname{Im} z_+ > 0$ and $\operatorname{Im} z_- < 0$.*
(iv) *$d_+(T) = d_-(T) = 0$.*

If the symmetric operator T is positive, these conditions are equivalent to
(v) *$(T - xI)\mathcal{D}(T)$ is dense in $\mathcal{H}$ for one (then for all) $x < 0$.*

Proof [Sm9, Propositions 3.8 and 3.15 and Theorem 13.10]. □

A *conjugation* is a mapping $J : \mathcal{H} \to \mathcal{H}$ such that for $\alpha, \beta \in \mathbb{C}$ and $x, y \in \mathcal{H}$,

$$J(\alpha x + \beta y) = \overline{\alpha}J(x) + \overline{\beta}J(y), \quad \langle Jx, Jy \rangle = \langle y, x \rangle, \quad \text{and} \quad J^2x = x. \tag{A.28}$$

Proposition A.43 *Let T be a densely defined symmetric operator on $\mathcal{H}$. If there exists a conjugation J such that $J\mathcal{D}(T) \subseteq \mathcal{D}(T)$ and $JTx = TJx$ for all $x \in \mathcal{D}(T)$, then T has a self-adjoint extension on $\mathcal{H}$.*

Proof [Sm9, Theorem 13.25]. □

Suppose that T is a self-adjoint operator. By the spectral theorem, there exists a unique projection-valued measure E_T on the Borel σ-algebra $\mathfrak{B}(\mathbb{R})$ such that $T = \int_{\mathbb{R}} \lambda dE_T(\lambda)$. The support of E_T is the spectrum $\sigma(T)$. For each Borel function f on the spectrum of T there exists an operator $f(T) = \int f(\lambda) \, dE_T(\lambda)$ defined by

$$f(T)\varphi = \int f(\lambda) \, dE_T(\lambda)\varphi, \quad \text{where} \ \varphi \in \mathcal{H} \ \text{and} \ \int |f(\lambda)|^2 d\langle E_T(\lambda)\varphi, \varphi \rangle < \infty.$$

The assignment $f \mapsto f(T)$ is the functional calculus of T, see [Sm9, Chapter 5].

Bibliography

[AM]   Agler, J. and J.E. McCarthy: *Pick Interpolation and Hilbert Function Spaces*, Amer. Math. Soc., Providence, R.I., 2002.

[Ak]   Akhiezer, N.I.: *The Classical Moment Problem*, Oliver and Boyd, Edinburgh and London, 1965.

[AG]   Akhiezer, N.I. and I.M. Glazman: *Theory of Linear Operators in Hilbert Space*, Ungar, New York, 1961.

[AK]   Akhiezer, N.I. and M.G. Krein: *Some Questions in the Theory of Moments*, Kharkov, 1938; Amer. Math. Soc. Providence, R. I., 1962.

[AL]   Anjos, M.A. and J.B. Lasserre (Editors): *Handbook on Semidefinite, Conic and Polynomial Optimization*, Internat. Ser. Oper. Res. and Management Sci. **166**, Springer-Verlag, New York, 2012.

[Bk1]   Bakan, A.: Codimension of polynomial subspaces in $L^2(\mathbb{R}^d, d\mu)$ for discrete indeterminate measures, Proc. Amer. Math. Soc. **130**(2002), 3545–3553.

[BR]   Bakan, A. and S. Ruscheweyh: Representation of measures with simultaneous polynomial denseness in $L_p(\mathbb{R}, \mu)$, $1 \le p < \infty$, Ark. Math. **43**(2005), 221–249.

[BW]   Bakonyi, M. and H.J. Woerdeman: *Matrix Completions, Moments, and Sums of Hermitean Squares*, Princeton Univ. Press, Princeton, NJ, 2011.

[Bv]   Barvinok, A.: *A course in Convexity*, Amer. Math. Sov., Providence, R.I., 2002.

[Ba]   Bauer, H.: *Measure and Integration Theory*, Walter de Gruyter, Berlin, 2001.

[BS]   Becker, E. and N. Schwartz: Zum Darstellungssatz von Kadison–Dubois, Arch. Math. **40**(1983), 42–428.

[BN]   Ben-Tal, A. and A. Nemirovsky: *Lectures on Modern Convex Optimization*, SIAM, Philadelpia, 2001.

[Bz]   Berezansky, Y.M.: *Expansions in Eigenfunctions of Self-adjoint Operators*, Amer. Math. Soc., Providence, R.I., 1968.

[Be]   Berg, C.: Markov's theorem revisited, J. Approx. Theory **78**(1994), 260–275.

[BC1]   Berg, C. and J.P.R. Christensen: Density questions in the classical theory of moments, Ann. Institut Fourier (Grenoble) **31**(1981), 99–114.

[BC2]   Berg, C. and J.P.R. Christensen: Exposants critiques dans le problem des moments, C.R. Acad. Sci. Paris **296**(1983), 661–663.

[BD]   Berg, C. and A.J. Duran: The index of determinacy for measures and the l^2-norm of orthonormal polynomials, Trans. Amer. Math. Soc. **347**(1995), 2795–2811.

[BP]   Berg, C. and H.L. Pedersen: On the order and the type of the entire functions associated with an indeterminate Hamburger moment problem, Ark. Mat. **32**(1994), 1–11.

© Springer International Publishing AG 2017
K. Schmüdgen, *The Moment Problem*, Graduate Texts in Mathematics 277,
DOI 10.1007/978-3-319-64546-9

[BS1]   Berg, C. and R. Szwarc: The smallest eigenvalue of Hankel matrices, Constr. Approx. 34(2011), 107–133.

[BS2]   Berg, C. and R. Szwarc: On the order of indeterminate moment problems, Adv. Math. 250(2014), 105–143.

[BS3]   Berg, C. and R. Szwarc: A determinant characterization of moment sequences with finitely many mass points, Linear Multilinear Algebra 3(2015), 1568–1576.

[BT]   Berg, C. and M. Thill: Rotation invariant moment problems, Acta Math. 167(1991), 207–227.

[BV]   Berg, C. and G. Valent: The Nevanlinna parametrization for some indeterminate Stieltjes moment problems associated with birth and death processes, Methods Appl. Anal. 1(1994), 169–209.

[BCI]   Berg, C., Y. Chen and M.E.H. Ismail: Small eigenvalues of large Hankel matrices: the indeterminate case, Math. Scand. 9(2002), 7–81.

[BCJ]   Berg, C., J.P.R. Christensen, and C.U. Jensen: A remark on the multidimensional moment problem, Math. Ann. 243(1979), 163–169.

[BCRl]   Berg, C., J.P.R. Christensen, and P. Ressel: *Harmonic Analysis on Semigroups*, Springer-Verlag, New York, 1984.

[Bn]   Bernstein, S.: Sur le representations des polynomes positifs, Soobshch. Kharkov matem. ob-va 14(1915), 227–228.

[Bl]   Billingsley, P.: *Probability and Measures*, Wiley, New York, 1995.

[Bi]   Bisgaard, T.M.: The two-sided complex moment problem, Ark. Mat. 27(1989), 23–28.

[Bx]   Bix, R.: *Conics and Cubics*, Springer-Verlag, New York, 1998.

[Bl1]   Blekherman, G.: Nonnegative polynomials and sums of squares, J. Amer. Math: Soc. 25(2012), 617–635.

[Bl2]   Blekherman, G.: Positive Gorenstein ideals, Proc. Amer. Math. Soc. 43(2015),69–86.

[BlLs]   Blekherman, G. and J.B. Lasserre: The truncated *K*-moment problem for closure of open sets, J. Funct. Anal. 23(2012), 3604–3616.

[BTB]   Blekherman, G., R.R. Thomas, and P.A. Parillo (Editors): *Semidefinite Optimization and Convex Algebraic Geometry*, MOS-SIAM Series Optimization 13, Philadelphia, 2012.

[BCRo]   Bochnak, J., M. Coste and M.-F. Roy: *Real Algebraic Geometry*, Springer-Verlag, Berlin, 1998.

[Bou]   Bourbaki, N.: *Integration*, Hermann, Paris, 1969.

[Br]   Bram J.: Subnormal operators, Duke Math. J. 22(1955), 75–94.

[BCa1]   Buchwalther, H. and G. Cassier: Measures canoniques dans le probleme classique des moments, Ann. Inst. Fourier (Grenoble) 34(1984), 45–52.

[BCa2]   Buchwalther, H. and G. Cassier: La parametrisation de Nevanlinna dans le probleme de moments de Hamburger, Expo. Math. 2(1984), 155–178.

[BGHN]   Bultheel, A., P. Gonzalez-Vera, E. Hendriksen, and O. Njastad: *Orthogonal Rational Functions*, Cambridge Univ. Press, Cambridge, 1999.

[BSS]   Burgdorf, S., C. Scheiderer, and M. Schweighofer: Pure states, nonnegative polynomials, and sums of squares, Comm. Math. Helvetici 87(2012), 113–140.

[Ca1]   Carathéodory, C.: Über den Variabilitätsbereich der Koeffizienten von Potenzreihen, die gegebene Werte nicht annehmen. Math. Ann. 64(1907), 95–115.

[Ca2]   Carathéodory, C.: Über den Variabilitätsbereich der Fourier'schen Konstanten von positiven harmonischen Funktionen, Rend. Circ. Mat. Palermo 32(1911), 193–217.

[Cl]   Carleman, T.: *Les fonctions quasi-analytiques*, Gauthier-Villars, Paris, 1926.

[Cs]   Cassier, G.: Problem de moments sur un compact de R^n et decomposition de polynomes a plusieurs variables, J. Funct. Anal. 58(1984), 254–266.

[Ch]   Chandler, J.D., Jr.: Rational moment problem for compact sets, J. Approx. Theory 79(1993), 72–88.

[CI]   Chen, Y. and M.E.H. Ismail: Some indeterminate moment problems and Freud-like weights, Constr. Appr. 14(1998), 439–458.

[Chi1]   Chihara, T.S.: *An Introduction to Orthogonal Polynomials*, Gordon and Breach, New York, 1978.

[Chi2]   Chihara, T.S.: On indeterminate symmetric moment problems, Math. Anal. Appl.
          85(1982), 331–346.
[Chil]   Chihara, T.S. and M.E.H. Ismail: Extremal measures for a system of orthogonal
          polynomials, J. Constructive Appr. 9(1993), 111–119.
[Chq]   Choquet, G.: Lectures on Analysis, vol. 3, Benjamin, Reading, 1969.
[CL]   Choi, M.D. and T.Y. Lam: Extremal positive semidefinite forms, Math. Ann. 231(1977),
          1–18.
[CLR]   Choi, M.D., T.Y. Lam and B. Reznick: Real zeros of positive semidefinite forms, Math.
          Z. 171(1987), 559–580.
[Chl]   Christoffel, E.B.: Über die Gaussische Quadratur und eine Verallgemeinerung derselben,
          J. reine angew. Math. 55(1858), 61–82.
[CMN]   Cimprič, J., M. Marshall, M., and T. Netzer: On the real multidimensional rational K-
          moment problem, Trans. Amer. Math. Soc 363(2011), 5773–5788.
[Cd]   Coddington, E.A.: Formally normal operators having no normal extensions, Canad. J.
          Math. 17(1965), 1030–1040.
[Cw]   Conway, J.B.: A Course in Functional Analysis, Springer-Verlag, New York, 1990.
[CLO]   Cox, D., J. Little and D. O'Shea: Ideals, Varieties, and Algorithms, Springer-Verlag,
          New York, 1992.
[CF1]   Curto, R. and L. Fialkow: Recursiveness, positivity, and truncated moment problems,
          Houston J. 17(1991), 603–635.
[CF2]   Curto, R. and L. Fialkow: Solutions of the Truncated Moment Problem for Flat Data,
          Memoirs Amer. Math. Soc. 119, Providence, R. I., 1996.
[CF3]   Curto, R. and L. Fialkow: Flat Extensions of Positive Moment Matrices: Recursively
          Generated Relations, Memoirs Amer. Math. Soc. 136, Providence, R. I., 1998.
[CF4]   Curto, R. and L. Fialkow: Solution of the truncated hyperbolic moment problem, Integral
          Equ. Operator Theory 52(2005), 181–218.
[CF5]   Curto, R. and L. Fialkow: Truncated K-moment problems in several variables, J.
          Operator Theory 54(2005), 189–226.
[CF5]   Curto, R. and L. Fialkow: An analogue of the Riesz-Haviland theorem for the truncated
          moment problem, J. Funct. Anal. 255(2008), 2709–2731.
[CFM]   Curto, R., L. Fialkow and H.M. Möller: The extremal truncated moment problem, Int.
          Equ. Operator Theory 60(2008), 177–200.
[DJ]   De Jeu, M.: Determinate multidimensional moment measures, the extended Carleman
          condition and quasi-analytic weights, Ann. Prob. 31(2007), 1205–1227.
[DM]   Derkach, V. A. and M.M. Malamud: The extension theory of hermitian operators and
          the moment problem, J. Math. Sci. 73(1995), 141–242.
[DS]   Dette, H. and W.J. Studden: The Theory of Canonical Moments with Applications in
          Statistics, Probability, and Analysis, Wiley, New York, 1997.
[Dv1]   Devinatz, A.: Integral representations of positive definite functions, II; Trans. Amer.
          Math. Soc. 77(1955), 455–480.
[Dv2]   Devinatz, A.: Two parameter moment problems, Duke Math. J. 24(1957), 48–498.
[DF]   Diaconis, P. and D. Freeedman: The Markov moment problem and de Finetti's theorem
          I, Math. Z. 247(2004), 183–199.
[DSm1]   di Dio, Ph. and K. Schmüdgen: On the truncated multidimensional moment problem:
          atoms, determinacy, and core variety, ArXiv:1703.01497, to appear in J. Funct. Analysis.
[DSm2]   di Dio, Ph. and K. Schmüdgen: On the truncated multidimensional moment problem:
          Carathéodory numbers, ArXiv:1703.01494, to appear in J. Math. Anal. Appl.
[Dn]   Donoghue, W.F. Jr.: Monotone Matrix Functions and Analytic Continuation, Springer-
          Verlag, Berlin, 1974.
[Do]   Douglas, R.: Extremal measures and subspace density, Michigan Math. J. 11(1964),
          243–246.
[DX]   Dunkl, C.F. and Y. Xu: Orthognal Polynomials of Several Variables, Cambridge Univ.
          Press, Cambridge, 2001.

[DK]   Dym, H. and V. Katznelson: Contributions of Issai Schur to analysis, in Studies in memory of Issai Schur, Progress Math. **20**, Birkhäuser, Boston, 2003, pp. xci–clxxxviii.

[EGH]   Eisenbud, D., M. Green and J. Harris: Cayley–Bacharach theorems and conjectures, Bull. Amer. Math. Soc. **33**(1996), 295–324.

[El1]   Ellison, W.J.: A "Waring problem" for homogeneous forms, Cambridge Phil. Soc. **65**(1969), 663–672.

[El2]   Ellison, W.J.: Waring's problem, Amer. Math. Monthly **78**(1971), 10–36.

[Es]   Eskin, G.I.: A sufficient condition for the solvability of a multidimensional problem of moments. Dokl. Akad. Nauk SSSR **133**(1960), 540–543.

[Fv]   Favard, J.: Sur le polynomes de Tchebycheff, C. R. Acad. Sci. Paris **200**(1935), 2052–2055.

[Fj]   Fejér, L.: Über trigonometrische Polynome, J. reine angew. Math. **146**(1916), 53–82.

[F1]   Fialkow, L.: The truncated K-moment problem: a survey, Preprint, 2015.

[F2]   Fialkow, L.: The core variety of a multisequence in the truncated moment problem, Preprint, 2015.

[FN1]   Fialkow, L. and J. Nie.: Positivity of Riesz functionals and solutions of quadratic and quartic moment problems, J. Funct. Analysis **258**(2010), 328–356.

[FN2]   Fialkow, L. and J. Nie: The truncated moment problem via homogenization and flat extensions, J. Funct. Analysis **23**(2012), 1682–1700.

[Fi]   Fischer, E.: Über das Carathéodory'sche Problem, Potenzreihen mit positivem reellen Teil betreffend, Rend. Circ. Mat. Palermo **32**(1911), 240–256.

[Fo]   Foias, C.: Décomposition en opérateurs et vectors propres. I. Rev. Roumaine Math. Pures Appl. **7**(1962), 241–282.

[Fr1]   Friedrich, J.: A note on the two dimensional moment problem, Math. Nachr. **85**(1984), 285–286.

[Fr2]   Friedrich, J.: Operator moment problems, Math. Nachr. **151**(1991), 273–293.

[FK]   Fritzsche, B. and B. Kirstein (Editors): *Ausgewählte Arbeiten zu den Ursprüngen der Schur-Analysis*, Teubner-Verlag, Stuttgart-Leipzig, 1991.

[Fu]   Fuglede, B.: The multidimensional moment problem, Expo. Math. **1**(1983), 47–65.

[Fn]   Fulton, W.: *Algebraic curves: an introduction to algebraic geometry*, Addison-Wesley, 1998.

[Gb]   Gabardo, J.-P.: A maximum entropy approach to the classical moment problem, J. Funct. Analysis **106**(1992), 80–94.

[Gn]   Gantmacher, F.R.: *Matrizenrechnung*, DVW, Berlin, 1986.

[Gr]   Garnett, J.B.: *Bounded Analytic Functions*, Springer-Verlag, New York, 2006.

[Ge]   Gebhardt, R.: Das eindimensionale Momentenproblem–Lösungen endlicher Ordnung und das Momentenproblem auf $[0, \infty)$, Diplomarbeit, Universität Leipzig, 2010.

[Gs]   Geronimus, Ya. L.: On polynomials orthogonal on the circle, on trigonometric moment problem and on allied Carathéodory and Schur functions (Russian), Mat. Sb. **15**(1944), 99–130.

[GLPR]   Gravin, N., J. Lasserre, D.V. Pasechnik, and D. Robins: The inverse moment problem for convex polytopes, Discrete Comput. Geom. **48**(2012),596–621.

[GHPP]   Gustafsson, B., C. He, M. Peyman, and M. Putinar: Reconstruction planar domains from their moments, Inverse Problems **1**(2000), 1053–1070.

[Hm]   Hamburger, H.L.: Über eine Erweiterung des Stieltjesschen Momentenproblems, Math. Ann. **81**(1920), 235–319, and **82**(1920), 120–164, 168–187.

[Hn]   Handelman, D.: Representing polynomials by positive linear functions on compact convex polyhedra, Pacific J. Math. **132**(1988), 35–62.

[Hs]   Hausdorff, F.: Summationsmethoden und Momentenfolgen, Math. Z. **9**(1921), 74–109 and 280–299.

[Ha]   Haviland, E. K.: On the momentum problem for distribution functions in more than one dimension, I. Amer. J. Math. **57**(1935), 562–582, and II. **58**(1936), 164–168.

[He]   Heine, E.: *Handbuch der Kugelfunktionen. Theorie und Anwendungen*, Vols. 1 and 2, G. Reimer, Berlin, 1878 and 1881.

[Hz]   Herglotz, G.: Über Potenzreihen mit positivem reellen Teil im Einheitskreis, Ber. Ver.
          Sächs. Ges. d. Wiss. Leipzig **63**(1911), 501–511.

[H1]   Hilbert, G.: Über die Darstellung definiter Formen als Summe von Formenquadraten,
          Math. Ann. **32**(1888), 342–350.

[H2]   Hilbert, G.: Beweis für die Darstellbarkeit der ganzen Zahlen durch eine feste Zahl n-ter
          Potenzen, Math. Ann. **7**(1909), 281–300.

[HS]   Hildebrandt, T.H. and I.J. Schoenberg: On linear functional operators and the moment
          problem for a finite interval in one or several dimensions, Ann. Math **34**(1933), 317–328.

[Hr]   Hörmander, L.: *The Analysis of Linear Partial Differential Operators I.*, Springer-Verlag,
          Berlin, 1973.

[Hu]   Hurwitz, A.: Über die Komposition der quadratischen Formen von beliebig vielen
          Variablen, Nachr. Königl. Ges. Wiss. Göttingen, 1898.

[If]   Infusino, M.: Quasi-analyticity and determinacy of the full moment problem from finite
          to infinite dimensions, In: *Stochastic and Infinite Dimensional Analysis*, Trends in Math.,
          Birkhäuser, 2016.

[Ii]   Isii, K.: The extrema of probability determined by generalized moments (1): bounded
          random variables, Annales Inst. Math. Statist. **12**(1960), 119–133.

[Is]   Ismail, M.E.H.: *Classical and Quantum Orthogonal Polynomials in One Variable*,
          Cambridge Univ. Press, Cambridge, 2004.

[IM]   Ismail, M.E.H. and D.R. Masson: q-Hermite polynomials, biorthogonal rational func-
          tions and q-beta integrals, Trans. Amer. Math Soc. **346**(1994), 63–116.

[Jac]   Jacobi, K.G.J.: Über Gauss neue Methode, die Werthe der Integrale näherungsweise zu
          finden, J. reine angew. Math. **1**(1826), 301–308.

[Jc]   Jacobi, T.: A representation theorem for certain partially ordered commutative rings,
          Math. Z. **23**(2001), 259–273.

[JP]   Jacobi, J. and A. Prestel: Distinguished representations of strictly positive polynomials,
          J. reine angew. Math. **532**(2001), 223–235.

[JT]   Jones, W.B. and W.J. Thron: *Continued Fractions. Analytic Theory and Applications*,
          Addison-Wesley, Reading, Mass., 1980.

[Kd]   Kadison, R.V.: *A Representation Theory for Commutative Topological Algebras*,
          Memoirs Amer. Math. Soc. **7**, Providence, R.I., 1951.

[Ka]   Karlin, S.: Representation theorems for positive functions, J. Math. Mech. **12**(1960),
          599–618.

[KSh]   Karlin, M. and L.S. Shapley: *Geometry of Moment Spaces*, Memoirs Amer. Math. Soc.
          12, Providence, R. I., 1953.

[KSt]   Karlin, M. and W. Studden: *Tchebycheff Systems: with Applications in Analysis and
          Statistics*, Interscience, New York, 1966.

[KNa]   Kelley, J.E. and I. Namioka: *Linear Topological Spaces*, Springer-Verlag, Berlin, 1976.

[Kp1]   Kemperman, J.H.B.: The general moment problem, Ann. Math. Stat. **39**(1968), 93–122.

[Kp2]   Kemperman, J.H.B.: Geometry of the moment problem, in: Landau, H.J. (Editor)
          Moments in Mathematics, pp. 110–124, Amer. Math. Soc., Providence, R.I., 1987.

[Kv]   Khrushchev, S.V.: Schur's algorithm, orthogonal polynomials and convergence of Wall's
          continued fractions in $L^2(\mathbb{T})$, J. Approx. Theory **108**(2001), 161–248.

[Ki]   Kilpi, Y.: Über das komplexe Momentenproblem, Ann. Acad. Sci. Fenn A Math.
          236(1957), 1–31.

[KlM]   Klebanov, L.B. and S.T. Mkrtchayn: Estimates of the closeness of distributions in terms
          of identical moments, in: Problems of stability of stochastic models, Proc. Fourth All-
          Union Sem., Moscow, 1980, 64–72; Translations: J. Soviet Math. **32**(1986), 54–60.

[Ks]   Koosis, P.: *The Logarithmic Integral* I, Cambridge Univ. Press, Cambridge, 1988.

[KMP]   Korenblum, B., A. Mascuilli and J. Panariell: A generalization of Carleman's uniqueness
          theorem and a discrete Phragmen–Lindelöf theorem, Proc. Amer. Math. Soc. **12**(1998),
          2025–2032.

[Kö]   Köthe, G.: *Topological Vector Spaces*, Springer-Verlag, Berlin, 1983.

[Kr1]   Krein, M.G.: On an extrapolation problem due to Kolmogorov, Doklady Akad. Nauk
         SSSR **46**(1945), 306–309.
[Kr2]   Krein, M.G.: The ideas of P.L. Cebysev and A.A. Markov in the theory of limiting values
         of integrals and their further developments, Uspehi Mat. Nauk **6**(1951), 2–120; Amer.
         Math. Transl., Amer. Math. Soc., RI, **12**(1951), 1–122.
[Kr3]   Krein, M.G.: The description of all solutions of the truncated power moment problem
         and some problems of operator theory, Mat. Issled. **2**(1967), 114–132. Amer. Mat. Soc.
         Transl. **95**(1970), 219–234.
[KN]   Krein, M.G. and A.A. Nudelman: *The Markov Moment Problem and Extremal Problems*,
         Amer. Math. Soc., Providence, R. I, 1977.
[Kv1]   Krivine, J.L.: Anneaux preordonnees, J. Analyse Math. **12**(1964), 307–326.
[Kv2]   Krivine, J.L.: Quelques properties des preordres dans les anneaux commutatifs unitaires,
         C. R. Acad. Sci. Paris **258**(1964), 3417–3418.
[KM]   Kuhlmann, S. and M. Marshall: Positivity, sums of squares and the multi-dimensional
         moment problem, Trans. Amer. Math. Soc. **354**(2002), 4285–4302.
[KMS]   Kuhlmann, S., M. Marshall, and N. Schwartz: Positivity, sums of squares and the multi-
         dimensional moment problem. II, Adv. Geom. **5**(2005), 583–606.
[L1]   Landau, H.J.: The classical moment problem: Hilbertian proofs, J. Funct. Anal.
         38(1980), 255–272.
[L2]   Landau, H.J. (Editor): *Moments in Mathematics*, Proc. Symposia Appl. Math. **37**, Amer.
         Math. Soc., Providence, R. I, 1987.
[Ls1]   Lasserre, J.B.: Global optimization with polynomials and the problem of moments,
         SIAM J. Optim. **11**(2001), 796–817.
[Ls2]   Lasserre, J.B.: *Moments, Positive Polynomials and Their Applications*, Imperial College
         London, 2010.
[Ls3]   Lasserre, J.B.: The K-moment problem for continuous linear functionals, Trans. Amer.
         Mat. Soc. **365**(2013), 2489–2504.
[Ls4]   Lasserre, J.B.: Borel measures with a density on a compact semi-algebraic set, Arch.
         Math. (Basel) **101**(2013), 361–371.
[La1]   Laurent, M.: Revisiting two theorems of Curto and Fialkov on moment matrices, Proc.
         Amer. Math. Soc. **133**(2005), 2965–2976.
[La2]   Laurent, M.: Sums of squares, moment matrices and optimization, In: Emerging
         Applications of Algebraic Geometry, M. Putinar and S. Sullivant (Editors), Springer,
         New York, 2009, 157–270.
[LM]   Laurent, M. and B. Mourrain: A generalized flat extension theorem for moment matrices,
         Archiv Math. **93**(2009), 87–98.
[Lin1]   Lin, G.D.: On the moment problem, Statist. Probab. Lett. **35**(1997), 85–90. Erratum
         50(2000), 205.
[Lin2]   Lin, G.D.: Recent developments on the moment problem, arXiv:1703.01027.
[Lu]   Lukacs, F. : Verschärfung des ersten Mittelwertsatzes der Integralrechnung für rationale
         Polynome, Math. Z. **2**(1918), 177–216.
[Ml]   Malyshev, V.A.: Cell structure of the space of real polynomials, St. Petersburg Math. J.
         15(2004), 191–248.
[MA]   Marcellan, F. and R. Alvarez-Nodarse: On the "Favard theorem" and its extension, J.
         Comput. Appl. Math. **37**(2001), 213–254.
[Mv1]   Markov, A.A.: Deux demonstrations de la convergence de fractions continues, Acta
         Math. **19**(1895), 235–319.
[Mv2]   Markov, A.A.: Lectures on functions deviating least from zero, Mimeographed Notes, St.
         Petersburg, 1906, 244–281. Reprinted in: Selected papers on continued fractions and the
         theory of functions deviating least from zero, OGIZ, Moscow, 1948, 244–281. (Russian)
[Ms1]   Marshall, M.: *Positive Polynomials and Sums of Squares*, Math. Surveys and Mono-
         graphs **146**, Amer. Math. Soc., Providence, R.I., 2008.
[Ms2]   Marshall, M.: Representations of non-negative polynomials, degree bounds and applica-
         tions to optimization, Canad. J. Math. **61**(2009), 205–221.

[Ms3]   Marshall, M.: Application of localization to the multivariate moment problem, Math. Scand. **115**(2014), 269–286.

[Mt]   Matzke, J.: Mehrdimensionale Momentenprobleme und Positivitätskegel, Dissertation A, Universität Leipzig, 1992.

[MS]   Mourrain, B. and K. Schmüdgen: Flat extensions in ∗-algebras, Proc. Amer. Math. Soc. **144**(2016), 4873–4885.

[Mo]   Motzkin, T.: The arithmetic-geometric inequality, In: Inequalities, O. Sisha (Editor), Academic Press, 1967, 205–224.

[Na]   Naimark, M.A.: Extremal spectral functions of a symmetric operator, Dokl. Akad. Nauk SSSR **54**(1946), 7–9.

[Nat]   Natanson, M.B.: *Additive number theory*, Springer-Verlag, New York, 1996.

[Nt]   Netzer, T.: An elementary proof of Schmüdgen's theorem on the moment problem of closed semi-algebraic sets, Proc. Amer. Math. Soc. **136**(2008), 529–537.

[Nv1]   Nevanlinna, R.: Asymptotische Entwicklungen beschränkter Funktionen und das Stieltjessche Momentenproblem, Ann. Acad. Sci. Fenn. A **18**(1922), 1–53.

[Nv2]   Nevanlinna, R.: Über beschränkte analytische Funktionen, Ann. Acad. Sci. Fenn. A **32** (1929), 1–75.

[Nie]   Nie, J.: Optimimality conditions and finite convergence of Lasserre's hierarchy, Math. Program. Ser. A **14**(2014), 97–121.

[Nu1]   Nussbaum, A.E.: Quasi-analytic vectors, Ark. Math. **6**(1965), 179–191.

[OW]   Overton, M. and R.S. Womersley: On the sum of the k largest eigenvalues of a symmetric matrix, SIAM J. Matrix Anal. Appl. **13**(1992), 41–45.

[Pa]   Parillo, P.A.: Semidefinite programming relaxations for semialgebraic problems, Math. Program. **9**(2003), 293–320.

[Pd1]   Pedersen, H. L.: La parametrisation de Nevanlinna at le problem des moments de Stieltjes indetermine, Expo. Math. **15**(1997), 273–278.

[Pd2]   Pedersen, H. L.: On Krein's theorem for indeterminacy of the classical moment problem, J. Approx. Theory **95**(1998), 90–100.

[Pt]   Petersen, L.C.: On the relation between the multidimensional moment problem and the one-dimensional moment problem, Math. Scand. **51**(1982), 361–366.

[PO]   Petrovskii, I.G. and O.A. Oleinik: On the topology of real algebraic surfaces, Isvestiya Akad. Nauk SSSR Ser. Mat. **13**(1949), 389–402; Amer. Math. Translation **1992**, No. 70.

[Pi]   Pick, G.: Über die Beschränkungen analytischer Funktionen, welche durch vorgegebene Funktionswerte bewirkt sind, Math. Ann. **77**(1916), 7–23.

[Pl]   Plaumann, D.: Sums of squares on reducible real curves, Math. Z. **265**(2009), 777–797.

[P]   Polya, G.: Über positive Darstellung von Polynomen, Vierteljschr. Naturforsch. Ges. Zürich **73**(1928), 141–145.

[PSz]   Polya, G. and G. Szegö: *Problems and Theorems in Analysis*. II, Springer-Verlag, New York, 1976.

[PR]   Powers, V. and B. Reznick: Polynomials that are positive on an interval, Trans. Amer. Math. Soc. **352**(2000), 4677–4692.

[PoS]   Powers, V. and C. Scheiderer: The moment problem for non-compact semi-algebraic sets, Adv. Geom.**1**(2001), 71–88.

[PD]   Prestel, A. and C.N. Delzell: *Positive Polynomials–from Hilbert's 17th Problem to Real Algebra*, Springer-Verlag, New York, 2001.

[Pu1]   Putinar, M.: The L-moment problem in two dimensions, J. Funct. Analysis **94**(1990), 288–307.

[Pu2]   Putinar, M.: Positive polynomials on compact semi-algebraic sets, Indiana Univ. Math. J. **42**(1994), 969–984.

[Pu3]   Putinar, M.: Extremal solutions of the two-dimensional L-problem of moments, J. Funct. Analysis **136**(1996), 331–364.

[Pu4]   Putinar, M.: A dilation approach to cubature formulas, Expo. Math **15**(1997), 183–192.

[Pu5]   Putinar, M.: Extremal solutions of the two-dimensional L-problem of moments. II, J. Approx. Theory **92**(1998), 38–58.

[PSr]   Putinar, V. and C. Scheiderer: Multivariate moment problems: geometry and indetermi-
          nateness, Ann. Scuola Norm. Sup. Pisa Cl. Sci. (5), V(2006), 1–21.
[PSm]   Putinar, N. and K. Schmüdgen: Multivariate determinateness, Indiana Univ. Math. J.
          57(2008), 2931–2968.
[PVs1]   Putinar, N. and F.-H. Vasilescu: Solving the moment problem by dimension extension,
          Ann. Math. 149(1999), 1087–1107.
[PVs2]   Putinar, N. and F.-H. Vasilescu: A uniqueness criterion in the multivariate moment
          problem, Math. Scand. 92(2003), 295–300.
[RW]   Rade, L. and B. Westergren: *Springers Mathematische Formeln*, Springer-Verlag, Berlin,
          1991.
[RS2]   Reed, M. and B. Simon: *Methods of Modern Mathematical Physics II. Fourier Analysis
          and Self-Adjointness*, Academic Press, New York, 1975.
[RRS]   Ren, Q., J. Richter-Gebert and B. Sturmfels: Cayley–Bacharach formulas, Amer. Math.
          Monthly 122(2015), 845–854.
[Re1]   Reznick, B.: *Sums of Even Powers of Linear Forms*, Memoirs Amer. Math. Soc. 9,
          Providence, R.I., 1982.
[Re2]   Reznick, B.: Uniform denominators in Hilbert's seventheenth problem, Math. Z.
          220(1995), 75–98.
[Re3]   Reznick, B.: Homogeneous polynomial solutions to constant PDE's, Adv. Math.
          117(1996), 179–192.
[Re4]   Reznick, B.: On Hilbert's construction of positive polynomials, arXiv:0707.2156.
[RiS]   C. Riener and M. Schweighofer: Optimization approaches to quadrature: new charac-
          terizations of Gaussian quadrature on the line and quadrature with few nodes on plane
          algebraic curves, on the plane and in higher dimensions, arXiv:1607.08404.
[Rz1]   Riesz, F.: Über ein Problem des Herrn Carathéodory, J. reine angew. Math. 146(1916),
          83–87.
[Rz2]   Riesz, M.: Sur le problème des moments. Premiere Note. Arkiv för Mat., Astr. och Fysik
          16(12) (1921), 23pp.; Deuxieme Note. Arkiv för Mat., Astr. och Fysik 16(19) (1922),
          21pp.; Troisieme Note. Arkiv för Mat., Astr. och Fysik 17(12) (1923), 52p.
[RzSz]   Riesz, F. and B. Sz.-Nagy: *Vorlesungen über Funltionalanalysis*, DVW, Berlin, 1956.
[Ri]   Richter, H.: Parameterfreie Abschätzung und Realisierung von Erwartungswerten, Bl.
          der Deutsch. Ges. Versicherungsmath. 3(1957), 147–161.
[Rb]   Robinson, R.M.: Some definite polynomials which are not sums of squares of real
          polynomials, in: Selected Questions of Algebra and Logic, Acad. Sci. USSR, 1973,
          264–282.
[Rf]   Rockafellar, R.T.: *Convex Analysis*, Princeton Univ. Press, Princeton, 1970.
[Rg]   Rogosinski, W.W.: Moments of non-negative mass, Proc. Roy. Soc. London Ser A
          245(1958), 1–27.
[Ru1]   Rudin, W.: *Functional Analysis*, Second Edition, McGraw–Hill Inc., New York, 1973.
[Ru2]   Rudin, W.: *Real and Complex Analysis*, Third Edition, McGraw–Hill Inc., New York,
          1987.
[Sr1]   Scheiderer, C.: Non-existence of degree bounds for weighted sums of squares represen-
          tations, J. Complexity 2(2005), 823–844.
[Sr2]   Scheiderer, C.: Sums of squares on real algebraic curves. Math. Z. 245(2003), 725–760.
[Sr3]   Scheiderer, C.: Positivity and sums of squares: A guide to recent results, In: *Emerging
          Applications of Algebraic Geometry*, M. Putinar and S. Sullivant (Editors), Springer-
          Verlag, Berlin, pp. 1–54, 2009.
[Sm1]   Schmüdgen, K.: Positive cones in enveloping algebras, Reports Math. Phys. 14(1978),
          385–404.
[Sm2]   Schmüdgen, K.: A positive polynomial which is not a sum of squares. A positive, but
          not strongly positive functional, Math. Nachr. 88(1979), 385–390.
[Sm3]   Schmüdgen, K.: A formally normal operator having no normal extension, Proc. Amer.
          Math. Soc. 98(1985), 503–504.

[Sm4]   Schmüdgen, K.: *Unbounded Operator Algebras and Representation Theory*, Birkhäuser-Verlag, Basel, 1990.

[Sm5]   Schmüdgen, K.: On determinacy notions for the two dimensional moment problem, Ark. Math. **29**(1991), 277–284.

[Sm6]   Schmüdgen, K.: The K-moment problem for compact semi-algebraic sets, Math. Ann. **289**(1991), 203–206.

[Sm7]   Schmüdgen, K.: On the moment problem of closed semi-algebraic sets, J. reine angew. Math. **558**(2003), 225–234.

[Sm8]   Schmüdgen, K.: Noncommutative real algebraic geometry–some basic concepts and first ideas, In: Emerging Applications of Algebraic Geometry, M. Putinar and S. Sullivant (Editors), Springer, New York, 2009, 325–350.

[Sm9]   Schmüdgen, K.: *Unbounded Self-adjoint Operators on Hilbert Space*, Springer-Verlag, New York, 2012.

[Sm10]   Schmüdgen, K.: On the multi-dimensional truncated moment problem: maximal masses, Methods Funct. Anal. Topology **21**(2015), 266–281.

[Sm11]   Schmüdgen, K.: A fibre theorem for moments problems and some applications, Israel J. Math. **28**(2016), 43–66.

[Sn]   Schneider, R.: *Convex Bodies: The Brunn–Minkowski Theory*, Cambridge Univ. Press, Cambridge, 1993.

[SSz]   Schoenberg, I.J. and G. Szegö: An extremum problem for polynomials, Compositio Math. **14**(1960), 260–268.

[Sw1]   Schweighofer, M.: An algorithmic approach to Schmüdgen's Positivstellensatz, J. Pure Applied Algebra **166**(2002), 307–319.

[Sw2]   Schweighofer, M.: Iterated rings of bounded elements and generalizations of Schmüdgen's Positivstellensatz, J. reine angew. Math. **554**(2003), 19–45.

[Sw3]   Schweighofer, M.: Optimization of polynomials on compact semi-algebraic sets, SIAM J. Optimization **15**(2005), 805–825.

[Su]   Schur, I.: Über Potenzreihen, die im Innern des Einheitskreises beschränkt sind, I. J. reine angew. Math. **147**(1917), 205–232, and II. **148**(1918), 122–145.

[Sh]   Shmuljan, J.L.: An operator Hellinger integral, Mat. Sb. **91**(1959), 381–430.

[SR]   Semple, J.G. and L. Roth: *Introduction to Algebraic Geometry*, Clarendon Press, Oxford, 1949.

[ST]   Shohat, J.A. and J.D. Tamarkin: *The Problem of Moments*, Amer. Math. Soc., Providence, R.I., 1943.

[Sim1]   Simon, B.: The classical moment problem as a self-adjoint finite difference operator, Adv. Math. **137** (1998), 82–203.

[Sim2]   Simon, B.: *Orthogonal Polynomials on the Unit Circle, Part I: Classical Theory*, Amer. Math. Soc., Providence, R.I., 2005.

[Sim3]   Simon, B.: *Szegö's Theorem and Its Descendants*, Princeton Univ. Press, Princeton, 2011.

[Ste1]   Stengle, G.: A Nullstellensatz and a Positivstellensatz in semialgebraic geometry, Math. Ann. **207**(1974), 87–97.

[Ste2]   Stengle, G.: Complexity estimates of Schmüdgen's Positivstellensatz, J. Complexity **12** (1996), 167–174.

[Stj]   Stieltjes, T.J.: Recherches sur les fractions continues, Ann. Fac. Sci. Toulouse **8**(1894), 1–122, and **9**(1895), 1–47.

[St1]   Stochel, J.: Moment functions on real algebraic sets, Ark. Mat. **30**(1992), 133–148.

[St2]   Stochel, J.: Solving the truncated moment problem solves the moment problem, Glasgow J. Math. **43**(2001), 335–341.

[StS1]   Stochel, J. and F.H. Szafraniec: On normal extrensions of unbounded operators. I, J. Operator Theory **14**(1985), 31–45.

[StS2]   Stochel, J. and F.H. Szafraniec: On normal extrensions of unbounded operators. II, Acta Sci. Math. (Szeged) **53**(1989), 153–177.

[StS3]   Stochel, J. and F.H. Szafraniec: On normal extrensions of unbounded operators. III, Publ. RIMS Kyoto Univ. **25**(1989), 105–139.

[StS4]   Stochel, J. and F.H. Szafraniec: The complex moment problem and subnormality: a polar decomposition approach. J. Funct. Anal. **159**(1998), 432–491.

[Stn]   Stone, M.H.: A theory of spectra I, Proc. Nat. Acad. Sci. USA **26** (1940), 280–283.

[Stv]   Stoyanov, J.: Krein's condition in probabilistic moment problems, Bernoulli **6**(2000), 939–949.

[SKv]   Stoyanov, J. and P. Kopanov: Lin's condition and moment determinacy of functions of random variables, Preprint, 2017.

[Sf]   Szafraniec, F.H.: Boundedness of the shift operator related to positive definite forms: An appplication to moment problems, Ark. Mat. **19**(1981), 251–259.

[Sz]   Szegö, G.: *Orthogonal Polynomials*, Amer. Math. Soc., Providence, R.I., 1939.

[SK]   Sz.-Nagy, B. and A. Koranyi: Relations d'un probleme de Nevanlinna et Pick avec la theorie des operateurs l'espace hilbertien, Acta Math. Acad. Sci. Hungar. **7**(1956), 295–303.

[Tch]   Tchakaloff, V.: Formules de cubatures mécanique à coefficients non négatifs, Bull. Sci. Math. **82**(1957), 123–134.

[To]   Toeplitz, O.: Über die Fouriersche Entwicklung positiver Funktionen, Rend. Circ. Mat. Palermo **32**(1911), 191–192.

[VA]   Van Assche, V.: Orthogonal polynomials, associated polynomials and functions of the second kind, J. Comp. Appl. Math. **37**(1991), 237–249.

[V]   Vandenberghe, L. (Editor): *Handbook of Semidefinite Programming: Theory, Algorithm, and Applications*, Kluwer Academic, Boston, 2005.

[VB]   Vandenberghe, L. and S. Boyd: Semidefinite programming, SIAM Rev. **38**(1996), 49–95.

[Vs1]   Vasilescu, F.: Hamburger and Stieltjes moment problems in several variables, Trans. Amer. Math. Soc. **354**(2001), 1265–1278.

[Vs2]   Vasilescu, F.: Spectral measures and moment problems, Theta Ser. Adv. Math. **2**, Bucharest, 2003.

[Vs3]   Vasilescu, F.: Dimensional stability in truncated moment problems, J. Math. Anal. Appl. **388**(2012), 219–230.

[Veg]   Vegter, G.: The apolar bilinear form in geometric modeling, Math. Comp. **69**(1999), 691–720.

[Ver]   Verblunsky, S.: On positive harmonic functions: A contribution to the algebra of Fourier series, Proc. London Math. Soc. **38**(1935), 125–157.

[Wr]   Walker, R.J.: *Algebraic curves*, Springer-Verlag, New York, 1978.

[Wl]   Wall, H.S.: *Analytic Theory of Continued Fractions*, Amer. Math. Soc., Providence, R.I., 2000.

[Wei]   Weidmann, J.: *Linear Operators in Hilbert Spaces*, Springer-Verlag, Berlin, 1987.

[Wen]   Wendroff, B.: On orthogonal polynomials, Proc. Amer. Math. Soc. **12**(1961), 554–555.

[Wö]   Wörmann, T.: Strict positive Polynome in der semialgebraischen Geometrie, Dissertation, Universität Dortmund, 1998.

[Y]   Young, N.: *An Introduction to Hilbert Space*, Cambridge Univ. Press, Cambridge, 1988.

[Zh]   Zhang, F.: *Matrix Theory*, Universitext, Springer-Verlag, Berlin, 1999.

Symbol Index

$\langle A, B \rangle$, 399, 505
$\overline{A}, A^T$, 504
$A_F^{[n]}, A_K^{[n]}$, 184
$A \succ 0, A \succ 0$, 504
a_n, 98
$A_n(x, z)$, 109
$A(y)$, 399
$A(z, w), A(z)$, 147
A, 18, 43, 415
$\mathsf{A}_{\mathbb{C}}$, 48
$\mathsf{A}_{\mathcal{D}}$, 332
A_h, 43
$\hat{\mathsf{A}}$, 18
A^2, 44
A^2, 425
$\alpha(n)$, 489
$\alpha_n(s), \alpha_n(\mu)$, 264

$B(\mathcal{K}(\mathsf{f}))$, 293
b_n, 98
$B_n(x, z)$, 109
$B(z, w), B(z)$, 147
$\mathbf{B}(\mathcal{H})$, 514
$\mathfrak{B}(\mathcal{X})$, 499
β_s, 191

$C_c(\mathcal{X}), C_c(\mathcal{X}; \mathbb{R}), C_c(\mathcal{X})_+, C_0(\mathcal{X})$, 499
C_f, 344
$c_L(x)$, 32, 457
$C\{m_n\}$, 81
$C_n(x, z)$, 109
$C^\wedge$, 511
$C(\mathcal{X}; \mathbb{R})$, 13

C_z, 151
$C(z, w), C(z)$, 147
$\mathsf{C}_{d,m}$, 482
c_{m+1}, 231, 266
$\mathsf{C}(s), \mathsf{C}(L), \mathsf{C}(\mathsf{A}, \mathcal{K})$, 449
$\mathbb{C}_+$, 502

$\underline{D}_m(s), \overline{D}_m(s)$, 230
$D_n, D_n(s)$, 93
$D_n(s)$, 63
$D_n(x, z)$, 109
$d_+(T), d_-(T)$, 515
$D(z, w), D(z)$, 147
d, 123
$\mathcal{D}_L$, 303, 425
$\mathcal{D}[\mathsf{s}]$, 177
$\mathcal{D}(T)$, 514
$\mathbb{D}$, 271
$\mathbf{D}(\mathbb{R}_d[\underline{x}])$, 332

E_+, 14
Es, 63
E^*, 511
$E_{\mathcal{T}}$, 516
$\overline{E}_t, E_t$, 464
$\mathrm{Exr}(C)$, 513
$\epsilon(t)$, 232

f_s, 218
$\hat{f}$, 230, 426
$f_z(x)$, 125
f_λ, 322

© Springer International Publishing AG 2017
K. Schmüdgen, *The Moment Problem*, Graduate Texts in Mathematics 277,
DOI 10.1007/978-3-319-64546-9

Index

© Springer International Publishing AG 2017
K. Schmüdgen, *The Moment Problem*, Graduate Texts in Mathematics 277,
DOI 10.1007/978-3-319-64546-9

Printed in the United States
By Bookmasters